OPTICAL HOLOGRAPHY

OPTICAL HOLOGRAPHY

Robert J. Collier
Christoph B. Burckhardt
Lawrence H. Lin
Bell Telephone Laboratories
Murray Hill, New Jersey

1983

Library of Congress Catalog Card Number: 74–137619

PRINTED IN THE UNITED STATES OF AMERICA

To
Susan, Anne-Marie, and Silvia

CONTENTS

LIST OF COLOR PLATES . . . xiii

PREFACE . . . xv

1. Introduction to Basic Concepts

1.1 Optical Holography . . . 3
1.2 Light Waves . . . 4
1.3 Interference Patterns . . . 6
1.4 Diffraction . . . 12
1.5 Hologram Formation . . . 14
1.6 Wavefront Reconstruction . . . 16
1.7 Plane and Volume Hologram-Formation Geometries . . . 19
1.8 Basic Holography Equations . . . 22
1.9 Partial Coherence . . . 26
References . . . 33

2. Early Holography

2.1 Bragg's X-Ray Microscope . . . 36
2.2 Holography for Electron Microscopy . . . 39
2.3 X-Ray Holography . . . 41
2.4 Beginnings of Optical Holography . . . 42
2.5 In-Line Holograms . . . 47
2.6 Off-Axis Holograms . . . 52
References . . . 56

3. **Geometric Analysis of Point-Source Holograms**

3.1 Computation of Subject–Reference Phase Differences 59
3.2 Reconstruction with a Point Source 68
3.3 Characteristics of the Images 71
3.4 Third-Order Aberrations 78
References 78

4. **The Fourier Transform**

4.1 Linear Space-Invariant Systems and the Fourier Transform 81
4.2 Correspondences and Transform Relations 84
4.3 The Convolution Operation 86
4.4 Other Operational Correspondences 91
4.5 Functional Correspondences 94
References 95

5. **Propagation and Diffraction**

5.1 The Wave Equation and Its Monochromatic Solution 97
5.2 The Plane Wave Solution 97
5.3 Diffraction from Periodic Objects 101
5.4 The General Diffraction Problem 104
5.5 The Relation to the Fresnel–Kirchhoff Integral 106
References 110

6. **Optical Systems with Spherical Lenses**

6.1 The Spherical Lens 112
6.2 A Simple Optical System 115
6.3 A More General Optical System 119
6.4 The Effect of Finite Lens Size 128
6.5 Coherent and Incoherent Transfer Functions 135
References 136

7. **Light Sources and Optical Technique**

7.1 Light Sources for Hologram Formation 137
7.2 Fringe Visibility in Hologram Recording 152
7.3 Illumination with an Expanded Laser Beam 164
7.4 Division and Attenuation of the Laser Beam 167
7.5 Mechanical Stability in Hologram Formation 169
7.6 Light Sources for Hologram Reconstruction 170
7.7 Simple Holographic Technique 174
References 177

8. Analysis of Plane Holograms

8.1 Off-Axis Holography with Nondiffuse Subject Light 180
8.2 Off-Axis Holography with a Diffuse Signal 194
8.3 Hologram-Forming Geometries 204
8.4 Effects of Resolution and Size of the Recording Medium 217
8.5 Maximum Efficiency of Plane Holograms 223
References 226

9. Diffraction from Volume Holograms

9.1 Holograms Formed with Two Plane Waves 228
9.2 Bragg's Law 231
9.3 Coupled Wave Theory 233
9.4 The Wave Equation 233
9.5 Solution of the Wave Equation 239
9.6 Transmission Holograms 244
9.7 Reflection Holograms 253
9.8 Discussion of Volume Hologram Properties 261
References 263

10. Hologram Recording Materials

10.1 Optical Changes in Photosensitive Materials 266
10.2 Exposure and Sensitivity 269
10.3 Recording Resolution 270
10.4 Persistence of the Hologram and Erasability 271
10.5 Noise and Recording Linearity 272
10.6 Ideal Wavefront Reconstruction and Ideal Recording Material . . . 273
10.7 Representative Exposure Characteristics for Real Materials . . . 276
10.8 Silver Halide Photographic Emulsions 280
10.9 Dichromated Gelatin Films 293
10.10 Photoconductor–Thermoplastic Films 298
10.11 Photochromic Materials 305
10.12 Ferroelectric Crystals 307
References 309

11. Pulsed-Laser Holography

11.1 The Multimode Ruby Laser 312
11.2 The Single-Frequency Ruby Laser 315
11.3 Coherence Length of the Single-Frequency Laser 316
11.4 The Ruby Amplifier 319
11.5 Protection of Optical Components 322

11.6 Arrangements for Forming Pulsed-Laser Holograms . . . 324
11.7 Maximum Allowable Subject Movement . . . 328
11.8 Safe Illumination of Human Subjects . . . 331
11.9 Hologram Recording Materials . . . 332
References . . . 335

12. Nonlinear Recording, Speckle, and Film Grain Noise

12.1 Effects of Nonlinear Recording . . . 337
12.2 Speckle Pattern . . . 345
12.3 Film Grain Noise . . . 351
References . . . 359

13. Real-Image Applications

13.1 Microscopy . . . 360
13.2 Analysis of Aerosol Particles . . . 365
13.3 Imaging through Phase-Distorting Media . . . 368
13.4 High-Resolution Projection Imaging . . . 375
13.5 Multiple Imaging . . . 377
References . . . 379

14. Holograms and Hologram Spatial Filters Formed with Spatially Modulated Reference Waves

14.1 Associative Storage . . . 382
14.2 The Fourier Transform Hologram with a Spatially Modulated Reference Wave . . . 387
14.3 Some Experiments with Fourier Transform Holograms . . . 391
14.4 Character Recognition . . . 394
14.5 Multiplexing and Coding . . . 409
14.6 Image Processing . . . 414
References . . . 416

15. Holographic Interferometry

15.1 Real-Time Interferometry . . . 420
15.2 Double-Exposure Interferometry . . . 423
15.3 Fringe Localization and Interpretation . . . 426
15.4 Interferometry of Vibrating Surfaces . . . 437
15.5 Contour Generation . . . 444
15.6 Applications and Improvements . . . 450
References . . . 452

16. **Information Storage**

16.1 Page-Organized Microimage Storage System 455
16.2 A Standing-Wave Optical Memory 458
16.3 Holographic Storage in Thick Media 461
16.4 A Holographic Flying-Spot Store 476
16.5 Write, Read, and Erase *in Situ* 489
References 493

17. **Color Holography**

17.1 Color Mixing with Laser Light 494
17.2 Recording Materials 500
17.3 Monochrome Images 500
17.4 Achromatic Images 501
17.5 Multicolor Images from Plane Holograms 504
17.6 Multicolor Images from Volume Holograms 514
References 521

18. **Composite Holograms**

18.1 Image Resolution and Element Size 523
18.2 Hologram Information Reduction 526
18.3 Hyper- and Hypostereoscopic Hologram Images 532
18.4 Wide-Angle Hologram Images 533
18.5 3D Image Synthesis from Photographs 535
References 540

19. **Computer-Generated Holograms**

19.1 The Sampling Theorem 543
19.2 The Discrete Fourier Transform and the Fast Fourier Transform . . 548
19.3 Binary Fourier Transform Holograms 551
19.4 Applications 558
19.5 The Kinoform 560
References 563

20. **Three Topics in Search of a Chapter: Replication, TV Transmission, and Incoherent-Light Holograms**

20.1 Hologram Replication 564
20.2 Television Transmission of Holograms 573
20.3 Holograms Formed with Spatially Incoherent Subject Light 579
References 582

Appendix I. **Equivalence of the Fresnel-Kirchhoff Integral and the Diffraction Formula in the Spatial Frequency Domain** 583

References 586

Appendix II. **Complex Representation of the Electric Field** 587

References 589

Appendix III. **Capacity-Speed Product of an Acoustic Beam Deflector** 590

References 593

INDEX 595

LIST OF COLOR PLATES

Following page 510

PLATE I. Achromatic Imaging

PLATE II. Color Cross-Modulation

PLATE III. Multicolor Imaging by Coding

PLATE IV. Multicolor Imaging by Spatial Sampling

PLATE V. The First Multicolor Image from a Volume Hologram

PLATE VI. Multicolor Pattern from a Reflection Hologram

PLATE VII. Multicolor 3D Image with White-Light Illumination

PREFACE

Electron microscopists, electrical engineers, opticists, physicists, and chemists have all had a hand in fostering the development of holography. Undoubtedly this interdisciplinary interest, coming soon after the invention of the laser, has given a broad scope to research into the properties of the hologram. At the same time, it has been responsible for a flood of publications written in diverse scientific languages and dispersed among a variety of journals. As might be expected, the level of prior knowledge required of the reader fluctuates markedly from one publication to the other. Counter to its purpose, this melange may well discourage communication between current investigators and potential users of holographic technique.

Our book is addressed to those who will learn for themselves whether holographic techniques can solve technical problems. Our intention is to provide a ground from which innovation can spring. The reader need have no more than an undergraduate training in science or engineering. An elementary introduction to optics, an acceptance of Maxwell's equations, and a sometime acquaintance with the Fourier transform should suffice.

The text concentrates on the properties of holograms formed with visible light. We begin by assuming that the reader is not a recent student of physical optics. Consequently the basic optical concepts which play a role in holography are reviewed first. Specific properties of simple holograms are then explained by means of geometrical analyses. Chapters 4 through 6 together form a short course in the mathematics and physics of Fourier optics. They lay the foundation for the theory of plane and volume holograms following in Chapters 8 and 9. Chapter 7 is interposed for the benefit of those who might prefer to try their hand at forming a hologram before encountering further analysis. The chapter establishes the reasons for acquiring a continuous-wave laser, sets forth the requisite optical technique, and describes a simple experiment which can be performed with modest equipment. To obtain even better results, the practical holographer must choose a recording material best suited to his purpose (Chapter 10). When the subject of the hologram cannot be immobilized, the holographer may progress to the pulsed-laser holography techniques of Chapter 11.

Aside from Chapter 19, describing computer-generated holograms, the second half of the book deals with potential applications of holography: high-resolution imaging, imaging through diffusing media, spatial filtering, character recognition and coding, displacement interferometry and contour generation, information storage, monochrome and multicolor 3D display, and 3D television. It also reveals problems which must be overcome. These include low diffraction efficiency, noise, distortion, nonlinearities, the low sensitivity characteristic of high-resolution recording material, and the disturbing speckle pattern observed when laser light is employed to illuminate a hologram.

The book is meant to be an instrument for learning rather than a catalog of contributions to holography. We therefore have selected from the substantial amount of literature on holography that which seemed suited to the purpose. Similarly, the reference list at the end of each chapter is limited to those papers which influenced the writing of that chapter. However, a short bibliography at the end of the preface offers an opportunity for wider reading and cites more complete lists of published contributions.

We appreciate the support extended to us by the Bell Telephone Laboratories and gratefuly acknowledge the individual help given to us by many of our colleagues. We are particularly indebted to Mr. H. L. Beauchamp and Mr. E. T. Doherty for their aid in obtaining the desired holograms, photographs and figures.

H. J. Caufield, and Sun Lu, *The Applications of Holography*. Wiley (Interscience), New York, 1970.

R. P. Chambers and J. S. Courtney-Pratt, "Bibliography on Holograms," *J. Soc. Motion Pict. Telev. Eng.* **75,** 373, 759 (1966).

R. P. Chambers and B. A. Stevens, "Bibliography on Holograms," *J. Soc. Motion Pict. Telev. Eng.* **76,** 392 (1967).

J. B. DeVelis and G. O. Reynolds, *Theory and Applications of Holography*. Addison-Wesley, Reading, Massachusetts, 1967.

M. Françon, *Holographie*. Masson, Paris, 1969.

J. W. Goodman, *Introduction to Fourier Optics*. McGraw-Hill, New York, 1968.

H. Kiemle and D. Röss, *Einführung in die Technik der Holographie*. Akademische Verlagsgesellschaft, Frankfurt am Main, 1969.

W. E. Kock, *Lasers and Holography: An Introduction to Coherent Optics*. Doubleday, New York, 1969.

J. N. Latta, "A Classified Bibliography on Holography and Related Fields," *J. Soc. Motion Pict. Telev. Eng.* **77**, 422, 540 (1968).

Yu. I. Ostrosky, *Holography*. Izdatel'stvo Nauka, Leningrad, 1970. (In Russian.)

H. M. Smith, *Principles of Holography*. Wiley (Interscience), New York, 1969.

G. W. Stroke, *An Introduction to Coherent Optics and Holography*, 2nd ed. Academic Press, New York, 1969.

Chapter 1

INTRODUCTION TO BASIC CONCEPTS

Invention of the laser has revitalized the ancient science of optics, injected practical significance into the study of coherent light, and aroused new hopes for some old ideas. Among these is the method of *wavefront reconstruction*, first invented by Dennis Gabor in 1948 [*1.1*]. Its renaissance is largely a product of the coherence properties of laser light. Wavefront reconstruction, or *holography* as it is now called, is a way to retrieve the likeness of a subject from a record of its unfocused diffraction pattern. Gabor invented the method to improve the resolution in images obtained with an electron microscope. Although unable to demonstrate the validity of his principle with electron waves, he was able to do so with visible light. At the time, however, absence of a source of coherent radiation with sufficient intensity, even in the visible region of the spectrum, seemed to relegate holography forever to the class of optical curiosities. In the face of many obstacles and with only marginal results, the method was abandoned by its early investigators. Holography after a decade seemed destined for obscurity. This bleak prospect brightened in the early 1960s with work carried out at the University of Michigan Institute of Science and Technology. In 1962, Leith and Upatnieks [*1.2*] began publication of a series of reports demonstrating a new approach to optical holography based on communications theory techniques. By 1964 they had convincing proof that holography was practical and that the use of laser light was an important factor. Indeed, the three-dimensional images they were able to generate by illuminating a photographic plate with light from a laser excited the imagination of all who saw them and stimulated many to initiate their own investigations into the potential of holography. Today holography is widely known as a practical means for storing wavefronts in a record from which the wavefronts may later be reconstructed. The record, formed in photosensitive material, is called a *hologram*.

Unlike photographic negatives, the featureless gray cast observed on the surface of a hologram plate (Fig. 1.1) portends little of its imaging capability. Standard photography accustoms us to expect visible evidence of the recording process to appear on plate or film in the form of a two-dimensional image. This requires a lens to gather light diverging from each subject point

FIG. 1.1. A photograph of a hologram.

and to focus the light to corresponding points in the image. The lack of image detail on a hologram surface suggests that holography has dispensed with the imaging lens, as is indeed the case. Only the photographic plate and a coherent reference wave are required to form a record of the unfocused light scattered from a laser illuminated subject. This unfocused light or diffraction pattern is a superposition of light waves from many subject points, a pattern often macroscopically uniform in intensity. Despite this seeming lack of information, when the proper record of the pattern, the hologram, is exposed, developed and illuminated, an image of a three-dimensional subject can form in a volume of space. The image, which may be located some distance from the hologram, possesses the depth and parallax properties normally associated with real objects.

As we shall see, holography is concerned with much more than pictorial three-dimensional imaging. A closer look at the method will reveal its potential role in information storage, interferometry, microscopy, and data processing. In this first chapter, concepts which are basic to understanding holography are informally introduced. Many of the topics covered in summary fashion will be taken up again in subsequent chapters, to be reexamined with greater rigor.

1.1 Optical Holography

Holography is an *interference* method of recording the light waves diffracted by a subject illuminated with coherent light. The diffracted waves are caused to interfere with a phase-related reference wave. If the waves are highly coherent, the relative phase between subject and reference wave remains constant in time producing an observable effect on the intensity distribution of the resulting interference pattern. The photographic record of this pattern, the hologram, contains sufficient information about both the phase and amplitude of the diffracted waves to permit their reconstruction. Wavefront reconstruction takes place, in a second step, when the hologram is illuminated with the reference wave.

Preservation of relative phase information in a retrievable form is the unique characteristic of the holographic process. In contrast, photography can preserve only the spatial distribution of light intensity in the subject scene. This is reproduced by the camera lens in the focused-image intensity and recorded by the amount of silver developed in the photographic emulsion exposed to that intensity. However, the intensity is a quantity averaged over all phases of the light wave and so contains no information about the original phase of the light at the subject. Holography, by means of a reference wave, encodes the unfocused subject light intensity with both amplitude and phase information prior to its recording.

The method of holography applies to all waves: to electron waves, X rays, light waves, microwaves, acoustic waves and seismic waves, providing the waves are coherent enough to form the required interference patterns. Holograms have in fact been formed with each of these waveforms. However, it is in the optical region of the electromagnetic spectrum that holography seems best suited. Progress in *optical* holography, using the laser, has been explosive while work at shorter wavelengths has been impeded by the lack of a coherent source. Toward the other end of the wavelength scale, active research is currently in progress in acoustic holography. New techniques for forming acoustic holograms are evolving, but results are still somewhat on the primitive side. We shall therefore confine our attention to visible light and optical holography.

Before proceeding further with an interpretation of the holographic method, let us review some relevant aspects of physical optics.

1.2 Light Waves

Propagation of the mechanical energy associated with elastic deformation, the propagation of sound and of light can all be described in terms of wave motion. In the case of mechanical energy propagating over the surface of the ocean, the transverse wave motion is quite apparent. Less easy to observe are compressional sound waves in air. Instead of observing the air molecules directly we must infer their motion by recording their effect on a diaphragm. Light wave propagation presents even greater observational difficulties. Since the oscillation frequencies of the electric and magnetic fields of a light wave approach 10^{15} Hz, there is no detector with response rapid enough to record their instantaneous values.

A demonstration that light does propagate as a wave was first provided by Thomas Young in 1802. Young inferred the wave properties from observations of the interference of light coming from separated points on a wavefront. He observed the overlap, on a screen, of the light from two such secondary sources and found that there was cancellation of light intensity as well as addition, a feat difficult to account for on a particle basis but easily accommodated by a wave theory. Young's experimental arrangement for securing interference of two light sources is shown in Fig. 1.2. A pinhole

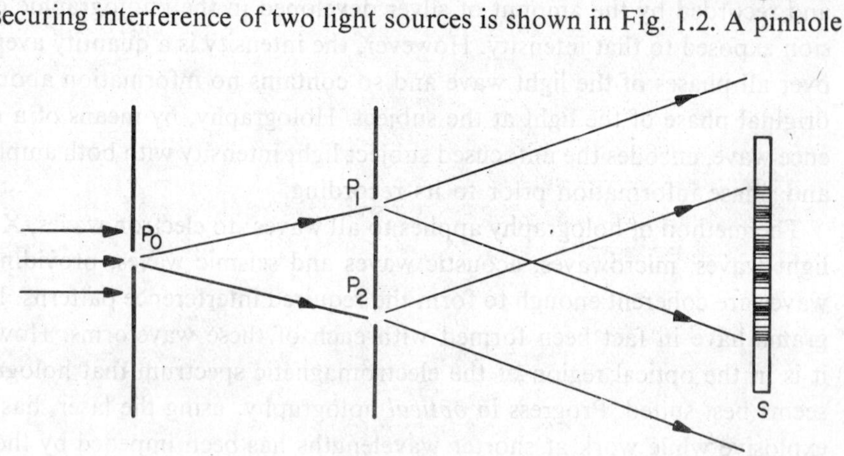

FIG. 1.2 Young's double-aperture experiment.

P_0, illuminated by collimated light, diffracts a spherical wave to an opaque screen some distance away containing two additional pinholes P_1 and P_2. These sample the spherical wavefront and diffract two secondary, phase-related, spherical waves. On an observation screen S, parallel to the first screen and placed where the waves overlap, an alternating bright and dark interference pattern is observed, with linear fringes running perpendicular to the direction of the line joining P_1 and P_2.

Observation of the wave properties of light and formation of a hologram are closely related. Both depend on recording the intensity of the bright and dark spatial patterns or fringes found wherever coherent light waves intersect. The spatial pattern is variously called an *interference pattern* or *standing wave pattern*, the latter referring to the fact that the spatial distribution of fringe intensity remains constant in time. This constancy in time allows the fringes obtained in Young's experiment to be observed and their intensity and spacing measured. The formation of the fringe pattern implies that light has a wave character, and measurement of the spacing and contrast of the fringe pattern can reveal properties of the wave such as wavelength and degree of coherence.

Although one cannot directly observe oscillation of the light wave fields, the success of the theory of Maxwell, which predicts electromagnetic or light waves, is practical proof of their existence. Maxwell's theory predicts the presence of two vector force fields in light waves: the electric and magnetic fields. These fields can propagate through space unsupported by any known medium, and only the time-averaged effects of their interaction with matter can be observed. Holography is concerned with the interaction of light waves with photosensitive matter, e.g., silver halide grains in photographic emulsion. Presumably one must deal with the presence of two force fields both of which may possibly interact in the recording medium. This concern is removed when one considers one of the results of a standing wave experiment carried out by Wiener (1890), Fig. 1.3. He was able to demonstrate that a light standing wave pattern blackened a photographic plate most at the *electric* field antinodal regions and not at all at the magnetic field antinodal regions. Thus, it is the electric field which is the light force component of major consequence in the forming of holograms. This holds true, not only for photographic emulsion, but for all the photosensitive media in which holograms have been formed. (The light interacts with electrons which are at rest or slowly moving relative to the velocity of light, and the contribution of the magnetic field to the Lorentz electromagnetic force is negligible.) We thus eliminate the magnetic field in hologram formation and proceed as though the electric field were the sole

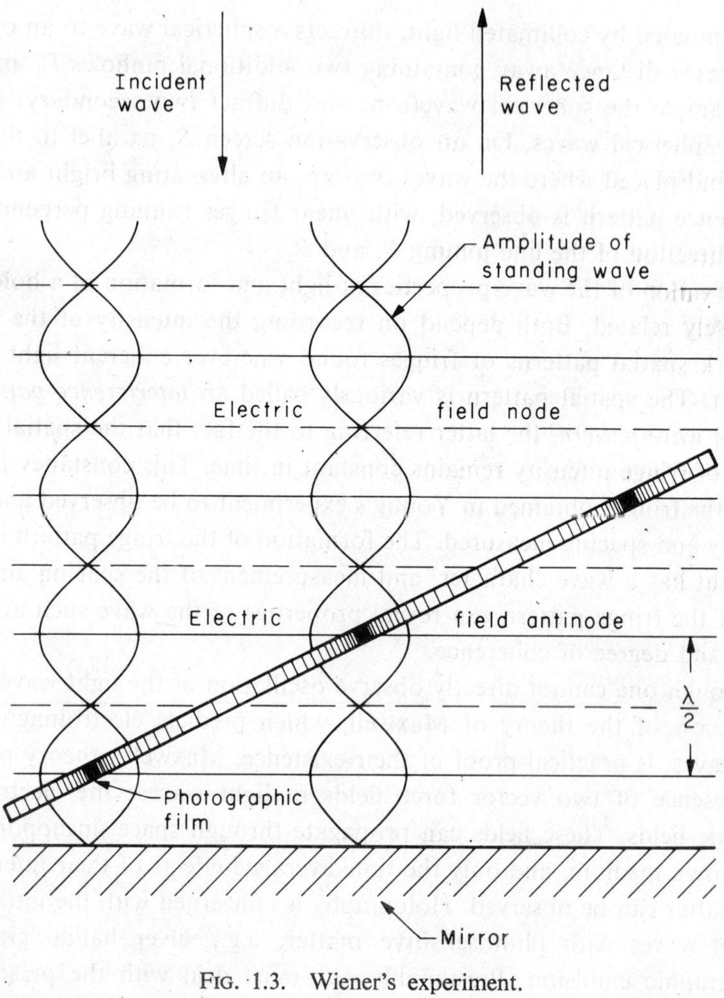

FIG. 1.3. Wiener's experiment.

component of the light wave. On reconstruction from holograms formed in nonmagnetic materials, it is again the interaction of the electric field that is important.

1.3 Interference Patterns

The recording of a hologram is essentially a measurement of the intensity of an interference pattern. If the relative phase between the interfering wave fields has some degree of constancy in time, then the spatial distribution of fringes of intensity in the interference pattern will also have some degree of constancy. In this section we shall examine the intensity of such patterns.

In doing so, we shall restrict our consideration to the interference of monochromatic waves of identical wavelength, derived from a single continuously oscillating source. That is, we assume the ideal case of perfectly coherent light. Relative phase and the interference pattern intensity are absolutely constant in time. The salient facts are more apparent with this simplification. Light with only partial coherence will be considered in Section 1.9.

Before proceeding further, we must first be more precise about the meaning of the interaction of light with photosensitive media. The darkening of a unit volume of photographic emulsion or the bleaching of a unit volume of photochromic material is a function of the energy absorbed by that volume averaged over a time long compared to the light vibration period. From the energy law of Maxwell's theory, we know that u the energy per unit volume or energy density in the electric field of a light wave is given in mks units by

$$u = \tfrac{1}{2}\varepsilon\vec{v}\cdot\vec{v},$$

where ε is the dielectric constant of the medium in which the wave is traveling and $\vec{v}$ is the electric field vector. We write the time average of u as

$$\begin{aligned}\langle u\rangle &= \frac{1}{2T}\int_{-T}^{T} u\,dt\\ &= \frac{1}{2}\varepsilon\cdot\frac{1}{2T}\int_{-T}^{T}\vec{v}\cdot\vec{v}\,dt\\ &= \frac{1}{2}\varepsilon\langle\vec{v}\cdot\vec{v}\rangle,\end{aligned}$$

where $2T$ is the time over which the average is taken and where the brackets $\langle\ \rangle$ are a symbol for the time-averaging process.

At any point in the light wave the Poynting vector may be interpreted as giving the magnitude and direction of the energy flow per unit time per unit area normal to the flow. It is common usage in classical optics to call the time average of the magnitude of this flow of power per unit cross section the *intensity* of the light at the point. If we label the intensity I_P, then

$$I_P = s\langle u\rangle = \tfrac{1}{2}s\varepsilon\langle\vec{v}\cdot\vec{v}\rangle,$$

where s is the speed of the light in the medium. In the mks system I_P is expressed in units of watts per square meter. On the other hand, in holography it is the custom to define intensity in an abbreviated form such that

$$I = 2\langle\vec{v}\cdot\vec{v}\rangle. \tag{1.1}$$

The discussion of monochromatic waves which follows will show that the *intensity I reduces to the square of the amplitude of a light wave* and is a very important parameter in the theory of holography. Although the choice of the term *intensity* for the square of the amplitude is rather unfortunate, there is generally no confusion as to which intensity is meant in a given situation. Furthermore, the proportionality between I and I_P allows us to express relative intensities equivalently in terms of I or I_P. Thus if $\vec{r}_1$ is the radius vector to one point in a light beam and $\vec{r}_2$ is the radius vector to another, the relative intensities at the two points are given by

$$\frac{I(\vec{r}_1)}{I(\vec{r}_2)} = \frac{I_P(\vec{r}_1)}{I_P(\vec{r}_2)}.$$

Using the definition of the intensity I_P, we may now describe the interaction of light with photosensitive material in terms of an *exposure* value E. Let us consider light passing with very little attenuation through a film of silver halide photographic emulsion. In any volume of the emulsion, the number of silver halide grains per unit time made developable by the light is a function of the average electric energy flowing into that volume per unit time. If we consider a volume whose cross section is of unit area and normal to the energy flow, the energy flow into it per unit time is the intensity I_P. After an exposure time τ_e, the total number of developable silver halide grains in the volume is given by some function of the exposure $E = I_P\tau_e \propto I\tau_e$. Consequently the darkening or loss in transmittance of the film after development of these grains can also be expressed as a function of the exposure E (see Section 2.5.1).

Insight into the interference process is gained by substituting expressions for relevant wave amplitudes into $I = 2\langle\vec{v}\cdot\vec{v}\rangle$, the intensity defined in Eq. (1.1). If the electric field $\vec{v}$ exists as a physical quantity, it must be a real function of space and time and if it represents a truly monochromatic wave, it must be a simple harmonic function of time. We let f be the frequency of the wave oscillation and write for the electric field

$$\vec{v} = \vec{a}\cos(2\pi ft + \varphi). \tag{1.2}$$

Here $\vec{a}$ is the amplitude, a function of spatial coordinates only, and φ is the phase function of spatial coordinates only. Substitution of Eq. (1.2) into Eq. (1.1) yields

$$I = \frac{2}{2T}\int_{-T}^{T}\frac{\vec{a}\cdot\vec{a}}{2}[1+\cos(4\pi ft + 2\varphi)]\,dt$$

$$= \vec{a}\cdot\vec{a} \quad \text{for} \quad T \gg 1/f \tag{1.3}$$

$$= a^2 = a_x^{\,2} + a_y^{\,2} + a_z^{\,2} \tag{1.4}$$

with a_x, a_y, and a_z representing the Cartesian components of the vector $\vec{a}$. Intensity is thus equal to the square of the amplitude of the electric field. As is evident from Eq. (1.4), measurement of the intensity of a single wave provides no information about the phase of the wave.

Interference patterns imply the simultaneous presence of more than one wave, and so we must consider how to add a number of interfering monochromatic waves and then apply Eq. (1.1). Each wave may be represented by $\vec{v}_i = \vec{a}_i \cos(2\pi ft + \varphi_i)$, where the frequency f has a single value, identical for each wave ($\vec{v}_i$ is the electric field vector in the region of interference). The sum of these sinusoidal functions is a sinusoid itself, and so we have

$$\vec{a}_1 \cos(2\pi ft + \varphi_1) + \vec{a}_2 \cos(2\pi ft + \varphi_2) + \cdots = \vec{a} \cos(2\pi ft + \varphi). \quad (1.5)$$

The above may be rewritten as

$$\begin{aligned} &\mathrm{Re}\big[\vec{a}_1 \exp[i(2\pi ft + \varphi_1)]\big] + \mathrm{Re}\big[\vec{a}_2 \exp[i(2\pi ft + \varphi_2)]\big] + \cdots \\ &\quad = \mathrm{Re}\big[\vec{a} \exp[i(2\pi ft + \varphi)]\big] \end{aligned} \quad (1.6)$$

where Re[ ] indicates the real part of the complex quantity within the brackets. Computations are made simpler by using complex notation, and we can facilitate its use by dropping the reminder that the wave functions are real.

At this point we distinguish several terms which we shall apply to the complex light wave function of space and time appearing on the right-hand side of Eq. (1.6). The complex quantity[1]

$$\vec{\mathbf{v}} = \vec{a} \exp(i\varphi) \exp(2\pi i f t)$$

containing the temporal phase factor varying at the oscillation frequency f is called the *complex electric field vector*; the complex quantity

$$\vec{\mathbf{a}} = \vec{a} \exp(i\varphi)$$

contains only amplitude and phase factors which do not vary at the frequency f and is called the *complex amplitude vector*; the real quantity $\vec{a}$ is simply the *amplitude vector*. When we deal with scalar waves (see the next section), the suffix "vector" will be dropped. (A general definition of the complex electric field is given in Appendix II.)

[1] Complex quantities are denoted by **boldface** type.

Dropping the symbol Re[ ] in Eq. (1.6) and dividing each term by $\exp(2\pi i f t)$, we obtain

$$\vec{a}_1 \exp(i\varphi_1) + \vec{a}_2 \exp(i\varphi_2) + \cdots = \vec{a} \exp(i\varphi) = \mathbf{\vec{a}}. \quad (1.7)$$

Thus the complex amplitude vector of a sum of monochromatic waves is obtained by adding the complex amplitude vectors of the individual waves according to the rules for adding complex numbers.

We can now write the intensity I in Eq. (1.3) in terms of $\mathbf{\vec{a}}$ by forming the product $\mathbf{\vec{a}} \cdot \mathbf{\vec{a}}^* = [\vec{a} \exp(i\varphi)] \cdot [\vec{a} \exp(-i\varphi)] = \vec{a} \cdot \vec{a}$ so that

$$I = \vec{a} \cdot \vec{a} = \mathbf{\vec{a}} \cdot \mathbf{\vec{a}}^* = [\vec{a}_1 \exp(i\varphi_1) + \vec{a}_2 \exp(i\varphi_2) + \cdots] \times [\vec{a}_1 \exp(-i\varphi_1) + \vec{a}_2 \exp(-i\varphi_2) + \cdots] \quad (1.8)$$

where the asterisk indicates the complex conjugate.

1.3.1 Interference of Two Waves

Holography is often concerned with the interference of *two* waves, a subject wave and a reference wave. In this case, the intensity I in Eq. (1.8) takes the form

$$I = \vec{a} \cdot \vec{a} = \vec{a}_1 \cdot \vec{a}_1 + \vec{a}_2 \cdot \vec{a}_2 + \vec{a}_1 \cdot \vec{a}_2 [\exp[i(\varphi_2 - \varphi_1)] + \exp[-i(\varphi_2 - \varphi_1)]]$$

or

$$I = I_1 + I_2 + 2\vec{a}_1 \cdot \vec{a}_2 \cos(\varphi_2 - \varphi_1). \quad (1.9)$$

Thus the intensity at any point in the interference pattern formed by two wave trains is the sum of the intensities of the individual waves plus an interference term. *Relative phase* information is contained in this time-independent interference term.

It should be noted that for a nonzero value of the interference term the vector $\vec{a}_2$ must have a component parallel to $\vec{a}_1$. Two light waves which are polarized in mutually perpendicular directions contribute only the sum of their individual intensities to I in Eq. (1.9) and can produce neither an interference pattern nor a hologram. On the other hand, if the two interfering waves are polarized parallel to one another, the resultant intensity I can be greater or less than the sum of their intensities. For example, if the waves have constant and equal amplitudes

$$|\vec{a}_1| = |\vec{a}_2| = (I_1)^{1/2} = (I_2)^{1/2},$$

then, from Eq. (1.9), the maximum value of the intensity I is four times the individual intensities I_1 or I_2, and the minimum value of I is 0.

Equation (1.9) tells us that only vector wave components parallel to one another produce interference patterns (or holograms). In analyzing the hologram process, then, we can consider the interacting wave amplitudes as scalar quantities. This, of course, results in a simplification of notation which is helpful to an understanding of the basic concepts. Scalar wave analysis will be used throughout much of the remainder of the book.

The particular shape of an interference or standing wave pattern is determined by the spatial variation of the amplitudes and phases of the interfering waves. An instructive example, and one to be employed in our further discussion of holography, is the interference of two plane waves. We assume they are derived from the same perfectly coherent source and intersect at an angle 2θ, as in Fig. 1.4. For such waves the loci of constant phase (the wavefronts) are planes. To simplify the figure, only the positive maximum amplitude or crest wavefront planes F_1 and F_2, normal to the figure plane and spaced by the wavelength λ, are represented.[2] They are indicated by the traces they make in the plane of the figure. Thus two sets of periodically spaced lines represent the succession of wavefronts in each wave train. Wave normals 1 and 2, perpendicular to the fronts and contained in the figure plane, indicate the directions of propagation.

The intersections of F_1 and F_2 in *lines perpendicular to the plane* of Fig. 1.4 (indicated by the heavy dots) represent addition of the wave crests. As the waves progress in the direction of their wave normals, these intersection lines move so as to generate *planes* of maximum resultant light amplitude which bisect the angle between the wave normals. The planes are normal to the figure and located where the vertical line density (in Fig. 1.4) is greatest. The time average of the square of the resultant amplitude, the intensity, is also maximum along these loci or fringes. More generally, such planes are the loci of all wave interference for which $\varphi_2 - \varphi_1 = 2n\pi$, $n = 0, 1, 2, \ldots$, in Eq. (1.9). When account is taken of all the other relative phases in which amplitude addition takes place, the intensity of the pattern takes on a sinusoidal distribution in the y direction indicated by the density of the remaining vertical lines. (An analysis of plane wave interference is given in Section 9.1.)

Examination of the heavy-lined triangle in Fig. 1.4 shows that the period of the sinusoidal intensity distribution is given by

$$2d \sin \theta = \lambda. \tag{1.10}$$

[2] Unless otherwise noted, the symbol λ represents the wavelength of light in the medium in which the light is propagating.

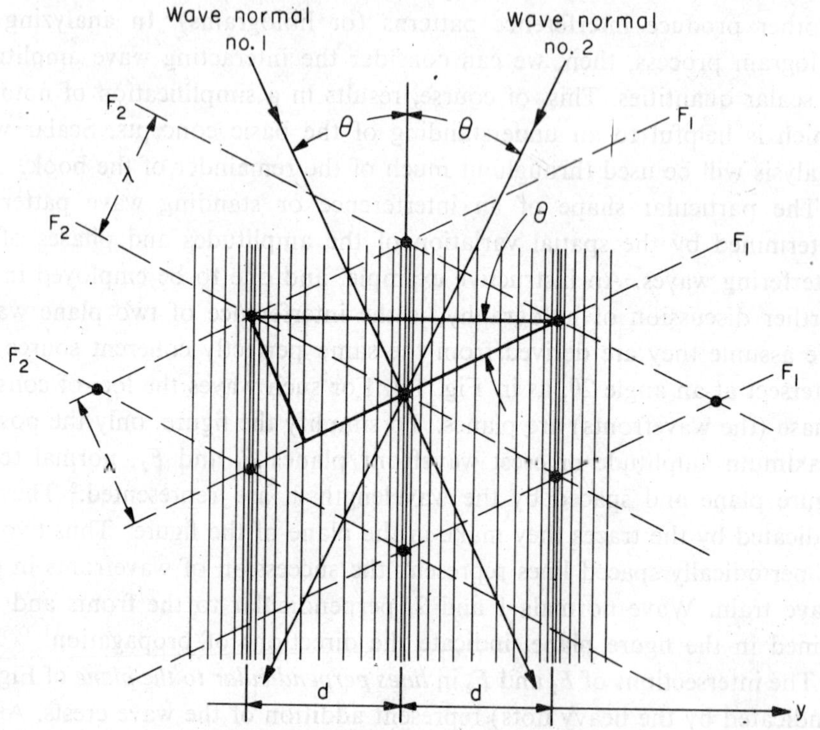

FIG. 1.4. Interference pattern of two intersecting plane waves.

Note that θ is the grazing angle made by each wave normal with the generated intensity planes. As the angle 2θ between the wave normals increases, the period d decreases. If one wishes to record the fringe pattern, as in forming a hologram, the photographic medium employed must be able to resolve approximately $1/d$ fringe pairs per unit distance.

1.4 Diffraction

The object of holography is to usefully record diffraction patterns. The term diffraction is applied to the bending of wave normals (rays) when they encounter obstacles whose optical transmission or reflection properties change significantly in distances approaching the wavelength of the illuminating light. Small holes in screens diffract light; large holes diffract only the light passing close to the edges of the hole but not that passing near the center. Diffraction will be the main topic of Chapter 5 and so we need not dwell on its complexities here. However, it will be useful to the discussion of the remainder of this chapter to set down two equations characterizing diffraction by plane and volume diffraction gratings.

A hologram is itself a diffracting object with some peculiar properties. Despite their complex microscopic structure, holograms can be classified as behaving like (1) plane diffraction gratings or (2) volume diffraction gratings. In Fig. 1.5, a plane diffraction grating is shown edge-on. The grating may consist of a set of periodically spaced transparent slits in an opaque screen.

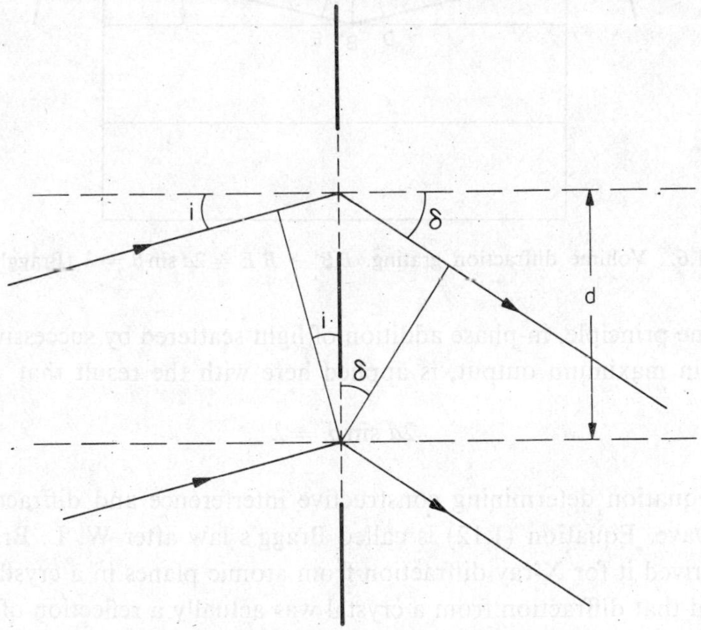

FIG. 1.5. Plane diffraction grating.

When a plane wave is incident on the grating, the condition determining the in-phase or constructive addition of diffracted light is the *grating equation*,

$$d(\sin i + \sin \delta) = \lambda \tag{1.11}$$

where we consider only first-order diffraction and where d is the grating spacing, i the angle of incidence, and δ the angle of diffraction. As may be seen from the figure, Eq. (1.11) states that the light amplitude diffracted by each slit adds in-phase to that diffracted by all the others, thus constructing a maximum-output diffracted plane wave. Negative- and higher-order diffraction are possible as well.

Figure 1.6 displays a *volume* diffraction grating consisting of periodically spaced scattering planes (seen edge-on), illuminated with a plane wave.

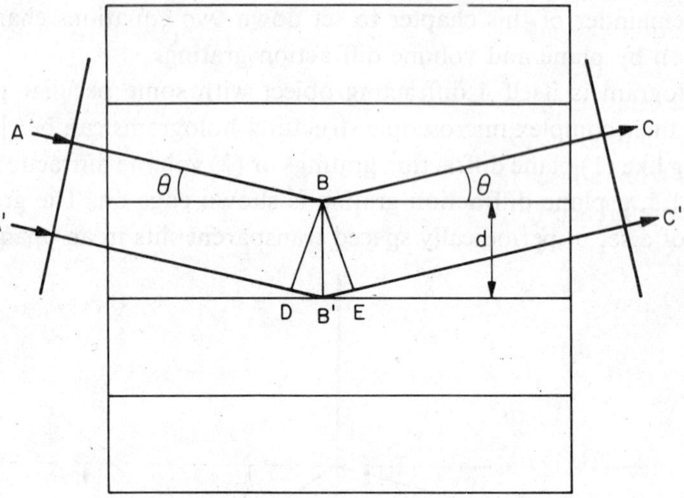

FIG. 1.6. Volume diffraction grating. $DB' + B'E = 2d \sin \theta = \lambda$ (Bragg's law).

The same principle, in-phase addition of light scattered by successive planes to obtain maximum output, is applied here with the result that

$$2d \sin \theta = \lambda \tag{1.12}$$

is the equation determining constructive interference and diffraction of a plane wave. Equation (1.12) is called Bragg's law after W. L. Bragg [*1.3*] who derived it for X-ray diffraction from atomic planes in a crystal. Bragg assumed that diffraction from a crystal was actually a reflection of incident waves from the crystal planes. Maximum diffraction occurs when the grazing angles of incidence and reflection are equal, as shown in Fig. 1.6, and given by θ in Eq. (1.12).

Comparison of Eqs. (1.11) and (1.12) reveals that the latter is a more selective criterion for observing maximum diffraction. For volume gratings, once the incident angle has been chosen, the wavelength and the diffraction angle are determined. This is not so for plane gratings. Equation (1.11) allows arbitrary choice of both angle of incidence and wavelength (within the limitation that $|\sin i|$ and $|\sin \delta| \leq 1$).

1.5 Hologram Formation

We can now turn our attention to some fundamental aspects of the holographic method itself and begin by describing the formation of the hologram. Gabor derived the name *hologram* from the Greek word ὅλος (holos) meaning whole, noting that a record of phase as well as amplitude information provided a rather complete description of a light wave. The method for preserving phase information comes out of Eq. (1.9) where we see that relative phase is preserved in two-beam interference patterns. The basic technique of hologram formation, then, is to divide the coherent light coming from a laser into two beams: one to illuminate a subject and one to act as a reference (Fig. 1.7).

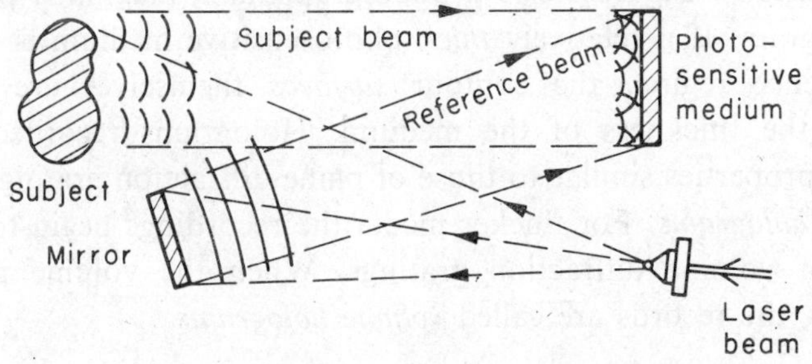

FIG. 1.7. Formation of a hologram.

Reference wavefronts are often (but not necessarily) unmodulated spherical or plane fronts. The reference beam is directed so as to intersect the light transmitted or reflected by the subject. Assuming the two beams to be perfectly coherent, an interference pattern will form in the volume of space where the beams overlap. A photosensitive medium, placed in the overlap region, will undergo certain chemical or physical changes due to exposure to the light intensity. After removal from the light and after any processing required to record these changes as an alteration of the optical transmission of the medium, the medium becomes the hologram.

When the recording medium is silver-halide photographic emulsion, the change in transmittance might be an increase in absorption due to conversion of silver halide to silver atoms during exposure and development. Under such circumstances an *absorption* hologram is obtained. If the same hologram is bleached, so as to convert the silver to a transparent compound whose index of refraction differs from that of gelatin, then the record is written in the resulting localized changes in the index of refraction of the emulsion. The hologram is then called a *phase* hologram. In the absorption hologram case, an exposure and development process is chosen to make the spatial variation in the absorption constant of the hologram plate correspond to the pattern of intensity of the incident light. In the case of a phase hologram, the spatial phase modulation imposed on a wave as it passes through the hologram is made to correspond to the incident intensity pattern.

The intensity of the interference pattern formed by simple unmodulated plane or spherical waves can, for the most part, be regarded as a three-dimensional contour map. The contour surfaces would represent regions of maximum light intensity corresponding to $\varphi_2 - \varphi_1 = 2\pi n$ in Eq. (1.9). If a very *thin* photosensitive medium is placed in the interference region and properly exposed to the light there, it will record the *line traces* of these maximum intensity contour surfaces (silver lines in the case of thin photographic emulsion). Figure 1.8 is a photograph of a magnified portion of such a hologram. If a relatively *thick* photosensitive medium is placed in the interference region, the contour *surfaces* themselves are recorded throughout the thickness of the medium. Holograms recorded in thin media have properties similar to those of plane diffraction gratings and are called *plane holograms.* For thicker media the recordings begin to take on properties of volume diffraction gratings. When the volume properties predominate, the records are called *volume holograms.*

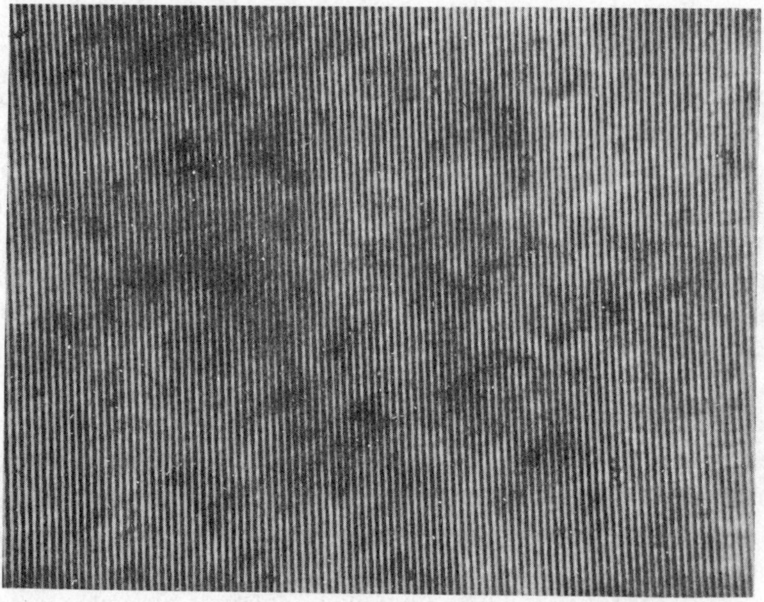

FIG. 1.8. Photograph of a magnified portion of a hologram.

1.6 Wavefront Reconstruction

The silver lines on plane holograms and the silver surfaces in volume holograms are very closely spaced and can therefore diffract light significantly. When the hologram is illuminated by the original reference beam, part of the light diffracted out of the reference beam is directed and shaped by the hologram into a re-creation of the light wavefronts originally coming from the subject. A reconstructed wave train proceeds out from the hologram exactly as did the original subject wave. An observer, viewing a wave identical with the original subject wave, quite naturally perceives it to diverge from a *virtual* image of the subject located precisely at the original subject location (Fig. 1.9). On the other hand, if the reference beam is

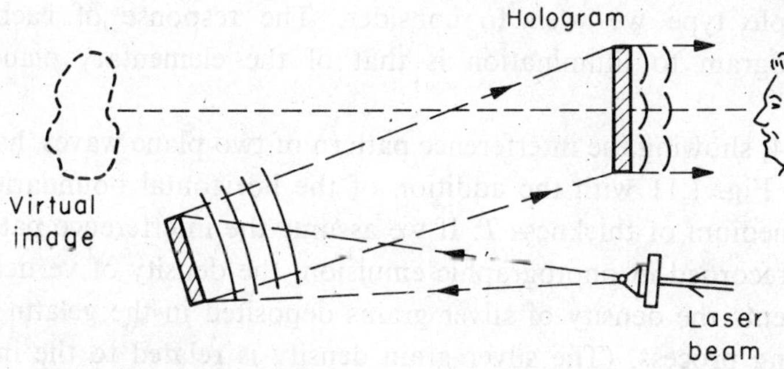

FIG. 1.9. Generation of a virtual image of the subject using the original reference beam to illuminate the hologram.

accurately retroreflected so that all rays of the reflected beam are opposite to the original reference, then such a *conjugate beam*, illuminating the back side of the hologram, produces a *real* image of the subject at the original subject location (Fig. 1.10). Because the light converges to the image, the real image can be directly detected with photographic film or photodetectors without need for a lens. Hence a hologram is a diffracting record of the interference of a particular subject and a particular reference beam. It acts as a combination record and projection system which provides an image of the original subject when illuminated by the reference and does so without the need for additional lenses.

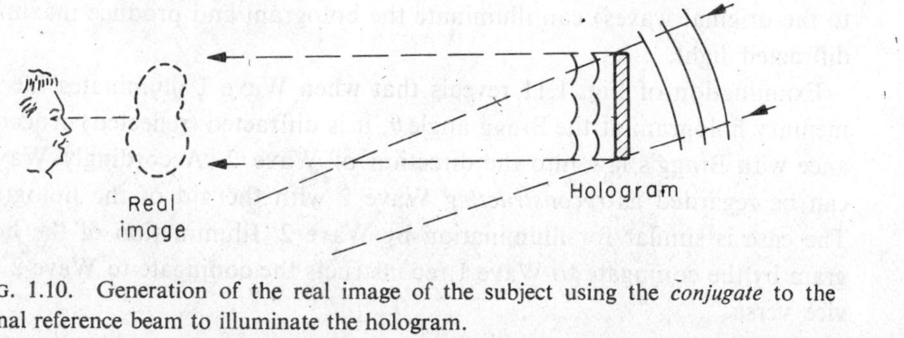

FIG. 1.10. Generation of the real image of the subject using the *conjugate* to the original reference beam to illuminate the hologram.

It is not hard to see how the wavefront reconstruction process works when *an elementary* hologram, formed by the intersection of two *plane waves*, is properly illuminated. Before considering the problem, we can, in a qualitative way, establish its relevance to the practical case where the subject beam is complex and carries information. To do this, consider an arbitrary subject illuminated by laser light. The arbitrary wavefront transmitted by the subject can be Fourier analyzed into a sum of plane waves. Each will interfere with a reference plane wave to form superimposed holograms of the simple type we wish to consider. The response of each component hologram to illumination is that of the elementary plane-wave hologram.

Figure 1.4, showing the interference pattern of two plane waves, has been redrawn as Fig. 1.11 with the addition of the horizontal boundaries of a recording medium of thickness T. If we assume the interference pattern to have been recorded in photographic emulsion, the density of vertical lines now represents the density of silver grains deposited in the gelatin during the recording process. (The silver grain density is related to the intensity of the interference pattern through an exposure curve, characteristic of the particular emulsion and its processing.) We can focus our attention on the loci of maximum density, as was done in discussing Fig. 1.4, and we again find a periodically spaced set of planes, this time a set of silver scattering planes. Their spacing is given by Eq. (1.10): $2d \sin \theta = \lambda$, where θ is the angle that the wave normal of each forming plane wave makes with the recorded planes.

Comparison of Fig. 1.11 with Fig. 1.6 shows that the hologram is a volume diffraction grating. Suppose a plane wave of wavelength λ is incident on the grating. If diffraction is to be significant, the grazing angle θ, which the incident wave normal makes with the grating planes, must obey Bragg's law, Eq. (1.12): $2d \sin \theta = \lambda$. Since Eqs. (1.10) and (1.12) are identical, the conclusion to be drawn is that each of the original forming plane waves as well as their conjugates (plane waves traveling in antiparallel directions to the original waves) can illuminate the hologram and produce maximum diffracted light.

Examination of Fig. 1.11 reveals that when Wave 1 illuminates the elementary hologram at the Bragg angle θ, it is diffracted (reflected) in accordance with Bragg's law into the direction of Wave 2. Accordingly Wave 1 can be regarded as *reconstructing* Wave 2 with the aid of the hologram. The case is similar for illumination by Wave 2. Illumination of the hologram by the conjugate to Wave 1 reconstructs the conjugate to Wave 2 and vice versa.

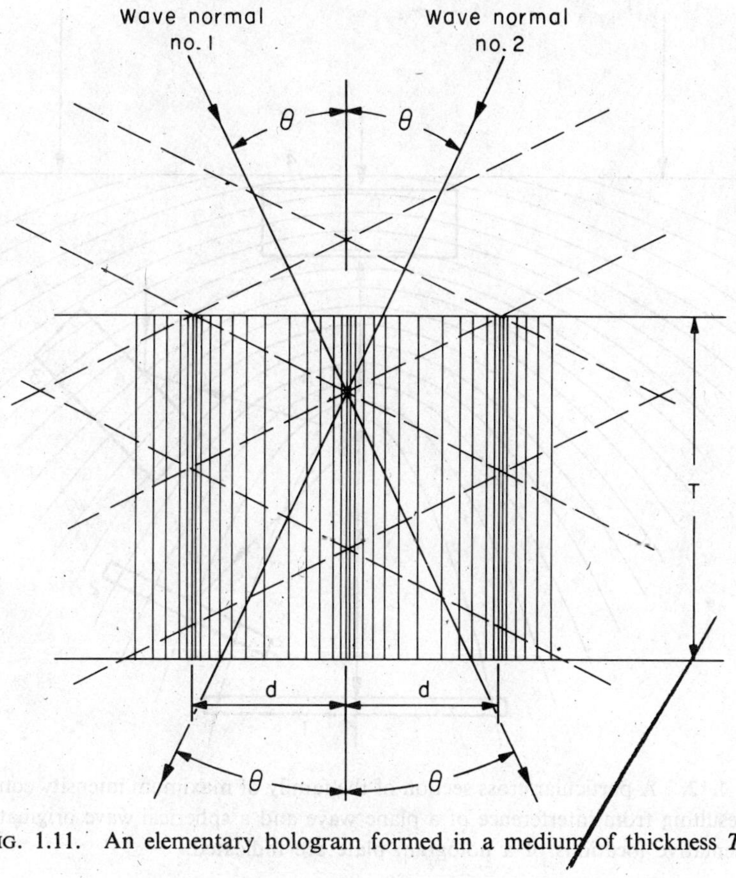

FIG. 1.11. An elementary hologram formed in a medium of thickness T.

1.7 Plane and Volume Hologram-Formation Geometries

To show how the hologram-formation *geometry* affects the diffraction properties of the hologram, let us consider the interference of a spherical wave, derived from a point source S a finite distance from the recording medium, and a reference plane wave traveling downward in Fig. 1.12. The point source is an elementary subject and the spherical wave is our subject wave. (As will become apparent later, more complicated subjects can be regarded as aggregates of elementary point sources; each of these sources interacts with the reference wave but their mutual interaction can for the moment be disregarded.) Figure 1.12 is a particular cross section of the family of maximum intensity contour surfaces resulting from the interference of the plane wave reference and the point source subject wave. In this cross section, which contains the point source and the plane wave normal, the traces of the contour surfaces are parabolas. If we consider any

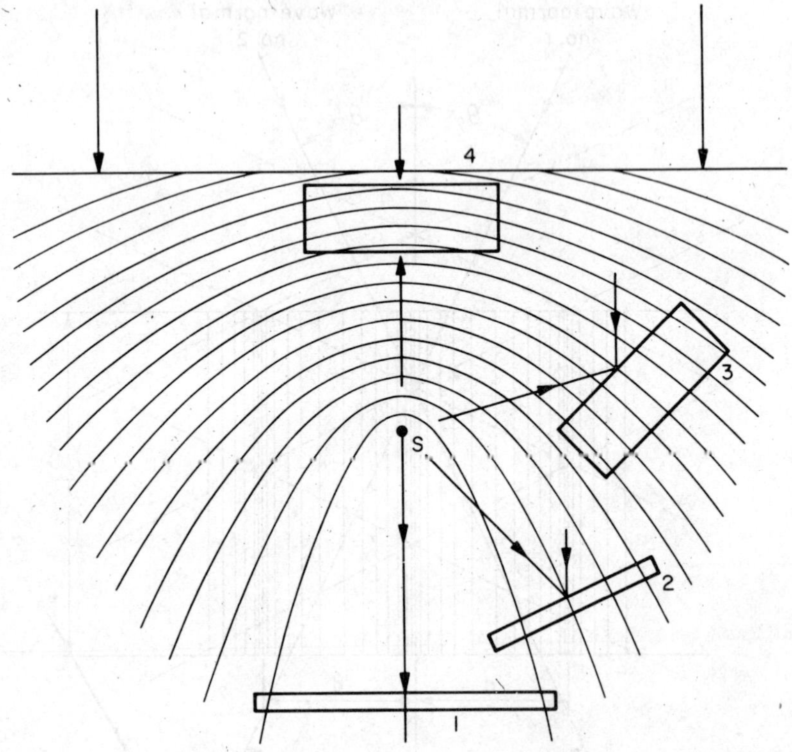

FIG. 1.12. A particular cross section of the family of maximum intensity contour surfaces resulting from interference of a plane wave and a spherical wave originating at S. Representative locations of a hologram plate are indicated.

small portion of the spherical subject wave interfering with the plane wave reference, the resulting interference pattern will be similar to that shown in Fig. 1.4. The appropriate ray from source S, the plane wave normal, and the spacing of the maximum intensity contours all satisfy Eq. (1.10).

Some representative locations of the hologram plate are indicated by Positions 1, 2, 3, and 4. Gabor (1948), unaided by the laser and forced to make the best of light sources with low coherence, formed his holograms in Position 1. Here the mean direction of the light from point S and the direction of the reference wave are collinear, and the holograms so formed have been called "in-line" holograms. A point scatterer placed in the plane wave at S can produce the desired spherical wave, and the remaining unscattered light is the plane reference wave. For interfering waves derived in this manner, Position 1 minimizes the difference in the paths taken by subject and reference light, making the best of low coherence. The relatively large spacing between contours eases the requirement on resolution capability of the recording medium.

Leith and Upatnieks (1962) [*1.2*] formed "off-axis" holograms with a beam arrangement generally equivalent to placing the hologram plate in Position 2. Using laser light, they were able to tolerate greater differences in path length for light traveling from source to hologram via the subject beam as against light traveling the path of the reference beam. The off-axis configuration overcomes problems associated with in-line holography, and the greater coherence of the laser light allows reconstruction of 3D images. It is, perhaps, this latter result of Leith's and Upatnieks' work which has most contributed to the appeal of holography and its revitalization. In Position 2 the mean direction of the light from the point source makes an acute angle with the reference wave direction. Providing the thickness of the recording medium is small compared with the spacing of the contours, the hologram formed in this position acts as a *plane* diffraction grating. An incident ray, in this case, may interact with only one contour before passing through the medium. The recording, therefore, is essentially that of a line pattern on a surface. The same is true, of course, of Gabor's in-line holograms.

For holograms formed in Position 3, the angle between the mean direction of the light from the point source and the plane wave direction is approximately 90°, and the spacing between the interference contour surfaces, according to Eq. (1.10), is smaller. Providing that the hologram thickness T is greater than that of the contour spacing, the recording must be considered as a *volume* diffraction grating. van Heerden (1963) [*1.4*] described the diffraction properties of a volume hologram and treated the hologram as a series of partially reflecting planes which respond selectively to incident light in accordance with Bragg's law. Holograms formed in Position 3 have been called "Bragg-effect" holograms; Pennington and Lin (1965) [*1.5*] used the selective Bragg diffraction properties to overcome the color cross-modulation problems of planar holography and form the first multicolor-imaging hologram.

In Position 4, where the hologram plate receives the plane wave from one side and the spherical wavefront from the other, the separation of the interference surfaces becomes nearly $\lambda/2$, and the surfaces approach planes parallel to the hologram plate surface. Denisyuk (1962) [*1.6*] first described hologram formation with this configuration. Recording of the interference pattern in silver-halide emulsion results in a stack of closely spaced, partially reflecting silver planes which acts as a resonant reflective filter. Even in emulsions which are but 12–15 μm thick, some 50–100 silver planes can form. Because the planes are similar to those formed with Lippmann color photography methods and because they diffract in accordance with Bragg's law, such volume holograms are called *Lippmann–Bragg* holograms. They are also called *reflection* holograms, since the illuminating wave seems to reflect into the subject wave direction during the wave-reconstruction process. Stroke and Labeyrie (1966) [*1.7*] have demonstrated that, as wavelength filters, they are selective enough to allow the use of white-light illumination for reconstruction of the subject wave. This has led to the term *white-light* holograms.

Holograms are characterized by features other than the reference-to-subject beam angle used in their formation. Subsequent chapters will analyze the properties of lens and lensless Fourier transform holograms, Fraunhofer holograms, image holograms, and other variations of the basic record.

1.8 Basic Holography Equations

The basic concepts of holography, which have been introduced in preceding sections by referring to specific waveforms and beam configurations, can be formulated in more general terms. Consider the situation in Fig. 1.13 where Objects 1 and 2 may, in the general case, reflect light diffusely. Each

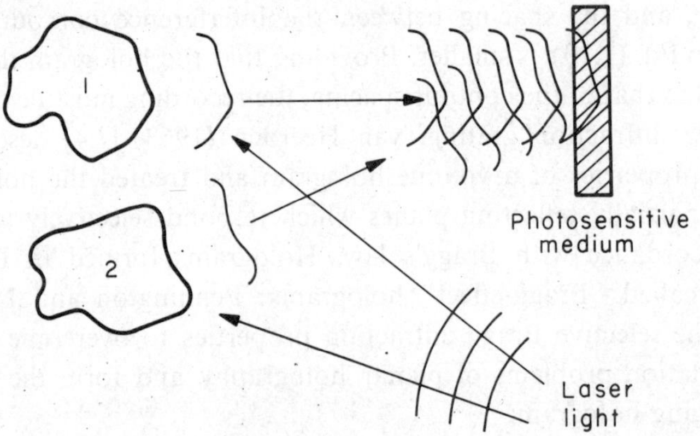

FIG. 1.13. A general hologram-forming situation.

is illuminated by coherent light from the same source. Their reflections interfere in a region of space where a photosensitive plate has been located. In both the formation and the reconstruction process we shall be concerned only with complex light amplitudes in the immediate vicinity of the photosensitive plate. The complex amplitude of the light arriving at the plate from Object 1 can be expressed as $\mathbf{a}_1 = a_1 \exp(i\varphi_1)$ where a_1 and φ_1 are both functions of the spatial coordinates at the plate. Similarly, the complex amplitude of the light arriving at the plate from Object 2 is $\mathbf{a}_2 = a_2 \exp(i\varphi_2)$. The complex conjugates of $\mathbf{a}_1$ and $\mathbf{a}_2$ will be designated $\mathbf{a}_1{}^*$ and $\mathbf{a}_2{}^*$.

If the exposure of the photosensitive plate to the interference pattern formed by $\mathbf{a}_1$ and $\mathbf{a}_2$ has been properly controlled and the plate properly developed, and if the hologram is of the absorption type, then we shall find (in Section 2.6.2) that the transmittance t of the completed hologram (the ratio of light amplitude transmitted by the hologram to that incident on it) contains a term t_{E} proportional to the exposure $E = I_{\mathrm{P}}\tau_{\mathrm{e}}$ and hence proportional to the intensity I. (The basic aspects of holography, which we wish to deal with here, can be revealed most simply by confining our attention to plane absorption holograms. Phase and volume holograms will be considered in Chapters 7–9.) Summing the amplitudes $\mathbf{a}_1$ and $\mathbf{a}_2$ and multiplying by the complex conjugate of the sum as in Eqs. (1.7) and (1.8), we may write for the intensity

$$I = (\mathbf{a}_1 + \mathbf{a}_2)(\mathbf{a}_1 + \mathbf{a}_2)^*. \tag{1.13}$$

We assume a linear relation between t and E, and consequently between t and I, of the form

$$t = \frac{\text{light amplitude transmitted}}{\text{light amplitude incident}} = t_0 - t_{\mathrm{E}} = t_0 - kI \tag{1.14}$$

where t_0 is the transmittance of the unexposed plate. Experimental t–E curves in Figs. 10.7 and 10.8 show this to be a reasonable assumption for a moderate range of exposure. An approximation containing nonlinear terms is given in Eq. (12.1). The uniform transmittance t_0 and the constant k are of no interest here; the essential part of the transmittance is that proportional to the spatially varying function I which will diffract incident light. Expanding Eq. (1.13),

$$\begin{aligned} I &= \mathbf{a}_1\mathbf{a}_1{}^* + \mathbf{a}_2\mathbf{a}_2{}^* + \mathbf{a}_1\mathbf{a}_2{}^* + \mathbf{a}_1{}^*\mathbf{a}_2 \\ &= I_1 + I_2 + \mathbf{a}_1\mathbf{a}_2{}^* + \mathbf{a}_1{}^*\mathbf{a}_2 \end{aligned} \tag{1.15}$$

where the symbols I_1 and I_2 have been substituted for the individual wave intensities.

Suppose we wish to reconstruct the wavefront which originally arrived at the plate from Object 2 by illuminating the hologram with the wavefront from Object 1 (Fig. 1.14). The complex light amplitude, transmitted to the back side of the hologram, is the product of $\mathbf{a}_1$, the illumination, times t, the hologram transmittance. The significant terms in the product are the complex amplitudes of the waves diffracted by the exposure-dependent transmittance t_E and given by

$$\mathbf{w} \propto \mathbf{a}_1 I = \mathbf{a}_1(I_1 + I_2) + \mathbf{a}_1\mathbf{a}_1\mathbf{a}_2{}^* + I_1\mathbf{a}_2 \tag{1.16}$$

where we have replaced an equal sign and the constant k [in Eq. (1.14)] by the proportionality symbol $\propto$. Each of the terms in Eq. (1.16) represents the complex amplitude of a wave emerging from the hologram. Of interest is the last term where, if I_1 is constant, the diffracted wave is a proportional reconstruction of that wave which arrived at the hologram from Object 2 during the forming of the hologram. Observation of the reconstructed diverging wave permits the viewer to see a virtual image of Object 2, provided the other diffracted waves represented in Eq. (1.16) do not interfere.

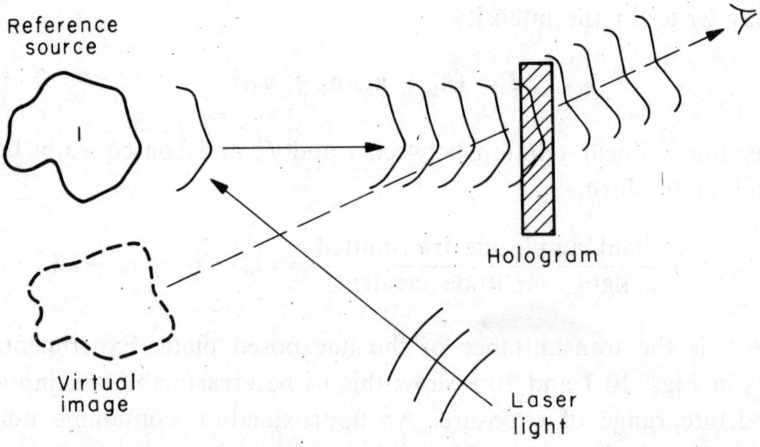

FIG. 1.14. Reconstruction of the wavefront from Object 2 of Fig. 1.13 by illuminating the hologram with the wavefront from Object 1. A virtual image of Object 2 can be seen.

To be able to view the image of Object 2 without overlap of light diffracted by the other waves of Eq. (1.16) was a major problem for early workers in holography. They were limited by the poor coherence of their light and found it difficult or impossible to employ solutions which were later made feasible with the advent of the laser. For example, lateral displacement of the two sources of the interference (Objects 1 and 2 in Fig. 1.13) and the use of light diffusers came about only after the revival of holography in the early 1960s.

If Objects 1 and 2 do indeed reflect light diffusely, we may regard the unfocused individual intensities of the reflected light at the hologram as macroscopically constant, i.e., the spatial variations of I_1 and I_2 in the hologram plane are so rapid as not to be resolved by the eye. I_1 and I_2 can then be considered real constants, and the first and last terms on the right of Eq. (1.16) represent proportional reconstructions of the original waves coming from Objects 1 and 2. Since Objects 1 and 2 do not overlap in Fig. 1.13, their virtual images will not overlap either. The second term on the right of Eq. (1.16) represents a diffuse wave which does not form an image. It can however contribute a uniform background to the images generated by the other waves, reducing their contrast.

Object 1 in Fig. 1.14 may be regarded as the reference source and $\mathbf{a}_1$ the complex amplitude of the reference wave. Rather than a diffuse wave, it is common and more practical to make the reference wave plane or spherical. This may be done by replacing Object 1 in both Figs. 1.13 and 1.14 with a plane mirror. The intensity of the light reflected to the hologram by the mirror, I_1, is essentially constant over the hologram plane, and the subject wave from Object 2 is again reconstructed. The practical feature stems from the fact that reflection of laser light from *any* plane mirror can provide an illuminating beam which duplicates the reference. Otherwise the original Object 1 must always be used to carry out the reconstruction. In Chapter 8 (see also Section 2.6), following Leith and Upatnieks [*1.2*], we show that when the angle between the unmodulated reference beam direction and the mean direction of the subject light is sufficiently great, an undisturbed virtual image of Object 2 can be obtained. Early workers, who placed both the source of the unmodulated reference wave and the subject on the same normal to the (in-line) hologram and who used *nondiffusing* subjects, found that the second term on the right of Eq. (1.16) generated a real image of the subject. Observation of either virtual or real image was disturbed by superposition of out-of-focus light coming from the other. Leith's and Upatnieks' off-axis or off-set-reference method eliminated this obstacle to the development of holography.

A real image of Object 2 can in the general case best be obtained by illuminating the hologram with a wave which is the conjugate to the reference wave. One wave is conjugate to another when in any plane its complex amplitude is the complex conjugate of the other and all its rays are directed antiparallel to the other's. The conjugate wave to the reference is easy to form if the reference is a plane wave. It is another plane wave directed antiparallel to the original. From Eq. (1.15) we have at the plane of the hologram in this case

$$\mathbf{w}' \propto \mathbf{a}_1{}^*I = \mathbf{a}_1{}^*(I_1 + I_2) + I_1\mathbf{a}_2{}^* + \mathbf{a}_1{}^*\mathbf{a}_1{}^*\mathbf{a}_2. \qquad (1.17)$$

The second term on the right is proportional to the complex amplitude of a wave *conjugate* to the original wave which diverged from Object 2; here it is a *converging* wave (with all rays oppositely directed to the corresponding rays in the original). The wave converges to a real image of Object 2, but because of the conjugate nature of the wave, the image is pseudoscopic (it has inverted depth and peculiar parallax properties, see Section 8.1). Whether the several diffracted waves overlap depends on the choice of angle between the mean directions of the subject and reference beams.

1.9 Partial Coherence

Thus far we have discussed the properties of light waves and holography under the assumption of perfectly coherent light. We have assumed a single point source emitting one infinitely long monochromatic wave train bearing spherical or plane wavefronts. Under these conditions the phase difference for two fixed points along a ray direction is time independent, or equivalently, the difference in the phase measured at a single point in space at the beginning and end of a fixed time interval Δt does not change with time. These are statements of perfect *temporal coherence*. Similarly, the phase difference for two fixed points in a plane normal to a ray direction is time independent. The latter is a statement of perfect *spatial* or *lateral coherence*.

Real light sources have only a degree of coherence. Despite this, the results we have obtained in previous sections, using the idealization of perfect coherence, are still relevant. However the practical limitation does mean we must reconsider the time-averaging process involved in expressing the intensity of an interference pattern. Of particular concern is the interference term [see Eq. (1.9) of Section 1.3.1]. The main effect of partial coherence is that the depth of modulation of the intensity in the interference pattern is reduced [the factors multiplying the cosine term in Eq. (1.9) are smaller] so that the minimum intensity no longer goes to zero. The consequence for holography is reduced diffraction efficiency over that obtainable with perfectly coherent light (see Chapter 7).

1.9.1 Coherence Length and Temporal Coherence

A useful concept in gauging temporal coherence is the *coherence length* of a light source. Suppose a source emits a monochromatic wave train of finite length l and suppose that instantaneous values of the wave amplitude can simultaneously be measured at two points z_1 and z_2 along the wave normal. If $\Delta z = z_2 - z_1$ is slightly less than l, then for a brief period the propagating monochromatic wave train can exist at both z_1 and z_2, and for that brief period it would appear that the source is temporally coherent. When the wave train moves on, so that it no longer spans the gap Δz, the evidence of temporal coherence vanishes. The separation $\Delta z = l$, for which one can just observe some degree of constancy of relative phase as a function of time, is a measure of the coherence length of the wave source. (A more relevant definition of *coherence length* will be given in Chapter 7.) The length l can be expressed as the product $l = c\,\Delta t$, where Δt is the time during which the source emits without interruption and c is the wave velocity. In holography, as we shall see, the coherence length places an upper limit on the allowable path difference between subject and reference beam. It, in effect, limits the subject depth.

Coherence length may be expressed in other terms by analyzing the single-frequency wave train of time duration Δt into its Fourier components. The complex electric field $\mathbf{v}$ of a pulse of duration Δt and frequency f_0, can be represented as

$$\begin{aligned} \mathbf{v} &= a_0 \exp(2\pi i f_0 t) \qquad \text{for} \quad -\Delta t/2 < t < \Delta t/2, \\ &= 0 \qquad \text{otherwise} \end{aligned}$$

where a_0 is a constant and t represents the time. (We have arbitrarily picked the phase φ to be zero.) The intensity spectrum of the light is proportional to the square of the Fourier transform of $\mathbf{v}$. We may write the latter as

$$\begin{aligned} |\mathbf{V}(f)|^2 &= \frac{1}{\Delta t}\left[\int_{-\Delta t/2}^{\Delta t/2} a_0 \exp(2\pi i f_0 t) \exp(-2\pi i f t)\, dt\right]^2 \\ &= (a_0^2\,\Delta t)\left[\frac{\sin \pi(f-f_0)\,\Delta t}{\pi(f-f_0)\,\Delta t}\right]^2. \end{aligned} \tag{1.18}$$

The first nulls of this well-known $(\sin x/x)^2$ function (Fig. 1.15) occur when $f - f_0 = \pm 1/\Delta t$. The half-width of the central maximum of the function can be taken as the bandwidth Δf, so that $\Delta f = 1/\Delta t$. If we substitute $\Delta t = l/c$ into the expression for the bandwidth Δf, we obtain a useful relation between coherence length and bandwidth,

$$l = c/\Delta f. \tag{1.19}$$

Since the monochromatic Fourier components are infinitely long wave trains, the notion of limited coherence implied in this approach has to do with waves of many different frequencies falling in and out of step with one another. This process will be important when the observation points z_1 and z_2 are separated by a large distance, so that appreciable phase shifts can accumulate between z_1 and z_2.

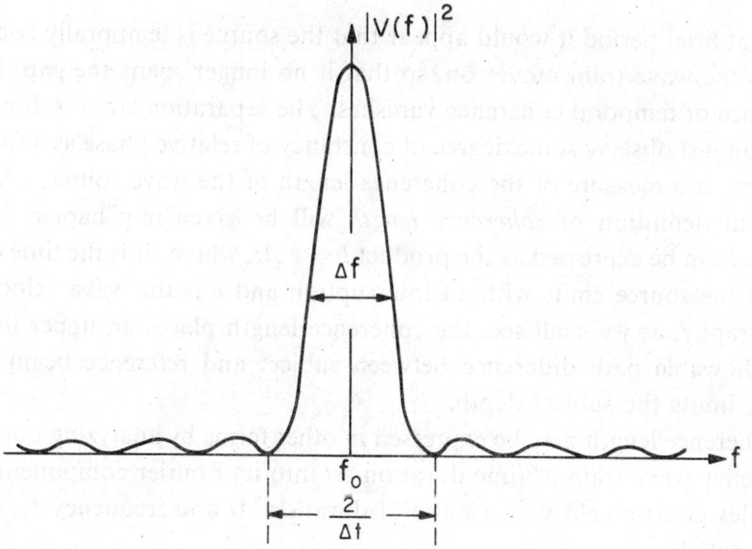

FIG. 1.15. Intensity spectrum of a single-frequency pulse of duration Δt. Bandwidth is defined as $\Delta f = 1/\Delta t$.

The concept of pulses of single frequency is somewhat more realistic than that of infinitely long wave trains. Thermal light sources consist of atomic oscillators emitting a series of finite wave trains of random length and at random times although their frequencies are altered by their thermal motions or by local fields. However, if the emission spectrum can be described by a bell-shaped curve of appropriately defined bandwidth Δf, then the relation $l \approx c/\Delta f$ turns out to be one practical definition of coherence length of thermal sources. From this definition it is clear that only those thermal light sources which emit over very narrow bandwidths Δf can have the long coherence lengths desirable for holography. (Coherence of laser sources is discussed in Chapters 7 and 11.)

Temporal coherence can be related to the fringe contrast of an interference pattern. Amplitude division methods of beam sampling (e.g., the Michelson interferometer using beam splitters and mirrors) allow phase comparison of the amplitude of a plane wave at various points along the direction of propagation (Fig. 1.16). Tilting one mirror facilitates the comparison by causing plane wave samples of the beam to intersect, as in Fig. 1.4, and produce the linear fringe system whose intensity in an observation plane is given by Eq. (1.15):

$$I = I_1 + I_2 + \mathbf{a}_1\mathbf{a}_2^* + \mathbf{a}_1^*\mathbf{a}_2 .$$

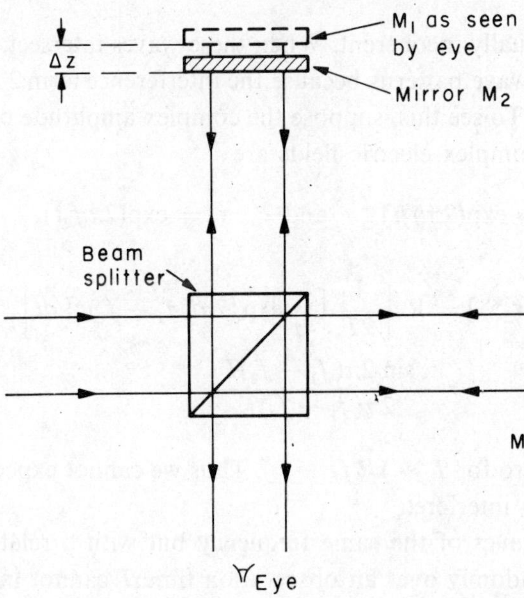

FIG. 1.16. Indicating now a Michelson interferometer is used to divide the amplitude of a beam and allow phase comparison of the wave at two positions along the beam separated by Δz.

A more general expression for the intensity, suitable for light having only a degree of coherence, is obtained by substituting the complex electric fields $\mathbf{v}_1$ and $\mathbf{v}_2$ for the complex amplitudes $\mathbf{a}_1$ and $\mathbf{a}_2$ and by adding the time-average brackets [as in Eq. (1.1)]. We then have

$$\begin{aligned} I &= I_1 + I_2 + \langle \mathbf{v}_1\mathbf{v}_2{}^* + \mathbf{v}_1{}^*\mathbf{v}_2 \rangle \\ &= I_1 + I_2 + 2\mathrm{Re}[\langle \mathbf{v}_1\mathbf{v}_2{}^* \rangle]. \end{aligned} \tag{1.20}$$

We expect the time-average process to produce different results in the case of partial coherence from that obtained in the case of perfect coherence. This will assert itself in the *visibility* V of the fringes defined by Michelson as

$$V = \frac{I_{\max} - I_{\min}}{I_{\max} + I_{\min}}. \tag{1.21}$$

When $I_1 = I_2$, a maximum value $V = 1$ indicates perfect coherence [see Eq. (1.9)]. Since the interference fringes form an intensity standing wave pattern, V is also an intensity standing wave ratio.

Before any further consideration of partial coherence, we may use Eq. (1.20) to show that two monochromatic waves of substantially different frequency are mutually incoherent. When these waves intersect, they do not produce standing wave patterns because the interference term $2\,\mathrm{Re}[\langle \mathbf{v}_1\mathbf{v}_2^*\rangle]$ is essentially zero. To see this, suppose the complex amplitude of each wave is unity and the complex electric fields are

$$\mathbf{v}_1 = \exp(2\pi i f_1 t) \qquad \text{and} \qquad \mathbf{v}_2 = \exp(2\pi i f_2 t).$$

Then

$$\mathrm{Re}[\langle \mathbf{v}_1\mathbf{v}_2^*\rangle] = \mathrm{Re}\left[\frac{1}{2T}\int_{-T}^{T} \exp\{2\pi i(f_1 - f_2)t\}\, dt\right]$$
$$= \frac{\sin 2\pi(f_1 - f_2)T}{2\pi(f_1 - f_2)T}$$

and approaches zero for $T \gg 1/(f_1 - f_2)$. Thus we cannot expect green and blue laser light to interfere.

Similarly two waves of the same frequency but with a relative phase φ which changes randomly over an observation time T cannot interfere. The product $\mathbf{v}_1\mathbf{v}_2^*$ becomes $\exp(2\pi i f_1 t + i\varphi)\cdot\exp(-2\pi i f_1 t) = \exp(i\varphi)$. If for short random periods during the observation period T the phase φ takes on all possible values between 0 and 2π, then the mean value of the phase factor $\exp(i\varphi)$ is zero (see random-walk statistics in Section 12.3.3). Thus, for example, light waves from two He–Ne lasers which are not phase related do not interfere.

1.9.2 Complex Degree of Coherence

The general subject of partial coherence is treated extensively in Chapter 10 of Born and Wolf [1.8]. In our Chapter 7 where we discuss more quantitatively the coherence properties of laser light sources as well as thermal sources, we require some of the results found in the reference. Rather than reproduce proofs of these results we simply introduce and summarize the necessary relations here as basic concepts. Their use is confined primarily to Chapters 7 and 11 where CW laser and pulsed-laser sources, respectively, are considered.

Suppose, as in Fig. 1.17, that light waves emerge from pinholes P_1 and P_2 in an opaque screen and interfere on an observation screen S. An extended partially coherent source illuminates the pinholes. The intensity at any point Q on S is represented by Eq. (1.20) where $\mathbf{v}_1$ and $\mathbf{v}_2$ are the complex electric fields of the waves from P_1 and P_2 arriving at Q. It is shown by Born and Wolf [1.8] that the time average of the interference term

$$\langle \mathbf{v}_1\mathbf{v}_2^* + \mathbf{v}_1^*\mathbf{v}_2\rangle = 2\,\mathrm{Re}[\langle \mathbf{v}_1\mathbf{v}_2^*\rangle]$$

can be written in terms of a *complex degree of coherence* $\gamma_{12}(\tau)$. $\gamma_{12}(\tau)$ relates the correlation of the electric fields at P_1 and P_2 to the time average of the interference term at Q. Since the latter is measurable in terms of the visibility, $\gamma_{12}(\tau)$ has the merit of expressing coherence as a measurable quantity. We now show this to be so.

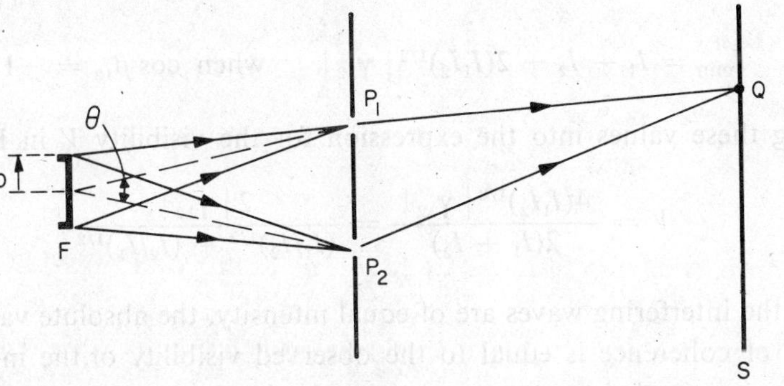

FIG. 1.17. Double-aperture experiment with an extended source.

Suppose $\mathbf{v}_{P_1}(t)$ and $\mathbf{v}_{P_2}(t)$ are the complex electric fields at P_1 and P_2 while $2\langle \mathbf{v}_{P_1}(t)\mathbf{v}^*_{P_1}(t)\rangle$ and $2\langle \mathbf{v}_{P_2}(t)\mathbf{v}^*_{P_2}(t)\rangle$ are the corresponding light intensities. Then the complex degree of coherence $\gamma_{12}(\tau)$ is defined by Born and Wolf as the normalized correlation of $\mathbf{v}_{P_1}(t)$ and $\mathbf{v}_{P_2}(t)$

$$\gamma_{12}(\tau) = \frac{\langle \mathbf{v}_{P_1}(t+\tau)\mathbf{v}^*_{P_2}(t)\rangle}{[\langle \mathbf{v}_{P_1}(t)\mathbf{v}^*_{P_1}(t)\rangle\langle \mathbf{v}_{P_2}(t)\mathbf{v}^*_{P_2}(t)\rangle]^{1/2}}$$

$$= \frac{\lim_{T\to\infty} \frac{1}{2T}\int_{-T}^{T} \mathbf{v}_{P_1}(t+\tau)\mathbf{v}^*_{P_2}(t)\,dt}{\left[\left(\lim_{T\to\infty}\frac{1}{2T}\int_{-T}^{T}\mathbf{v}_{P_1}(t)\mathbf{v}^*_{P_1}(t)\,dt\right)\left(\lim_{T\to\infty}\frac{1}{2T}\int_{-T}^{T}\mathbf{v}_{P_2}(t)\mathbf{v}^*_{P_2}(t)\,dt\right)\right]^{1/2}}. \tag{1.22}$$

They then relate γ_{12} to 2 $\mathrm{Re}[\langle \mathbf{v}_1\mathbf{v}_2{}^*\rangle]$ through the equation

$$\begin{aligned} 2\,\mathrm{Re}[\langle \mathbf{v}_1\mathbf{v}_2{}^*\rangle] &= 2(I_1I_2)^{1/2}\,\mathrm{Re}[\gamma_{12}(\tau)] \\ &= 2(I_1I_2)^{1/2}\,|\gamma_{12}(\tau)|\cos\beta_{12}(\tau) \end{aligned} \tag{1.23}$$

where I_1 and I_2 represent the individual intensities of light arriving at Q from P_1 and P_2, respectively, τ is the difference in light transit time between P_1 and Q as compared to that between P_2 and Q, and β_{12} is the phase of $\gamma_{12}(\tau)$. The quantity $\gamma_{12}(\tau)$ is a measure of the coherence between light at P_1 and light at P_2 and encompasses both temporal and spatial aspects. We can now substitute Eq. (1.23) into Eq. (1.20) to get the maximum and minimum values of I:

$$I_{\max} = I_1 + I_2 + 2(I_1 I_2)^{1/2} |\gamma_{12}| \qquad \text{when } \cos\beta_{12} = 1$$

and

$$I_{\min} = I_1 + I_2 - 2(I_1 I_2)^{1/2} |\gamma_{12}| \qquad \text{when } \cos\beta_{12} = -1.$$

Placing these values into the expression for the visibility V in Eq. (1.21)

$$V = \frac{4(I_1 I_2)^{1/2} |\gamma_{12}|}{2(I_1 + I_2)} = \frac{2|\gamma_{12}|}{(I_1/I_2)^{1/2} + (I_2/I_1)^{1/2}}. \tag{1.24}$$

When the interfering waves are of equal intensity, the absolute value of the degree of coherence is equal to the observed visibility of the interference fringes.

1.9.3 Spatial Coherence

When $\tau \to 0$, the visibility of the fringes obtained in the double-aperture arrangement of Fig. 1.17 measures essentially the degree of *spatial* coherence between P_1 and P_2. If the extended source F consists of a number of uncorrelated spatially separated oscillators, then each oscillator illuminating the double hole produces an independent interference pattern of periodic fringes on the observation screen S. Whether a maximum, minimum, or intermediate value of the fringe intensity is observed at Q for a given pattern depends on the phase of the light radiated to P_1 compared to the phase of the light radiated to P_2 by the same oscillator. This in turn depends on the oscillator location within the source and on the angle θ subtended at the source by the distance $\overline{P_1P_2}$. Thus it is possible for some oscillators to produce at Q fringes whose maxima fall on minima of patterns produced by other oscillators. For these cases the resultant intensity is uniform and the visibility zero. Hence, an extended aggregate of uncorrelated oscillators is a poor source for the purposes of interferometry and holography, both of which depend on fringe recording.

The van Cittert–Zernike theorem (see Chapter 10 of Born and Wolf [*1.8*]) formally relates the degree of spatial coherence to the lateral extent of a source through a Fourier transform relationship. (In optics we are generally concerned with Fourier transforms which relate the signal distribution over spatial coordinates to the distribution over the spatial frequencies or angular directions. The topic of Fourier transforms will be covered in Chapter 4.) We shall merely state the theorem here and defer further discussion. For the case of an extended source containing mutually incoherent oscillators, but with a narrow spectral bandwidth $\Delta\nu$, the van Cittert–Zernike theorem can be expressed as follows: *When a small source illuminates two closely spaced points located in a plane a long distance from the source, the degree of coherence between the complex electric fields at the two points is given by the magnitude of a normalized Fourier transform of the intensity distribution of the source.* For a circular source of radius r_0 and of uniform intensity, the degree of spatial coherence $|\mu_s|$ can be plotted as in Fig. 1.18, the

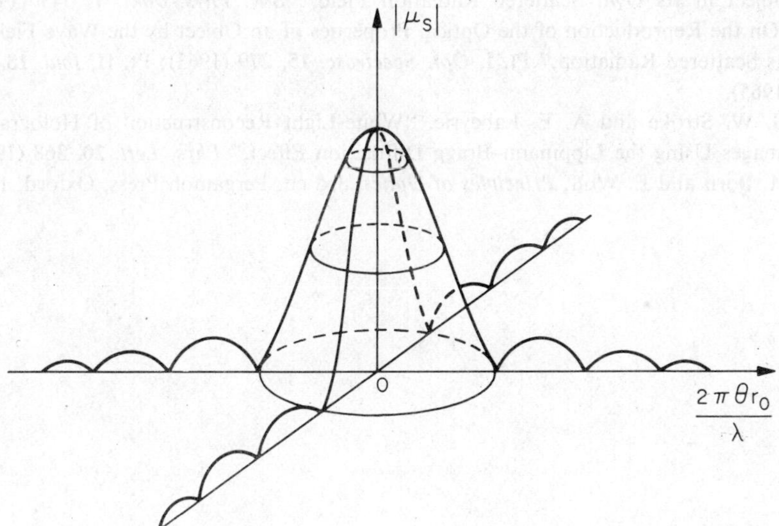

FIG. 1.18. The degree of spatial coherence $|\mu_s|$ as a function of source radius and the angle at the source subtended by the sampling points.

parameters being θ, the angle subtended at the source by the two points, and r_0, the radius of the source. We see that in the region of the central maximum, the smaller the source and the farther away the source (i.e., the smaller θ), the greater the degree of spatial coherence.

REFERENCES

1.1. D. Gabor, "A New Microscopic Principle," *Nature* **161**, 777 (1948); "Microscopy by Reconstructed Wavefronts," *Proc. Roy. Soc.* **A197**, 454 (1949); "Microscopy by Reconstructed Wavefronts: II," *Proc. Phys. Soc.* **B64**, 449 (1951).

1.2. E. N. Leith and J. Upatnieks, "Reconstructed Wavefronts and Communication Theory," *J. Opt. Soc. Amer.* **52**, 1123 (1962); "Wavefront Reconstruction with Continuous-Tone Objects," *J. Opt. Soc. Amer.* **53**, 1377 (1963); "Wavefront Reconstruction with Diffused Illumination and Three-Dimensional Objects," *J. Opt. Soc. Amer.* **54**, 1295 (1964).

1.3. W. L. Bragg, "The Diffraction of Short Electromagnetic Waves by a Crystal," *Proc. Cambridge Phil. Soc.* **17**, 43 (1912).

1.4. P. J. van Heerden, "Theory of Optical Information Storage in Solids," *Appl. Opt.* **2**, 393 (1963).

1.5. K. S. Pennington and L. H. Lin, "Multicolor Wavefront Reconstruction," *Appl. Phys. Lett.* **7**, 56 (1965).

1.6. Yu. N. Denisyuk, "Photographic Reconstruction of the Optical Properties of an Object in Its Own Scattered Radiation Field," *Sov. Phys.-Dokl.* **7**, 543 (1962); "On the Reproduction of the Optical Properties of an Object by the Wave Field of Its Scattered Radiation," Pt. I, *Opt. Spectrosc.* **15**, 279 (1963); Pt. II, *Ibid.* **18**, 152 (1965).

1.7. G. W. Stroke and A. E. Labeyrie, "White-Light Reconstruction of Holographic Images Using the Lippmann–Bragg Diffraction Effect," *Phys. Lett.* **20**, 368 (1966)

1.8. M. Born and E. Wolf, *Principles of Optics*, 3rd ed. Pergamon Press, Oxford, 1964.

Chapter 2

EARLY HOLOGRAPHY

Holography was intended as a tool for the electron microscopist to enable him to form images of subjects having atomic dimensions. Just prior to the invention of holography, in 1947, the resolving power of an electron microscope objective lens was the order of 10 Å, and it was felt that spherical aberration would set a theoretical limit at about 5 Å. It occurred to Gabor that the aberrated image produced by the lens still preserved all of the subject information, albeit in a somewhat coded form. If the aberrated image could somehow be decoded, the resolution limit of the electron microscope might be reduced to perhaps 1 Å, and atomic structure might be made observable.

Gabor decided to dispense with the electron microscope objective altogether and perform a decoding operation on a photographic record of unfocused electron waves diffracted by the subject. The electron wave record, or hologram, is decoded by illuminating it with coherent visible light whereupon part of the illuminating wave is diffracted by the hologram. Waves arising out of this second diffraction are the optical equivalent of the unfocused electron waves. They yield a magnified optical image of the original subject. To get the desired result, the light beam illuminating the hologram must be an accurately scaled imitation of the electron beam. The scale factor is given by the ratio of the wavelength of light to that of the electron waves. If the hologram itself is similarly enlarged, a *distortion-free* optical image is obtained with magnification equal to the wavelength ratio.

To test his theory, Gabor *formed* the first hologram with visible light instead of electron waves. Although this move into the optical range proved to be a step in the right direction, its impact could be fully appreciated only after laser sources became available.

2.1 Bragg's X-Ray Microscope

The inspiration for this two-step, lensless, wave-reconstruction microscopy came from the field of X-ray crystallography where investigation of the atomic arrangement in crystals had been pursued for some time despite the lack of an X-ray lens. As early as 1929 Sir Lawrence Bragg had been able to produce an optical image of atomic structure in the crystal *diopside* [*2.1*]. His method was one of optical Fourier synthesis. As suggested in Section 1.6, any spatial distribution of light—in particular, the optical image of atoms in a crystal—can be analyzed into Fourier components, i.e., into sets of sinusoidal fringes. The superposition of the fringe sets is the image. Bragg was aware that normally gathered X-ray diffraction data contained the amplitudes and the orientations of these sinusoidal fringes. His task then was to form the fringes in accordance with the data and superimpose them on a photographic plate. The work led to what he called an "X-ray microscope" [*2.2, 2.3*]. Gabor, in his first paper on holography [*2.4*] acknowledges this X-ray microscope as a starting point for his own thinking. A brief consideration of the method will reveal the basis upon which Gabor built holography.

Bragg's first success in producing a discernible image of atomic structure resulted from a straightforward application of the sinusoidal fringe superposition principle. A photograph of a set of opaque cylindrical rods (with their axes parallel to the plane of the photograph and spaced by a distance twice their diameters) was projected, out of focus, onto a screen. The resulting pattern on the screen was an approximation to a sinusoidal fringe pattern. On the screen was placed a suitably shuttered "photographic paper." This was sequentially exposed to 40 such fringe patterns whose spacing and orientation were determined by the observed X-ray spectral data for diopside. When the multiply exposed photographic paper was developed, an image of the atomic structure in the crystal was observed on the paper. The multiple exposure resulted in a rather low-contrast image, and Bragg sought a method which could simultaneously form the entire set of required fringes on the image plane. He selected Young's double-aperture arrangement for forming sinusoidal fringes by light interference. An opaque plate, perforated with appropriately drilled holes and illuminated with coherent light, produced the desired multiple-fringe superposition and indeed improved the image.

An important aspect of the work was the recognition by Bragg of the fundamental relation between his experimental efforts and the imaging principle of Abbé. Abbé, in 1873, described the formation of an optical image by a lens in terms of a double diffraction process. His theory is most easily explained when the subject to be imaged is a one-dimensional grating (Fig. 2.1). In the first step of the process, a plane wave illuminates the grating, and the lens produces in its back focal plane an array of bright point foci S_i, the far-field or Fraunhofer diffraction pattern of the grating. The second step of the process results from considering these foci as sources of a second diffraction pattern formed at the image plane. Light diverging

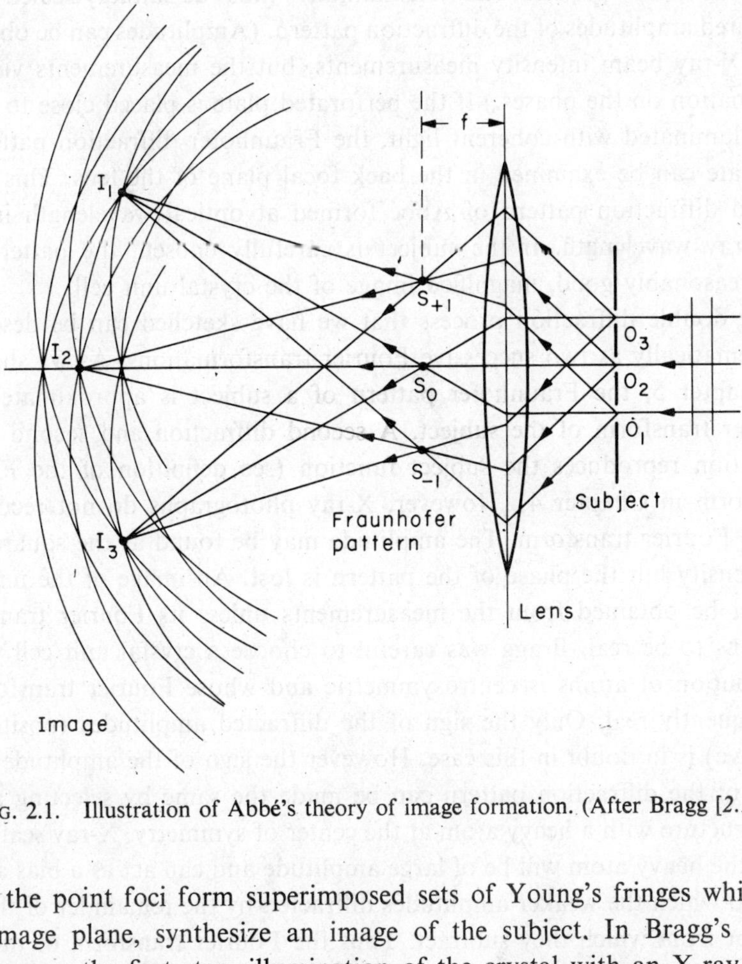

FIG. 2.1. Illustration of Abbé's theory of image formation. (After Bragg [*2.1*].)

from the point foci form superimposed sets of Young's fringes which, at the image plane, synthesize an image of the subject. In Bragg's X-ray microscopy the first step, illumination of the crystal with an X-ray beam to acquire the spectral data, also leads to an array of points. If the crystal is properly oriented in the incident beam and the diffracted X rays selectively detected, then the X rays may appear to scatter from a projection of atoms in the unit cell onto a plane. In this case the far-field X-ray pattern of the structure of the unit cell is a set of point foci on a plane. The connection with Abbé's theory is made by perforating the opaque plate in a manner which replicates this far-field X-ray pattern. Coherent, visible-light illumination of the plate (the second step) then allows the holes to act as secondary sources of a second diffraction pattern which, according to Abbé, forms the image at optical wavelengths.

To actually accomplish image formation requires holes to be accurately drilled in positions corresponding to those of the points of the X-ray Fraunhofer diffraction pattern. The hole diameters must be suitably scaled to the measured amplitudes of the diffraction pattern. (Amplitudes can be obtained from X-ray beam intensity measurements, but the measurements yield no information on the phases.) If the perforated plate is placed close to a lens and illuminated with coherent light, the Fraunhofer diffraction pattern of the plate can be examined in the back focal plane of the lens. This is the second diffraction pattern of Abbé formed at optical wavelength instead of X-ray wavelength. If the subject is carefully chosen, the pattern will be a reasonably good, magnified image of the crystal unit cell.

The double diffraction process that we have sketched can be described mathematically as two successive Fourier transformations. As we shall see in Chapter 5, the Fraunhofer pattern of a subject is approximately the Fourier transform of the subject. A second diffraction and second transformation reproduces the subject function (see definition of the Fourier transform in Chapter 4). However, X-ray photographs do not record all of the Fourier transform. The amplitude may be found as the square root of intensity but the phase of the pattern is lost. An image of the unit cell cannot be obtained from the measurements unless its Fourier transform happens to be real. Bragg was careful to choose a crystal unit cell whose distribution of atoms is centrosymmetric and whose Fourier transform is consequently real. Only the sign of the diffracted amplitudes (positive or negative) is in doubt in this case. However the sign of the amplitude in all parts of the diffraction pattern can be made the same by selecting a unit cell structure with a heavy atom at the center of symmetry. X-ray scattering from the heavy atom will be of large amplitude and can act as a bias amplitude to which the weaker amplitudes diffracted by the remainder of the cell add, or from which they subtract. Thus the Fourier transform of the unit cell modulates a uniform background amplitude but with a depth insufficient to change sign. Diopside satisfies the requirements for a heavy center of symmetry. For more general subjects that do not have a center of symmetry, the Fourier transform has a spatially varying phase factor. Loss of phase information in recording the first diffraction pattern can no longer be tolerated if an image is to result from the second diffraction.

2.2 Holography for Electron Microscopy

Gabor added to Bragg's two-wavelength, double diffraction process the means for simultaneously recording both the phase and the amplitude distribution of the diffraction pattern. With the new concept came a more general interpretation of the imaging process and the possibility of application to more general subjects. Yet even the new principle, at least in the form Gabor deemed necessary for electron microscopy, placed restrictions on the shape of the subject. Gabor required a subject with large transparent areas so that, when electron waves coherently illuminated the subject from the rear, a strong undiffracted component passed through to the recording plate. Interference of this undiffracted or background wave with waves diffracted by the edges of opaque areas of the subject produces the intensity pattern to be recorded as a hologram. The undiffracted wave provides a reference phase to which the spatial phase variation of the diffracted waves can be compared. The relative phase is encoded in the cosine factor of the interference term [see Eq. (1.9)]. Optical decoding or wave reconstruction (the second diffraction) takes place upon illumination of the magnified hologram with a suitably scaled optical facsimile of the undiffracted wave.

Gabor was able to accompany the first report of his "New Microscopic Principle" [*2.4*] with an all-optical verification of it. Figure 2.2 shows some early results. Visible light was used in *both* the recording and reconstructing stages. However, despite considerable subsequent effort, mainly by Haine *et al.* [*2.5*, *2.6*], the application of holography to electron microscopy (electron wave formation and optical reeonstruction) proved to be an exercise in futility. The best resolution obtained with the method was inferior to the then current state of the art. Haine and Mulvey [*2.6*], in the early 1950s, experimented first with the original hologram-forming configuration of Gabor called the *projection method* and then with a more advantageous

FIG. 2.2. Results of some early optical holography by Gabor. Left: The original subject. Center: The hologram. Right: The reconstruction. [Courtesy D. Gabor, *Research* **4**, 107 (1951).]

scheme called the *transmission method* [2.5]. With either method one attempts to obtain simultaneously (1) a high degree of spatial coherence so that interference fringe visibility will be high, (2) a fringe spacing which is easily resolved in the recording, and (3) a small separation between hologram and source to keep exposure time to a minimum.

The projection method, shown schematically in Fig. 2.3, requires the subject to be placed very close to the electron wave source and relatively distant from the recording plate. As we shall show in Chapter 3 [see Eq. (3.9) in Section 3.1.1], the fringe spacing is maximized (the spatial frequency of the fringes is minimized) with this arrangement. However, a position too close to the source means that the angle θ subtended by the subject at the source is large. If spatial coherence is to be high the product of the source radius and the angle θ must be small (see Fig. 1.18). An effective source for resolving 1 Å is achieved by demagnifying an ordinary electron beam source by a factor of 10^6 using electron lenses. The penalty is high, however; severe distortions of the beam shape take place which cannot be compensated for by optical imitation in the reconstruction process. These distortions limit the usable field of fringes and thus the aperture of the resulting hologram. The limited aperture of the hologram, in turn, limits its image resolution capability.

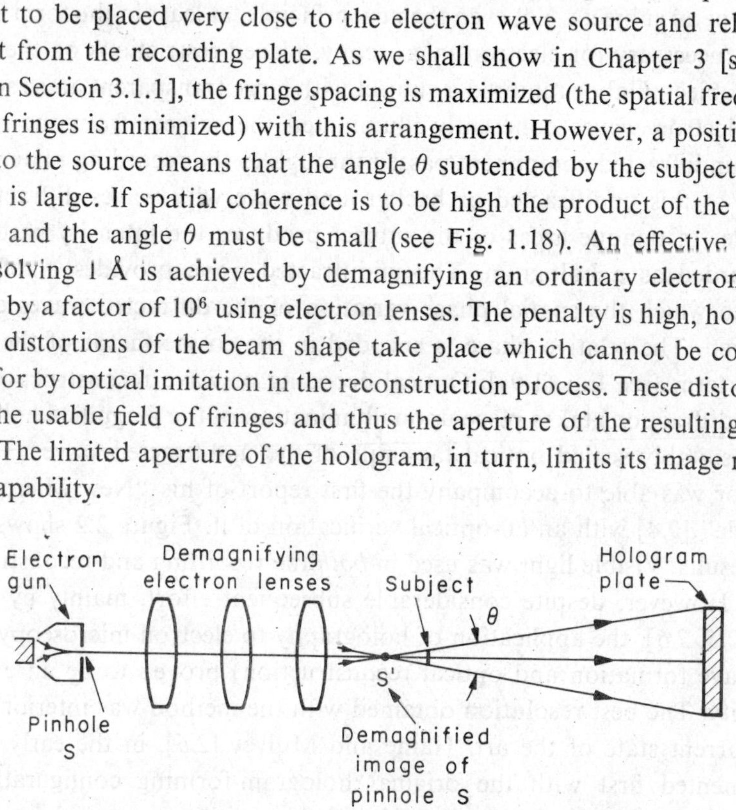

FIG. 2.3. The projection method of forming an electron-wave hologram.

The transmission method (Fig. 2.4) uses a larger effective source. To achieve sufficient spatial coherence, the source and subject are separated by a distance large enough to make θ small. At the same time, the hologram is desired close to the subject so as to minimize exposure time. However, the fringes in the near-field or Fresnel diffraction pattern are closely spaced and require magnification by an electron lens system before the photographic film can record them. In essence, the arrangement is a defocused transmission electron microscope with the subject in the normal position in front of the objective. A slit system at the electron source is added to control the spatial

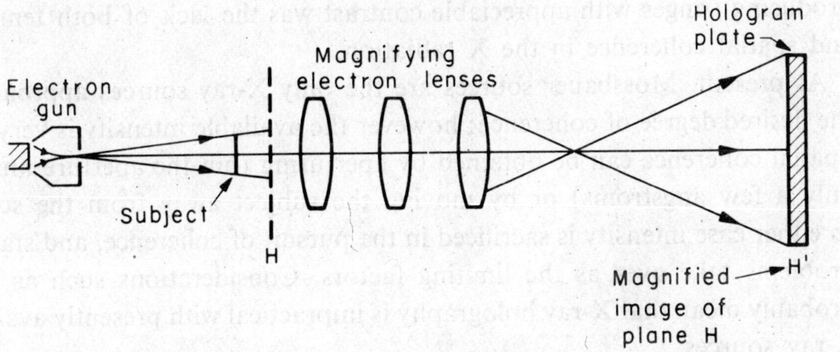

FIG. 2.4. The transmission method of forming an electron-wave hologram.

coherence. The main advantage of the transmission method is that it avoids the severe beam distortions consequent to demagnification. Despite this advantage, it is still the achievable degree of spatial coherence that is the major obstacle to attaining high resolution. To gain the desired spatial coherence the source must be apertured down to about 1 μm^2. This results in low beam intensity and requires exposure times as long as 2 hr. Instabilities due to voltage drift, thermal motions, vibrations, and subject contamination then combine to limit hologram fringe resolution and consequently image resolution.

2.3 X-Ray Holography

Gabor did not suggest that holography should be applied to X-ray microscopy, because it seemed even more difficult to get an intense source of coherent X rays than to obtain a suitably coherent electron source. Others, e.g., Baez, El-Sum, and Kirkpatrick, motivated by the lack of an X-ray lens, did carry out X-ray hologram experiments. El Sum and Kirkpatrick (1952) were able to generate a visible image of a thin wire by illuminating an X-ray diffraction pattern which had been recorded by Kellstrom 20 years earlier [*2.7*, *2.8*]. This decidedly two-step wavefront reconstruction process proved to be their only significant experimental success with X-ray holograms. Attempts in the 1950s to demonstrate the utility of wavefront reconstruction X-ray microscopy failed to produce X-ray diffraction patterns with more than one useful interference fringe. (A single fringe is insufficient for reconstructing the original wavefront. Kellstrom had obtained four of five good contrast fringes on either side of a central unusable area.) Primarily responsible for the difficulty in producing fringes with appreciable contrast was the lack of both temporal and spatial coherence in the X radiation.

At present, Mossbauer sources are the only X-ray sources approaching the desired degree of coherence; however the available intensity is very low. Spatial coherence can be obtained by aperturing (but the aperture must be only a few angstroms) or by moving the subject away from the source. In either case intensity is sacrificed in the pursuit of coherence, and stability problems take over as the limiting factors. Considerations such as these probably mean that X-ray holography is impractical with presently available X-ray sources.

2.4 Beginnings of Optical Holography

First to investigate *optical* holography for its own sake was Rogers [*2.9*]. Working for the most part with a high-pressure mercury arc lamp, Rogers, in 1952, reported a series of experiments which were forerunners of much of the holographic investigation carried out more than a decade later with the laser. Some of the more interesting results were:

1. A hologram made of a hologram (a method now used to copy holograms).
2. Three-dimensional image generation (a possibility recognized by Gabor).
3. Image subtraction by superposition of "negative" and "positive" holograms.
4. Relief, phase holograms (for high diffraction efficiency).
5. An unsuccessful attempt at forming a multicolor-imaging, composite hologram (using color selective dyes).
6. Initiation of work on a computed hologram (a forerunner of recent activity in computer-generated holograms).

Rogers, moreover, pursued Gabor's suggestion that the hologram of a point source is closely related to a Fresnel zone plate. The zone-plate analogy is a useful way of understanding the imaging properties of a hologram and can be derived through simple geometric considerations ([*2.9*], see also El-sum [*2.8*]).

2.4.1 Geometric Analysis of an Elementary Gabor Hologram

As discussed in Section 2.2, Gabor required hologram subjects to be small opaque areas on a relatively large transparent background. In the analysis to follow, the subject is an idealization of Gabor's subjects, namely a single point scatterer of light. The temporal and spatial coherence properties of the illuminating source are also idealized by assuming a point source of monochromatic spherical waves. Finally, the photosensitive medium recording the hologram is assumed to be thin. With these assumptions the arrangement analogous to Gabor's experiment is shown in Fig. 2.5. A point source S, a distance v from the hologram plane H, illuminates the scattering center P which is a distance u from H. We wish to derive an expression for the intensity of the light pattern on the hologram plane H due to interference of light scattered by P with the background light.

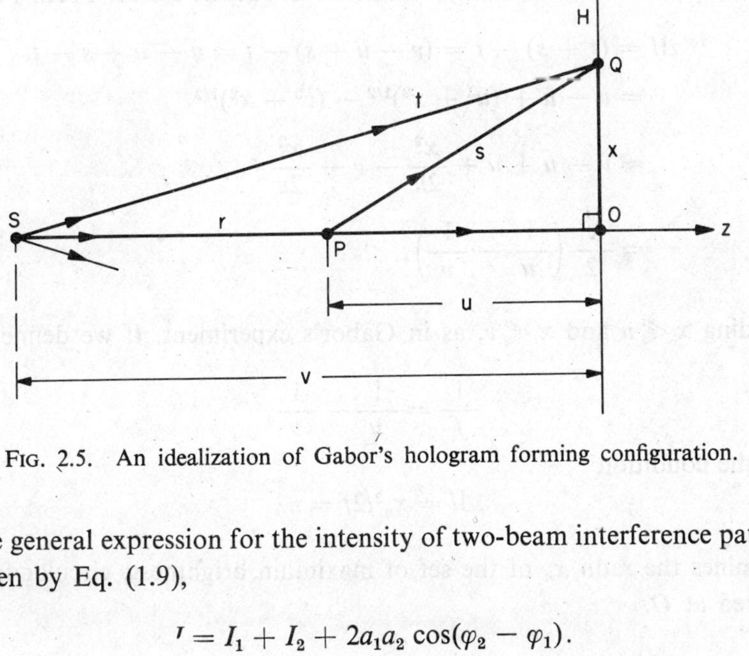

FIG. 2.5. An idealization of Gabor's hologram forming configuration.

The general expression for the intensity of two-beam interference patterns is given by Eq. (1.9),

$$I = I_1 + I_2 + 2a_1a_2 \cos(\varphi_2 - \varphi_1).$$

We assume a hologram is formed whose amplitude transmittance is proportional to I. Diffraction of light illuminating the hologram will result from the spatial variation of the relative phase $\Delta\varphi = \varphi_2 - \varphi_1$ (a_1 and a_2 are nearly constant over the hologram plane). At some point Q on the hologram, $\Delta\varphi$ can be expressed in terms of the difference in path taken by light traveling from S to Q directly (reference wave) compared to that traveling from S to Q via P (subject wave). Assume a continuously oscillating source producing a light wave (of wavelength λ and frequency F) whose absolute phase $\Phi = 2\pi Ft$ is a linear function of the time. A wavefront arriving at

Q at time t_Q has an absolute phase proportional to the time when it was emitted. If the wave velocity is c, then the phase of the wavefront arriving at Q at time t_Q via path SPQ is $\Phi_s = 2\pi F(t_Q - (SPQ/c))$. Similarly, the phase of the wavefront simultaneously arriving at Q via SQ is $\Phi_r = 2\pi F(t_Q - (SQ/c))$. Since $SPQ > SQ$, $\Phi_r > \Phi_s$ and

$$\Phi_r - \Phi_s = \frac{2\pi F}{c}(SPQ - SQ) = \varphi_2 - \varphi_1 = \Delta\varphi = \frac{2\pi\, \Delta l}{\lambda}.$$

Whenever Δl equals $n\lambda$, where $n = 1, 2, 3, \ldots$, $\cos \Delta\varphi = 1$, and the interference fringe has its maximum intensity. The analysis can be confined to the xz plane of Fig. 2.5 since, with S and P each located on the z axis, the intensity pattern is rotationally symmetric about the z axis. From Fig. 2.5

$$\begin{aligned}\Delta l &= (r + s) - t = (v - u + s) - t = v - u + s - t \\ &= v - u + (u^2 + x^2)^{1/2} - (v^2 + x^2)^{1/2} \\ &\approx v - u + u + \frac{x^2}{2u} - v - \frac{x^2}{2v} \\ &= \frac{x^2}{2}\left(\frac{1}{u} - \frac{1}{v}\right),\end{aligned}$$

providing $x \ll u$ and $x \ll v$, as in Gabor's experiment. If we define

$$\frac{1}{f} = \frac{1}{u} - \frac{1}{v}, \tag{2.1}$$

then the condition

$$\Delta l = x_n^2/2f = n\lambda \tag{2.2}$$

determines the radii x_n of the set of maximum brightness, circular fringes centered at O.

2.4.2 The Zone Plate

Solving for x_n in Eq. (2.2), the bright fringes have radii

$$x_n = (f\lambda)^{1/2} \cdot (2n)^{1/2}, \tag{2.3}$$

which are proportional to the square roots of the even integers. The condition, Eq. (2.3), is identical to that defining the transmitting zones of a Fresnel zone plate [*2.10*]. [A zone plate (see Fig. 2.6) can be constructed simply by drawing on white paper concentric circles whose radii are proportional to the square roots of consecutive integers 1, 2, 3,.... These form annular zones, every other one of which is to be blackened. This done, the figure is photographically reduced to a suitable size, and the resulting transparency is the zone plate [2.10].] Since Eq. (2.3) governs the periodicity of the zone plate as well as that of the point source hologram, one would expect their light-diffracting properties to be similar. This is true except that the zones of the zone plate are a square-wave version of the sinusoidal fringes of the hologram grating. A sinusoidal grating diffracts the incident beam only into the +1 and −1 orders, the fundamental directions, while the square-wave grating diffracts the beam into higher orders, or harmonics, as well. Thus the diffraction properties of the Gabor hologram can be inferred from the known properties of the zone plate [2.11] providing we confine our interest to first-order diffraction.

FIG. 2.6. Drawing of a Fresnel zone plate.

The zone plate is a diffraction grating with focusing properties. It is at once both a positive and negative lens. The quantity f in Eq. (2.2) is the focal length of the zone-plate lens and also of the hologram. Equation (2.1) is a *focal equation* defining the lens-to-image distance u in terms of f and the lens-to-subject distance v. As in Fig. 2.7 when a point source S illuminates a zone plate and only first-order diffraction is considered, two images are formed: One is a virtual image P from which the +1 order diffracted waves appear to diverge and the other is a real image P' to which the -1 order diffracted waves converge. By analogy a point-subject hologram also behaves as a diffraction grating with focusing properties and as a negative and positive lens producing virtual and real images. Suppose the hologram is illuminated by the point source S in its original position, a distance v from the hologram. According to Eq. (2.1) the hologram will form a virtual image of S at a distance u from the hologram. The point image can be considered a virtual image of S or better, a virtual image of the original scattering center P, since the image is located at the original site of P. If we choose the latter view, then the spherical wave diffracted from the hologram and appearing to diverge from P can be considered a reconstruction of the wave scattered from P.

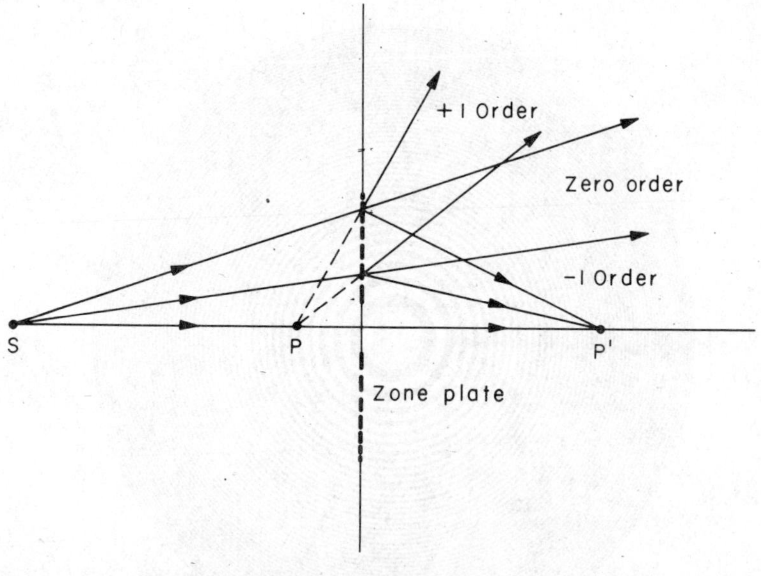

FIG. 2.7. Focusing properties of a zone plate.

The zone-plate behavior of point-subject holograms can be used to explain the imaging properties of more general holograms. A normal, extended subject for a hologram may be thought of as an aggregate of point subjects. Scattered light from each of the points interferes with a reference wave to produce a superposition of many zone-plate-like holograms. (It is assumed that the scattered waves are weak relative to the reference so their mutual interaction can be neglected.) When the entire hologram is illuminated with the reference, each individual hologram generates a virtual image of its associated point subject and, in the process, they combine to image the extended subject.

2.5 In-Line Holograms

Despite a suggestion by Gabor that in the optical regime "where beam splitters are available, methods can be found for providing the coherent background which will allow better separation of object planes and more effective elimination of the twin wave," optical holography in the 1950s continued to be carried out with his original in-line arrangement. Lack of a good source of coherent light probably discouraged experimentation. Source and subject were placed on an optic axis, defined by the normal to the photographic plate. To reveal some of the difficulties and limitations of the method, we return to the analysis initiated by Gabor and used in Section 1.8. There attention is centered on the amplitudes of the arriving and departing waves at the hologram plane.

Suppose we consider an on-axis subject, suitable for a Gabor hologram, to be illuminated with coherent light. The *total* complex amplitude $\mathbf{u}$ of the light striking the photographic plate at the hologram plane can be expressed as a complex function of spatial coordinates where $\mathbf{u} = u_0 \exp(i\varphi_u)$. Part of $\mathbf{u}$ is the undiffracted background or reference wave $\mathbf{r} = r_0 \exp(i\varphi_r)$, and part is the light diffracted by the subject $\mathbf{a} = a_0 \exp(i\varphi_a)$. Thus

$$\mathbf{u} = \mathbf{r} + \mathbf{a}, \tag{2.4}$$

and the intensity I at the hologram plane is

$$\begin{aligned} I &= \mathbf{u}\mathbf{u}^* = (\mathbf{r} + \mathbf{a})(\mathbf{r} + \mathbf{a})^* \\ &= r_0^2 + a_0^2 + \mathbf{r}\mathbf{a}^* + \mathbf{r}^*\mathbf{a} \\ &= r_0^2 + a_0^2 + 2r_0a_0 \cos(\varphi_r - \varphi_a). \end{aligned} \tag{2.5}$$

2.5.1 Response of Photographic Emulsion

Hologram formation on photographic plates requires that photographic emulsion be exposed to the light of intensity I, developed, fixed, and then illuminated by the reference light to produce an image. We inquire as to the response of emulsion to these operations. Hurter and Driffield in 1890 characterized the photographic plate by a curve (called the H&D curve) which is a plot of the *optical density* of the developed plate versus the logarithm of the exposure. They defined the optical density D as

$$D = \log(1/\mathcal{T})$$

where $\mathcal{T}$ is the intensity transmittance, the ratio of the intensity of light transmitted by the emulsion to that incident on it. The exposure E is defined as

$$E = I_P \tau_e \propto I \tau_e$$

(see Section 1.3) where τ_e is the exposure time and I_P or its abbreviated form I [defined in Eq. (1.1)] is the intensity of the light pattern to which the plate had been exposed. A general form of the H&D curve is shown in Fig. 2.8. Although well suited to photography and used by early holographers, the curve is not the best way to characterize emulsion response for hologram formation (see Section 2.6.2).

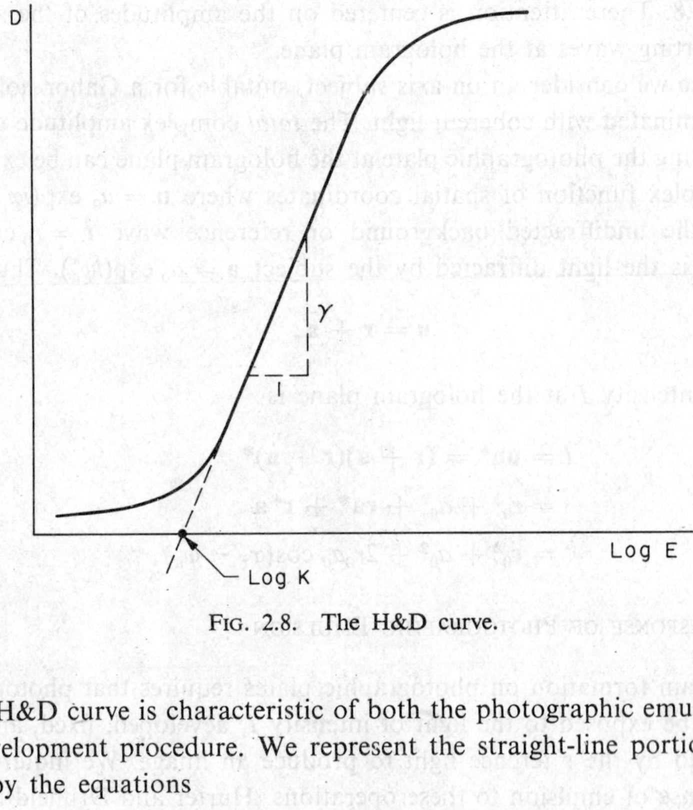

FIG. 2.8. The H&D curve.

The H&D curve is characteristic of both the photographic emulsion and the development procedure. We represent the straight-line portion of the curve by the equations

$$D = \log(1/\mathcal{T}) = \gamma(\log E - \log K),$$
$$\log(1/\mathcal{T}) = \log(E/K)^{\gamma}, \tag{2.6}$$

or

$$\mathcal{T} = K^{\gamma} E^{-\gamma} \propto E^{-\gamma} \propto I^{-\gamma}.$$

In the above, γ is the slope and $\log K$ is the intercept of the linear portion of the H&D curve extended to the log E axis. For coherent-light illumination we are generally interested in the *amplitude* transmittance t rather than $\mathscr{T}$, where $t = \sqrt{\mathscr{T}}$, so that

$$t \propto I^{-\gamma/2}. \tag{2.7}$$

Suppose the photographic emulsion, exposed to the intensity I, is developed as a negative. Then the amplitude transmittance is given by

$$t_n \propto I^{-\gamma_n/2}.$$

If, on the other hand, the negative is printed as a positive, the printing illumination producing the positive exposure is proportional to $\mathscr{T}_n$, the intensity transmittance of the negative, and the resulting amplitude transmittance of the print is

$$t_p = (\mathscr{T}_p)^{1/2} \propto (\mathscr{T}_n)^{-\gamma_p/2} \propto (I^{-\gamma_n})^{-\gamma_p/2} \propto I^{\Gamma/2}, \quad \text{where} \quad \gamma_n\gamma_p = \Gamma. \tag{2.8}$$

The subscript "p" refers the previously defined parameters to the positive print.

2.5.2 The Reconstruction

Gabor actually did make a positive hologram. He then illuminated the positive with the original reference wave $\mathbf{r}$. The complex amplitude of the light just after passing through the hologram is

$$\mathbf{w} = \mathbf{r}t_p \propto \mathbf{r}I^{\Gamma/2}.$$

If by proper development Γ is made equal to 2, then

$$\mathbf{w} \propto \mathbf{r}I = \mathbf{r}(r_0{}^2 + a_0{}^2 + \mathbf{ra}^* + \mathbf{r}^*\mathbf{a}) = r_0{}^2\mathbf{r} + \mathbf{r}a_0{}^2 + \mathbf{rra}^* + r_0{}^2\mathbf{a}$$

$$\mathbf{w} \propto r_0{}^2(\mathbf{r} + \mathbf{a}) + r_0{}^2\big((a_0{}^2/r_0)\exp(i\varphi_r)\big) + r_0{}^2\big(\exp(i2\varphi_r)\mathbf{a}^*\big) \tag{2.9}$$

where $\mathbf{rr}^* = r_0{}^2$ and $\mathbf{r} = r_0 \exp(i\varphi_r)$. Providing the reference wave amplitude is uniform over the hologram plate ($r_0{}^2 =$ constant), the first term on the right of Eq. (2.9) represents a wavefront whose complex amplitude is proportional to that of the original wave $\mathbf{u}$ in Eq. (2.4). If, further, the reference wave is so strong that $a_0{}^2/r_0 \ll 1$, then the second term can be neglected. Finally for a reference wave whose phase is nearly constant over the hologram plane (as, for example, in the transmission method of forming Gabor holograms), the third term is proportional to the conjugate of the subject wave complex amplitude. It generates a conjugate twin image of the subject.

If the transmission method (Fig. 2.4) is used, the twin image is real (see Section 3.3.1). In this case viewing through the hologram toward the illuminating source, one sees the illuminating source, a virtual image of the subject, and a conjugate real image of the subject. (Recall that an in-line hologram behaves as a set of superimposed zone plates, and the two waves generating the virtual and real images can be reconstructed simultaneously.) If the viewer focuses on the virtual image, its twin, the real image, appears out of focus. On the other hand, if a white screen is placed in the plane where the diffracted rays converge to focus the real image, then out-of-focus light from the virtual image is found to be present. This overlap of light from the twin image in the viewing direction is a handicap of the in-line method which Gabor and his successors strove to eliminate.

2.5.3 Contrast in In-Line Holograms

Before considering some of the methods employed to eliminate the twin image problem, let us inquire as to the effect on the reconstruction of choosing a value for gamma other than $\Gamma = 2$. The complex amplitude of the diffracted waves at the hologram plane can be represented in general form as

$$\begin{aligned}\mathbf{w} \propto \mathbf{r}I^{\Gamma/2} &= \mathbf{r}(r_0^2 + a_0^2 + \mathbf{r}\mathbf{a}^* + \mathbf{r}^*\mathbf{a})^{\Gamma/2}\\ &= \mathbf{r}r_0^{\Gamma}\left(1 + \frac{a_0^2}{r_0^2} + \frac{\mathbf{r}}{r_0^2}\mathbf{a}^* + \frac{\mathbf{r}^*}{r_0^2}\mathbf{a}\right)^{\Gamma/2}\\ &\approx \mathbf{r}r_0^{\Gamma}\left(1 + \frac{\Gamma}{2}\frac{a_0^2}{r_0^2} + \frac{\Gamma}{2}\frac{\mathbf{r}}{r_0^2}\mathbf{a}^* + \frac{\Gamma}{2}\frac{\mathbf{r}^*}{r_0^2}\mathbf{a} + \cdots\right).\end{aligned}$$

Assuming $r_0^2 \gg a_0^2$ so that the second term is negligible and multiplying through by the factor $\mathbf{r}$, we have

$$\mathbf{w} \propto r_0^{\Gamma}\left(\mathbf{r} + \frac{\Gamma}{2}\mathbf{a}\right) + \text{the conjugate real-image term.}$$

Focusing on the virtual image and neglecting the unfocused real-image light,

$$\mathbf{w} \propto \mathbf{r} + \frac{\Gamma}{2}\mathbf{a}. \tag{2.10}$$

Equation (2.10) tells us that the value of gamma controls the contrast, i.e., it controls the ratio of the signal amplitude to the background. For example, when $\Gamma = +2$ (positive hologram) then $\mathbf{w}$ is a reconstruction of the original wave $\mathbf{u} = \mathbf{r} + \mathbf{a}$ where the subject light adds to the background;

when $\Gamma = -2$ (negative hologram) then $\mathbf{w} = \mathbf{r} - \mathbf{a}$ and the subject light is subtracted from the background. In the latter case the result is a negative image. Values of Γ between $+2$ and -2 produce varying degrees of contrast.

2.5.4 Elimination of the Twin Image Problem

We have seen that in-line holograms, as formed by early holographers, require: (1) subjects consisting of small opaque objects on a large transparent background so that $r_0^2 \gg a_0^2$, (2) a positive print to be made of the original exposure, and (3) the overall gamma to be $+2$ to maintain original contrast in the subject. While these requirements were restrictive, it was the overlap of one unfocused image onto its twin focused image that seemed most hampering. Several early attempts were made to eliminate the unwanted image light while still maintaining the in-line geometry. The first of such attempts was the two-hologram method of Bragg and Rogers [*2.12*]. Their procedure is straightforward: A subject, suitable for an in-line hologram, is illuminated with parallel coherent light and its diffraction pattern recorded in Hologram 1. If, after development, the hologram is replaced and the subject removed, illumination by the original source generates a virtual image of the subject at the original subject position $u = -f$ [see Eq. (2.1) for $v = \infty$]. A real image is also formed at a distance f on the far side of the hologram. However, in the plane of the real image is also the unwanted diffraction pattern produced by the wave appearing to diverge from the virtual image. Since the virtual image and the subject occupied identical positions, $2f$ units distant from the real image plane, it was clear to Bragg and Rogers that the unwanted diffraction pattern was identically that of the subject observed at a distance of $2f$. They suggested that a *negative* photographic record of the subject diffraction pattern be made at a distance $2f$ from the subject. This, then, is the second hologram.

To observe the real image and simultaneously eliminate disturbance from its twin, one must illuminate Hologram 1 in its original position and register Hologram 2 with the disturbing pattern in the real image plane. The negative record on Hologram 2 should just cancel the actual pattern received from the virtual image; remaining should be the real image on a uniform background. In practice the cancellation is imperfect, and the method at best only partially successful. Moreover it introduces noise and applies only to the real image.

El-Sum [*2.7*] suggested other methods for eliminating the twin image, and Lohmann [*2.13*] proposed a method involving filtering the Fraunhofer diffraction pattern. Neither these nor the Bragg–Rogers method proved entirely effective nor were they very convenient or flexible. The problem remained one of major importance. Before considering Leith's and Upatnieks' solution to the twin image problem, we note that, in certain cases, the in-line hologram is capable of producing the desired image without undue interference from the twin image. Using the in-line method, Thompson *et al.* [*2.14*] were able to form pulsed-laser holograms of moving aerosol particles. (The available pulsed lasers lacked sufficient coherence length, so the in-line method seemed a good choice.) Aerosol particles are so small that a hologram made in the near field of the ensemble of particles is still in the far field of any individual particle. If one examines the real image of a particle, its virtual image appears to lie in a plane a long distance away. Hence, at the real image plane, the diffraction pattern of the point-like virtual image appears as an unobtrusive uniform background, a spherical wave of constant amplitude. The real image may therefore be clearly observed.

2.6 Off-Axis Holograms

By breaking with the in-line geometry concept and introducing the reference beam at an angle to the subject beam, Leith and Upatnieks devised the most general and successful method of separating the twin image (as well as undiffracted light) from the desired image. Figure 2.9 indicates the arrangement initially employed by Leith and Upatnieks [*2.15*] with a mercury discharge light source. Two lines from the Fraunhofer diffraction pattern of a line grating are selected as phase-related secondary sources.

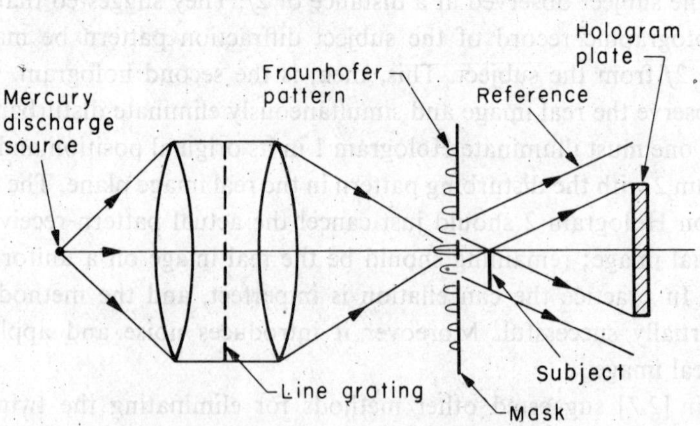

FIG. 2.9. Initial arrangement used by Leith and Upatnieks to form an off-axis hologram. [After E. Leith and J. Upatnieks, *J. Opt. Soc. Amer.* **52**, 1123 (1962).]

One of these acts as the reference wave source while the other transilluminates a subject. The hologram can be formed anywhere in the region of beam overlap. The resemblance of the experimental arrangement to that of Young is apparent.

While it is perhaps more appropriate to delay the explanation of the off-axis method till after the spatial frequency domain concept has been introduced (Chapter 5), some of the obvious benefits can be appreciated by returning to a zone-plate picture. Consider Fig. 2.10 which indicates a zone-plate pattern formed by interference of a plane wave reference with the spherical wave scattered by a point scatterer at P. The in-line procedure entails placing a small photographic plate (the dashed-line rectangle) at the center of the pattern such that its normal passes through P and is parallel to the direction of propagation of the reference plane wave. When nonlaser sources are used, the portion of the interference pattern which can be recorded as a hologram is generally confined to a small radius about the center, because of the limited coherence length of the light. As the light

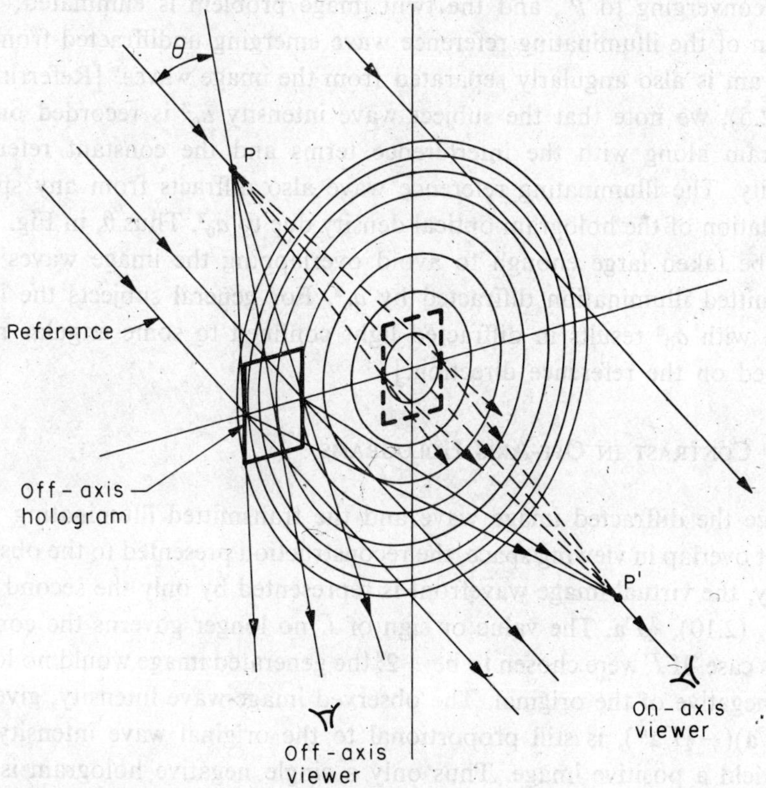

FIG. 2.10. A simple hologram recorded off axis.

path from source to hologram plane via the subject increases over the direct path via the reference wave and the difference approaches the coherence length of the source, the interference fringe visibility becomes zero and the record of the interference no longer diffracts with useful efficiency. One can see from Fig. 2.10 that the path difference increases as one proceeds out from the center of the interference pattern. The small hologram that can be formed is illuminated by the plane wave reference and diffracts a wave appearing to diverge from a virtual image at P and a second wave converging to a real image at P'. On-axis observation of either image is disturbed by the out-of-focus light from the other and by the undiffracted illumination as well.

Suppose, however, that the coherence length of the light is sufficient to produce useful-fringe visibility over a large portion of the zone-plate pattern such that the small photographic plate can be moved so far off axis that it intercepts no axial rays from P. Then, as indicated in the figure, the rays diverging from the virtual image at P are all angularly separated from those converging to P', and the twin image problem is eliminated. That portion of the illuminating reference wave emerging undiffracted from the hologram is also angularly separated from the image waves. [Referring to Eq. (2.5), we note that the subject wave intensity a_0^2 is recorded on the hologram along with the interference terms and the constant reference intensity. The illuminating reference wave also diffracts from any spatial modulation of the hologram optical density due to a_0^2. Thus θ, in Fig. 2.10, must be taken large enough to avoid overlapping the image waves with transmitted illumination diffracted by a_0^2. For general subjects the interaction with a_0^2 results in diffracted light confined to some angular range centered on the reference direction.]

2.6.1 Contrast in Off-Axis Holograms

Since the diffracted image wave and the transmitted illuminating wave do not overlap in viewing space, the reconstruction presented to the observer of, say, the virtual-image wavefront is represented by only the second term in Eq. (2.10), $\frac{1}{2}\Gamma\mathbf{a}$. The value or sign of Γ no longer governs the contrast in this case. If Γ were chosen to be -2, the generated image would no longer be a negative of the original. The observed image-wave intensity, given by $(-\frac{1}{2}\Gamma\mathbf{a})(-\frac{1}{2}\Gamma\mathbf{a}^*)$, is still proportional to the original wave intensity and will yield a positive image. Thus only a single negative hologram is necessary to produce a positive image whose contrast is that of the original subject.

2.6.2 LINEAR RESPONSE

With the advent of the off-axis method it became clear that characterization of photographic emulsion by the slope of the linear region of its H&D curve is inconvenient [*2.16*]. Of course any method of characterization must be directed toward achieving a recording which faithfully reconstructs the subject wave when illuminated by the reference wave. Fidelity is obtained when the *amplitude transmittance* of the developed hologram is linearly proportional to the intensity of the interference pattern. Then the virtual image wave, as given in Eq. (2.9), is $r_0^2\mathbf{a}$ and proportional to $\mathbf{a}$, providing r_0^2 is constant over the hologram. Therefore an exposure characteristic of significance to holography is the plot of the amplitude transmittance t versus exposure E as shown in Fig. 2.11a. A proper exposure for holography

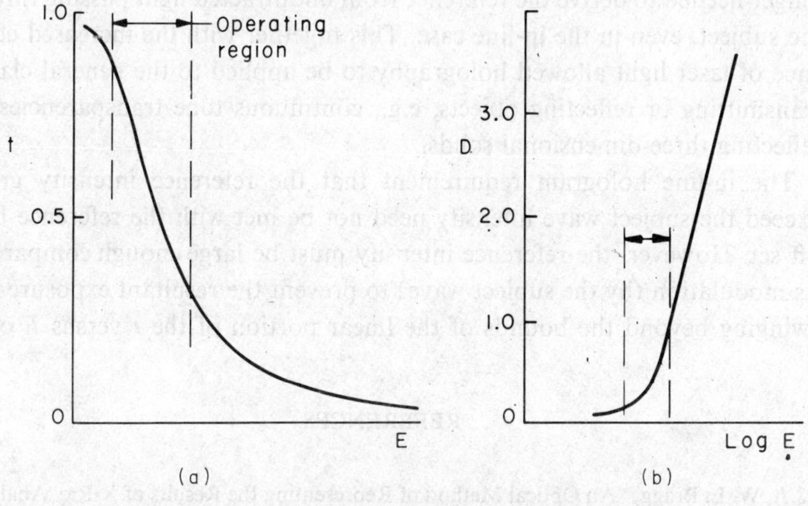

FIG. 2.11. (a) Amplitude transmittance versus exposure. (b) The H&D curve showing the location of the linear region of the curve in (a). [After H. Kogelnik, "Reconstructing Response and Efficiency of Hologram Gratings," *Proc. Symp. Mod. Opt.*, New York, 1967, p. 605. Polytechnic Press, Brooklyn, New York, 1967.]

will produce an amplitude transmittance having values within the limits of the linear portion of the curve. If the maxima or minima of the intensity pattern, to which the hologram plate is exposed, produce transmittance values corresponding to the nonlinear portion of the curve, distortions will be observed in the reconstruction. Figure 2.11b shows that the linear region of the t versus E plot for the commonly used Eastman Kodak 649F emulsion lies at the toe of the H&D curve. In this case the latter provides little guidance in selecting the exposure limits. It is interesting to note that the average exposure value, corresponding to the center of the linear region of the t versus E curve, is considerably less than that corresponding to the center of the linear region of the H&D curve. Hence, for a given emulsion, hologram exposures are generally shorter than those for normal photographs and the plates are less dense (see Fig. 1.1).

2.6.3 Further Effects of the Off-Set Reference and Increased Coherence

Introduction of the off-axis technique had the effect of freeing optical holography from restraints originating in the early work with electron waves and X rays. It became evident that by using beam splitters one no longer needed to derive the reference from undiffracted light passing through the subject, even in the in-line case. This together with the increased coherence of laser light allowed holography to be applied to the general class of transmitting or reflecting objects, e.g., continuous tone transparencies and reflecting three-dimensional solids.

The in-line hologram requirement that the reference intensity greatly exceed the subject wave intensity need not be met with the reference beam off set. However, the reference intensity must be large enough compared to its modulation (by the subject wave) to prevent the resultant exposure from swinging beyond the bounds of the linear portion of the t versus E curve.

REFERENCES

2.1. W. L. Bragg, "An Optical Method of Representing the Results of X-Ray Analyses," *Z. Kristallogr. Kristallgeometrie Kristallphys. Kristallchem.* **70**, 475 (1929).

2.2. W. L. Bragg, "A New Type of 'X-Ray Microscope,' " *Nature* **143**, 678 (1939).

2.3. W. L. Bragg, "The X-Ray Microscope," *Nature* **149**, 470 (1942).

2.4. D. Gabor, "A New Microscopic Principle," *Nature* **161**, 777 (1948).

2.5. M. E. Haine and J. Dyson, "A Modification to Gabor's Proposed Diffraction Microscope," *Nature* **166**, 315 (1950).

2.6. M. E. Haine and T. Mulvey, "The Formation of the Diffraction Image with Electrons in the Gabor Diffraction Microscope," *J. Opt. Soc. Amer.* **42**, 763 (1952).

2.7. H. M. A. El-Sum and P. Kirkpatrick, "Microscopy by Reconstructed Wave-Fronts," *Phys. Rev.* **85**, 763 (1952).

2.8. H. M. A. El-Sum, "Reconstructed Wavefront Microscopy," Ph. D. Thesis, Stanford Univ. (1952). (Available from Univ. Microfilm Inc., Ann Arbor, Michigan.)

2.9. G. L. Rogers, "Experiments in Diffraction Microscopy," *Proc. Roy. Soc.* (*Edinburgh*) **63A**, 193 (1952).

2.10. O. E. Myers, Jr., "Studies of Transmission Zone Plates," *Amer. J. Phys.* **19**, 359 (1951).

2.11. M. Sussman, "Elementary Diffraction Theory of Zone Plates," *Amer. J. Phys.* **28**, 394 (1960).

2.12. W. L. Bragg and G. L. Rogers, "Elimination of the Unwanted Image in Diffraction Microscopy," *Nature* **167**, 190 (1951).

2.13. A. Lohmann, "Optische Einseitenbandübertragung angewandt auf das Gabor-Mikroskop." *Opt. Acta* **3**, 97 (1956).

2.14. B. J. Thompson, J. Ward, and W. Zinky, "Application of Hologram Techniques for Particle-Size Determination," *J. Opt. Soc. Amer.* **55**, 1566A (1965); *Appl. Opt.* **6**, 519 (1967).

2.15. E. N. Leith and J. Upatnieks, "Reconstructed Wavefronts and Communication Theory," *J. Opt. Soc. Amer.* **52**, 1123 (1962).

2.16. A. Kozma, "Photographic Recording of Spatially Modulated Coherent Light," *J. Opt. Soc. Amer.* **56**, 428 (1966).

Chapter 3

GEOMETRIC ANALYSIS OF POINT-SOURCE HOLOGRAMS

In Chapter 2 the Gabor in-line hologram and the Leith–Upatnieks off-axis hologram were described in terms of the interference of light coming from two point sources. Despite the simplifications involved, such point-source holograms clearly illustrate many of the basic features of holography. From them much can be learned about the spacing of the interference fringes to be recorded, about properties of the virtual and real images that are generated, and about the magnification obtainable in the reconstruction process.

There are no point sources of light in the physical world, but the extended sources and illuminated subjects we do encounter can be thought of as collections of point sources. Let us suppose that $\mathbf{a}_1$, $\mathbf{a}_2$, . . ., etc. represent the complex amplitudes of light waves arriving at the hologram plane from one such collection of subject point sources. If $\mathbf{r}$ represents the reference light complex amplitude arriving at the hologram, then the total complex amplitude at the hologram is

$$\mathbf{a}_1 + \mathbf{a}_2 + \cdots + \mathbf{r}.$$

The quantity important to the holographic recording process is the total intensity

$$\begin{aligned} I &= (\mathbf{a}_1 + \mathbf{a}_2 + \cdots + \mathbf{r})(\mathbf{a}_1{}^* + \mathbf{a}_2{}^* + \cdots + \mathbf{r}^*) \\ &= \mathbf{a}_1\mathbf{a}_1{}^* + \mathbf{a}_2\mathbf{a}_2{}^* + \cdots + \mathbf{r}\mathbf{r}^* + (\mathbf{a}_1\mathbf{a}_2{}^* + \mathbf{a}_2\mathbf{a}_1{}^* + \cdots) \\ &\quad + \mathbf{r}(\mathbf{a}_1{}^* + \mathbf{a}_2{}^* + \cdots) + \mathbf{r}^*(\mathbf{a}_1 + \mathbf{a}_2 + \cdots). \end{aligned}$$

Apart from the intermodulation terms (the terms $\mathbf{a}_1\mathbf{a}_2^* + \mathbf{a}_2\mathbf{a}_1^* + \cdots$), which indicate interference among the subject wave components, the subject point sources behave independently. Concern with effects of the intermodulation terms on reconstruction can be removed in the ways touched on in Chapter 2. For the case of in-line holography, the reference wave amplitude is made much larger than the subject wave amplitude. The intermodulation terms then become negligible. In the case of off-axis holography, the angle between subject and reference waves is chosen large enough to angularly separate the image waves from the waves diffracted by the intermodulation terms. (The latter waves propagate in directions close to that of the illuminating beam.)

Since the subject point sources can for our purposes be regarded as independent, we restrict our attention here to a *single* subject point source and to the holograms formed with a reference point source. We assume all waves travel from left to right. We further assume that (1) all illumination is perfectly coherent, (2) the holograms are exposed and developed to produce an amplitude transmittance proportional to the intensity of the interference pattern, and (3) the holograms behave as plane diffraction gratings.

The reader who anticipates making his own holograms and seeks to observe the results and properties described in this chapter should be forewarned regarding the choice of recording media. Many of the effects to be discussed, e.g., simultaneous observation of real and virtual images and the results of illuminating the hologram at a different angle or with a different wavelength than that used to form it, can best be observed when the hologram truly behaves as a plane diffraction grating. The high-resolution photographic emulsions normally used in holography range from 6 to 15 μm in thickness; therefore small reference-to-subject beam angles must be employed to avoid the angular and spectral selective properties of volume-grating behavior. One recording material thin enough to behave essentially as a plane diffraction grating is thermoplastic ([*3.1*], see also Chapter 10). Holograms formed in it will exhibit the properties we shall analyze.

3.1 Computation of Subject–Reference Phase Differences

In the analysis to follow, a subject, reference, or illuminating wave arriving at any point Q on the hologram plane in Fig. 3.1 is to be represented by the difference in its phase at Q over its phase at a fixed point of origin O. (It is assumed that the amplitude of the spherical wave arising from each point source is approximately uniform across the hologram plane and therefore plays no part in the essential process.) If we assume that the space on either side of the hologram has the same index of refraction and if the hologram can be regarded as very thin, then relative phase can be computed from geometrical light-path differences. We shall employ the basic analysis introduced by Meier [*3.2*]. It involves making paraxial approximations. (For a nonparaxial analysis see Champagne [*3.3*].)

Let $\mathbf{a} = a_0 \exp(i\varphi_a)$ be the complex amplitude of the light arriving at the hologram plane from the subject point source and let $\mathbf{r} = r_0 \exp(i\varphi_r)$ be the complex amplitude of the reference wave at the hologram. Then, as in Section 2.5 (but without restriction to the in-line configuration), the intensity recorded at the hologram plane is

$$I = a_0^2 + r_0^2 + \mathbf{ra}^* + \mathbf{r}^*\mathbf{a}. \tag{3.1}$$

We are primarily interested in the interference terms

$$\mathbf{ra}^* + \mathbf{r}^*\mathbf{a} \doteq 2a_0r_0 \cos(\varphi_r - \varphi_a) \tag{3.2}$$

which describe the periodic spatial variation of intensity in the interference fringes. The periodicity of the hologram fringes is governed by the argument of the cosine, i.e., by the phase difference $\varphi_r - \varphi_a$.

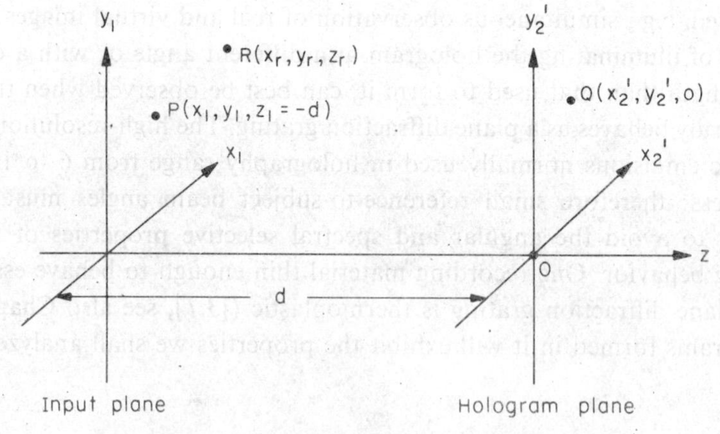

FIG. 3.1 Parameters required for computation of $\varphi_r - \varphi_a$.

Consider now hologram formation as indicated in Fig. 3.1. The subject point source P is in the x_1y_1 plane, $z_1 = -d$ units from the origin O of the z axis. The origin O is located in the hologram $x_2'y_2'$ plane. (The primes will be carried until we reach that stage of the analysis where enlargement of the hologram plane is considered.) A reference point source R is located in some arbitrary plane $x_r y_r$, a distance z_r from the hologram plane. If R is to the left of the hologram plane and the reference wave diverges from R, z_r is negative (as shown); if R is to the right of the hologram plane and the reference wave converges to R, z_r is positive. We wish to compute $\varphi_r - \varphi_a$ at an arbitrary point Q in the hologram plane. We can differentiate $(1/2\pi)(\varphi_r - \varphi_a)$ with respect to a spatial coordinate to determine the number of cycles of intensity variation per unit distance along the coordinate axis. We then know how many fringes per unit distance in that direction must be recorded by the photographic emulsion as a function of the arrangement of P, R, and the recording plate.

The starting phases of the waves emanating from P and R are perfectly arbitrary. Let us assume that these have been adjusted so that each wave has the same phase at the point O in the hologram plane. We can call that phase value zero. Since P and R are point sources, they each emit a spherical wave whose phase at any point in space is proportional to the radial distance of that point from the source. Then, by computing the path difference $PQ - PO$, we obtain the phase φ_a of the light complex amplitude at Q coming from P. By similar computation we obtain the phase φ_r of the light complex amplitude at Q coming from R. The sign of the phase at Q relative to that at O must be carefully considered. The two diagrams of Fig. 3.2

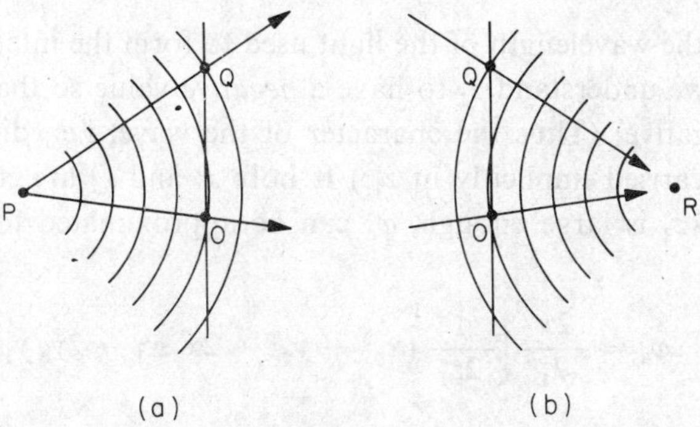

FIG. 3.2. Aid in determination of the signs of relative phase at Q and O for (a) a diverging subject wave and (b) a converging reference wave.

are useful in discussing this point. The magnitude of the phase difference φ_a, corresponding to the path difference $PQ - PO$, is $|\varphi_a| = (2\pi/\lambda)\,|(PQ - PO)|$ where λ represents the wavelength. If P is a real point source of diverging spherical waves and if $PQ > PO$, then the wavefront arriving at Q was emitted at an earlier time than that simultaneously arriving at O (see Fig.

3.2a). Therefore its phase at Q must be less than that of the front at O (assuming phase to increase with time), and consequently $\varphi_a = -(2\pi/\lambda) \times (PQ - PO)$. In Fig. 3.2b a converging reference wave is indicated. Here R exists as an effective point focus on the opposite side of the hologram from P. For $RQ > RO$, the phase of the wavefront at Q is greater than at O since the wavefront at Q was emitted later. Thus, for a *converging* reference wave, $\varphi_r = +(2\pi/\lambda)(RQ - RO)$, while for the usual case of a *diverging* reference wave, $\varphi_r = -(2\pi/\lambda)(RQ - RO)$.

We can now return to the computation of $\varphi_r - \varphi_a$ for the case where P and R are each sources of diverging light on the same side of the hologram. For the phase of the subject wave at Q, $\varphi_a(x_2', y_2')$,

$$\varphi_a = -\frac{2\pi}{\lambda_1}(PQ - PO)$$

$$= -\frac{2\pi}{\lambda_1}\{[(x_2' - x_1)^2 + (y_2' - y_1)^2 + z_1^2]^{1/2} - [x_1^2 + y_1^2 + z_1^2]^{1/2}\}$$

$$= +\frac{2\pi}{\lambda_1} z_1 \left\{\left[1 + \frac{(x_2' - x_1)^2 + (y_2' - y_1)^2}{z_1^2}\right]^{1/2} - \left[1 + \frac{x_1^2 + y_1^2}{z_1^2}\right]^{1/2}\right\},$$

where λ_1 is the wavelength of the light used to form the interference pattern and where we understand z_1 to have a *negative* value so that the sign of φ_a remains negative. (Thus the character of the wave, i.e., diverging or converging, is carried implicitly in z_1.) If both P and Q are not far off the z axis, and if z_1 is large enough, φ_a can be approximated to the first order in $1/z_1$ by

$$\varphi_a \approx \frac{2\pi}{\lambda_1}\left[\frac{1}{2z_1}(x_2'^2 + y_2'^2 - 2x_2'x_1 - 2y_2'y_1)\right]. \tag{3.3}$$

The next higher-order terms in the binomial expansion are third-order in $1/z_1$. [The first-order approximation will be satisfactory for most of the purposes of this chapter. We shall point out when an expression derived in this chapter with the aid of Eq. (3.3) differs, because of the approximation, from that obtained through other considerations.] $\varphi_r(x_2', y_2')$, the reference phase at Q, obtained in a fashion analogous to that used for φ_a, is

$$\varphi_r \approx \frac{2\pi}{\lambda_1}\left[\frac{1}{2z_r}(x_2'^2 + y_2'^2 - 2x_2'x_r - 2y_2'y_r)\right]. \tag{3.4}$$

The subject–reference phase difference at Q is then given by

$$\varphi_r - \varphi_a = \frac{2\pi}{\lambda_1}\left[(x_2'^2 + y_2'^2)\left(\frac{1}{2z_r} - \frac{1}{2z_1}\right) - x_2'\left(\frac{x_r}{z_r} - \frac{x_1}{z_1}\right) - y_2'\left(\frac{y_r}{z_r} - \frac{y_1}{z_1}\right)\right]$$

$$= \frac{2\pi}{\lambda_1}\Delta l. \tag{3.5}$$

The quantity in the brackets is the path difference Δl for light traveling to Q from P as against that traveling to Q from R.

3.1.1 The In-Line Hologram

Both subject and reference point sources are on-axis in the case of the in-line hologram so that in Eq. (3.5), x_1, y_1, x_r, and y_r are all zero. If we relabel $z_1 = -u$ and $z_r = -v$ to correspond with the notation of Section 2.4.1, then the path difference in Eq. (3.5) becomes

$$\Delta l = (x_2'^2 + y_2'^2)\left(\frac{1}{2}\right)\left(\frac{1}{z_r} - \frac{1}{z_1}\right)$$

$$= (x_2'^2 + y_2'^2)\left(\frac{1}{2}\right)\left(\frac{1}{u} - \frac{1}{v}\right) = \frac{(x_2'^2 + y_2'^2)}{2f} = \frac{\varrho^2}{2f} \tag{3.6}$$

where we have used Eq. (2.1), $f^{-1} = u^{-1} - v^{-1}$, and where ϱ is the radial distance from the origin O at the hologram plane. The bright fringes of the interference pattern occur whenever $\Delta l = n\lambda_1$ where n is an integer. Since Δl is symmetric about the origin, the fringes are circular and form the zone plate pattern given by

$$\Delta l = \frac{x_2'^2 + y_2'^2}{2}\left(\frac{1}{u} - \frac{1}{v}\right) = \frac{\varrho^2}{2f} = n\lambda_1. \tag{3.7}$$

As stated in Eq. (3.2) the spatial variation of the interference pattern intensity is governed by a cosine function, $\cos(\varphi_r - \varphi_a) = \cos(2\pi\,\Delta l/\lambda_1)$. If Δl were linearly dependent on the spatial variable, then a constant frequency could be ascribed to the cosinusoidal intensity variation. This is generally not the case, but a *local* spatial frequency, the fringe frequency $\nu(\varrho)$, can be defined. (Here ϱ is the spatial variable measured along a direction perpendicular to the crests of the interference pattern, and ν is considered to be a function of ϱ.) We define ν as the spatial rate of change of the phase of the intensity pattern at Q divided by 2π radians:

$$\nu(\varrho) = \frac{\partial(\varphi_r - \varphi_a)}{\partial\varrho} \cdot \frac{1}{2\pi} = \frac{\partial}{\partial\varrho}\left(\frac{\Delta l}{\lambda_1}\right). \tag{3.8}$$

For Δl given in Eq. (3.7),

$$\nu(\varrho) = \varrho/f\lambda_1. \tag{3.9}$$

Thus, proceeding radially outward from the hologram center, the fringe frequency increases linearly with ϱ. At some value of ϱ, ν may exceed the resolution capability ν_m of the photosensitive medium. Such a radius defines the limiting aperture and the image resolution of the hologram.

When comparing the recording resolution requirements of in-line holograms with other hologram-forming configurations to be analyzed in this chapter, we shall find it convenient and sufficient to consider only ξ' the fringe frequency component of ν along the x_2' direction. As a further simplification we consider R to be at infinity (a plane wave reference with $z_r = \infty$). For this case [see Eq. (3.6)]

$$\xi' = \frac{\partial(\varphi_r - \varphi_a)}{2\pi\, \partial x_2'} = -\frac{x_2'}{z_1\lambda_1}. \tag{3.10}$$

The farther the subject is from the hologram, the coarser and the more easily recorded are the fringes. Gabor attempted to apply this fact in his "projection method" (Section 2.2). Unfortunately image resolution decreases as well if x_2' is limited.

3.1.2 The Off-Axis Hologram

The relation obtained by setting $\Delta l = n\lambda_1$ in Eq. (3.5),

$$\begin{aligned} \Delta l = (x_2'^2 + y_2'^2)\left(\frac{1}{2}\right)\left(\frac{1}{z_r} - \frac{1}{z_1}\right) \\ - x_2'\left(\frac{x_r}{z_r} - \frac{x_1}{z_1}\right) - y_2'\left(\frac{y_r}{z_r} - \frac{y_1}{z_1}\right) = n\lambda_1, \end{aligned} \tag{3.11}$$

is the general expression for a circle whose center has the coordinates

$$x_2' = \frac{z_1 x_r - z_r x_1}{z_1 - z_r}, \qquad y_2' = \frac{z_1 y_r - z_r y_1}{z_1 - z_r}, \tag{3.12}$$

and whose radius ϱ is given by

$$\varrho^2 = \left(\frac{z_1 x_r - z_r x_1}{z_1 - z_r}\right)^2 + \left(\frac{z_1 y_r - z_r y_1}{z_1 - z_r}\right)^2 + \frac{2n\lambda_1 z_1 z_r}{z_1 - z_r}. \tag{3.13}$$

Suppose we consider the off-axis intensity pattern formed by the interference of an axial plane wave reference ($x_r = y_r = 0$, $z_r = \infty$) with a spherical subject wave diverging from an off-axis point (x_1, $y_1 = 0$, z_1). The center of the set of circular fringes, whose radii correspond to integral values of n in Eq. (3.13), is given by Eq. (3.12) as $x_2' = x_1$ and $y_2' = 0$. A zone-plate pattern is thus centered at the foot of the perpendicular dropped from P to the hologram plane. The situation is indicated in Fig. 3.3. If the

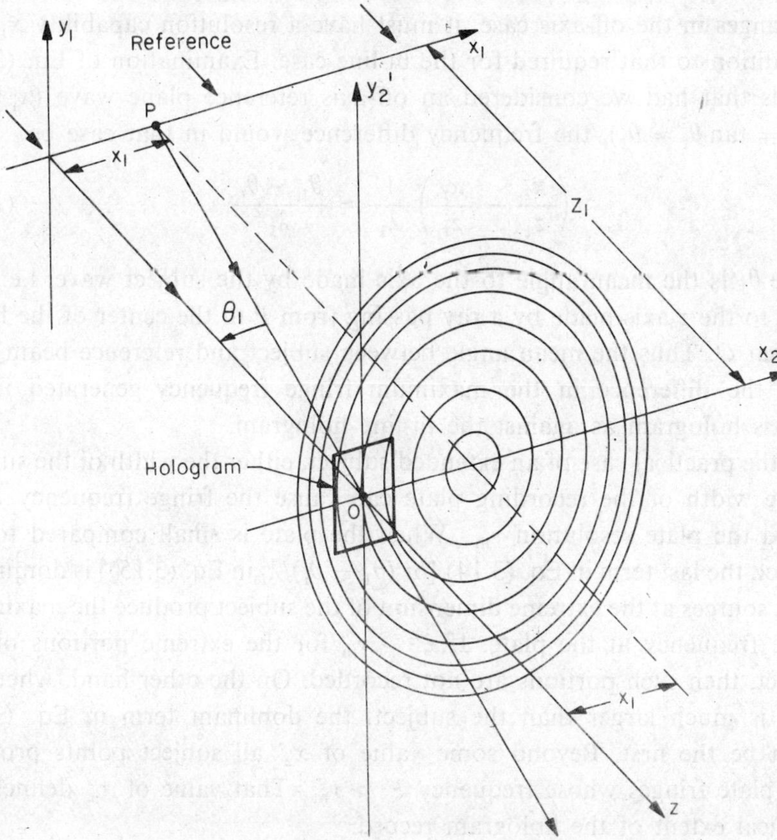

FIG. 3.3. Hologram produced by off-axis subject P and axial plane wave reference.

photographic plate is centered at O so that it can record only off-center portions of the interference pattern, a Leith–Upatnieks hologram is obtained. The fringe frequency ξ' in the x_2' direction can be found by differentiating $\Delta l/\lambda_1$ in Eq. (3.11) under the conditions $x_r = y_r = y_1 = 0$ and $z_r = \infty$. The result is

$$\xi' = -\frac{x_2'}{z_1\lambda_1} + \frac{x_1}{z_1\lambda_1}. \tag{3.14}$$

Assuming the hologram plate to be centered at O, let us compare ξ' in Eq. (3.14) with that found for the in-line configuration, Eq. (3.10). At the hologram center ($x_2' = 0$) the in-line fringe frequency is zero while the off-axis fringe frequency is $x_1/z_1\lambda_1$. As one proceeds outward in the negative x_2' direction (see Fig. 3.3), the frequency of each fringe system increases linearly with x_2' and this frequency difference is maintained. At the edge of the hologram are generated the highest frequencies to be recorded. If the photosensitive medium on the hologram plate is to record the fringes in the off-axis case, it must have a resolution capability $x_1/z_1\lambda_1$ in addition to that required for the in-line case. Examination of Eq. (3.11) reveals that had we considered an off-axis reference plane wave ($x_r \neq 0$, $x_r/z_r = \tan\theta_r \approx \theta_r$), the frequency difference would in that case be

$$\left(\frac{x_1}{z_1} - \frac{x_r}{z_r}\right)\frac{1}{\lambda_1} \approx \frac{\theta_1 - \theta_r}{\lambda_1} \tag{3.15}$$

where θ_1 is the mean angle to the axis made by the subject wave, i.e., the angle to the z axis made by a ray passing from P to the center of the hologram at O. Thus the mean angle between subject and reference beam provides the difference in the maximum fringe frequency generated in an off-axis hologram as against the in-line hologram.

In the practical case of an extended subject, either the width of the subject or the width of the recording plate can cause the fringe frequency ξ' to exceed the plate resolution ν_m. When the plate is small compared to the subject, the last term in Eq. (3.14) [or $(\theta_1 - \theta_r)/\lambda_1$ in Eq. (3.15)] is dominant. Point sources at the extreme dimension of the subject produce the maximum fringe frequency at the plate. If $\xi' > \nu_m$ for the extreme portions of the subject, then such portions are not recorded. On the other hand, when the plate is much larger than the subject, the dominant term in Eq. (3.14) might be the first. Beyond some value of x_2' all subject points produce zone-plate fringes whose frequency $\xi' > \nu_m$. That value of x_2' defines the practical extent of the hologram record.

3.1.3 Lensless Fourier Transform Hologram

We consider now the arrangement shown in Fig. 3.4 where subject and reference points are in the same plane. The coordinates of the subject point P are (x_1, $y_1 = 0$, z_1) and those of the reference point R are (x_r, $y_r = 0$, $z_r = z_1$). The phase difference $\varphi_r - \varphi_a$ in Eq. (3.5) becomes

$$\varphi_r - \varphi_a = -\frac{2\pi}{\lambda_1}\left(\frac{x_r}{z_1} - \frac{x_1}{z_1}\right)x_2'. \tag{3.16}$$

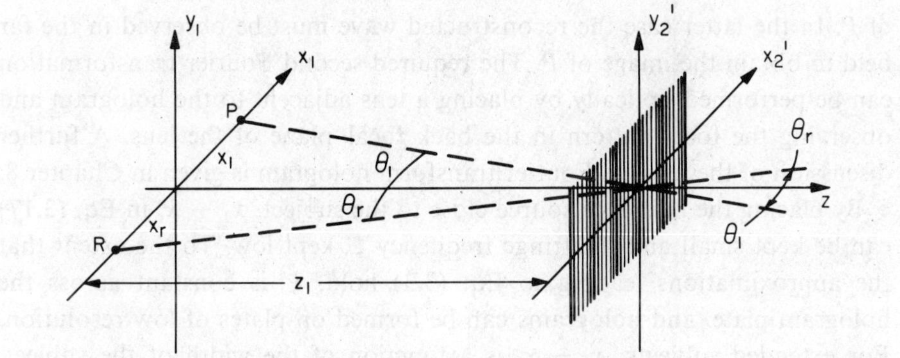

FIG. 3.4. Lensless Fourier-transform hologram configuration. (*Note*: θ_1 is a negative angle while θ_r is positive.)

Differentiation of $(\varphi_r - \varphi_a)/2\pi$ with respect to x_2' yields the *constant* fringe frequency

$$\xi' = \frac{x_1 - x_r}{z_1 \lambda_1}. \tag{3.17}$$

Since the intensity of the interference pattern is independent of y_2', as is apparent from Eq. (3.16), the fringes are vertical, uniformly spaced, *linear* fringes. Their intensity varies cosinusoidally in the x_2' direction. (The physical arrangement and results are equivalent to those of Thomas Young's experiment. The method was suggested by Winthrop and Worthington for X-ray holography [*3.4*] and by Stroke for optical holography [*3.5*].)

As may be seen in Fig. 3.4, the ratio $x_r/z_1 = \tan\theta_r \approx \theta_r$ in our first-order approximation, and similarly the ratio $x_1/z_1 \approx \theta_1$. We can write $\varphi_r - \varphi_a$ in Eq. (3.16) as

$$\varphi_r - \varphi_a = \frac{2\pi}{\lambda_1}(\theta_1 - \theta_r)x_2', \tag{3.18}$$

an expression depending only on the angle subtended at the hologram by the distance separating P and R. If points P and R are at an infinite distance from the hologram ($z_r = z_1 = \infty$, while $x_r/z_r \approx \theta_r$ and $x_1/z_1 \approx \theta_1$ are still finite), Eq. (3.18) continues to hold. The waves arriving at the hologram from point sources at infinity are plane waves. They are the far-field pattern or Fourier transforms of the point sources. Hence the linear fringe system of Eq. (3.18) can be regarded as the interference of a plane wave reference with the Fourier transform of the subject point source P. Not only can the hologram formed as in Fig. 3.4 be illuminated with the original reference point R to produce an image of P, but it can equally well be illuminated by a plane wave to reconstruct a plane wave which is the Fourier transform of P. In the latter case the reconstructed wave must be observed in the far field to obtain the image of P. The required second Fourier transformation can be performed optically by placing a lens adjacent to the hologram and observing the focal pattern in the back focal plane of the lens. A further discussion of the lensless Fourier transform hologram is given in Chapter 8.

By placing the reference source close to the subject, $x_1 - x_r$ in Eq. (3.17) can be kept small and the fringe frequency ξ' kept low. To the extent that the approximations leading to Eq. (3.3) hold, ξ' is constant across the hologram plate, and holograms can be formed on plates of low resolution. For extended subjects, $x_1 - x_r$ is a function of the width of the subject. Fringes produced by extreme portions of the subject may yet exceed the plate resolution, thus preventing these portions from being recorded. However, for small subjects the lensless Fourier transform hologram configuration produces a uniform, low-frequency fringe system over a large plate area. The result is a wide-aperture hologram which images with high resolution.

Equation (3.18) may be used to express the fringe separation, $d = 1/\xi'$, for the interference pattern formed by the intersection of two plane waves:

$$d = \lambda_1/(\theta_1 - \theta_r). \tag{3.19}$$

Suppose that as in Section 1.3.1, $\theta_r = -\theta_1$. Substituting into Eq. (3.19) we obtain

$$2\theta_1 d = \lambda_1 \tag{3.20}$$

which is a small-angle approximation to Eq. (1.10)

$$2d \sin\theta = \lambda.$$

That the lensless Fourier transform hologram is equivalent to the hologram record of two intersecting plane waves is understandable when one notes that in $(\varphi_r - \varphi_a)$ are phase expressions for two spherical waves of equal but opposite curvature. Phase contributions due to curvature of the wavefronts cancel leaving only those contributions due to the difference in mean directions of the waves.

3.2 Reconstruction with a Point Source

Having considered various hologram-forming arrangements, let us now investigate the reconstruction process. We assume that it is possible to magnify or demagnify the hologram after formation and before reconstruction. To take account of this, the hologram plane coordinates are relabeled $x_2 = mx_2'$ and $y_2 = my_2'$, where m is the linear magnification. We also assume that the reconstructing wavelength λ_2 need not be the same as the forming wavelength λ_1; their ratio is given by $\mu = \lambda_2/\lambda_1$. The reconstructing or illuminating wave originates from a point source $C(x_c, y_c, z_c)$ as in Fig. 3.5. We do not require C to be the original reference source; it may be the source of a diverging wave or the focus of a converging wave.

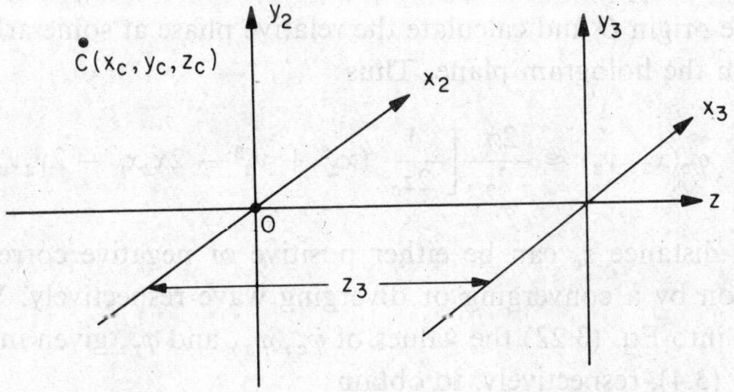

FIG. 3.5. Illumination of a hologram in plane x_2y_2 with a point source $C(x_c, y_c, z_c)$. The image plane can be separated from the hologram by a positive distance z_3, as shown (in which case the image is real), or by a negative distance (virtual image).

When the hologram is properly recorded in photographic emulsion, its amplitude transmittance t is proportional to the intensity I given by Eq. (3.1) (see also Section 1.8), where the spherical wave intensities $I_1 = a_0^2$ and $I_2 = r_0^2$ are approximately uniform over the hologram plane. Hence for holograms of point sources, diffraction results only from illumination of the spatially varying interference terms in the transmittance proportional to

$$\mathbf{ra}^* + \mathbf{r}^*\mathbf{a}.$$

The complex amplitudes of the diffracted waves at the hologram plane are proportional to the products of the illumination complex amplitude $\mathbf{c}$ times the above transmittance terms,

$$\mathbf{cra}^* + \mathbf{cr}^*\mathbf{a},$$

where $\mathbf{c} = c_0 \exp(i\varphi_c)$. In Section 1.8 it is said that the first term above, containing $\mathbf{a}^*$, yields a real image while the second, containing $\mathbf{a}$, produces a virtual image. As we shall see, this is not always the case. Nevertheless the phase of the diffracted wave

$$\mathbf{cra}^* = c_0 r_0 a_0 \exp[i(\varphi_c + \varphi_r - \varphi_a)]$$

is to be labeled

$$\varphi_R = \varphi_c + \varphi_r - \varphi_a, \tag{3.21}$$

while the phase of the wave **cr*a** is to be labeled

$$\varphi_V = \varphi_c - \varphi_r + \varphi_a. \tag{3.22}$$

As in the computation of φ_a, we can cause the phase φ_c of wave **c** to be zero at the origin O and calculate the relative phase at some arbitrary point (x_2, y_2) on the hologram plane. Thus

$$\varphi_c(x_2, y_2) \approx \frac{2\pi}{\lambda_2}\left[\frac{1}{2z_c}(x_2^2 + y_2^2 - 2x_2x_c - 2y_2y_c)\right]. \tag{3.23}$$

The axial distance z_c can be either positive or negative corresponding to illumination by a converging or diverging wave respectively. We can now substitute into Eq. (3.22) the values of φ_c, φ_a, and φ_r, given in Eqs. (3.23), (3.3), and (3.4), respectively, to obtain

$$\begin{aligned}\varphi_V = &\left(\frac{2\pi}{\lambda_2}\right)\left(\frac{1}{2}\right)\left(\frac{x_2^2 + y_2^2 - 2x_2x_c - 2y_2y_c}{z_c}\right)\\ &+ \frac{2\pi}{\lambda_1}\left(\frac{1}{2}\right)\left(\frac{x_2'^2 + y_2'^2 - 2x_2'x_1 - 2y_2'y_1}{z_1}\right)\\ &- \frac{2\pi}{\lambda_1}\left(\frac{1}{2}\right)\left(\frac{x_2'^2 + y_2'^2 - 2x_2'x_r - 2y_2'y_r}{z_r}\right).\end{aligned}$$

Introducing $x_2 = mx_2'$, $y_2 = my_2'$, and $\mu = \lambda_2/\lambda_1$,

$$\begin{aligned}\varphi_V(x_2, y_2) = \frac{\pi}{\lambda_2}\Bigg[&(x_2^2 + y_2^2)\left(\frac{1}{z_c} + \frac{\mu}{m^2z_1} - \frac{\mu}{m^2z_r}\right)\\ &- 2x_2\left(\frac{x_c}{z_c} + \frac{\mu x_1}{mz_1} - \frac{\mu x_r}{mz_r}\right) - 2y_2\left(\frac{y_c}{z_c} + \frac{\mu y_1}{mz_1} - \frac{\mu y_r}{mz_r}\right)\Bigg].\end{aligned} \tag{3.24}$$

Similarly

$$\begin{aligned}\varphi_R(x_2, y_2) = \frac{\pi}{\lambda_2}\Bigg[&(x_2^2 + y_2^2)\left(\frac{1}{z_c} - \frac{\mu}{m^2z_1} + \frac{\mu}{m^2z_r}\right)\\ &- 2x_2\left(\frac{x_c}{z_c} - \frac{\mu x_1}{mz_1} + \frac{\mu x_r}{mz_r}\right) - 2y_2\left(\frac{y_c}{z_c} - \frac{\mu y_1}{mz_1} + \frac{\mu y_r}{mz_r}\right)\Bigg].\end{aligned} \tag{3.25}$$

the hologram is indeed to image the point-source subject P, the phases the reconstructed waves at the hologram, φ_V and φ_R, must correspond to those of spherical waves. A first-order approximation to a spherical-wave phase distribution over the hologram can be written, as in Eq. (3.3),

$$\varphi(x_2, y_2) = \frac{2\pi}{\lambda_2}\left[\frac{1}{2z_3}(x_2{}^2 + y_2{}^2 - 2x_2x_3 - 2y_2y_3)\right]. \tag{3.26}$$

In the above equation, z_3 is the hologram-to-image plane separation while x_3 and y_3 represent the coordinates of the image point P in the image plane (see Fig. 3.5). We must try to arrange φ_V and φ_R to have the same form as φ. If this can be done, then the image waves are, to first order, spherical and converge or diverge according to the signs of φ_V and φ_R. They represent the case of perfect first-order imaging of the point source. The higher-order terms, neglected in the expansion of φ, φ_V, and φ_R, however, may differ and so represent aberrations (see Section 3.4).

3.3 Characteristics of the Images

By factoring out the coefficient of $(x_2{}^2 + y_2{}^2)$ in φ_V and φ_R, one can produce the desired form, indicating perfect first-order imaging. The image coordinates (x_{3V}, y_{3V}, z_{3V}) for φ_V and (x_{3R}, y_{3R}, z_{3R}) for φ_R can be identified as

$$\begin{aligned} z_{3V} &= \left(\frac{1}{z_c} + \frac{\mu}{m^2 z_1} - \frac{\mu}{m^2 z_r}\right)^{-1} = \frac{m^2 z_c z_1 z_r}{m^2 z_1 z_r + \mu z_c z_r - \mu z_c z_1}, \\ x_{3V} &= \frac{m^2 x_c z_1 z_r + \mu m x_1 z_c z_r - \mu m x_r z_c z_1}{m^2 z_1 z_r + \mu z_c z_r - \mu z_c z_1}, \\ y_{3V} &= \frac{m^2 y_c z_1 z_r + \mu m y_1 z_c z_r - \mu m y_r z_c z_1}{m^2 z_1 z_r + \mu z_c z_r - \mu z_c z_1}, \end{aligned} \tag{3.27}$$

and

$$\begin{aligned} z_{3R} &= \left(\frac{1}{z_c} - \frac{\mu}{m^2 z_1} + \frac{\mu}{m^2 z_r}\right)^{-1} = \frac{m^2 z_c z_1 z_r}{m^2 z_1 z_r - \mu z_c z_r + \mu z_c z_1}, \\ x_{3R} &= \frac{m^2 x_c z_1 z_r - \mu m x_1 z_c z_r + \mu m x_r z_c z_1}{m^2 z_1 z_r - \mu z_c z_r + \mu z_c z_1}, \\ y_{3R} &= \frac{m^2 y_c z_1 z_r - \mu m y_1 z_c z_r + \mu m y_r z_c z_1}{m^2 z_1 z_r - \mu z_c z_r + \mu z_c z_1}. \end{aligned} \tag{3.28}$$

Along with the relations defining the image location, we can define expressions for the lateral magnification M_{lat} as

$$M_{lat} \equiv \frac{dx_3}{dx_1} = \frac{dy_3}{dy_1},$$

from which we obtain

$$M_{\text{lat,V}} = m\left(1 + \frac{m^2 z_1}{\mu z_c} - \frac{z_1}{z_r}\right)^{-1},$$
$$M_{\text{lat,R}} = m\left(1 - \frac{m^2 z_1}{\mu z_c} - \frac{z_1}{z_r}\right)^{-1}, \tag{3.29}$$

and the angular magnification M_{ang} as

$$M_{\text{ang}} = \frac{d(x_3/z_3)}{d(x_1/z_1)},$$

from which we obtain

$$| M_{\text{ang}} | = \mu/m. \tag{3.30}$$

3.3.1 In-Line Hologram Images

The letters V and R refer to the reconstructed waves **cr*a** and **cra***, respectively. Whether the images produced by these waves are actually virtual or real depends upon their divergence or convergence, i.e., upon the sign of $z_{3\text{V}}$ and $z_{3\text{R}}$. A negative value implies a diverging wave and a virtual image; a positive value implies a converging wave and a real image. (*Note*: z_1, the subject-to-hologram distance, has a negative value if the subject is a real object.) We shall begin our analysis of image characteristics with Gabor's in-line holography where reference, subject, and illuminating sources are all *on axis* so that

$$x_r = x_1 = x_c = 0.$$

Only the x and z components of the image coordinates will be analyzed, since no new information is obtained by considering the y component.

Gabor's "projection method" required that the subject be placed close to the source, i.e., $z_r = z_1 + \Delta$ where Δ is a negative distance and where $\Delta/z_1 \ll 1$ (see Section 2.2). Suppose $\mu = m = 1$ and $z_c = z_r$. (This corresponds to Gabor's all-optical verification of his invention.) With the above x and z values Eqs. (3.27) and (3.28) give

$$x_{3\text{V}} = 0, \qquad z_{3\text{V}} = z_1 = z_c - \Delta,$$
$$x_{3\text{R}} = 0, \qquad z_{3\text{R}} = \left(\frac{2}{z_c} - \frac{1}{z_1}\right)^{-1} \approx z_c + \Delta. \tag{3.31}$$

The images lie close to and symmetric about the illuminating source. Since

z_c is negative, *both* images are *virtual*. To photograph these, Gabor needed a lens to form real images at a photographic plate.

For $\mu = m = 1$, the lateral magnification in Eq. (3.29) becomes

$$M_{\text{lat,V}} = \left[1 + z_1\left(\frac{1}{z_c} - \frac{1}{z_r}\right)\right]^{-1},$$
$$M_{\text{lat,R}} = \left[1 - z_1\left(\frac{1}{z_c} + \frac{1}{z_r}\right)\right]^{-1}. \tag{3.32}$$

When the illuminating source is located at the original reference source position $z_c = z_r$, then $M_{\text{lat,V}} = 1$ and $M_{\text{lat,R}} \simeq -1$, remembering that $z_1/z_r \approx 1$. (The -1 magnification in the case of the virtual image produced by φ_R indicates an inverted image.) When the illuminating wavefront has *less curvature* at the hologram than the reference wave, i.e., $|z_c| > |z_r|$, the lateral magnification increases. It becomes large, approaching z_1/Δ as $z_c \to \infty$ and the illumination becomes a plane wave. However the corresponding value of the image plane distance z_3 becomes large as well. The angular magnification, $M_{\text{ang}} = \mu/m = 1$, remains constant.

Gabor's original plan was to form the hologram at electron wavelengths and illuminate it at optical wavelengths. For this case $\mu = \lambda_2/\lambda_1 \simeq 10^5$. To avoid aberrations, he planned to scale the hologram by a factor $\mu = m$ and to place the illuminating source at distance $z_c = mz_r$ from the hologram. The lateral magnification under these conditions becomes

$$M_{\text{lat}} = \pm m = \pm\mu, \tag{3.33}$$

but the angular magnification remains unity. Consequently the distance from hologram to image plane is μ times the distance between object and hologram.

The essential feature of the "*transmission method*" of Haines and Dyson (see Section 2.2) was to place the subject close to the hologram so that $|z_1| \ll |z_r|$. Again with $x_r = x_1 = x_c = 0$, $\mu = m = 1$, and $z_c = z_r$, we obtain from Eqs. (3.27) and (3.28)

$$x_{3V} = 0, \qquad z_{3V} = z_1,$$
$$x_{3R} = 0, \qquad z_{3R} = \left(\frac{2}{z_r} - \frac{1}{z_1}\right)^{-1}. \tag{3.34}$$

If the reference is a plane wave so that $z_r \to \infty$, then a virtual image is found at z_1 and a real image at $-z_1$, the images symmetric this time about the hologram. Both images are upright since in this case $M_{\text{lat,V}} = M_{\text{lat,R}} = +1$.

A result of general consequence, not confined to on-axis holography, can be obtained by considering the hologram to be illuminated by a plane wave ($z_c = \infty$) of wavelength different from that used to form it ($\mu > 1$). With no scaling ($m = 1$), the *lateral magnification* $M_{lat} = (1 - z_1/z_r)^{-1}$ depends on the ratio z_1/z_r but *is independent of the wavelength change.* If, in addition, the reference wave had been a plane wave ($z_r = \infty$), then no magnification whatsoever is obtainable in the reconstruction process. Of course scaling up the dimensions of the hologram can produce large lateral magnification even with plane reference and illuminating waves. However, optical enlargement of the hologram is an impractical and undesirable step in an otherwise lensless imaging process. It also places the image plane at a considerable distance from the hologram. For example, when $z_c = z_r = \infty$ and $\mu = m$, the axial distance of the real image is $z_{3R} = -mz_1 = -\mu z_1$.

3.3.2 Leith–Upatnieks Off-Axis Hologram Images

Leith's and Upatnieks' method allows off-axis positions for the subject, reference, and illuminating sources, and there is no need to restrict the reference source to the z axis as we did in Section 3.1.2. Effects of illuminating the hologram with a wavelength and mean angle different from those of the reference beam can be illustrated easily by assuming plane waves for both reference and illuminating beams. In this section and throughout the remainder of the chapter we shall consider that the hologram dimensions are kept constant so that $m = 1$. Again we confine our attention to the x and z image coordinates. With these simplifications Eqs. (3.27) and (3.28) reduce to

$$x_{3V} = x_1 + \left(\frac{x_c}{z_c}\right)\frac{z_1}{\mu} - \left(\frac{x_r}{z_r}\right)z_1 = x_1 + z_1\left(\frac{\theta_c}{\mu} - \theta_r\right), \qquad z_{3V} = \frac{z_1}{\mu},$$

$$x_{3R} = x_1 - z_1\left(\frac{\theta_c}{\mu} + \theta_r\right), \qquad z_{3R} = -\frac{z_1}{\mu} \tag{3.35}$$

where $\theta_c \approx \tan\theta_c = x_c/z_c$ and $\theta_r \approx \tan\theta_r = x_r/z_r$ are the angles that the illuminating and reference beams make with the positive direction of the z axis (see Fig. 3.6).

When the illuminating wave is identical to the reference, $\mu = 1$ and $\theta_c = \theta_r$. A virtual image then appears at the location of the original subject source (x_1, z_1), and a real image appears in a plane z_1 distant on the other side of the hologram from the illumination sources. Both images are upright.

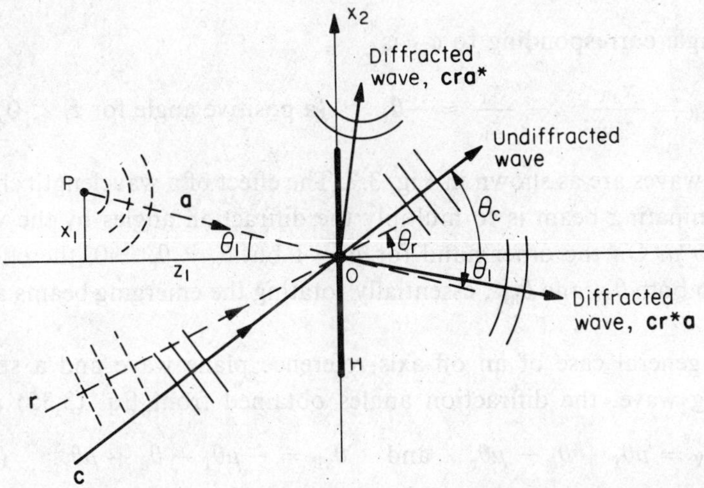

FIG. 3.6. Illumination of Leith–Upatnieks hologram with plane wave **c**. Hologram had been made with spherical wave **a** and plane wave reference **r**. (*Note*: θ_r and θ_c are positive angles while θ_1 is negative.)

As in Section 1.4, it is customary to describe the response of a diffraction grating to incident illumination in terms of the angles of incidence and diffraction. Since holograms discussed in this chapter are similar to plane diffraction gratings, it is appropriate to calculate the diffraction angles corresponding to illumination of a hologram with a plane wave making an angle θ_c to the hologram plane. We can simplify matters further by considering $\theta_c = \theta_r = 0$ as in Fig. 3.7. Then from Eq. (3.35) the diffraction angle corresponding to the φ_V diffracted wave is

$$\theta_{3V} = \frac{x_{3V}}{z_{3V}} = \frac{x_1}{z_1} \approx \theta_1 \qquad \text{(a negative angle for } z_1 < 0),$$

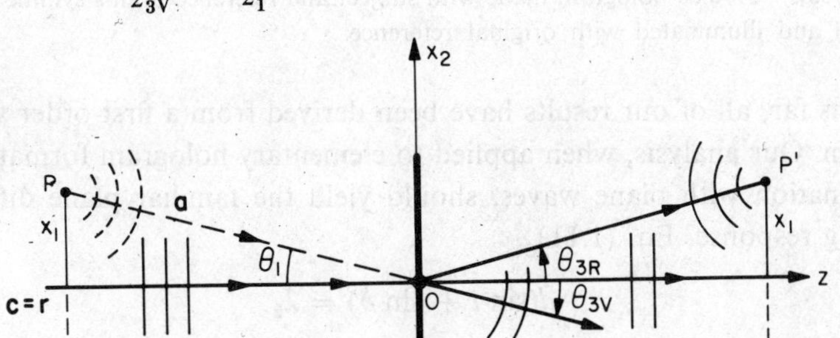

FIG. 3.7. Illumination of off-axis hologram with a plane wave **c** where $\theta_c = \theta_r = 0$.

and the angle corresponding to φ_R is

$$\theta_{3R} = \frac{x_{3R}}{z_{3R}} = -\frac{x_1}{z_1} \approx -\theta_1 \quad \text{(a positive angle for } z_1 < 0).$$

The image waves are as shown in Fig. 3.7. The effect of a wavelength change in the illuminating beam is to multiply the diffraction angles by the wavelength ratio μ. On the other hand for $\mu = 1$ but $\theta_c \neq \theta_r = 0$, the effect is to add θ_c to both θ_{3V} and θ_{3R}, essentially rotating the emerging beams about the y axis.

For the general case of an off-axis reference plane wave and a similar illuminating wave, the diffraction angles obtained from Eq. (3.35) are

$$\theta_{3V} = \mu\theta_1 + \theta_c - \mu\theta_r \quad \text{and} \quad \theta_{3R} = -\mu\theta_1 + \theta_c + \mu\theta_r. \tag{3.36}$$

A choice of angles often used in making holograms is $\theta_1 = -\alpha$, $\theta_r = +\alpha$' and $\theta_c = +\alpha$. If μ is kept equal to 1, the diffraction angles $\theta_{3V} = -\alpha$ and $\theta_{3R} = +3\alpha$ are as shown in Fig. 3.8.

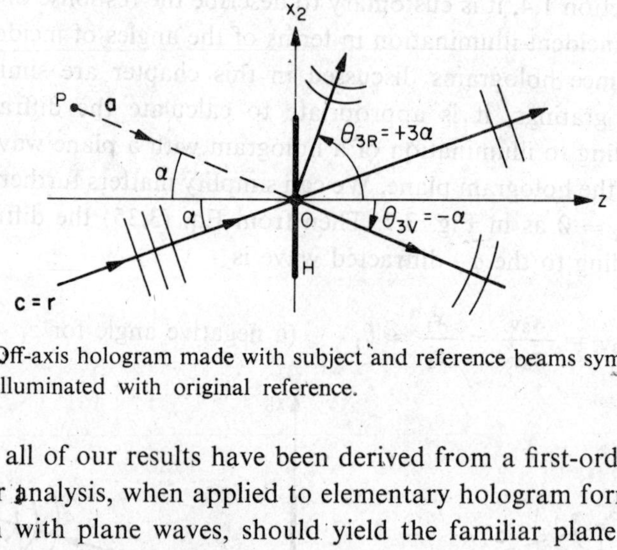

FIG. 3.8. Off-axis hologram made with subject and reference beams symmetric about normal and illuminated with original reference.

Thus far, all of our results have been derived from a first-order approximation. Our analysis, when applied to elementary hologram formation and illumination with plane waves, should yield the familiar plane diffraction grating response, Eq. (1.11)

$$d(\sin i + \sin \delta) = \lambda_2.$$

However the first-order analysis allows only a small angle approximation to Eq. (1.11) to be obtained. To illustrate, let us first restate Eq. (1.11) in a form pertinent to holographically formed gratings. Consider a grating to have been formed by interference of two plane waves on a photographic emulsion as in Fig. 1.4. For the case shown there, each forming angle can be set equal to θ_1 so that $d = \lambda_1/(2 \sin \theta_1)$. Substituting d into Eq. (1.11) and assuming that an illuminating plane wave is incident at an angle $i = \theta_1$, we obtain for the angle of diffraction δ (which is also the angle θ_3 in the notation of this chapter)

$$\begin{aligned} \sin \delta \equiv \sin \theta_3 &= 2(\lambda_2/\lambda_1) \sin \theta_1 - \sin \theta_1 \\ &= \sin \theta_1 (2\mu - 1). \end{aligned} \tag{3.37}$$

Now let us evaluate the diffraction angle θ_{3V}, in Eq. (3.36), under similar conditions. The conditions are met by setting $\theta_1 = -\theta_r$ and $\theta_c = \theta_r$. We obtain

$$\theta_{3V} = \theta_1(2\mu - 1), \tag{3.38}$$

the first-order approximation to Eq. (3.37). Thus, results obtained through first-order geometric considerations hold only so long as $\sin \theta \approx \tan \theta \approx \theta$.

3.3.3 Images When All Sources Are Equidistant from the Hologram

Suppose the subject and reference point sources lie in the same plane so that $z_1 = z_r$, and suppose that the illuminating wave is identical to the reference ($x_c = x_r$, $z_c = z_r$). Equations (3.27) and (3.28) reduce to

$$\begin{aligned} x_{3V} &= x_r(1 - \mu) + \mu x_1, & z_{3V} &= z_1, \\ x_{3R} &= x_r(1 + \mu) - \mu x_1, & z_{3R} &= z_1. \end{aligned} \tag{3.39}$$

The configuration forms the lensless Fourier transform hologram, and in this case both images are virtual, lying in the original subject plane. If the reference source lies on the z axis, then the images are symmetric about the z axis. The image at (x_{3R}, z_{3R}) is inverted.

Any of the waves can be converging to a point which is a positive distance from the hologram plane. Suppose the subject beam is such a converging beam so that z_1 is positive and suppose $z_r = z_c = -z_1$. Then

$$z_{3V} = z_1/(2\mu - 1),$$

and the image produced by φ_V is real, for $2\mu > 1$. On the other hand the image produced by φ_R is characterized by

$$z_{3R} = -z_1/(2\mu + 1)$$

and is virtual. If, instead, it is the illuminating wave that is chosen to be converging to a point a positive distance from the hologram, then inspection of z_{3V} and z_{3R} in Eqs. (3.27) and (3.28) reveals that both images are real; in this case the image at (x_{3V}, z_{3V}) is inverted.

3.4 Third-Order Aberrations

Equation (3.26) representing a spherical-wave phase distribution over the hologram is only a first-order approximation in $1/z_3$. The next order of approximation consists of a number of terms all multiplying the factor $(1/z_3)^3$. This is also true, of course, of the expressions for φ_a, φ_r, and φ_c in Eqs. (3.3), (3.4), and (3.23), respectively. The third-order terms of Eqs. (3.3), (3.4), and (3.23) are to be added together in accordance with Eq. (3.21) or (3.22) to give the third-order terms in φ_V or φ_R. The phase differences between the third-order expansion of Eq. (3.26) and the third-order expansion of φ_V (or φ_R) are the aberrations.

Meier [3.2] has calculated hologram aberrations in terms of their usual classifications: spherical aberration, coma, astigmatism, field curvature, and distortion. He finds that one or the other of the waves diffracted by the hologram yields an image free of all aberrations providing the illuminating wave is identical with the reference. For this case the magnification is unity. Magnification is achieved (1) by illuminating the hologram with a spherical wave whose curvature differs from the reference, keeping $\mu = m = 1$; (2) by illuminating with a wavelength differing from that used to form the hologram, $\mu \neq 1$; and (3) by scaling the hologram, $m \neq 1$. The first method cannot be employed without producing aberrations. If plane waves are used for reference and illumination, then the condition $\mu = m$ and $\theta_c = \theta_r$ produces an aberration-free image from φ_V, while $\mu = m$, $\theta_c = -\theta_r$ produces an aberration-free image from φ_R. (The scaling, however, requires a lens which may degrade the image.) When reference source and subject sources are equidistant from the hologram (the lensless Fourier transform configuration, $z_1 = z_r$), magnification can be achieved without optical scaling and with zero spherical aberration. The magnification is obtained by making $\mu > 1$. However at least one of the other aberrations will be present.

REFERENCES

3.1. J. C. Urbach and R. W. Meier, "Thermoplastic Xerographic Holography," *Appl. Opt.* **5**, 666 (1966).

3.2. R. W. Meier, "Magnification and Third-Order Aberrations in Holography," *J. Opt. Soc. Amer.* **55**, 987 (1965).

3.3. E. B. Champagne, "Nonparaxial Imaging, Magnification, and Aberration Properties in Holography," *J. Opt. Soc. Amer.* **57**, 51 (1967).

3.4. J. T. Winthrop and C. R. Worthington, "X-Ray Microscopy by Successive Fourier Transformation," *Phys. Lett.* **15**, 124 (1965).

3.5. G. W Stroke, "Lensless Fourier-Transform Method for Optical Holography," *Appl. Phys. Lett.* **6**, 201 (1965).

Chapter 4

THE FOURIER TRANSFORM

The analysis of point-source holograms, presented in Chapter 3, is based on the determination of differences in geometrical path to the hologram taken by simple spherical or plane waves issuing from point sources. For such simple waves, the complex light amplitude in the immediate vicinity of the hologram can be easily specified; fundamental aspects of hologram formation and wavefront reconstruction are then describable in terms of this complex amplitude. When, on the other hand, a complicated distribution of light complex amplitude exists in an input plane, and it is desired to know how this distribution is modified by passage through free space, optical components, holograms, and other obstacles, then consideration must be given to more general light wave propagation.

Electromagnetic waves can be temporally modulated or, as is generally the case at optical wavelengths, they can be spatially modulated. When temporally modulated, the wave propagation may be analyzed in either of two domains: a temporal domain or a temporal-frequency domain. Analogously, spatially modulated wave propagation, which is our concern here, may be analyzed in either a spatial domain or a spatial-frequency domain. In the spatial domain the light complex amplitude $\mathbf{a}(x, y)$ is expressed as a function of the x, y spatial coordinates of an observation plane through which the light propagates. The same complex amplitude distribution can be expressed in terms of orthogonal spatial frequencies ξ and η as well. According to the basic theorem of Fourier analysis, as applied to light distributions, any two-dimensional complex amplitude pattern can be considered as a discrete or continuous set of sinusoidally varying patterns (periodic components). The reciprocal of the spatial period of any of the components of the set, measured in a given direction on the observation plane, is called the *spatial frequency* of that component in the assigned direction. By resolving spatial periods into two orthogonal components along the x and y directions we can obtain the corresponding spatial frequency components ξ and η. Thus we may express the light complex amplitude distribution $\mathbf{a}(x, y)$ in the spatial domain as another function $\mathbf{A}(\xi, \eta)$ in the spatial frequency domain. The function $\mathbf{A}(\xi, \eta)$ is given by the two-dimensional Fourier transform of $\mathbf{a}(x, y)$, $\mathscr{F}[\mathbf{a}(x, y)]$, where

$$\mathscr{F}[\mathbf{a}(x, y)] = \int_{-\infty}^{\infty}\int_{-\infty}^{\infty} \mathbf{a}(x, y) \exp(2\pi i\xi x) \exp(2\pi i\eta y)\, dx\, dy$$

$$= \mathbf{A}(\xi, \eta). \tag{4.1}$$

Equation (4.1), which states that the Fourier transform of $\mathbf{a}(x, y)$ is $\mathbf{A}(\xi, \eta)$, may be represented symbolically as $\mathbf{a}(x, y) \supset \mathbf{A}(\xi, \eta)$. Conversely, $\mathbf{a}(x, y)$ is the *inverse* Fourier transform of $\mathbf{A}(\xi, \eta)$, $\mathscr{F}^{-1}[\mathbf{A}(\xi, \eta)]$, where

$$\mathscr{F}^{-1}[\mathbf{A}(\xi, \eta)] = \int_{-\infty}^{\infty}\int_{-\infty}^{\infty} \mathbf{A}(\xi, \eta) \exp(-2\pi i\xi x) \exp(-2\pi i\eta y)\, d\xi\, d\eta$$

$$= \mathbf{a}(x, y). \tag{4.2}$$

That the inverse Fourier transform of $\mathbf{A}(\xi, \eta)$ is $\mathbf{a}(x, y)$ may be be expressed symbolically by $\mathbf{A}(\xi, \eta) \subset \mathbf{a}(x, y)$. Note that with the designation of $\supset$ for Fourier transformation and $\subset$ for inverse Fourier transformation, the statement $\mathbf{a}(x) \supset \mathbf{A}(\xi, \eta)$ reads correctly in either direction, The two functions $\mathbf{A}(\xi, \eta)$ and $\mathbf{a}(x, y)$ are said to form a Fourier transform pair.[1]

Fourier transformation from the spatial domain to the spatial frequency domain can often provide physical insight into the working of optical systems. *The transformation can be regarded as a decomposition of a general light wave into many plane waves whose direction cosines correspond to the spatial frequencies.* Analysis of the propagation and diffraction of the plane wave is relatively straightforward and can reveal the essential physics of a light propagation process.

Although the Fourier transform and its inverse are defined by infinite integrals, in Eqs. (4.1) and (4.2), it is possible in many cases to approximate the infinite integrals by finite integrals and to perform the transformations

[1] The signs in the exponentials of Eqs. (4.1) and (4.2) have been chosen to be consistent with our definition of plane waves in Chapter 5, Eq. (5.7). An amplitude distribution in the spatial domain describing a plane wave propagating in positive x and y directions is then represented by positive spatial frequencies in the spatial frequency domain.

optically. In Chapter 6, for example, it is shown that light complex amplitude spatial distributions in the front and back focal planes of a spherical lens form a Fourier transform pair. This allows us to form Fourier transform holograms which have useful properties associated with the Fourier transform. In this chapter we review the basic features of the Fourier transform assuming the reader to be generally familiar with the theory. For a more complete development of the subject he is referred to Bracewell [*4.1*], Papoulis [*4.2*], and Jennison [*4.3*].

4.1 Linear Space-Invariant Systems and the Fourier Transform

Let us consider the optical system in Fig. 4.1 as a "black box," by which we mean that we are not concerned with what is inside the box, but only want to know how it performs. We want to know the *output* function in plane P_2 for a given *input* function in plane P_1.

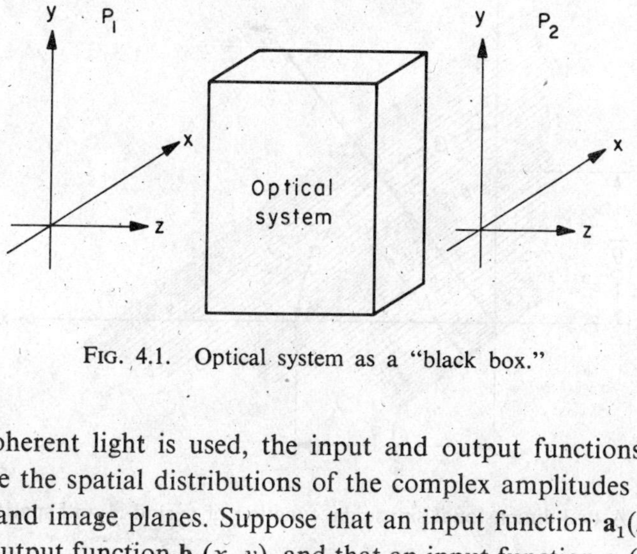

FIG. 4.1. Optical system as a "black box."

When coherent light is used, the input and output functions may, for example, be the spatial distributions of the complex amplitudes of light in the object and image planes. Suppose that an input function $\mathbf{a}_1(x, y)$ gives rise to an output function $\mathbf{b}_1(x, y)$, and that an input function $\mathbf{a}_2(x, y)$ gives rise to an output function $\mathbf{b}_2(x, y)$. The system is said to be *linear* if the law of superposition holds, i.e., if the input function $c_1\mathbf{a}_1(x, y) + c_2\mathbf{a}_2(x, y)$ gives rise to the output function $c_1\mathbf{b}_1(x, y) + c_2\mathbf{b}_2(x, y)$ regardless of the values of the constants c_1 and c_2 and the form of the functions $\mathbf{a}_1(x, y)$ and $\mathbf{a}_2(x, y)$. The system is said to be *space invariant* if the input $\mathbf{a}_1(x - u, y - u)$ gives rise to the output $\mathbf{b}_1(x - u, y - v)$, regardless of $\mathbf{a}_1(x, y)$. Here u and v are constants, and we have assumed the output plane to be scaled such that the magnification is unity. It should be emphasized that optical systems are very often *not* space invariant over their whole input and output planes. They are, however, usually space invariant over sufficiently small areas, which are called *isoplanatic* patches. The system is then treated as linear and space invariant for any particular isoplanatic patch.

Linear and space invariant systems have the property that a sinusoidal input gives rise to a sinusoidal output of the same frequency. A sinusoidal two-dimensional function is indicated in Fig. 4.2. The amplitude variation as a function of x and y and the period Λ is

$$a(x, y) = A(\vec{n}, \Lambda) \cos 2\pi\left(\frac{\vec{r} \cdot \vec{n}}{\Lambda}\right), \tag{4.3}$$

where $A(\vec{n}, \Lambda)$ is the amplitude of the cosine function, $\vec{r} = \vec{i}x + \vec{j}y$ is the position vector, $\vec{i}$ and $\vec{j}$ are unit vectors in the directions of the x and y

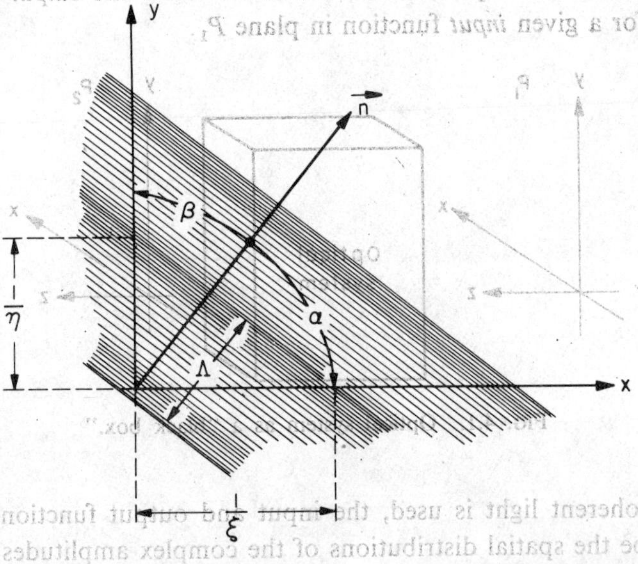

FIG. 4.2. A two-dimensional sinusoidal function of period Λ and spatial frequencies ξ and η.

axes, and $\vec{n}$ is a unit vector in the direction defining the period Λ. From the figure, $\vec{n} = \vec{i} \cos \alpha + \vec{j} \cos \beta$ and

$$a(x, y) = A(\xi, \eta) \cos 2\pi(\xi x + \eta y), \tag{4.4}$$

where the spatial frequencies

$$\xi = (\cos \alpha)/\Lambda, \qquad \eta = (\cos \beta)/\Lambda$$

are the reciprocals of the spatial periods measured along the x and y axes respectively. We can express $a(x, y)$ as $\mathrm{Re}[\mathbf{a}(x, y)]$ and then drop the reminder that $a(x, y)$ is real (as in Section 1.3) to obtain

$$\begin{aligned} \mathbf{a}(x, y) &= \mathbf{A}(\xi, \eta) \exp[-2\pi i(\xi x + \eta y)] \\ &= \mathbf{A}(\xi, \eta) \exp(-2\pi i \xi x) \exp(-2\pi i \eta y). \end{aligned} \tag{4.5}$$

Then for a linear space-invariant system the output $\mathbf{b}(x, y)$ corresponding to an input $\mathbf{a}(x, y)$ has the same spatial frequency as $\mathbf{a}(x, y)$, and

$$\mathbf{b}(x, y) = \mathbf{S}(\xi, \eta)\mathbf{A}(\xi, \eta) \exp(-2\pi i \xi x) \exp(-2\pi i \eta y), \tag{4.6}$$

where $\mathbf{S}(\xi, \eta)$ is called the *frequency transfer function* (see, e.g., Bracewell [*4.1*, p. 186]).

This simple relation between a sinusoidal input and output invites description of a linear and space invariant optical system by its frequency transfer function $\mathbf{S}(\xi, \eta)$. Input functions to optical systems are usually not sinusoidal functions, but may be decomposed into sinusoids of the form described by the Fourier transform pair

$$\mathbf{A}(\xi, \eta) = \int_{-\infty}^{\infty} \int_{-\infty}^{\infty} \mathbf{a}(x, y) \exp(2\pi i \xi x) \exp(2\pi i \eta y)\, dx\, dy$$

and

$$\mathbf{a}(x, y) = \int_{-\infty}^{\infty} \int_{-\infty}^{\infty} \mathbf{A}(\xi, \eta) \exp(-2\pi i \xi x) \exp(-2\pi i \eta y)\, d\xi\, d\eta,$$

as given in Eqs. (4.1) and (4.2). $\mathbf{A}(\xi, \eta)$ is often called the *spectrum* of $\mathbf{a}(x, y)$. Suppose that $\mathbf{a}(x, y)$ in Eq. (4.2) is the input function to a linear and space invariant system, and we want to know the output function $\mathbf{b}(x, y)$. According to Eq. (4.6) we must multiply every Fourier component by the appropriate frequency transfer coefficient $\mathbf{S}(\xi, \eta)$. Carrying out the process, we obtain for the output function $\mathbf{b}(x, y)$,

$$\mathbf{b}(x, y) = \int_{-\infty}^{\infty} \int_{-\infty}^{\infty} A(\xi, \eta) S(\xi, \eta) \exp(-2\pi i \xi x) \exp(-2\pi i \eta y)\, d\xi\, d\eta. \tag{4.7}$$

Equation (4.7) states that *the output function of a linear, space invariant system is the Fourier transform of the product of the spectrum of the input function and the frequency transfer function.* Expressing the same result in the spatial frequency or Fourier transform domain: *The spectrum of the output function of a linear space invariant system is the product of the spectrum of the input function and the frequency transfer function,*

$$\mathbf{B}(\xi, \eta) = \mathbf{A}(\xi, \eta)\mathbf{S}(\xi, \eta). \tag{4.8}$$

4.2 Correspondences and Transform Relations

Equation (4.8) specifies the input–output relation of a linear, space-invariant system as a multiplication operation in the spatial frequency domain. As we shall see, there is a corresponding operation in the spatial domain which also specifies the input–output relation. The correspondence is a consequence of the general properties of Fourier analysis and is only one of a number of such correspondences between the domains. There are, in general, two types of interdomain correspondences. The first type concerns *operations*. To every operation in the spatial domain, e.g., addition or multiplication of two functions, there corresponds an operation in the spatial frequency domain. The corresponding operation in the spatial frequency domain may or may not be the same operation as that in the spatial domain. The second type of correspondence concerns *functions*. To every function in the spatial domain there usually corresponds another function in the spatial frequency domain. (There are highly irregular functions which do not have a Fourier transform, but they need not concern us here.)

While input and output functions in optics are usually two-dimensional, the essential features of optical problems can often be revealed by one-dimensional analysis. Reduction to one lateral dimension simplifies the necessary mathematical expressions and corresponding figures. Moreover, a two-dimensional function, expressed in an appropriate coordinate system, is very often separable into a product of two one-dimensional functions. The Fourier transform of such a function can then be obtained as the product of two one-dimensional Fourier transforms. The one-dimensional Fourier transform pair has the form

$$\mathbf{A}(\xi) = \int_{-\infty}^{\infty} \mathbf{a}(x) \exp(2\pi i \xi x)\, dx, \tag{4.9}$$

$$\mathbf{a}(x) = \int_{-\infty}^{\infty} \mathbf{A}(\xi) \exp(-2\pi i \xi x)\, d\xi. \tag{4.10}$$

Certain functions which are two dimensional in rectangular coordinates are more naturally expressed as a function of one dimension in polar coordinates. Examples which are of interest to a discussion of holography are the Gaussian function and the circular aperture function. The Gaussian function $\exp(-\pi g r^2)$, where g is a constant and $r^2 = x^2 + y^2$, is separable into the product

$$\exp(-\pi g x^2) \exp(-\pi g y^2)$$

so that its Fourier transform is obtained by a double application of Eq. (4.9). The one-dimensional integration proceeds as follows:

$$\begin{aligned}
\mathbf{A}(\xi) &= \int_{-\infty}^{\infty} \exp(-\pi g x^2)\exp(2\pi i \xi x)\,dx \\
&= \int_{-\infty}^{\infty} \exp[-\pi(gx^2 - 2i\xi x)]\,dx \\
&= \exp\left(-\frac{\pi\xi^2}{g}\right)\int_{-\infty}^{\infty} \exp\left[-\pi\left(\sqrt{g}x - \frac{i\xi}{\sqrt{g}}\right)^2\right]dx \\
&= \frac{1}{\sqrt{g}}\exp\left(-\frac{\pi\xi^2}{g}\right)\int_{-\infty}^{\infty} \exp\left[-\pi\left(\sqrt{g}x - \frac{i\xi}{\sqrt{g}}\right)^2\right]\sqrt{g}\,dx \\
&= \frac{1}{\sqrt{g}}\exp\left(-\frac{\pi\xi^2}{g}\right),
\end{aligned}$$

where the infinite integral has the value unity. We can then write for $A(\nu)$, the Fourier transform of $\exp(-\pi g r^2)$,

$$\mathbf{A}(\nu) = \frac{1}{\sqrt{g}}\exp\left(-\frac{\pi\xi^2}{g}\right)\cdot\frac{1}{\sqrt{g}}\exp\left(-\frac{\pi\eta^2}{g}\right) = \frac{1}{g}\exp\left(-\frac{\pi\nu^2}{g}\right),$$

where

$$\nu^2 = \xi^2 + \eta^2.$$

The circular aperture function $\mathrm{rect}(r/2c)$ represents a unit amplitude constant over a circular area of radius c and a zero amplitude for $r > c$. To obtain its transform we must express Eq. (4.1) in cylindrical coordinates. Let

$$\begin{aligned}
x &= r\cos\theta, & y &= r\sin\theta \\
\xi &= \nu\cos\varphi, & \eta &= \nu\sin\varphi
\end{aligned}$$

$$\begin{aligned}
\mathbf{a}(x, y) = \mathrm{rect}(r/2c) &= 1 \quad \text{for} \quad r < c, \\
&= 0 \quad \text{for} \quad r > c.
\end{aligned}$$

Then Eq. (4.1) becomes

$$\begin{aligned}
\mathbf{A}(\nu) &= \int_0^c \left[\int_0^{2\pi} \exp[2\pi i r\nu\cos(\theta - \varphi)]\,d\theta\right] r\,dr \\
&= \int_0^c [2\pi J_0(2\pi r\nu)] r\,dr \\
&= \frac{1}{2\pi\nu^2}\int_0^c (2\pi\nu r)J_0(2\pi\nu r)\,d(2\pi\nu r) \\
&= \frac{2\pi\nu c}{2\pi\nu^2}J_1(2\pi\nu c) = (\pi c^2)\frac{J_1(2\pi\nu c)}{\pi\nu c}.
\end{aligned}$$

In the computation of $\mathbf{A}(\nu)$ we have used the relations

$$2\pi J_0(x) = \int_0^{2\pi} \exp(ix \cos \beta)\, d\beta \qquad \text{and} \qquad \int xJ_0(x)\, dx = xJ_1(x)$$

(see Jahnke and Emde [*4.4*], pp. 146, 149), where J_0 is the Bessel function of the first kind and zero order while J_1 is the Bessel function of the first kind, first order. The function

$$\frac{J_1(2\pi\nu c)}{\pi\nu c}$$

has a maximum value of 1 at $\nu = 0$ so that the maximum of $A(\nu)$ occurs at the origin of the spatial frequency plane and has a value πc^2. $A(\nu)$ is plotted in Fig. 4.7. Its width is generally taken to be $\nu_0 = 0.61/c$, half the width of the central lobe.

Extensive tables of Fourier transform relations have been prepared (see, e.g., Campbell and Foster [*4.5*]) and these will be referred to at times. However, it will be helpful to review some of the basic operations and list in our own notation some of the commonly used functional correspondences. We shall use lower-case letters for functions in the spatial domain, upper-case letters for functions in the spatial frequency domain and the sign $\supset$ for correspondence via Fourier transformation between the two domains. All correspondences written with the sign $\supset$ hold as well for the inverse transformation designated by $\subset$ except the two correspondences (4.20) and (4.21), involved in shift operations, and the correspondence (4.33).

4.3 The Convolution Operation

Let us begin our list of corresponding operations with

$$\mathbf{b}(x) = \int_{-\infty}^{\infty} \mathbf{a}(u)\mathbf{s}(x-u)\, du \supset \mathbf{A}(\xi)\mathbf{S}(\xi) = \mathbf{B}(\xi). \tag{4.11}$$

The integral on the left, called the convolution integral, is often written as

$$\int_{-\infty}^{\infty} \mathbf{a}(u)\mathbf{s}(x-u)\, du = \mathbf{a}(x) * \mathbf{s}(x)$$

where $*$ symbolizes the convolution operation. Correspondence (4.11) is the very important convolution theorem which states that the *Fourier transform of the convolution of two functions is the product of their Fourier* *transforms*. The proof of correspondence (4.11) is obtained simply by using the Fourier transform definitions Eqs. (4.9) and (4.10):

$$\int_{-\infty}^{\infty} \mathbf{a}(u)\mathbf{s}(x-u)\,du = \int_{-\infty}^{\infty} \mathbf{a}(u) \int_{-\infty}^{\infty} \mathbf{S}(\xi) \exp[-2\pi i(x-u)\xi]\,d\xi\,du$$

$$= \int_{-\infty}^{\infty} \mathbf{S}(\xi) \left[\int_{-\infty}^{\infty} \mathbf{a}(u) \exp(2\pi iu\xi)\,du\right] \exp(-2\pi ix\xi)\,d\xi$$

$$= \int_{-\infty}^{\infty} \mathbf{A}(\xi)\mathbf{S}(\xi) \exp(-2\pi ix\xi)\,d\xi,$$

where the last line is equivalent to $\supset \mathbf{A}(\xi, \eta)\mathbf{S}(\xi, \eta)$. We note that there is nothing in the last integral to distinguish the function which is shifted in the convolution integral. Hence the convolution is commutative, i.e., $\mathbf{a}(x)*\mathbf{s}(x) = \mathbf{s}(x)*\mathbf{a}(x)$.

A convolution relation of importance in Chapter 14 is concerned with the consequence of shifting one of the convolved functions. If $\mathbf{a}(x)$ is shifted a distance c units from its original position, the convolution $\mathbf{a}(x-c)*\mathbf{s}(x)$ can be expressed in terms of a shifted convolution of the original functions $\mathbf{a}(x)$ and $\mathbf{s}(x)$. Let

$$\mathbf{a}(x) * \mathbf{s}(x) = \int_{-\infty}^{\infty} \mathbf{a}(u)\mathbf{s}(x-u)\,du = \mathbf{h}(x).$$

Then

$$\mathbf{a}(x-c) * \mathbf{s}(x) = \int_{-\infty}^{\infty} \mathbf{a}(u-c)\mathbf{s}(x-u)\,du$$

$$= \int_{-\infty}^{\infty} \mathbf{a}(v)\mathbf{s}[(x-c)-v]\,dv = \mathbf{h}(x-c), \qquad (4.12)$$

where $v = u - c$.

The meaning of the convolution integral can be understood with the help of Fig. 4.3, where the real functions $a(u)$, $s(u)$, and its inversion with respect to the origin, $s(-u)$, are displayed. The integral in Eq. (4.11) requires us to invert $s(u)$, shift the inverse $s(-u)$ along the u axis to the right by an amount x, multiply the inverted, shifted function $s(x-u)$ by $a(u)$, and compute the area under the curve $a(u)s(x-u)$. This process gives one value of the function $b(x)$. If the operation is repeated for various values of the shift x, $b(x)$ can be plotted as a function of x.

Figure 4.4 illustrates the convolution process for two simple rectangular functions. The shifted function $s(x-u)$ scans the function $a(u)$ (upper plot). A nonzero value of the convolution is obtained only for those values

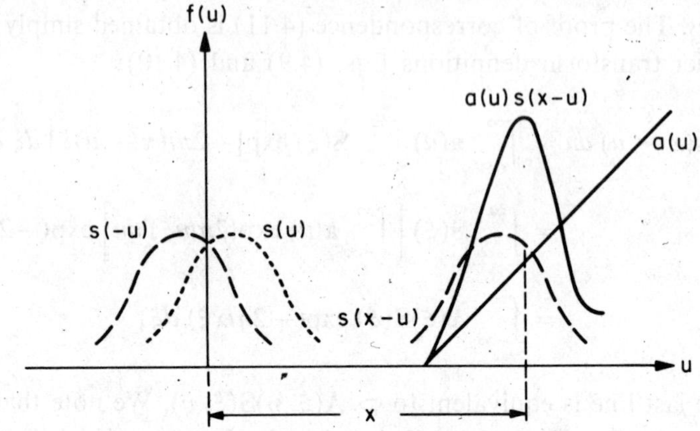

FIG. 4.3. Functions illustrating the convolution operation. The area under the curve $a(u)s(x - u)$ represents one point on the curve of the convolution, $b(x)$ versus x.

of the shift x which cause overlap of the two functions. The width of the convolution, plotted as a function of x (bottom plot), is the sum of the widths of the two functions being convolved. This is true irrespective of the shape of the functions.

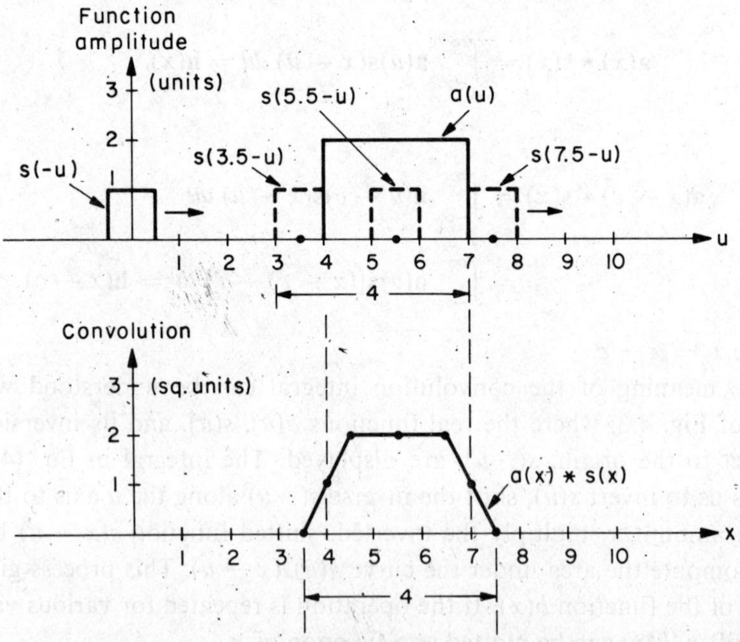

FIG. 4.4. Convolution of two rectangular functions. The scanning of one function with the other is indicated in the upper plot. The lower shows that the convolution, plotted as a function of x, has a width equal to the sum of the widths of the functions being convolved.

If we regard $\mathbf{A}(\xi)$, on the right hand side of correspondence (4.11), as the frequency spectrum of the input function for a linear, space-invariant system and $\mathbf{S}(\xi)$ as the frequency transfer function of the system, then $\mathbf{B}(\xi)$, according to Eq. (4.8), is the frequency spectrum of the output function $\mathbf{b}(x)$. The latter, expressed as a convolution integral, must be a function of the input function and the response characteristics of the system. Since $\mathbf{a}(x)$ in correspondence (4.11) can be identified as the input function [whose transform is $\mathbf{A}(\xi)$], it remains for us to determine the nature of $\mathbf{s}(x)$. To do this we first consider some useful properties of the Dirac delta function $\delta(x)$, viz.,

$$\delta(x) = 0 \qquad \text{wherever} \quad x \neq 0 \tag{4.13a}$$

$$\delta(x) = \delta(-x) \tag{4.13b}$$

$$\int_{-\infty}^{\infty} \delta(x)\, dx = 1 \tag{4.13c}$$

$$\delta(ax) = (1/|a|)\, \delta(x) \tag{4.13d}$$

$$\int_{-\infty}^{\infty} f(x)\, \delta(x-a)\, dx = f(a) \qquad \text{(sifting property).} \tag{4.13e}$$

The last property implies that convolution of a function with a delta function yields the function itself.

Suppose we choose the input function $\mathbf{a}(x)$ to be an impulse, i.e., the delta function. Substituting into Eq. (4.11) we have

$$\begin{aligned} \mathbf{b}(x) &= \int_{-\infty}^{\infty} \delta(u)\mathbf{s}(x-u)\, du = \int_{-\infty}^{\infty} \mathbf{s}(u)\, \delta(x-u)\, du \\ &= \int_{-\infty}^{\infty} \mathbf{s}(u)\, \delta(u-x)\, du = \mathbf{s}(x), \end{aligned} \tag{4.14}$$

where we have used Eq. (4.13e), and the facts that the convolution is commutative [see the proof of Eq. (4.11)] and the delta function is symmetric. We see that $\mathbf{s}(x)$ is the output response when an impulse is presented to the input of the system. The impulse response $\mathbf{s}(x)$ is called the *spread function* in optics, where its width is a measure of the distribution of the complex amplitude of light in an output plane corresponding to the presence of a point source (a delta function) in an input plane. As defined in correspondence (4.11),

$$\mathbf{s}(x) \supset \mathbf{S}(\xi), \tag{4.15}$$

meaning that *for a linear space invariant system, the Fourier transform of the spread function is the frequency transfer function.* Moreover (4.11) can be interpreted to read that *the output function of a linear space invariant system is the convolution of the input function with the spread function.*

Another way of understanding the convolution operation and its relation to a linear space invariant system can be gained by considering the function

$$I(x) = (1/\Delta u)\ \mathrm{rect}(x/\Delta u), \tag{4.16}$$

a narrow rectangular function centered at the origin, with a base extending from $-\Delta u/2$ to $+\Delta u/2$ and height equal to $1/\Delta u$ (Fig. 4.5a). Suppose $I(x)$ to be an input to the system and let the associated output be $s(x)$ (Fig. 4.5b). We wish to describe an arbitrary input $a(x)$ in terms of $I(x)$. The real input function $a(x)$, shown in Fig. 4.5c, is divided into a sum of rectan-

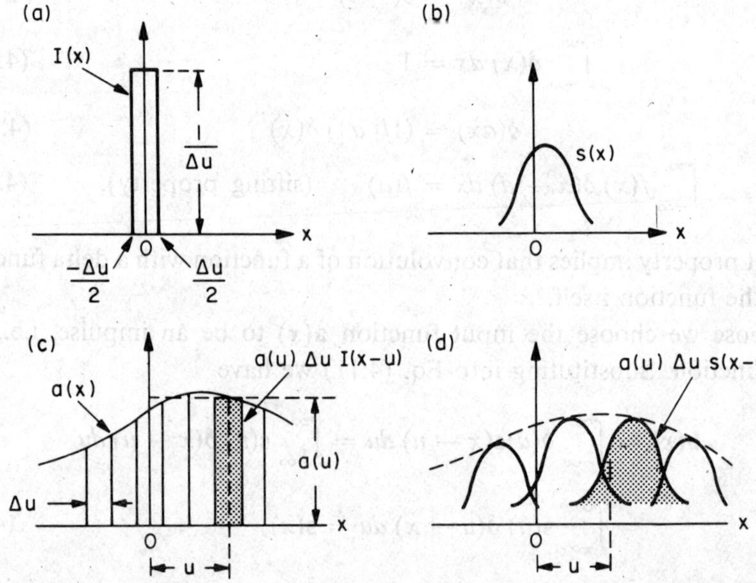

FIG. 4.5. Convolution of an input function with the response to a narrow rectangular function. (a) The rectangular function $I(x)$. (b) The response to the input $I(x)$ of a linear, space invariant system. (c) An input function divided into rectangular functions. (d) The convolution at any value of x is the sum of the ordinates of all the response curves for that value of x.

gular functions of width Δu. At a particular value of x, $x = u$, the height of the rectangular function is $a(u)$ and it is shifted to the right a distance u from the center of $I(x)$. The height $a(u)$ is a factor $a(u)\ \Delta u$ times the height of $I(x)$. Hence the rectangular function at $x = u$ can be represented by

$$a(u) I(x - u)\ \Delta u.$$

Invoking the property of linearity, the output corresponding to this input is $a(u)\,\Delta u$ greater than that corresponding to $I(x)$. Space invariance of the system implies that the shift in the input u produces no change in the shape of the output function $s(x)$ but only a corresponding shift u. Hence the output function corresponding to $a(u)I(x-u)\,\Delta u$ is

$$a(u)s(x-u)\,\Delta u.$$

Similar outputs, resulting from each of the input rectangular functions composing $a(x)$, will add linearly to form the sum

$$\sum a(u)s(x-u)\,\Delta u$$

as indicated in Fig. 4.5d.

Now let $\Delta u \to du$, an infinitesimal quantity. The summation then goes over into the integral found in correspondence (4.11)

$$\mathbf{b}(x) = \int_{-\infty}^{\infty} \mathbf{a}(u)\mathbf{s}(x-u)\,du, \tag{4.17}$$

where, in general, $\mathbf{a}(x)$ and $\mathbf{s}(x)$ are complex. We see that the linear and space-invariant properties again lead to an output in the form of a convolution between the input function and the response to a narrow impulse function.

4.4 Other Operational Correspondences

We list, below, a number of operations in the spatial domain and their corresponding operations in the spatial frequency domain. Proof of the first of these, the correlation correspondence, follows that used for the convolution theorem. The remainder of the proofs are equally straightforward and may be found in Bracewell [*4.1*].

a. *Correlation Operation*

$$\int_{-\infty}^{\infty} \mathbf{a}^*(u)\mathbf{s}(x+u)\,du \supset \mathbf{A}^*(\xi)\mathbf{S}(\xi). \tag{4.18}$$

The integral on the left is called the cross correlation of $\mathbf{a}(x)$ with $\mathbf{s}(x)$ and can be written

$$\mathbf{c}(x) = \int_{-\infty}^{\infty} \mathbf{a}^*(u)\mathbf{s}(x+u)\,du = \mathbf{a}^*(x) \star \mathbf{s}(x), \tag{4.19}$$

where the pentacle denotes the correlation operation. Note that correlation is in general not a commutative operation; it differs from the convolution operation in that the complex conjugate of $\mathbf{a}(x)$ is involved and the function $\mathbf{s}(x)$ is not inverted. Correspondence (4.18) states that *the Fourier transform of the cross correlation of two functions is the product of the complex conjugate of the Fourier transform of one function and the Fourier transform of the other.* If $\mathbf{a}(x) = \mathbf{s}(x)$, $\mathbf{c}(x)$ in Eq. (4.19) becomes the *autocorrelation.*

b. *Shift Operation*

A shift of a function in the spatial domain causes *not* a shift of the corresponding function in the spatial frequency domain (the Fourier transform) but a multiplication of the transform of the unshifted function by a phase factor whose phase is a linear function of frequency:

$$\mathbf{a}(x - c) \supset \mathbf{A}(\xi) \exp(2\pi i \xi c). \tag{4.20}$$

Correspondence (4.20) is useful in optical pattern recognition schemes. If we now consider a shift of a function in the spatial frequency domain, we find the corresponding function in the spatial domain (the inverse Fourier transform) to have been multiplied by a phase factor whose phase is a linear function of the spatial coordinates:

$$\mathbf{A}(\xi - c) \subset \mathbf{a}(x) \exp(-2\pi icx). \tag{4.21}$$

Note the difference in the sign of the phase relative to that in correspondence (4.20).

c. *Similarity Theorem*

If the scale of a function in the spatial domain is compressed, then its Fourier transform expands:

$$\mathbf{a}(cx) \supset (1/|c|)\, \mathbf{A}(\xi/c). \tag{4.22}$$

d. *Addition and Multiplication by a Constant*

$$\mathbf{a}(x) + \mathbf{b}(x) \supset \mathbf{A}(\xi) + B(\xi) \tag{4.23}$$

and

$$c\mathbf{a}(x) \supset c\mathbf{A}(\xi). \tag{4.24}$$

e. *Inversion*

Inversion of the function in the spatial domain causes inversion in the spatial frequency domain:

$$\mathbf{a}(-x) \supset \mathbf{A}(-\xi). \tag{4.25}$$

f. *Forming the Complex Conjugate*

$$\mathbf{a}^*(\pm x) \supset \mathbf{A}^*(\mp \xi). \tag{4.26}$$

If the function in the spatial domain is *real and even*, then $a(x) = a(-x)$, and from correspondence (4.25) $\mathbf{A}(\xi) = \mathbf{A}(-\xi)$. From correspondence (4.26) $\mathbf{A}(\xi) = \mathbf{A}^*(\xi)$. Hence *for a real and even function in the spatial domain the Fourier transform is also real and even.* Aware of this theorem, Bragg selected centrosymmetric (real and even) subjects for his "X-ray microscope" experiments (see Section 2.1).

When $a(x) \supset \mathbf{A}(\xi)$ and $a(x)$ is *real*, then, from correspondence (4.26), $a(x) \supset \mathbf{A}^*(-\xi)$ and $\mathbf{A}(\xi) = \mathbf{A}^*(-\xi)$. The last is a statement of the hermitian property of $\mathbf{A}(\xi)$. We can say that, in general, a real function in the spatial domain has a transform which is hermitian. This implies that the Fourier transform of a real signal is completely determined by a knowledge of its positive frequencies, and, in analyzing problems dealing with real signals, e.g., electrical waveforms, one need consider only these frequencies.

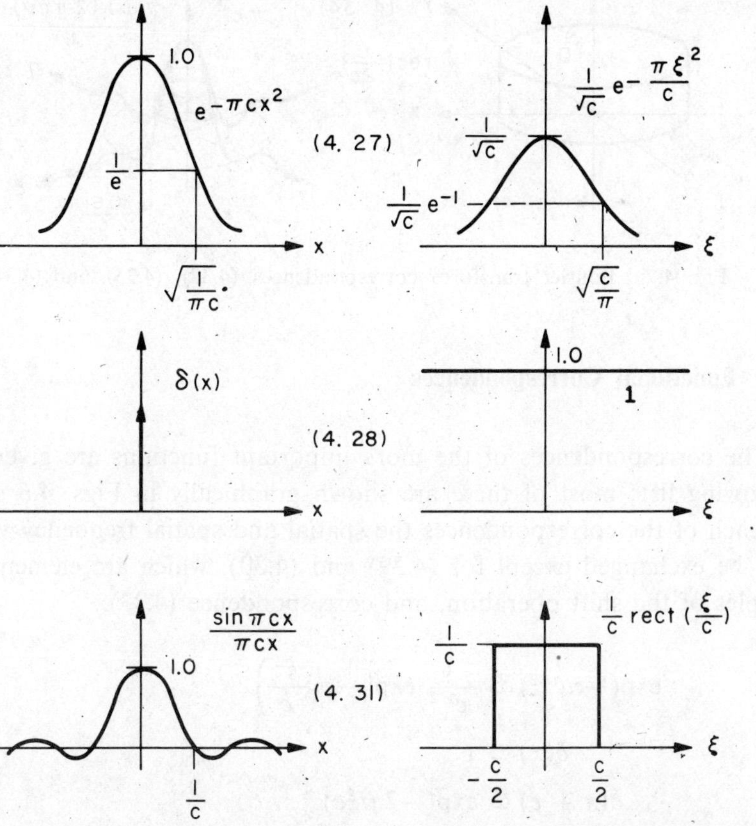

FIG. 4.6. Fourier transform correspondences (4.27), (4.28), and (4.31).

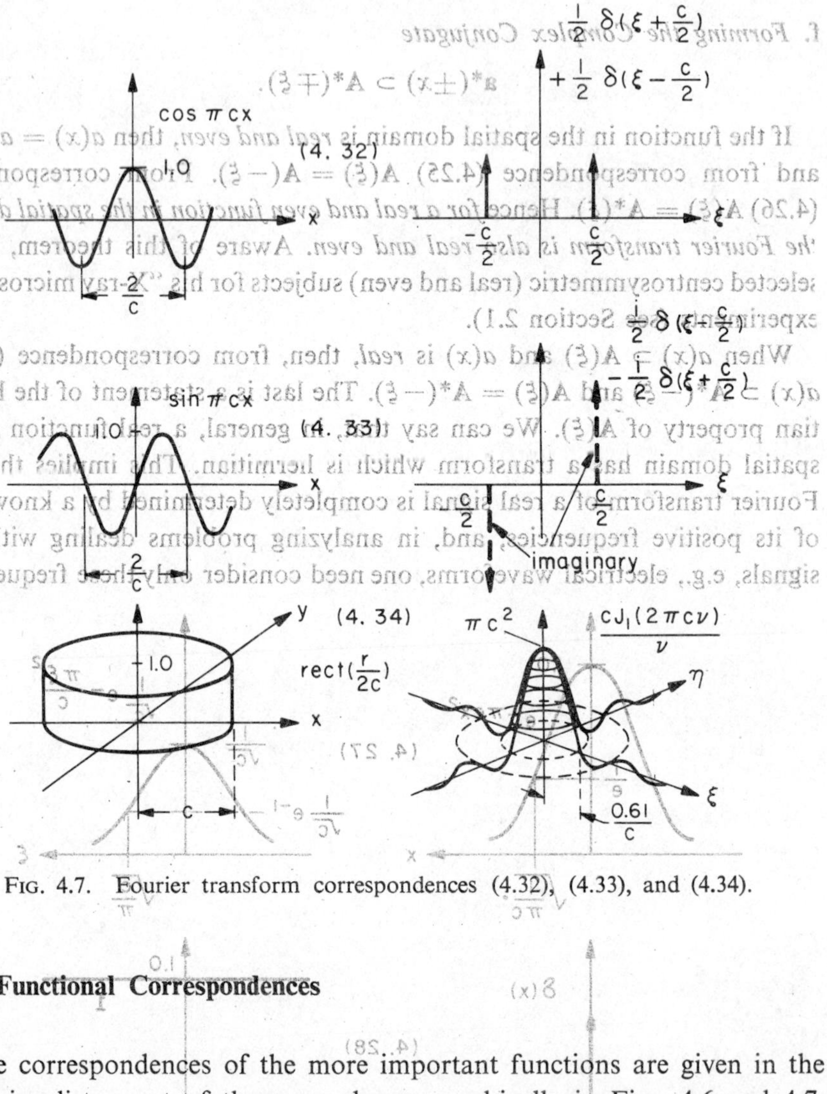

FIG. 4.7. Fourier transform correspondences (4.32), (4.33), and (4.34).

4.5 Functional Correspondences

The correspondences of the more important functions are given in the following list; most of these are shown graphically in Figs. 4.6 and 4.7. In each of the correspondences the spatial and spatial frequency variables can be exchanged except for (4.29) and (4.30), which are elementary examples of the shift operation, and correspondence (4.33).

$$\exp(-\pi c x^2) \supset \frac{1}{c^{1/2}} \exp\left(-\pi \frac{\xi^2}{c}\right) \tag{4.27}$$

$$\delta(x) \supset 1 \tag{4.28}$$

$$\delta(x + c) \supset \exp(-2\pi i \xi c) \tag{4.29}$$

$$\exp(2\pi i c x) \supset \delta(\xi + c) \tag{4.30}$$

$$\frac{\sin \pi cx}{\pi cx} \supset \frac{1}{c}\,\text{rect}\left(\frac{\xi}{c}\right) \tag{4.31}$$

$$\cos \pi cx \supset \frac{1}{2}\,\delta\left(\xi + \frac{c}{2}\right) + \frac{1}{2}\,\delta\left(\xi - \frac{c}{2}\right) \tag{4.32}$$

$$\sin \pi cx \supset \frac{i}{2}\,\delta\left(\xi - \frac{c}{2}\right) - \frac{i}{2}\,\delta\left(\xi + \frac{c}{2}\right) \tag{4.33}$$

$$\text{rect}\left(\frac{r}{2c}\right) \supset \frac{cJ_1(2\pi c\nu)}{\nu}. \tag{4.34}$$

In correspondence (4.34), $\text{rect}(r/2c)$ and $cJ_1(2\pi c\nu)/\nu$ are each rotationally symmetric functions with $r^2 = x^2 + y^2$ and $\nu^2 = \xi^2 + \eta^2$. J_1 is the Bessel function of the first kind and first order.

REFERENCES

4.1. R. Bracewell, *The Fourier Transform and Its Applications*, McGraw-Hill, New York, 1965.

4.2. A. Papoulis, *The Fourier Integral and Its Applications*, McGraw-Hill, New York, 1962.

4.3. R. C. Jennison, *Fourier Transforms*, Pergamon Press, Oxford, 1961.

4.4. E. Jahnke and F. Emde, *Tables of Functions*, 4th ed., Dover, New York 1945.

4.5. G. A. Campbell and R. M. Foster, *Fourier Integrals for Practical Applications*, Van Nostrand, Princeton, New Jersey, 1961.

Chapter 5

PROPAGATION AND DIFFRACTION

To form a hologram it is generally necessary to insert a number of obstacles into the path of light issuing from a source. The obstacles may be beam splitters, mirrors, microscope objectives and other lenses, masks, subjects, and finally a photographic plate. Each of these modifies the light incident on it. Because of their finite size they may intercept only a portion of the propagating light and, in the process, lose optical information.

Modification of light waves is not limited to diffraction by obstacles. Mere propagation of light through space alters the light complex amplitude distribution as well. Illustrative of this is the discussion in Chapter 6 of those properties of thin lenses which make possible the display of the Fourier transform of an original light amplitude distribution. We find there that propagation of the light through distances equal to the lens focal length is just as important in achieving the optical Fourier transform as is passage through the glass lens itself. An understanding of the process and, indeed, of the imaging properties of the hologram itself can be gained by application of diffraction theory.

In this chapter we treat the propagation of a plane wave and the diffraction of plane waves first from simple and then from more general obstacles. The complex light amplitude distribution in the plane of the obstacle will be related to that in some other plane farther along the direction of propagation. We have chosen to carry out the analysis in the spatial frequency domain. Although this approach differs from that in many optics textbooks, we find that it develops naturally from elementary concepts. The usual method of analysis, in the spatial domain, results in the Fresnel–Kirchhoff integral relation between the light amplitudes in two planes. We shall demonstrate the equivalence of the two methods.

5.1 The Wave Equation and Its Monochromatic Solution

Maxwell's equations relate the space and time derivatives of the vector quantities which describe the electromagnetic field. For propagation in free space, they can be combined to give the wave equation

$$\nabla^2\vec{v}(x, y, z, t) = \frac{1}{c^2}\frac{\partial^2\vec{v}(x, y, z, t)}{\partial t^2} \tag{5.1}$$

(see Ramo and Whinnery [*5.1*], p. 272 ff). Consistent with the discussion of Section 1.2, we consider only the electric field vector $\vec{v}$. The quantity c is the velocity of light propagation, t stands for time, and ∇^2 is the Laplacian operator, and we have assumed a Cartesian coordinate system x, y, z. The conditions for interference set forth in Sections 1.3.1 imply that, for our purposes, we can work with the scalar version of Eq. (5.1)

$$\nabla^2 v(x, y, z, t) = \frac{1}{c^2}\frac{\partial^2 v(x, y, z, t)}{\partial t^2} \tag{5.2}$$

where $v(x, y, z, t)$, one of the two mutually perpendicular components of the electric field, oscillates in a plane perpendicular to the direction of propagation.

If, as in past chapters, we consider only monochromatic light of frequency f, a solution to Eq. (5.2) is the sinusoidal scalar field

$$v(x, y, z, t) = a(x, y, z)\cos[2\pi ft + \varphi(x, y, z)]. \tag{5.3}$$

The above may be rewritten, as in Eq. (1.6),

$$v(x, y, z, t) = \mathrm{Re}[\mathbf{a}(x, y, z)\exp(2\pi ift)] \tag{5.4}$$

where $\mathbf{a}(x, y, z)$, the complex amplitude or phasor describing both phase and amplitude of the wave, is

$$\mathbf{a}(x, y, z) = a(x, y, z)\exp[i\varphi(x, y, z)]. \tag{5.5}$$

Mathematically it is convenient to drop the label Re[ ] and consider v in Eq. (5.2) as complex $\mathbf{v}$. In doing so we must bear in mind that the physical value of the electric field is real.

5.2 The Plane Wave Solution

A plane wave is a wave whose amplitude and phase, at any instant of time, are constant over some plane represented by the equation

$$\vec{r}\cdot\vec{n} = \text{constant} \tag{5.6}$$

where $\vec{r}$ is the position vector of a point in space and $\vec{n}$ is a unit vector normal to the plane (see Fig. 5.1). Let us consider the form of the following complex electric field **v** as a possible solution to the wave equation:

$$\mathbf{v}(x, y, z, t) = a_1 \exp(-ik\vec{r}\cdot n) \exp(i2\pi ft) \tag{5.7}$$

where a_1 is the constant crest amplitude of the wave and k is a constant to be determined. If $\vec{r} \cdot \vec{n}$ is indeed constant over some plane, then the phase of the light wave in Eq. (5.7) at any instant, is constant over that plane. At a particular value of $r = r_1$ and $t = t_1$, the phase of the wave is $2\pi ft_1 - k\vec{r}_1\cdot\vec{n} = \varphi_1(r_1, t_1)$. At some later time $t_2 > t_1$, the same value of phase φ_1 can be found at a greater distance $\vec{r}_2\cdot\vec{n} > \vec{r}_1\cdot\vec{n}$, while the phase at the original plane takes on a greater value. Thus, with time, the planes of constant phase propagate through space, and solutions to the wave equation of the form (5.7) are *plane waves*. The direction $\vec{n}$, normal to the planes of constant phase, is the direction of propagation. If $\cos\alpha$, $\cos\beta$, and $\cos\gamma$ are the direction cosines of $\vec{n}$ (see Fig. 5.1), Eq. (5.7) takes the form

$$\mathbf{v}(x, y, z, t) = a_1 \exp[-ik(x\cos\alpha + y\cos\beta + z\cos\gamma)] \exp(i2\pi ft) \tag{5.8}$$

where x, y, and z are the Cartesian components of $\vec{r}$.

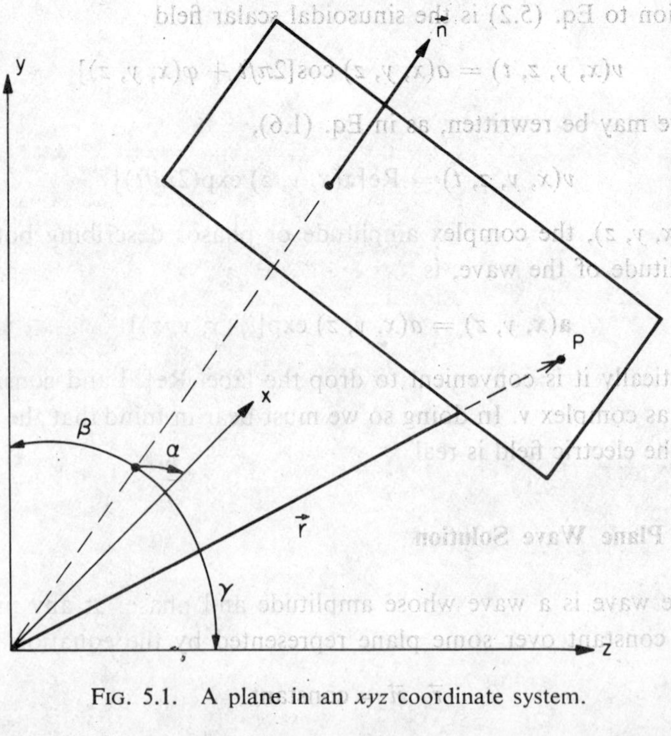

FIG. 5.1. A plane in an *xyz* coordinate system.

Insertion of the trial solution Eq. (5.8) into the wave equation Eq. (5.2) yields

$$-k^2(\cos^2 \alpha + \cos^2 \beta + \cos^2 \gamma) = -4\pi^2 f^2/c^2 = -4\pi^2/\lambda^2 \tag{5.9}$$

where λ is the wavelength of the light. Since the geometric relation

$$\cos^2 \alpha + \cos^2 \beta + \cos^2 \gamma = 1 \tag{5.10}$$

must hold, we conclude that $\mathbf{v}$ is a solution to the wave equation providing

$$k = 2\pi/\lambda. \tag{5.11}$$

Here, k is called the *propagation number*. Equation (5.8) can then be written

$$\begin{aligned} \mathbf{v}(x, y, z, t) &= a_1 \exp\left[-2\pi i\left(x\frac{\cos\alpha}{\lambda} + y\frac{\cos\beta}{\lambda} + z\frac{\cos\gamma}{\lambda}\right)\right]\exp(i2\pi f t) \\ &= a_1 \exp[-2\pi i(\xi x + \eta y + \zeta z)]\exp(i2\pi f t) \\ &= \mathbf{a}(x, y, z)\exp(i2\pi f t). \end{aligned} \tag{5.12}$$

Throughout this chapter we consider only monochromatic light. In the analysis to follow, the factor $\exp(i2\pi f t)$ will therefore be dropped, and the complex amplitude $\mathbf{a}(x, y, z)$ will represent the wave electric field. The expressions

$$\xi = \cos \alpha/\lambda, \tag{5.13a}$$

$$\eta = \cos \beta/\lambda, \tag{5.13b}$$

$$\zeta = \cos \gamma/\lambda, \tag{5.13c}$$

are called *spatial frequencies*. They represent the reciprocal of the spatial wave periods in the x, y, and z direction, respectively. The unit of spatial frequency is cycles per millimeter (cycles/mm).

It should be noted that the spatial frequencies can assume positive and negative values. If the direction of propagation forms an angle of less than 90° with the respective axis, the spatial frequency is positive; if it forms an angle of more than 90°, the spatial frequency is negative. By orienting the x, y, z axes to make one axis, say the z axis, coincide with the direction of propagation ($\xi = \eta = 0$, $\zeta = 1/\lambda$), one can see that the phase in Eq. (5.12), at any instant, decreases with increasing distance from the source in the direction of propagation. (The reader should be cautioned that some books choose the opposite sign convention, which is, of course, as valid as ours when applied consistently.)

Spatial frequencies ξ, η, and ζ are often expressed in terms of the angles $\theta_1 = 90° - \alpha$, $\theta_2 = 90° - \beta$, and $\theta_3 = 90° - \gamma$. With these angles

$$\xi = \sin\theta_1/\lambda, \tag{5.14a}$$

$$\eta = \sin\theta_2/\lambda, \tag{5.14b}$$

$$\zeta = \sin\theta_3/\lambda. \tag{5.14c}$$

Figure 5.2 illustrates the situation for a plane wave whose direction of propagation lies in the *yz* plane. It is seen that θ_2 or θ_3 is the angle the direction of propagation makes with the *xz* or *xy* plane, respectively.

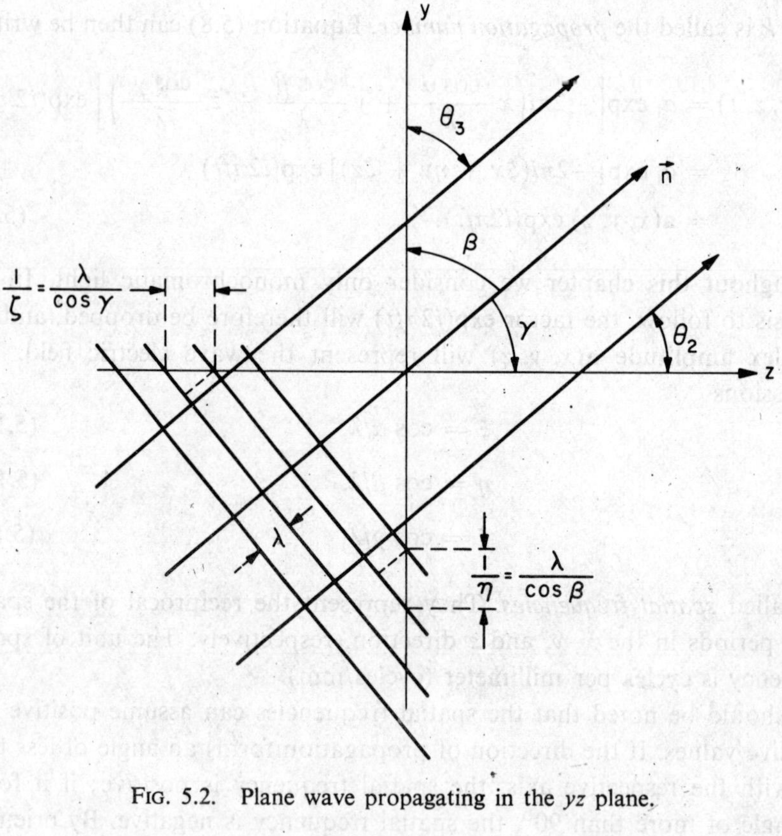

FIG. 5.2. Plane wave propagating in the *yz* plane.

ξ, η, and ζ are not independent but are related through Eq. (5.10). When Eqs. (5.13a–c) are inserted there, we have

$$\lambda^2\xi^2 + \lambda^2\eta^2 + \lambda^2\zeta^2 = 1 \tag{5.15}$$

or

$$\zeta = \pm(1/\lambda)(1 - \lambda^2\xi^2 - \lambda^2\eta^2)^{1/2} \tag{5.16}$$

where the sign is determined by the direction of propagation [see discussion following Eq. (5.13)]. We can now write the complex amplitude, $\mathbf{a}(x, y, z)$, of the plane wave represented by Eq. (5.12) as

$$\begin{aligned}\mathbf{a}(x, y, z) &= a_1 \exp\left[-2\pi i\left(x\frac{\cos\alpha}{\lambda} + y\frac{\cos\beta}{\lambda} + z\frac{\cos\gamma}{\lambda}\right)\right] \\ &= a_1 \exp[-2\pi i(\xi x + \eta y)] \cdot \exp\left[-\left(\frac{2\pi i}{\lambda}\right)z(1 - \lambda^2\xi^2 - \lambda^2\eta^2)^{1/2}\right] \\ &= \mathbf{a}(x, y, 0)\exp\left[-i\left(\frac{2\pi}{\lambda}\right)z(1 - \lambda^2\xi^2 - \lambda^2\eta^2)^{1/2}\right]. \qquad (5.17)\end{aligned}$$

Equation (5.17) is very useful for treating diffraction problems. It states that the value of the complex amplitude of a plane wave at any distance z is given by the product of the complex amplitude at $z = 0$ times a z-dependent exponential.

5.3 Diffraction from Periodic Objects

We are now ready to discuss what happens to a light wave upon encountering an obstacle. In order to obtain an exact solution to the diffraction problem, the wave equation, Eq. (5.2) must be solved with the appropriate boundary conditions on the obstacle. Unfortunately, such a straightforward approach is feasible only for very simple obstacles. Even then, the solution is quite involved. Therefore diffraction problems of practical interest are usually treated in an approximate way. For many problems in optics these approximate solutions are quite adequate. The reasons will become clear as we proceed.

Let us begin by assuming a plane wave propagating in the direction of the positive z axis with a crest amplitude a_1 and incident on a transparency in the plane $z = 0$. The transparency has the amplitude transmittance $t(x, y)$ shown in Fig. 5.3, where

$$t(x, y) = t_0 + t_1 \cos 2\pi\eta y \qquad (5.18)$$

is simply periodic in y with a spatial frequency η, and t_0 and t_1 are real-valued constants.[1] [It is assumed that $t(x, y)$ is real, i.e., that no phase shift

[1] As implied in Eqs. (4.3) and (4.4), the term spatial frequency applies to the sinusoidal components of any spatial distribution, e.g., transmittance or reflectance, and not only to those of a complex light amplitude distribution. The spatial frequency of a sinusoidal spatial component is $1/2\pi$ times the spatial rate of change of the phase of the component in a given direction.

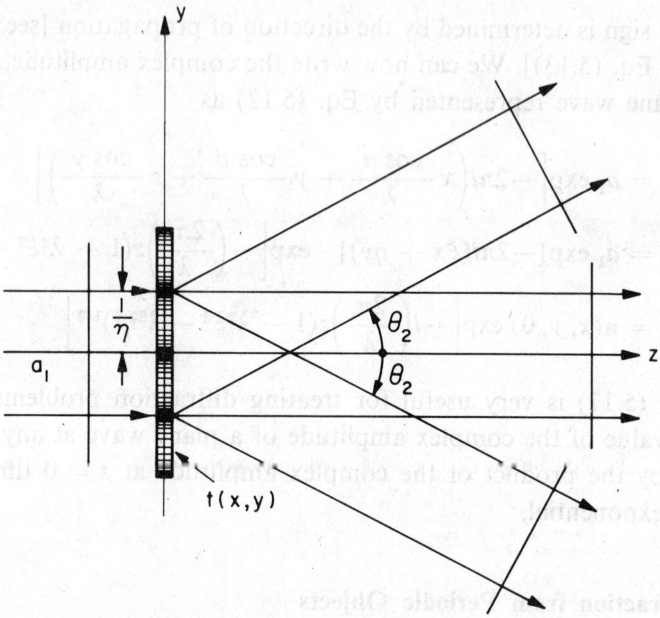

FIG. 5.3. Plane wave of amplitude a_1 incident on a transparency with an amplitude transmittance which varies cosinusoidally in the y direction. Three plane waves emerge from the transparency.

occurs in the transparency.] Directly behind the transparency the amplitude of the wave, $\mathbf{a}(x, y, 0)$, is the product of the incident light amplitude a_1 and the transmittance t:

$$\begin{aligned}\mathbf{a}(x, y, 0) = a_1 t(x, y) &= a_1 t_0 + a_1 t_1 \cos 2\pi\eta y \\ &= a_1 t_0 + \tfrac{1}{2} a_1 t_1 \exp(2\pi i \eta y) + \tfrac{1}{2} a_1 t_1 \exp(-2\pi i \eta y).\end{aligned} \tag{5.19}$$

We note that the second term in Eq. (5.19) has an xy dependence equivalent to that of the plane wave in Eq. (5.12), under the conditions $\xi = 0$ and $\eta < 0$. The second term may therefore be regarded as the representation of a plane wave whose direction of propagation is normal to the x axis ($\alpha = 90°$) and parallel to the yz plane and is headed downward, making a *negative* angle θ_2 with the z axis. From Eq. (5.14b), $\sin \theta_2 = \lambda\eta$. Similarly the xy dependence of the third term in Eq. (5.19) corresponds to that of a plane wave whose direction of propagation is parallel to the yz plane and headed upwards. The first term has no xy dependence because $\xi = \eta = 0$, and the term may be regarded as representing a plane wave propagating in the z direction. We therefore conclude that a plane wave propagating in the z direction and incident on a transparency sinusoidally periodic in the y direction causes three plane waves to emerge behind the transparency: One, of amplitude $a_1 t_0$, propagates in the z direction (the undiffracted wave); one, of amplitude $\frac{1}{2}a_1 t_1$, propagates *downward* in the yz plane, the direction of propagation forming an angle $|\theta_2| = \arcsin(\lambda\eta)$ with the z axis (the negative first-order diffracted wave); the remainder, of amplitude $\frac{1}{2}a_1 t_1$, propagates *upward* making the same angle $|\theta_2|$ with the z axis (the positive first-order diffracted wave).

The phenomenon we have been discussing is a special case of *diffraction*, and transparencies with periodic amplitude transmittances are simple *diffraction gratings*. Under many conditions of formation, holograms can be considered to be transparencies with modulated periodic transmittances, and thus we expect holograms to respond to illumination in a manner closely related to the response of the simple diffraction grating we have just discussed.

Suppose we continue our consideration of diffraction of a plane wave incident on a sinusoidal transmittance $t(x, y)$ situated at $z = 0$ and ask what the complex light amplitude is in the xy plane at $z = d$. Immediately behind the transparency we have three plane waves whose complex amplitudes in the plane $z = 0$ are given by Eq. (5.19). Equation (5.17) may be applied to each of these to obtain the individual complex amplitudes at $z = d$. The resultant complex amplitude at $z = d$, $\mathbf{a}(x, y, d)$ is their sum

$$\begin{aligned}\mathbf{a}(x, y, d) = {} & a_1 t_0 \exp\left(-i\frac{2\pi d}{\lambda}\right) \\ & + \tfrac{1}{2} a_1 t_1 \exp(i2\pi\eta y) \exp\left[-i\frac{2\pi d}{\lambda}(1 - \lambda^2\eta^2)^{1/2}\right] \\ & + \tfrac{1}{2} a_1 t_1 \exp(-i2\pi\eta y) \exp\left[-i\frac{2\pi d}{\lambda}(1 - \lambda^2\eta^2)^{1/2}\right].\end{aligned} \tag{5.20}$$

[Equation (5.17) reduces to the first term for $\xi = \eta = 0$ and to the second and third terms for $\xi = 0$.] As long as the z-dependent exponential factors [evaluated at $z = d$ in Eq. (5.20)] have imaginary exponents, all three terms in Eq. (5.20) represent propagating waves. However, the exponents do not remain imaginary for all values of the illuminating wavelength λ. As $\lambda\eta \to 1$, the angle of diffraction $\theta_2 = \arcsin \lambda\eta$ becomes large, approaching 90°. At longer wavelengths such that

$$\lambda^2\eta^2 > 1, \tag{5.21}$$

$(1 - \lambda^2\eta^2)^{1/2}$ itself becomes imaginary. If we take the negative sign in front of the square root, the exponential factors containing the square root take the form

$$\exp\left[-i\frac{2\pi d}{\lambda}(-i)(\lambda^2\eta^2 - 1)^{1/2}\right] = \exp(-bd) \quad (5.22)$$

where $b > 0$ and real. The second and third terms of Eq. (5.20), corresponding to the first-order diffraction, now represent exponentially damped or *evanescent* waves. (The choice of sign thus corresponds to the physically realizable case.) If we rewrite the inequality (5.21) as $\lambda > 1/\eta$, we see that when a grating is illuminated with a wavelength exceeding the grating period $1/\eta$, evanescent waves are generated. Their amplitude, as a function of distance d from the grating, approaches zero as $d \gg \lambda$ [see Eq. (5.22)]. The condition for evanescent waves can be restated in terms of spatial frequency to read, $\eta > 1/\lambda$. Light distributions in planes separated from the transparency by a distance $d \gg \lambda$ will contain no information about spatial frequencies in the transparency greater than $1/\lambda$.

5.4 The General Diffraction Problem

We now want to treat diffraction from more general objects. Suppose the object amplitude transmittance to be a periodic function of y only but not in the simple cosinusoidal form of Eq. (5.18). The transparency might, for instance, consist of opaque horizontal strips separated by transparent ones. We could then write the amplitude transmittance as a Fourier series. In the more general case, where the amplitude transmittance is periodic in two dimensions x and y and complex, it can be written as a sum of terms each of which has the form $\exp(-i2\pi\xi x)\exp(-i2\pi\eta y)$ given by Eq. (4.5). Assigning to each term a proper coefficient, we have for the complex amplitude transmittance $\mathbf{t}(x, y)$, which is periodic in x and y but otherwise arbitrary, the following Fourier series:

$$\mathbf{t}(x, y) = \sum_l \sum_k \mathbf{t}_{lk} \exp(-i2\pi\xi_l x)\exp(-i2\pi\eta_k y). \quad (5.23)$$

The summation extends over all the terms necessary to describe the two-dimensional function. Suppose $\mathbf{t}(x, y)$ is located in the plane $z = 0$ and a plane wave of amplitude a_1, propagating in the direction of the positive z axis, is incident on it. Applying Eq. (5.17) to the variously directed plane waves implied in Eq. (5.23), the amplitude $\mathbf{a}_2(x, y, d)$ in the plane $z = d$ becomes

$$\mathbf{a}_2(x, y, d) = a_1 \sum_l \sum_k \Big[\mathbf{t}_{lk} \exp(-i2\pi\xi_l x) \exp(-i2\pi\eta_k y) \times \exp\Big(-i\frac{2\pi d}{\lambda}(1 - \lambda^2\xi_l^2 - \lambda^2\eta_k^2)^{1/2}\Big)\Big]$$

$$= a_1 \sum_l \sum_k \Big[\mathbf{t}_{lk} \exp\Big(-i\frac{2\pi d}{\lambda}(1 - \lambda^2\xi_l^2 - \lambda^2\eta_k^2)^{1/2}\Big) \times \exp(-i2\pi\xi_l x) \exp(-i2\pi\eta_k y)\Big]. \tag{5.24}$$

If we now allow $\mathbf{t}(x, y)$ to be a nonperiodic function, the Fourier series goes over into the Fourier integral [5.2], the coefficients $\mathbf{t}_{lk}$ become the product $\mathbf{T}(\xi, \eta)\, d\xi\, d\eta$, where $\mathbf{t}(x, y) \supset \mathbf{T}(\xi, \eta)$, and Eq. (5.24) takes the form

$$\mathbf{a}_2(x, y, d) = a_1 \iint \Big[\mathbf{T}(\xi, \eta) \exp\Big(-i\frac{2\pi d}{\lambda}(1 - \lambda^2\xi^2 - \lambda^2\eta^2)^{1/2}\Big)\Big] \times \exp(-i2\pi\xi x) \exp(-i2\pi\eta y)\, d\xi\, d\eta \tag{5.25}$$

where the integration is taken over all ξ and η for which $(\xi^2 + \eta^2) \leq 1/\lambda^2$. Inspection of the above Fourier transformation yields the following result:

If a plane wave of amplitude a_1 propagating in the z direction is incident on an amplitude transmittance $\mathbf{t}(x, y)$ at $z = 0$, the spectrum $\mathbf{A}_2(\xi, \eta)$ of the complex amplitude in the plane $z = d$ then is

$$\mathbf{A}_2(\xi, \eta)\,|_{z=d} = a_1 \mathbf{T}(\xi, \eta) \exp\Big[-i\frac{2\pi d}{\lambda}(1 - \lambda^2\xi^2 - \lambda^2\eta^2)^{1/2}\Big]. \tag{5.26}$$

If in Eq. (5.26) we consider paraxial rays so that $\xi, \eta \ll 1/\lambda$, the square root can be expanded in the form

$$(1 - \lambda^2\xi^2 - \lambda^2\eta^2)^{1/2} \approx 1 - \tfrac{1}{2}\lambda^2\xi^2 - \tfrac{1}{2}\lambda^2\eta^2, \tag{5.27}$$

and Eq. (5.26) approximated by

$$\mathbf{A}_2(\xi, \eta)\,|_{z=d} \approx a_1 \mathbf{T}(\xi, \eta) \exp[i\pi\lambda d(\xi^2 + \eta^2)]. \tag{5.28}$$

A phase factor $\exp(-i2\pi d/\lambda)$ constant in the xy plane has been neglected in Eq. (5.28). (Neglecting a phase factor constant over a plane is equivalent to shifting the time origin.) Because the phase in Eq. (5.28) $\varphi = \pi\lambda d(\xi^2 + \eta^2)$

$= \pi\lambda d\nu^2$ is a parabolic function of the coordinates ξ, η, the approximation made in obtaining Eq. (5.28) is called the parabolic approximation. Since we shall use it often, some comments as to its validity are appropriate. Suppose that $\eta = 0$, and we want to know the upper limit on the spatial frequency ξ for which the parabolic approximation is still valid. The next term (left out) in the expansion Eq. (5.27) is $(\lambda^4\xi^4/8)$. We must decide what error in phase we are going to tolerate. A well-known rule due to Rayleigh (see Born and Wolf [*1.8*], p. 468) states that an optical imaging system, which is essentially perfect, must not distort a wavefront by more than $\pi/2$ radians of phase. Taking this criterion as a guide, we require that

$$(2\pi d/\lambda)(\lambda^4\xi^4/8) < \pi/2, \tag{5.29}$$

or

$$\xi^4 < 2/(d\lambda^3). \tag{5.30}$$

A numerical example is illustrative. Assume $d = 10$ cm and $\lambda = 0.5\ \mu$m. Equation (5.30) gives $\xi = 113$ cycles/mm as the upper spatial-frequency limit for which the parabolic approximation can be considered valid.

5.5 The Relation to the Fresnel–Kirchhoff Integral

The spatial domain solution to the diffraction problem is expressed in the Fresnel–Kirchhoff integral as follows: If a plane wave of amplitude a_1, traveling in the direction of the positive z axis, is incident on an object with amplitude transmittance $\mathbf{t}(x_1, y_1)$ in the plane normal to the z axis at $z = 0$, the light complex amplitude $\mathbf{a}_2(x_2, y_2, d)$, in the plane $z = d$, is

$$\mathbf{a}_2(x_2, y_2, d) = \frac{ia_1}{\lambda}\int_{x_1=-\infty}^{+\infty}\int_{y_1=-\infty}^{+\infty}\mathbf{t}(x_1, y_1) \times \frac{\exp\{-i(2\pi/\lambda)[d^2 + (x_2 - x_1)^2 + (y_2 - y_1)^2]^{1/2}\}}{[d^2 + (x_2 - x_1)^2 + (y_2 - y_1)^2]^{1/2}} \times \cos\theta\, dx_1\, dy_1. \tag{5.31}$$

A derivation of the Fresnel–Kirchhoff integral can be found in O'Neill [*5.3*, p. 72 ff]. The angle θ is formed by the positive z axis and the straight line connecting the points $(x_1, y_1, 0)$ and (x_2, y_2, d). Cos θ is called the *obliquity factor*. The geometry is illustrated in Fig. 5.4. It should be mentioned that slightly different assumptions in the boundary conditions lead to different forms of the obliquity factor. The one in Eq. (5.31) is used by Sommerfeld

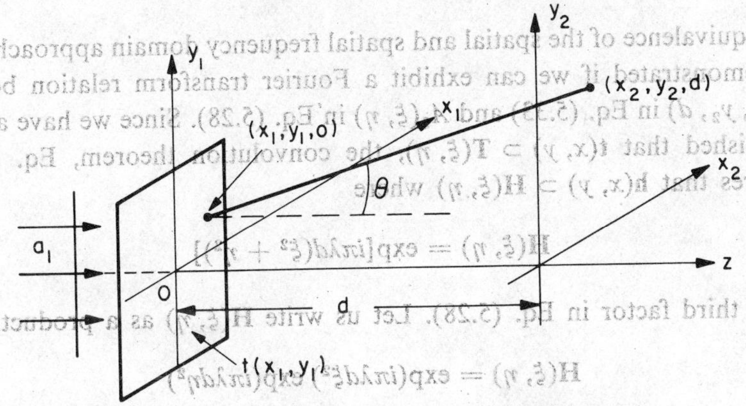

FIG. 5.4. Geometry for Fresnel–Kirchhoff integral.

whereas the one due to Kirchhoff reads $(1 + \cos\theta)/2$. As long as θ is not too large there is only a small difference between the two.

Equation (5.31) has the form of a convolution integral. In order to find the complex amplitude at $z = d$ we must *convolve* the amplitude transmittance $\mathbf{t}(x, y)$ with another function. This corresponds to the fact that in Eq. (5.26) we had to *multiply* the Fourier transform of $\mathbf{t}(x, y)$ with another function of spatial frequency. It can be shown that the Fresnel–Kirchhoff integral Eq. (5.31) and the formulation in the frequency domain, Eq. (5.26) are completely equivalent. Since the calculation is somewhat lengthy, it is given in Appendix I. Here we shall demonstrate this equivalence only for the parabolic approximation Eq. (5.28) and an approximate form of Eq. (5.31) which we shall now derive. Suppose that in Eq. (5.31), $(x_2 - x_1) \ll d$ and $(y_2 - y_1) \ll d$. Then $\cos\theta \approx 1$. We now expand the argument of the exponential in Eq. (5.31), obtaining

$$[d^2 + (x_2 - x_1)^2 + (y_2 - y_1)^2]^{1/2} \approx d + \frac{(x_2 - x_1)^2}{2d} + \frac{(y_2 - y_1)^2}{2d}, \tag{5.32}$$

and approximate the denominator in Eq. (5.31) by d. With these approximations the complex light amplitude at $z = d$ is

$$\mathbf{a}_2(x_2, y_2, d) = \frac{ia_1}{\lambda d} \int_{x_1=-\infty}^{+\infty} \int_{y_1=-\infty}^{+\infty} \mathbf{t}(x_1, y_1) \times \exp\left\{-\frac{i\pi}{\lambda d} [(x_2 - x_1)^2 + (y_2 - y_1)^2]\right\} dx_1\, dy_1. \tag{5.33}$$

A factor constant over the plane $z = d$ has been neglected. We see that the function $\mathbf{t}(x_1, y_1)$ is convolved with a function

$$\mathbf{h}(x, y) = \frac{i}{\lambda d} \exp\left[-\frac{i\pi}{\lambda d}(x^2 + y^2)\right]. \tag{5.34}$$

The equivalence of the spatial and spatial frequency domain approaches will be demonstrated if we can exhibit a Fourier transform relation between $\mathbf{a}_2(x_2, y_2, d)$ in Eq. (5.33) and $A_2(\xi, \eta)$ in Eq. (5.28). Since we have already established that $\mathbf{t}(x, y) \supset \mathbf{T}(\xi, \eta)$, the convolution theorem, Eq. (4.11), requires that $\mathbf{h}(x, y) \supset \mathbf{H}(\xi, \eta)$ where

$$\mathbf{H}(\xi, \eta) = \exp[i\pi\lambda d(\xi^2 + \eta^2)] \tag{5.35}$$

is the third factor in Eq. (5.28). Let us write $\mathbf{H}(\xi, \eta)$ as a product,

$$\mathbf{H}(\xi, \eta) = \exp(i\pi\lambda d\xi^2)\exp(i\pi\lambda d\eta^2) \tag{5.36}$$

and compute its inverse Fourier transform. We can do this in two steps. First we transform with respect to ξ, keeping η constant, and then we transform with respect to η, keeping x constant. Using correspondence (4.27) we obtain for the inverse Fourier transform of $\mathbf{H}(\xi, \eta)$

$$\begin{aligned} \mathscr{F}^{-1}\mathbf{H} &= \frac{1}{(-i\lambda d)^{1/2}}\exp\left(\frac{\pi x^2}{i\lambda d}\right)\frac{1}{(-i\lambda d)^{1/2}}\exp\left(\frac{\pi y^2}{i\lambda d}\right) \\ &= \frac{i}{\lambda d}\exp\left[-\frac{i\pi}{\lambda d}(x^2 + y^2)\right] \\ &= \mathbf{h}(x, y), \end{aligned}$$

as required.

In Section 2.1 of Chapter 2 we remarked that (the complex amplitude of) the far-field or Fraunhofer diffraction pattern of a subject transmittance is approximately the Fourier transform of that transmittance. Equation (5.33) allows us to verify the statement for the case of a transparency $\mathbf{t}(x_1, y_1)$ illuminated by a plane wave. Suppose we expand the exponential factor in Eq. (5.33) in the form

$$\begin{aligned} &\exp\left[-\frac{i\pi}{\lambda d}(x_2^2 + y_2^2)\right]\exp\left[-i\pi\frac{(x_1^2 + y_1^2)}{\lambda d}\right] \\ &\quad\times\exp\left\{i2\pi\left[x_1\left(\frac{x_2}{\lambda d}\right) + y_1\left(\frac{y_2}{\lambda d}\right)\right]\right\}. \end{aligned}$$

The first factor is independent of the variables of integration, x_1 and y_1, and can be removed from under the integral sign. If we define the far-field condition to correspond to a distance d so large compared to the dimensions of the subject that

$$(x_1^2 + y_1^2)/\lambda \ll d \qquad \text{(far-field condition)}, \tag{5.37}$$

then the second factor may be approximated by unity. Substituting

$$\xi = x_2/\lambda d \qquad \text{and} \qquad \eta = y_2/\lambda d, \tag{5.38}$$

we then have for $\mathbf{a}_2(x_2, y_2, d)$

$$\begin{aligned}\mathbf{a}_2 &= \frac{i a_1}{\lambda d} \exp\left[-\frac{i\pi}{\lambda d}(x_2{}^2 + y_2{}^2)\right] \\ &\quad \times \int_{-\infty}^{\infty}\int_{-\infty}^{\infty} \mathbf{t}(x_1, y_1) \exp[i2\pi(\xi x_1 + \eta y_1)]\, dx_1\, dy_1 \\ &= \frac{i a_1}{\lambda d} \exp\left[-\frac{i\pi}{\lambda d}(x_2{}^2 + y_2{}^2)\right] \mathbf{T}(\xi, \eta)\end{aligned} \tag{5.39}$$

where the spherical phase factor varies slowly over the $x_2 y_2$ plane and where we have used the definition for the Fourier transform, Eq. (4.1). If we multiply Eq. (5.39) by its complex conjugate, we find that the intensity in the far field of $\mathbf{t}$ is exactly the square of the absolute value of the Fourier transform of $\mathbf{t}$.

The definitions of ξ and η in Eq. (5.38), under the far-field condition [Eq. (5.37)], are equivalent to those in Eqs. (5.14a, b) where $\xi = (\sin\theta_1)/\lambda$ and $\eta = (\sin\theta_2)/\lambda$. Figure 5.5 illustrates a plane wave illuminating a subject transparency in a plane $x_1 y_1$. The area of the subject is small compared to the separation d of the plane $x_1 y_1$ from an observation plane $x_2 y_2$. Light rays diffracted by the subject may be regarded as bundles with equivalently small cross sections traveling in directions characteristic of the spatial frequencies of the subject transmittance. One such bundle traveling at an angle θ_2 to the z axis is shown in the figure. It intercepts the $x_2 y_2$ plane in a relatively small area around the coordinate y_2. If d is large enough so that

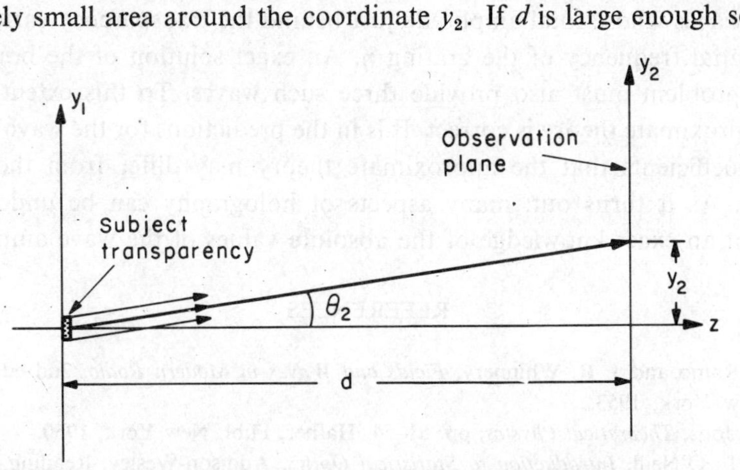

FIG. 5.5. Geometry illustrating the far-field condition.

y_2 is large compared with the intercept area, we can set

$$\lambda\eta = \sin\theta_2 \approx y_2/d \quad \text{or} \quad \eta \approx y_2/\lambda d.$$

Similarly.

$$\xi \approx x_2/\lambda d.$$

This section should not end without a word on the basic assumptions of the Fresnel–Kirchhoff theory. As implied in Section 5.3, the Fresnel–Kirchhoff integral, as well as our equivalent formulation in the frequency plane, is *not* an exact solution of the boundary value problem. The basic approximation of this theory is illustrated by considering a plane wave incident on an opaque screen containing a hole. In our derivation we would associate an amplitude transmittance of 1 with the hole and of 0 with the screen. This is certainly valid as long as we are far away from the edge of the hole. In that region of the hole close to the edge, the field is perturbed by the optical properties of the screen material. It is this perturbation that is neglected in the Fresnel–Kirchhoff theory. We should expect the theory to hold only for diffraction from objects which are large compared to a wavelength of light. This condition is satisfied for many problems in optics.

Sometimes, especially in holography, the Fresnel–Kirchhoff integral, or its spatial-frequency-plane equivalent, is used when the details of the object are no longer large compared to a wavelength. Surprisingly, it still gives results which are at least qualitatively correct. This feature is illustrated in the example of the sinusoidal grating transparency represented in Eq. (5.18). We need not suppose that the spatial period $1/\eta$ is much larger than λ. Our theory nevertheless predicts, in accordance with physical fact, three plane waves whose total amplitude just behind the transparency varies with the spatial frequency of the grating η. An exact solution of the boundary value problem must also provide three such waves. To this extent, then, the approximate theory is correct. It is in the predictions for the wave amplitude coefficients that the approximate theory may differ from the exact theory. As it turns out, many aspects of holography can be understood without an exact knowledge of the absolute values of the wave amplitude.

REFERENCES

5.1. S. Ramo and J. R. Whinnery, *Fields and Waves in Modern Radio*, 2nd ed. Wiley, New York, 1953.

5.2. G. Joos, *Theoretical Physics*, pp. 51–54, Hafner, Publ. New York, 1950.

5.3. E. L. O'Neill, *Introduction to Statistical Optics*. Addison-Wesley, Reading, Massachusetts, 1963.

Chapter 6

OPTICAL SYSTEMS WITH SPHERICAL LENSES

A spherical lens can form not only an image of a light amplitude distribution but also a pattern which is the Fourier transform of that distribution. With a simple lens, therefore, one can cause the subject light at a hologram plane to be the Fourier transform of some original pattern. The transform, stored in a hologram, has properties which are important for optical pattern recognition and attractive for optical information storage.

As an image-forming device the lens may be used in holography to construct an *image* hologram. The lens focuses the image of the subject onto the hologram plate where it interferes with a reference wave. This method of forming a hologram considerably reduces the degree of coherence required of the reconstructing light. Image holograms, properly made, can be illuminated with an ordinary frosted incandescent bulb.

These reasons and others related to providing suitably shaped beams for hologram formation make it desirable to consider some of the properties of thin-lens optical systems. In this chapter we shall derive the conditions under which a lens forms:

(a) the Fourier transform and
(b) the image of an input complex amplitude distribution.

Although the condition for an image can be obtained from geometrical-optics considerations (neglecting diffraction), the Fourier transform condition cannot and must be derived from diffraction theory. We therefore consider both image and Fourier transform conditions from the point of view of physical optics, taking into account the finite wavelength of light and the consequent diffraction effects.

6.1 The Spherical Lens

A simple spherical lens consists of transparent material bounded by two spherical surfaces. Light propagates in the lens material n times slower than in free space, n being the index of refraction of the lens material. Figure 6.1 shows such a lens, centered on the z axis of an x, y, z coordinate system. Suppose that a plane wave of wavelength λ, traveling from left to right in the direction of the z axis, is incident on the lens. We wish to compute $\mathbf{a}_r$ the complex light amplitude in a plane normal to the z axis and tangent to the right surface of the lens. We shall express $\mathbf{a}_r$ in terms of $\mathbf{a}_l$, the complex amplitude in a similar plane, tangent to the left surface of the lens. Assuming no absorption, this amounts to deriving a phase factor to multiply $\mathbf{a}_l$. To obtain that factor, we must calculate the change in the phase of the wave as it passes between the planes $z = z_2$ and $z = z_3$ in Fig. 6.1. We make the further assumption that $d = z_3 - z_2$ is so small that the planes at z_2 and z_3 are nearly coincident, and the lens can be considered to be *thin*. Under this condition, a light ray, incident at a point (x_0, y_0) on the left surface of the lens intercepts the right surface at approximately the same position coordinates (x_0, y_0). Consequently the phase modulation of an incident wave by the thin lens can be considered to have been

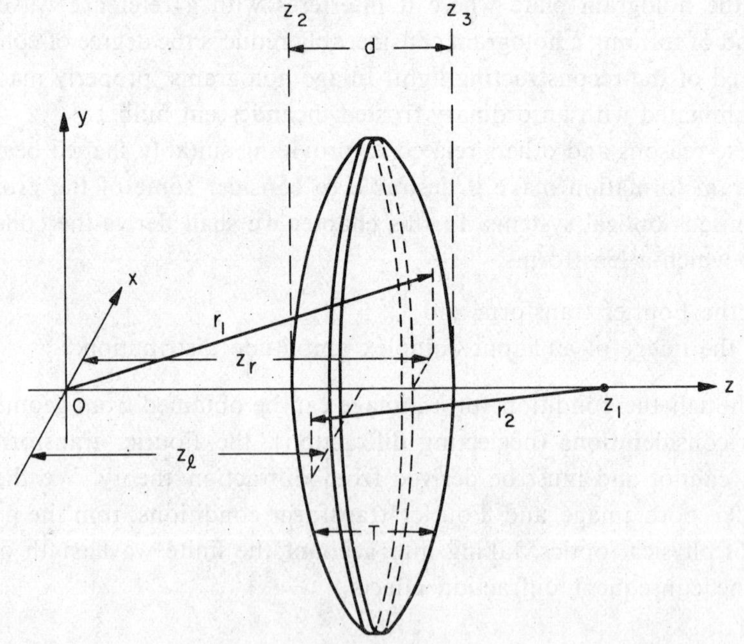

FIG. 6.1. A spherical lens.

produced by a transparency with a transmittance $\mathbf{t}(x, y) = \exp[i\,\Delta\varphi(x, y)]$ whose location is in the xy plane centered in the lens.

The right surface of the lens is represented by the equation of a sphere of radius r_1

$$x^2 + y^2 + z_r^2 = r_1^2.$$

Here, z_r is the axial distance between the xy plane at $z = 0$ and any point on the right surface of the lens. Solving for z_r, we have

$$z_r = (r_1^2 - x^2 - y^2)^{1/2}. \tag{6.1}$$

Similarly the left surface is represented by the sphere of radius r_2

$$x^2 + y^2 + (z_l - z_1)^2 = r_2^2$$

where z_l is the axial separation of a point on the left surface from the xy plane at the origin, z_1 is the coordinate of the center of curvature of the left surface, and

$$z_l = z_1 - (r_2^2 - x^2 - y^2)^{1/2}. \tag{6.2}$$

The thickness of lens material through which the wave travels varies with x and y and is given by

$$T(x, y) = z_r - z_l = (r_1^2 - x^2 - y^2)^{1/2} - z_1 + (r_2^2 - x^2 - y^2)^{1/2}. \tag{6.3}$$

Corresponding to the distance T traveled through the lens material, the light undergoes a phase shift

$$\Delta\varphi_1 = -(2\pi/\lambda')T = -(2\pi/\lambda)nT \tag{6.4}$$

where λ' is the wavelength in the lens material, n is the index of refraction of the lens, and $\lambda = n\lambda'$ is the wavelength in air. (The negative sign is consistent with the concept that phase decreases with increasing distance from the source.)

The wave passing between $z = z_2$ and $z = z_3$ travels a distance in air $d - T$; the corresponding phase shift is

$$\Delta\varphi_2 = -(2\pi/\lambda)(d - T) \tag{6.5}$$

where $d = z_3 - z_2$. The total phase shift of the wave in passing from z_2 to z_3 is the sum

$$\Delta\varphi = \Delta\varphi_1 + \Delta\varphi_2 = -(2\pi/\lambda)(n - 1)T - (2\pi/\lambda)d. \tag{6.6}$$

We can neglect the last term in Eq. (6.6) since it represents a phase change independent of x and y (constant over the xy plane at $z = z_3$). Equation (6.6) then becomes

$$\Delta\varphi = -(2\pi/\lambda)(n - 1)T(x, y). \tag{6.7}$$

We now insert $T(x, y)$, represented by Eq. (6.3), into Eq. (6.7) to obtain

$$\Delta\varphi = -(2\pi/\lambda)(n - 1)[(r_1^2 - x^2 - y^2)^{1/2} + (r_2^2 - x^2 - y^2)^{1/2}]. \tag{6.8}$$

Again a phase change $+(2\pi/\lambda)z_1$, independent of x and y, has been neglected. To obtain the desired relation between $\mathbf{a}_r$ and $\mathbf{a}_l$, we approximate the square roots in Eq. (6.8) by a series expansion, keeping only first-order terms, so that

$$(r_1^2 - x^2 - y^2)^{1/2} \approx r_1\left(1 - \frac{x^2 + y^2}{2r_1^2}\right), \tag{6.9}$$

$$(r_2^2 - x^2 - y^2)^{1/2} \approx r_2\left(1 - \frac{x^2 + y^2}{2r_2^2}\right). \tag{6.10}$$

This *paraxial* approximation is valid when $(x^2 + y^2) \ll r_1^2$ or r_2^2. Once more neglecting phase shifts independent of x and y, Eq. (6.8) becomes

$$\begin{aligned}\Delta\varphi &= +\frac{2\pi}{\lambda}(n - 1)\left(\frac{x^2 + y^2}{2r_1} + \frac{x^2 + y^2}{2r_2}\right)\\ &= \frac{\pi}{\lambda}(n - 1)\left(\frac{1}{r_1} + \frac{1}{r_2}\right)(x^2 + y^2).\end{aligned} \tag{6.11}$$

The product $(n - 1)(1/r_1 + 1/r_2)$ is related to the focal length f of a thin lens by the "lens-makers' formula" (see e.g., Jenkins and White [6.1], p. 51)

$$\frac{1}{f} = (n - 1)\left(\frac{1}{r_1} + \frac{1}{r_2}\right), \tag{6.12}$$

and the phase shift now takes the form

$$\Delta\varphi = (\pi/\lambda f)(x^2 + y^2). \tag{6.13}$$

Assuming a given lens is sufficiently thin and modifies only the phase of the light incident on it, we may as a result of Eq. (6.13) associate with the lens a transmittance $\mathbf{t}(x, y)$, a two-dimensional distribution over an xy plane centered in the lens, where

$$\mathbf{t}(x, y) = \exp(i\,\Delta\varphi) = \exp[(i\pi/\lambda f)(x^2 + y^2)]. \tag{6.14}$$

The complex light amplitude $\mathbf{a}_r$ immediately to the right of the lens is given by the product of $\mathbf{t}(x, y)$ and the complex light amplitude $\mathbf{a}_l$ incident on the left surface of the lens:

$$\mathbf{a}_r = \mathbf{a}_l \exp[(i\pi/\lambda f)(x^2 + y^2)]. \tag{6.15}$$

When the phase modulation $\Delta\varphi$, expressed as a function of x and y in Eq. (6.13), is compared with phase distributions in Eq. (3.3), (3.4), or (3.26), it is seen to be the first-order approximation to a spherical wave converging to a point on the z axis a distance f from the lens ($f > 0$).

6.2 A Simple Optical System

We are now in a position to treat optical systems consisting of thin lenses and the free spaces between them. A varicty of optical systcms, such as thc magnifier, microscope, and telescope do indeed contain no other elements than lenses and free space. (Labeling free space as an optical element should not come as a surprise to the reader of Chapter 5.) We first consider a very simple optical system which nevertheless, is capable of performing the Fourier transform operation. This will prepare us for more complicated systems to be taken up in the next section. The system to be analyzed is shown in Fig. 6.2. It consists of a spherical lens of focal length f centered in the plane $z = 0$ and a transparency with a complex amplitude transmittance $\mathbf{t}(x_1, y_1)$ immediately behind it. A plane wave propagating in the direction of the positive z axis is incident on the lens. Its complex amplitude in a plane immediately to the left of the lens is $\mathbf{a}_l$. We want to know the complex amplitude in the plane $z = f$.

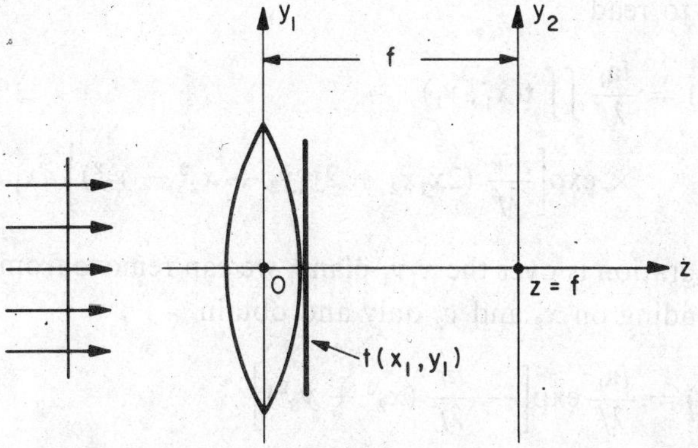

FIG. 6.2. A simple optical system for the Fourier transform operation,

Using Eq. (6.15), the complex amplitude $\mathbf{a}_r(x_1, y_1)$ immediately to the right of the lens is

$$\mathbf{a}_r(x_1, y_1) = \mathbf{a}_l \exp[(i\pi/\lambda f)(x_1^2 + y_1^2)]. \tag{6.16}$$

The wave then passes through the transparency, and the complex wave amplitude just to the right of the transparency is the product

$$\begin{aligned}\mathbf{a}_t(x_1, y_1) &= \mathbf{a}_r(x_1, y_1)\mathbf{t}(x_1, y_1)\\ &= \mathbf{a}_l\mathbf{t}(x_1, y_1) \exp[(i\pi/\lambda f)(x_1^2 + y_1^2)].\end{aligned} \tag{6.17}$$

[For a *thin* lens it does not matter whether the transparency is to the right or left of the lens. The same factors appear in the product of Eq. (6.17).] To the right of the transparency the wave travels through free space. Calculation of the complex amplitude in the plane $z = f$ in terms of the complex amplitude in the plane $z = 0$ can be carried out either in the spatial domain or the spatial frequency domain, as discussed in Chapter 5. Suppose we choose the spatial domain and employ Eq. (5.33). The complex amplitude $\mathbf{a}_2(x_2, y_2)$, in the plane $z = f$, is represented by

$$\begin{aligned}\mathbf{a}_2(x_2, y_2) &= \frac{i}{\lambda f} \iint \mathbf{a}_t(x_1, y_1)\\ &\quad \times \exp\left\{-\frac{i\pi}{\lambda f}[(x_2 - x_1)^2 + (y_2 - y_1)^2]\right\} dx_1\, dy_1\\ &= \frac{i\mathbf{a}_l}{\lambda f} \iint \mathbf{t}(x_1, y_1) \exp\left[\frac{i\pi}{\lambda f}(x_1^2 + y_1^2)\right]\\ &\quad \times \exp\left\{-\frac{i\pi}{\lambda f}[(x_2 - x_1)^2 + (y_2 - y_1)^2]\right\} dx_1\, dy_1.\end{aligned} \tag{6.18}$$

The integration extends over the surface of the lens. Equation (6.18) can be simplified to read

$$\begin{aligned}\mathbf{a}_2(x_2, y_2) &= \frac{i\mathbf{a}_l}{\lambda f} \iint \mathbf{t}(x_1, y_1)\\ &\quad \times \exp\left[\frac{i\pi}{\lambda f}(2x_1x_2 + 2y_1y_2 - x_2^2 - y_2^2)\right] dx_1\, dy_1.\end{aligned} \tag{6.19}$$

Since the integration is over the x_1y_1 plane, we can remove from the integral a factor depending on x_2 and y_2 only and obtain

$$\begin{aligned}\mathbf{a}_2(x_2, y_2) &= \frac{i\mathbf{a}_l}{\lambda f} \exp\left[-\frac{i\pi}{\lambda f}(x_2^2 + y_2^2)\right]\\ &\quad \times \iint \mathbf{t}(x_1, y_1) \exp\left[\frac{i2\pi}{\lambda f}(x_1x_2 + y_1y_2)\right] dx_1\, dy_1.\end{aligned} \tag{6.20}$$

If we let

$$\xi = x_2/\lambda f \tag{6.21}$$

and

$$\eta = y_2/\lambda f \tag{6.22}$$

and substitute these into Eq. (6.20), the complex amplitude at $z = f$ can be expressed as

$$\mathbf{a}_2(x_2, y_2) = \frac{i\mathbf{a}_1}{\lambda f} \exp\left[-\frac{i\pi}{\lambda f}(x_2{}^2 + y_2{}^2)\right]$$
$$\times \iint \mathbf{t}(x_1 y_1) \exp[i2\pi(x_1\xi + y_1\eta)]\, dx_1\, dy_1. \tag{6.23}$$

The integral in Eq. (6.23) is recognized as a two-dimensional Fourier transform integral, providing the function $\mathbf{t}(x_1, y_1)$ is zero outside the lens area. The latter provision allows the limits of integration to be extended to $+\infty$ and $-\infty$, as required in the Fourier transform. Multiplying the integral is a factor proportional to the transmittance which would be attributed to a thin diverging lens of focal length $-f$ and located at $z = f$. The exponential is a spherical phase factor; in this case it has the $x_2 y_2$ phase distribution of a spherical wave diverging from an axial source. We therefore conclude that a *thin lens, with a transparency adjacent to it and illuminated with a plane wave, produces in its back focal plane a complex amplitude distribution which is proportional to the product of a spherical phase factor and the Fourier transform of the transparency transmittance.*

Equations (6.21) and (6.22) are definitions which relate the spatial frequencies ξ, η of light diffracted from the transparency to the coordinates (x_2, y_2) of the Fourier transform of the transparency displayed in the lens focal plane. They should of course be equivalent to the spatial frequency definitions in Chapter 5. For small angles consistent with previous approximations the relations do satisfy the original definitions of ξ and η given in Eqs. (5.14a, b). This agreement is illustrated in Fig. 6.3 where, according to Eq. (5.14b), the plane wave diffracted from the transparency at an angle θ to the z axis is characterized by the spatial frequency $\eta = (\sin\theta)/\lambda$. The ray passing through the center of the lens undeviated (in the case of a thin lens) meets the refracted rays in the $x_2 y_2$ focal plane at a distance $+y_2$ from the z axis. For small angles $y_2/f \approx \theta \approx \sin\theta = \eta\lambda$ so that $\eta \approx y_2/\lambda f$. Similar remarks hold for ξ and x_2.

In Section 5.3 we show that the spatial frequencies diffracted by a subject pattern are those of the two-dimensional Fourier components of the subject. Thus, if the maximum spatial frequency in the subject is known, the

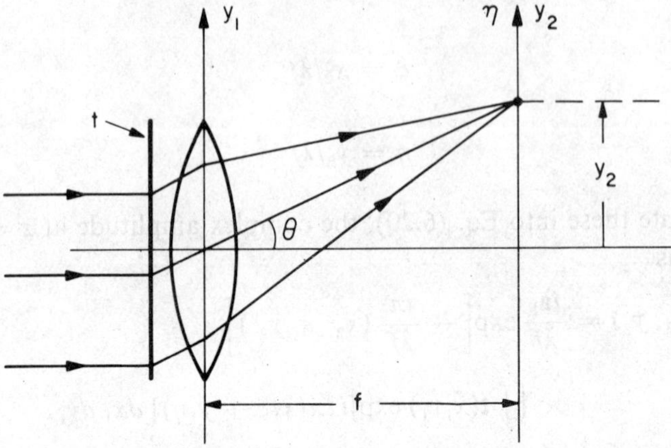

FIG. 6.3. Geometry indicating the relation between spatial frequencies and focal plane coordinates.

maximum extent of its Fourier transform displayed in the back focal plane of a given lens can be computed from Eqs. (6.21) and (6.22). If we consider only a single dimension, assume a moderate value of $|\xi_{\max}| = 10$ cycles/mm for the maximum subject spatial frequency, and assign values of $f = 500$ mm, and $\lambda = 0.5\ \mu\text{m} = 5 \times 10^{-4}$ mm, we find the maximum extent of the transform in the positive x direction, $x_{2,\max} = 2.5$ mm, to be quite small.

For some applications, where intensity is the only concern, the spherical phase factor in Eq. (6.23) is of no significance. In others it may be necessary to eliminate it. If the spherical phase factor is undesirable it can be removed by placing a converging lens of positive focal length f into the plane $z = f$.

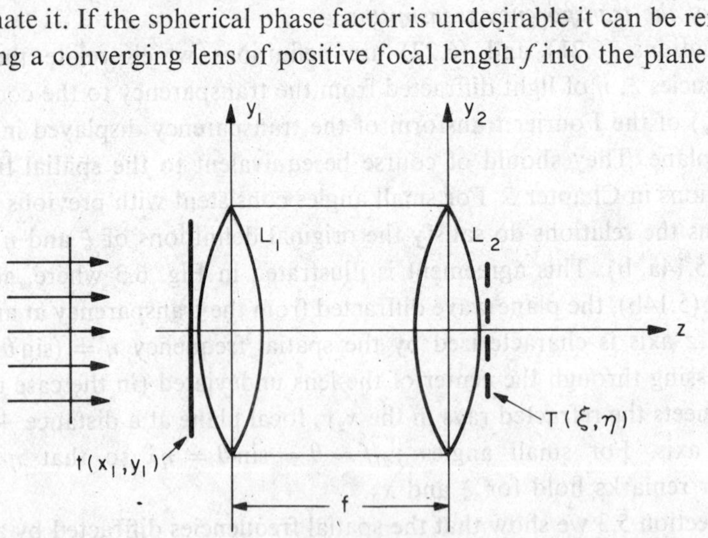

FIG. 6.4. Optical system for performing an exact Fourier transformation. L_1 and L_2 have the same focal length f.

It is evident from Eqs. (6.15) and (6.23) that immediately behind this second lens we then have the exact Fourier transform without the spherical phase factor. The optical system which results is shown in Fig. 6.4.

Let us return to the system shown in Fig. 6.2 and consider the case where the transparency is uniformly transparent, i.e., $\mathbf{t}(x_1, y_1) = 1$. The amplitude in the plane $z = f$, according to Eq. (6.23), is then

$$\mathbf{a}_2(x_2, y_2) = \frac{i\mathbf{a}_1}{\lambda f} \exp\left[\left(-\frac{i\pi}{\lambda f}\right)(x_2{}^2 + y_2{}^2)\right] \times \iint \exp[i2\pi(x_1\xi + y_1\eta]\, dx_1\, dy_1. \qquad (6.24)$$

If, for the moment, we assume that the lens is infinitely large so that the limits on the integration extend to infinity, then the integral can be considered to be the Fourier transform of unity. Correspondence (4.30), with $c = 0$, tells us that the value of the integral is $\delta(\xi) \cdot \delta(\eta) = \delta(\xi, \eta) = \delta(x_2/\lambda f, y_2/\lambda f)$, which is zero everywhere except at $x_2 = y_2 = 0$. Hence Eq.(6.24) becomes

$$\mathbf{a}_2(x_2, y_2) = (i\mathbf{a}_1/\lambda f)\, \delta(x_2/\lambda f, y_2/\lambda f), \qquad (6.25)$$

and we find that a plane wave, incident on a lens with positive focal length f, converges to a mathematical point in a plane one focal length away from the lens. The focus to a mathematical point is a consequence of our assuming an infinitely large lens. A lens of finite dimension produces a spot of finite size centered at $x_2 = y_2 = 0$. The consequences of finite lens size will be considered in Section 6.4.

6.3 A More General Optical System

There are other optical systems, in addition to those shown in Figs. 6.2 and 6.4, which can perform the Fourier transform operation. This will become evident now as we analyze a more general optical system. In this section we shall obtain not only the condition for forming the Fourier transform but also the condition for forming an image. The optical system is shown in Fig. 6.5. A spherical wave is incident on a transparency of complex amplitude transmittance $\mathbf{t}(x_1, y_1)$. The radius of curvature of the spherical wave is d_1, i.e., the wave diverges from a point at a distance d_1 to the left of $\mathbf{t}(x_1, y_1)$. At a distance d_2 to the right of the transparency there is a spherical lens of focal length f. We want to know the complex amplitude in a plane a distance d_3 to the right of the lens.

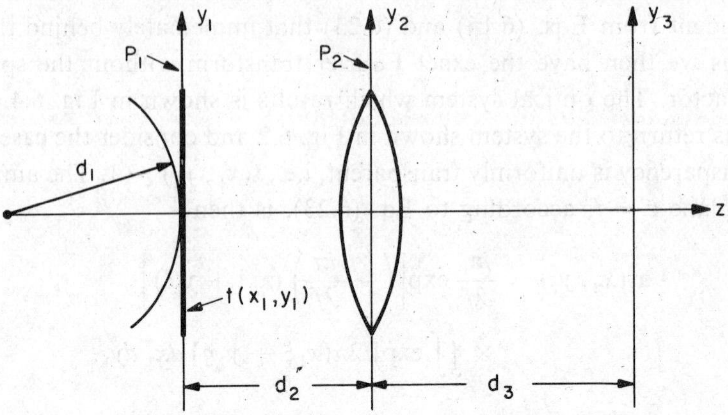

FIG. 6.5. A more general optical system. The lens has a focal length *f*.

We shall carry out the analysis using the approximate formula for propagation in free space Eq. (5.33) (the convolution form in the spatial domain) and the approximate lens transmittance Eq. (6.15). It would be simpler, and therefore preferable, to be able to use only the multiplicative forms for analyzing the passage of light through the optical system of free space and lenses. We can readily see, however, that this is not possible. If we begin with free space and work in the spatial frequency domain, the multiplicative process of Eq. (5.28) can be used. However when we consider light passing through a lens, we find that the multiplicative form, as exhibited in Eq. (6.17), applies to the spatial domain, and we must convert Eq. (6.17) into a convolution in the spatial frequency domain. On the other hand if we treat the problem in the spatial domain, we have convolutions for propagation in free space and multiplications for passage through lenses. The choice is arbitrary, and we choose to work in the spatial domain.

The analysis of the system of Fig. 6.5 involves two multiplications and two convolutions. To keep things manageable, we shall, in this section, adopt an operational notation and approach introduced by Vander Lugt [*6.2*]. The notation is a shorthand way of writing down and manipulating the relations we have already derived.

6.3.1 THE OPERATIONAL NOTATION

Equation (6.14) states that a thin lens is a transmission element with a transmittance

$$g(x, y) = \exp[(i\pi/\lambda f)(x^2 + y^2)]. \tag{6.26}$$

The form of $\mathbf{g}(x, y)$ is very similar to that of $\mathbf{h}(x, y)$ in Eq. (5.34), the quantity which is convolved with an input transmittance in the spatial-domain treatment of propagation in free space,

$$\mathbf{h}(x, y) = (i/\lambda d) \exp[-(i\pi/\lambda d)(x^2 + y^2)]. \tag{6.27}$$

The similarity between Eqs. (6.26) and (6.27) prompts one to define a function ψ as

$$\psi(x, y; p) = \exp[-(i\pi p/\lambda)(x^2 + y^2)] \tag{6.28}$$

where p stands for any parameter. Passage of a wave through a spherical lens of focal length f can now be described by multiplying the complex amplitude arriving at the lens by $\psi^*(x, y; F)$. The asterisk signifies the complex conjugate and

$$F = 1/f. \tag{6.29}$$

Propagation of a wave through a distance d in space can be described by convolving the complex amplitude with $(i/\lambda)D \cdot \psi(x, y; D)$ where[2]

$$D = 1/d. \tag{6.30}$$

We list some properties of $\psi(x, y; p)$ which will be helpful later. The following equations can be verified by inserting Eq. (6.28):

$$\psi(x, y; p) = \psi^*(x, y; -p). \tag{6.31}$$

$$\psi(-x, -y; p) = \psi(x, y; p). \tag{6.32}$$

$$\psi(x, y; p_1)\psi(x, y; p_2) = \psi(x, y; p_1 + p_2). \tag{6.33}$$

$$\begin{aligned} \psi(x, y; p_1)\psi^*(x, y; p_2) &= \psi(x, y; p_1 - p_2) \\ &= \psi^*(x, y; p_2 - p_1). \end{aligned} \tag{6.34}$$

$$\psi(cx, cy; p) = \psi(x, y; c^2 p). \tag{6.35}$$

$$\psi(x - u, y - v; p) = \psi(x, y; p)\psi(u, v; p) \exp[(+i2\pi p/\lambda)(ux + vy)]. \tag{6.36}$$

Notice that the relation

$$\psi^*(x, y; 0) = 1 \tag{6.37}$$

is a statement of the fact that a lens of infinite focal length does not change the amplitude distribution incident on it.

[2] To the reader consulting Vander Lugt's paper it should be pointed out that he uses upper case F and D for our lower case f and d and vice versa. His definition of time dependence, furthermore, is opposite to ours, as discussed in Sections 3.1 and 5.2.

This operational notation will now be applied to the analysis of the optical system shown in Fig. 6.5. The diverging spherical wave incident on the transparency $\mathbf{t}(x_1, y_1)$ in plane P_1 is represented by $\psi(x_1, y_1; D_1)$ [see discussion of Eq. (6.23)], and the light transmitted to the back of the transparency is the product

$$\mathbf{a}_t(x_1, y_1) = \psi(x_1, y_1; D_1)\mathbf{t}(x_1, y_1). \tag{6.38}$$

Convolution of $\mathbf{a}_t$ with $(iD_2/\lambda)\psi(x, y; D_2)$ gives the amplitude distribution on the left surface of the lens

$$\mathbf{a}_l(x_2, y_2) = \frac{iD_2}{\lambda}\int_{P_1}\int \mathbf{a}_t(x_1, y_1)\psi(x_2 - x_1, y_2 - y_1; D_2)\,dx_1\,dy_1, \tag{6.39}$$

and multiplication of $\mathbf{a}_l$ by the lens transmittance $\psi^*(x_2, y_2; F)$ produces the complex amplitude at the right surface

$$\mathbf{a}_r(x_2, y_2) = \mathbf{a}_l(x_2, y_2)\psi^*(x_2, y_2; F). \tag{6.40}$$

Finally we obtain the complex amplitude $\mathbf{a}(x_3, y_3)$ in an xy plane a distance d_3 from the lens by convolving $\mathbf{a}_r$ with $(iD_3/\lambda)\psi(x, y; D_3)$:

$$\mathbf{a}(x_3, y_3) = \frac{iD_3}{\lambda}\int_{P_2}\int \mathbf{a}_r(x_2, y_2)\psi(x_3 - x_2, y_3 - y_2; D_3)\,dx_2\,dy_2. \tag{6.41}$$

Equation (6.41) can be put into a more desirable form by (1) separating the ψ functions of Eqs. (6.39) and (6.41) into factors each dependent on only one coordinate plane [through the use of Eq. (6.36)], (2) substituting Eqs. (6.39) and (6.40) into Eq. (6.41), and (3) collecting factors which are functions of the same x, y coordinates [through the use of Eqs. (6.31)–(6.34)]. The result is

$$\begin{aligned}\mathbf{a}(x_3, y_3) = &-\left(\frac{D_2D_3}{\lambda^2}\right)\psi(x_3, y_3; D_3)\\ &\times \int_{P_1}\int\int_{P_2}\int \psi(x_1, y_1; D_1+D_2)\mathbf{t}(x_1, y_1)\psi(x_2, y_2; D_2 - F+D_3)\\ &\times \exp\left\{\frac{i2\pi}{\lambda}[x_2(D_2x_1 + D_3x_3) + y_2(D_2y_1 + D_3y_3)]\right\}\\ &\times dx_1\,dy_1\,dx_2\,dy_2.\end{aligned} \tag{6.42}$$

6.3.2 The Imaging Condition

We shall first show that the output function $\mathbf{a}(x_3, y_3)$ in Eq. (6.42) has the form of the input function $\mathbf{t}(x_1, y_1)$, and hence is the image of $\mathbf{t}(x_1, y_1)$, under the condition for imaging obtained from geometrical optics. The latter condition, written for the parameters of the optical system of Fig. 6.5, is

$$\frac{1}{d_2} + \frac{1}{d_3} = \frac{1}{f} \tag{6.43}$$

or, in the notation we have adopted in this chapter

$$D_2 + D_3 = F. \tag{6.44}$$

When Eq. (6.44) is substituted into the second ψ factor under the integral of Eq. (6.42), we have that $\psi(x_2, y_2; D_2 - F + D_3) = 1$ and the integral over P_2 becomes

$$\int_{P_2}\int 1 \cdot \exp\left\{\frac{i2\pi}{\lambda}\,[x_2(D_2x_1 + D_3x_3) + y_2(D_2y_1 + D_3y_3)]\right\} dx_2\, dy_2$$
$$= \delta\left(\frac{D_2x_1 + D_3x_3}{\lambda}, \frac{D_2y_1 + D_3y_3}{\lambda}\right) \tag{6.45}$$

where we have used correspondence (4.30) with $c = 0$. Writing the delta function in the form

$$\delta\left[\frac{D_2}{\lambda}\left(x_1 + \frac{D_3}{D_2}\,x_3\right), \frac{D_2}{\lambda}\left(y_1 + \frac{D_3}{D_2}\,y_3\right)\right],$$

we employ a two-dimensional version of Eq. (4.13d), i.e., $\delta(ax, by) = (1/|\,ab\,|)\,\delta(x, y)$, to obtain

$$\delta\left(\frac{D_2x_1 + D_3x_3}{\lambda}, \frac{D_2y_1 + D_3y_3}{\lambda}\right) = \frac{\lambda^2}{D_2{}^2}\,\delta\left(x_1 + \frac{D_3}{D_2}\,x_3, y_1 + \frac{D_3}{D_2}\,y_3\right). \tag{6.46}$$

Substituting the above back into Eq. (6.42) produces

$$\mathbf{a}(x_3, y_3) = -\frac{D_3}{D_2}\psi(x_3, y_3; D_3)$$
$$\times \int_{P_1}\int \psi(x_1, y_1; D_1 + D_2)\mathbf{t}(x_1, y_1)$$
$$\times\, \delta\left(x_1 + \frac{D_3}{D_2}\,x_3, y_1 + \frac{D_3}{D_2}\,y_3\right) dx_1\, dy_1$$
$$= -\frac{D_3}{D_2}\psi\left[x_3, y_3; D_3 + \left(\frac{D_3}{D_2}\right)^2 (D_1 + D_2)\right]$$
$$\times\, \mathbf{t}\left(-\frac{D_3x_3}{D_2}, -\frac{D_3y_3}{D}\right) \tag{6.47}$$

where we have used the property that the convolution of a function with the delta function is the function itself [see Eq. (4.13e)] and have used Eqs. (6.32)–(6.35) to make the expression more compact. The ψ function in Eq. (6.47) is a spherical phase factor which in many imaging application is of little consequence. Generally the image is recorded as intensity, $\mathbf{aa}^*$ and the phase factor cancels out. Under such conditions the imaging operation is not affected by D_1, the curvature of the illuminating beam. The amplitude function remaining in Eq. (6.47) is

$$\mathbf{t}\left(-\frac{D_3}{D_2}x_3, -\frac{D_3}{D_2}y_3\right) \tag{6.48}$$

an inverted, scaled image of $\mathbf{t}(x_1, y_1)$ with a magnification

$$M = -D_2/D_3 = -d_3/d_2. \tag{6.49}$$

6.3.3 The Fourier Transform Condition

Let us now return to Eq. (6.42) and determine the conditions under which the output complex amplitude $\mathbf{a}(x_3, y_3)$ displayed in the plane x_3y_3 is the Fourier transform of the input transmittance $\mathbf{t}(x_1, y_1)$. Since the desired Fourier transform relation must hold for the complex amplitudes displayed in the planes x_1y_1 and x_3y_3, we must eliminate from Eq. (6.42) amplitudes expressed in the x_2, y_2 coordinates of the plane P_2. To emphasize this point we rewrite Eq. (6.42) as

$$\mathbf{a}(x_3, y_3) = \left(-\frac{D_2D_3}{\lambda^2}\right)\psi(x_3, y_3; D_3) \times \int_{P_1}\int \psi(x_1, y_1; D_1 + D_2)\mathbf{t}(x_1, y_1)\mathbf{I}_2\, dx_1\, dy_1 \tag{6.50}$$

where

$$\mathbf{I}_2 = \int_{P_2}\int \psi(x_2, y_2; D_2 - F + D_3)\exp\left\{i2\pi\left[x_2\left(\frac{D_2x_1 + D_3x_3}{\lambda}\right) + y_2\left(\frac{D_2y_1 + D_3y_3}{\lambda}\right)\right]\right\} dx_2\, dy_2 \tag{6.51}$$

$$= \int_{P_2}\int \psi(x_2, y_2; D_2 - F + D_3)\exp\{i2\pi(x_2\xi + y_2\eta)\}\, dx_2\, dy_2 \tag{6.52}$$

and

$$\xi = \left(\frac{D_2x_1 + D_3x_3}{\lambda}\right), \quad \eta = \left(\frac{D_2y_1 + D_3y_3}{\lambda}\right). \tag{6.53}$$

Variables x_2 and y_2 can be eliminated by carrying out the integration of the Fourier integral in Eq. (6.52). The function $\psi(x_2, y_2; D_2 - F + D_3)$ is a two-dimensional Gaussian whose transform $\mathbf{I}_2(x_1, x_3)$ can be computed with the help of Eq. (4.27) and manipulated, using the properties of the ψ functions, into the form

$$\mathbf{I}_2(x_1, x_3) = \frac{\lambda}{i(D_2 - F + D_3)} \times \psi^*\left(x_1 + \frac{D_3}{D_2} x_3, y_1 + \frac{D_3}{D_2} y_3; \frac{D_2{}^2}{D_2 - F + D_3}\right). \tag{6.54}$$

A final manipulation of $\mathbf{I}_2$ is accomplished through application of Eqs. (6.36) and (6.35) with the result:

$$\mathbf{I}_2 = \frac{\lambda}{i(D_2 - F + D_3)} \times \psi^*\left(x_1, y_1; \frac{D_2{}^2}{D_2 - F + D_3}\right)\psi^*\left(x_3, y_3; \frac{D_3{}^2}{D_2 - F + D_3}\right) \times \exp\left\{\frac{i2\pi}{\lambda}\left(\frac{D_2 D_3}{D_2 - F + D_3}\right)(x_1x_3 + y_1y_3)\right\}. \tag{6.55}$$

We can now insert $\mathbf{I}_2$ into Eq. (6.50), collect ψ functions of the same coordinates using Eq. (6.34), and arrive at the following representation of the complex amplitude at a distance d_3 from the lens:

$$\mathbf{a}(x_3, y_3) = \frac{iD_2D_3}{\lambda(D_2 - F + D_3)}\psi\left(x_3, y_3; D_3 - \frac{D_3{}^2}{D_2 - F + D_3}\right) \times \int_{P_1}\int \psi\left\{x_1, y_1; D_1 + D_2 - \frac{D_2{}^2}{D_2 - F + D_3}\right\}\mathbf{t}(x_1, y_1) \times \exp\left\{i2\pi\left[\frac{D_2D_3}{\lambda(D_2 - F + D_3)}\right](x_1x_3 + y_1y_3)\right\} dx_1\, dy_1. \tag{6.56}$$

The integral over P_1 in Eq. (6.56) has the form of a Fourier transform providing the ψ function under the integral is unity. This is so when

$$D_1 + D_2 - \frac{D_2{}^2}{D_2 - F + D_3} = 0. \tag{6.57}$$

Let us suppose that the transparency $\mathbf{t}(x_1, y_1)$ is illuminated by a plane wave so that $D_1 = 1/d_1 = 0$ and $d_1 = \infty$. Then, in this case,

$$D_3 = F \quad \text{or} \quad d_3 = f, \tag{6.58}$$

and Eq. (6.56) becomes

$$\mathbf{a}(x_3, y_3) = \frac{iF}{\lambda}\psi\left(x_3, y_3; F - \frac{F^2}{D_2}\right)$$
$$\times \int_{P_1}\int \mathbf{t}(x_1, y_1) \exp\left[\frac{i2\pi F}{\lambda}(x_1x_3 + y_1y_3)\right] dx_1\, dy_1. \quad (6.59)$$

We see that, apart from a spherical phase factor, plane wave illumination of a transparency $\mathbf{t}(x_1, y_1)$ situated in front of a lens produces in the back focal plane of the lens a complex amplitude distribution which has the form of the Fourier transform of $\mathbf{t}(x_1, y_1)$. This holds independent of the distance d_2 separating lens and transparency. The spherical phase factor can be reduced to unity by causing

$$D_2 = F, \quad (6.60)$$

i.e., placing the transparency $t(x_1, y_1)$ in the front focal plane of the lens. This system, shown in Fig. 6.6, is the usual method of displaying the Fourier

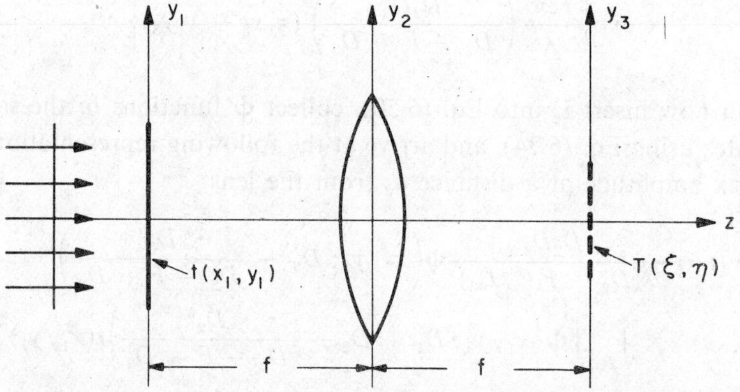

FIG. 6.6. Another optical system for performing the exact Fourier transformation. The lens has a focal length f.

transform of an input transparency. With the substitution of Eq. (6.60), Eq. (6.59) becomes

$$\mathbf{a}(x_3, y_3) = \frac{i}{\lambda f}\int_{P_1}\int \mathbf{t}(x_1, y_1) \exp[i2\pi(x_1\xi + y_1\eta)]\, dx_1\, dy_1 \quad (6.61)$$

where

$$\xi = x_3/f\lambda \quad \text{and} \quad \eta = y_3/f\lambda \quad (6.62)$$

are the coordinates of the spatial frequency plane.

It should be noted that with ξ and η chosen to be positive, the sign in the exponential factor under the integral of Eq. (6.61) is correct for the transformation of a spatial distribution to a spectral distribution but is not correct for the opposite transformation. The analog transformations performed by optical systems, as in Fig. 6.6, are always characterized by a plus sign in the exponential if ξ and η are chosen positive. To bring the optical transforms into line with what one would compute from Eqs. (4.1) and (4.2), the coordinates on any lens back focal plane which displays the transform from the spatial frequency to the spatial distribution are inverted with respect to those on the front focal plane displaying the spatial frequency distribution. On the other hand if the front focal plane displays a spatial distribution, the coordinates in the back focal plane need not be inverted with respect to this plane. The convention is illustrated in Fig. 6.7.

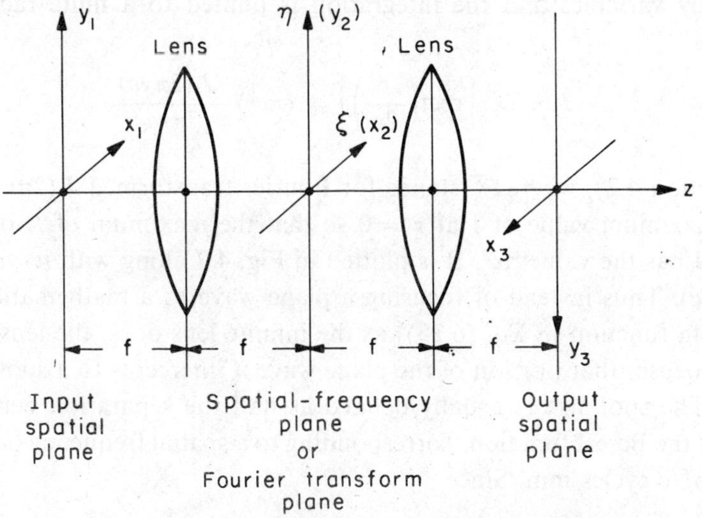

FIG. 6.7. Orientation of the coordinate axes of planes displaying optical Fourier transformations.

If the input transparency $\mathbf{t}(x_1, y_1)$ is illuminated by a spherical wave ($D_1 \neq 0$), it is apparent from Eq. (6.56) that the Fourier transform forms somewhere other than in the back focal plane [D_3 is given by Eq. (6.57)]. Furthermore, since D_3 will no longer equal F, the scale factor of the Fourier transform

$$\frac{D_2 D_3}{\lambda(D_2 - F + D_3)}$$

is a function of D_2. This offers the possibility of making a variable-scale Fourier transform system [*6.2*].

6.4 The Effect of Finite Lens Size

6.4.1 Effect on the Spatial Frequency Plot

Our discussion of optical Fourier transformation in Section 6.2 requires the assumption of an infinitely large lens so that the lens transmittance can be considered as a pure phase factor over infinite limits. Suppose now that the lens has a finite radius c, and let us reconsider the Fourier integral in Eq. (6.24) for the case $\mathbf{t}(x_1, y_1) = 1$,

$$I_1 = \iint_{-\infty}^{\infty} \exp[i2\pi(x_1\xi + y_1\eta)]\, dx_1\, dy_1.$$

When I_1 is expressed in cylindrical coordinates for both spatial and spatial frequency variables and the integration is limited to a finite radius c, I_1 becomes

$$I_1 = \mathscr{F}\left[\operatorname{rect}\left(\frac{r}{2c}\right)\right] = (\pi c^2)\frac{J_1(2\pi\nu c)}{\pi\nu c} \tag{6.63}$$

(see Section 4.2), where $\mathscr{F}$ stands for Fourier transform. $[J_1(2\pi\nu c)]/(\pi\nu c)$ has a maximum value of 1 at $\nu = 0$ so that the maximum of I_1 occurs on axis and has the value πc^2. It is plotted in Fig. 4.7 along with its transform $\operatorname{rect}(r/2c)$. Thus instead of focusing a plane wave to a mathematical point [the delta function in Eq. (6.25)] as the infinite lens does, the lens of finite radius focuses that portion of the plane wave it intercepts to a spot of finite width. The spot size is usually defined as half the separation between the zeros of the Bessel function, corresponding to a spatial frequency bandwidth $\nu_0 = 0.61/c$ cycles/mm. Since

$$\nu = (\xi^2 + \eta^2)^{1/2} = (1/\lambda f)(x_2{}^2 + y_2{}^2)^{1/2} = r/\lambda f, \tag{6.64}$$

from Eqs. (6.21) and (6.22), we can convert bandwidth to spatial distance and obtain for Δ the width (diameter) of the spot in the x_2y_2 plane

$$\Delta = 0.61\lambda f/c. \tag{6.65}$$

We may consider the width Δ in Eq. (6.65) a measure of the degree of uncertainty with which a point in the x_2y_2 spatial frequency plane represents the spatial frequency of the axial plane wave incident on the lens of finite radius c. The uncertainty in spatial frequency is a consequence of the inability of the finite lens to gather all the spatial light information.

Let us now consider a transparency with a transmittance which is no longer unity and which is represented in polar coordinates by $\mathbf{t}(r, \theta)$. The Fourier integral in Eq. (6.23) produced by the optical Fourier transform system of Fig. 6.2 becomes

$$\mathbf{I}_t = \mathscr{F}\left[\mathbf{t}(r, \theta)\,\mathrm{rect}\left(\frac{r}{2c}\right)\right]$$

$$= \mathbf{T}(\nu, \varphi) * \left[(\pi c)^2 \frac{J_1(2\pi\nu c)}{\pi\nu c}\right] \tag{6.66}$$

where $\mathbf{t}(r, \theta) \supset \mathbf{T}(\nu, \varphi)$. As noted in Section 4.3 the convolution of two functions represents the scanning of one function by the other. The function $\mathbf{T}(\nu, \varphi)$, the Fourier transform of $\mathbf{t}(r, \theta)$ displayed by an infinitely large lens, can be considered as composed of ideal points or delta functions. Each of the latter when convolved with the spot $(\pi c)^2[J_1(2\pi\nu c)/\pi\nu c]$ of width $\Delta = 0.61\lambda f/c$ broadens to the width Δ.

6.4.2 Effect on the Choice of Imaging and Transform Systems

Let us next consider the effects of the finite lens size on an *imaging system*. For perfect imaging, the lens must gather all the light information coming from the subject. An imaging system demonstrably poor in this respect is the one illustrated in Fig. 6.8, where the subject transparency $\mathbf{t}(x_1, y_1)$ is illuminated by a plane wave. We confine our attention to propagation in the *yz* plane. The input transparency $\mathbf{t}$ can be imagined to be decomposed into its Fourier sinusoidal-grating components. One of these might be characterized by the spatial frequency η. The situation is similar to that illustrated in Fig. 5.3, with the exception that the cross section of the light beam incident on the grating is limited by the spatial extent of $\mathbf{t}(x_1, y_1)$. In the case of a plane wave incident on a grating of infinite extent, as in Section 5.3, the diffracted waves are plane waves whose complex amplitudes just to the right of $\mathbf{t}(x_1, y_1)$ are given by Eq. (5.19)

$$\mathbf{a}(x_1, y_1) = a_1 t_0 + \tfrac{1}{2} a_1 t_1 \exp(2\pi i \eta y_1) + \tfrac{1}{2} a_1 t_1 \exp(-2\pi i \eta y_1) \tag{6.67}$$

and whose diffraction angles are

$$\theta_2 = \pm \arcsin \lambda\eta\,.$$

If we neglect those diffraction effects caused by the edge of the transparency, we can consider every small bundle of rays incident on any small area of the grating to diffract in the directions θ_2.

First consider a bundle of rays centered on the z axis (Fig. 6.8). The maximum angle through which these rays can be diffracted and still be intercepted by the lens of radius r_2 is limited by the ratio $r_2/d_2 = \tan\theta_2$. To simplify the results suppose $r_2/d_2 = \tan\theta_2 \approx \sin\theta_2$. For central rays, then, we see that the maximum spatial frequency $\eta_{\max}$ in the input plane that can be transferred to the image by the lens is given by the condition

$$\sin\theta_2 = \lambda\eta_{\max} \leq r_2/d_2$$

or

$$\eta_{\max} \leq r_2/d_2\lambda. \tag{6.68}$$

Now suppose that the transparency **t** is circular with radius r_1 and consider a bundle of rays incident near the outer edge $y_1 = r_1$. If the lens is to intercept all the rays from this part of the transparency, the diffraction angle θ_2' must not exceed $(r_2 - r_1)/d_2$, i.e.,

$$\tan\theta_2' \approx \sin\theta_2' = \lambda\eta'_{\max} \leq (r_2 - r_1)/d_2$$

or

$$\eta'_{\max} \leq (r_2 - r_1)/d_2\lambda. \tag{6.69}$$

Thus the maximum spatial frequency transferred to the image by the lens decreases linearly as a function of subject radius. Information is lost unless the subject spatial frequencies are small enough to obey Eq. (6.69).

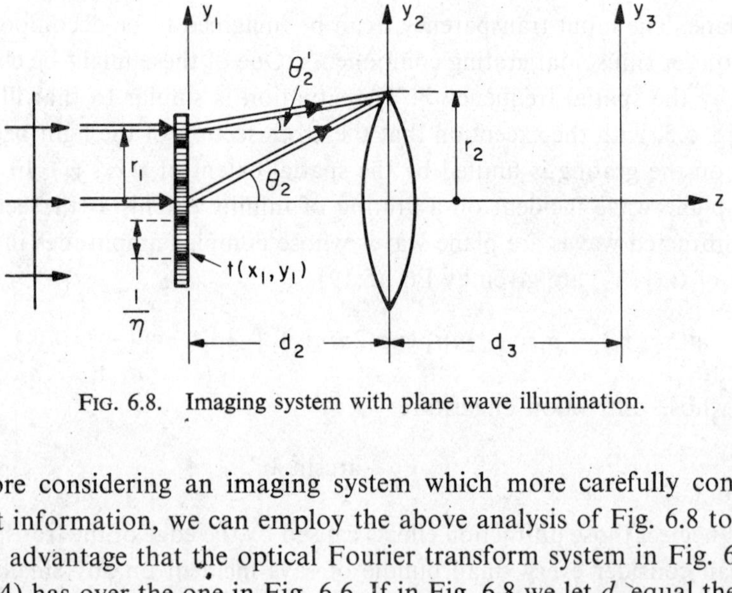

FIG. 6.8. Imaging system with plane wave illumination.

Before considering an imaging system which more carefully conserves subject information, we can employ the above analysis of Fig. 6.8 to point out an advantage that the optical Fourier transform system in Fig. 6.2 (or Fig. 6.4) has over the one in Fig. 6.6. If in Fig. 6.8 we let d_3 equal the focal length of the lens, we obtain an optical Fourier transform system which, under the assumption of an infinitely large lens, will display $\mathbf{T}(\xi, \eta)$ in the x_3y_3 plane regardless of the value of d_2. [The spherical phase factor in Eq. (6.59), which we have neglected here, does depend on d_2.] However for the physically realizable case of a finite lens of radius r_2, Eqs. (6.68) and (6.69) tell us that the maximum spatial frequencies intercepted by the lens are inversely proportional to the distance d_2 separating the transparency $\mathbf{t}(x_1, y_1)$ and the transforming lens. When $d_2 = 0$ as in Fig. 6.2 or 6.4, all spatial frequencies of the subject diffraction are accepted by the lens, and the *complete* spectrum of $\mathbf{t}(x_1, y_1)$ can be plotted in the back focal plane. In this respect such systems are superior to that of Fig. 6.6, where $d_2 = f$.

An imaging system which is efficient in conserving subject light information is that illustrated in Fig. 6.9. The first lens L_1, adjacent to the subject

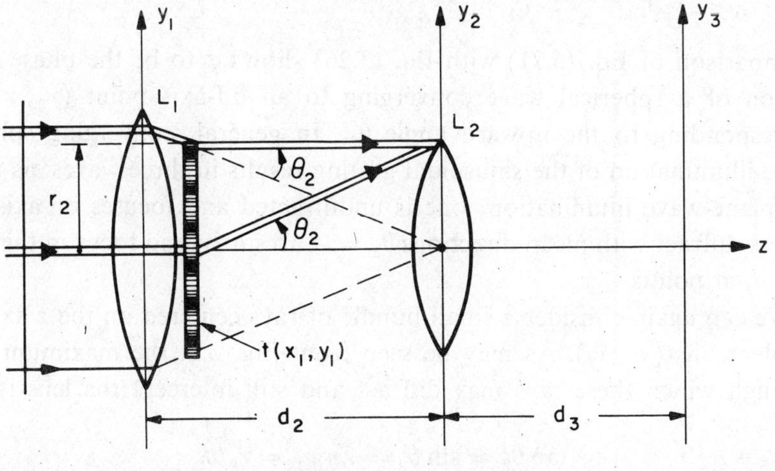

FIG. 6.9. Imaging system with spherical wave illumination. The focal length of lens L_1 is d_2 and that of lens L_2 is f.

transparency $\mathbf{t}(x_1, y_1)$, provides spherical-wave illumination which converges to an on-axis focus at a second lens L_2. Since the latter is located in the back focal plane of the first lens, the Fourier transform or spectrum of $\mathbf{t}(x_1, y_1)$ is plotted over the surface of L_2. This pattern is in turn transformed by L_2 into the image of $\mathbf{t}(x_1, y_1)$ in the plane x_3y_3. Once again consider the Fourier sinusoidal-grating component of $\mathbf{t}(x_1, y_1)$ with spatial frequency η, and let us confine the analysis to the yz plane. The sinusoidal component this time is illuminated by a spherical wave. If we substitute for a_1, in Eq. (6.67), a spherical phase factor corresponding to a spherical wave converging to a point focus at a distance d_2 from $\mathbf{t}(x_1, y_1)$, the third term on the right of Eq. (6.67) becomes

$$\frac{1}{2} a_1 t_1 \exp(-2\pi i \eta y_1) \propto \exp\left[i \frac{2\pi}{\lambda}\left(\frac{1}{2d_2}\right)(x_1{}^2 + y_1{}^2)\right]$$
$$\times \exp\left[-i \frac{2\pi}{\lambda}(\lambda\eta) y_1\right]$$
$$= \exp\left\{i \frac{2\pi}{\lambda}\left(\frac{1}{2d_2}\right)[x_1{}^2 + y_1{}^2 - 2(d_2 \sin\theta_2) y_1]\right\}. \tag{6.70}$$

For small angles $d_2 \sin\theta_2 \approx d_2\theta_2 \approx y_2$. With the substitution of y_2 the phase in Eq. (6.70) is

$$\varphi = \frac{2\pi}{\lambda}\left(\frac{1}{2d_2}\right)[x_1{}^2 + y_1{}^2 - 2y_2 y_1]. \tag{6.71}$$

Comparison of Eq. (6.71) with Eq. (3.26) shows φ to be the phase distribution of a spherical wave converging to an off-axis point $(y_2, z = d_2)$ corresponding to the upward angle θ_2. In general, converging spherical wave illumination of the sinusoidal grating results in three waves, as is true for plane-wave illumination; one is undiffracted and focuses on-axis; the others diffract with mean directions $\theta_2 = \pm\arcsin \lambda\eta$ and focus in the plane $z = d_2$ at points $\pm y_2$.

We can again consider a small bundle of rays centered on the z axis and incident on $\mathbf{t}(x_1, y_1)$. As may be seen from Fig. 6.9, the maximum angle through which these rays may diffract and still intercept the lens is

$$\tan\theta_2 \approx \sin\theta_2 = \lambda\eta_{\max} = r_2/d_2. \tag{6.72}$$

However this same condition holds (within the range of our approximations) for the bundle of rays incident on the extremities of the subject. If we generalize the analysis to include sinusoidal grating components of arbitrary orientation, then Eq. (6.72) can be expressed as

$$\lambda\nu_{\max} = r_2/d_2$$

or

$$\nu_{\max} = r_2/d_2\lambda \tag{6.73}$$

where $\nu = (\xi^2 + \eta^2)^{1/2}$ as in Eq. (6.64). Equation (6.73) states that the lens, in the optical system of Fig. 6.9, transfers all available light information to the image for spatial frequency components of the subject below ν_{max} and does this independent of position in the input plane. For spatial frequency components of the subject with frequencies greater than ν_{max}, no information is transferred. If we describe these characteristics in terms of a frequency transfer function [see definition in Eq. (4.6)], we see that the latter is constant up to ν_{max} and zero beyond ν_{max}. This function, sometimes called a "top-hat" function, is plotted in Fig. 6.10. For an imaging system employing coherent light it is also called the *coherent transfer function*.

6.4.3 Effect on Image Resolution

To find the spread function $\mathbf{s}(x_3, y_3)$ of the imaging system of Fig. 6.9 we must, in accordance with Eq. (4.15), find the inverse Fourier transform of the frequency transfer function $\mathbf{S}(\nu)$. The latter, plotted in Fig. 6.10, is the function rect $(\nu/2\nu_{max})$ whose transform is of the form given by Eq. (4.34). In the present case the variables of the transform are in the spatial domain, and the spread function (normalized to a maximum value of unity) is

$$s(r) = \frac{J_1(2\pi\nu_{max}r)}{\pi\nu_{max}r} \tag{6.74}$$

where $r = (x_3^2 + y_3^2)^{1/2}$. A plot of $s(r)$ versus r is given in Fig. 6.11.

We have seen that when the illuminating beam focuses in the plane of the imaging lens, as in the optical system of Fig. 6.9, the frequency transfer function is the same for every position in the input plane and ν_{max} is not a function of the input-plane coordinates. Consequently, the image resolution, as defined by the spread function, is independent of position in the input. In accordance with Section 4.3 the output or image of a linear, space invariant optical system is the convolution of the input with the spread function. In the convolution process each delta function point of the input is broadened in the output image to a spot of width $\Delta = 0.61/\nu_{max} = 0.61 d_2\lambda/r_2$ The width Δ is that of the smallest resolvable spot in the image. [Here we consider a magnification of unity. Otherwise Δ is multiplied by d_3/d_2, see Eq. (6.49).] A single transfer function cannot describe the optical system of Fig. 6.8 where plane wave illumination is used, nor can it describe systems where spherical illuminating waves do not focus in the plane of the imaging lens. Such systems where resolution varies as a function of position in the input plane are called *space variant*.

In this section we have discussed some limitations imposed on the performance of an optical system by the finite size of physically realizable

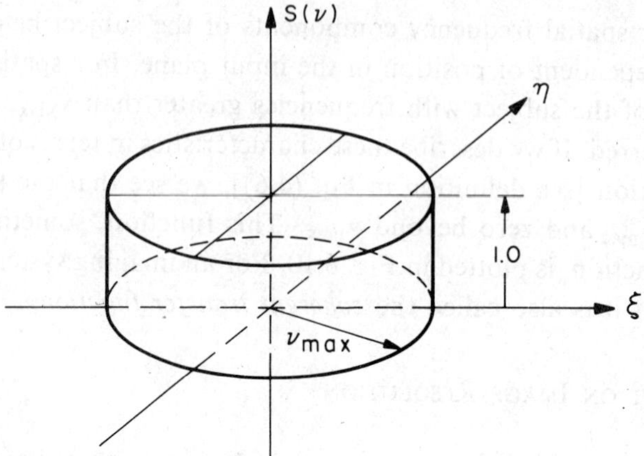

FIG. 6.10. Coherent transfer function, rect($\nu/2\nu_{max}$), of the optical system in Fig. 6.9.

lenses. If the limit on performance imposed by the lens is due only to its finite size, then we say the lens is diffraction limited. Our analysis contains many instances where "small-angle approximations" are made. When light rays are incident on thin lenses at angles which exceed the values for which these approximations are valid, degradation from optimum diffraction-limited performance becomes apparent. This occurs as we depart from the use of small, on-axis subjects and small ratios of lens diameter to focal length.

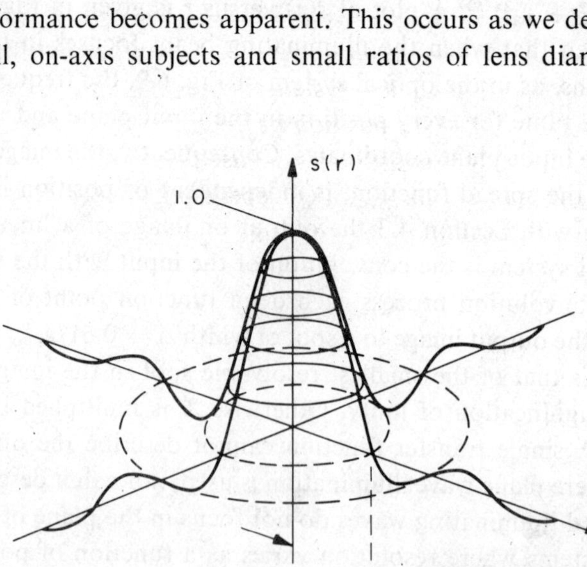

FIG. 6.11. Spread function of the imaging system of Fig. 6.9.

6.5 Coherent and Incoherent Transfer Functions

Although incoherent light imaging is normally not important in holography, it is convenient at this point to make a brief excursion and compare the frequency transfer function of a coherently illuminated system with that of a system illuminated incoherently. Suppose that a coherently illuminated optical system responds to an input complex amplitude $\mathbf{a}_1(x, y)$ or $\mathbf{a}_2(x, y)$ by producing an output complex amplitude $\mathbf{b}_1(x, y)$ or $\mathbf{b}_2(x, y)$, respectively. If the system is linear, then an input complex amplitude $\mathbf{a}_1(x, y) + \mathbf{a}_2(x, y)$ must produce an output complex amplitude $\mathbf{c}(x, y) = \mathbf{b}_1(x, y) + \mathbf{b}_2(x, y)$. The output intensity $I = \mathbf{cc}^*$ expands into

$$\begin{aligned} I &= \mathbf{b}_1\mathbf{b}_1{}^* + \mathbf{b}_2\mathbf{b}_2{}^* + \mathbf{b}_1\mathbf{b}_2{}^* + \mathbf{b}_1{}^*\mathbf{b}_2 \\ &= I_1 + I_2 + \mathbf{b}_1\mathbf{b}_2{}^* + \mathbf{b}_1{}^*\mathbf{b}_2 \end{aligned} \tag{6.75}$$

where I_1 is the output intensity due to $\mathbf{a}_1$ alone and I_2 the output intensity due to $\mathbf{a}_2$ alone. All four terms of Eq. (6.75) are in general nonzero for the case of coherent illumination. The system is therefore not linear in intensity. However for incoherent illumination, there are only the individual intensities

$$I = I_1 + I_2,$$

showing that *incoherently illuminated systems are linear in intensity.* They can be characterized by a spread function and frequency transfer function for intensity. The latter is known as the *optical transfer function* and its modulus is called the *modulation transfer function.*

The spread function for intensity is the distribution of light intensity over the output plane corresponding to an impulse function in the input plane and, as such, is the square of the absolute value of the complex spread function for amplitude. For the imaging system of Fig. 6.9 this is given by

$$s_{\mathrm{I}}(r) = \left[\frac{J_1(2\pi\nu_{\max}r)}{\pi\nu_{\max}r}\right]^2. \tag{6.76}$$

Paralleling the definition for the coherent system, the optical transfer function is the Fourier transform of $s_{\mathrm{I}}(r)$ and, according to the correspondences of Eqs. (4.18) and (4.34), it is the autocorrelation of the coherent transfer function $\operatorname{rect}(\nu/2\nu_{\max})$. Both the optical transfer function and the coherent transfer function are rotationally symmetric, depending only on ν. Cross sections of the magnitudes of these functions are compared in Fig. 6.12.

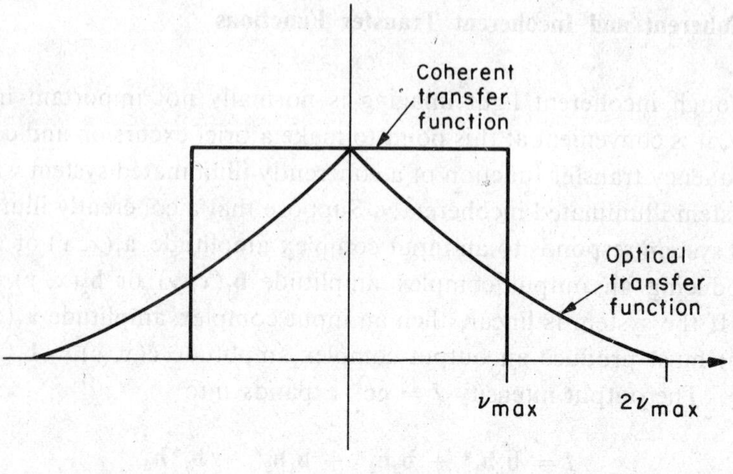

FIG. 6.12. Cross sections of coherent transfer function and optical transfer function for the system of Fig. 6.9.

REFERENCES

6.1. F. A. Jenkins and H. E. White, *Fundamentals of Optics*, 3rd ed. McGraw-Hill, New York, 1957.

6.2. A. Vander Lugt, "Operational Notation for the Analysis and Synthesis of Optical Data-Processing Systems," *Proc. Inst. Elec. Electron. Eng.* **54**, 1055 (1966).

Chapter 7

LIGHT SOURCES AND OPTICAL TECHNIQUE

As laser sources became available for forming holograms, holography began its transformation to a practical science. Before the laser, the standard sources were mercury discharge lamps which represented at best a compromise between coherence and output intensity. Spatial coherence could be acquired only by reducing the area of the source and thereby drastically reducing the useful output. In contrast, laser radiation can be both highly coherent and intense. Yet, even with a laser source, holography can be difficult and unrewarding if the holographer is unaware of laser properties and requisite optical technique.

In this chapter we examine the coherence properties of continuous wave light sources as they relate to hologram formation and to wavefront reconstruction. Methods of deploying laser beams to obtain the desired subject and reference illumination are discussed, and we present a simple example of an optical arrangement and procedure employed in practical holography.

7.1 Light Sources for Hologram Formation

The requirements placed on light sources to be used for hologram formation depend on parameters determined by the subject and the arrangement of necessary optical components. Figures 7.1a and 7.1b illustrate two common methods of deriving subject and reference waves from a single source. In Fig. 7.1a they are derived by *wavefront division* and in Fig. 7.1b by *amplitude division.* In each case the waves reaching the hologram originate in a source wavefront whose angular extent is $2\theta_0$. As indicated in Fig. 1.18, the degree of spatial or lateral coherence $|\mu_s|$ of a source depends not only on the source size but also on θ_0. If we expect the relative phase for all of the subject and reference wavefronts to remain substantially time inde-

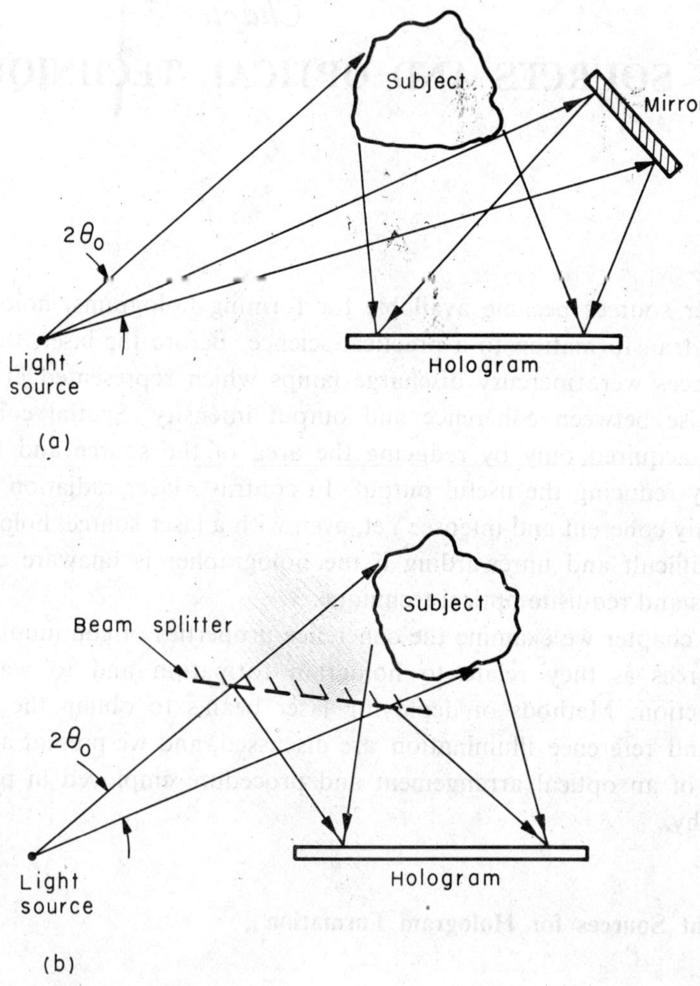

FIG. 7.1. Methods of hologram formation employing (a) wavefront division, (b) amplitude division.

pendent (permitting the recording of an interference pattern), the source emission must be sufficiently spatially coherent over $2\theta_0$. It must also have sufficient temporal or longitudinal coherence. More precisely, the maximum optical path difference for light rays traveling between source and hologram (via reflections from the subject surface or reference mirrors) must be less than the coherence length.

We state in Section 1.9.2, Eq. (1.22) that the complex degree of coherence $\gamma_{12}(\tau)$ encompasses both spatial and temporal coherence. By letting $\tau \to 0$ in Eq. (1.22) we imply that we are concerned only with the lateral aspect of the source coherence. If the correlation of $\mathbf{v}_{P_1}$ and $\mathbf{v}_{P_2}$ in Eq. (1.22) is a function of τ alone, we imply concern only with the longitudinal coherence. This is a somewhat artificial division but useful in discussing the limitations of conventional thermal sources and the advantages of lasers. Our discussion of laser sources will be confined here to continuous-wave gas discharge devices and to He–Ne and argon lasers in particular. Hologram formation using pulsed lasers is treated in Chapter 11. Semiconductor lasers are at present not very useful for holography as they emit only weakly in the visible range, and their coherence properties are poor.

7.1.1 Spatial Coherence Requirements

The coherence properties of a laser beam are intimately related to the oscillation mode-structure of the laser. A comprehensive review of the mode-structure can be found in Kogelnik and Li [*7.1*]. So far as holography is concerned, the radiation from a laser oscillating in any *one transverse mode* is spatially coherent. It is preferable that the mode of oscillation be the lowest order mode, the TEM_{00}, since more uniform illumination can be obtained from this mode than from higher order modes. Higher order modes are also inherently less stable, there being a tendency to oscillate in two or more simultaneously. Almost all commercially available lasers oscillate in only the lowest order mode or can be so adjusted.

In contrast, a conventional nonlaser source is far from being spatially coherent. To increase the spatial coherence of a source of arbitrary cross section, all but a small fraction of its radiating surface must be masked off. If we set some meaningful criterion for minimum spatial coherence, we can use the van Cittert–Zernike theorem (see Section 1.9.3) to estimate the maximum spatial extent of a nonlaser source suitable for forming holograms with the methods illustrated in Fig. 7.1a, b.

For this purpose consider the geometry shown in Fig. 7.2 where a small pinhole S, located in the $x'y'$ plane, allows radiation from only a small area of the source to pass to an xy plane a distance R away. We define two sampling positions P_1 and P_2 on the xy plane with coordinates $(0, 0)$ and (x, y), respectively. The angle subtended at the pinhole by the distance r between P_1 and P_2 is θ. Figure 7.2 describes a situation similar to that

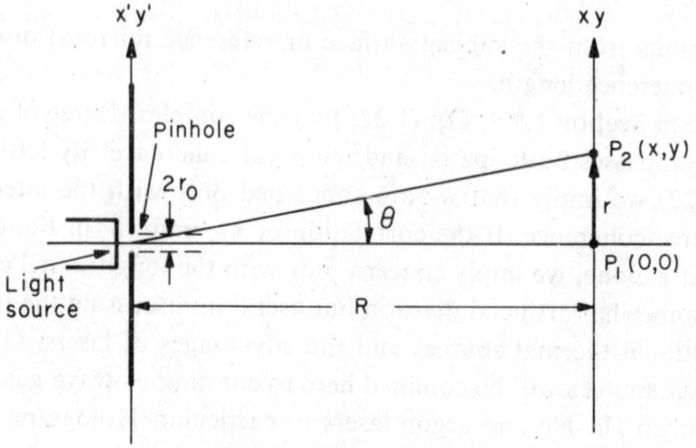

FIG. 7.2. Geometry related to discussion of the spatial coherence of a thermal source.

illustrated in Fig. 1.17; Eq. (1.22) defining the complex degree of coherence $\gamma_{12}(\tau)$ applies:

$$\gamma_{12}(\tau) = \frac{\lim_{T\to\infty} \frac{1}{2T}\int_{-T}^{T} \mathbf{v}_{P_1}(t+\tau)\mathbf{v}_{P_2}^*(t)\,dt}{\left[\left(\lim_{T\to\infty} \frac{1}{2T}\int_{-T}^{T} \mathbf{v}_{P_1}(t)\mathbf{v}_{P_1}^*(t)\,dt\right)\left(\lim_{T\to\infty} \frac{1}{2T}\int_{-T}^{T} \mathbf{v}_{P_2}(t)\mathbf{v}_{P_2}^*(t)\,dt\right)\right]^{1/2}}$$

where $\mathbf{v}_{P_1}$ and $\mathbf{v}_{P_2}$ represent the complex electric fields at P_1 and P_2 and τ is the transit time difference for light traveling from P_1 and P_2 to an observation point Q (see Fig. 1.17). For the case $\tau = 0$ and with the points P_1 and P_2 located as in Fig. 7.2, $\gamma_{12}(0) \equiv \boldsymbol{\mu}_s(x, y)$ can be written as

$$\boldsymbol{\mu}_s(x, y) = \frac{\int_{-\infty}^{\infty} \mathbf{v}(0, 0, t)\mathbf{v}^*(x, y, t)\,dt}{\left[\int_{-\infty}^{\infty} \mathbf{v}(0, 0, t)\mathbf{v}^*(0, 0, t)\,dt \int_{-\infty}^{\infty} \mathbf{v}(x, y, t)\mathbf{v}^*(x, y, t)\,dt\right]^{1/2}}. \tag{7.1}$$

The quantity $\boldsymbol{\mu}_s$ is the *complex degree of spatial (lateral) coherence* of the apertured source as measured in the *xy* plane.

By means of the van Cittert–Zernike theorem [7.2] we can express $|\boldsymbol{\mu}_s|$, the *degree of spatial coherence*, as the magnitude of a normalized Fourier transform of the intensity distribution over the pinhole:

$$|\boldsymbol{\mu}_s(x, y)| = \left| \frac{\int_{-\infty}^{\infty}\int_{-\infty}^{\infty} I(x', y') \exp[2\pi i(\xi x' + \eta y')\,dx'\,dy'}{\int_{-\infty}^{\infty}\int_{-\infty}^{\infty} I(x', y')\,dx'\,dy'} \right| \tag{7.2}$$

where $\xi = x/\bar{\lambda}R$, $\eta = y/\bar{\lambda}R$ [as in Eq. (5.38)], and $\bar{\lambda}$ is the mean wavelength of the radiation emitted by the source. The validity of the theorem depends on the following assumptions:

1. The radiation from the source is quasi monochromatic; i.e., the mean wavelength $\bar{\lambda}$ is much greater than the deviation $\Delta\lambda$.
2. The separation R between the pinhole and the xy plane is much greater than the extent of the pinhole or the distance r.
3. The radiation inside the pinhole is spatially incoherent.
4. The coherence length of the source radiation $c/\Delta f = \bar{\lambda}^2/\Delta\lambda$ is much greater than the maximum difference in optical paths between either of the sampling points and any point in the source. (Here, f is the temporal frequency.)

The assumptions can all be satisfied in practice; the bandwidth $\Delta\lambda$ can be made as small as necessary by passing the source radiation through a narrow-band filter.

If we assume the intensity of the source to be uniform over the circular pinhole of radius r_0, integration of Eq. (7.2) yields

$$|\mu_s| = \frac{(\pi r_0^2)}{(\pi r_0^2)} \frac{J_1(2\pi\nu r_0)}{\pi\nu r_0} = \frac{J_1(2\pi\nu r_0)}{\pi\nu r_0} \tag{7.3}$$

where we have employed correspondence (4.34). As indicated in Chapter 4, $J_1(2\pi\nu r_0)/\pi\nu r_0$ has a maximum value of one. The spatial frequency ν can be expressed as

$$\nu = (\xi^2 + \eta^2)^{1/2} = (1/\bar{\lambda}R)(x^2 + y^2)^{1/2} = r/\bar{\lambda}R \approx \theta/\bar{\lambda}$$

where the angle θ is that indicated in Fig. 7.2, and $|\mu_s|$, the degree of spatial coherence, then becomes

$$|\mu_s| = \left| \frac{J_1(2\pi r_0\theta/\bar{\lambda})}{\pi r_0\theta/\bar{\lambda}} \right|. \tag{7.4}$$

Figure 7.3 is a plot of $|\mu_s|$ versus $r_0\theta/\bar{\lambda}$.

Suppose we desire to form a hologram using one of the methods of Fig. 7.1 and require $2\theta_0 = 30°$ or $\theta_0 = \pi/12$ rad. If we also require the degree of spatial coherence to be no less than $|\mu_s| = 1/\sqrt{2} = 0.707$ (an arbitrary criterion which will be justified later) and if we set $\bar{\lambda} = 0.5$ μm, then from Fig. 7.3 we find that $(r_0\theta_0/\bar{\lambda})_{\max} = 0.25$, and the maximum allowed diameter for the pinhole is $2r_0 \approx 1$ μm. When we use a pinhole whose diameter is as small as one micrometer to obtain sufficient spatial coherence from a thermal

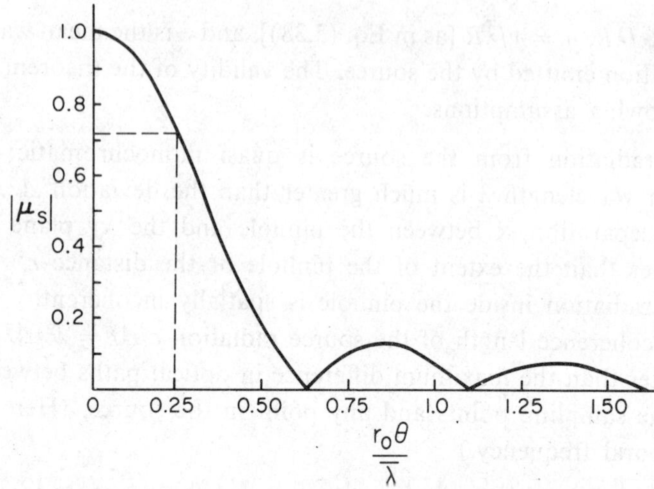

FIG. 7.3. Degree of spatial coherence versus $r_0\theta/\bar{\lambda}$.

source, we greatly reduce the useful power. Among the nonlaser sources, the high-pressure mercury arc lamp radiates the most power per unit area of its emitting surface. Typically this is about 100 W/cm² at $\bar{\lambda} = 5461$ Å, with $\Delta\lambda \approx 50$ Å. A pinhole 1 μm in diameter would limit the useful power of the source to 1 μW. Use of a wavelength filter to increase the temporal coherence would cause a further reduction. In contrast, useful power of 100 mW is not difficult to obtain from a continuous-wave laser.

7.1.2 Temporal Coherence of Conventional Sources

The temporal coherence of a light source is ultimately determined by the spectral purity of its radiation. When a laser oscillates in only one axial or longitudinal mode (and, of course, one transverse mode), it possesses almost perfect temporal coherence so far as holography is concerned. While most commercially available lasers are adjusted to oscillate in the lowest order transverse mode, they are, however, not constructed for single-longitudinal-mode (single-frequency) operation. In fact, the temporal coherence property of a typical (multiple-mode) gas laser is not much better than that of the *spontaneous* emission from the same atomic transition under a similar gas discharge condition. In this section we first set up a criterion for temporal coherence and then examine the temporal coherence properties of some conventional gas discharge sources.

Equation (7.4) tells us that the degree of spatial coherence between light at points P_1 and P_2 on the xy plane of Fig. 7.2 depends on the position coordinates of P_1 and P_2 only as a function of the angle θ between light rays passing from the extended source to each point. Let us consider now the case where P_1 and P_2 are spaced along the same ray coming from a point source so that $\theta = 0$, $|\mu_s| = 1$, and the complex degree of coherence depends only on τ. To signify this we replace $\gamma_{12}(\tau)$ in Eq. (1.22) by $\mu_T(\tau)$, the *complex degree of temporal coherence*, and replace $\mathbf{v}_{P_1}(t) = \mathbf{v}_{P_2}(t)$ by $\mathbf{v}(t)$ and $\mathbf{v}_{P_1}(t+\tau)$ by $\mathbf{v}(t+\tau)$ so that

$$\mu_T(\tau) = \frac{\lim\limits_{T\to\infty} \dfrac{1}{2T}\displaystyle\int_{-T}^{T} \mathbf{v}(t+\tau)\mathbf{v}^*(t)\,dt}{\lim\limits_{T\to\infty} \dfrac{1}{2T}\displaystyle\int_{-T}^{T} \mathbf{v}(t)\mathbf{v}^*(t)\,dt} = \frac{\displaystyle\int_{-\infty}^{\infty} \mathbf{v}(t+\tau)\mathbf{v}^*(t)\,dt}{\displaystyle\int_{-\infty}^{\infty} \mathbf{v}(t)\mathbf{v}^*(t)\,dt} \tag{7.5}$$

where $\mathbf{v}(t)$ is the complex electric field at P_2, $\mathbf{v}(t+\tau)$ is the complex electric field at P_1 (a point on the same ray between the source and P_2 but closer to the source). Equation (7.5) holds when a wave is amplitude-divided (Fig. 7.1b) and light from the same original ray travels different paths to an observation point. In this latter arrangement, τ, the transit time difference, can be finite though P_1 and P_2 are coincident. The transit time difference τ can be expressed in terms of an optical path difference $\Delta L = c\tau$ where c is the speed of light. We now define a coherence length $\Delta L_H = c\tau_H$ such that $|\mu_T(\tau_H)| = 1/\sqrt{2}$. We shall show in the next section that this criterion is meaningful for holography. For successful hologram formation we require that the difference between any pair of optical paths from light source to any point on the recording medium be less than ΔL_H. To determine ΔL_H for a given source we must be able to plot $|\mu_T(\tau)|$ as a function of τ. This may be done by rewriting Eq. (7.5) in terms of temporal frequency f and introducing the power spectrum of the desired source.

Equation (7.5) can be transformed into

$$\mu_T(\tau) = \frac{\displaystyle\int_{-\infty}^{\infty} \Phi(f)\exp(2\pi i f\tau)\,df}{\displaystyle\int_{-\infty}^{\infty} \Phi(f)\,df} \tag{7.6}$$

where $\Phi(f) = \mathbf{V}(f)\mathbf{V}^*(f)$ is the power spectrum and $\mathbf{V}(f)$ is the temporal Fourier transform of $\mathbf{v}(\tau)$. [See Appendix II for a general definition of $\mathbf{v}(\tau)$.] The equivalence of Eqs. (7.5) and (7.6) may be demonstrated simply by considering the result of equating the numerators and denominators separately. Equality of the numerators is merely a statement of the autocorrelation correspondence which holds equally well for temporal variables as for spatial [see correspondence (4.18)]. Equality of the denominators represents the special case of the autocorrelation correspondence at $\tau = 0$. Paralleling the spatial Fourier transform relation between $|\boldsymbol{\mu}_s(x, y)|$ and $I(x', y')$ in Eq. (7.2), the two functions $|\boldsymbol{\mu}_T(\tau)|$ and $\Phi(f)$ form a temporal Fourier transform pair.

Using Eq. (7.6) let us calculate the degree of temporal coherence $|\boldsymbol{\mu}_T(\tau)|$ for some conventional nonlaser light sources. Gas discharge lamps are the most coherent nonlaser sources. An appropriate filter or monochromator can be used to isolate a single spectral line from the discharge. The power spectrum $\Phi(f)$ corresponding to a spectral line emitted by a *low-pressure* discharge lamp is approximated by the Gaussian expression [*7.3*]

$$\Phi(f) = \Phi(\bar{f}) \exp\left\{-\left[2\frac{(\ln 2)^{1/2}(f-\bar{f})}{\Delta f_D}\right]^2\right\} \tag{7.7}$$

where $\bar{f}$ is the center frequency of the spectral line and Δf_D is the full (Doppler) width of the line at half intensity. Upon substitution of Eq. (7.7) into Eq. (7.6) we find in the numerator of Eq. (7.6) the shifted Fourier transform of a Gaussian function. This may be evaluated with the help of temporal versions of correspondences (4.21) and (4.27). The denominator of Eq. (7.6) may be looked up in a table of integrals with the result that

$$\boldsymbol{\mu}_T(\tau) = \exp(-2\pi i\bar{f}\tau) \exp\left[-\left(\frac{\pi\tau\,\Delta f_D}{2(\ln 2)^{1/2}}\right)^2\right]$$

and

$$|\boldsymbol{\mu}_T(\tau)| = \exp\left[-\left(\frac{\pi\tau\,\Delta f_D}{2(\ln 2)^{1/2}}\right)^2\right]. \tag{7.8}$$

A *high-pressure* lamp emits a spectral line whose power spectrum is better approximated by a Lorentzian [*7.3*]

$$\Phi(f) = \Phi(\bar{f})\left[1 + \left\{\frac{2(f-\bar{f})}{\Delta f_L}\right\}^2\right]^{-1} \tag{7.9}$$

where Δf_L is the Lorentz line width due to collisions of the various gas and charged particles. Substitution of Eq. (7.9) into Eq. (7.6) and application of (1) the shift theorem, (2) correspondence 444 of Campbell and Foster [*7.4*], and (3) a table of integrals yields

$$\boldsymbol{\mu}_T(\tau) = \exp(-2\pi i\bar{f}\tau) \exp(-\pi\tau\,\Delta f_L)$$

and

$$|\boldsymbol{\mu}_T(\tau)| = \exp(-\pi\tau\,\Delta f_L). \tag{7.10}$$

We can now determine $\Delta L_H = c\tau_H$ from Eqs. (7.8) and (7.10) or Fig. 7.4, where $|\mu_T|$ is plotted as a function of either $\tau\,\Delta f_D$ or $\tau\,\Delta f_L$. When $|\mu_T| = 1/\sqrt{2} = 0.707$, $\tau_H\,\Delta f_D = 0.32$ while $\tau_H\,\Delta f_L = 0.11$. The corresponding coherence lengths are then

$$\Delta L_H = \frac{c\sqrt{2}\ln 2}{\pi\,\Delta f_D} = \frac{0.32c}{\Delta f_D}, \quad \text{low-pressure discharge,} \qquad (7.11)$$

and

$$\Delta L_H = \frac{c\ln 2}{2\pi\,\Delta f_L} = \frac{0.11c}{\Delta f_L}, \quad \text{high-pressure discharge.} \qquad (7.12)$$

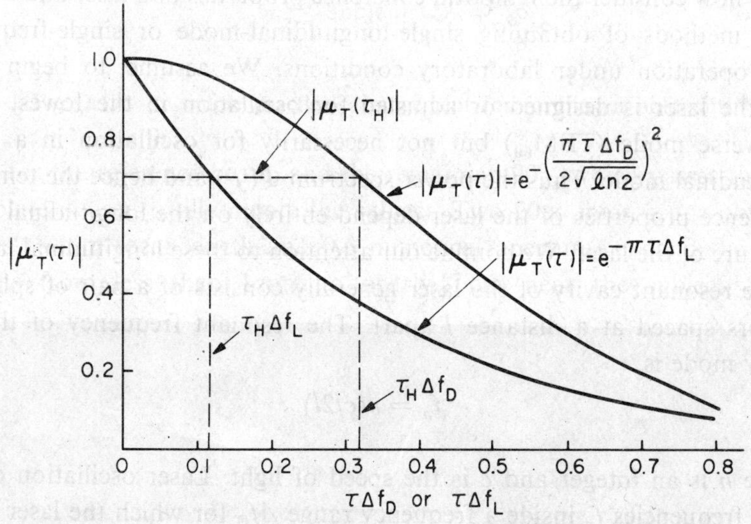

FIG. 7.4. Degree of temporal coherence of a thermal source with a Gaussian or Lorentzian spectral profile.

One of the most coherent of the nonlaser sources is the 6058 Å orange line of ^{86}Kr. The Doppler width Δf_D of this line can be made as narrow as 4.5×10^8 Hz [*7.5*] and its coherence length is $\Delta L_H \approx 21$ cm according to Eq. (7.11). If it were not for the low output power per unit area of the source and the small degree of spatial coherence which results when the source is extended, such a source would be useful for holography. A much brighter source is the 5461 Å green line from a high-pressure mercury arc lamp. However, its line width Δf_L is about 5×10^{12} Hz ($\Delta\lambda \approx 50$ Å) yielding a coherence length $\Delta L_H \approx 8\ \mu$m. To use the light for holography, the line must be narrowed further by passing it through a wavelength filter.

Let us now calculate the coherence lengths of the spontaneous emissions at 6328 Å and at 4880 Å from neon and argon gas discharges, respectively.

We will later compare these with coherence lengths for laser lines at the same wavelengths. The Doppler widths of the two lines are approximately 1.5×10^9 Hz at 6328 Å and 7.5×10^9 Hz at 4880 Å under gas discharge conditions common to helium–neon lasers and argon ion lasers. According to Eq. (7.11) the corresponding coherence lengths are approximately 6.4 cm for the 6328 Å red line of atomic neon and 1.5 cm for the 4880 Å blue line of the argon ion.

7.1.3 Temporal Coherence of the Gas Laser

We now consider the temporal coherence properties of a laser and discuss some methods of obtaining single-longitudinal-mode or single-frequency laser operation under laboratory conditions. We assume, to begin with, that the laser is designed or adjusted for oscillation in the lowest order transverse mode (TEM_{00}) but not necessarily for oscillation in a single longitudinal mode. Thus the power spectrum $\Phi(f)$ and hence the temporal coherence properties of the laser depend entirely on the longitudinal mode structure of the laser. We confine our attention to these longitudinal modes.

The resonant cavity of the laser generally consists of a pair of spherical mirrors spaced at a distance l apart. The resonant frequency of the nth cavity mode is

$$f_n = n(c/2l) \tag{7.13}$$

where n is an integer, and c is the speed of light. Laser oscillation occurs at all frequencies f_n inside a frequency range Δf_G for which the laser active medium has sufficient gain to overcome the power loss (including the loss due to the output coupling) of the cavity. This oscillation frequency range Δf_G can be greater or less than the Doppler width Δf_D. The width Δf_M of each mode is determined by the loss and the mechanical and thermal stability of the cavity structure. Typically $\Delta f_M \approx 10^5$ Hz (over an observation period of a few minutes). Let us assume that only one longitudinal mode is allowed to oscillate and that the power spectrum $\Phi(f)$ for the mode can be approximated by a Gaussian having a half-intensity width of $\Delta f_M = 10^5$ Hz. The coherence length of such a single-frequency laser, according to Eq. (7.11), is $\Delta L_H \approx 1 \times 10^5$ cm or 1 km. The optical path lengths encountered in most holographic applications are certainly much less than 1 km. We are therefore justified in considering that $\Delta L_H \to \infty$ and that the power spectrum may be written as a δ function. If more than one longitudinal mode can oscillate, $\Phi(f)$ can then be represented by series of δ functions.

What is the coherence length of the laser if more than one longitudinal mode can oscillate? We consider first the case where oscillation can be supported in just two modes. This will happen when the frequency range Δf_G, within which the laser has sufficient gain for oscillation, spans twice the mode spacing $f_{n+1} - f_n$, i.e.,

$$\Delta f_G/2 = f_{n+1} - f_n = c/2l.$$

The power spectrum of this two-mode laser source can be represented by two delta functions

$$\Phi(f) = I_n\,\delta(f - f_n) + I_{n+1}\,\delta(f - f_{n+1})$$

where the I_n's represent the total power in an individual mode. Substituting $\Phi(f)$ into Eq. (7.6) we have

$$\mu_T(\tau) = \frac{1}{(I_n + I_{n+1})} \cdot [I_n \exp(2\pi i f_n \tau) + I_{n+1} \exp(2\pi i f_{n+1} \tau)]$$

$$= \exp(2\pi i f_n \tau)\left[a_n + a_{n+1} \exp\left(2\pi i \frac{c}{2l}\,\tau\right)\right] \tag{7.14}$$

where we have used the expression for the mode spacing $f_{n+1} - f_n = c/2l$ and where

$$a_n = \frac{I_n}{I_n + I_{n+1}}, \qquad a_{n+1} = \frac{I_{n+1}}{I_n + I_{n+1}}, \qquad \text{and} \qquad a_n + a_{n+1} = 1.$$

The degree of coherence $|\mu_T(\tau)|$ is given by

$$|\mu_T(\tau)| = |a_n + a_{n+1} \exp(i\pi c\tau/l)|$$

$$= |[a_n{}^2 + a_{n+1}^2 + 2a_n a_{n+1} \cos(\pi c\tau/l)]^{1/2}| \tag{7.15}$$

and is plotted in Fig. 7.5 for two values of the parameter $b = |a_{n+1} - a_n|$. The parameter b represents the minimum value of $|\mu_T(\tau)|$. When $b = 0$, $a_n = a_{n+1} = \frac{1}{2}$ and

$$|\mu_T(\tau)| = |[\tfrac{1}{2} + \tfrac{1}{2}\cos(\pi c\tau/l)]^{1/2}| = |\cos(\pi c\tau/2l)|.$$

When the power in one mode is twice that in the other, $b = \frac{1}{3}$, $a_n = \frac{1}{3}$, $a_{n+1} = \frac{2}{3}$, and

$$|\mu_T(\tau)| = |(\tfrac{5}{9})^{1/2}[1 + \tfrac{4}{5}\cos(\pi c\tau/l)]^{1/2}|.$$

Regardless of the value of b, $|\mu_T(\tau)|$ for the two-mode case is *periodic* in τ with a period $2l/c$. The condition $|\mu_T(\tau)| = 1/\sqrt{2}$ is satisfied by multiple values of τ as shown in Fig. 7.5. We define the smallest of these as τ_H from which we obtain the coherence length $\Delta L_H = c\tau_H$. For $b = 0$, $\tau_H = \frac{1}{2}(l/c)$, and $\Delta L_H = \frac{1}{2}l$. For path length values between 0 and $2\,\Delta L_H$, the degree of temporal coherence varies monotonically with path difference as in the case of the nonlaser sources indicated in Fig. 7.4. The length of a typical gas laser is $l = 2\,\Delta L_H = 1$ m. Path differences required for hologram formation can generally be kept within the limit $\Delta L_H = l/2 = 0.5$ m.

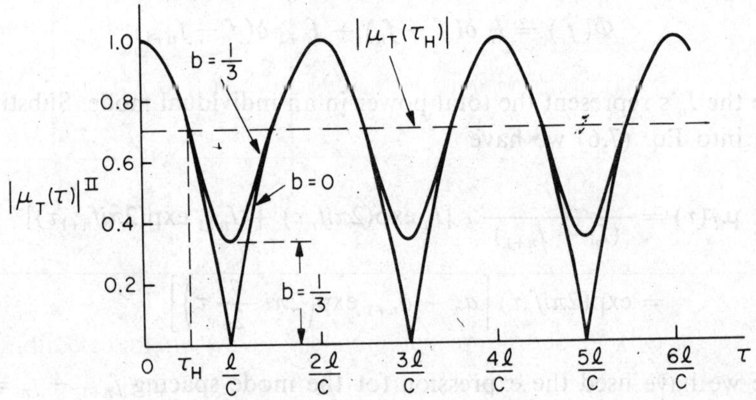

FIG. 7.5. Degree of temporal coherence of a laser operating in two longitudinal modes (for two values of the parameter b).

Due to thermal instabilities in the cavity length l, the value of b can take on a range of values. When, however, the gain required for oscillation can be obtained only over a frequency range Δf_G much narrower than the Doppler line width Δf_D, the oscillating modes are nearly equal in power and $b \approx 0$. Figure 7.5 indicates that even when one mode has twice the power of the other ($b = \frac{1}{3}$), the value of τ_H (corresponding to $|\mu_T(\tau)| = 0.707$) differs very little from that corresponding to $b = 0$. We therefore shall restrict ourselves in further discussion to the case where the powers of the modes are equal.

We note that for the case of two oscillating modes and a cavity length of 1 m, the coherence length $\Delta L_H = 0.5$ m represents a drastic reduction from the 1 km computed for the case of single-mode oscillation.

When N modes of oscillation are permitted, we can write the power spectrum of the laser radiation as

$$\Phi(f) = I_n\,\delta(f - f_n) + I_{n+1}\,\delta(f - f_{n+1}) + \cdots + I_{n+N-1}\,\delta(f - f_{n+N-1}).$$

Substitution into Eq. (7.6) yields the degree of coherence

$$|\mu_T(\tau)| = \frac{1}{N}|\exp(2\pi i f_n \tau) + \exp(2\pi i f_{n+1}\tau) + \cdots + \exp(2\pi i f_{n+N-1}\tau)|, \tag{7.16}$$

where we have used the condition that $I_n = I_{n+1} = \cdots = I_{n+N-1}$. We can write Eq. (7.16) in terms of the mode spacing $c/2l$ by extracting the factor $\exp(2\pi i f_{n-1}\tau)$ so that

$$\begin{aligned} |\mu_T(\tau)| &= \Bigg| \frac{\exp(2\pi i f_{n-1}\tau)}{N} \\ &\quad \times [\exp(\pi i c\tau/l) + \exp(2\pi i c\tau/l) + \cdots + \exp(N\pi i c\tau/l)] \Bigg| \\ &= \left| \left[\frac{\exp(2\pi i f_{n-1}\tau)}{N} \right] \left\{ \frac{[\exp(\pi i c\tau/l)][\exp(N\pi i c\tau/l) - 1]}{[\exp(\pi i c\tau/l) - 1]} \right\} \right| \\ &= \left| \frac{\exp(2\pi i f_n \tau)}{N} \left[\frac{\exp(N\pi i c\tau/l) - 1}{\exp(\pi i c\tau/l) - 1} \right] \right| \end{aligned}$$

where we have used the formula for the sum of a geometric progression of ratio $\exp(\pi i c\tau/l)$. If we multiply the quantity inside the absolute value sign by its complex conjugate, take its square root, and then take the absolute value of the result, $|\mu_T(\tau)|$ becomes

$$|\mu_T(\tau)| = \left| \frac{1}{N} \left(\frac{1 - \cos(N\pi c\tau/l)}{1 - \cos(\pi c\tau/l)} \right)^{1/2} \right| = \left| \frac{\sin(N\pi c\tau/2l)}{N \sin(\pi c\tau/2l)} \right|. \tag{7.17}$$

Again, as in the case of two-mode oscillation, the degree of coherence $|\mu_T(\tau)|$ is periodic in τ. It is plotted in Fig. 7.6 for the cases of a laser oscillating in one, two, three, and four modes. The principal maxima (equal to unity) occur at equal intervals of $\tau = 2l/c$ regardless of the value of N. Zeros occur whenever $\tau = (m/N)(2l/c)$, where $m = 1, 2, 3, \ldots$, but $m/N \neq 0, 1, 2, 3, \ldots$.

From the above discussion it is clear that single-frequency operation of a laser is highly desirable. A single-frequency laser offers nearly unlimited coherence length while even a few simultaneously oscillating modes can reduce the coherence length severely (see Fig. 7.6). Short coherence length in hologram formation limits the permissible subject depth and necessitates the often troublesome equalization of the mean optical path lengths of the subject and reference beams. Unfortunately, stable single-frequency operation of a laser is not easily obtained, and its achievement inevitably results in a reduction of output power.

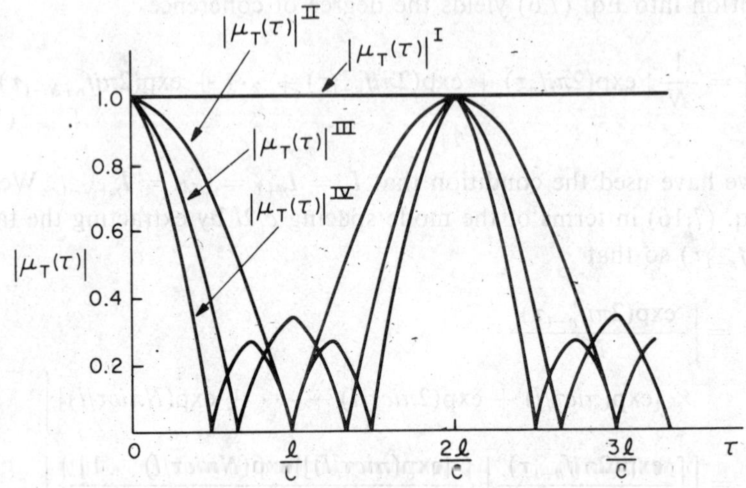

7.6. Degree of temporal coherence for laser operating in one (I), two (II), three (III), and four (IV) longitudinal modes.

7.1.4 Methods of Achieving Single-Frequency Operation

The most obvious method of achieving single-frequency operation of a laser is to reduce the cavity length l such that

$$f_{n+1} - f_n = \frac{c}{2l} = \Delta f_G.$$

For this case the gain is sufficient to sustain oscillation only over a band of frequencies Δf_G equal to the mode spacing. Hence only a single mode can be supported. If we take Δf_G to be approximately the Doppler linewidth $\Delta f_D = 1.5 \times 10^9$ Hz, the cavity length corresponding to single-mode operation of a He–Ne laser at 6328 Å is $l = c/(2\Delta f_D) = 10$ cm. Only a small volume of the active medium can be enclosed in so short a cavity; consequently only a fraction of a milliwatt of output power can be obtained from the laser [7.5]. The single-frequency cavity length for the argon ion laser operating at 4880 Å is but 2 cm, making it very difficult to build such a laser.

A better approach to single-frequency operation is to couple together two resonant cavities so that only one mode, common to both cavities, can have sufficient gain to sustain laser oscillation. It is frequently desirable to make the optical length l_1 of one cavity long, to enclose a large volume of active medium while the optical length l_2 of the second cavity is kept short to widely separate the frequencies of common modes [see Eq. (7.13)]. If the length l_2 is limited to 10 cm for the He–Ne laser and to 2 cm for the argon ion laser so that the shorter cavity can support only one mode, then single-frequency operation is obtained.

Examples of the two-cavity method are shown in Fig. 7.7a,b. The first configuration [*7.6, 7.7*] has the virtue of simplicity, but the second configuration [*7.8–7.11*] is more effective at high power levels. In the former case (Fig. 7.7a) the mode-selecting second cavity is simply a fused-quartz etalon. Its physical length t is determined by the relation between it and the required optical length $l_2 = tn_i$, where n_i is the index of refraction of fused quartz. A slight tilting of the etalon is usually required to make l_2 an integral number of half-wavelengths and satisfy Eq. (7.13). Coating of the surfaces of the etalon, to increase its finesse, may not even be necessary in many cases. In Fig. 7.7b the mode-selecting second cavity is formed by the mirrors M_2 and M_3 together with the beam splitter. The configuration produces high

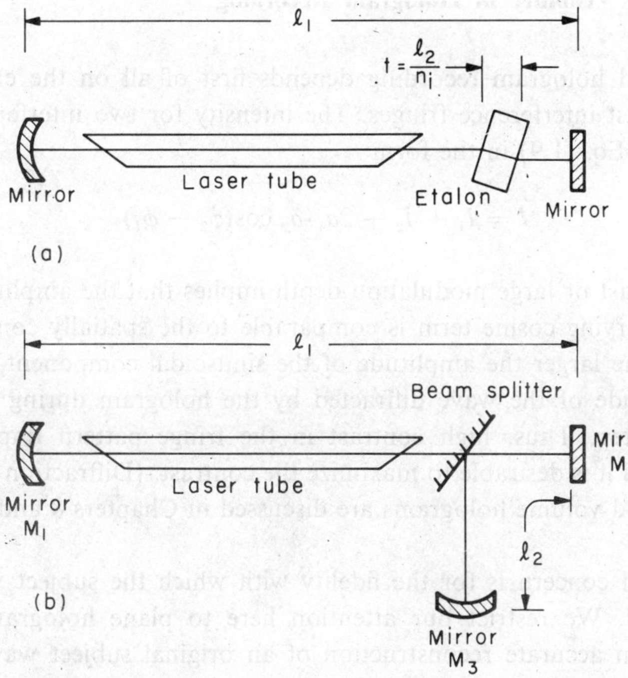

FIG. 7.7. Laser configurations for two-cavity, single-longitudinal-mode operation: (a) An intracavity etalon. (b) An interferometer resonator.

transmission loss at all frequencies except the frequency of the common modes. The curvatures of the mirrors M_1 and M_3 are chosen to match properly wavefronts at the beam splitter.

Highest reported output powers from single-frequency lasers are about 50 mW at 6328 Å (He–Ne) and near 1 W at 4880 Å (argon ion). The main difficulty in achieving single-frequency operation with two-cavity methods is maintaining the dimensional stability of both cavities. With the increased temperatures accompanying high power levels the difficulty increases. If $l_1 \gg l_2$, only l_2 need be critically adjusted during the operation of the laser. This adjustment can be made by changing the tilt of the etalon in Fig. 7.7a or the position of the mirror M_3 in Fig. 7.7b. One can also servo-control the dimension l_2 by a feedback system which automatically locks the oscillation frequency to the center of the line $\bar{f}$. In practice it is not too difficult to maintain stable operation of the laser for a few minutes without any feedback system. This is usually long enough to expose a hologram.

7.2 Fringe Visibility in Hologram Recording

Successful hologram recording depends first of all on the existence of high contrast interference fringes. The intensity for two interfering beams is given by Eq. (1.9) in the form

$$I = I_1 + I_2 + 2a_1 \cdot a_2 \cos(\phi_2 - \phi_1).$$

High contrast or large modulation depth implies that the amplitude of the spatially varying cosine term is comparable to the spatially constant term $I_1 + I_2$. The larger the amplitude of the sinusoidal component, the larger the amplitude of the wave diffracted by the hologram during the reconstruction step. Thus, high contrast in the fringe pattern implies bright images, and it is desirable to maximize the contrast. (Diffraction efficiencies of plane and volume holograms are discussed in Chapters 8 and 9, respectively.)

A second concern is for the fidelity with which the subject wave is reconstructed. We restrict our attention here to plane holograms. If, for example, an accurate reconstruction of an original subject wave is to be obtained from an absorption hologram, the hologram recording must have an amplitude transmittance which is linearly related to the intensity of the interference pattern. Let us see how (1) high contrast in the fringe pattern and (2) linear recording of the subject wave can be obtained.

7.2.1 Optimizing Fringe Visibility

Contrast is quantitatively measured by the fringe visibility or standing wave ratio V given in Eq. (1.21) by

$$V = \frac{I_{\max} - I_{\min}}{I_{\max} + I_{\min}} \tag{7.18}$$

where $I_{\max}$ and $I_{\min}$ denote the maximum and minimum intensities of the interference fringes in an observation plane. The fringe visibility may vary over the observation plane; it is defined over any small area larger than a fringe spacing. We shall find that the visibility depends on the degree of coherence $|\gamma_{12}(\tau)|$ between the interfering beams, the angle Ω between the directions of polarization of the beams, and the ratio R of the intensities of the two beams all measured in the plane of observation. To facilitate obtaining a useful relation between V and the parameters $|\gamma_{12}(\tau)|$, Ω, and R, we impose the condition that the latter are all constant over any small area of the observation plane, and we consider the optical arrangement indicated in Fig. 7.8, where both reference and subject wave are derived from the same laser source. Each wave is spherical, the reference diverging from the point source R while the signal appears to diverge from a virtual point source P. We are to observe the interference over a very small area A of the observation plane. The area A is large enough for the visibility to be well defined but small enough that the signal and reference waves can be considered plane and that $|\gamma_{12}(\tau)|$ and Ω can be considered constants.

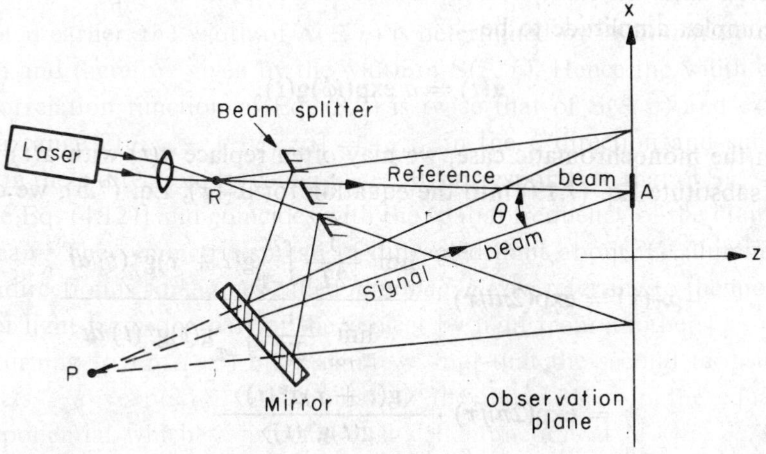

FIG. 7.8. Ray directions for the essentially plane wavefronts interfering over the small area A.

The analysis will be carried out for light propagation vectors lying in the xz plane only.

We suppose that the laser is operating in a single transverse mode but in several longitudinal modes. Hence, spatial coherence can be considered essentially perfect, and the degree of coherence $|\gamma_{12}(\tau)|$ can be replaced by the degree of temporal coherence $|\boldsymbol{\mu}_T(\tau)|$. Before beginning the analysis of the interference of the waves indicated in Fig. 7.8, let us rewrite the expression for $\boldsymbol{\mu}_T(\tau)$ in Eq. (7.5) in a form which separates out the rapidly varying temporal phase factor contained in the complex electric field $\mathbf{v}$. Recall that in Section 1.3 the complex electric field for a monochromatic light wave was defined as

$$\mathbf{v} = a \exp(i\phi) \exp(2\pi i f t).$$

For light emission at several frequencies in a narrow band, $\mathbf{v}$ may be written

$$\begin{aligned}\mathbf{v} &= a \exp(i\phi)[c_0 \exp(2\pi i f t) + c_1 \exp\{2\pi i(f + \varepsilon_1)t\} \\ &\qquad + c_2 \exp\{2\pi i(f + \varepsilon_2)t\} + \cdots] \\ &= a \exp(i\phi) \exp(2\pi i f t)[c_0 + c_1 \exp(2\pi i \varepsilon_1 t) \\ &\qquad + c_2 \exp(2\pi i \varepsilon_2 t) + \cdots]\end{aligned}$$

or

$$\mathbf{v}(t) = a \exp(i\phi)\mathbf{g}(t) \exp(2\pi i f t) \tag{7.19}$$

where $\mathbf{g}(t) = \sum c_i \exp(2\pi i \varepsilon_i t)$ varies at a much slower rate than $\exp(2\pi i f t)$ providing $\varepsilon_i \ll f$. We see that the complex electric field for narrow-band emission takes the same form as that for monochromatic light if we regard the complex amplitude to be

$$\mathbf{a}(t) = a \exp(i\phi)\mathbf{g}(t). \tag{7.20}$$

As in the monochromatic case, we may often replace $\mathbf{v}(t)$ with $\mathbf{a}(t)$. If we now substitute Eq. (7.19) into the equation for $\boldsymbol{\mu}_T(\tau)$, Eq. (7.5), we obtain

$$\begin{aligned}\boldsymbol{\mu}_T(\tau) &= \exp(2\pi i f \tau) \cdot \frac{\lim\limits_{T\to\infty} \dfrac{1}{2T} \displaystyle\int_{-T}^{T} \mathbf{g}(t+\tau)\mathbf{g}^*(t)\, dt}{\lim\limits_{T\to\infty} \dfrac{1}{2T} \displaystyle\int_{-T}^{T} \mathbf{g}(t)\mathbf{g}^*(t)\, dt} \\ &= \exp(2\pi i f \tau) \cdot \frac{\langle \mathbf{g}(t+\tau)\mathbf{g}^*(t)\rangle}{\langle \mathbf{g}(t)\mathbf{g}^*(t)\rangle}\end{aligned}$$

or

$$\boldsymbol{\mu}_T'(\tau) = \boldsymbol{\mu}_T(\tau) \exp(-2\pi i f \tau) = \frac{\langle \mathbf{g}(t+\tau)\mathbf{g}^*(t)\rangle}{\langle \mathbf{g}(t)\mathbf{g}^*(t)\rangle}. \tag{7.21}$$

Let us return now to the interference of the partially coherent reference and signal waves of Fig. 7.8. For the complex amplitude of the reference wave over the area A we write

$$\mathbf{r} = r \exp(i\phi)\mathbf{g}(t+\tau) = r \exp(2\pi i \xi_r x)\mathbf{g}(t+\tau) \tag{7.22}$$

where the spatial frequency ξ_r is defined for the mean wavelength $\bar{\lambda}$. The parameter τ is defined by the quantity $c\tau$, where c is the velocity of light and where $c\tau$ is the difference in optical path from laser source to the area A when reference and signal routes are compared. The intensity of the reference beam at A is

$$I_r = \langle \mathbf{r}\mathbf{r}^* \rangle = r^2 \langle \mathbf{g}(t+\tau)\mathbf{g}^*(t+\tau) \rangle. \tag{7.23}$$

We can divide the signal wave into two components. One is polarized parallel to the reference wave polarization direction and has a complex amplitude at A given by

$$\mathbf{a}_{\parallel} = a \exp(2\pi i \xi_a x)\mathbf{g}(t) \cos \Omega \tag{7.24}$$

and an intensity

$$I_{a_{\parallel}} = \langle \mathbf{a}_{\parallel}\mathbf{a}_{\parallel}^* \rangle = a^2 \cos^2 \Omega \, \langle \mathbf{g}(t)\mathbf{g}^*(t) \rangle. \tag{7.25}$$

Here Ω is the angle between the direction of polarization of reference and signal beams and ξ_a is the spatial frequency of the signal wave defined for the mean wavelength. The other component polarized perpendicular to the reference wave has a complex amplitude at A,

$$\mathbf{a}_{\perp} = a \exp(2\pi i \xi_a x)\mathbf{g}(t) \sin \Omega \tag{7.26}$$

with an intensity

$$I_{a_{\perp}} = \langle \mathbf{a}_{\perp}\mathbf{a}_{\perp}^* \rangle = a^2 \sin^2 \Omega \, \langle \mathbf{g}(t)\mathbf{g}^*(t) \rangle. \tag{7.27}$$

We now desire (1) to substitute Eqs. (7.22)–(7.27) into an expression for the intensity of the interference pattern formed by $\mathbf{r}$, $\mathbf{a}_{\parallel}$, and $\mathbf{a}_{\perp}$; (2) to express the intensity in terms of the degree of temporal coherence $|\boldsymbol{\mu}_T(\tau)|$; and (3) to compute the maximum and minimum values of the intensity and thereby the visibility V of Eq. (7.18). Equation (1.20), the expression for the intensity of the interference of *two* waves

$$I = I_1 + I_2 + 2\,\mathrm{Re}[\langle \mathbf{v}_1\mathbf{v}_2^* \rangle] \tag{7.28}$$

can be written in terms of complex amplitudes and generalized to express the intensity of the fringe pattern produced by **r**, $\mathbf{a}_{\parallel}$, and $\mathbf{a}_{\perp}$. We note that the total subject wave intensity $I_a = I_{a_\parallel} + I_{a_\perp} = a^2\langle \mathbf{g}(t)\mathbf{g}^*(t)\rangle$ [from Eqs. (7.25) and (7.27)], and that only the subject wave component polarized parallel to the reference wave polarization will produce an interference term. The intensity of the pattern over A is then the sum of the individual wave intensities plus the interference term

$$\begin{aligned} I &= I_r + I_{a_\parallel} + I_{a_\perp} + 2\,\mathrm{Re}[\langle \mathbf{r}\mathbf{a}_\parallel^*\rangle] \\ &= r^2\,\langle \mathbf{g}(t+\tau)\mathbf{g}^*(t+\tau)\rangle + a^2\,\langle \mathbf{g}(t)\mathbf{g}^*(t)\rangle + 2\,\mathrm{Re}[\langle \mathbf{r}\mathbf{a}_\parallel^*\rangle]. \end{aligned}$$

The quantity $\langle \mathbf{r}\mathbf{a}_\parallel^*\rangle$ is given by Eqs. (7.22) and (7.24) as

$$\begin{aligned} \langle \mathbf{r}\mathbf{a}_\parallel^*\rangle &= ra\cos\Omega\,\exp[2\pi i(\xi_r - \xi_a)x]\,\langle \mathbf{g}(t+\tau)\mathbf{g}^*(t)\rangle \\ &= ra\cos\Omega\,\exp[2\pi i(\xi_r - \xi_a)x]\boldsymbol{\mu}_T'(\tau)\langle \mathbf{g}(t)\mathbf{g}^*(t)\rangle \\ &= ra\cos\Omega\,\exp[2\pi i(\xi_r - \xi_a)x]\cdot|\boldsymbol{\mu}_T(\tau)|\,\exp[i\zeta(\tau)]\langle \mathbf{g}(t)\mathbf{g}^*(t)\rangle \end{aligned} \tag{7.29}$$

where we have (1) used Eq. (7.21), (2) noted that $|\boldsymbol{\mu}_T'(\tau)| = |\boldsymbol{\mu}_T(\tau)|$, and (3) expressed $\boldsymbol{\mu}_T'(\tau)$ as the product of its modulus $|\boldsymbol{\mu}_T(\tau)|$ and its phase factor $\exp[i\zeta(\tau)]$. Substituting these into the expression for I, we have for the intensity over A

$$\begin{aligned} I &= r^2\langle \mathbf{g}(t+\tau)\mathbf{g}^*(t+\tau)\rangle + a^2\langle \mathbf{g}(t)\mathbf{g}^*(t)\rangle \\ &\quad + 2\,\mathrm{Re}\{ra\cos\Omega\,|\boldsymbol{\mu}_T(\tau)|\,\exp[2\pi i(\xi_r - \xi_a)x + i\zeta(\tau)]\langle \mathbf{g}(t)\mathbf{g}^*(t)\rangle\}. \end{aligned}$$

The quantities in the time-average brackets are spatially constant and equal [see Eq. (7.19)] so that to a multiplicative constant, which need not concern us,

$$\begin{aligned} I &= r^2 + a^2 + 2ra\,|\boldsymbol{\mu}_T(\tau)|\cos\Omega\cos[2\pi(\xi_r - \xi_a)x + \zeta(\tau)] \\ &= r^2 + a^2 + 2ra\,|\boldsymbol{\mu}_T(\tau)|\cos\Omega\cos\beta(x,\tau) \end{aligned} \tag{7.30}$$

where

$$\beta(x,\tau) = 2\pi(\xi_r - \xi_a)x + \zeta(\tau).$$

The maximum intensity of the fringe pattern I_{max} is obtained when $\cos\beta(x,\tau) = +1$, and the minimum value I_{min} is obtained when $\cos\beta(x,\tau) = -1$. Substituting the extreme values of intensity into Eq. (7.18), the visibility becomes

$$V = \frac{2ra\,|\boldsymbol{\mu}_T(\tau)|\cos\Omega}{r^2 + a^2} = \frac{2\,|\boldsymbol{\mu}_T(\tau)|\sqrt{R}\cos\Omega}{R+1} \tag{7.31}$$

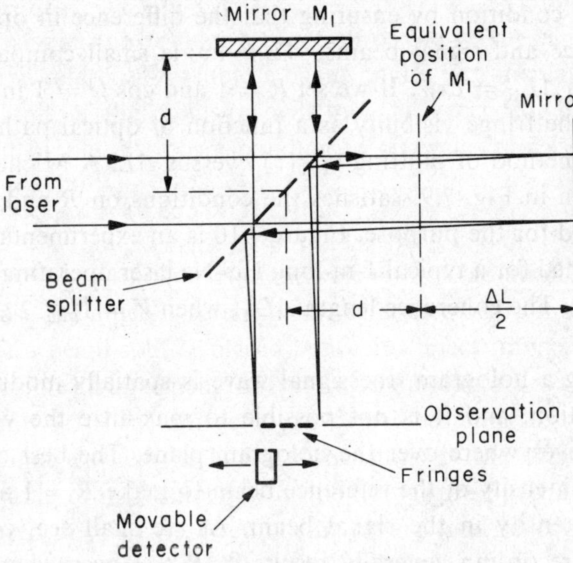

FIG. 7.9. Michelson interferometer.

where the beam ratio $R = (r/a)^2$. Note that the visibility is independent of the spatial frequencies of either beam and that $V(R) = V(1/R)$. The visibility has a maximum value of unity when $|\mu_T|$, R, and $\cos \Omega$ are each unity.

The degree of temporal coherence $|\mu_T(\tau)|$ of the light from a laser operating in several longitudinal modes is given in Eq. (7.17) and plotted in Fig. 7.6 as a function of τ. We see from the figure that $|\mu_T(\tau)|$ approaches unity for small values of τ. In particular, τ should be small compared to τ_H, the value for which the degree of temporal coherence has the value 0.707. We

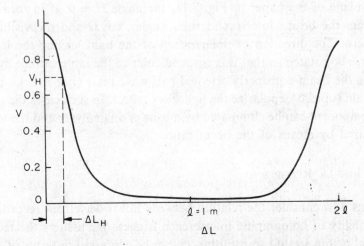

FIG. 7.10. Measured curve of visibility V versus optical path difference ΔL.

can achieve this condition by ensuring that the difference in optical paths between reference and signal beams, $\Delta L = c\tau$, is small compared to the coherence length $\Delta L_H = c\tau_H$. If we set $R = 1$ and $\cos \Omega = 1$ in Eq. (7.31), observation of the fringe visibility as a function of optical path difference offers a simple method of plotting $|\mu_T(\tau)|$ versus ΔL. A Michelson interferometer, shown in Fig. 7.9, satisfies the conditions on R and $\cos \Omega$ and may be employed for the purpose. Figure 7.10 is an experimental plot of V versus ΔL obtained for a typical 1-m-long He–Ne laser operating at a wavelength of 6328 Å. The coherence length ΔL_H, when $V = 1/\sqrt{2}$, is approximately 12 cm.

When forming a hologram, the signal wave is spatially modulated with subject information, and it is not possible to maximize the visibility by making $R = 1$ everywhere over the hologram plane. The best one can do is to adjust the intensity of the reference beam to make $R = 1$ at locations of maximum intensity in the signal beam. As we shall see, concern for linearity in the recording generally results in $R > 1$ everywhere over the hologram plane.

Light from a typical laser is linearly polarized and it is generally not difficult to maintain $\cos \Omega = 1$ and so maximize visibility. Reference to Fig. 7.11, however, shows that it is desirable that both reference and signal beam at the hologram plane be polarized in directions perpendicular to the plane formed by their propagation vectors. (When the beams are kept parallel to an optical table, the desirable polarization direction is perpendicular to the table.) For this case $\cos \Omega = 1$ regardless of the angle θ between the propagation vectors. On the other hand, when the laser light is polarized parallel to the plane of the propagation vectors, e.g., parallel to the plane of the paper in Fig. 7.11, the angle $\Omega = \theta$. If in forming the hologram the beams intersect at right angles, $\cos \Omega$ and the visibility are both zero. The direction of polarization of the light leaving the laser can of course be rotated so that it is perpendicular to the table surface by inserting into the beam a properly oriented half-wave plate (Fig. 7.12).

Certain subjects depolarize the light they reflect. In such cases the undesirable component can be eliminated by means of an analyzer and the visibility maximized by means of the beam ratio.

7.2.2 Linear Recording

Let us now consider the relation between linear wavefront recording and the visibility of holographic interference fringes. We assume the recording material is thin so that transmittances can be expressed in terms of x and y

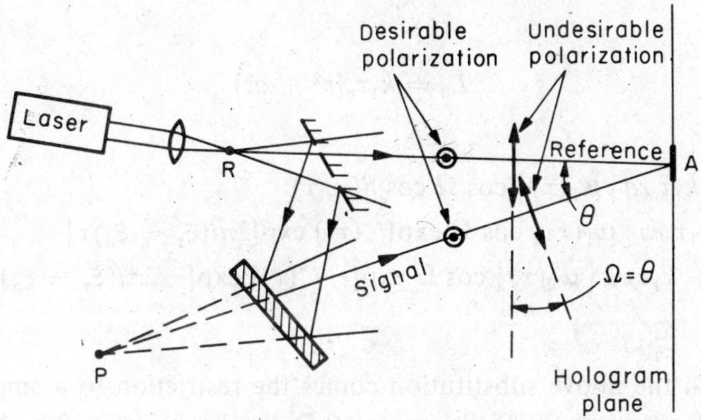

FIG. 7.11. Desirable and undesirable orientations for polarization direction.

only. Suppose the medium is given an exposure (see Section 2.5.1)

$$E(x, y) = I_P(x, y)\tau_e = k_1 I(x, y)\tau_e \tag{7.32}$$

where τ_e is the exposure time, k_1 is the constant of proportionality between the intensities I_P and I defined in Section 1.3, and I, the intensity of the interference pattern, is given by Eq. (7.30). The amplitude transmittance of the exposed and developed record (the hologram) will be dependent in some manner on the exposure E. Suppose that the exposure-dependent part of the transmittance is represented by $\mathbf{t}_E$. Here we allow the possibility of phase as well as amplitude modulation by the hologram. When I from Eq. (7.30) is inserted into Eq. (7.32), we have

$$\begin{aligned} E(x) &= k_1\tau_e[r^2 + a^2 + 2ra\,|\,\mu_T(\tau)\,|\cos\Omega\cos\beta(x, \tau)] \\ &= E_0 + E_1(x) \end{aligned} \tag{7.33}$$

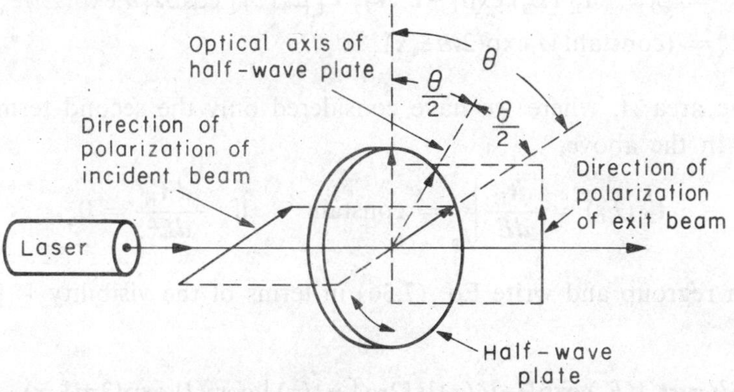

FIG. 7.12. Polarization vector orientation with a half-wave plate.

where

$$E_0 = k_1\tau_e[r^2 + a^2]$$

and

$$\begin{aligned}E_1(x) &= 2k_1\tau_e ra\,|\,\mu_T(\tau)\,|\cos\Omega\cos\beta(x, \tau)\\ &= k_1\tau_e ra\,|\,\mu_T(\tau)\,|\cos\Omega\exp[i\zeta(\tau)]\exp[2\pi i(\xi_r - \xi_a)x]\\ &\quad + k_1\tau_e ra\,|\,\mu_T(\tau)\,|\cos\Omega\exp[-i\zeta(\tau)]\exp[-2\pi i(\xi_r - \xi_a)x].\end{aligned} \tag{7.34}$$

Along with the above substitution comes the restriction to a small area A of the hologram plane. If we ensure that the beam ratio $R > 1$, then E_1 will always be less than E_0 and we can express the amplitude transmittance $\mathbf{t}_E$ over the area A as a Taylor series

$$\mathbf{t}_E = \mathbf{t}_E(E_0) + E_1\frac{d\mathbf{t}_E}{dE}\bigg|_{E_0} + \frac{1}{2}E_1^2\frac{d^2\mathbf{t}_E}{dE^2}\bigg|_{E_0} + \cdots. \tag{7.35}$$

When the hologram of transmittance $\mathbf{t}_E$ is illuminated by the original reference wave $\mathbf{r} = r\exp(2\pi i\xi_r x)$ (assumed monochromatic at this stage), the complex amplitude of the modulated light at the back side of area A is

$$\mathbf{w} = \mathbf{r}\mathbf{t}_E.$$

If, in Eq. (7.35), the coefficients of second and higher order terms are negligible, a wave can be reconstructed whose amplitude is linearly proportional to the original subject wave amplitude $\mathbf{a} = a\exp(2\pi i\xi_a x)$. Under this condition the reconstructed wave of interest becomes

$$\begin{aligned}\mathbf{w}_V &= \mathbf{r}E_1\mathbf{t}_E'(E_0)\\ &= \{k_1\tau_e r^2\mathbf{t}_E'(E_0)\exp[-i\zeta(\tau)]\cdot|\,\mu_T(\tau)\,|\cos\Omega\}a\exp(2\pi i\xi_a x)\\ &= (\text{constant})a\exp(2\pi i\xi_a x)\end{aligned} \tag{7.36}$$

over the area A, where we have considered only the second term in Eq. (7.34). In the above,

$$\mathbf{t}_E'(E_0) = \frac{d\mathbf{t}_E}{dE}\bigg|_{E_0} = \text{constant} \quad \text{if} \quad \frac{d^2\mathbf{t}_E}{dE^2} = 0.$$

We can regroup and write Eq. (7.36) in terms of the visibility V [see Eq. (7.31)] as

$$\begin{aligned}\mathbf{w}_v &= \tfrac{1}{2}k_1\tau_e r\mathbf{t}_E'(E_0)\exp[-i\zeta(\tau)]\cdot[2ra\,|\,\mu_T(\tau)\,|\cos\Omega]\exp(2\pi i\xi_a x)\\ &= \tfrac{1}{2}r\mathbf{t}_E'(E_0)\exp[-i\zeta(\tau)]\cdot E_0 V\exp(2\pi i\xi_a x).\end{aligned} \tag{7.37}$$

The intensity of the reconstructed wave at the hologram is

$$I_v = \tfrac{1}{4}[r\mathbf{t}_E'(E_0)E_0V]^2. \qquad (7.38)$$

Since $V \propto |\mu_T(\tau)|$, we find that our choice of $|\mu_T(\tau_H)| = 1/\sqrt{2}$ in Section 7.1.2 as a minimum tolerable degree of temporal coherence for hologram formation implies that we wish the intensity of the reconstructed wave to remain within 50% of the value it would have for perfect coherence.

When the complex amplitude of the reconstructed wavefront is a function proportional to that of the original signal wavefront over the entire hologram plane, the recording is called *linear and space invariant* or simply *linear.* The condition for linear recording is that the derivative of the complex transmittance of the hologram $\mathbf{t}$ with respect to the exposure E must be

$$d\mathbf{t}/dE = \text{constant} \qquad (7.39)$$

whose value remains independent of the entire range of exposure values found over the hologram plane during formation of the hologram. The range of exposure, $E_{min} < E < E_{max}$, can be obtained from Eqs. (7.33) and (7.31), where E_{min} corresponds to $\cos\beta = -1$ and E_{max} corresponds to $\cos\beta = +1$. Expressing the exposure range in terms of the maximum visibility V_{max} found on any area of the hologram plane, we have

$$E_0(1 - V_{max}) < E < E_0(1 + V_{max}). \qquad (7.40)$$

Whether the condition $d\mathbf{t}/dE =$ constant can be satisfied over the range in (7.40) depends on the exposure characteristic of the photosensitive material used to record the hologram (see Chapter 10).

We can make some general observations on the conditions for linear recording, without reference to specific materials, by considering separately plane absorption and plane phase holograms. The complex transmittance of a photosensitive medium suitable for forming a plane hologram can be characterized by

$$\mathbf{t} = t \exp(i\phi)$$

and by its derivative with respect to exposure

$$\frac{d\mathbf{t}}{dE} = \frac{dt}{dE}\exp(i\phi) + it\exp(i\phi)\frac{d\phi}{dE}. \qquad (7.41)$$

When the material is exposed to light, the modulus t and the phase ϕ of the transmittance may be altered. While to some degree photosensitive materials exhibit both of these changes, the more important recording materials respond with a significant alteration of only one or the other. If dt/dE is finite and $d\phi/dE = 0$, the material forms an absorption hologram. If $d\phi/dE$ is finite and $dt/dE = 0$, the material forms a phase hologram.

When an absorption hologram is to be formed in a material such as photographic emulsion, the condition for a linear recording, Eq. (7.39), becomes

$$dt/dE = \text{constant}.$$

To be assured of linearity we must first obtain a curve of transmittance versus exposure (t–E) for the particular emulsion chosen. The exposure range $E_0(1 - V_{\max}) < E < E_0(1 + V_{\max})$ must remain within the (approximately) linear portion of the t–E curve, Fig. 7.13. One obtains the curve experimentally by measuring the intensity transmittance $\mathcal{T}$ as a function of E and then computing $t = \sqrt{\mathcal{T}}$. Some t–E curves corresponding to emulsions commonly used in holography are found in Chapter 10.

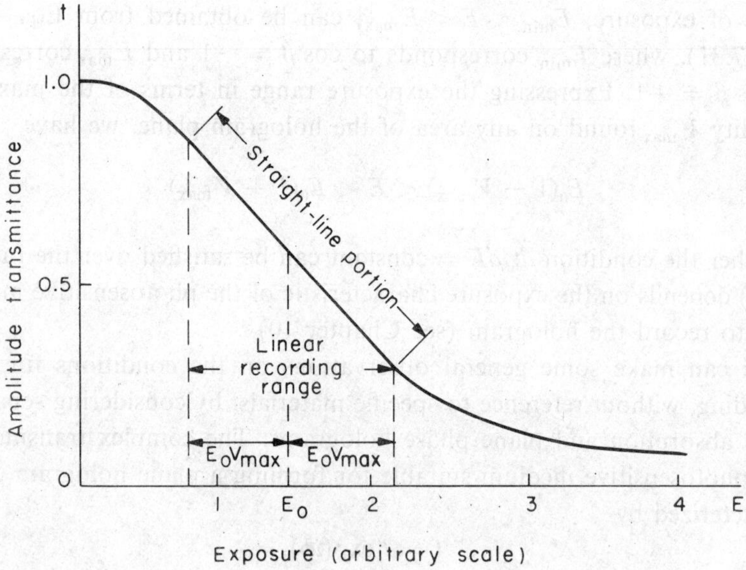

FIG. 7.13. Amplitude transmittance versus exposure.

When a phase hologram is formed in an appropriate material such as thermoplastic, Eqs. (7.39) and (7.41) require that

$$\exp(i\phi)\,\frac{d\phi(E)}{dE} = \text{constant} \tag{7.42}$$

for linear recording [$\phi(E)$ is the exposure-dependent portion of the transmittance phase associated with the recording material]. However, Eq. (7.42) has only the trivial solution: $\phi =$ constant and $d\phi/dE = 0$, indicating that strictly linear recording cannot be achieved in phase holograms. For a small but finite $\phi(E)$ the exponential in Eq. (7.42) becomes $\exp(i\phi) \approx 1$ and we obtain as a linearity condition

$$\frac{d\phi(E)}{dE} = \text{constant}, \qquad \phi(E) \ll 1. \tag{7.43}$$

A given recording material might have a $\phi(E) - E$ characteristic as indicated in Fig. 7.14. Only for a small exposure range in the straight-line portion of the curve is $\phi(E)$ small and the recording essentially linear. For a larger exposure range and correspondingly larger value of $\phi(E)$ the recording becomes nonlinear. In practice, phase holograms yield high diffraction efficiency, indicative of appreciable exposure range and moderate values of $\phi(E)$, without intolerable nonlinearity effects in the recording.

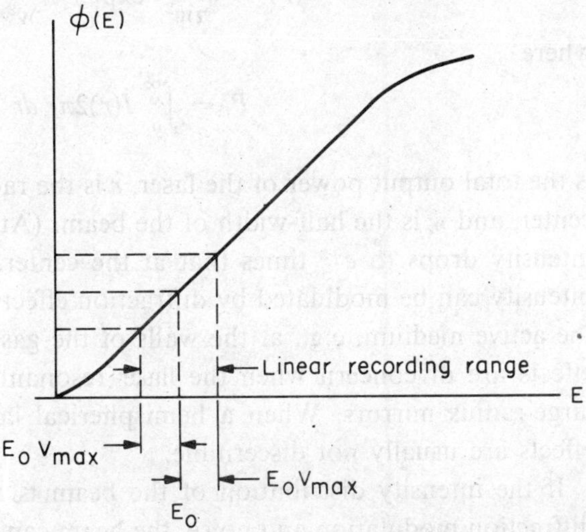

FIG. 7.14. $\phi(E)$ versus E.

For many holographic materials, e.g., photographic emulsions, the straight-line portion of the exposure characteristic does not extend to $E = 0$. Since the minimum value of the exposure range is $E_0(1 - V_{\max})$, the visibility should not be permitted to reach its maximum value of unity. By ensuring that the intensity of the reference beam at the hologram is everywhere greater than the signal intensity, the minimum value of the beam ratio $R_{\min}$ is everywhere greater than unity and the maximum value of visibility $V_{\max}$ is thus kept less than unity.

7.3 Illumination with an Expanded Laser Beam

The beam of light emerging from a laser is typically 1 mm in diameter. For most holographic applications, however, we must illuminate a subject and a hologram recording medium of much larger size. We also desire the illumination to be as uniform as possible. Expansion of a laser beam can be accomplished with one or more lenses or spherical mirrors without significant loss of power or serious alteration of the intensity distribution in the beam. *Uniform* illumination can be obtained, however, only by sacrificing a portion of the laser output power.

The intensity distribution in a laser beam depends on its transverse mode structure. Except for the lowest order TEM_{00} mode, the intensity vanishes at one or more places over the cross section of the beam. Uniform illumination, therefore, cannot be achieved with higher order modes. In the absence of diffraction effects due to the finite extent of the active laser medium, the radial distribution of intensity in the TEM_{00} mode has the Gaussian form

$$I(r) = \frac{2P_t}{\pi w^2} \exp\left(-\frac{2r^2}{w^2}\right) \tag{7.44}$$

where

$$P_t = \int_0^\infty I(r) 2\pi r \, dr$$

is the total output power of the laser, r is the radial distance from the beam center, and w is the half-width of the beam. (At the radius $r = w$, the beam intensity drops to e^{-2} times that at the center.) This smooth variation in intensity can be modulated by diffraction effects originating at the edges of the active medium, e.g., at the walls of the gas discharge tube. Diffraction effects are of concern when the laser resonant cavity is formed with two large-radius mirrors. When a hemispherical laser cavity is employed, the effects are usually not discernible.

If the intensity distribution of the beam is a smooth Gaussian, free of diffraction modulation and noise, the beam can be expanded very simply by reflecting it from a front-surface concave or convex mirror. The main advantage of this method is that multiple reflections are avoided. The mirror, however, must be used in an off-axis configuration, introducing aberrations into the expanded beam. To minimize the aberrations the radius of curvature of the mirror should be as large as possible consistent with obtaining the desired beam cross section in a given distance.

If the beam intensity distribution is not a smooth Gaussian but contains variations which we can consider to be noise, the latter can be effectively removed by spatial frequency filtering. The laser beam is brought to a focus by a lens, and a pinhole is placed at the focal point. A simple microscope objective lens can serve both to focus and expand the beam (Fig. 7.15). The pinhole passes only the very lowest spatial frequencies characteristic of the slowly varying Gaussian distribution but blocks the higher spatial frequencies characteristic of the noise. Emerging from the pinhole is an essentially smooth Gaussian intensity distribution superimposed on spherical wavefronts. (A collimating lens can be used to obtain essentially plane wavefronts if desired.)

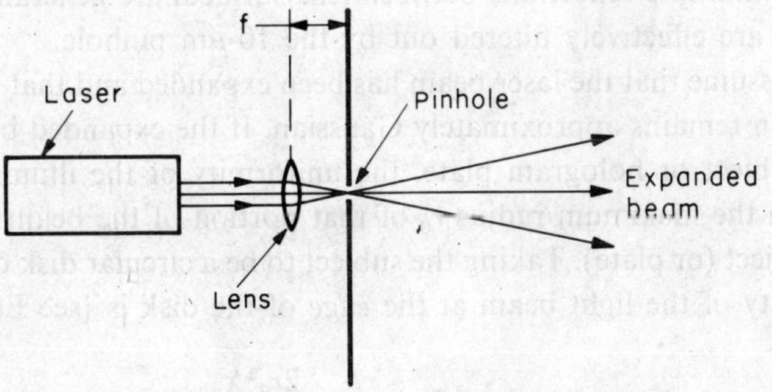

FIG. 7.15. Beam expander and spatial filter.

Let us now estimate the size of the pinhole required for effective spatial filtering by considering the incident beam to be Gaussian. If the diameter of the focusing lens is sufficiently large relative to the incident Gaussian beam width, we can neglect the effect of lens-edge diffraction on the Gaussian beam profile. The complex amplitude of the light incident on the lens is the square root of Eq. (7.44) and proportional to $\exp(-r_1^2/w_1^2)$, where w_1 is the radius or the half-width of the incident beam at the front focal plane of the lens and r_1 is the radial distance in that plane. Its Fourier transform is displayed by the lens in the plane of the pinhole and (from the discussion in Section 4.2) given by the correspondence

$$\exp\left(-\frac{r_1^2}{w_1^2}\right) \supset \text{const} \cdot \exp(-\pi^2 w_1^2 \nu^2)$$
$$\supset \text{const} \cdot \exp\left(-\frac{\pi^2 w_1^2 r_2^2}{\lambda^2 f^2}\right) \qquad (7.45)$$

where $\nu = r_2/\lambda f$, f is the focal length of the lens, and r_2 is the radial distance in the plane of the pinhole. The intensity of the Gaussian beam in the back focal plane is the square of the right-hand side of correspondence (7.45)

$$I_2 \propto \exp\left(-\frac{2\pi^2 w_1^2}{\lambda^2 f^2} r_2^2\right)$$

with a beam half-width

$$w_2 = \lambda f/\pi w_1. \tag{7.46}$$

Substituting typical values of $\lambda = 0.63\ \mu m$, $f = 16$ mm, and $w_1 = 1$ mm, we find for the full width of the laser beam near the focus, $2w_2 = 6.4\ \mu m$. If a 10-μm-diameter pinhole is used as the spatial filter, it will easily pass the 6.4 μm laser beam. However, spatial frequencies in excess of $\nu = r_2/f\lambda = 5\ \mu m/(16\ mm \times 0.63\ \mu m) = 0.5$ cycles/mm are stopped. The spatial frequencies associated with diffraction modulation, scattering from dust particles, and multiple reflections between lens surfaces are generally of higher value and are effectively filtered out by the 10-μm pinhole.

Let us assume that the laser beam has been expanded and that its intensity distribution remains approximately Gaussian. If the expanded beam illuminates a subject or hologram plate, the uniformity of the illumination will depend on the maximum radius r_0 of that portion of the beam intercepted by the subject (or plate). Taking the subject to be a circular disk of radius r_0, the intensity of the light beam at the *edge* of the disk is [see Eq. (7.44)]

$$I_e = I_0 \exp\left(-\frac{2r_0^2}{w^2}\right) \tag{7.47}$$

where $I_0 = 2P_t/\pi w^2$ is the light intensity at the center of the disk and P_t and w are the total power and half-width of the laser beam at the plane of the disk. The laser power which actually contributes to the illumination of the disk is obtained by integrating Eq. (7.44) over the radius r

$$P_i = \frac{2P_t}{\pi w^2}\int_0^{r_0} \exp\left(-\frac{2r^2}{w^2}\right) 2\pi r\, dr$$

$$= P_t\left[1 - \exp\left(-\frac{2r_0^2}{w^2}\right)\right].$$

Substituting Eq. (7.47) for the exponential, we have

$$\frac{P_i}{P_t} + \frac{I_e}{I_0} = 1. \tag{7.48}$$

If, for example we require that the light intensity at the edge of the disk be at least 50% of that at the center, the laser power which can be used for the purpose is only 50% of the total available power.

If we desire to illuminate *uniformly* a disk of radius r_0 with a Gaussian beam we must pass the beam through an absorption filter which reduces the intensity everywhere in the beam to that at the edge. The intensity transmittance $\mathcal{T}$ of the filter should be

$$\mathcal{T} = \exp\left[\frac{2(r^2 - r_0^2)}{w^2}\right] \quad \text{for} \quad r \leq r_0$$

$$= 0 \quad \text{otherwise.}$$

The laser power P_i which can be used for illumination is now given by the product of $I(r)$ [from Eq. (7.44)], the transmittance $\mathcal{T}$, and the area πr_0^2:

$$P_i = \frac{2P_t}{\pi w^2}(\pi r_0^2)\exp\left(-\frac{2r_0^2}{w^2}\right) = 2P_t\left(\frac{r_0}{w}\right)^2 \exp\left(-\frac{2r_0^2}{w^2}\right). \tag{7.49}$$

The maximum value of P_i can be found by differentiating Eq. (7.49) with respect to r_0. That value is

$$P_{i,\max} = P_t/e = 0.37P_t$$

when $r_0^2 = w^2/2$.

7.4 Division and Attenuation of the Laser Beam

To form a hologram, the output power from a laser must be divided into subject illumination and a reference wave (see Fig. 7.1). If the subject requires illumination from more than one angle (to eliminate shadows), several divisions of the laser beam may be necessary. Amplitude division (Fig. 7.1b) rather than wavefront division is generally preferred since it results in more uniform illumination intensity and requires less beam expansion. The beam may be divided either before or after it is expanded. Two advantages of division before expansion are (1) the aperture of the beam splitter can be small, and (2) diffraction effects related to dust and surface defects on the beam splitter can be spatially filtered with the lens–pinhole combination used for the subsequent beam expansion. An advantage of division after expansion is that only one set of beam expanding optical components is needed.

Partial reflection from a partially silvered (or dielectric-coated) mirror is the simplest method for dividing a laser beam into two parts. The unsilvered surface is antireflection-coated to minimize the interference between reflections from the two surfaces. A variation of the partially reflecting mirror is a cube-type beam splitter made of two right-angle glass prisms cemented together at their hypotenuse surfaces. One of the two hypotenuse surfaces is partially silvered before the cementing. Cube-type beam splitters are advantageous because they are less likely to be damaged in cleaning and produce no lateral translation of the transmitted beam. On the other hand, the two output beams are fixed approximately perpendicular to each other.

The beams emerging from the beam splitter should have an intensity ratio which optimizes the fringe visibility at the hologram plane. Of course the optimum ratio varies with the circumstances of hologram formation. Therefore a variable-ratio lossless beam splitter is highly desirable, especially in applications of holography to real-time interferometry where the beam ratio must be changed to a new optimum during the reconstruction process.

Perhaps the simplest variable-ratio beam splitter is a partially silvered mirror whose reflectivity varies continuously in an annular ring. A laser beam incident on a selected portion of the ring can thus be divided into two beams having the desired intensity ratio. An obvious undesirable feature of this device is its gradient of reflectivity resulting in nonuniform division across the beam.

A superior variable-ratio beam splitter is a polarization beam splitter made of doubly refracting material such as calcite or crystal quartz. An example [*7.12*] of this type of beam splitter using a Foster–Seely prism is shown in Fig. 7.16. Half-wave plate *A* changes the direction of polarization of a linearly polarized laser beam to any arbitrary direction before the laser beam enters the Foster–Seely prism *B*. The prism is cut and silvered in such a way that the extraordinary ray, polarized perpendicular to the paper, passes through the prism unhindered; the ordinary ray polarized parallel to the paper, however, suffers a refraction and a reflection leaving the prism at 90° to the extraordinary ray. By rotating the half-wave plate *A*, we can therefore obtain any desired intensity ratio between the two output

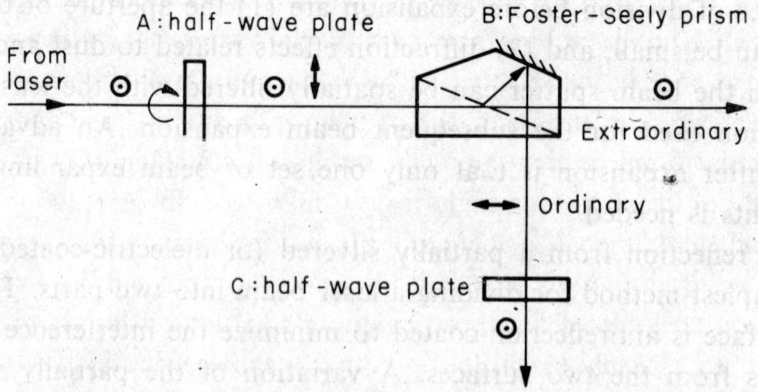

FIG. 7.16. Polarization beam splitter using a Foster–Seely prism.

beams. A fixed half-wave plate *C* changes the direction of polarization of the emerging ordinary ray so that both output beams are polarized perpendicular to the paper as desired in most cases. Of course, compensators or electro-optical crystals may be used instead of the two half-wave plates. These allow use of the beam splitter at any wavelength. Electro-optical crystals offer the possibility of an electrically alterable intensity ratio.

When only one of two output beams is used and the other discarded, a beam splitter also acts as an attenuator. It is generally more economical, however, to use a neutral density filter as a fixed attenuator and a pair of polarizing prisms or sheets as a variable-ratio attenuator.

There are two types of neutral density filters and both are commercially available. In one type the unwanted energy is absorbed by the filter and in the other type unwanted energy is reflected. The absorption type of neutral density filter is superior at low power levels, say below 1 W/cm^2. Light multiply reflected between the surfaces of the filter is absorbed, thus avoiding unwanted interference. At higher levels, however, an absorption filter can be damaged by the heat generated from the absorbed energy.

7.5 Mechanical Stability in Hologram Formation

The highest spatial frequency in the interference fringe pattern to be recorded determines the degree to which the position of the hologram-recording medium must remain fixed relative to the pattern. We assume, for the time being, that the fringe pattern is stationary. The smallest fringe spacing encountered when forming most holograms is approximately the wavelength of the forming light. We must therefore ensure that the hologram recording medium does not move more than a small fraction of a wavelength during its exposure. Under laboratory environment it is not difficult to achieve this relative stability if the holder for the recording medium is carefully designed. On the other hand, the task of maintaining a stable fringe pattern is a more formidable one and is a familiar problem in precision optical interferometry.

To ensure a steady fringe pattern, all optical components as well as the subject, the light source, and the recording medium must be carefully and securely mounted on a massive optical bench or table. Tables made of granite, concrete, steel, or aluminum weighing tons are not uncommon. The mass is used to make the mechanical resonance frequency of the table (about 1 Hz) much lower than the frequencies of the building vibration. Isolation of the optical table from the building vibration is adequately achieved if the table is supported in a sand bath or by a pneumatic suspension system. For example, one inexpensive suspension system consists of many rubber inner tubes filled with air at a low pressure. An air bearing is another, perhaps more luxurious, method.

Airborne disturbances, both acoustic and thermal, can also cause an unstable interference fringe pattern and are generally more difficult to eliminate than mechanical vibrations. Air-conditioning units and ventilating fans are best turned off during the exposure of a hologram. Sources of heat such as lamps and some electrical equipment should be kept away from the area of the optical table immediately before and during the exposure. The effects of airborne disturbances are minimized if all optical path lengths, especially those between the beam splitter and the hologram recording medium, are made as short as possible.

The problems of both mechanical and airborne disturbances are lessened, of course, if the exposure time can be reduced. It is therefore important to fully utilize the output power of a laser by using the optimum beam intensity ratio and by minimizing the number of optical components and their associated optical loss. An electronic servo-control technique can be devised to stabilize the interference fringe pattern [*7.13*, *7.14*]. This technique is applicable, however, only when the mechanical or airborne disturbance of the fringes can be compensated by a change in the path length of reference beam or signal beam or by a change in the oscillation frequency of the laser.

7.6 Light Sources for Hologram Reconstruction

If an original wavefront recorded in a hologram is to be reconstructed with minimum aberration, the direction and radius of curvature of the reconstructing beam must duplicate those of the reference beam used to form the hologram. The resolution of the image produced by the wave reconstructed under this condition is limited only by the size of the hologram and the coherence properties of the reconstructing light source.

To obtain the highest achievable resolution in the image, the light source for wavefront reconstruction should be as coherent as the light source for hologram formation. However, in many cases, this is neither necessary nor desirable. A powerful multilongitudinal-mode laser can often be used successfully to illuminate a hologram made with a single-frequency laser and, moreover, to produce a brighter image than the single-mode laser without appreciable loss in image quality. In some cases an arc lamp or an incandescent lamp may be used in the reconstruction step with a not intolerable loss in image resolution. In this section we shall consider how the degree of coherence of the illuminating light source affects the image resolution.

We have seen that the degree of spatial coherence of a nonlaser light source depends on its radial extent according to Eq. (7.4). When the source is a point so that in Eq. (7.4) the radius $r_0 = 0$, the degree of spatial coherence $|\mu_s|$ is a maximum of unity. Suppose, for the purposes of analysis, that we now have a more realizable source of finite extent and thus of lesser spatial coherence, but one which nonetheless has a very high degree of temporal coherence. We consider the extended source to illuminate a hologram which has been formed with a perfect reference point source and inquire as to the effect on the resulting image of this lesser degree of spatial coherence. The effect may be determined for the case of a small source by examining the expressions for the virtual image coordinates in Eq. (3.27),

$$x_{3V} = \frac{x_c z_1 z_r + \mu x_1 z_c z_r - \mu x_r z_c z_1}{z_1 z_r + \mu z_c z_r - \mu z_c z_1}, \qquad z_{3V} = \frac{z_c z_1 z_r}{z_1 z_r + \mu z_c z_r - \mu z_c z_1} \tag{7.50}$$

where the scaling factor m in Eq. (3.27) has been set equal to unity, (x_c, z_c) are the coordinates of the illuminating source, $\mu = \lambda_2/\lambda_1$, and the remaining parameters are defined in Fig. 7.17. (*Note*: The results of Chapter 3 used in the remainder of this section are valid for *plane* holograms.) Figure 7.17 illustrates the reconstruction situation. Illumination of the hologram H by the original reference point source R, located a distance z_r from H, yields the subject image P in its original location (x_1, z_1). Consider now a second illuminating point source R' also located a distance $z_c = z_r$ from the hologram and radiating light of the same wavelength as R [$\mu = 1$ in

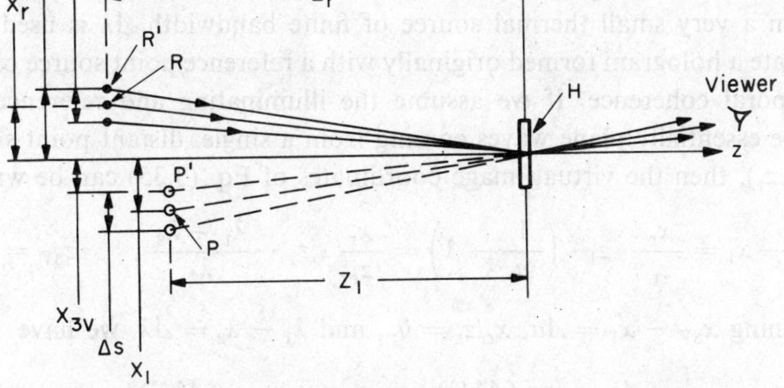

FIG. 7.17. Illustrating the effect of an illuminating source of finite extent on an image point *P*.

Eq. (7.50)]. From Eq. (7.50) we find that another virtual image P' is located at

$$x_{3V} = x_1 + (z_1/z_r)(x_c - x_r), \qquad z_{3V} = z_1. \tag{7.51}$$

Let us define $\Delta s/2 = x_{3V} - x_1$ and $\Delta r/2 = x_c - x_r$. Imagine the nonlaser source of finite extent to be composed of point sources whose extreme coordinates are $(x_c = x_r \pm \Delta r/2, z_c = z_r)$. The total span of possible image positions of the point P along the x axis, corresponding to the source of width Δr, is then given by Eq. (7.51) as

$$\Delta s = 2(x_{3V} - x_1) = 2(z_1/z_r)(x_c - x_r)$$

or

$$\Delta s = (z_1/z_r)\,\Delta r \tag{7.52}$$

Thus the degree of spatial coherence of the illuminating source, determined by the source diameter Δr, fixes the minimum resolvable image dimension Δs.

As an example, suppose a mercury arc lamp, whose radiation is passed through a narrow-band wavelength filter and whose diameter is 0.3 mm, is used as the illuminating source. If both source and image are at distance z_r from the hologram, the smallest detail we can expect to resolve in the image is about $\Delta s = 0.3$ mm. At $z_1 = z_r = 1$ m the resolution approaches that of the eye situated at the hologram and is quite sufficient for many visual displays. Equation (7.52) implies that the closer the image is to the hologram, the less is the demand for spatial coherence in the illuminating source. If the central plane of the image lies in the hologram plane, the requirement on spatial coherence is least.

We can also employ Eq. (3.35) to demonstrate the effect on the image when a very small thermal source of finite bandwidth $\Delta\lambda$ is used to illuminate a hologram formed originally with a reference point source of perfect temporal coherence. If we assume the illuminating and reference beams to be essentially plane waves coming from a single, distant point source at (x_r, z_r), then the virtual image coordinates of Eq. (3.35) can be written as

$$x_{3V} - x_1 = \frac{x_r}{z_r} \cdot z_1 \cdot \left(\frac{1}{\mu} - 1\right) = \frac{x_r}{z_r} \cdot z_1 \cdot \frac{\lambda_1 - \lambda_2}{\lambda_2}, \qquad z_{3V} = \frac{z_1}{\mu}.$$

Defining $x_{3V} - x_1 = \Delta\sigma$, $x_r/z_r = \theta_r$, and $\lambda_1 - \lambda_2 = \Delta\lambda$, we have

$$\Delta\sigma = \theta_r z_1(\Delta\lambda/\lambda), \qquad z_{3V} - z_1 = (\Delta\lambda/\lambda)z_1. \tag{7.53}$$

Suppose we consider the subject to be on axis so that the small change in

z position does not contribute to loss of resolution in a plane transverse to the viewing direction. Since x_1 corresponds to the original subject position and also to the image position when the original reference wavelength λ_1 illuminates the hologram, $\Delta\sigma$ represents a displacement from x_1 due to illumination with a source whose wavelength $\lambda_2 = \lambda_1 - \Delta\lambda$. Thus a point image would spread to the dimension $\Delta\sigma$ given in Eq. (7.53) if the hologram were illuminated by a very small nonmonochromatic source of bandwidth $\Delta\lambda$.

If we once again consider a high-pressure mercury arc lamp with a bandwidth of 50 Å centered at 5461 Å and assign some typical values $\theta_r = 15°$ and $z_1 = 100$ mm in Eq. (7.53), the point image spread $\Delta\sigma$ becomes 0.24 mm. Note that the bandwidth of the source, representing a relatively low degree of coherence, places a limit on the depth of the image over which one can display small detail and a limit on the offset angle θ_r of the reference beam. If the central plane of a 3D image lies in the hologram plane ($z_1 = 0$) and if the angle θ_r is small, white light illumination is possible. A volume of the image surrounding the hologram plate will then appear achromatic, i.e., color dispersion will be negligible (see Section 17.4).

Source size and bandwidth are primary determinants of the smallest resolvable dimension in the image when nonlaser sources are employed to illuminate a hologram, but this is often not the case when a laser source is used. A multilongitudinal-mode He–Ne laser can have sufficient temporal coherence that when it illuminates a hologram formed on a standard 4 × 5 inch high-resolution photographic plate, the image resolution is essentially diffraction limited (limited only by the angular aperture of the hologram). Suppose we consider the formation of the real image of a point as indicated in Fig. 7.18. We compute the maximum difference in optical path ΔL taken by rays passing from the source (at infinity) to the image via the hologram. We assume that the degree of temporal coherence of the source places no limit on image resolution provided ΔL does not exceed the source coherence length ΔL_H. If we write ΔL in terms of the extent of the hologram h, then we can obtain the maximum value h_m allowed under the condition that the source temporal coherence plays no significant part in determining image resolution. From Fig. 7.18 we have

$$\Delta L = h \sin\theta_r + (h^2 + z_i^2)^{1/2} - z_i. \tag{7.54}$$

Common values for the parameters θ_r and z_i might be $\theta_r = 30°$ and $z_i = 2h$. These yield $\Delta L = 0.736h$. Equating ΔL and ΔL_H, we find that

$$h_m = \Delta L_H / 0.736.$$

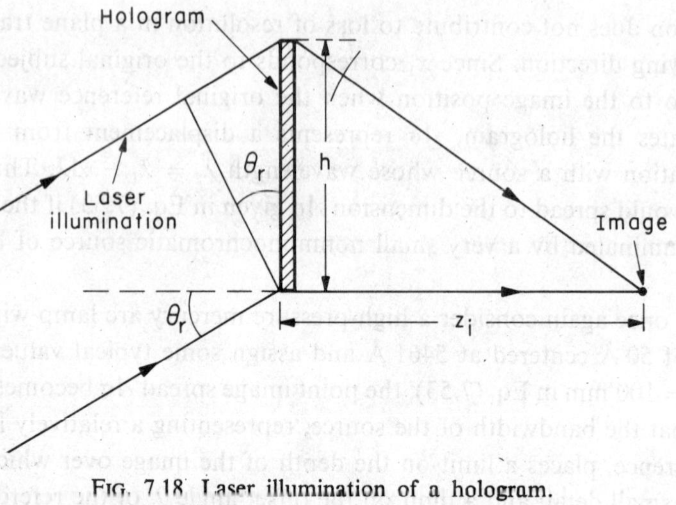

FIG. 7.18 Laser illumination of a hologram.

A 1-m-long He–Ne laser has a coherence length $\Delta L_H \approx 10$ cm permitting h_m to be somewhat over 13 cm. (Note that a multilongitudinal-mode argon laser with $\Delta L_H \approx 2$ cm restricts h_m to $\frac{1}{5}$ this value.)

7.7 Simple Holographic Technique

In the preceding sections of this chapter we have specified properties of light sources and optical components which are desirable for general application in holography. The requirements placed on the components, of course, vary with the application. In this section we present a simple example of holographic technique which can form good holograms and reconstruct wavefronts yielding high-quality 3D images. Only a minimum number of inexpensive optical components are employed.

The optical arrangement for forming the hologram is shown as a schematic drawing in Fig. 7.19 and as an actual photograph in Fig. 7.20. All optical components are securely fastened to a massive steel optical table which is supported on a number of inflated airplane-tire inner tubes. In the case of the beam splitter and mirrors, tacky wax is used to attach these components to a metal rod which in turn is locked into the collar of a stand and the latter bolted to the table. We show as a light source a helium–neon laser of sufficient length to support several longitudinal modes. It has an output power of 1.8 mW at 6328 Å. A glass plate, approximately 5 mm thick, serves as the beam splitter B and reflects about 5% of the incident power from each (front and back) surface. The front surface reflection is

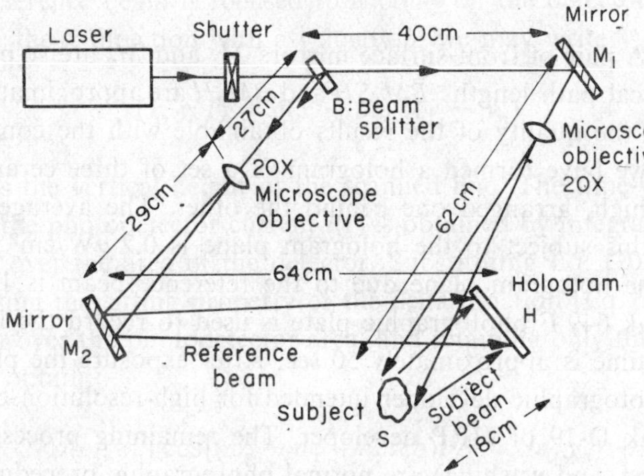

Fig. 7.19. A simple practical arrangement for forming a hologram.

used for the reference beam, and the remaining 90% which passes through the plate is used to illuminate the subject. Two 20× microscope objective lenses (8-mm focal length and 0.5 numerical aperture) expand the beam widths from less than 1 mm to nearly 7 cm near the subject S and

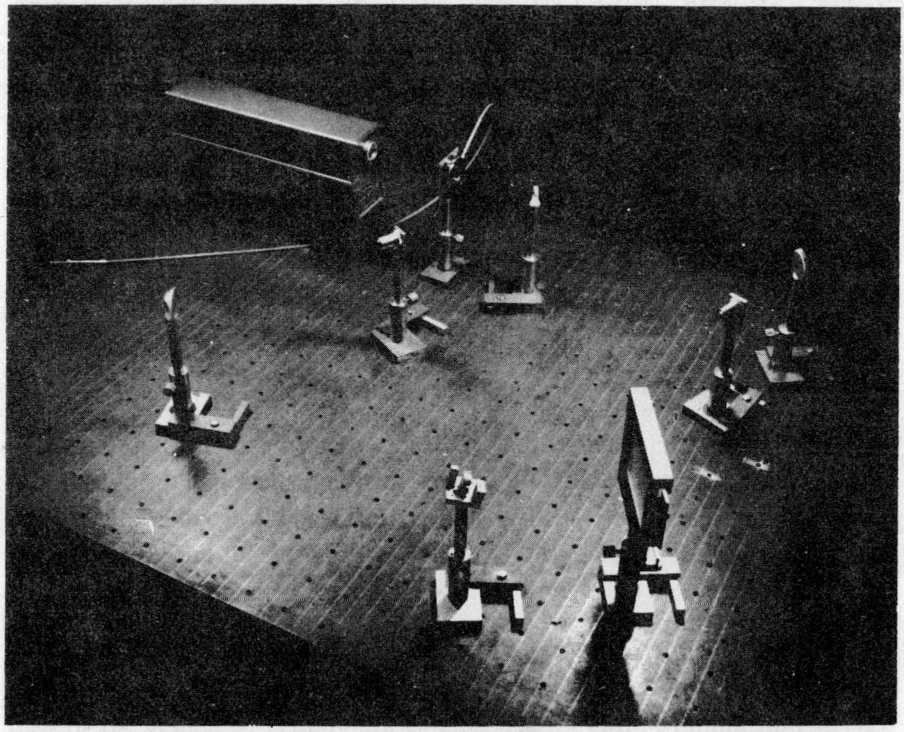

FIG. 7.20. Photograph of the actual arrangement corresponding to Fig. 7.19.

hologram H. A pair of front surface mirrors M_1 and M_2 are so placed that the mean optical path lengths BM_1SH and BM_2H are approximately equal.

To indicate the quality of the results obtainable with the configuration of Fig. 7.20, we have formed a hologram of a set of three ceramic block letters, 2 cm high, arranged one behind the other. The average intensity imparted by this subject to the hologram plane is 0.2 $\mu W/cm^2$ while the intensity at the hologram plane due to the reference beam is 1 $\mu W/cm^2$. When a Kodak 649 F photographic plate is used to record the hologram, the exposure time is approximately 50 sec. After exposure the plate is developed in photographic developer intended for high-resolution emulsions, such as Kodak D-19 or HRP developer. The remaining processing steps (fixing, clearing, and washing) are normal photographic procedures which are set down in detail in Chapter 10. Figure 7.21 is a photograph of the hologram itself. Note the coarse diffraction patterns characteristic of scattering from dust particles. These are probably located on the microscope objective lenses used to expand the beam (the fringes relevant to the holographic process are, of course, too small to be observed without magnification). Dust particle fringes can, if desired, be eliminated by spatial frequency filtering with a 10-μm pinhole placed near the focus of the microscope objective (see Section 7.3). However, their presence does little to degrade the image produced by the hologram.

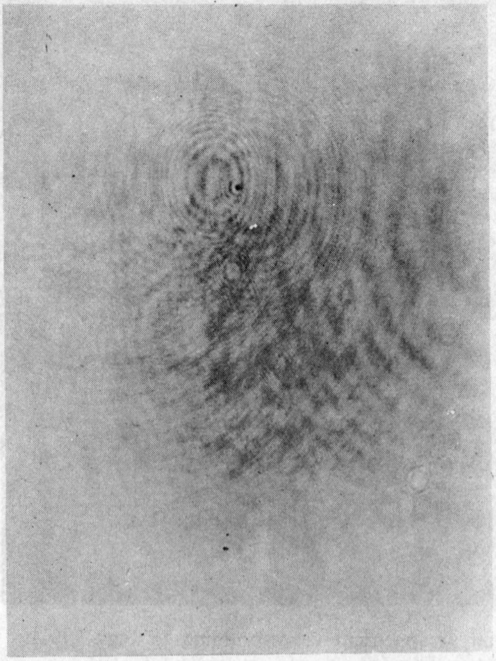

FIG. 7.21. Photograph of hologram formed with the arrangement of Fig. 7.20.

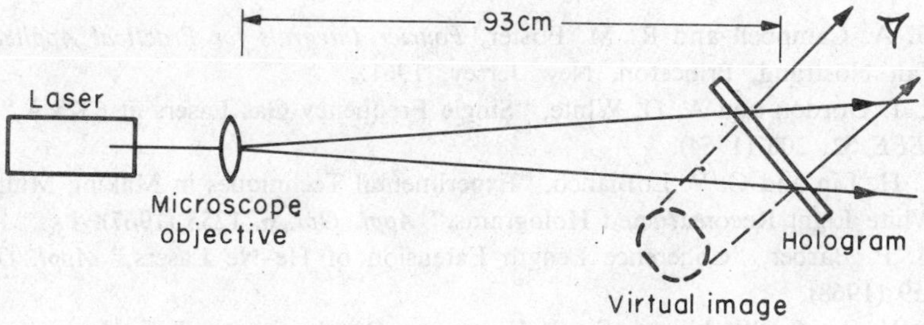

FIG. 7.22. Arrangement for reconstruction.

A simple arrangement for reconstructing the original subject wavefront is shown in Fig. 7.22. A laser and a beam-expanding microscope objective duplicate the original reference wavefront at the hologram. The wavefront reconstructed in this manner generates for the observer a virtual image whose size and position, relative to the hologram, are identical to those of the original subject. Figure 7.23 is a photograph of the image.

FIG. 7.23. Image produced by hologram of Fig. 7.21.

REFERENCES

7.1. H. Kogelnik and T. Li, "Laser Beams and Resonators," *Appl. Opt.* **5**, 1550 (1966).

7.2. M. Born and E. Wolf, *Principles of Optics*, 3rd ed., Chapter X. Pergamon Press, Oxford, 1965.

7.3. A. C. G. Mitchell and M. W. Zemansky, *Resonance Radiation and Excited Atoms*, second printing, Chapter III. Cambridge Univ. Press, London and New York, 1961.

7.4. G. A. Campbell and R. M. Foster, *Fourier Integrals for Practical Applications.* Van Nostrand, Princeton, New Jersey, 1961.

7.5. E. I. Gordon and A. D. White, "Single Frequency Gas Lasers at 6328 Å," *Proc. IEEE* **52**, 206 (1964).

7.6. L. H. Lin and C. V. LoBianco, "Experimental Techniques in Making Multicolor White Light Reconstructed Holograms," *Appl. Opt.* **6**, 1255 (1967).

7.7. H. P. Barber, "Coherence Length Extension of He–Ne Lasers," *Appl. Opt.* **7**, 559 (1968).

7.8 P. W. Smith, "Stabilized, Single-Frequency Output from a Long Laser Cavity," *IEEE J. Quantum Electron.* **1**, 343 (1965).

7.9. P. W. Smith, "On the Stabilization of a High-Power Single-Frequency Laser," *IEEE J. Quantum Electron.* **2**, 666 (1966).

7.10. P. Zory, "Measurements of Argon Single-Frequency Laser Power and the 6328 Å Neon Isotope Shift Using an Interferometer Laser," *J. Appl. Phys.* **37**, 3643 (1966).

7.11. W. W. Rigrod and A. M. Johnson, "Resonant Prism Mode Selector for Gas Lasers," *IEEE J. Quantum Electron.* **3**, 644 (1967).

7.12. H. J. Caulfield and W. J. Beyen, "Birefringent Beam Splitting for Holography," *Rev. Sci. Instrum.* **38**, 977 (1967).

7.13. D. B. Neumann and H. W. Rose, "Improvement of Recorded Holographic Fringes by Feedback Control," *Appl. Opt.* **6**, 1097 (1967).

7.14. H. W. Rose and H. D. Pruett, "Stabilization of Holographic Fringes by FM Feedback," *Appl. Opt.* **7**, 87 (1968).

Chapter 8

ANALYSIS OF PLANE HOLOGRAMS

The fringes recorded by small, in-line holograms with nondiffuse light are coarsely spaced relative to the thickness of photographic emulsion. A ray in the wavefront illuminating such a hologram interacts with only one recorded fringe before passing through the hologram. Consequently the hologram response is closely approximated by that of a plane diffraction grating—with focusing properties. Gabor analyzed these properties by considering the hologram to be strictly two dimensional. Predictions of the analysis using the two-dimensional model were in good agreement with the experimental observations.

The off-axis technique, introduced by Leith and Upatnieks, led to holograms whose fringe frequency exceeded that of the in-line hologram by a term proportional to the angle between subject and reference beams [see Eq. (3.15)]. A typical value of the fringe spacing is obtained by considering the interference of two plane waves. Equation (1.10) $2d \sin \theta = \lambda$, relates the half-angle θ between beam directions and the wavelength λ to the fringe spacing d. For $\theta = 15°$ and $\lambda = 0.5\ \mu\text{m}$ (green light), $d = 1\ \mu\text{m}$. Photographic emulsions used to record off-axis holograms are often 15 μm in thickness, and the hologram formed therein can no longer be realistically regarded as two dimensional. Nevertheless, a two-dimensional analysis, restated in terms of communication theory, was extended to the off-axis case by Leith and Upatnieks [*8.1*, *8.2*]. Despite the discrepancy between the facts of photographic emulsion and the assumptions of the planar model, it provided a useful framework for the further development of holography. However its application to holograms which might be better described as volume diffraction gratings leads to partially fulfilled predictions and leaves many of the observed properties of holograms unexplained.

It is therefore important to remember that the results of plane hologram analysis are applicable with accuracy only to holograms formed on suitably thin media. A good example is thermoplastic film material whose thickness may be made comparable to the wavelength of light. The observed properties of holograms formed in thermoplastic are correctly predicted by planar analysis.

With analytic tools developed from the diffraction theory of Chapters 5 and 6, we can now treat those properties of plane holograms which are not revealed by the geometric analysis of Chapter 3. Diffraction theory allows us to discuss the Fourier transform hologram. We shall derive the condition for separation of the image-forming diffracted waves of off-axis holograms, consider factors affecting the quality of the image, and compute maximum values of diffraction efficiency for absorption and phase holograms.

8.1 Off-Axis Holography with Nondiffuse Subject Light

Hologram formation with a reference wave directed to interfere with a subject wave at some angle was described in Section 2.6 as the most successful method for separating the twin images. Analysis of the method in terms of spatial frequencies leads to the concept of a carrier wave (the reference) whose spatial frequency is modulated by subject information. Hence the name *carrier-frequency hologram* is an appropriate and descriptive alternate to *off-axis hologram*. The carrier-frequency method eliminates the need to derive a strong reference wave from the subject transmission. As a consequence Gabor's restriction to transmission subjects with large transparent areas is no longer necessary.

Figure 8.1 represents a simple wavefront division method of illuminating a transmission subject with a coherent plane wave and deriving an off-axis plane reference wave from the same source. As subject we might choose a continuous-tone transparency. Suppose that the complex amplitude of the subject wave at the plane of the hologram is $\mathbf{a}(x, y)$ while that of the reference plane wave is $\mathbf{r} = r \exp(2\pi i \xi_r x)$. The reference wave spatial frequency $\xi_{\text{Ref}} = -\xi_r = -(\sin\theta)/\lambda$ corresponds to the *downward* direction of the reference wave propagation vector and to the angle θ which it makes in the xz plane with the z axis. We assume, as in Section 1.8, that we form an absorption hologram and that the amplitude transmittance of the fully processed hologram exposed to the interference pattern produced by $\mathbf{a}(x, y)$ and $\mathbf{r}$ is

$$t = t_0 - kI, \tag{8.1}$$

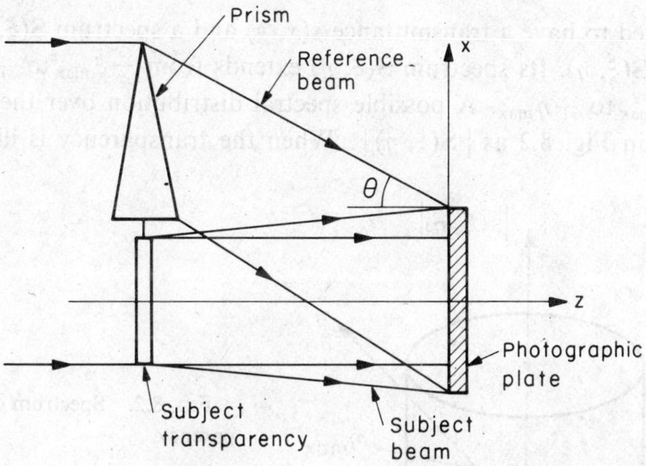

FIG. 8.1. A simple arrangement for forming an off-axis hologram.

where t_0 is the transmittance of the unexposed (but developed) plate, k is a constant, and I is the intensity of the interference pattern. The intensity I, as in Eq. (1.15), is

$$\begin{aligned} I &= \mathbf{aa}^* + \mathbf{rr}^* + \mathbf{ar}^* + \mathbf{a}^*\mathbf{r} \\ &= \mathbf{aa}^* + r^2 + \mathbf{a}r \exp(-2\pi i \xi_r x) + \mathbf{a}^* r \exp(2\pi i \xi_r x). \end{aligned} \tag{8.2}$$

When the hologram is illuminated by the original reference wave, the complex amplitude just behind the hologram takes the form

$$\begin{aligned} \mathbf{w}(x, y) = \mathbf{r}t = t_0 r \exp(2\pi i \xi_r x) \\ - k[\mathbf{aa}^* r \exp(2\pi i \xi_r x) \\ + r^3 \exp(2\pi i \xi_r x) + \mathbf{a}r^2 + \mathbf{a}^* r^2 \exp(4\pi i \xi_r x)]. \end{aligned} \tag{8.3}$$

8.1.1 Separation of the Diffracted Waves

In Section 1.8 we state without proof that a properly directed reference wave would separate the desired reconstructed wave from the remaining waves diffracted by the hologram. Figure 2.10 of Section 2.6 indicates geometrically that the mean angle between subject and reference beam must be sufficiently large if this is to be satisfactorily accomplished. We now employ a spatial frequency analysis of Eq. (8.3) in order to relate the condition for angular separation of the diffracted waves to the maximum spatial frequency of the subject transmittance [*8.1*]. The subject transparency is considered to have a transmittance $\mathbf{s}(x, y)$ and a spectrum $\mathbf{S}(\xi, \eta)$ where $\mathbf{s}(x, y) \supset \mathbf{S}(\xi, \eta)$. Its spectrum $\mathbf{S}(\xi, \eta)$ extends from $-\xi_{\max}$ to $+\xi_{\max}$ and from $-\eta_{\max}$ to $+\eta_{\max}$. A possible spectral distribution over the $\xi\eta$ plane is plotted in Fig. 8.2 as $|\mathbf{S}(\xi, \eta)|$. When the transparency is illuminated

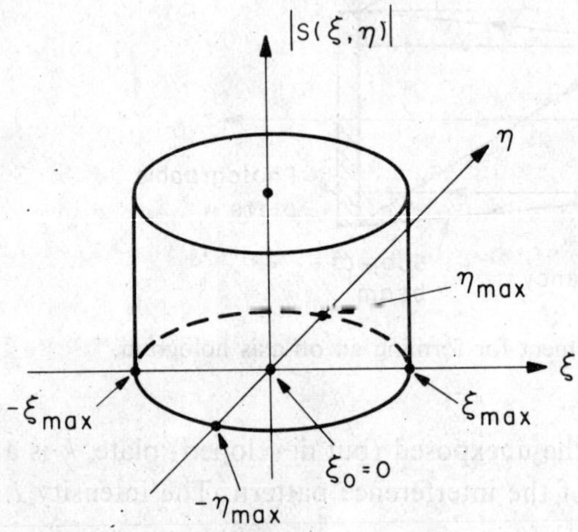

FIG. 8.2. Spectrum of the transparency.

by a z-directed plane wave, the complex amplitude of the subject wave received at the hologram plane is $\mathbf{a}(x, y)$, and the associated spectrum is given by Eq. (5.26) as

$$\mathbf{A}(\xi, \eta) = a_1 \mathbf{S}(\xi, \eta) \exp[-(2\pi d/\lambda)(1 - \lambda^2\xi^2 - \lambda^2\eta^2)^{1/2}] \tag{8.4}$$

where a_1 is the constant amplitude of the plane wave incident on the transparency and d is the distance separating the transparency from the hologram. Note that the maximum extent or width of $\mathbf{A}(\xi, \eta)$ in the spatial frequency plane is determined by the extent over which the spectrum $\mathbf{S}(\xi, \eta)$ has nonzero values. Equation (8.3) contains not only $\mathbf{a}(x, y)$ but also its conjugate $\mathbf{a}^*(x, y)$ whose spectrum is

$$\begin{aligned}\mathbf{A}'(\xi, \eta) &= \mathbf{A}^*(-\xi, -\eta) \\ &= a_1 \mathbf{S}^*(-\xi, -\eta) \exp[+i(2\pi d/\lambda)(1 - \lambda^2\xi^2 - \lambda^2\eta^2)]^{1/2}\end{aligned} \tag{8.5}$$

where we have used the correspondence (4.26). We now employ Eqs. (8.4) and (8.5) to plot the absolute value of the spectrum of $\mathbf{w}(x, y)$ in Eq. (8.3).

The first term on the right in Eq. (8.3), $t_0 r \exp(2\pi i \xi_r x)$, represents undiffracted light proceeding in the direction of the illuminating wave. For normal absorption (low efficiency) holograms this usually will be a very strong component. According to correspondence (4.30), its spectrum is the delta function centered at $(-\xi_r, 0)$. We may write the transform relation as

$$t_0 r \exp(2\pi i \xi_r x) \supset t_0 r\, \delta(\xi + \xi_r) \tag{8.6}$$

and plot the spectrum in Fig. 8.3 as a large vertical arrow.

The second term in Eq. (8.3) (the first in the brackets), $-kaa^*r \times \exp(2\pi i \xi_r x)$, transforms into frequency space, according to correspondences (4.18) and (4.21), as a shifted autocorrelation function of the spectrum $\mathbf{A}(\xi, \eta)$. We are primarily interested in the maximum extent of the autocorrelation function over the spatial frequency plane; this will be a major determinant of the spatial frequency assigned to the reference wave. For this purpose we recall that the correlation integral like the convolution integral represents a scanning of one function by another (see Section 4.3, Fig. 4.4). The range of variables for which the integral is nonzero is given by the sum of the widths of the scanning and scanned functions; in the case of autocorrelation the maximum extent is twice the width of the function being autocorrelated. If we neglect constant multipliers, we have for the transform of the second term in Eq. (8.3)

$$\begin{aligned}\mathbf{aa}^* \exp(2\pi i \xi_r x) &\supset \mathbf{A}^*(\xi, \eta) \star \mathbf{A}(\xi + \xi_r, \eta)\\ &= \left[\mathbf{S}^*(\xi, \eta) \exp\left(i \frac{2\pi d}{\lambda} (1 - \lambda^2\xi^2 - \lambda^2\eta^2)^{1/2}\right)\right]\\ &\quad \star \left[\mathbf{S}(\xi + \xi_r, \eta)\right.\\ &\quad \left.\times \exp\left(-i \frac{2\pi d}{\lambda} (1 - \lambda^2(\xi + \xi_r)^2 - \lambda^2\eta^2)^{1/2}\right)\right].\end{aligned} \tag{8.7}$$

As noted earlier, the width of $\mathbf{A}(\xi, \eta)$ is determined by its *amplitude* distribution and therefore given by the width of $\mathbf{S}(\xi, \eta)$. Hence the width of the autocorrelation function of Eq. (8.7) is twice that of $\mathbf{S}(\xi, \eta)$ and extends over a range $2[\xi_{\max} - (-\xi_{\max})] = 4\xi_{\max}$ in the ξ direction and similarly $4\eta_{\max}$ in the η direction. Its center $(-\xi_r, 0)$ corresponds to that of $\mathbf{S}(\xi + \xi_r, \eta)$ [see Eq. (4.12)] and coincides with the spatial frequency of the illuminating beam. The symmetric spread of diffracted light about the illumination beam direction is sometimes called *intermodulation*, referring to the modulation of light from one part of the subject by light from another.

Returning to Eq. (8.3) once again we find that the second term in the brackets $-kr^3 \exp(2\pi i \xi_r x)$ is similar to the initial term in the equation, an exponential which transforms to a delta function at $(\xi = -\xi_r, 0)$. It generally has a lesser strength than that corresponding to the initial term.

All three terms which we have thus far discussed are called zero-order terms because they represent light waves emerging from the hologram with their mean directions undeviated from the propagation axis of the illuminating wave.

The third term in the bracketed portion of Eq. (8.3) is proportional to the original wave $\mathbf{a}(x, y)$ which arrived at the hologram from the subject. The absolute value of its spectrum $|\mathbf{A}(\xi, \eta)|$ is proportional to $|\mathbf{S}(\xi, \eta)|$ as is evident from Eq. (8.4). We assume that $|\mathbf{S}(\xi, \eta)|$ is a symmetric distribution about a center spatial frequency $\xi_0 = 0$, $\eta_0 = 0$ with limits $\pm\xi_{\max}$ and $\pm\eta_{\max}$ (see Fig. 8.2), and therefore the same distribution represents $\mathbf{a}(x, y)$ in the spectral plot of Fig. 8.3.

The final term in Eq. (8.3), $-k\mathbf{a}^*r^2 \exp(4\pi i\xi_r x)$ represents the conjugate to the complex amplitude of the subject wavefront at the hologram, modulated onto a high carrier frequency. (Note that the corresponding diffracted wave is not antiparallel to the original subject wave and not its conjugate as defined in Section 1.8.) It has a spectrum given by the correspondence

$$-k\mathbf{a}^*r^2 \exp(4\pi i\xi_r x) \supset -kr^2\mathbf{A}^*[-(\xi + 2\xi_r), -\eta].$$

Its absolute value according to Eq. (8.5) is proportional to $|\mathbf{S}^*[-(\xi + 2\xi_r), -\eta]|$, a distribution similar to that shown in Fig. 8.2 but inverted and shifted in the $-\xi$ direction by $2\xi_r$, and its limits are $\xi = -2\xi_r \pm \xi_{\max}$ and $\eta = \pm\eta_{\max}$.

In Fig. 8.3 are plotted the absolute values of the spectra of the waves emerging from the back side of the hologram. It is clear that the employment

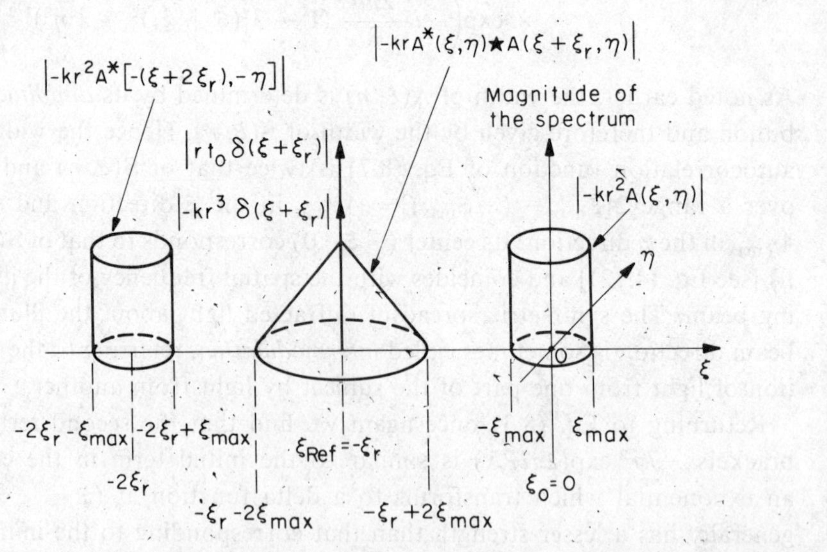

FIG. 8.3. Spectrum of the light emerging from an illuminated off-axis hologram.

of a reference wave with a suitably high spatial frequency or large angle θ insures angular separation of the image-forming waves. To avoid angular overlap of the image-forming waves by the zero-order waves, Fig. 8.3 indicates that the reference spatial frequency ξ_{Ref} must satisfy

$$|\xi_{\text{Ref}} - \xi_0| = |-\xi_r - \xi_0| = \xi_r + \xi_0 \geq 3\xi_{\max} \tag{8.8}$$

where ξ_0 is the center spatial frequency of the subject spectrum (assumed to be 0). To obtain even the minimum condition for angular separation $|\xi_{\text{Ref}} - \xi_0| = 3\xi_{\max}$, however, requires a recording medium with high resolution in the x direction as may be seen by reconsidering the effective hologram transmittance in Eq. (8.2). Let us substitute for the complex amplitude of the subject light at the hologram, $\mathbf{a}(x, y)$, its highest spatial frequency component in the $+x$ direction

$$A(\xi_0 + \xi_{\max}) \exp[-2\pi i(\xi_0 + \xi_{\max})x].$$

Equation (8.2) can then be written

$$\begin{aligned} t(x) \propto I = A^2 + r^2 &+ A(\xi_0 + \xi_{\max})r \exp[-2\pi i(\xi_0 + \xi_{\max} + \xi_r)x] \\ &+ A(\xi_0 + \xi_{\max})r \exp[2\pi i(\xi_0 + \xi_{\max} + \xi_r)x] \\ &= I_0 + (\text{constant}) \cos[2\pi(\xi_r + \xi_0 + \xi_{\max})x]. \end{aligned}$$

In the argument of the cosine is the fringe frequency $\xi_r + \xi_0 + \xi_{\max}$ which the medium must record. Taken with the condition $\xi_r + \xi_0 = 3\xi_{\max}$, from Eq. (8.8), this amounts to a spatial frequency of $4\xi_{\max}$, four times the highest spatial frequency in the subject. The high resolution needed in the recording medium is the price of twin image separation by means of the off-axis method. In terms of image quality and of flexibility in the choice of subject, the benefits outweigh this disadvantage, especially when high resolution photographic emulsion is available. Furthermore, some overlap of intermodulation diffraction can be tolerated since the amplitude of this term falls off rapidly away from its center frequency (see Fig. 8.3). It is possible to further reduce the effect of the overlap by making the amplitude of the reference wave large in comparison to that of the subject wave. The first term in the brackets of Eq. (8.3) is then small compared to the third and fourth terms.

Although Eq. (8.8) is the condition for no overlap in the spatial frequency domain (no angular overlap), it does not guarantee that in the spatial plane of the image formed by one of the diffracted waves there will be no unwanted light from the other waves. This is evident from Fig. 8.4 which shows the formation of a real image in a plane relatively close to the hologram plane (Position 1). Figure 8.4 indicates the cones of light emitted by two points of the illuminated hologram. Although there is angular separation of the diffracted waves, the site of the real image in Position 1 is illuminated by undesirable light from the zero-order waves as well as by the desired real image wave. One solution, the usual one, is to choose the reference wave spatial frequency sufficiently high and the distance of the subject from the hologram sufficiently large that there is no overlap in the image plane (Position 2).

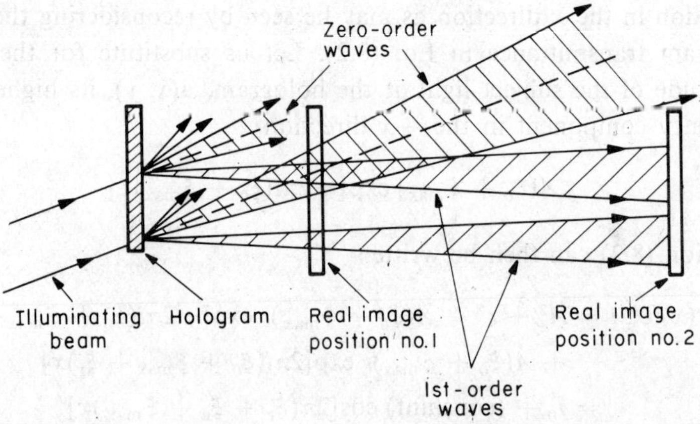

FIG. 8.4. Overlap on the image plane in Position 1 despite angular separation. No overlap in Position 2.

A second, more complicated solution is to filter out the undesired spatial frequency components. This may be done by using a lens to display in its back focal plane the frequency spectrum of the complex amplitude just behind the hologram and then blocking out all but the useful portion of the spectrum. The complexity of the method relegates its use to situations where the resolution capacity of the recording medium is very low. As an example, let us suppose we wish to form a hologram on Kodak Tri-X panchromatic film of a subject transparency whose center spatial frequency is $\xi_0 = 0$ and whose maximum spatial frequency is $\xi_{\max} = \eta_{\max} = 20$ cycles/mm (corresponding to subject detail 0.025 mm in dimension). If we wish to record an off-axis hologram as described in this section, the absolute value of minimum spatial frequency of the reference wave required for complete angular separation of the diffracted waves is $|\xi_{\text{Ref}}| = 3\xi_{\max} = 60$ cycles/mm, and the highest fringe frequency to be recorded is $4\xi_{\max} = 80$ cycles/mm. This is just within the resolving power of the emulsion. Although further increase in the angular separation of the waves can be achieved by increasing the reference-to-subject beam angle, i.e., increasing $|\xi_{\text{Ref}}|$, the resolution limit of the emulsion will then be exceeded. The angle in the arrangement of Fig. 8.1 corresponding to 60 cycles/mm is given by

$$\sin\theta \approx \theta = \lambda\xi_{\text{Ref}} = (0.633 \times 10^{-3})(6 \times 10^{1}) = 3.79 \times 10^{-2} \text{ rad} = 2.16^{\circ},$$

where $\lambda = 0.633 \times 10^{-3}$ mm is the wavelength of radiation from the He–Ne laser. Because the angle is so small, the reference beam must be brought in by means of a beam splitter placed between subject transparency and hologram. A possible arrangement for forming the hologram is shown in Fig. 8.5. Since θ is only 2.16° the diffracted waves will certainly overlap in the plane of the real image when the lateral dimensions of the subject transparency are comparable to its distance from the hologram. Spatial frequency filtering is required in this case. Methods involving total reflection of unwanted waves have been suggested [*8.3, 8.4*].

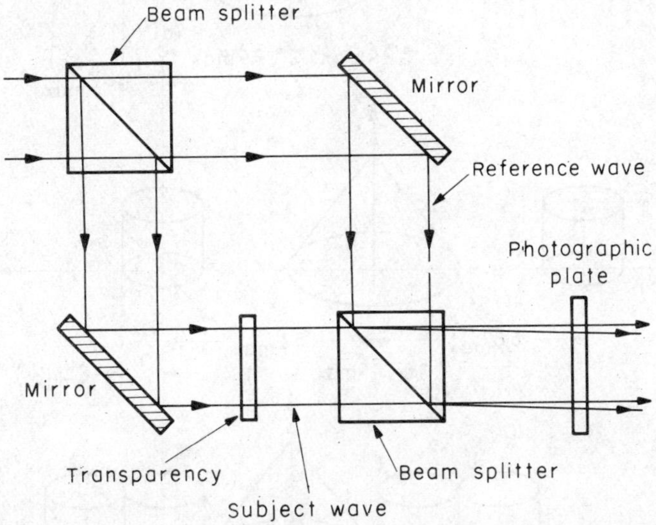

FIG. 8.5. Arrangement for forming an off-axis hologram with a small angle between subject beam and reference beam.

It should be noted in passing that when a number of holograms are superimposed on the same photographic plate, the proper choice of the reference waves employed in their formation can produce an angular separation of the reconstructed image-forming waves. Suppose we consider *two* superimposed holograms. Each is formed with a subject located on the z-axis and illuminated by an axial plane wave. Each subject has a mean spatial frequency $\xi_0 = 0$, $\eta_0 = 0$ and a spatial frequency bandwidth $\pm\xi_{max}$, $\pm\eta_{max}$. One hologram is formed with a plane reference wave of spatial frequency $\xi_{Ref} = -\xi_r = -3\xi_{max}$ and $\eta_{Ref} = 0$; the other is formed with a plane reference wave of spatial frequency $\xi_{Ref} = -\xi_r - 2\xi_{max} = -5\xi_{max}$ and $\eta_{Ref} = 0$. Based on analysis similar to that leading to Fig. 8.3, we find that the spectral response of each hologram taken individually and illuminated by its respective reference wave is as shown in Fig. 8.6a,b. When the two superimposed holograms are illuminated by a plane wave of spatial

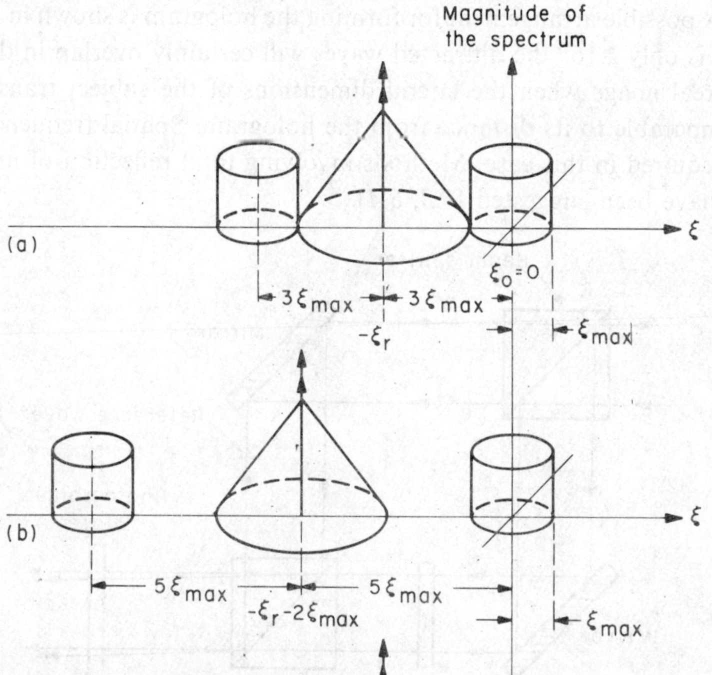

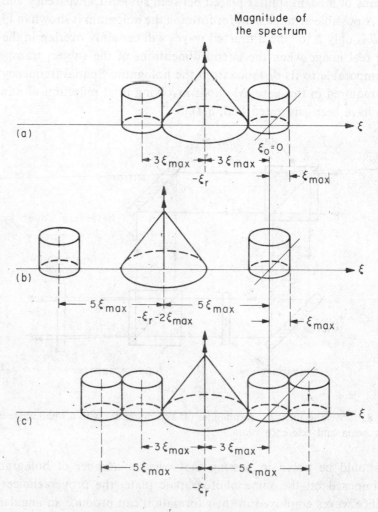

FIG. 8.6. Spectrum from a hologram made and illuminated with a reference wave of spatial frequency (a) $-\xi_r = -3\xi_{max}$ and (b) $-\xi_r - 2\xi_{max}$. (c) Spectrum from the two holograms in (a) and (b) superimposed and illuminated with plane wave of spatial frequency $-\xi_r$.

frequency $\xi = -\xi_r$, the spectrum is as given by Fig. 8.6c. The image-forming waves are angularly separated, and the condition for achieving this separation is

$$\Delta\xi_{Ref} \geq 2\xi_{max} \tag{8.9}$$

where $\Delta\xi_{Ref}$ is the difference in the reference-wave spatial frequencies and $2\xi_{max}$ is the spatial frequency bandwidth of the subject in the x direction. Choice of the illuminating wave spatial frequency is arbitrary.

8.1.2 Generation of the Real Image

If one illuminates a transparency with a plane wave, the diffraction pattern in the near field is a projection of the subject transparency. Each small area of the hologram which records such a pattern contains information about only a small part of the subject. Suppose the hologram formed as in Fig. 8.1 is now illuminated by the original reference plane wave. Figure 8.7 shows the illuminating and reconstructed waves as well as the on-axis *virtual* image. (A simple subject transparency consisting of three small transparent holes in an opaque background has been assumed.) Because the subject is neither illuminated by diffuse light nor does it itself diffuse the illuminating plane wave, the observer receives light from only one of the virtual image spots at a time. He can learn about all three spots by moving his head so as to scan the image, a procedure tiresome and impractical. It is therefore preferable in the case of nondiffuse subject light to view the projected *real* image. This may be done by allowing the image to form on a ground-glass diffusing screen placed in the real-image plane. Diffusion by the screen permits viewing the entire image from a single position. (A less convenient viewing method is to convert the virtual image to a real image by means of a lens.)

When the hologram is illuminated by the *original* reference wave, as in Fig. 8.7, the virtual image appears on-axis in the position of the original subject, and it is free of spherical and off-axis aberrations (see Section 3.4, [*3.2*]). The real image, formed off axis, is not. Let us now analyze two methods of illuminating the hologram to obtain aberration-free, on-axis real images. We assume the hologram has been formed with the arrangement of Fig. 8.1, where the subject is centered on the z axis and where the reference plane wave propagates at an angle $-\theta$ to the z axis. In the first method the hologram is illuminated by a plane wave proceeding in a direction $+\theta$ to the z axis, as shown in Fig. 8.8. Illumination of the hologram in this manner corresponds to multiplying the hologram transmittance, as expressed in Eq. (8.2), by $r \exp(-2\pi i\xi_r x)$ where $\xi_r = (\sin\theta)/\lambda$. Of all the

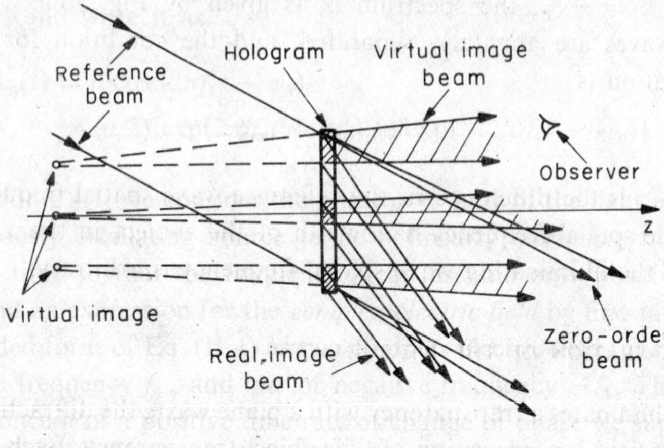

FIG. 8.7. Observation of the virtual image of a subject illuminated with nondiffuse light.

waves diffracted, we shall be concerned only with the real-image forming wave at the hologram plane, $\mathbf{a}^* r^2$. This wave has a complex amplitude conjugate to that of the original subject wave at the same hologram plane. Its properties are best analyzed in the spatial frequency plane. The frequency spectrum of $\mathbf{a}^*(x, y)$ is given in Eq. (8.5) as

$$\mathbf{A}'(\xi, \eta) = \mathbf{A}^*(-\xi, -\eta)$$
$$= a_1 \mathbf{S}^*(-\xi, -\eta) \exp\left[+i \frac{2\pi d}{\lambda} (1 - \lambda^2\xi^2 - \lambda^2\eta^2)^{1/2}\right].$$

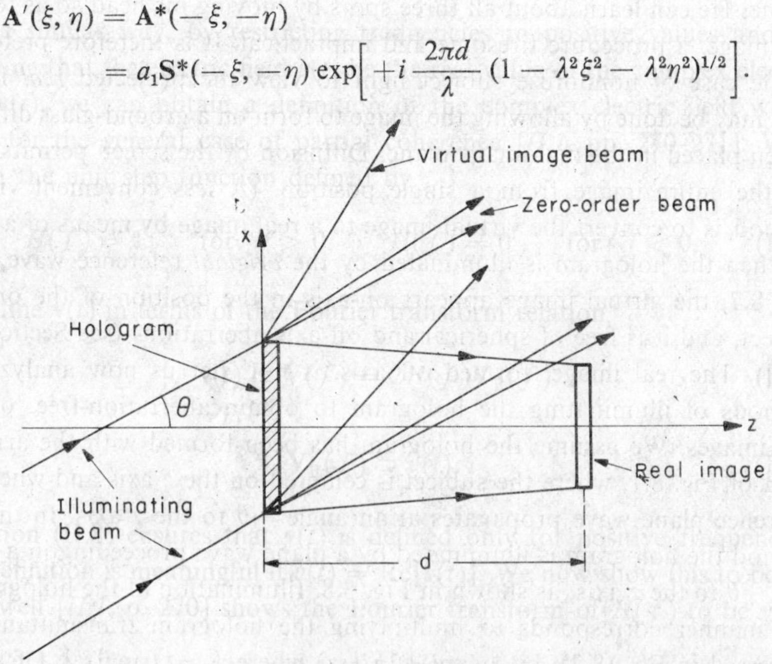

FIG. 8.8. Generation of an on-axis real image.

Suppose we consider the wave corresponding to $\mathbf{a}^*(x, y)$ to propagate to the right of the hologram along the z axis for a distance d. According to Eq. (5.26) the spatial frequency spectrum of the complex amplitude of the wave at d is

$$\begin{aligned} \mathbf{A}''(\xi, \eta) &= \mathbf{A}'(\xi, \eta) \exp\left[-i\frac{2\pi d}{\lambda}(1 - \lambda^2\xi^2 - \lambda^2\eta^2)^{1/2}\right] \\ &= a_1\mathbf{S}^*(-\xi, -\eta). \end{aligned} \tag{8.10}$$

If we take the inverse Fourier transform of both sides of Eq. (8.10) and use correspondence (4.26) we have that the complex amplitude in a plane a distance d from the hologram is proportional to the complex conjugate of the transmittance of the original subject transparency $\mathbf{s}^*(x, y)$. The intensity at d is then proportional to that in the original subject plane. Thus the hologram generates an image of the original transparency which is real (the image-forming wave propagates *to* the image) and which is located on the z axis. The image plane lies to the right of the hologram the same distance d as the original subject had been to the left. A photographic plate placed in the image plane at d will record the image without the need for a lens.

A second method for generating a real image is indicated in Fig. 8.9. Note that the illuminating beam is incident on the right side of the hologram (whereas the original reference illuminated the left surface) and is antiparallel to the original reference. It is therefore the conjugate to the original reference beam. However, in the plane of the hologram the complex amplitude

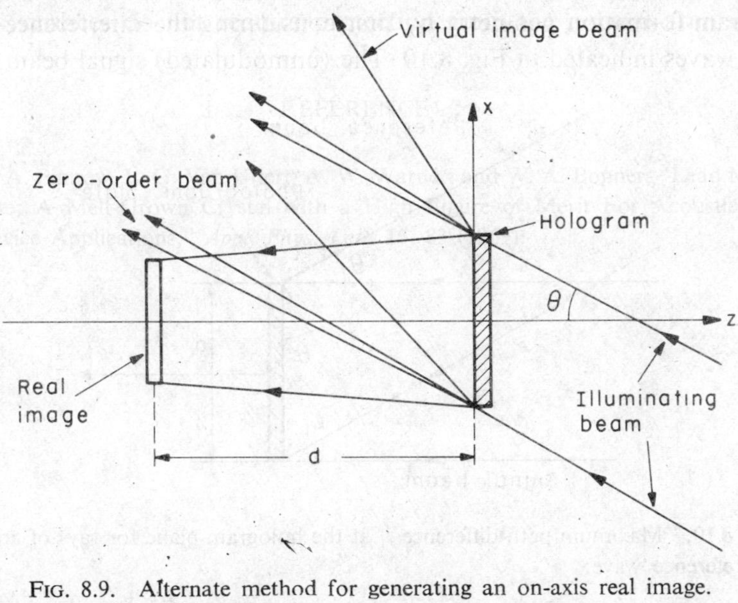

FIG. 8.9. Alternate method for generating an on-axis real image.

of the illuminating wavefront is $r \exp(-i2\pi\xi_r x)$, identical to that of the illuminating beam shown in Fig. 8.8. In contrast to the latter beam, the illuminating beam in Fig. 8.9 travels from right to left. Since the complex amplitude in the plane of the hologram is the same for both arrangements, our previous discussion applies to Fig. 8.9. The only difference is that the real image now forms on axis a distance d to the *left* of the hologram, consistent with the direction of propagation of the illuminating beam, and the image coincides with the original subject location. To accomplish this, the image-forming wave must be antiparallel to the original subject and therefore must be its conjugate.

The arrangements of Figs. 8.8 and 8.9 for generating the real image are equally good providing the hologram is thin. For thick holograms the arrangement of Fig. 8.9 is superior, since Bragg's law, the criterion for significant diffraction from a volume grating, is satisfied by the conjugate to the original reference wave as well as the original (see Section 1.6). When the recording medium cannot be characterized as strictly plane, Figs. 8.7 and 8.9 indicate the preferred arrangement for generating the virtual and real images respectively.

8.1.3 Coherence Requirements for Off-Axis Holograms

The coherence length required of a laser which is used, as in Fig. 8.1, to form an off-axis hologram is greater than required of a laser used to form an in-line hologram. We can relate the necessary coherence length to the hologram-formation geometry by first considering the interference of two plane waves indicated in Fig. 8.10. The (unmodulated) signal beam is axial

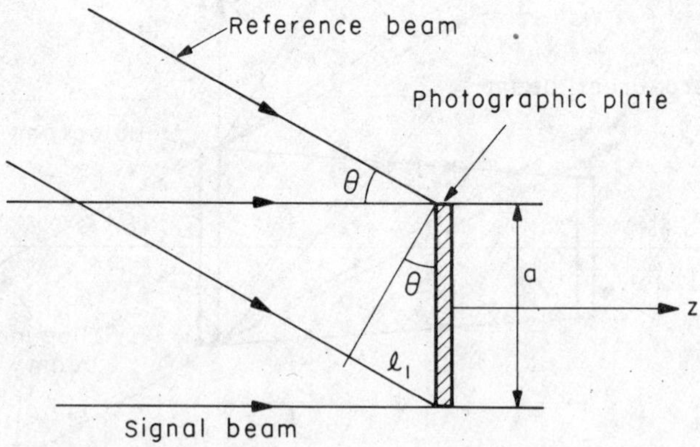

Fig. 8.10. Maximum path difference l_1 at the hologram plane for rays of an off-axis plane reference wave.

and the reference beam makes an angle θ with it. Hence the paths from source to photographic plate taken by all rays in the signal are equal while those for the reference beam rays differ by as much as l_1. We may set reference and signal paths equal for the central ray of the reference beam in which case the maximum signal-to-reference beam path difference is given by

$$\frac{l_1}{2} = \frac{a}{2} \sin |\theta| = \frac{a}{2} \lambda \xi_r \tag{8.11}$$

where a is the length of the plate, $|\theta|$ is the magnitude of the angle the reference beam makes with the z axis, and ξ_r is the corresponding reference wave spatial frequency. Thus, even with no information in the signal beam, the coherence length must at least equal $l_1/2$.

Next consider the added path length in the signal beam due to diffraction by an information-bearing transparency as in Fig. 8.11. The longest subject-

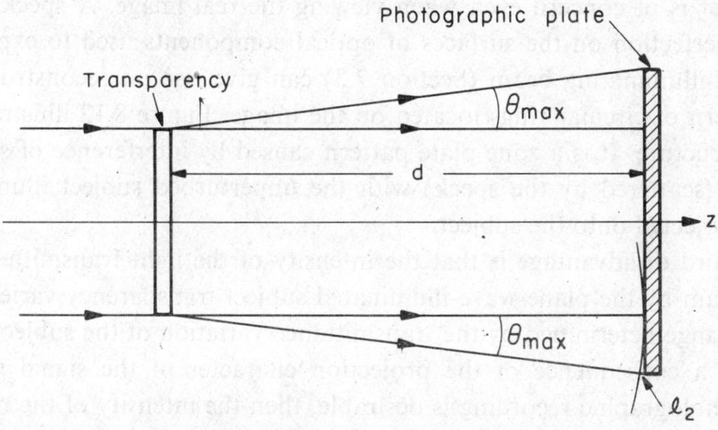

FIG. 8.11. Maximum path difference l_2 for a signal beam with subject information.

ray path differs from the unmodulated signal-beam path by

$$l_2 = \frac{d}{\cos \theta_{\max}} - d \approx \frac{d}{2}\, \theta_{\max}^2 \approx \frac{d}{2}\, \lambda^2 \xi_{\max}^2 \tag{8.12}$$

for small values of $\theta_{\max}$. Here d is the distance separating transparency and photographic plate, $\theta_{\max}$ is the maximum angle through which the plane wave incident on the transparency is diffracted, and $\xi_{\max}$ is the maximum spatial frequency of the transparency. Thus, if the path length of the central ray of the reference beam is set equal to that of the central ray of the signal beam, the total requirement on the coherence length of the laser light is

$$\Delta L_{\mathrm{H}} > \frac{l_1}{2} + l_2 = \frac{a}{2}\lambda\xi_{\mathrm{r}} + \frac{d}{2}\lambda^2\xi_{\max}^2. \tag{8.13}$$

In the Gabor in-line arrangement $\xi_{\mathrm{r}} = 0$ and the required coherence length L_{H} need only equal or exceed l_2. Several techniques exist for forming off-axis holograms in a manner requiring this lesser coherence length. The geometries are specialized, however, and intended for use only where laser light is not available [*8.5*, *8.6*] or where holograms are formed with multi-mode pulsed lasers (see Section 11.6.1).

8.2 Off-Axis Holography with a Diffuse Signal

Difficulty in observing the virtual image has already been cited as one disadvantage of illuminating subject transparencies with a uniform wave. Another is of concern even when viewing the real image. A speck of dust or imperfection on the surfaces of optical components used to expand the subject-illuminating beam (Section 7.3) can give rise on reconstruction to a pattern of circular rings located on the image. Figure 8.12 illustrates this ring structure. It is a zone plate pattern caused by interference of spherical waves (scattered by the speck) with the unperturbed subject illumination and projected onto the subject.

A third disadvantage is that the intensity of the light transmitted to the hologram by the plane-wave-illuminated subject transparency varies over a large range determined by the transmittance variation of the subject. Again this is a consequence of the projection character of the signal wave. If linear holographic recording is desirable, then the intensity of the reference beam must exceed that of the signal over the entire hologram (see Section 7.2.2). Where the signal is weak, the beam ratio is too high, and the diffraction efficiency is consequently poor.

These disadvantages can be avoided by using diffuse light to illuminate the subject transparency [*8.7*]. This is usually accomplished by inserting a diffusing screen such as a sheet of ground glass between the illuminating laser source and the transparency. Since the diffusing screen scatters light over a wide angular range, the transparency can now be viewed comfortably from a single position of the viewer's head. The same is true for viewing the virtual image generated by the reconstructed subject wave, at least over the angular range of subject rays received and recorded by the hologram.

FIG. 8.12. Ringlike noise in the real image of a nondiffusely illuminated transparency. (Courtesy Leith and Upatnieks [8.2].)

Although the phase of the diffuse subject light is a rapidly varying spatial function of the hologram coordinates, the light at the hologram plane can still be coherent. For this to be so, (1) the initial wave illuminating the diffusing screen must be spatially coherent over the full area of the screen, (2) the maximum path length from source to hologram via the diffusing screen must not differ from that via the reference beam by more than the coherence length, and (3) the screen must be motionless. An arrangement for forming a hologram of a diffusely illuminated transparency is shown schematically in Fig. 8.13a and the viewing of the virtual image is indicated in Fig. 8.13b.

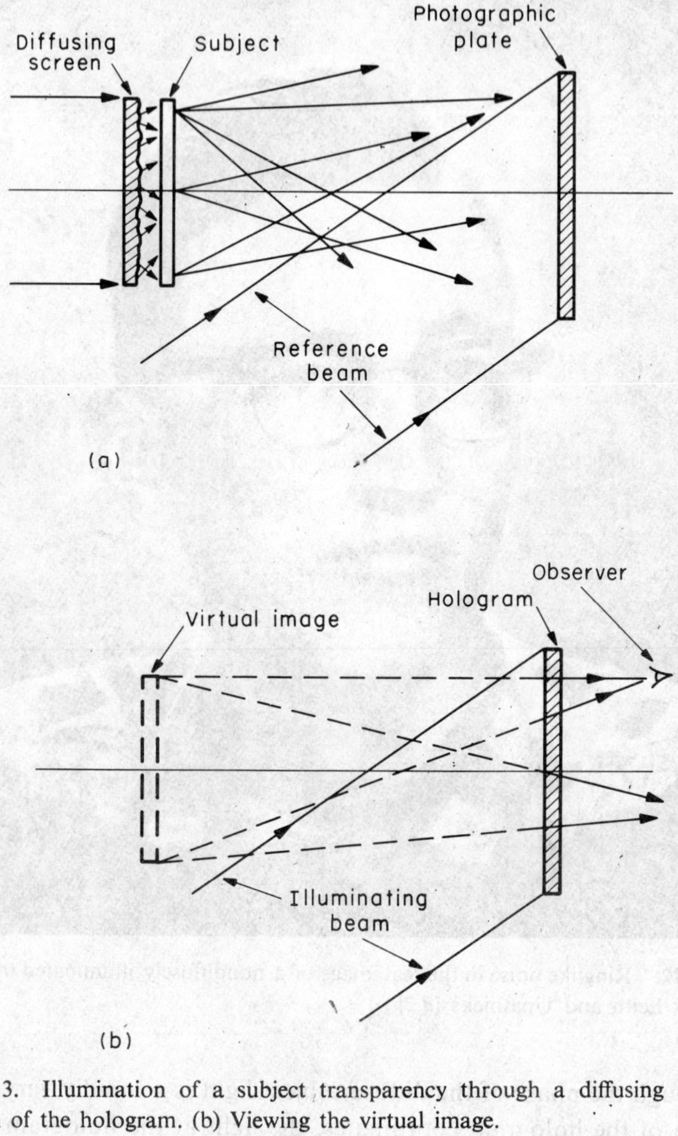

FIG. 8.13. Illumination of a subject transparency through a diffusing screen. (a) Formation of the hologram. (b) Viewing the virtual image.

The hologram of a diffusely illuminated subject has a number of unusual properties. Since the diffusing screen has a much broader spectrum than the transparency, it scatters light over a wide angular range, and every point on the hologram plate receives light from every point of the transparency. In the reconstruction step the entire virtual image of the subject can be observed by looking through any part of the hologram. As the viewing position is changed the image is seen from a different aspect. If the subject is a two-dimensional transparency and one wishes to display the real image on some observation plane, a complete image can be obtained even if the hologram is broken or partially destroyed, with but a fragment remaining. Of course, the resolution in the image decreases (as with any lens of finite size) as the fragment becomes smaller. Figure 8.14 displays photographs of three real images obtained from the same hologram illuminated by a laser beam of decreasing diameter. This property of the diffuse light hologram to record information in a nonlocalized way makes it attractive for information storage. In contrast to storing information by means of microimages where a scratch or blemish entirely removes or obliterates information, information stored in a hologram made with diffuse subject light is relatively immune to such imperfections of the recording medium.

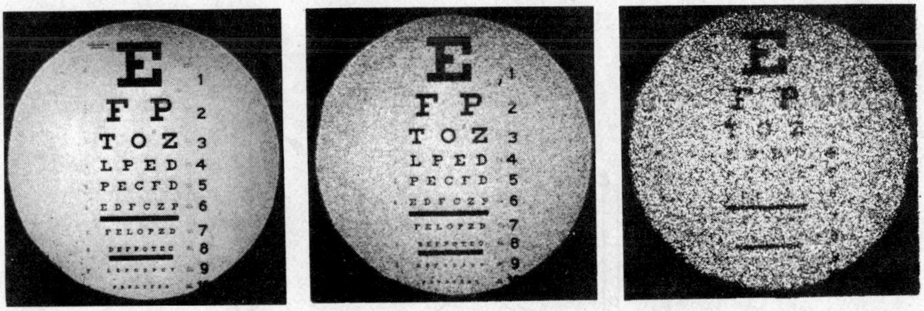

FIG. 8.14. Photographs of real images of a diffusely illuminated subject obtained by illuminating the hologram with a beam of decreasing diameter. The beam diameter from left to right was 0.81 cm, 0.26 cm, and 0.08 cm.

Diffuse illumination places an increased demand on the resolution capability of the recording medium. The maximum fringe frequency which must be recorded by a hologram is determined by the maximum angle which light rays from the subject make with the reference beam direction. Figure 8.15a illustrates an angle ψ formed with a plane reference wave and nondiffused light passing through a transparency. When a diffusing screen is interposed between illuminating source and transparency, a substantial increase in ψ results (Fig. 8.15b).

8.2.1 Three-Dimensional Images

Most three-dimensional objects reflect light more or less diffusely so that holograms made with such subjects exhibit the properties we have just discussed. Apart from these are properties related to the three-dimensional

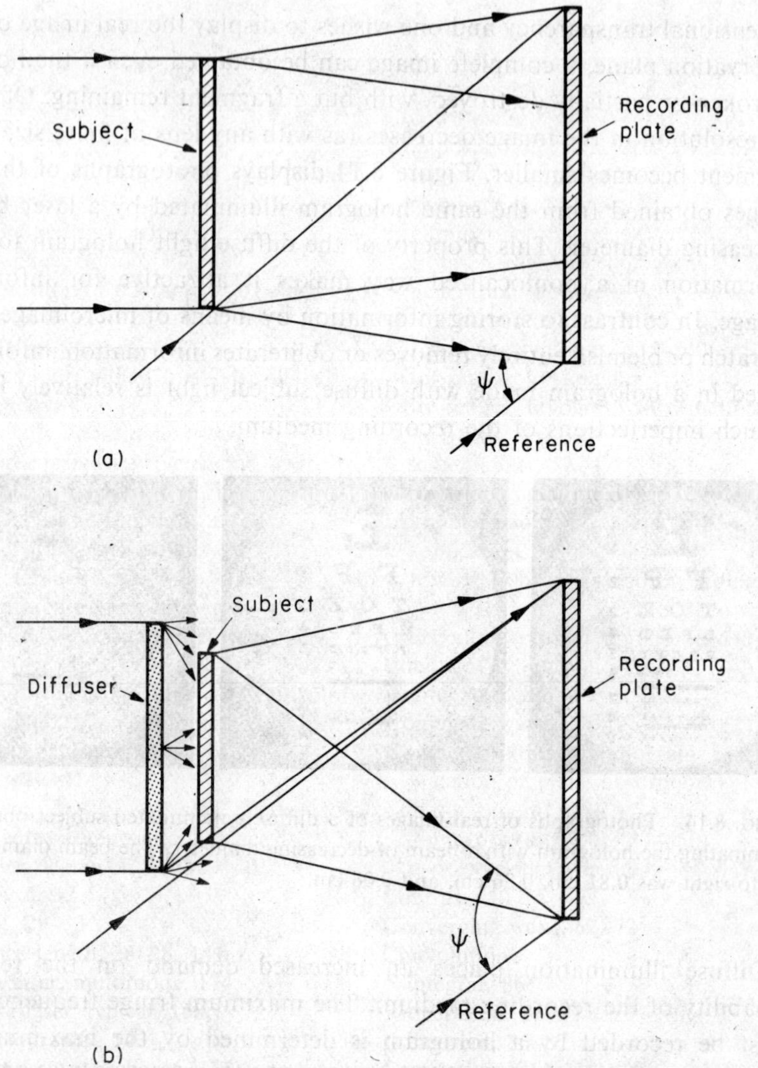

FIG. 8.15. Maximum angles between subject and reference rays reaching the hologram in the case (a) where the subject is illuminated nondiffusely and (b) where the subject is diffusely illuminated.

nature of the subject. Since the hologram can reconstruct a wave which is an exact replica of the original subject wave, the virtual image from which the reconstructed wave appears to diverge contains all the depth and parallax properties of the subject. Each eye of the observer views the virtual image through a different area of the hologram and thus sees the image from a different aspect (Fig. 8.16). These aspects, moreover, are identical to those seen by an observer of the original subject looking through an

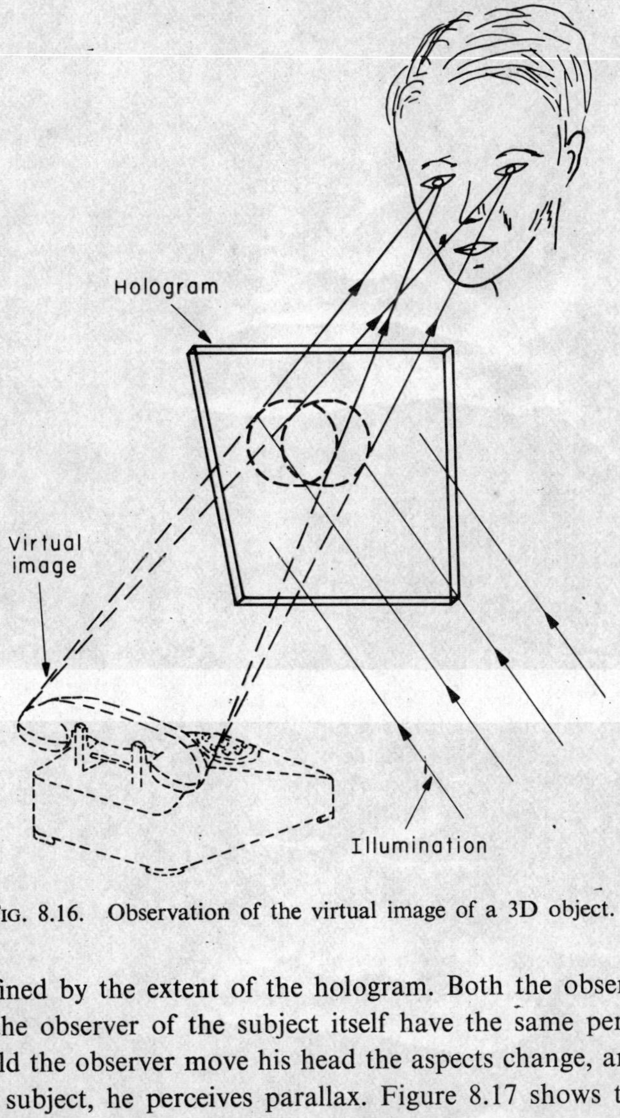

FIG. 8.16. Observation of the virtual image of a 3D object.

aperture defined by the extent of the hologram. Both the observer of the image and the observer of the subject itself have the same perception of depth. Should the observer move his head the aspects change, and, as with the original subject, he perceives parallax. Figure 8.17 shows two photographs of the same virtual image taken from different angles to exhibit the parallax. Of course, the 3D nature of the image can be fully appreciated only by actually viewing through the hologram.

The *real* image of a three-dimensional object generated by a hologram usually has the curious property that its depth is inverted. We say that the image is *pseudoscopic*. Consider the formation of a hologram with a simple three-dimensional subject consisting of two separated point sources (Fig. 8.18a) and the subsequent generation of a real image by illuminating the hologram with the conjugate to the reference wave (Fig. 8.18b). In the formation step the point P_1 is closer to the hologram plane than the point P_2.

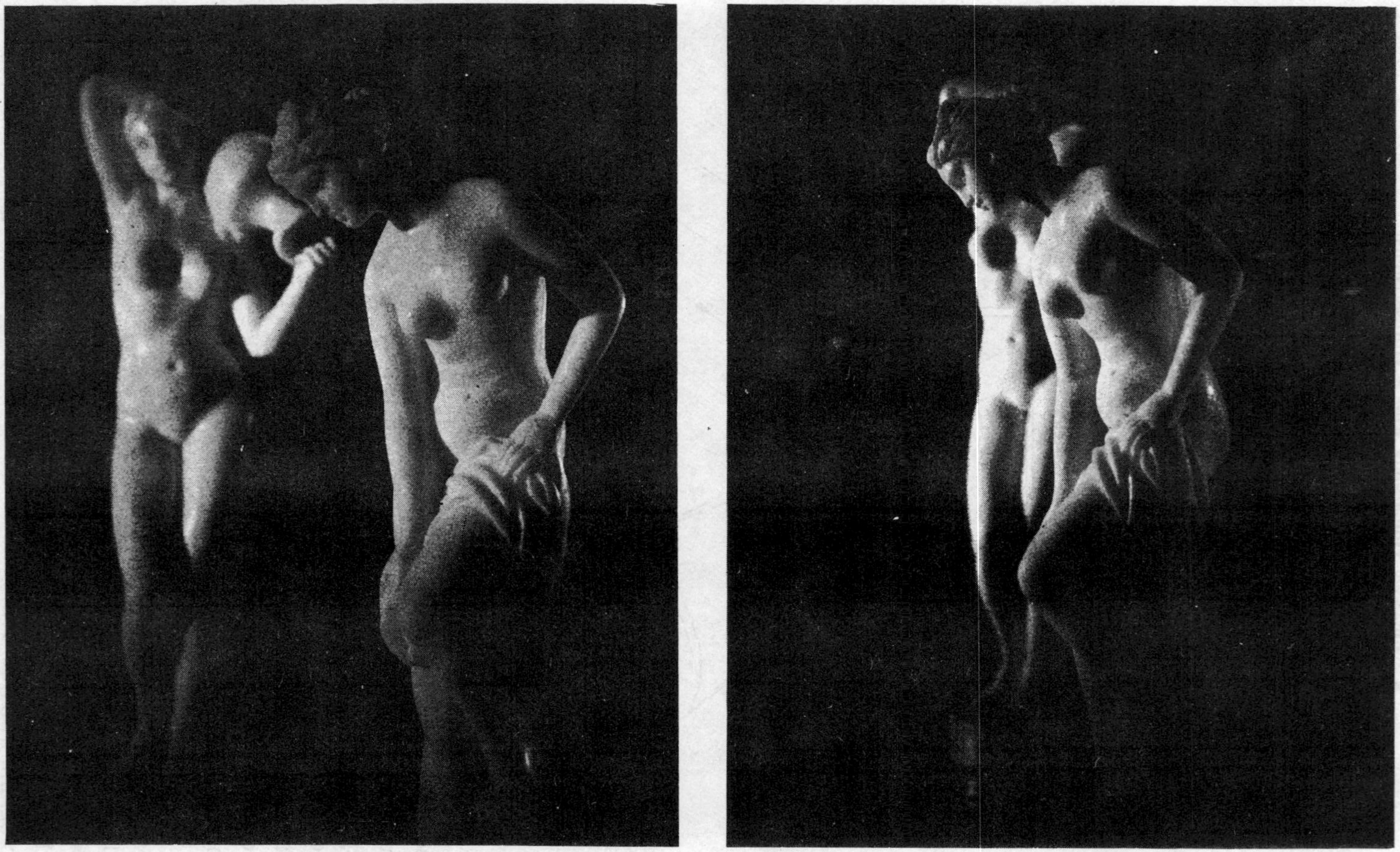

FIG. 8.17. Two photographs of the same virtual image of a 3D object taken from different angles.

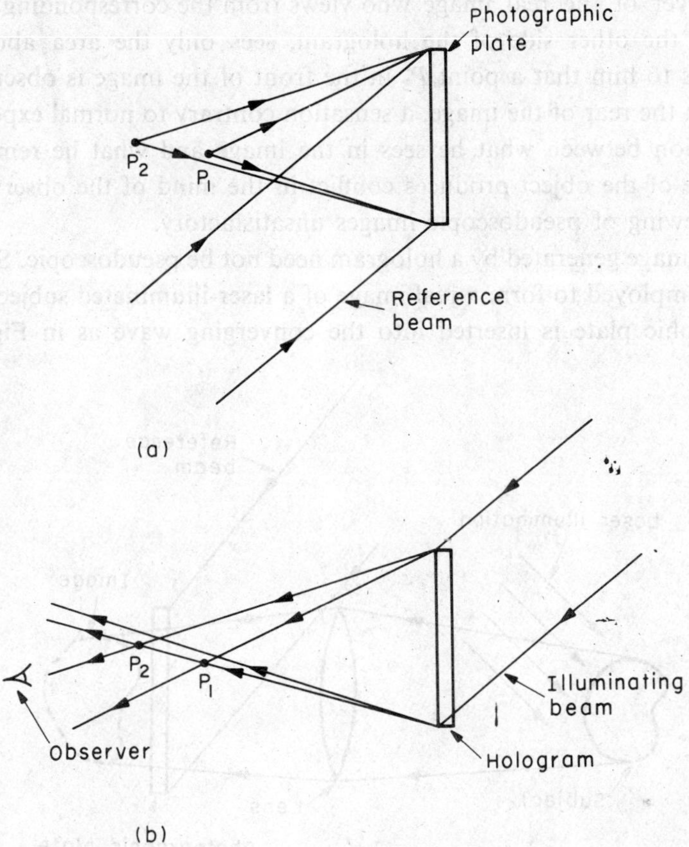

FIG. 8.18. (a) Formation of the hologram of a simple 3D subject. (b) Generation of the real image.

The reconstruction step causes real images of P_1 and P_2 to form at their original locations (as discussed in Section 8.1.2). However the observer of the real image must be positioned as in Fig. 8.18b if he is to receive the light diffracted by the hologram from right to left. To him P_2 is the closer point. Thus he observes an *inverted* depth relation relative to his normal view of the original sources, where light propagates from left to right through the photographic plate. This depth inversion leads to a number of confusing visual clues. Suppose that P_1 and P_2 are points on the surface of some three-dimensional subject and that there is a range of viewing angles over which the surface in the vicinity of P_1 obstructs the view of P_2. Over this angular range the hologram records information only about the surface in the vicinity of P_1 and records no information about P_2. Upon reconstruction, the observer of the real image who views from the corresponding angles, but from the other side of the hologram, sees only the area about P_1. It appears to him that a point P_2 in the front of the image is obscured by an area in the rear of the image, a sensation contrary to normal experience. Competition between what he sees in the image and what he remembers to be true of the object produces conflict in the mind of the observer and makes viewing of pseudoscopic images unsatisfactory.

A real image generated by a hologram need not be pseudoscopic. Suppose a lens is employed to form a real image of a laser-illuminated subject and a photographic plate is inserted into the converging wave as in Fig. 8.19.

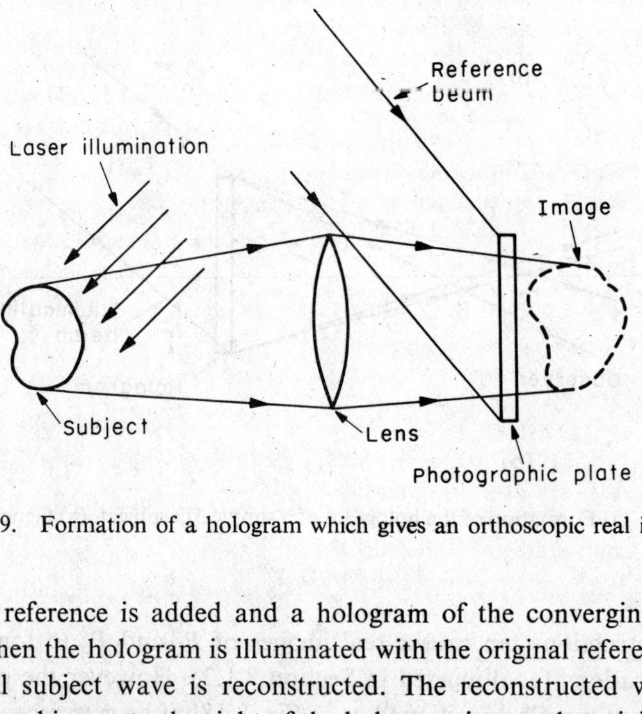

FIG. 8.19. Formation of a hologram which gives an orthoscopic real image.

A suitable reference is added and a hologram of the converging wave is formed. When the hologram is illuminated with the original reference wave, the original subject wave is reconstructed. The reconstructed wave converges to a real image to the right of the hologram just as the original wave, and the image is therefore normal in its depth properties or *orthoscopic*. On the other hand, if the conjugate to the reference wave traveling from right to left illuminates the hologram, the conjugate to the original subject wave will appear to diverge from the original subject location. That is, a virtual image of the image produced by the lens is generated at the original image plane of the lens. Looking through the hologram from the left, points formerly in the rear of the image produced by the lens now appear closer to the observer, so that it is the virtual image which is pseudoscopic [*8.8*].

If, with holography, one forms a pseudoscopic image of an image which is already pseudoscopic relative to some original subject, then the resulting holographic image is orthoscopic with respect to that subject. Hence if the subject wave for a second hologram is the pseudoscopic image-forming wave emerging from a first hologram, then illumination of the second hologram with the conjugate to its original reference will yield an orthoscopic image [*8.9*]. The same results can be obtained with autocollimating devices which do not require a second recording to convert pseudoscopic images into orthoscopic ones and vice versa [*8.10*].

8.2.2 Speckle Pattern

Since the images generated by holograms made with diffuse-light signals are replicas of the original laser-illuminated subjects, they suffer from a problem common to laser-illuminated, diffusely scattering objects. An observer of either image or object sees an annoying, granular *speckle pattern* which gives the observed surface a spangled or scintillating appearance. The grain size depends on the smallest aperture of the recording or viewing instrument (e.g., the iris of the eye) and on its distance from the observed surface [*8.11*]. In any area of the surface smaller than can be resolved by the observer there are many scattering centers which impose random phases on the complex amplitude of the light they scatter. Because the light is coherent and the relative phases constant in time, these amplitude components add. The observer, unable to resolve the individual scattering centers, sees the light emitted from the small area as a small spot of uniform intensity whose value depends on the net phase resulting from the amplitude addition. Even when the surface is uniformly illuminated, the observed reflected intensity will seem to vary randomly from spot to spot, since at each spot the intensity depends on the net amplitude resulting from the coherent addition of the randomly phased, scattered amplitudes. Motion of the observer so as to view the surface from a new aspect requires that the light rays traveling from surface to viewer proceed along a new path. The phases of the net amplitude of light from the minimum resolvable spots under observation change accordingly and, along with these, the intensity also changes. A continuous motion of the head tends to average out the intensity variation contributing to the speckle pattern and thereby tends to improve perception. However this is not very convenient when examining detail, and the presence of speckle pattern must be regarded as detrimental both to perception and viewing comfort. (For a more detailed discussion of speckle see Chapter 12.)

8.3 Hologram-Forming Geometries

8.3.1 Fresnel Holograms

When the photosensitive medium or plate for recording a hologram is placed in the near-field or Fresnel diffraction region of the subject and at an arbitrary distance from the reference source, the record formed is called a *Fresnel hologram.* It is a natural way to form a hologram since no lenses or other imaging devices are required either in formation or in the reconstruction. Except for aspeets of the lenless Fourier transform hologram discussed in Chapter 3, all of the hologram properties discussed thus far are those exhibited by the Fresnel hologram. It is the common hologram formed with the arrangements indicated in Figs. 7.1, 7.19, and 7.20. The hologram itself appears as in Fig. 7.21; it can be illuminated with the simple apparatus of Fig. 7.22 and is generally used to produce images of 3D objects such as those in Fig. 7.23. Figure 2.4 of Section 2.2 shows an early Fresnel hologram-forming arrangement with an on-axis reference wave.

8.3.2 Image Holograms

When special geometries and sometimes lenses or other imaging devices are used in the formation of the hologram, a number of useful properties are realized. Suppose that the photographic plate in Fig. 8.19 is moved until it lies in the central plane of the image formed by the lens, as in Fig. 8.20. Addition of the reference beam then allows formation of an *image*

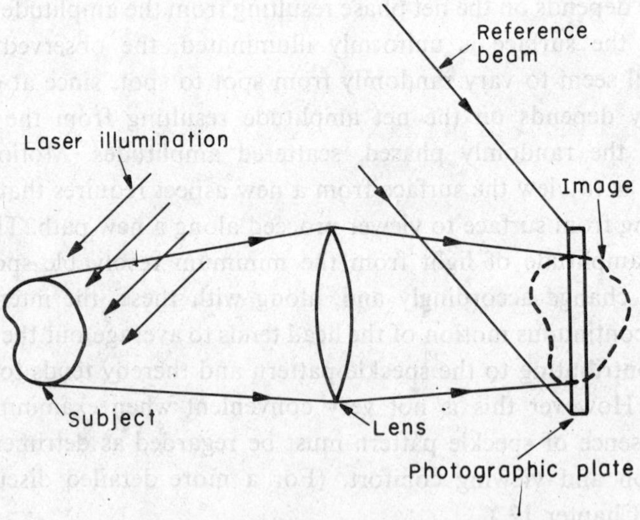

FIG. 8.20. Formation of an image hologram.

hologram [*8.12*, *8.13*]. On reconstruction with the original reference wave, part of the image generated by the hologram is virtual and part real. To the observer there is no marked difference between this image and those generated from lensless Fresnel holograms. However the angle from which the image can be viewed is limited by the aperture of the lens and the 3D image appears centered about the hologram plate. The technique has the merit of minimizing the requirements on the hologram-illuminating source. This advantage was pointed out in Section 7.6 where the distance z_1 of the image from the hologram was related to the requirements for spatial and temporal coherence of the source. Equation (7.52),

$$\Delta s = (z_1/z_r)\,\Delta r\,,$$

describes how the degree of spatial coherence, represented by the extent of the source Δr and the distance of the source from the hologram z_r, determine the resolution of the image Δs. As $z_1 \to 0$ (the case of the image hologram) even large sources with little spatial coherence can be used to generate hologram images with sufficient resolution. The consequence is

FIG. 8.21. The image obtained from an image hologram illuminated with an extended source.

that image holograms can be brightly illuminated by extended sources. Of course, $z_1 = 0$ only for a single plane of the subject, and image resolution on either side of the plane will degrade when a broad source is used (Fig. 8.21).

Equation (7.53) relates the resolution $\Delta\sigma$ in the image generated by the illuminated hologram to the bandwidth $\Delta\lambda$ of the illuminating source,

$$\Delta\sigma = \theta_r z_1 (\Delta\lambda/\lambda)$$

where z_1 again is the distance between hologram and image, λ is the center wavelength, and θ_r is the angle the reference beam makes with the normal to the hologram plate (the subject and image are considered to be on-axis). We see that when both θ_r and z_1 are small, the illuminating source bandwidth may be large and yet have small effect on the minimum resolvable image spot. One can even use a white-light source; the central plane of the image, located on the hologram, will then appear achromatic. However image points out of this plane will exhibit color dispersion and blurring which degrade the resolution.

8.3.3 Fourier Transform Holograms

In the next three sections we shall distinguish between several methods of producing holograms which generate at the hologram plane wave amplitudes which are either the exact Fourier transform of the subject or the Fourier transform multiplied by a slowly varying phase factor. Common to all of the methods is the restriction that the reference source must effectively lie in the same (input) plane as the subject. As a consequence, the analysis is intended to apply strictly to planar subjects (e.g., transparencies) and is less applicable as the subject extends out of the input plane. We generally require the subject to be illuminated with a plane wave. In some of the hologram-forming arrangements to be discussed a lens is used. When the lens precedes the subject, it is the lens which is illuminated with the plane wave. When the lens follows the input plane, it is assumed to operate on light from both the subject and the reference source.

We call a *Fourier transform hologram* one which records the interference of two waves whose complex amplitudes at the hologram are the Fourier transforms of both the subject and reference source. As we shall see in Chapter 14, such holograms are employed as spatial filters for pattern recognition, and the properties of the Fourier transform provide the basis for the recognition process. There the transmittance of the reference source is spatially modulated (an extended source). Here we confine our analysis to point reference sources.

The Fourier transform of a two-dimensional subject can be displayed in the back focal plane of a lens as discussed in Section 6.3.3 (see Fig. 6.6). An arrangement for forming Fourier transform holograms in the manner of Vander Lugt [*8.14*] is shown in Fig. 8.22. If $\mathbf{s}(x, y)$ is the transmittance

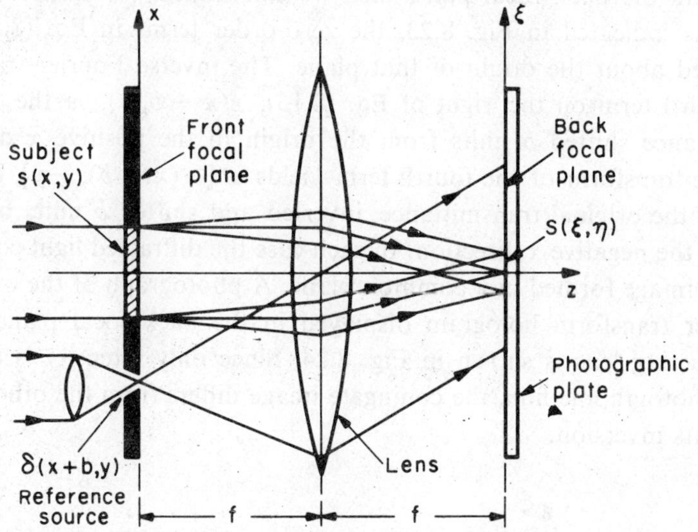

FIG. 8.22. An arrangement for forming a Fourier transform hologram. (*Note*: The lens is not drawn to scale.)

of the transparency in the front focal plane of the lens, the subject amplitude at the hologram located in the back focal plane is $\mathbf{S}(\xi, \eta)$ where $\mathbf{s}(x, y) \supset \mathbf{S}(\xi, \eta)$. Also located in the front focal plane is a point source $\delta(x + b, y)$ whose transform, a plane wave amplitude given by $\exp(-2\pi i \xi b)$, acts as the reference wave and illuminates the back focal plane along with $\mathbf{S}(\xi, \eta)$. The intensity of the interference pattern formed by the two transforms is

$$\begin{aligned} I &= [\exp(-2\pi i \xi b) + \mathbf{S}(\xi, \eta)][\exp(2\pi i \xi b) + \mathbf{S}^*(\xi, \eta)] \\ &= 1 + |\mathbf{S}(\xi, \eta)|^2 + \mathbf{S}(\xi, \eta)\exp(2\pi i \xi b) + \mathbf{S}^*(\xi, \eta)\exp(-2\pi i \xi b). \quad (8.14) \end{aligned}$$

We assume the developed hologram has a transmittance $t(x, y) \propto I$. If the hologram is illuminated with a plane wave propagating along the z axis with constant amplitude r_0, the product $r_0 t(x, y)$ represents the complex amplitude **W** of the diffracted light just behind the hologram, where

$$\mathbf{W} \propto r_0 t(x, y) \propto I = 1 + |\mathbf{S}|^2 + \mathbf{S} \exp(2\pi i \xi b) + \mathbf{S}^* \exp(-2\pi i \xi b). \quad (8.15)$$

A lens placed immediately before or after the hologram (Fig. 8.23) will display in its back focal plane the product of the inverse Fourier transform of **W** and a spherical phase factor (see Section 6.2). If we detect only the intensity in the back focal plane then we can neglect the spherical phase factor. As indicated in Fig. 8.23, the zero-order terms in Eq. (8.15) will be focused about the origin of that plane. The inverse Fourier transform of the third term on the right of Eq. (8.15), $\mathbf{s}(x - b, y)$, is the original transmittance shifted b units from the origin in the positive x direction while the transform of the fourth term yields $\mathbf{s}^*[-(x + b), -y]$, the conjugate of the original transmittance, inverted and shifted b units from the origin in the negative x direction. In each case the diffracted light converges to a real image formed in a common plane. A photograph of the output of a Fourier transform hologram displayed in the back focal plane of the reconstructing lens is shown in Fig. 8.24. Since only intensity is detected by the photographic film, the conjugate image differs from the other image only by its inversion.

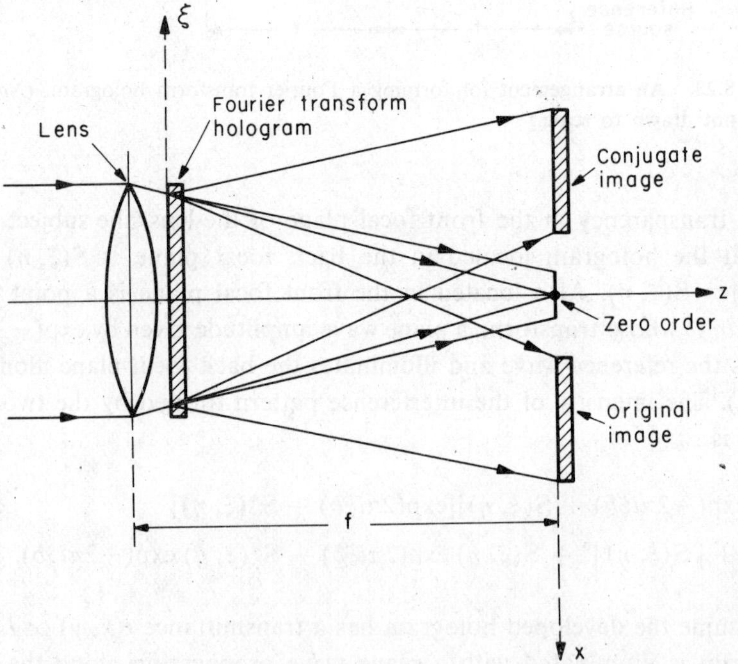

FIG. 8.23. Generation of two real images from a Fourier transform hologram.

FIG. 8.24. Photograph of the image plane of a Fourier transform hologram.

A useful property of the Fourier transform hologram formed with a plane reference wave is that the generated image remains stationary when the hologram is translated. This, for example, would allow holograms recorded on a reel of film to project stationary images while the film was moving. To demonstrate the invariance of image position to hologram translation, consider the displacement of the complex amplitude in the third term of Eq. (8.15) by ξ_0 units in the $+\xi$ direction so that it is now represented by

$$\mathbf{S}(\xi - \xi_0, \eta) \exp[2\pi i(\xi - \xi_0)b].$$

When the hologram is illuminated as before and the corresponding diffracted wave is transformed, we then have for the complex amplitude at the back focal plane

$$\begin{aligned}\mathscr{F}^{-1}&[\mathbf{S}(\xi - \xi_0, \eta) \exp(2\pi i \xi b)] \exp(-2\pi i \xi_0 b) \\ &= [\mathbf{s}(x, y) \exp(-2\pi i \xi_0 x) * \delta(x - b)] \exp(-2\pi i \xi_0 b) \\ &= \mathbf{s}(x - b, y) \exp(-2\pi i \xi_0 x)\end{aligned}$$

where $\mathscr{F}^{-1}$ signifies the inverse Fourier transformation and where we have used correspondences (4.11), (4.21), (4.29), and the consequences of convolving a function with a delta function. The phase factor $\exp(-2\pi i \xi_0 x)$ drops out of the expression for intensity and the image intensity

$$\mathbf{s}(x - b, y)\mathbf{s}^*(x - b, y)$$

is identical to that observed with no hologram displacement.

Holograms of the Fourier transform of a transparency must record the large intensity variations present in the transform. Light passing through the transparency undeviated (zero-order) is focused by the lens to a high-intensity peak at the origin of the frequency or hologram plane. By comparison, the higher spatial frequencies diffracted by the transparency and focused to other regions of the frequency plane are much less intense. A reference beam intensity great enough to linearly record the low frequency peak may be too great for the weak high-frequency intensities. Diffraction efficiency is then very poor for these high frequencies. When their complex amplitudes in the reconstruction are no greater than the amplitudes of light scattered by the hologram as noise, then subject information is lost.

8.3.4 Quasi-Fourier Transform Holograms

We designate a *quasi-Fourier transform hologram* as one formed under conditions where (1) the hologram plate is located in the back focal plane of a lens and (2) the subject transparency and reference point source are coplanar in a plane which is in front or in back of the lens but not in the front focal plane. Either the subject or the lens, whichever precedes, is illuminated with a plane wave.

A specific arrangement with the transparency and reference point source adjacent to the lens is shown in Fig. 8.25. We know (Section 6.2) that the subject wave amplitude at the hologram is the Fourier transform of the transparency times a spherical phase factor, and we inquire as to the effect of this factor. As in Eq. (6.23) the complex amplitude of the subject wave at the hologram plane can be represented by

$$\mathbf{a}(x_2, y_2) = \frac{ia}{\lambda f} \exp\left[-\frac{i\pi}{\lambda f}(x_2^2 + y_2^2)\right]\mathbf{S}(\xi, \eta)$$

$$= \mathbf{c} \exp\left[-\frac{i\pi}{\lambda f}(x_2^2 + y_2^2)\right]\mathbf{S}(\xi, \eta) \tag{8.16}$$

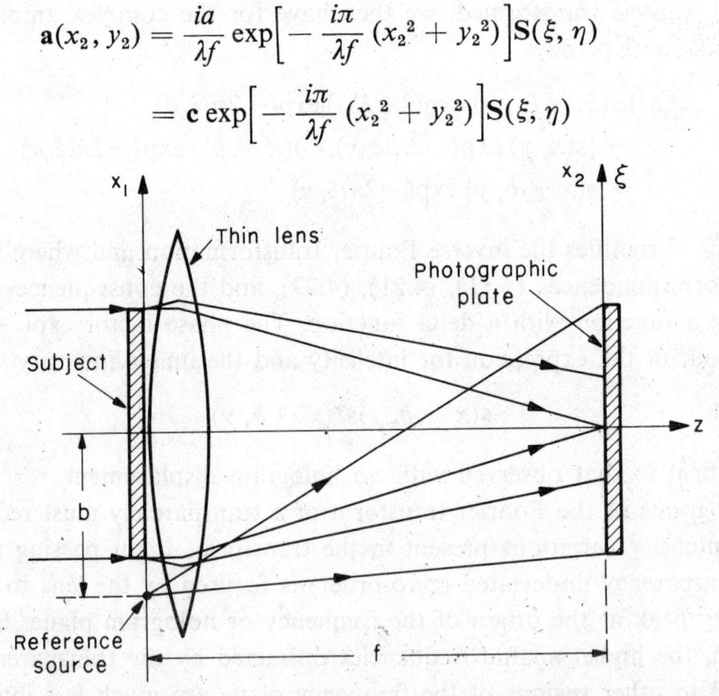

FIG. 8.25. Formation of a quasi-Fourier transform hologram with the input plane adjacent to the lens.

where $\mathbf{c} = ia/\lambda f$, $\xi = x_2/\lambda f$, $\eta = y_2/\lambda f$, $\mathbf{S}(\xi, \eta) \subset \mathbf{s}(x_1, y_1)$, and $\exp[-(i\pi/\lambda f)(x_2^2 + y_2^2)]$ is the spherical phase factor. Since $\exp[-(i\pi/\lambda f) \times (x_2^2 + y_2^2)]$ may be recognized as the transmittance of a thin *diverging* lens of focal length $-f$ [see Eq. (6.15)], $\mathbf{a}(x_2, y_2)$ may be considered to represent the combined transmittance formed by placing a transparency of transmittance $\mathbf{S}(\xi, \eta)$ adjacent to a diverging lens of focal length $-f$. Similarly $\mathbf{a}^*(x_2, y_2)$ is equivalent to the combined transmittance formed by placing a transparency, whose transmittance is $\mathbf{S}^*(\xi, \eta)$, adjacent to a converging lens of focal length f.

The hologram recording process will remove the spherical phase factor in $\mathbf{a}(x_2, y_2)$ if the reference point source is coplanar with the subject. Figure 8.25 shows a point source located $-b$ units below the origin of the x_1y_1 plane on the x axis. It produces a spherical wave whose phase variation over the x_2y_2 plane, relative to an arbitrarily assigned phase of zero at $x_2 = 0$, $y_2 = 0$, has the form of Eq. (3.3), viz.,

$$\varphi(x_2, y_2) = \frac{\pi}{\lambda z_1}(x_2^2 + y_2^2 + 2x_2 b).$$

(The *thin* lens with small curvature to its surface will contribute only a nearly constant phase factor to the light emitted by a point adjacent to it.) Recall that in Eq. (3.3) z_1 is a negative quantity. In the arrangement of Fig. 8.25 $z_1 = -f$. The complex amplitude of the spherical reference wave at the hologram is therefore

$$\mathbf{r} = r_0 \exp\left[-\frac{i\pi}{\lambda f}(x_2^2 + y_2^2 + 2x_2 b)\right]. \tag{8.17}$$

Interference between $\mathbf{r}$ and $\mathbf{a}(x_2, y_2)$ from Eq. (8.16) produces a pattern whose intensity is

$$\begin{aligned} I &= r_0^2 + \mathbf{aa}^* \\ &\quad + \mathbf{c}r_0\mathbf{S}(\xi, \eta) \exp\left[-\frac{i\pi}{\lambda f}(x_2^2 + y_2^2)\right] \exp\left[\frac{i\pi}{\lambda f}(x_2^2 + y_2^2 + 2x_2 b)\right] \\ &\quad + \mathbf{c}^* r_0 \mathbf{S}^*(\xi, \eta) \exp\left[\frac{i\pi}{\lambda f}(x_2^2 + y_2^2)\right] \exp\left[-\frac{i\pi}{\lambda f}(x_2^2 + y_2^2 + 2x_2 b)\right] \\ &= r_0^2 + \mathbf{aa}^* + \mathbf{c}r_0\mathbf{S}(\xi, \eta) \exp\left(i2\pi \frac{x_2}{\lambda f} b\right) + \mathbf{c}^* r_0 \mathbf{S}^*(\xi, \eta) \exp\left(-i2\pi \frac{x_2}{\lambda f} b\right) \\ &= r_0^2 + |\mathbf{cS}|^2 + \mathbf{c}r_0\mathbf{S}(\xi, \eta) \exp(2\pi i \xi b) + \mathbf{c}^* r_0 \mathbf{S}(\xi, \eta) \exp(-2\pi i \xi b). \end{aligned} \tag{8.18}$$

This is equivalent to the intensity in Eq. (8.14) characterizing the Fourier transform hologram and hence the hologram produced with the arrangement of Fig. 8.25 has the properties of the Fourier transform hologram.

We can verify that these same properties are obtained for quasi-Fourier transform holograms formed such that the plane of the subject and reference source is not adjacent to the lens. Let us return to Chapter 6 and to Eq. (6.59) as an aid. There we have arriving at the (hologram) plane in the back focal plane of a lens, the complex amplitude of light coming from a transmittance $\mathbf{t}(x_1, y_1)$. The latter is illuminated by a plane wave and is located some arbitrary distance d in front of the lens. We find the complex amplitude to be a product of the Fourier transform of $\mathbf{t}(x_1, y_1)$ times a spherical phase factor

$$\exp\left[-\frac{i\pi}{\lambda}\left(\frac{1}{f}-\frac{d}{f^2}\right)(x_2^2+y_2^2)\right]$$

which is independent of the form of $\mathbf{t}(x_1, y_1)$. Equation (6.59) holds for both subject and reference waves. The phase factor is independent of whether we consider the subject transmittance or the reference source to be $\mathbf{t}(x_1, y_1)$ since each is the same distance d from the lens. Consequently in the products $\mathbf{ar}^*$ and $\mathbf{a}^*\mathbf{r}$ arising from the recording process, the phase factors cancel out leaving only the Fourier transforms of subject and reference source as in Eq. (8.18).

Suppose now that subject and reference source are *behind* the lens in a plane a distance d from the back focal plane and suppose the lens to be illuminated with a plane wave. Equations (8.16)–(8.18) and the associated conclusions still apply if we substitute d for f. In this case the converging light falling on the subject can be considered to derive from parallel light illuminating a lens of focal length d adjacent to the subject.

8.3.5 Lensless Fourier Transform Holograms

Suppose that the lens is removed from the hologram-forming arrangement of Fig. 8.25, but the reference point source is maintained in the plane of the subject transparency (Fig. 8.26). The subject wavefront to be recorded at the hologram plane is now the near-field or Fresnel diffraction pattern (Fresnel transform) of the subject transparency. Nevertheless we shall see that the hologram formed with the arrangement of Fig. 8.26 has a transmittance resembling that of the Fourier transform hologram. The image-forming terms are again products of a Fourier transform and a phase factor linear in the coordinate of the hologram plane. Hence the name *lensless Fourier transform hologram* is applied to holograms formed without

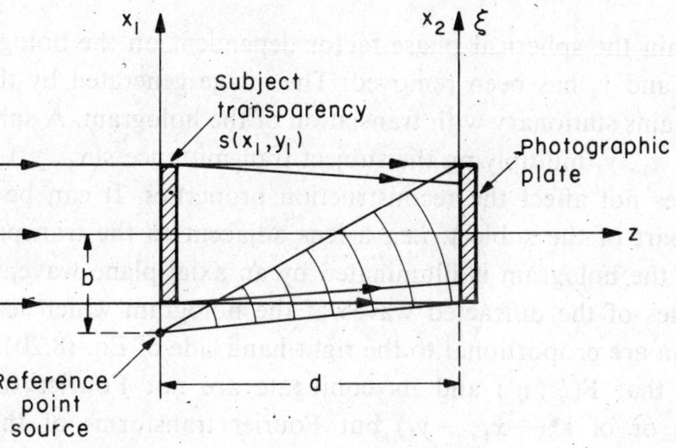

FIG. 8.26. Arrangement for forming a lensless Fourier transform hologram.

lenses but with a point reference source in the plane of the subject. As before, the subject is illuminated with a plane wave.

According to Eq. (5.33) and the discussion following in Section 5.5, the complex amplitude of the subject light in the x_2y_2 hologram plane of Fig. 8.26 can be written as

$$\begin{aligned}\mathbf{a}(x_2, y_2) &= \frac{ia_1}{\lambda d} \exp\left[-\frac{i\pi}{\lambda d}(x_2^2 + y_2^2)\right] \\ &\quad \times \iint \left[\mathbf{s}(x_1, y_1) \exp\left\{-\frac{i\pi}{\lambda d}(x_1^2 + y_1^2)\right\}\right] \\ &\quad \times \exp[2\pi i(\xi' x_1 + \eta' y_1)]\, dx_1\, dy_1 \\ &= \mathbf{c} \exp\left[-\frac{i\pi}{\lambda d}(x_2^2 + y_2^2)\right] \mathbf{F}(\xi', \eta') \qquad (8.19)\end{aligned}$$

where

$$\mathbf{F}(\xi', \eta') \subset \mathbf{s}(x_1, y_1) \exp\left[-\frac{i\pi}{\lambda d}(x_1^2 + y_1^2)\right], \qquad \mathbf{c} = \frac{ia_1}{\lambda d},$$

d is the separation between the x_1y_1 and x_2y_2 planes, and $\xi' = x_2/\lambda d$, $\eta' = y_2/\lambda d$. When a hologram is formed with a reference wave $\mathbf{r} = r_0 \exp[-(i\pi/\lambda d)(x_2^2 + y_2^2)] \exp(-2\pi i \xi' b)$ [of the form given by Eq. (8.17) but with $f = d$], the image-forming terms of the hologram transmittance are represented as in Eq. (8.18) by

$$\begin{aligned}&\mathbf{a}(x_2, y_2)\mathbf{r}^* + \mathbf{a}^*(x_2, y_2)\mathbf{r} \\ &\quad = \mathbf{c} r_0 \mathbf{F}(\xi', \eta') \exp(2\pi i \xi' b) + \mathbf{c}^* r_0 \mathbf{F}^*(\xi', \eta') \exp(-2\pi i \xi' b). \qquad (8.20)\end{aligned}$$

Once again the spherical phase factor dependent on the hologram coordinates x_2 and y_2 has been removed. The image generated by the hologram thus remains stationary with translation of the hologram. A spherical phase factor in x_1, y_1 multiplying the subject transmittance $\mathbf{s}(x_1, y_1)$ in the transform does not affect the reconstruction properties. It can be regarded as merely part of the subject, i.e., a lens adjacent to the transparency.

When the hologram is illuminated by an axial plane wave, the complex amplitudes of the diffracted waves at the hologram which lead to image formation are proportional to the right-hand side of Eq. (8.20). Apart from the fact that $\mathbf{F}(\xi', \eta')$ and its conjugate are not Fourier transforms of $\mathbf{s}(x_1, y_1)$ or of $\mathbf{s}^*(-x_1, -y_1)$ but Fourier transforms of their products with spherical phase factors, the hologram transmittance in Eq. (8.20) is similar to that obtained with Fourier transform holograms. Plane wave illumination requires a lens, as in Fig. 8.23 to carry out a Fourier transformation of the hologram transmittance. [We note that ξ' and η' in Eq. (8.20) differ by a constant scale factor $\alpha = d/f$ from the spatial frequencies $\xi = x_2/\lambda f$ and $\eta = y_2/\lambda f$ defined in Eq. (8.16) for Fourier transformation by a lens of focal length f. Let us write ξ' and η' in the form $\xi' = \xi/\alpha$, $\eta' = \eta/\alpha$ and employ the similarity theorem, correspondence (4.22). We find that transformation of the terms of Eq. (8.20) by a lens of focal length f divides the spatial dimensions of the image by the factor α as compared to those obtained when $d = f$.] Carrying out the transformation, we obtain complex amplitudes in the image plane x_3y_3, located at the back focal plane of the lens which are proportional to

$$\mathbf{s}\,[(\alpha x_3 - b), \alpha y_3] \exp\left\{-\frac{i\pi}{\lambda d}[(\alpha x_3 - b)^2 + \alpha^2 y_3^2]\right\}$$

and

$$\mathbf{s}^*[-(\alpha x_3 + b), -\alpha y_3] \exp\left\{\frac{i\pi}{\lambda d}[(\alpha x_3 + b)^2 + \alpha^2 y_3^2]\right\}$$

times the spherical phase factors introduced by the reconstructing lens. All the phase factors drop out when the intensities of the images are observed. The image intensities $|\mathbf{s}[(\alpha x_3 - b), \alpha y_3]|^2$ and $|\mathbf{s}[-(\alpha x_3 + b), -\alpha y_3]|^2$ represent one upright image centered at $(b/\alpha, 0)$ and one inverted image centered at $(-b/\alpha, 0)$ in the x_3y_3 plane.

While it is of interest to show that the hologram formed with the lensless Fourier transform method can be illuminated with a plane wave so as to produce an image display similar to that of the Fourier transform hologram, in practice it is often simpler to illuminate the hologram with the original reference point source. As with most holograms the virtual image then coincides with the original subject position.

Basic to the lensless Fourier transform method is the requirement that the curvature in the spherical phase factor associated with the reference wave be the same as that associated with the subject light. Translation invariance has been effectively used to produce composite holograms which reduce information content in pictorial holograms to the minimum necessary and allow alteration of the stereoscopic aspects of a 3D image (see Chapter 18).

8.3.6 Lensless Fraunhofer Holograms

In Section 2.5.4 it was noted that in-line holograms formed in the far field of a subject permitted the examination of one image without confusion from the other. Thompson *et al.* [*8.15*] employed the method to measure the size and shape of dynamic aerosol particles, perhaps the first practical application of holography. The Fraunhofer or far-field diffraction pattern of a subject can be recorded on a photographic plate a distance d from the subject provided

$$(x_1^2 + y_1^2)/\lambda \ll d, \qquad \text{as in Eq. (5.37).} \tag{8.21}$$

Here x_1 and y_1 are the coordinates of any point on the subject and λ is the wavelength of the illuminating light.

Suppose a subject transmittance $\mathbf{s}(x_1, y_1)$, whose Fourier transform is $\mathbf{S}(\xi, \eta)$, is illuminated with a plane wave traveling in a direction normal

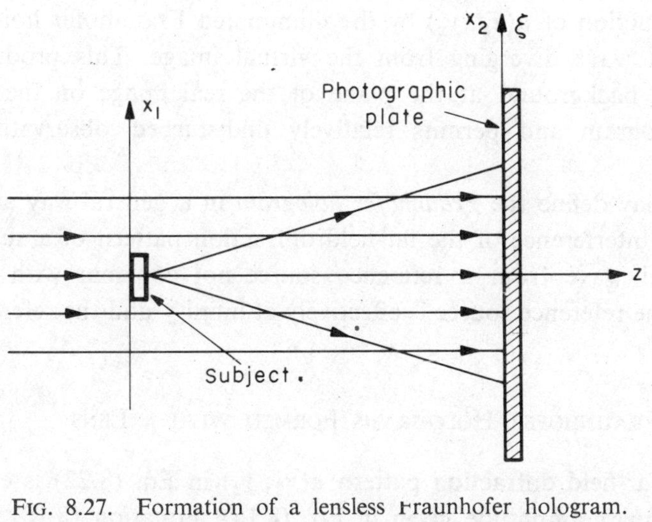

FIG. 8.27. Formation of a lensless Fraunhofer hologram.

to the x_1y_1 plane (see Fig. 8.27). Equation (5.39) expresses the complex amplitude $\mathbf{a}(x_2, y_2)$ of the light arriving at a plane x_2y_2 separated from x_1y_1 by a "far-field" distance d:

$$\mathbf{a}(x_2, y_2) = \frac{ia}{\lambda d} \exp\left[-\frac{i\pi}{\lambda d}(x_2^2 + y_2^2)\right]\mathbf{S}(\xi, \eta). \qquad (8.22)$$

As employed by Thompson *et al.* [*8.15*], a *Fraunhofer hologram* is formed by illuminating a subject with plane coherent light, placing a photographic plate in a plane a far-field distance from the subject and properly exposing it. Undiffracted light passing through the subject acts as the reference wave. As noted in Chapter 2 and in common with other in-line holograms, the image possesses the original contrast of the subject providing a positive hologram is properly made. In the reconstruction step, the hologram is illuminated by a plane wave similar to that which illuminated the subject.

Since the plane reference wave does not remove the spherical phase factor in Eq. (8.22) during formation of the hologram, the image generated by the Fraunhofer hologram does not have a position which is invariant to hologram translation. We may consider that the hologram has a built-in lens, represented by the spherical phase factor, which translates with the hologram and in turn translates the image.

When the subject becomes so small that it approaches an on-axis point-scatterer, as in the case of an aerosol particle, Fig. 8.27 becomes equivalent to Fig. 2.5 and a zone plate interference pattern is formed. If we can approximate the small subject by a delta function $\delta(x) \supset \mathbf{S}(\xi, \eta) = 1$, then $\mathbf{a}(x_2, y_2)$ in Eq. (8.22) approaches an unmodulated spherical wave. The reconstruction of $\mathbf{a}(x_2, y_2)$ by the illuminated Fraunhofer hologram is an identical wave diverging from the virtual image. This produces only a uniform background at the plane of the real image on the far side of the hologram and permits relatively undisturbed observation of that image.

We may define the *Fraunhofer hologram* in a general way as one which records interference of the far-field diffraction pattern of a subject with a spherical wave from a reference source not coplanar with the subject. Often the reference source is effectively at infinity and the reference wave is plane.

8.3.7 Fraunhofer Holograms Formed with a Lens

The far-field diffraction pattern $\mathbf{a}(x_2, y_2)$ in Eq. (8.22) is equivalent to the complex amplitude given in Eq. (8.16). Equation (8.16) applies to a coherently illuminated subject placed up against a lens of focal length f; the light amplitude is observed in the back focal plane. It is apparent then that a *Fraunhofer hologram formed with a lens* will result when (1) a subject adjacent to a lens is illuminated with a plane wave, (2) a plane reference wave is introduced (e.g., off-axis), and (3) subject and reference waves are made to interfere over a hologram plate placed in the back focal plane of the lens. Properties similar to those of the lensless Fraunhofer hologram will be obtained. Figure 8.28 indicates the arrangement for forming such a Fraunhofer hologram.

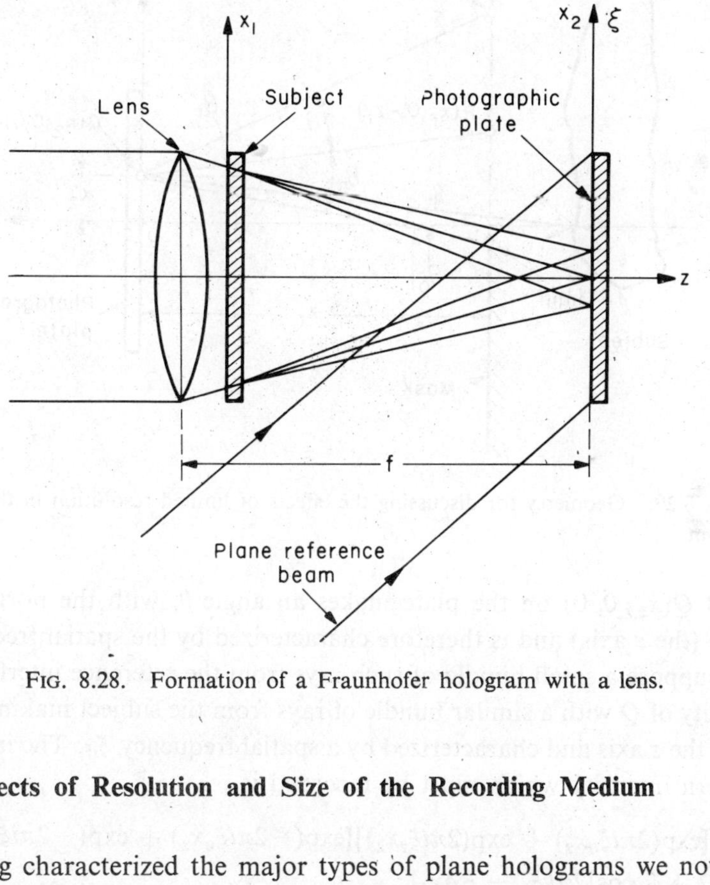

FIG. 8.28. Formation of a Fraunhofer hologram with a lens.

8.4 Effects of Resolution and Size of the Recording Medium

Having characterized the major types of plane holograms we now turn to some limitations placed on their imaging properties by the recording medium. The resolution required of recording media for several types of plane holograms has been discussed in Section 3.1. In the following section we present a simple, approximate model for treating the effects of limited resolution. The point of view taken is similar to that of Lukosz [*8.16*] and Urbach and Meier [*8.17*].

8.4.1 Limited Resolution in the Recording Medium

Let us initially carry out our analysis in the xz plane of Fig. 8.29 where a subject, a reference point source $R(x_r, 0, -z_r)$, and a recording plate are indicated. The plane normal to the axis and containing the reference point source is separated from the recording plate by the axial distance z_r. As indicated in Fig. 8.29 a ray from the reference source to an arbitrary

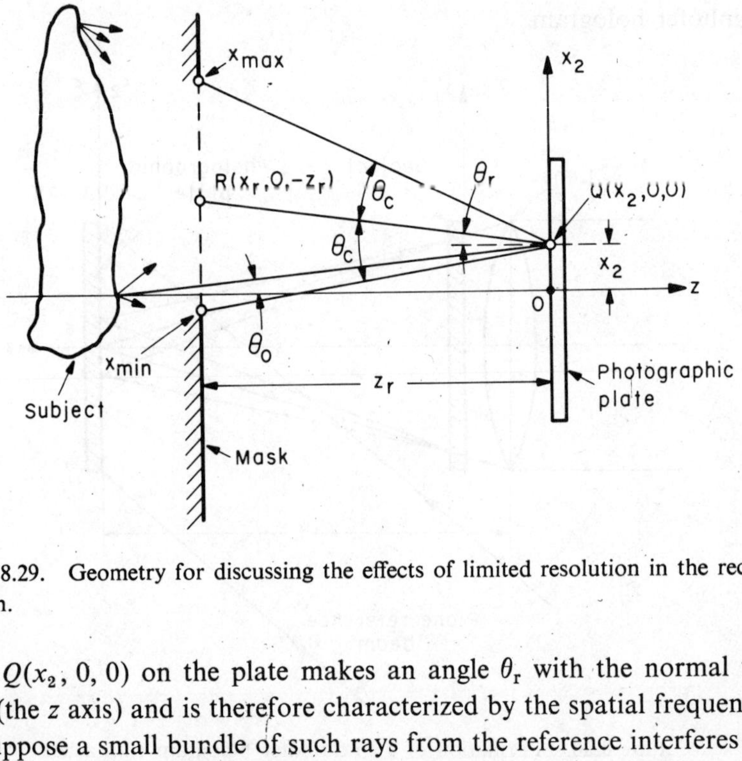

FIG. 8.29. Geometry for discussing the effects of limited resolution in the recording medium.

point $Q(x_2, 0, 0)$ on the plate makes an angle θ_r with the normal to the plate (the z axis) and is therefore characterized by the spatial frequency ξ_r. We suppose a small bundle of such rays from the reference interferes in the vicinity of Q with a similar bundle of rays from the subject making an angle θ_0 to the z axis and characterized by a spatial frequency ξ_0. The interference pattern intensity which must be recorded is

$$\begin{aligned} I &= [\exp(2\pi i\xi_0 x_2) + \exp(2\pi i\xi_r x_2)][\exp(-2\pi i\xi_0 x_2) + \exp(-2\pi i\xi_r x_2)] \\ &= 2 + 2\cos[2\pi(\xi_0 - \xi_r)x_2] \end{aligned} \tag{8.23}$$

where we have assumed unit amplitude waves and where the frequency of the cosine term $\xi_0 - \xi_r$ is the *fringe frequency*. For small angles, we approximate the fringe frequency of the pattern by

$$\xi_0 - \xi_r = \frac{\sin\theta_0 - \sin\theta_r}{\lambda} \approx \frac{\theta_0 - \theta_r}{\lambda}. \tag{8.24}$$

Suppose now that the recording medium is perfectly capable of recording all fringe frequencies below a cutoff frequency ξ_c but is totally incapable of recording fringe frequencies exceeding ξ_c. For a fixed value of ξ_r and a subject which scatters light over a broad band of spatial frequencies ξ_0, this means that there will be limits to the subject spatial frequencies that can be stored in the hologram. That is, if rays from the subject are to be stored in the hologram, their angles with the z axis must lie within certain limits. In Fig. 8.29 one such limiting ray passing from the subject to an arbitrary point Q on the hologram plane intersects the plane of the reference source at the point x_{max}. We call this ray a *marginal ray*; its spatial frequency $\xi_{0,max}$ satisfies

$$\xi_{0,max} - \xi_r = \xi_c. \tag{8.25a}$$

If the angle the marginal ray makes with the z axis is $\theta_{0,max}$, then we can write, with the help of Eq. (8.24), that

$$\frac{\theta_{0,max} - \theta_r}{\lambda} \approx \xi_c = \frac{\theta_c}{\lambda} \tag{8.25b}$$

or

$$\theta_{0,max} - \theta_r \approx \theta_c. \tag{8.25c}$$

The point x_{max} can be obtained by examining Fig. 8.29 and is given by

$$\frac{x_{max} - x_2}{z_r} = \tan(\theta_r + \theta_c) \approx \theta_r + \theta_c \approx \frac{x_r - x_2}{z_r} + \xi_c\lambda$$

or

$$x_{max} \approx x_r + z_r\lambda\xi_c. \tag{8.26}$$

In the same manner we obtain for x_{min}, the intersection coordinate of the other marginal ray,

$$x_{min} \approx x_r - z_r\lambda\xi_c. \tag{8.27}$$

Note that the marginal rays intersect the plane of the reference source at a coordinate *independent of the point* Q in the recording plane.

We now define a hypothetical opaque mask located in a plane transverse to the z axis and containing the reference source. It has an aperture which is centered at the source R and extends between limits given by the points $(x_{max}, 0, -z_r)$ and $(x_{min}, 0, -z_r)$. Through the mask pass all subject rays whose interference with reference rays at some point Q on the hologram plate will produce fringes capable of being resolved and recorded by the photosensitive medium. *If a ray from the subject to an arbitrary point* Q

on the hologram plate, or an extension of that ray, is intercepted by the hypothetical mask, then that ray will not be recorded by the hologram. Several such rays are indicated in Fig. 8.30. The rule as stated holds for subjects in front of as well as behind the reference source plane. We can extend the analysis to the case of a two-dimensional mask by considering spatial frequencies ν, where $\nu^2 = \xi^2 + \eta^2$, and a cutoff frequency ν_c; the aperture in the mask then becomes a circle of radius $z_r \lambda \nu_c$ centered at R.

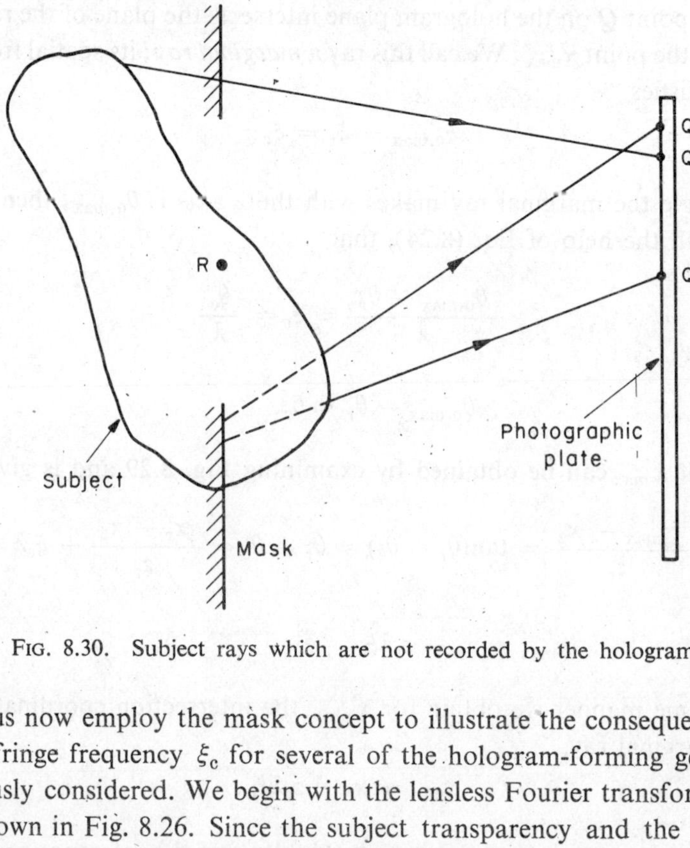

FIG. 8.30. Subject rays which are not recorded by the hologram.

Let us now employ the mask concept to illustrate the consequences of a cutoff fringe frequency ξ_c for several of the hologram-forming geometries previously considered. We begin with the lensless Fourier transform geometry shown in Fig. 8.26. Since the subject transparency and the reference point source are in the same plane, the plane of the mask, only the *size* of the subject which can be recorded is restricted by the mask. To the extent that the small-angle approximations used in the analysis are valid, rays of all spatial frequencies coming from the subject are faithfully recorded by the hologram. The same is true for the arrangement in Fig. 8.25 using a lens. These two geometries therefore allow the holographic recording of high spatial frequency subjects on low resolution media.

Next consider the geometry for forming in-line Fresnel holograms with an axial plane wave illuminating the subject transparency. The reference

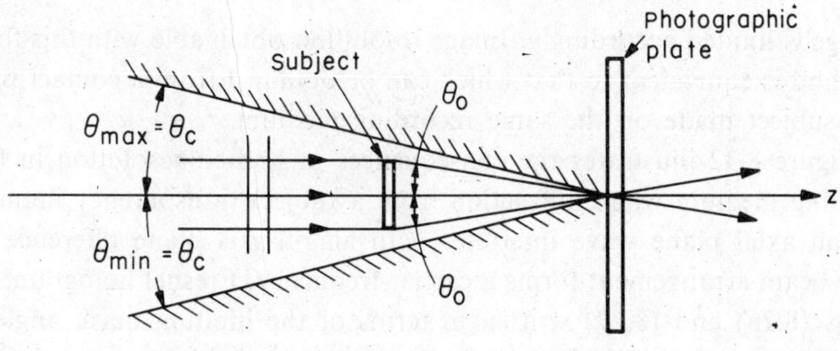

FIG. 8.31. Effect of limited recording resolution with in-line hologram geometry.

source R lies on the z axis at infinity ($x_r = 0$, $z_r = \infty$). The effect of the mask at infinity can best be described in terms of the limitation on angle which it imposes. If we define $\theta_{max} = x_{max}/z_r$ and $\theta_{min} = x_{min}/z_r$, Eqs. (8.26) and (8.27) become

$$\theta_{max} \approx \theta_c, \quad \theta_{min} \approx \theta_c. \tag{8.28}$$

These relations indicate that the extreme angles θ_{max} and θ_{min} which subject rays can make with the z axis and still be recorded are directly determined by the cutoff angle θ_c (Fig. 8.31). The subject plane wave with ray direction $|\theta_c|$ and spatial frequency $|\xi_c| = |\theta_c|/\lambda$ corresponds to a spatial frequency component of the subject transmittance which also has the value $|\xi_c|$ (see Section 5.3). Hence the hologram will not record spatial frequency components of the subject transmittance greater than $|\xi_c|$; the resolution in the

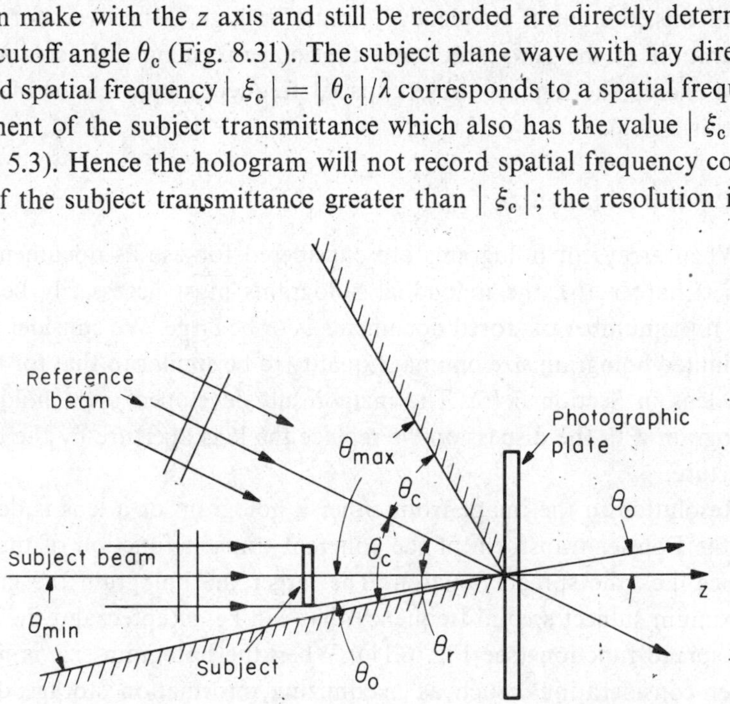

FIG. 8.32. Effect of limited recording resolution with off-axis geometry and a plane reference wave.

image is limited accordingly. Image resolution obtainable with this forming method is equivalent to that which can be obtained from a contact print of the subject made on the same recording medium.

Figure 8.32 illustrates the consequences of limited resolution in the recording medium when diffraction from a subject transparency illuminated by an axial plane wave interferes with an off-axis plane reference wave. The beam arrangement forms a carrier-frequency, Fresnel hologram. Equations (8.26) and (8.27) written in terms of the limiting mask angles θ_{max} and θ_{min} now take the form

$$\theta_{max} = \theta_r + \theta_c, \qquad \theta_{min} = \theta_r - \theta_c. \tag{8.29}$$

We see from the figure that unless θ_c is large enough, only a limited range of positive spatial frequencies diffracted by the subject will pass to the plate without intercepting the angular mask. One such ray from the subject making an angle θ_0 to the z axis is shown. The negative subject spatial frequencies are far less limited (the case would be reversed if the reference wave had a positive spatial frequency). This uneven recording of positive and negative spatial frequencies can reduce resolution in the image. To avoid this, the cutoff frequency ξ_c must be high. Moreover, for complete separation of the zero order and the image-forming diffracted waves, ξ_c must exceed the highest subject spatial frequency by a factor of four (see Section 8.1.1).

8.4.2 Hologram Size

When arrays of holograms are considered for use as document storage files (Chapter 16), the individual holograms must necessarily be small in size if the number of stored documents is to be large. We consider the effect of limited hologram size on image quality to be similar to that for the finite-size lens in Section 6.4.2. The main result developed there holds for the hologram if in the discussion we replace the lens aperture by the hologram aperture.

Resolution in the image from either a hologram or a lens is determined by the Fourier transform of the coherent transfer function of the imaging device, i.e., the spread function. The larger the hologram the greater the maximum subject spatial frequency that can be accepted and the narrower the spread function (see Fig. 6.11). When the hologram size is limited by other considerations, such as maximizing information storage density, it is important to optimize the use of the limited area. If we assume that the recording medium has sufficiently high resolution to record all spatial frequencies coming from the subject, then the goal is to record uniformly all frequency components from all parts of the subject.

In Section 6.4.2 it is shown that a lens imaging system, where the subject is illuminated with an axial plane wave, yields maximum image resolution only for subject points near the optic axis. Resolution in the image falls off linearly with distance from the axis. A better way to illuminate the subject, shown in Fig. 6.9, is to use a converging spherical wave which focuses at the lens. With this method the highest possible image resolution consistent with the fixed size of the lens is obtained for all subject points. The imaging system is said to be space-invariant and characterized by the top-hat transfer function of Fig. 6.10. The analogous, space-invariant, hologram-forming arrangements are shown in Figs. 8.25 and 8.28. These methods make optimum use of the limited hologram area by causing the small hologram to receive the same range of spatial frequencies from each subject point and consequently produce an image which is uniformly well resolved over its entire extent. Since a cone of rays from each subject point fills the hologram area (see Fig. 6.9) information about every subject point is stored at every point of the hologram. Holograms formed as in Fig. 8.25 or 8.28 are thus resistant to dust and scratches, as are holograms formed with diffuse subject light, but do not have the problem of speckle. The main problem associated with the hologram-forming arrangements of Figs. 8.25 and 8.28 is the large range of subject light intensity that must be recorded. As previously noted, there is a large intensity peak at zero spatial frequency. Some useful compromises between the redundancy with which information is stored and this intensity range will be discussed in Chapter 16.

8.5 Maximum Efficiency of Plane Holograms

If bright reconstructions are desirable, it is useful to know the maximum diffraction efficiency obtainable from various hologram types. Here we compute the efficiencies [*8.18*] of plane absorption and phase holograms formed by interference of an off-axis reference plane wave $\mathbf{r} = r \exp(2\pi i \xi x)$ with an axial, unmodulated plane wave signal of amplitude a.

We begin with the absorption hologram. If we assume that the exposure-dependent term t_E of the amplitude transmittance of the developed hologram is proportional to the intensity of the interference pattern to which the hologram was exposed, then

$$\begin{aligned} t_E \propto I &= [a + r \exp(2\pi i \xi x)][a + r \exp(-2\pi i \xi x)] \\ &= a^2 + r^2 + 2ar \cos(2\pi \xi x) \end{aligned}$$

or

$$t_E = t_{E_0} + t_1 \cos(2\pi\xi x). \tag{8.30}$$

For the plane absorption hologram the total amplitude transmittance t is defined in Eq. (1.14) as $t = t_0 - t_E$ where t_0 is the transmittance of the unexposed plate. At most, t can vary between 0 and 1 (when $t_0 = 1$). This maximum range is achieved when $t_{E_0} = \frac{1}{2}$ and $t_1 = \frac{1}{2}$. Under these conditions

$$\begin{aligned} t &= t_0 - t_E \\ &= 1 - \tfrac{1}{2} - \tfrac{1}{2}\cos(2\pi\xi x) \\ &= \tfrac{1}{2} - \tfrac{1}{2}\cos(2\pi\xi x) \\ &= \tfrac{1}{2} - \tfrac{1}{4}\exp(2\pi i\xi x) - \tfrac{1}{4}\exp(-2\pi i\xi x). \end{aligned} \tag{8.31}$$

Suppose the hologram is illuminated by an axial plane wave of unit amplitude. The amplitude of the light emerging from the hologram is then t in Eq. (8.31). We see that light is diffracted only into the zero order and into the $+1$ and -1 orders. Since each first-order diffracted wave receives $\frac{1}{4}$ of the incident light amplitude, each first-order wave has $\frac{1}{16}$ the intensity of the incident light. Diffraction efficiency is defined as the ratio of power diffracted into one first-order wave to the power illuminating the hologram. In the present case where the hologram is uniformly exposed, we may substitute *intensity* for *power* in the definition. Thus the efficiency is $\frac{1}{16}$ or 6.25%. Practical recording media are not usually linear over the full range of exposure required to make the transmittance t vary from 0 to 1. Therefore the maximum efficiency of 6.25% cannot be achieved if it is also necessary to reconstruct a wavefront proportional to the original signal wavefront.

A somewhat higher diffraction efficiency can be obtained if the transmittance t_E varies as a square-wave function of x, e.g., a Ronchi grating. Computer-generated holograms may be of this nature. In this case t_E is 0 in $\frac{1}{2}$ of the square-wave period and 1 in the remainder. The first two terms of the Fourier series representation of the square wave are [*8.19*]

$$\begin{aligned} t_E &= \tfrac{1}{2} + (2/\pi)\cos(2\pi\xi x) - \cdots + \cdots \\ &= \tfrac{1}{2} + (1/\pi)[\exp(2\pi i\xi x) + \exp(-2\pi i\xi x)] - \cdots. \end{aligned} \tag{8.32}$$

When $t = 1 - t_E$, the diffraction efficiency is $(1/\pi)^2 = 10.1\%$. Square-wave gratings also diffract appreciably into higher orders.

In Section 7.2.2 the *complex* transmittance of a photosensitive medium is expressed as

$$\mathbf{t} = t\exp[i\varphi(x)].$$

For a lossless, phase hologram we can consider t to be constant and equal to unity so that

$$\mathbf{t} = \exp[i\varphi(x)]. \tag{8.33}$$

Although we note in Section 7.2.2 that phase holograms are capable of linear recording only when φ is very small, let us, for the purpose of computing maximum possible efficiency, remove this restriction on φ. The photosensitive medium is exposed so that the phase shift $\varphi(x)$ imposed on a plane wave traveling through the developed hologram is proportional to the intensity of the exposure. Thus

$$\varphi(x) \propto a^2 + r^2 + 2ar \cos 2\pi\xi x = \varphi_0 + \varphi_1 \cos(2\pi\xi x). \tag{8.34}$$

The transmittance of the hologram $\mathbf{t}$ becomes

$$\begin{aligned} \mathbf{t} &= \exp(i\varphi_0) \exp[i\varphi_1' \cos(2\pi\xi x)] \\ &\propto \exp[i\varphi_1 \cos(2\pi\xi x)]. \end{aligned} \tag{8.35}$$

When we neglect the constant phase factor $\exp(i\varphi_0)$, Eq. (8.35) may be represented by the Fourier series [*8.19*]

$$\mathbf{t} = \exp[i\varphi_1 \cos(2\pi\xi x)] = \sum_{n=-\infty}^{+\infty} i^n J_n(\varphi_1) \exp(in2\pi\xi x),$$

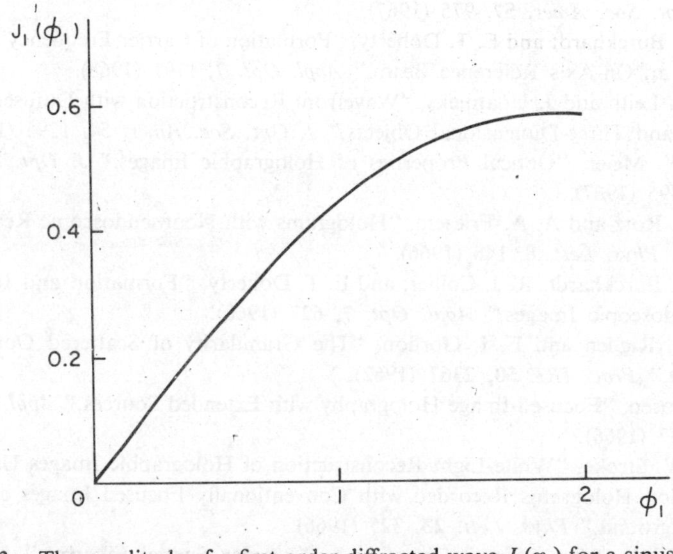

FIG. 8.33. The amplitude of a first-order diffracted wave $J_1(\varphi_1)$ for a sinusoidal phase grating.

where J_n is the Bessel function of the first kind and nth order. When **t** is illuminated with an axial plane wave of unit amplitude, the diffracted amplitude in the +1 order is $J_1(\varphi_1)$, a function plotted in Fig. 8.33. Its maximum amplitude is 0.582 and the maximum efficiency is 33.9%.

Somewhat more light can be diffracted into the first order if the phase modulation varies as a square-wave function of x and assumes the value $\varphi = 0$ during half the square-wave period and $\varphi = \pi$ during the remainder. The transmittance **t** in Eq. (8.33) is then +1 when $\varphi = 0$ and −1 when $\varphi = \pi$. This is analogous to the square-wave, absorption hologram where now the amplitude of the incident light diffracted into the first order is twice that for amplitude modulation. Hence efficiency is four times as large or 40.4%.

REFERENCES

8.1. E. N. Leith and J. Upatnieks, "Reconstructed Wavefronts and Communication Theory," *J. Opt. Soc. Amer.* **52**, 1123 (1962).

8.2. E. N. Leith and J. Upatnieks, "Wavefront Reconstruction with Continuous-Tone Objects," *J. Opt. Soc. Amer.* **53**, 1377 (1963).

8.3. L. H. Lin, "Increase of Hologram Image Separation by Total Reflection," *Appl. Opt.* **6**, 2004 (1967).

8.4. K. A. Stetson, "Holography with Total Internally Reflected Light," *Appl. Phys. Lett*, **11**, 225 (1967).

8.5. E. N. Leith and J. Upatnieks, "Holography with Achromatic-Fringe Systems," *J. Opt. Soc. Amer.* **57**, 975 (1967).

8.6. C. B. Burckhardt and E. T. Doherty, "Formation of Carrier Frequency Holograms with an On-Axis Reference Beam," *Appl. Opt.* **7**, 1191 (1968).

8.7. E. N. Leith and J. Upatnieks, "Wavefront Reconstruction with Diffused Illumination and Three-Dimensional Objects," *J. Opt. Soc. Amer.* **54**. 1295 (1964).

8.8. R. W. Meier, "Optical Properties of Holographic Images," *J. Opt. Soc. Amer.* **57**, 895 (1967).

8.9. F. B. Rotz and A. A. Friesem, "Holograms with Nonpseudoscopic Real Images," *Appl. Phys. Lett.* **8**, 146 (1966).

8.10. C. B. Burckhardt, R. J. Collier, and E. T. Doherty, "Formation and Inversion of Pseudoscopic Images," *Appl. Opt.* **7**, 627 (1968).

8.11. J. D. Rigden and E. I. Gordon, "The Granularity of Scattered Optical Maser Light," *Proc. IRE* **50**, 2367 (1962).

8.12. L. Rosen, "Focused-Image Holography with Extended Sources," *Appl. Phys. Lett.* **9**, 337 (1966).

8.13. G. W. Stroke, "White-Light Reconstruction of Holographic Images Using Transmission Holograms Recorded with Conventionally Focused Images and In-Line Background," *Phys. Lett.* **23**, 325 (1966).

8.14. A. Vander Lugt, "Signal Detection by Complex Spatial Filtering," *IEEE Trans. Inform. Theor.* IT–**10**, 139 (1964).

8.15. B. J. Thompson, J. H. Ward, and W. Zinky, "Application of Hologram Techniques for Particle-Size Analysis," *Appl. Opt.* **6**, 519 (1967).

8.16. W. Lukosz, "Equivalent-Lens Theory of Holographic Imaging," *J. Opt. Soc. Amer.* **58**, 1084 (1968).

8.17. J. C. Urbach and R. W. Meier, "Properties and Limitations of Hologram Recording Materials," *Appl. Opt.* **8**, 2269 (1969).

8.18. H. Kogelnik, "Reconstruction Response and Efficiency of Hologram Gratings," in *Proc. Symp. Modern Opt.* (J. Fox, ed.), pp. 605–617. Polytechnic Press, Brooklyn, New York, 1967.

8.19. H. P. Westman, ed., *Reference Data for Radio Engineers*, 4th ed. p. 1019. Internat. Telephone and Telegraph Corp., 1956.

Chapter 9
DIFFRACTION FROM VOLUME HOLOGRAMS

Coupled-wave theory can predict the response of elementary volume holograms to coherent illumination. Before applying the theory, let us employ the spatial frequency concepts of Chapter 5 to examine the formation of elementary volume holograms and to derive analytic expressions characterizing the geometry of their diffracting periodic structures.

9.1 Holograms Formed with Two Plane Waves

Consider two plane waves of unit amplitude propagating in the yz plane. They enter a recording medium and interfere as in Fig. 9.1. Snell's law requires that

$$\sin \Omega_S/\sin \psi_S = \sin \Omega_R/\sin \psi_R = n \tag{9.1}$$

where n is the index of refraction of the recording medium. Here Ω_S and Ω_R are the angles the two waves make with the z axis in air; ψ_S and ψ_R are the angles they make with the z axis in the medium. Our discussion begins, as in Section 3.1, with the summing of the complex amplitudes of the plane waves in the medium and then multiplying the net complex amplitude by its conjugate to obtain the intensity. Thus the net complex amplitude in the medium is

$$\mathbf{a}(y, z) = \exp[-i2\pi(\eta_S y + \zeta_S z)] + \exp[-i2\pi(\eta_R y + \zeta_R z)],$$

and

$$\mathbf{aa}^* = I = 2 + 2 \cos 2\pi[(\eta_S - \eta_R)y + (\zeta_S - \zeta_R)z] \tag{9.2}$$

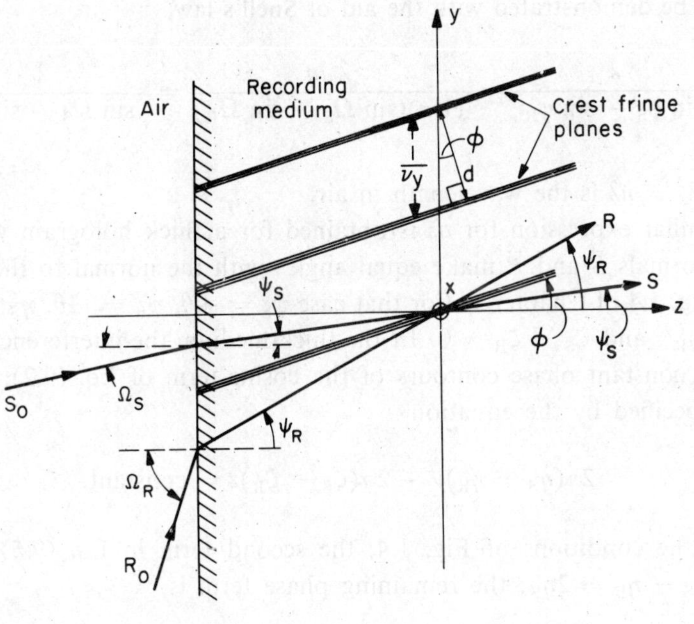

FIG. 9.1. Formation of an elementary hologram.

where

$$\eta_S = (\sin \psi_S)/\lambda, \qquad \eta_R = (\sin \psi_R)/\lambda \qquad \text{[from Eq. (5.14b)]}$$

$$\zeta_S = (1 - \lambda^2 \eta_S{}^2)^{1/2}/\lambda, \qquad \zeta_R = (1 - \lambda^2 \eta_R{}^2)^{1/2}/\lambda \qquad \text{[from Eq. (5.16)]}$$

and where λ is the wavelength in the medium.

Suppose for the moment that the medium is photosensitive only in one plane $z = z_1 =$ constant. Over this recording plane, the second term $(\zeta_S - \zeta_R)z_1$ in the argument of the cosine in Eq. (9.2) is merely a constant-phase contribution and can be set equal to zero. Remaining is the phase term containing the fringe frequency in the y direction

$$\nu_y = 1/d_y = \eta_S - \eta_R = (\sin \psi_S - \sin \psi_R)/\lambda \tag{9.3}$$

where d_y is the periodic spacing of the crest value of the cosinusoidal intensity fringes in the direction of the y axis. Since the intensity pattern expressed in Eq. (9.2) is independent of x, the crest values of the fringes are located on a series of lines in the plane $z = z_1$ parallel to the x axis and spaced by d_y. Equation (9.3) gives d_y in terms of wavelength and angles in the medium. The same form holds for the related parameters outside the medium in air, as may be demonstrated with the aid of Snell's law,

$$d_y = \frac{\lambda}{\sin \psi_S - \sin \psi_R} = \frac{\lambda_a/n}{(1/n)(\sin \Omega_S - \sin \Omega_R)} = \frac{\lambda_a}{\sin \Omega_S - \sin \Omega_R} \tag{9.4}$$

where $\lambda_a = n\lambda$ is the wavelength in air.

A similar expression for d_y is obtained for a thick hologram when the wave normals $\vec{S}$ and $\vec{R}$ make equal angles with the normal to the surface as in Fig. 1.4 of Chapter 1. For that case $\psi_S = +\theta$, $\psi_R = -\theta$, $\eta_S = -\eta_R$, $\eta_S^2 = \eta_R^2$, and $\zeta_S - \zeta_R = 0$. In the thick medium the interference fringes are the constant phase contours of the cosine term of Eq. (9.2), i.e., *surfaces* specified by the equation

$$2\pi(\eta_S - \eta_R)y + 2\pi(\zeta_S - \zeta_R)z = \text{constant}. \tag{9.5}$$

Under the conditions of Fig. 1.4, the second term in Eq. (9.5) is zero. With $\eta_S - \eta_R = 2\eta_S$, the remaining phase term is

$$2\pi(2\eta_S)y = \text{constant}$$

and the fringe contour equation reduces to

$$y = \text{constant}.$$

The fringe contours are thus *planes* parallel to the xz plane and crest values of intensity are spaced by a distance d_y in the y direction. When $\psi_S = -\psi_R = \theta$, the spacing is given by

$$d_y = 1/\nu_y = 1/2\eta_S \qquad \text{or} \qquad d_y = \lambda/(2 \sin \theta) \tag{9.6}$$

in agreement with Eq. (1.10). Since the planes are parallel to the z axis, θ refers to the angle that each interfering plane wave makes with the fringe planes in the medium.

Returning to Eq. (9.5) and the more general expression for the contours of constant phase, we note that the relation specifies planes perpendicular to the yz plane. (The traces of the surfaces in the yz plane are straight lines and there is no x dependence.) The angle ϕ that the planes make with the z axis is determined by differentiating their trace in the yz plane

$$(\eta_S - \eta_R)\frac{dy}{dz} + (\zeta_S - \zeta_R) = 0$$

or

$$\tan\phi = \frac{dy}{dz} = -\frac{\zeta_S - \zeta_R}{\eta_S - \eta_R}. \tag{9.7}$$

Using the definitions of the spatial frequencies given in Eq. (9.2), we obtain

$$\tan\phi = -\frac{(1-\sin^2\psi_S)^{1/2} - (1-\sin^2\psi_R)^{1/2}}{\sin\psi_S - \sin\psi_R} = -\frac{\cos\psi_S - \cos\psi_R}{\sin\psi_S - \sin\psi_R}$$

$$= \tan\left(\frac{\psi_S + \psi_R}{2}\right)$$

or

$$\phi = \frac{\psi_S + \psi_R}{2}. \tag{9.8}$$

From Fig. 9.1 we see that Eq. (9.8) implies that *the fringe planes of an elementary hologram bisect the angle between the wave normals of the interfering plane waves.* We can, of course, rotate the yz axes of Fig. 9.1 through the angle ϕ about the x axis to make the z axis coincide with the fringe plane. If now we define θ to be the *angle each beam makes with the fringe plane in the medium*, we find that the form of Eq. (1.10) holds regardless of orientation of the beams with respect to the hologram normal, i.e.,

$$2d\sin\theta = \lambda. \tag*{[1.10]}$$

Figure 9.1 also illustrates that the crest values of intensity are spatially separated by the distance

$$d = d_y\cos\phi = (\eta_S - \eta_R)^{-1}\cos\phi. \tag{9.9}$$

9.2 Bragg's Law

Figure 9.2 is a scale cross-sectional drawing which indicates an elementary hologram formed in photographic emulsion. A typical value of 15 μm has been assigned to the emulsion thickness. Horizontal lines in the figure represent traces of the planes of maximum silver density. These correspond to the crest value intensity planes of the interference pattern present during exposure of the emulsion. Suppose the hologram had been formed as shown in the figure, with an interbeam angle in air of 30° and light of wavelength $\lambda_a = 0.633$ μm. Substitution of these values into Eq. (9.4) yields the grating spacing $d = 1.22$ μm. A ray in the original reference beam now illuminating such a grating must pass through at least three maximum density planes

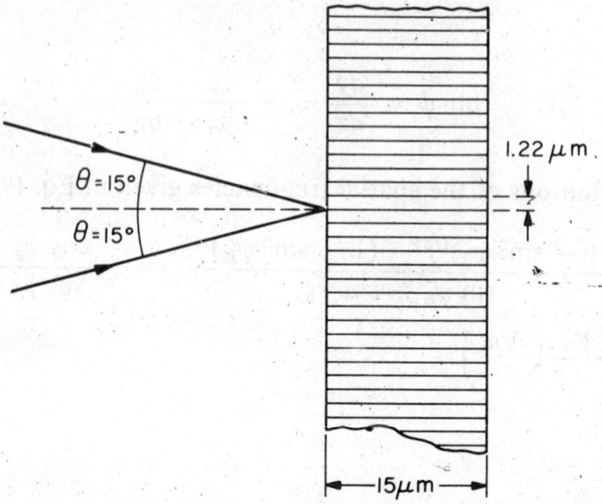

FIG. 9.2. Scale drawing of an interference pattern recorded in a Kodak 649F emulsion. Angle between the two beams 30°, wavelength 6328 Å, emulsion thickness 15 μm.

before emerging from the emulsion. It would indeed be surprising if plane hologram analysis, as given in Chapter 8, could account for all the properties of this volume hologram. Moreover, a thickness of 15 μm can be considered modest when compared to that possible with other recording media, e.g., photochromic crystals. Possibilities of larger thickness emphasize the need for a theory of diffraction from holograms in which a single illuminating ray can scatter successively from a large number of periodically spaced fringe planes. By requiring successively scattered waves to add in phase (constructively interfere) so that the net diffracted wave amplitude will be maximum, a restrictive relation is imposed on the wavelength of the illumination λ, the angle θ the illuminating beam makes to the scattering planes, and the spacing d between these planes. The relation is Bragg's law, given by Eq. (1.12), which can be written in terms of the wavelength in air λ_a and the average index of refraction of the hologram medium n_0 in the form

$$2d \sin \theta = \lambda_a / n_0. \tag{9.10}$$

Here, θ is the angle both illuminating and diffracted beams make with the scattering planes *in the hologram medium.*

Bragg's law determines the angle of incidence if the wavelength and spacing between planes are given, or it fixes the wavelength if the angle of incidence and the grating spacing are chosen independently. Volume holograms, whose properties are governed by Bragg's law, thus exhibit a selective response to illumination. In this chapter we address ourselves to the problem of obtaining a functional relation between the amplitude of the diffracted wave and the angle or wavelength of the reconstructing beam. We shall also compute the maximum efficiency possible under conditions specified for elementary transmission and reflection holograms. The results differ significantly from those obtained in Section 8.5 for elementary plane holograms.

9.3 Coupled Wave Theory

Most analyses of volume holograms [*9.1–9.6*] succeed in predicting the observed angular and wavelength selectivity. However, only those theories taking into account the depletion of the illuminating wave as it passes through the hologram are valid when diffraction efficiency is high. Computer solutions to the analytical problem of the volume hologram problem [*9.2, 9.3*] indicate that in fact high efficiencies are possible; these predictions are currently being borne out by experiment. Thus a linear theory [*9.1, 9.6*] is not sufficient. In contrast, *coupled wave analysis* is capable of not only predicting the selective response of the volume hologram but also carrying these predictions into the high-diffraction-efficiency regime. The theory predicts efficiencies which in certain cases approach 100%, implying nearly total depletion of the illuminating wave. The course of the remainder of this chapter follows Kogelnik's adaptation of the coupled wave theory [*9.4, 9.5*]. His approach has the merit of providing analytic as well as numerical results and is sufficiently flexible to describe the properties of a wide variety of volume holograms with and without absorption.

We are to be concerned only with the analysis of a hologram formed with two plane waves, a sinusoidal recording. Justification for limiting our consideration to this simple hologram follows the argument given in Section 1.6. That is, an arbitrary spatial function (a hologram in this case) can be Fourier analyzed into a sum of sinusoidal gratings any one of which will respond to incident light in the manner to be revealed by the theory.

9.4 The Wave Equation

Figure 9.3 indicates the geometry to be analyzed. Boundaries of the hologram are designated by the vertical traces in the y direction, $z = 0$ and $z = T$. Inside the hologram we assume that as a consequence of exposure and development either dielectric constant or absorption constant varies sinusoidally as a function of y and z, but not of x. For the purposes of

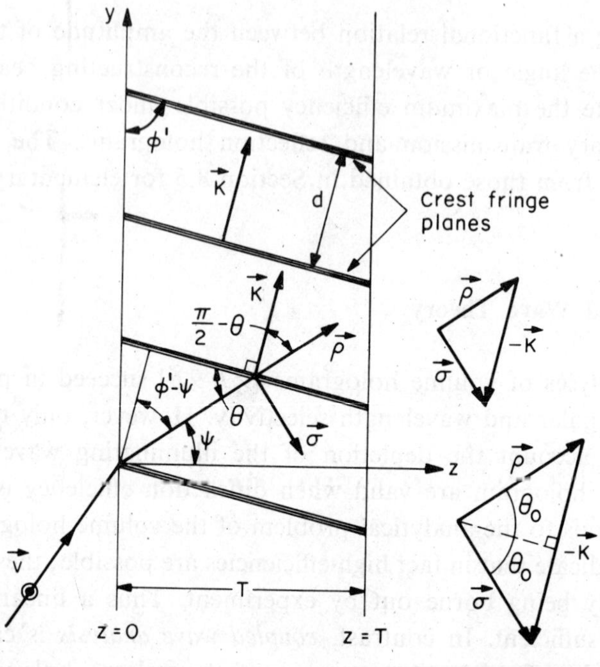

FIG. 9.3. Geometry for the analysis of volume holograms.

analysis we can consider the hologram to be formed so that the planes which are loci of constant dielectric constant or absorption constant are oriented perpendicular to the *yz* plane (the plane of the paper). The value of dielectric constant or absorption constant varies as a cosine function in the direction of the *grating vector* $\vec{K}$ lying in the *yz* plane normal to these isophase planes. Shown in the figure are traces of those planes which have crest values of the sinusoidal variation; they are separated by the periodic distance d and make an angle ϕ' with the boundaries. The grating vector $\vec{K}$, normal to the planes, is considered to have the absolute value

$$K = |\vec{K}| = 2\pi/d. \tag{9.11}$$

Incident on the hologram from the left is a monochromatic plane wave of light whose electric field vector $\vec{v}$ is polarized perpendicular to the plane of incidence, i,e., in the x direction perpendicular to the plane of the paper. Once the wave is *inside* the hologram boundaries, its direction of propagation, represented by the vector $\vec{\varrho}$, makes an angle ψ with the normal to the hologram (the z axis). Representing the direction of the diffracted wave is the vector $\vec{\sigma}$. (Kogelnik shows that with suitable modifications the theory can also apply to polarization of the electric field vector *in* the plane of incidence.)

Let us begin the analysis of wave propagation in the thick hologram with Maxwell's equations for a nonmagnetic material of permeability $\mu = 1$. They relate the electric field vector $\vec{v}$, the magnetic field vector $\vec{H}$, and the displacement vector $\vec{D}$ in the medium. Written in the mks system of units, they appear as

$$\operatorname{curl} \vec{v} = -\mu_0 \frac{\partial \vec{H}}{\partial t}, \tag{9.12a}$$

$$\operatorname{curl} \vec{H} = \varepsilon_0 \varepsilon \frac{\partial \vec{v}}{\partial t} + \sigma \vec{v}, \tag{9.12b}$$

$$\operatorname{div} \vec{D} = 0 \qquad \text{(no volume charges)}, \tag{9.12c}$$

$$\operatorname{div} \vec{H} = 0 \tag{9.12d}$$

where μ_0 is the permeability of free space, ε_0 is the dielectric constant of free space, ε is the relative dielectric constant of the hologram medium, and σ is the conductivity of the medium. If we take the curl of Eq. (9.12a)

$$\operatorname{curl}\operatorname{curl} \vec{v} = -\mu_0 \operatorname{curl} \frac{\partial \vec{H}}{\partial t} = -\mu_0 \frac{\partial}{\partial t} (\operatorname{curl} \vec{H}) \tag{9.13}$$

and differentiate Eq. (9.12b)

$$\frac{\partial}{\partial t} (\operatorname{curl} \vec{H}) = \varepsilon_0 \varepsilon \frac{\partial^2 \vec{v}}{\partial t^2} + \sigma \frac{\partial \vec{v}}{\partial t}, \tag{9.14}$$

we obtain

$$\operatorname{curl}\operatorname{curl} \vec{v} = -\mu_0 \left(\sigma \frac{\partial \vec{v}}{\partial t} + \varepsilon_0 \varepsilon \frac{\partial^2 \vec{v}}{\partial t^2} \right). \tag{9.15}$$

A well-known vector identity (see, e.g., Ramo and Whinnery [9.7], p. 114) gives us another expression for curl curl $\vec{v}$, viz.,

$$\operatorname{curl}\operatorname{curl} \vec{v} = \operatorname{grad}\operatorname{div} \vec{v} - \nabla^2 \vec{v} \tag{9.16}$$

where ∇^2 is the Laplacian operator. Before equating Eq. (9.15) and (9.16) we can show that the first term on the right of Eq. (9.16) vanishes. Consider Eq. (9.12c)

$$\operatorname{div} \vec{D} = \operatorname{div}(\varepsilon_0 \varepsilon \vec{v}) = 0.$$

The above may be expressed with the aid of another vector identity [9.7,

p. 114] which can be written for our variables as

$$\operatorname{div}(\varepsilon_0\varepsilon\vec{v}) = \vec{v}\cdot\operatorname{grad}(\varepsilon_0\varepsilon) + \varepsilon_0\varepsilon \operatorname{div}\vec{v} = 0. \tag{9.17}$$

Since $\vec{v}$ is parallel to the x axis while the variation of the dielectric constant occurs only in the yz plane, the scalar product

$$\vec{v}\cdot\operatorname{grad}(\varepsilon_0\varepsilon) = 0 \tag{9.18}$$

and from Eq. (9.17)

$$\operatorname{div}\vec{v} = 0. \tag{9.19}$$

Combining Eqs. (9.19), (9.16), and (9.15) we have

$$\nabla^2\vec{v} - \mu_0\sigma\frac{\partial\vec{v}}{\partial t} - \mu_0\varepsilon_0\varepsilon\frac{\partial^2\vec{v}}{\partial t^2} = 0. \tag{9.20}$$

We can now substitute for $\vec{v}$, ε, and σ in Eq. (9.20) expressions which specify the restrictions placed on these variables by our particular problem. As in Section 5.1, we may write the electric field vector $\vec{v}$, polarized in the x direction, as a scalar

$$v(y, z, t) = \operatorname{Re}[\mathbf{a}(y, z)\exp(i\omega t)] \tag{9.21}$$

which is independent of x and which oscillates at a single angular frequency ω. Employing complex notation, i.e., dropping Re[ ], we proceed to solve Eq. (9.20) for the complex quantity $\mathbf{a}(y, z)$. When Eq. (9.21) is substituted into Eq. (9.20), the result is

$$\nabla^2\mathbf{a} - i\omega\mu_0\sigma\mathbf{a} + \omega^2\mu_0\varepsilon_0\varepsilon\mathbf{a} = 0. \tag{9.22}$$

The relative dielectric constant ε may be considered to be composed of an average component ε_0[1] and a sinusoidally varying component of amplitude ε_1. Suppose $\vec{r} = \vec{i}x + \vec{j}y + \vec{k}z$ is a position vector from the origin to any point in the medium. ($\vec{i}, \vec{j}, \vec{k}$ are unit vectors directed along the x, y, z axes, respectively.) In the assumed spatial distribution of dielectric constant the loci of constant ε are planes which can be expressed as in Eq. (5.6) by

$$\vec{r}\cdot\vec{n} = \text{constant}$$

where $\vec{n}$, the normal to the planes, is equal to $\vec{K}/K$. The value of ε over any such plane is determined by the spatial phase $2\pi(\vec{r}\cdot\vec{n}/d) = \vec{K}\cdot\vec{r}$, where $\vec{r}\cdot\vec{n}$

[1] Note that ε_0 represents the average component of the dielectric constant of the hologram medium while ε_0 represents the dielectric constant of free space.

is the distance separating the plane of constant ε from the origin and where d is the spacing (measured along the normal) corresponding to an increment of 2π radians of phase. Similar remarks apply to the conductivity, which controls absorption, and we can express the spatially varying parameters by the relations

$$\varepsilon = \varepsilon_0 + \varepsilon_1 \cos \vec{K}\cdot\vec{r} \tag{9.23}$$

$$\sigma = \sigma_0 + \sigma_1 \cos \vec{K}\cdot\vec{r}. \tag{9.24}$$

Substituting the above into Eq. (9.22) we obtain the wave equation in the form

$$\nabla^2\mathbf{a} + \mathbf{q}^2\mathbf{a} = 0 \tag{9.25}$$

where

$$\mathbf{q}^2 = k^2\varepsilon_0 - i\omega\mu_0\sigma_0 + (k^2\varepsilon_1 - i\omega\mu_0\sigma_1)\cos \vec{K}\cdot\vec{r}. \tag{9.26}$$

In Eq. (9.26)

$$k = \omega(\mu_0\varepsilon_0)^{1/2} = \omega/c = 2\pi/\lambda_a,$$

$c = (\varepsilon_0\mu_0)^{-1/2}$ is the speed of light in vacuum (nearly the same in air), and λ_a is the wavelength in air. We can make our calculations easier if we write

$$\mathbf{q}^2 = k(\varepsilon_0)^{1/2}\left[k(\varepsilon_0)^{1/2} - \frac{2i\omega\mu_0\sigma_0}{2k(\varepsilon_0)^{1/2}} + 2\left(\frac{k\varepsilon_1}{2(\varepsilon_0)^{1/2}} - \frac{i\omega\mu_0\sigma_1}{2k(\varepsilon_0)^{1/2}}\right)\cos \vec{K}\cdot\vec{r}\right] \tag{9.27}$$

$$= \beta[\beta - 2i\alpha + 2\varkappa\{\exp(i\vec{K}\cdot\vec{r}) + \exp(-i\vec{K}\cdot\vec{r})\}]$$

$$= \beta^2 - 2i\alpha\beta + 2\varkappa\beta[\exp(i\vec{K}\cdot\vec{r}) + \exp(-i\vec{K}\cdot\vec{r})] \tag{9.28}$$

where

$$\beta = k(\varepsilon_0)^{1/2}, \tag{9.29}$$

$$\alpha = \frac{\omega\mu_0\sigma_0}{2k(\varepsilon_0)^{1/2}}, \tag{9.30}$$

$$\varkappa = \frac{1}{2}\left(k\frac{\varepsilon_1}{2(\varepsilon_0)^{1/2}} - i\frac{\omega\mu_0\sigma_1}{2k(\varepsilon_0)^{1/2}}\right) = \frac{1}{2}\left(k\frac{\varepsilon_1}{2(\varepsilon_0)^{1/2}} - i\alpha_1\right), \tag{9.31}$$

and

$$\alpha_1 = \frac{\omega\mu_0\sigma_1}{2k(\varepsilon_0)^{1/2}}. \tag{9.32}$$

The coupling parameter $\varkappa$ will be shown to be of central importance in the coupled wave theory. It describes the coupling between incident and diffracted waves. For $\varkappa = 0$, there is no coupling and no diffraction.

It is customary to describe the optical properties of material in terms of the index of refraction rather than the dielectric constant. To do so we first consider propagation in a *homogeneous* lossy dielectric medium. A solution of the wave equation corresponding to plane wave propagation through the medium in the z direction has the form

$$\mathbf{f} = A \exp(-\gamma_0 z)$$

where A is a constant amplitude and where the complex propagation constant γ_0 is given by

$$\gamma_0 = i(\varepsilon_0 \varepsilon_0 \mu_0 \omega^2 - i\mu_0 \omega \sigma_0)^{1/2}$$

(see, e.g., Ramo and Whinnery [9.7], p. 306). (Note that ε_1 and σ_1 are zero in the homogeneous medium and we have taken $\mu = 1$.) We can assign a meaning to β in Eq. (9.29) and α in Eq. (9.30) by writing γ_0 in terms of these parameters. We find

$$\gamma_0 = i(\beta^2 - 2i\alpha\beta)^{1/2} \approx i\beta + \alpha$$

providing $\alpha \ll \beta$, which is usually the case. The wave function $\mathbf{f}$ in the homogeneous medium then becomes

$$\mathbf{f} = A \exp(-i\beta z) \exp(-\alpha z)$$

where β is called the propagation constant and α the absorption constant. Since β and α are defined in terms of the average dielectric constant and average conductivity of the hologram, we may consider β and α to be propagation constant and absorption constant in an equivalent homogeneous medium where $\varepsilon = \varepsilon_0$ and $\sigma = \sigma_0$. The labels are meaningful under the condition $\alpha \ll \beta$.

Now let n_0 be the *average* index of refraction of the hologram and the index of refraction of the equivalent homogeneous medium. Recall that the index is defined as the ratio of the speed of light in vacuum to that in the medium. Since the latter is given by $(\varepsilon_0 \varepsilon_0 \mu_0)^{-1/2}$ when $\mu = 1$,

$$n_0 = \frac{(\varepsilon_0 \varepsilon_0 \mu_0)^{1/2}}{(\varepsilon_0 \mu_0)^{1/2}} = (\varepsilon_0)^{1/2} \tag{9.33}$$

Substituting into Eq. (9.29), we have that

$$\beta = 2\pi n_0 / \lambda_a . \tag{9.34}$$

From (9.34) we see that the restriction $\alpha \ll \beta$ implies that

$$\alpha \ll 2\pi n_0/\lambda_a. \tag{9.35}$$

We note that the factor $\exp(-\alpha z)$ in the wave function $\mathbf{f}$ represents attenuation as long as $\alpha > 0$. We assume that in the recording of an absorption hologram α is modulated, but that the recording process cannot reduce the absorption constant beyond zero. By means of Eqs. (9.30) and (9.31) one can rewrite Eq. (9.24) in terms of α and the modulation amplitude α_1 and show that unless $\alpha_1 \leq \alpha$, the net absorption can be negative (for $\cos \vec{K}\cdot\vec{r} = -1$). We therefore have the additional restriction

$$\alpha_1 \ll 2\pi n_0/\lambda_a. \tag{9.36}$$

In keeping with Eq. (9.33) we define the refractive index n of the hologram as

$$n^2 = \varepsilon. \tag{9.37}$$

Expressing n in terms of an average value n_0 and a modulation amplitude n_1 and using Eq. (9.23) we write

$$n^2 = (n_0 + n_1 \cos \vec{K}\cdot\vec{r})^2 = \varepsilon_0 + \varepsilon_1 \cos \vec{K}\cdot\vec{r}. \tag{9.38}$$

Assuming

$$n_1 \ll n_0, \tag{9.39}$$

we can expand the square of the index to obtain $n_0 = (\varepsilon_0)^{1/2}$ as in Eq. (9.33) and

$$n_1 = \varepsilon_1/2n_0 = \varepsilon_1/2(\varepsilon_0)^{1/2}. \tag{9.40}$$

The condition (9.39) is usually satisfied for volume holograms. If we substitute Eq. (9.40) into Eq. (9.31) we can express the coupling constant $\varkappa$ in the form

$$\varkappa = (\pi n_1/\lambda_a) - (i\alpha_1/2). \tag{9.41}$$

9.5 Solution of the Wave Equation

We now must solve the partial differential equation Eq. (9.25), the wave equation. To render the problem tractable we make some simplifying assumptions. First, we shall solve Eq. (9.25) only for angles of incidence close to those satisfying Bragg's law. It is in this region that appreciable diffraction is obtained. Second, we assume that only two waves propagate in the hologram. These are the incident wave and that diffracted wave which closely satisfies Bragg's law. This last assumption places a lower limit on the hologram thickness for which the theory is valid. We shall consider the consequence of this restriction at the end of the chapter.

Within the hologram, the complex amplitude of the incident plane wave $\mathbf{a}_i$ can be written in the form

$$\mathbf{a}_i = \mathbf{R}(z) \exp(-i\vec{\varrho}\cdot\vec{r}) \tag{9.42}$$

where $\vec{\varrho}$ has the direction of propagation of the wave, as indicated in Fig. 9.3. Here the phase factor $\exp(-i\vec{\varrho}\cdot\vec{r})$ is representative of the incident plane wave propagating in a medium with no variation in dielectric constant and no absorption. From Eqs. (5.7) and (5.11) and from our discussion in Section 9.4 of propagation in homogeneous media, it is clear that we should set

$$\varrho = |\vec{\varrho}| = \beta \tag{9.43}$$

where β is given by Eq. (9.34). Contained in this phase factor are the fast phase variations associated with any traveling wave. On the other hand the complex amplitude factor $\mathbf{R}(z)$ accounts for the slower variations of phase and amplitude incurred as the wave progresses through the thickness of the hologram (i.e., as a function of z). These variations are imparted by the spatial variation of dielectric constant and absorption.

Analogously we may write the complex amplitude of the wave diffracted by the hologram in the form

$$\mathbf{a}_d = \mathbf{S}(z) \exp(-i\vec{\sigma}\cdot\vec{r}). \tag{9.44}$$

$\vec{\sigma}$ is indicated in Fig. 9.3. When light illuminating the hologram is incident at the Bragg angle, the vector relation (which we choose to set)

$$\vec{\sigma} = \vec{\varrho} - \vec{K} \tag{9.45}$$

between incident and diffracted wave propagation vectors and the grating vector $\vec{K}$ takes on particular significance. Equation (9.45) is graphically illustrated by the vector diagram at the upper right of Fig. 9.3.

In the case of Bragg incidence, the propagation vector $\vec{\varrho}$ of the incident wave and the propagation vector $\vec{\sigma}$ of the diffracted wave each form an angle θ_0 (Bragg angle) with the crest planes of the sinusoidal hologram grating. As defined in Section 9.4, the vector $\vec{K}$ lies in the yz plane perpendicular to the crest planes. Since the electric field vector of the incident wave is polarized in the x direction, its propagation vector $\vec{\varrho}$ is also constrained to lie in the yz plane. Equation (9.45) then determines that $\vec{\sigma}$ is contained in the same plane. The triangle formed by the three coplanar vectors, for the case where $\vec{\varrho}$ and $\vec{\sigma}$ each form the angle θ_0 with the scattering planes, is shown on the lower right of Fig. 9.3. Because it is an isosceles triangle, $\varrho = \sigma = \beta$ whence

$$K/2 = \varrho \sin \theta_0 . \tag{9.46}$$

Using Eqs. (9.11) and (9.34) we obtain

$$\pi/d = \beta \sin \theta_0 = (2\pi n_0/\lambda_a) \sin \theta_0 , \tag{9.47}$$

which may be rearranged to read

$$2d \sin \theta_0 = \lambda_a/n_0 \qquad \text{(Bragg's law).}$$

Equation (9.45) is then an expression of Bragg's law for light incident at the Bragg angle.

Returning now to the problem of solving the wave equation Eq. (9.25), we can write the complex amplitude **a** of the electric field at any point in the hologram as the sum of the incident wave $\mathbf{a}_i$ and the diffracted wave $\mathbf{a}_d$

$$\mathbf{a} = \mathbf{a}_i + \mathbf{a}_d = \mathbf{R}(z) \exp(-i\vec{\varrho}\cdot\vec{r}) + \mathbf{S}(z) \exp(-i\vec{\sigma}\cdot\vec{r}). \tag{9.48}$$

Equation (9.48) is to be inserted into Eq. (9.25). In carrying out the partial differentiation, which is lengthy, use is made of the relations,

$$\vec{\varrho}\cdot\vec{r} = \varrho_y y + \varrho_z z , \qquad \vec{\sigma}\cdot\vec{r} = \sigma_y y + \sigma_z z ; \tag{9.49}$$

$$\varrho^2 = \varrho_y{}^2 + \varrho_z{}^2, \qquad \sigma^2 = \sigma_y{}^2 + \sigma_z{}^2 . \tag{9.50}$$

The terms resulting from the operations on **a** indicated in Eq. (9.25) can be grouped as coefficients of either $\exp(-i\vec{\varrho}\cdot\vec{r})$ or $\exp(-i\vec{\sigma}\cdot\vec{r})$. In considering the second term in Eq. (9.25), $\mathbf{q}^2\mathbf{a}$, terms with exponential factors $\exp[-i(\vec{K} + \vec{\varrho})\cdot\vec{r}]$ and $\exp[-i(\vec{\sigma} - \vec{K})\cdot\vec{r}]$ are neglected since waves propagating with the vectors $\vec{K} + \vec{\varrho}$ or $\vec{\sigma} - \vec{K}$ do not satisfy the Bragg vector relation Eq. (9.45). To ensure that $\nabla^2\mathbf{a} + \mathbf{q}^2\mathbf{a}$ vanishes as required by Eq. (9.25) and yet allow Eq. (9.45) to hold for arbitrary $\vec{K}$, the coefficients of $\exp(-i\vec{\varrho}\cdot\vec{r})$ and of $\exp(-i\vec{\sigma}\cdot\vec{r})$ must separately equal zero. Hence two equations emerge:

$$\mathbf{R}'' - 2i\varrho_z\mathbf{R}' - \varrho^2\mathbf{R} + \beta^2\mathbf{R} - 2i\alpha\beta\mathbf{R} + 2\varkappa\beta\mathbf{S} = 0, \tag{9.51}$$

$$\mathbf{S}'' - 2i\sigma_z\mathbf{S}' - \sigma^2\mathbf{S} + \beta^2\mathbf{S} - 2i\alpha\beta\mathbf{S} + 2\varkappa\beta\mathbf{R} = 0. \tag{9.52}$$

The primes denote differentiation with respect to z.

By making some reasonable assumptions and introducing useful notation we can simplify Eqs. (9.51) and (9.52). Previously it was noted that the fast variations in the wave functions of Eqs. (9.42) and (9.44) are contained in the phase factors while $\mathbf{R}(z)$ and $\mathbf{S}(z)$ change relatively slowly. We now assume $\mathbf{R}(z)$ and $\mathbf{S}(z)$ change slowly enough that $\mathbf{R}''$ and $\mathbf{S}''$ can be neglected. Later we shall be able to check the validity of this approximation. Equation (9.51) is reduced somewhat by observing that the sum of the third and fourth terms is zero [see Eq. (9.43)]. Now let us consider the sum of the third and fourth terms of Eq. (9.52), $\mathbf{S}(\beta^2 - \sigma^2)$, and evaluate the factor $\beta^2 - \sigma^2$ for the case where the angle of incidence θ deviates from the Bragg angle θ_0 by only a small angle δ. That is

$$\theta = \theta_0 + \delta. \tag{9.53}$$

Using Eq. (9.45) we have that

$$\begin{aligned} \beta^2 - \sigma^2 &= \beta^2 - (\vec{\varrho} - \vec{K})^2 \\ &= \beta^2 - \varrho^2 + 2\vec{\varrho}\cdot\vec{K} - K^2 \\ &= 2\varrho K \cos(\tfrac{1}{2}\pi - \theta) - K^2 \\ &= 2\varrho K \sin\theta - K^2 \end{aligned} \tag{9.54}$$

where the angle between $\vec{\varrho}$ and $\vec{K}$ is given in Fig. 9.3. Expanding $\sin\theta$ in terms of Eq. (9.53)

$$\begin{aligned} \sin\theta = \sin(\theta_0 + \delta) &= \sin\theta_0 \cos\delta + \sin\delta\cos\theta_0 \\ &\approx \sin\theta_0 + \delta\cos\theta_0 \\ &\approx (K/2\beta) + \delta\cos\theta_0 \end{aligned} \tag{9.55}$$

where we have set $\sin\delta \approx \delta$, $\cos\delta \approx 1$, and used Eqs. (9.46) and (9.43) to obtain

$$\sin\theta_0 = K/2\beta. \tag{9.56}$$

Continuing,

$$\begin{aligned} \beta^2 - \sigma^2 &\approx 2\varrho K[(K/2\beta) + \delta\cos\theta_0] - K^2 \\ &\approx 2\varrho K\delta\cos\theta_0 \\ &\approx 2\varrho 2\beta \sin\theta_0\,\delta\cos\theta_0 \end{aligned} \tag{9.57}$$

$$\beta^2 - \sigma^2 \approx 2\beta^2\delta\sin 2\theta_0. \tag{9.58}$$

We designate

$$\Gamma = \beta\delta \sin 2\theta_0 \tag{9.59}$$

so that

$$\beta^2 - \sigma^2 = 2\beta\Gamma. \tag{9.60}$$

Then by neglecting $\mathbf{R}''$ and $\mathbf{S}''$, employing Eq. (9.60), and introducing the symbols

$$\begin{aligned} c_{\mathrm{R}} &= \varrho_z/\beta = (\varrho \cos \psi)/\beta = \cos \psi, \\ c_{\mathrm{S}} &= \sigma_z/\beta, \end{aligned} \tag{9.61}$$

we can arrange Eqs. (9.51) and (9.52) in the form

$$c_{\mathrm{R}}\mathbf{R}' + \alpha\mathbf{R} = -i\varkappa\mathbf{S} \tag{9.62}$$

$$c_{\mathrm{S}}\mathbf{S}' + (\alpha + i\Gamma)\mathbf{S} = -i\varkappa\mathbf{R} \tag{9.63}$$

where $\mathbf{R}' = d\mathbf{R}/dz$ and $\mathbf{S}' = d\mathbf{S}/dz$.

The physics of the diffraction process is revealed by the above coupled-wave equations. For every increment of distance dz representing progress of the incident and diffracted waves through the thickness of the hologram, the wave complex amplitudes change by $d\mathbf{R}$ or $d\mathbf{S}$. Changes are caused by absorption, as indicated by the terms $\alpha\mathbf{R}$ and $\alpha\mathbf{S}$, or by coupling of one wave to be other through the coupling terms $\varkappa\mathbf{S}$ and $\varkappa\mathbf{R}$. As we shall see, the term $i\Gamma\mathbf{S}$ in Eq. (9.63) produces an additional phase factor in the diffracted wave. If the incident wave direction deviates too much from the Bragg angle, the factor Γ will be large. Then accumulation of this extra phase by the diffracted wave tends to force it out of synchronism with the incident wave and to cause interaction to cease.

Equations (9.62) and (9.63) represent a system of two linear first-order differential equations. Substituting Eq. (9.62) into (9.63) produces one second-order differential equation for $\mathbf{R}$

$$\mathbf{R}'' + \left(\frac{\alpha}{c_{\mathrm{R}}} + \frac{\alpha}{c_{\mathrm{S}}} + \frac{i\Gamma}{c_{\mathrm{S}}}\right)\mathbf{R}' + \frac{(\alpha^2 + i\Gamma\alpha + \varkappa^2)\mathbf{R}}{c_{\mathrm{R}}c_{\mathrm{S}}} = 0. \tag{9.64}$$

A general solution of a second-order differential equation with constant coefficients has the form

$$\mathbf{R}(z) = \exp(\gamma z). \tag{9.65}$$

Inserting Eq. (9.65) into Eq. (9.64) gives a quadratic equation for γ,

$$\gamma^2 + \left(\frac{\alpha}{c_{\mathrm{R}}} + \frac{\alpha}{c_{\mathrm{S}}} + \frac{i\Gamma}{c_{\mathrm{S}}}\right)\gamma + \left(\frac{\alpha^2 + i\Gamma\alpha + \varkappa^2}{c_{\mathrm{R}}c_{\mathrm{S}}}\right) = 0, \tag{9.66}$$

which when solved yields the roots

$$\gamma_{1,2} = -\frac{1}{2}\left(\frac{\alpha}{c_R} + \frac{\alpha}{c_S} + \frac{i\Gamma}{c_S}\right) \pm \frac{1}{2}\left[\left(\frac{\alpha}{c_R} - \frac{\alpha}{c_S} - \frac{i\Gamma}{c_S}\right)^2 - \frac{4\varkappa^2}{c_R c_S}\right]^{1/2} \tag{9.67}$$

where the subscript 1 refers to the plus sign in front of the square root and the subscript 2 refers to the minus sign. Particular solutions to Eq. (9.64) are $\exp(\gamma_1 z)$ and $\exp(\gamma_2 z)$, and a complete solution is the sum

$$\mathbf{R}(z) = \mathbf{R}_1 \exp(\gamma_1 z) + \mathbf{R}_2 \exp(\gamma_2 z) \tag{9.68}$$

where $\mathbf{R}_1$ and $\mathbf{R}_2$ are constants which must be evaluated from the boundary conditions. Substituting Eq. (9.68) into Eq. (9.62), we obtain a similar equation for $\mathbf{S}(z)$, viz.,

$$\mathbf{S}(z) = \mathbf{S}_1 \exp(\gamma_1 z) + \mathbf{S}_2 \exp(\gamma_2 z). \tag{9.69}$$

In the next two sections we shall evaluate the constants $\mathbf{R}_1$, $\mathbf{R}_2$, $\mathbf{S}_1$, and $\mathbf{S}_2$ for transmission and reflection holograms. In the case of the transmission hologram the grating is assumed to be "unslanted," i.e., the grating planes are normal to the surface of the hologram medium. For reflection holograms (see Section 1.7, Fig. 1.12, Position 4) the grating planes are assumed to be parallel to the surface. Our discussion will cover only *lossless dielectric* or *phase* gratings (where the index of refraction varies and the loss is zero) and *pure absorption* gratings (where the absorption varies but the index remains constant). Kogelnik [9.5] treats slanted gratings, lossy dielectric gratings, and mixed gratings as well.

Before going on we note that Eqs. (9.65) and (9.67) afford us the opportunity to check the validity of neglecting $\mathbf{R}''$ in comparison to $\varrho_z\mathbf{R}'$ in Eq. (9.51). From Eq. (9.65) $\mathbf{R}'' = \gamma^2 \exp(\gamma z)$ and from Eq. (9.61) $\varrho_z\mathbf{R}' = \gamma\beta \cos\psi \exp(\gamma z)$. Thus for $\psi < 90°$, $\mathbf{R}'' \ll \varrho_z\mathbf{R}'$ implies that $\gamma \ll \beta$. Examining Eq. (9.67), the expression for γ shows that $\gamma \ll \beta$, providing Γ (proportional to $\delta\beta$) is very small, and providing the inequalities (9.35), (9.36), and (9.39) are satisfied. Similar results hold for the neglect of $\mathbf{S}''$ in comparison to $\sigma_z\mathbf{S}'$.

9.6 Transmission Holograms

In Fig. 9.4 we suppose an illuminating wave from the left to be incident on a transmission hologram. Both illuminating wave and diffracted wave propagate from left to right through the hologram. Let us normalize the

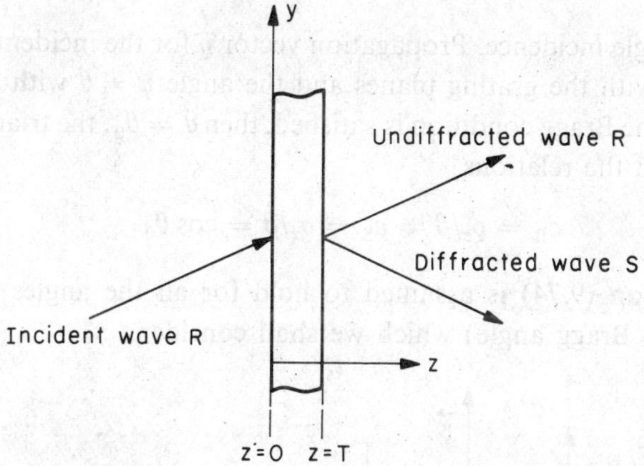

FIG. 9.4. The transmission hologram.

amplitude of the incident wave $\mathbf{R}(z)$ such that $\mathbf{R}(0) = 1$ at $z = 0$. Initially the amplitude of the diffracted wave is zero so that $\mathbf{S}(0) = 0$ at $z = 0$. Equations (9.68) and (9.69) evaluated at $z = 0$ then yield the boundary conditions

$$\mathbf{R}(0) = \mathbf{R}_1 + \mathbf{R}_2 = 1, \tag{9.70}$$

$$\mathbf{S}(0) = \mathbf{S}_1 + \mathbf{S}_2 = 0. \tag{9.71}$$

Using Eqs. (9.70) and (9.71) and the additional relation derived from Eq. (9.69), viz.,

$$\mathbf{S}'(0) = \gamma_1 \mathbf{S}_1 + \gamma_2 \mathbf{S}_2,$$

we can solve for $\mathbf{S}_1 = -\mathbf{S}_2$ in the coupled-wave equation (9.63) and obtain (when $z = 0$)

$$c_S(\gamma_1 \mathbf{S}_1 + \gamma_2 \mathbf{S}_2) = -i\varkappa$$

or

$$\mathbf{S}_1 = -\mathbf{S}_2 = -i\varkappa/[c_S(\gamma_1 - \gamma_2)]. \tag{9.72}$$

With these values of $\mathbf{S}_1$ and $\mathbf{S}_2$, the amplitude of the diffracted wave at the other boundary of the hologram $z = T$ can be obtained from Eq. (9.69). (T is the hologram thickness.) Thus

$$\mathbf{S}(T) = i[\varkappa/\{c_S(\gamma_1 - \gamma_2)\}]\,[\exp(\gamma_2 T) - \exp(\gamma_1 T)]. \tag{9.73}$$

In accordance with our plan, we shall restrict the orientation of the grating planes to lie perpendicular to the hologram surface. As a consequence the grating vector $\vec{K}$ is parallel to the surface. Figure 9.5 illustrates the geometry and indicates the vector triangle relation $\vec{\sigma} = \vec{\varrho} - \vec{K}$ between $\vec{\varrho}$, $\vec{K}$, and $\vec{\sigma}$

for Bragg angle incidence. Propagation vector $\vec{\varrho}$ for the incident wave forms the angle θ with the grating planes and the angle $\psi = \theta$ with the z axis as well. When the Bragg condition is satisfied, then $\theta = \theta_0$, the triangle becomes isosceles, and the relations

$$c_R = \varrho_z/\beta = c_S = \sigma_z/\beta = \cos\theta_0 \tag{9.74}$$

hold. Equation (9.74) is assumed to hold for all the angles of incidence (close to the Bragg angle) which we shall consider.

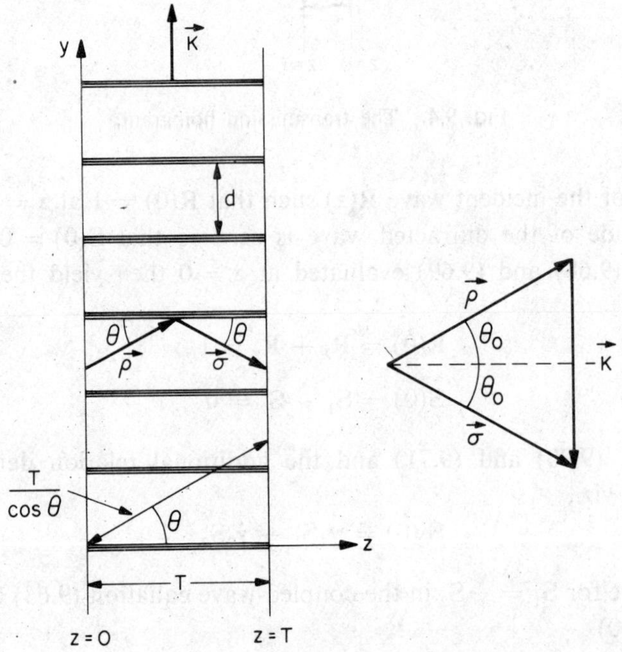

FIG. 9.5. Geometry for the transmission hologram with no slant.

9.6.1 Phase Transmission Holograms

Our purpose now is to express the amplitude of the diffracted wave emerging from the hologram, S(T) in Eq. (9.73), in terms appropriate for the case of a lossless dielectric or phase grating where $\alpha = \alpha_1 = 0$. In carrying out the computation it is convenient to define the parameters ξ and ν as

$$\xi = \delta\beta T \sin\theta_0 = \frac{\Gamma T}{2\cos\theta_0} \tag{9.75}$$

$$\nu = \frac{\varkappa T}{\cos\theta_0} = \frac{\pi n_1 T}{\lambda_a \cos\theta_0}. \tag{9.76}$$

We see from Eq. (9.73) that it is necessary first to express $(\gamma_1 - \gamma_2)$, $\gamma_1 T$, and $\gamma_2 T$ in terms of ξ and ν. From Eq. (9.67) we have that

$$\gamma_1 - \gamma_2 = (2i/T)(\xi^2 + \nu^2)^{1/2} \tag{9.77}$$

$$\gamma_{1,2}T = -i\xi \pm i(\xi^2 + \nu^2)^{1/2}. \tag{9.78}$$

Inserting Eqs. (9.77) and (9.78) into Eq. (9.73) yields

$$\mathbf{S}(T) = -i\,\frac{\exp(-i\xi)\sin(\xi^2 + \nu^2)^{1/2}}{(1 + \xi^2/\nu^2)^{1/2}}. \tag{9.79}$$

Since we have assumed a unit amplitude for the incident wave at $z = 0$, the efficiency of the hologram

$$\eta = |\,\mathbf{S}(T)\,|^2/|\,\mathbf{R}(0)\,|^2 = |\,\mathbf{S}(T)\,|^2.$$

First consider the angle of incidence to be exactly the Bragg angle so that $\delta = 0$ and $\xi = 0$. The efficiency will be 100% when $\sin \nu = 1$ or when

$$\nu = \frac{\pi n_1 T}{\lambda_a \cos\theta_0} = \frac{\pi}{2}. \tag{9.80}$$

Equation (9.80) can be rearranged to read

$$\frac{n_1 T}{\cos\theta_0} = \frac{\lambda_a}{2}. \tag{9.81}$$

The left-hand side of Eq. (9.81) is equivalent to the increment which would add to the optical path of an incident ray if an average index change n_1 were imposed on the hologram medium (see Fig. 9.5). When the equivalent path length change equals half the wavelength (in air) of the incident light, then the efficiency becomes 100%. Thus in forming a hologram in a lossless dielectric even a small index variation can yield 100% diffraction efficiency providing T, the medium thickness, is sufficiently large to satisfy Eq. (9.81).

When the angle of incidence differs from the Bragg angle, 100% efficiency cannot be achieved. Figure 9.6 plots results obtainable from Eq. (9.79) in normalized form. Along the vertical axis is plotted efficiency η divided by the efficiency η_0 obtainable when the illuminating beam is incident at the Bragg angle. The parameter ξ, proportional to the angular deviation δ from Bragg incidence, is plotted along the horizontal axis. Three curves corresponding to three values of the parameter ν are shown. For a constant thickness T and a constant arrangement of hologram-forming beams, ν is proportional to the amplitude of the index variation produced by exposure and processing of the hologram. Attached to each curve in Fig. 9.6 is the maximum value of efficiency η_0 obtainable when $\xi = 0$. Note that for $\nu = \pi/2$, $\eta_0 = 100\%$, while for $\nu = \pi/4$ or $3\pi/4$, $\eta_0 = 50\%$. Figure 9.6 serves to compare angular response of the hologram for three values of the modulation parameter ν.

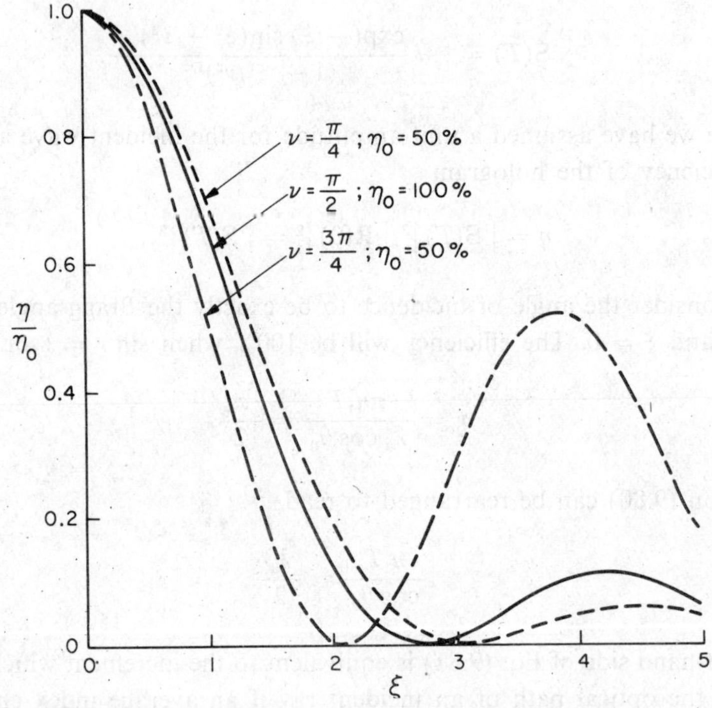

FIG. 9.6. Relative efficiency η/η_0 of the lossless dielectric transmission hologram versus $\xi = \delta(2\pi n_0/\lambda_a)T \sin\theta_0$ for various values of the parameter $\nu = \pi n_1 T/\lambda_a \cos\theta_0$. (After Kogelnik [9.5]. Copyright 1969 by American Telephone and Telegraph Company. Reprinted by permission.)

As a practical example of the use of the theory developed so far, let us compute the deviation δ from the Bragg angle of incidence which causes the efficiency to drop to zero. Suppose a hologram has been formed in dichromated gelatin (see Chapter 10) as a result of the interference of two plane waves whose directions in air make an angle of 60°. The Bragg angle in air is 30°, and within the emulsion $\theta_0 = 19.2°$ (taking $n_0 = 1.52$ for the index of refraction of the gelatin). Assume that the wavelength in air $\lambda_a = 4880$ Å, the thickness $T = 15\,\mu$m, and that the grating diffracts with

100% efficiency at Bragg angle incidence ($\nu = \pi/2$). From Eq. (9.75) we have

$$\xi = (\delta 2\pi n_0/\lambda_a)T \sin\theta_0 = 96.5\delta .$$

Since the $\nu = \pi/2$ plot in Fig. 9.6 goes to zero for $\xi = 2.7$, $\delta = 2.7/96.5 = 0.028$ rad $= 1.6°$ (inside the gelatin). In air δ is 2.45°.

The curves of Fig. 9.6 can also be used to determine the variation of efficiency as a function of deviation from the wavelength satisfying Bragg's law. Suppose a hologram is formed by two plane waves of wavelength λ_a meeting at an angle $2\theta_0$ in a recording medium of index n_0. When illuminated with a plane wave of wavelength λ_a satisfying Bragg's law, Eq. (9.10),

$$2n_0 d \sin\theta_0 = \lambda_a , \qquad [9.10]$$

the hologram diffracts with maximum efficiency. Now let the wavelength of the illumination be changed to $\lambda_a + \Delta\lambda$ where $\Delta\lambda/\lambda_a \ll 1$. Maximum efficiency is no longer obtained at the illumination angle θ_0 but is instead given by a new Bragg angle $\theta_0' = \theta_0 + \delta$. If we maintain the illumination at the original angle $\theta_0 = \theta_0' - \delta$, diffraction efficiency is reduced according to the deviation $-\delta$ from the Bragg angle θ_0' and to the corresponding *negative* value of the parameter $\xi = -\delta\beta T \sin\theta_0$. The curves $\eta/\eta_0 = |\mathbf{S}(T)|^2$ of Fig. 9.6 are symmetric in $\xi(\delta)$ [see Eq. (9.79)], so that knowing δ we may predict the reduced efficiency due to $\pm\xi$. We may express δ in terms of $\Delta\lambda$ by rewriting Eq. (9.10) for the new Bragg parameters $\theta_0 + \delta$ and $\lambda_a + \Delta\lambda$,

$$2n_0 d \sin(\theta_0 + \delta) = \lambda_a + \Delta\lambda . \qquad (9.82)$$

Setting $\sin\delta \approx \delta$ and $\cos\delta \approx 1$ and using Eq. (9.10), we obtain

$$\delta = \frac{\Delta\lambda}{2n_0\, d \cos\theta_0}, \qquad (9.83)$$

or

$$\delta = \frac{\Delta\lambda}{\lambda_a} \tan\theta_0 . \qquad (9.84)$$

The parameter ξ now becomes (for positive $\Delta\lambda$)

$$\xi = -\frac{\Delta\lambda}{\lambda_a} \tan\theta_0 \left(\frac{2\pi n_0}{\lambda_a}\right) T \sin\theta_0 . \qquad (9.85)$$

Let us compute the value of $\Delta\lambda$ which makes the efficiency of the dichromated gelatin hologram (in our first example) zero. Substituting $\xi = 2.7$,

$n_0 = 1.52$, $T = 15\ \mu$m, $\lambda_a = 0.488\ \mu$m, and $\theta_0 = 19.2°$ into Eq. (9.85) we find $|\Delta\lambda| = 0.0393\ \mu$m $= 393$ Å. [For small $\Delta\lambda$ the variation of ν can be neglected. This becomes evident when one differentiates Eq. (9.76) with respect to λ and examines Fig. 9.6.]

A simple rule-of-thumb for the angular sensitivity of a phase hologram can be extracted from the curves of Fig. 9.6. When $\xi \approx 3$, diffraction efficiency is essentially zero (at least over the range $\pi/2 \geq \nu \geq \pi/4$). Using Eq. (9.75) we compute the value δ_0 corresponding to $\xi = 3$,

$$\delta_0 = \frac{3}{\beta T \sin\theta_0} = \frac{3\lambda_a}{2\pi n_0 T \sin\theta_0} \approx \frac{\lambda_a}{2n_0 T \sin\theta_0} \tag{9.86}$$

or

$$\delta_0 \approx d/T \tag{9.87}$$

using Eq. (9.10). Inserting Eq. (9.87) into Eq. (9.84) we obtain a rule-of-thumb for wavelength sensitivity

$$\frac{\Delta\lambda_0}{\lambda_a} \approx \frac{d \cot\theta_0}{T} \tag{9.88}$$

where the deviation $\Delta\lambda_0$ leads to zero efficiency.

9.6.2 Absorption Transmission Holograms

When a transmission hologram such as that illustrated in Fig. 9.5 is characterized by $\varepsilon_1 = 0$ with α and α_1 finite, its diffraction is caused purely by the variation in absorption constant. In the analysis to follow, we can continue to use the parameter ξ as defined in Eq. (9.75) but it is convenient to define a new parameter ν_a. With the coupling constant $\varkappa$ [see Eq. (9.41)] now given by

$$\varkappa = -i\alpha_1/2, \tag{9.89}$$

we define ν_a as

$$\nu_a = \frac{\alpha_1 T}{2\cos\theta_0}. \tag{9.90}$$

We again, as in Section 9.6.1, wish to evaluate Eq. (9.73). As before we must first express $\gamma_1 - \gamma_2$ and $\gamma_{1,2}T$ in terms of ξ and ν_a. From Eq. (9.67) we now have

$$\gamma_1 - \gamma_2 = \left[-\left(\frac{\Gamma}{c_S}\right)^2 + \frac{\alpha_1^2}{c_S c_R}\right]^{1/2} = \left[\left(\frac{\alpha_1}{\cos\theta_0}\right)^2 - \left(\frac{\Gamma}{\cos\theta_0}\right)^2\right]^{1/2}$$
$$= \frac{2}{T}(\nu_a^2 - \xi^2)^{1/2} \tag{9.91}$$

and

$$\gamma_{1,2}T = -\frac{\alpha T}{\cos\theta_0} - i\xi \pm (\nu_a{}^2 - \xi^2)^{1/2}. \tag{9.92}$$

Inserting these values into Eq. (9.73) we have for $\mathbf{S}(T)$, the amplitude of the wave diffracted from a pure-absorption transmission hologram,

$$\mathbf{S}(T) = -\exp\left(-\frac{\alpha T}{\cos\theta_0}\right)\exp(-i\xi)\frac{\sinh(\nu_a{}^2 - \xi^2)^{1/2}}{(1 - \xi^2/\nu_a{}^2)^{1/2}}. \tag{9.93}$$

At Bragg incidence ($\xi = 0$) the diffracted wave amplitude becomes

$$S(T) = -\exp\left(-\frac{\alpha T}{\cos\theta_0}\right)\sinh\left(\frac{\alpha_1 T}{2\cos\theta_0}\right). \tag{9.94}$$

Its absolute value is plotted in Fig. 9.7 as a function of $\alpha_1 T/\cos\theta_0$ for various ratios α/α_1. For an incident wave of unit amplitude, efficiency is obtained by squaring $|S(T)|$. As α_1 is increased $|S(T)|$ increases; however we exclude the possibility of gain (negative absorption) and so with increasing

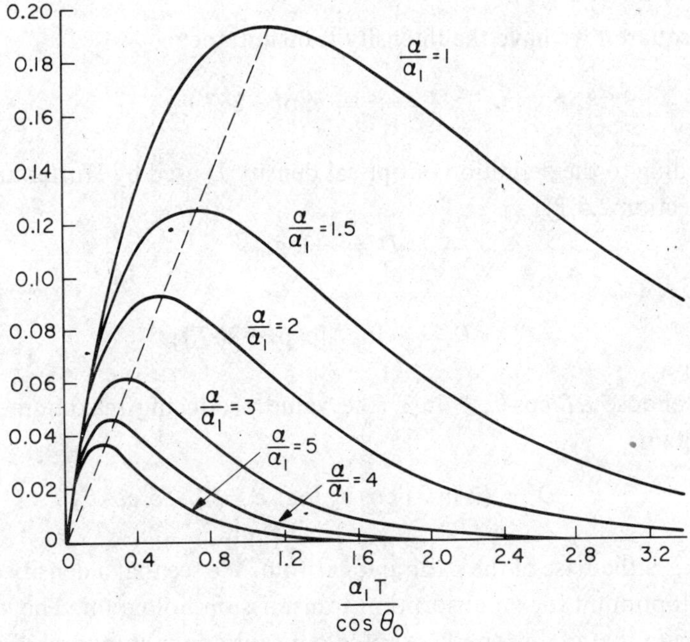

FIG. 9.7. Absolute value of the diffracted amplitude versus $\alpha_1 T/\cos\theta_0$ for the absorptive transmission hologram with various values of the parameter α/α_1. (After Kogelnik [9.5]. Copyright 1969 by American Telephone and Telegraph Company. Reprinted by permission.)

α_1, α must increase to maintain $\alpha \geq \alpha_1$. As may be seen from Eq. (9.94), α acts to attenuate $|S(T)|$. If we let α_1 assume its maximum allowable value, $\alpha_1 = \alpha$, insert this into Eq. (9.94), and maximize with respect to $\alpha_1 T/\cos\theta_0$, we obtain the maximum diffracted wave amplitude $|S(T)| = (3\sqrt{3})^{-1}$ corresponding to $\alpha_1 T/\cos\theta_0 = \ln 3$. Squaring $S(T)$ shows that the maximum efficiency $\eta_{max} = 1/27 = 3.7\%$ is little more than half that for a thin absorption hologram (see Section 8.5).

It is of interest to compute the optical density of an absorption hologram plate which diffracts with maximum efficiency. Suppose that its average absorption constant is α and its thickness is T. The measured density of the plate is essentially that of a homogeneous uniformly exposed photographic plate of thickness T having the same absorption constant α. We can define the amplitude transmittance t to be the ratio of the amplitude of the light transmitted normally through the plate to that incident on the plate. For a homogeneous medium the (real) wave amplitude has the form $A\exp(-\alpha z)$, as in Section 9.4, so that

$$t = \exp(-\alpha T).$$

If we square t we have the intensity transmittance

$$\mathcal{T} = t^2 = \exp(-2\alpha T).$$

According to the definition of optical density D used by Hurter and Driffield (see Section 2.5.1)

$$D = -\log_{10}\mathcal{T}$$

or

$$D = -\log_{10}[\exp(-2\alpha T)]. \tag{9.95}$$

If we choose $\alpha T/\cos\theta_0 = \ln 3$ (the value producing maximum efficiency), we obtain

$$D = (2\ln 3)\cos\theta_0 \log_{10} e = 0.955\cos\theta_0$$

where e is the base of the natural logarithm. We see that a density of less than one is optimum for an absorption transmission hologram. The value is low relative to that of a properly exposed conventional photograph, but in agreement with the experimental observations that good absorption holograms have the density of underexposed photographs. In Fig. 9.7 we show the curve connecting all the points corresponding to a density $D = 0.955\cos\theta_0$ and it is seen that this density is optimum for all ratios α/α_1.

Figure 9.8 is a plot of the *relative* efficiency η/η_0 versus ξ, obtained from Eq. (9.93) for the absorption transmission hologram. Each curve is drawn for $\alpha = \alpha_1$. The *shapes* of the curves are similar to those in Fig. 9.6 for $\nu \leq \pi/2$ exhibiting very little dependence on the parameter $\alpha_1 T/(2 \cos\theta_0)$.

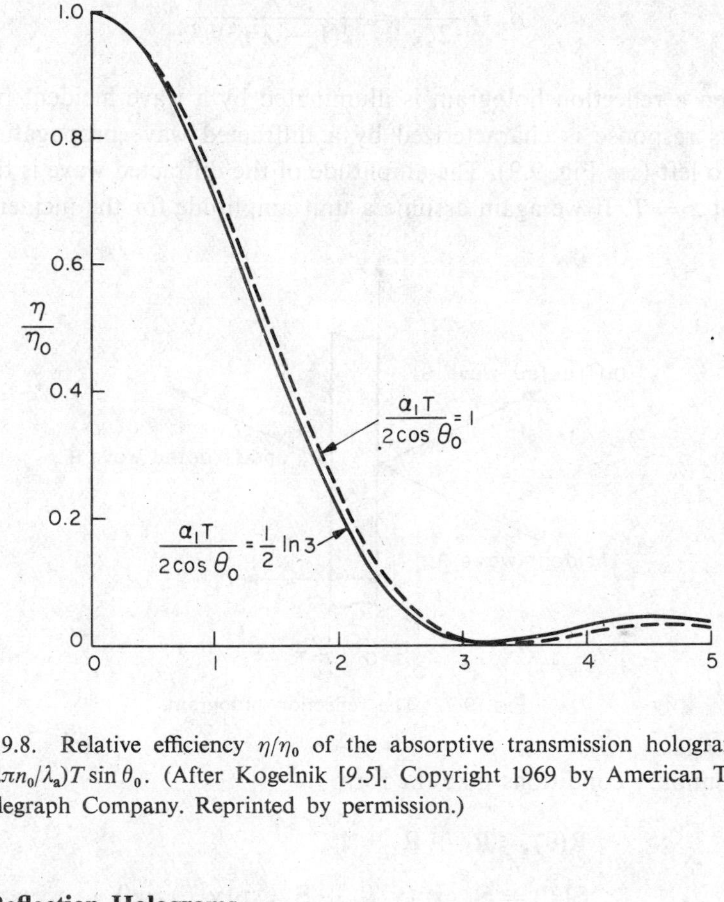

FIG. 9.8. Relative efficiency η/η_0 of the absorptive transmission hologram versus $\xi = \delta(2\pi n_0/\lambda_a)T \sin\theta_0$. (After Kogelnik [9.5]. Copyright 1969 by American Telephone and Telegraph Company. Reprinted by permission.)

9.7 Reflection Holograms

Reflection holograms with fringe planes parallel to the surface of the hologram are formed by two plane waves oppositely directed with respect to the z axis. We see from Eq. (9.5), viz.,

$$2\pi(\eta_S - \eta_R)y + 2\pi(\zeta_S - \zeta_R)z = \text{constant},$$

that when $\eta_S = \eta_R$ and $\zeta_S = -\zeta_R$, the fringe planes are given by the equation

$$2\pi(2\zeta_S)z = \text{constant}$$

or

$$z = \text{constant}.$$

They are planes parallel to the xy plane. Their spacing is

$$d_z = \frac{1}{2\zeta_S} = \frac{\lambda}{2(1 - \lambda^2\eta^2)^{1/2}}.$$

When a reflection hologram is illuminated by a wave incident from the left, its response is characterized by a diffracted wave propagating from right to left (see Fig. 9.9). The amplitude of the diffracted wave is therefore zero at $z = T$. If we again assume a unit amplitude for the incident wave,

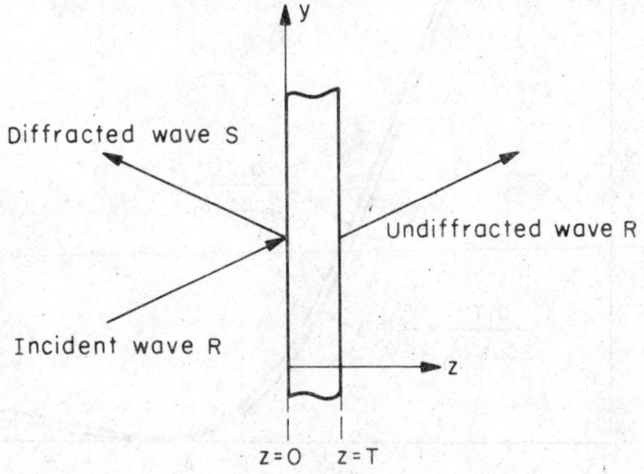

FIG. 9.9. The reflection hologram.

the boundary conditions take the form

$$\mathbf{R}(0) = \mathbf{R}_1 + \mathbf{R}_2 = 1 \tag{9.96}$$

$$\mathbf{S}(T) = \mathbf{S}_1 \exp(\gamma_1 T) + \mathbf{S}_2 \exp(\gamma_2 T) = 0. \tag{9.97}$$

We shall employ these boundary conditions in the coupled wave equations and so solve for the diffracted wave amplitude $\mathbf{S}(0)$ emerging from the hologram at $z = 0$. We can do this by substituting Eqs. (9.96) and (9.97) into Eq. (9.63), obtaining for $z = 0$,

$$-i\varkappa(\mathbf{R}_1 + \mathbf{R}_2) = -i\varkappa = c_S(\gamma_1\mathbf{S}_1 + \gamma_2\mathbf{S}_2) + (\alpha + i\Gamma)(\mathbf{S}_1 + \mathbf{S}_2). \tag{9.98}$$

Since $\mathbf{S}(0) = \mathbf{S}_1 + \mathbf{S}_2$, we must express the first term on the right of Eq. (9.98) as a function of $\mathbf{S}_1 + \mathbf{S}_2$, whereupon Eq. (9.98) provides the desired solution. Some manipulation of the boundary condition Eq. (9.97) allows this. First write Eq. (9.97) as

$$-\mathbf{S}_1 \exp(\gamma_1 T) = \mathbf{S}_2 \exp(\gamma_2 T). \tag{9.99}$$

Now in one case add $\mathbf{S}_1 \exp(\gamma_2 T)$ to both sides of Eq. (9.99) to give

$$\mathbf{S}_1[\exp(\gamma_2 T) - \exp(\gamma_1 T)] = (\mathbf{S}_1 + \mathbf{S}_2)\exp(\gamma_2 T), \tag{9.100}$$

and in a second case multiply both sides of Eq. (9.99) by -1 and add this time $\mathbf{S}_2 \exp(\gamma_1 T)$ to both sides with the result

$$\mathbf{S}_2[-\exp(\gamma_2 T) + \exp(\gamma_1 T)] = (\mathbf{S}_1 + \mathbf{S}_2)\exp(\gamma_1 T). \tag{9.101}$$

Hence the first term on the right of Eq. (9.98) becomes

$$c_S(\gamma_1 \mathbf{S}_1 + \gamma_2 \mathbf{S}_2) = c_S \frac{\gamma_1(\mathbf{S}_1 + \mathbf{S}_2)\exp(\gamma_2 T) - \gamma_2(\mathbf{S}_1 + \mathbf{S}_2)\exp(\gamma_1 T)}{\exp(\gamma_2 T) - \exp(\gamma_1 T)} \tag{9.102}$$

and

$$\mathbf{S}(0) = \mathbf{S}_1 + \mathbf{S}_2 = -i\varkappa\left(\alpha + i\Gamma + c_S\left[\frac{\gamma_1 \exp(\gamma_2 T) - \gamma_2 \exp(\gamma_1 T)}{\exp(\gamma_2 T) - \exp(\gamma_1 T)}\right]\right)^{-1}. \tag{9.103}$$

We shall now evaluate Eq. (9.103) under conditions where the grating vector $\vec{K}$ is perpendicular to the surface of the grating as illustrated in Fig. 9.10. Vector $\vec{\varrho}$ forms the angle θ with the grating planes, and the vector diagram at the right of the figure expresses Eq. (9.45), $\vec{\sigma} = \vec{\varrho} - \vec{K}$, for Bragg angle incidence. When θ equals the Bragg angle θ_0, the triangle is isosceles and we have

$$c_R = \varrho_z/\beta = -c_S = -\sigma_z/\beta = \cos\psi_0 \tag{9.104}$$

where ψ_0, the angle between $\vec{\varrho}$ and the z axis for Bragg incidence, is given in Fig. 9.10. In the following we assume that the angle of incidence θ is close enough to θ_0 so that Eq. (9.104) holds to a good approximation.

9.7.1 Phase Reflection Holograms

A phase reflection hologram is characterized by $\alpha = \alpha_1 = 0$. Here we find it convenient to employ the parameters

$$\xi_r = \frac{\Gamma T}{2\cos\psi_0} = \frac{\beta T\delta \sin 2\theta_0}{2\cos\psi_0} = \delta\beta T\cos\theta_0 \tag{9.105a}$$

$$\nu_r = \frac{\varkappa T}{\cos\psi_0} = \frac{\pi n_1 T}{\lambda_a \cos\psi_0} \tag{9.105b}$$

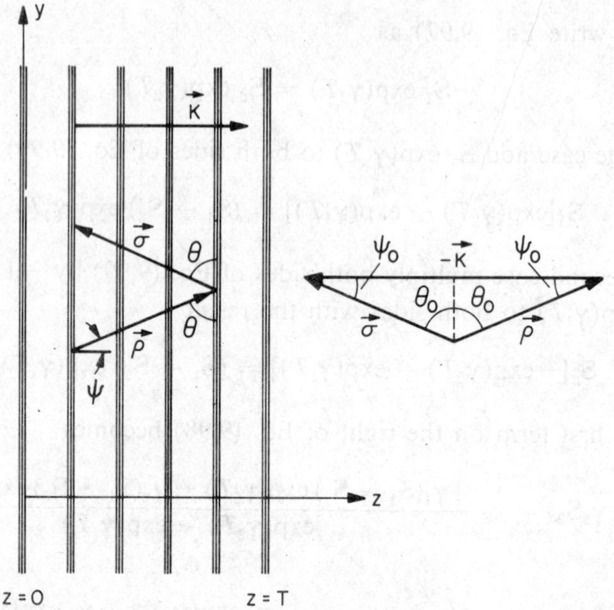

FIG. 9.10. Geometry for the reflection hologram with no slant.

In Eq. (9.105a) we have used the definition of Γ, Eq. (9.59), and the fact $\psi_0 = (\pi/2) - \theta_0$. To express $\mathbf{S}(0)$ in terms of ξ_r and ν_r we must substitute Eq. (9.104), (9.105a), and (9.105b) into the expression for $\gamma_{1,2}$ [Eq. (9.67)]. We find that

$$\gamma_{1,2}T = i\xi_r \pm (\nu_r^2 - \xi_r^2)^{1/2}. \tag{9.106}$$

The straightforward but lengthy procedure of substituting Eqs. (9.105a), (9.105b), and (9.106) into Eq. (9.103) yields the amplitude of the diffracted wave at $z = 0$,

$$\mathbf{S}(0) = \frac{-i}{(i\xi_r/\nu_r) + [1 - (\xi_r/\nu_r)^2]^{1/2} \coth(\nu_r^2 - \xi_r^2)^{1/2}}. \tag{9.107}$$

For Bragg incidence $\xi = 0$, and 100% efficiency is possible, as it was for the lossless dielectric transmission hologram. Unlike the transmission hologram where η_0 can be 100% for a particular product of thickness and coupling constant, the maximum possible efficiency is approached asymptotically with increasing ν_r. Figure 9.11 illustrates the difference with experimental curves of efficiency versus exposure obtained from dichromated gelatin holograms [9.10]. Curves A and B, for transmission gratings, exhibit a peak efficiency as a function of exposure (and thus of ν) while

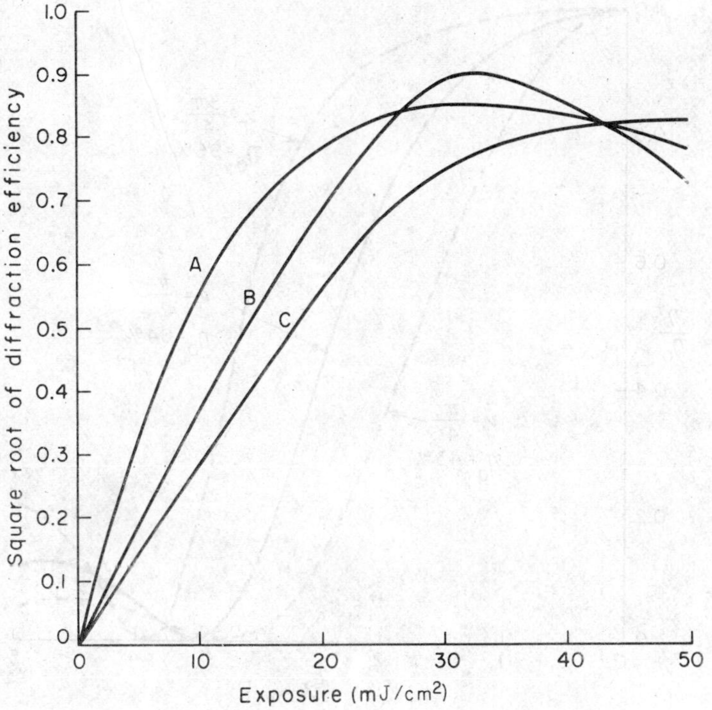

FIG. 9.11. Experimental exposure curves for gelatin dichromate gratings. A and B are for transmission gratings, C is for a reflection grating (from Lin [9.10].)

curve C, for the reflection grating, continues to increase with increasing exposure.

Relative efficiency η/η_0 is plotted in Fig. 9.12 as a function of ξ_r for three values of the parameter ν_r. For $\nu_r = \pi/4$, $\eta_0 = 43\%$; for $\nu_r = \pi/2$, $\eta_0 = 84\%$; and for $\nu_r = 3\pi/4$, $\eta_0 = 96\%$. Note that considerable broadening of the angular response of the reflection hologram accompanies increasing values of ν_r. Suppose we consider the $\nu_r = \pi/2$ curve and compute the wavelength sensitivity of the reflection hologram. From Eqs. (9.84) (here we take δ to be negative) and (9.105a) we have

$$\xi_r = -\frac{\Delta\lambda}{\lambda_a}\tan\theta_0\,\beta T\cos\theta_0 = -\frac{\Delta\lambda}{\lambda_a}\beta T\sin\theta_0$$

$$= -\frac{\Delta\lambda}{\lambda_a}\left(\frac{2\pi n_0}{\lambda_a}\right)T\sin\theta_0. \tag{9.108}$$

According to the curve corresponding to $\nu_r = \pi/2$ in Fig. 9.12, $\eta/\eta_0 = 0$ when $\xi_{r0} = 3.5$. If we assume $\lambda = 0.488\ \mu$m, $T = 15\ \mu$m, $n_0 = 1.52$, and

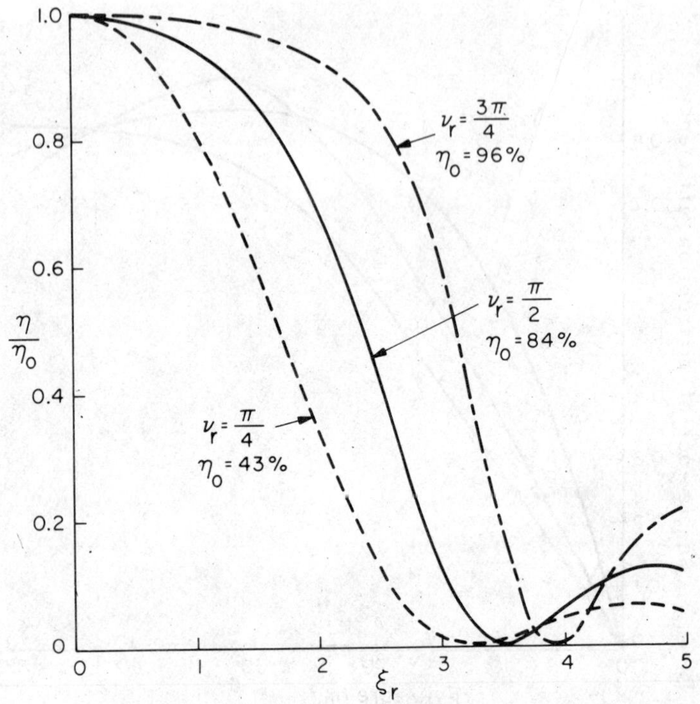

FIG. 9.12. Relative efficiency η/η_0 of the lossless dielectric reflection hologram versus $\xi_r = \delta(2\pi n_0/\lambda_a)T\cos\theta_0$ for various values of the parameter $\nu_r = \pi n_1 T/\lambda_a \cos\psi_0$. (After Kogelnik [9.5]. Copyright 1969 by American Telephone and Telegraph Company. Reprinted by permission.)

$\theta_0 = 80°$ within the gelatin, then solving for $\Delta\lambda$ in Eq. (9.108) yields $|\Delta\lambda| = 0.0059\ \mu m = 59$ Å. This high degree of wavelength selectivity permits illumination of the reflection hologram with a white-light source (see Chapter 17).

9.7.2 ABSORPTION REFLECTION HOLOGRAMS

Here again, as in Section 9.6.2, $\varepsilon_1 = 0$ while α and α_1 remain finite, and the coupling constant $\varkappa = -i\alpha_1/2$. We define the parameter ν_{ra} as

$$\nu_{ra} = \frac{\alpha_1 T}{2\cos\psi_0} \tag{9.109a}$$

and the parameter ξ_{ra} as

$$\xi_{ra} = \frac{\alpha T}{\cos\psi_0} + \frac{i\Gamma T}{2\cos\psi_0}. \tag{9.109b}$$

We then express $\gamma_{1,2}T$ in Eq. (9.67) in terms of $\nu_{\rm ra}$ and $\xi_{\rm ra}$ and obtain

$$\gamma_{1,2}T = \frac{i\Gamma T}{2\cos\psi_0} \pm (\xi_{\rm ra}^2 - \nu_{\rm ra}^2)^{1/2} \tag{9.110}$$

where we have used the condition characterizing the reflection hologram

$$c_{\rm R} = -c_{\rm S} = \cos\psi_0. \qquad [9.104]$$

If we insert $\gamma_{1,2}T$ into $\mathbf{S}(0)$ in Eq. (9.103), we then have for the diffracted wave amplitude emerging from the hologram

$$\mathbf{S}(0) = -\left\{\frac{\xi_{\rm ra}}{\nu_{\rm ra}} + \left[\left(\frac{\xi_{\rm ra}}{\nu_{\rm ra}}\right)^2 - 1\right]^{1/2} \coth(\xi_{\rm ra}^2 - \nu_{\rm ra}^2)^{1/2}\right\}^{-1}. \tag{9.111}$$

At Bragg incidence, $\Gamma = 0$,

$$\xi_{\rm ra}/\nu_{\rm ra} = 2\alpha/\alpha_1,$$

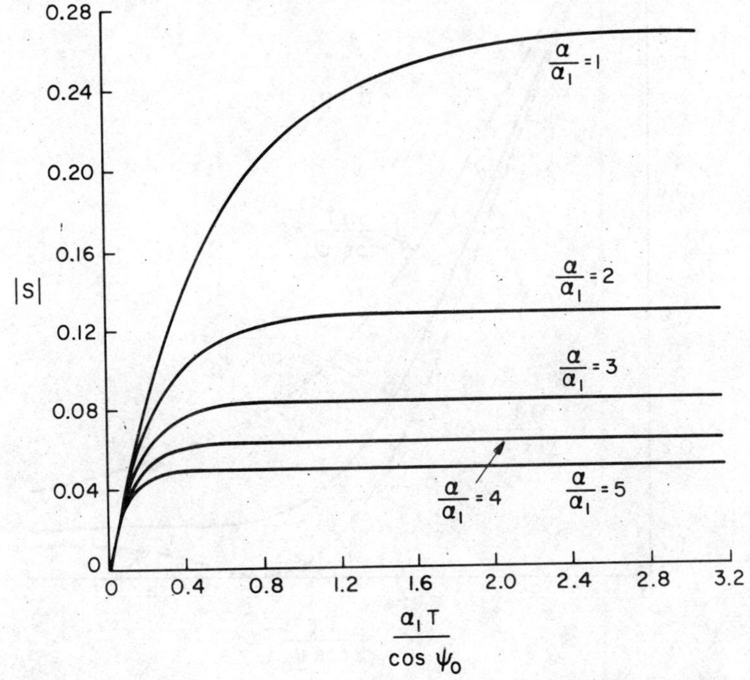

FIG. 9.13. Absolute value of the diffracted amplitude versus $\alpha_1 T/\cos\psi_0$ for the absorptive reflection hologram with various values of the parameter α/α_1. (After Kogelnik [9.5]. Copyright 1969 by American Telephone and Telegraph Company. Reprinted by permission.)

and S(0) becomes

$$S(0) = -\left\{\frac{2\alpha}{\alpha_1} + \left(\frac{4\alpha^2}{\alpha_1{}^2} - 1\right)^{1/2} \coth\left[\frac{T}{\cos\psi_0}\left(\alpha^2 - \frac{\alpha_1{}^2}{4}\right)^{1/2}\right]\right\}^{-1} \tag{9.112}$$

When α_1 takes on its maximum value $\alpha_1 = \alpha$,

$$|S(0)| = \left[2 + \sqrt{3}\coth\left(\frac{T}{\cos\psi_0}\frac{\sqrt{3}\alpha}{2}\right)\right]^{-1}$$

is optimized. As $\alpha_1 = \alpha \to \infty$, the maximum efficiency $\eta_{\max} = |S(0)|^2 \to (2 + \sqrt{3})^{-2} = 7.2\%$. This reflects the experimentally observed fact that the highest efficiencies for absorptive reflection holograms are achieved if the plate is quite dark. In Fig. 9.13 we plot $|S(0)|$ as a function of $\alpha_1 T/\cos\psi_0$ for several values of α/α_1. We see that the diffracted wave ampli-

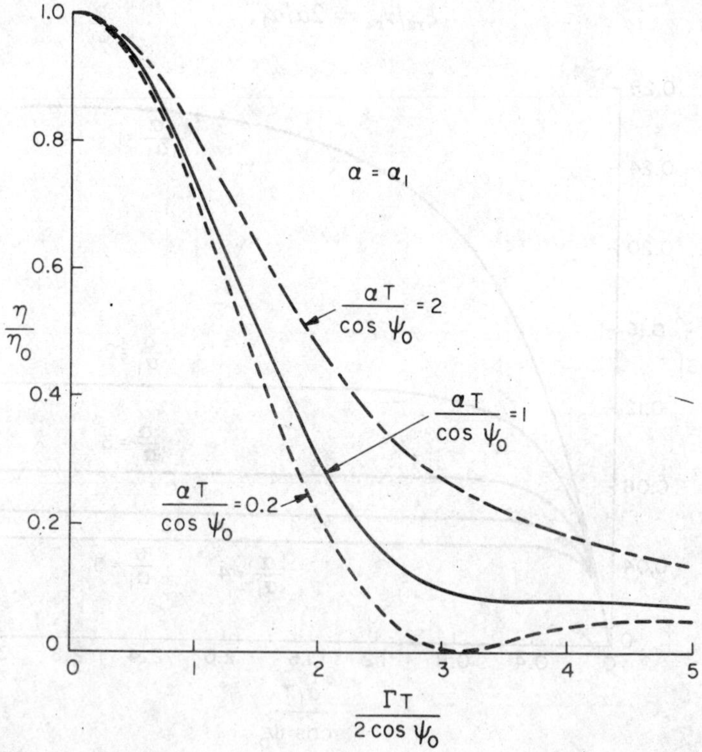

FIG. 9.14. Relative efficiency η/η_0 of the absorptive reflection hologram versus $\Gamma T/(2\cos\psi_0)$ for various values of $\alpha T/\cos\psi_0$ and for $\alpha_1 = \alpha$. (After Kogelnik [9.5]. Copyright 1969 by American Telephone and Telegraph Company. Reprinted by permission.)

tude is reasonably close to its asymptotic maximum when $\alpha_1 T/\cos\psi_0 = 2$. According to Eq. (9.95) the optical density of the plate would then be $D \simeq 1.7$. The value represents the minimum density to which the plate should be exposed if nearly maximum diffraction efficiency is desired.

In Fig. 9.14 we plot relative efficiency as a function of $\Gamma T/(2\cos\psi_0)$ for $\alpha = \alpha_1$ and various values of $\alpha T/\cos\psi_0$.

9.8 Discussion of Volume Hologram Properties

Table 9.1 compares theoretical and observed maximum efficiencies for plane and volume holograms under the conditions specified in this chapter. With the exception of the absorption reflection hologram the maximum predicted efficiencies are close to those observed. Since some of these predictions are made with the theory presented in this chapter and some with the results of the plane analysis in Section 8.5, we need some criterion for distinguishing when a hologram is thick enough for volume theory to apply. Klein [*9.12*] gave a lower limit to the thickness in terms of a parameter Q defined as

$$Q = 2\pi\lambda_a T/nd^2. \tag{9.113}$$

The coupled wave theory begins to give good results when $Q \geq 10$ [*9.5*]. Suppose we choose some typical values $T = 15\ \mu\text{m}$, $\lambda_a = 0.633\ \mu\text{m}$, $n = 1.52$ (for gelatin), and $Q = 10$ and solve for the fringe spacing d. We find

TABLE 9.1

MAXIMUM EFFICIENCY FOR VARIOUS HOLOGRAM TYPES

Hologram medium:	Thin		Thick		Thick	
Mode of diffraction:	Transmission		Transmission		Reflection	
Property modulated:	Amplitude transmittance	Phase shift	Absorption constant	Refractive index	Absorption constant	Refractive index
Maximum theoretical efficiency (%):	6.25	33.9	3.7	100	7.2	100
Maximum efficiency obtained experimentally:	6.0	32.6	3.0	90	3.8	80
Reference:	[*11.29*]	[*9.8*]	[*9.9*]	[*9.10*]	[*9.11*]	[*9.10*]

$d = 1.98\ \mu m$. Substituting this value into Eq. (1.10): $2d \sin\theta = \lambda$, we obtain the minimum interbeam angle 2θ required to form a volume hologram in 15-μm-thick gelatin. If $d = 1.98\ \mu m$ is the fringe spacing in the medium, then the interbeam angle in the hologram medium is given by setting $\lambda = \lambda_a/1.52 = 0.416\ \mu m$. For this case $2\theta \approx 12°$. As noted in Section 9.1, when λ is set equal to λ_a, then Eq. (1.10) gives the interbeam angle in air. Thus, the minimum interbeam angle in air required to form a volume hologram in 15-μm-thick photographic emulsion with He–Ne radiation is $2\theta = 18.4°$.

Holograms formed with this minimum interbeam angle show a moderate degree of angular sensitivity as exhibited (for phase holograms) by

$$\delta_0 \approx \frac{\lambda_a}{2n_0T \sin\theta_0} \quad [9.86]$$

$$\approx d/T \quad [9.87]$$

$$\approx 0.13 \text{ rad} \approx 7.5°$$

in the medium. However they exhibit very little wavelength sensitivity. For the parameters we have used above and $\theta_0 = 6°$,

$$\Delta\lambda_0 \approx \frac{d \cot\theta_0}{T} \lambda_a \approx 8000 \text{ Å}. \quad [9.88]$$

They, in fact, diffract with greater than 50% efficiency over the whole visible spectrum [*9.13*].

As θ_0 increases, the hologram response becomes more sensitive to angle of incidence. For example, when $\theta_0 = 30°$ while T, n_0, and λ_a remain as before, $\delta \approx 1.6°$ in the medium. With this increased angular sensitivity many holograms can be superimposed on the same plate in a way which allows the viewer to see only one image at a time. Each hologram may be formed with a subject beam having the same mean spatial frequency but with a reference beam whose direction is at least δ_0 away from the directions of any of the other reference beams. When the processed hologram plate is illuminated by any one of the original reference beams (satisfying the Bragg angle for only one of the superimposed holograms), then only the subject wave associated with that reference beam will reconstruct (see Chapter 16). Another method permits many subject beams to be recorded on the same hologram plate and reconstructed unambiguously even though each interferes with the same reference beam. In this case the plate is rotated by δ_0 after each exposure.

The increased wavelength sensitivity accompanying larger Bragg angles θ_0 in Eq. (9.88) ($\Delta\lambda_0 \approx 300$ Å when $\theta_0 = 30°$ in the medium) permits formation of multicolor holograms [*9.14*] which reconstruct multicolor waves without color cross-modulation (see Chapter 17). As the interbeam angle is increased further to form reflection holograms, the wavelength sensitivity is given by Eq. (9.108). One finds the reflection configuration a convenient way of achieving a sensitivity high enough to permit acceptable reconstruction with a white-light illuminating source [*9.15*]. Reflection holograms can generate multicolor images when illuminated with white light [*9.16*]. However, their angular selectivity decreases with increasing θ_0 because of the $\cos\theta_0$ factor in Eq. (9.105a).

Kogelnik [*9.5*], whose work has been closely followed here, has also considered the implications of an incident beam polarized in the plane of incidence. He finds that all previous results hold if only one modification is made, that being to multiply the coupling constant $\boldsymbol{\varkappa}$ by the cosine of the angle (in the medium) between incident and diffracted beam. When that angle is 90°, the diffraction amplitude goes to zero.

REFERENCES

9.1. E. N. Leith, A. Kozma, J. Upatnieks, J. Marks, and N. Massey, "Holographic Data Storage in Three-Dimensional Media," *Appl. Opt.* **5**, 1303 (1966).

9.2. C. B. Burckhardt, "Diffraction of a Plane Wave at a Sinuscidally Stratified Dielectric Grating," *J. Opt. Soc. Amer.* **56**, 1502 (1966).

9.3. C. B. Burckhardt, "Efficiency of a Dielectric Grating," *J. Opt. Soc. Amer.* **57**, 601 (1967).

9.4. H. Kogelnik, "Reconstructing Response and Efficiency of Hologram Gratings," in *Proc. Symp. Modern Opt.* (J. Fox, ed.), pp. 605–617. Polytechnic Press, Brooklyn, New York, 1967.

9.5. H. Kogelnik, "Coupled Wave Theory for Thick Hologram Gratings, *Bell. Syst. Tech. J.* **48**, 2909 (1969).

9.6. D. Gabor and G. W. Stroke, "The Theory of Deep Holograms," *Proc. Roy. Soc.* **A304**, 275 (1968).

9.7. S. Ramo and J. R. Whinnery, *Fields and Waves in Modern Radio*, 2nd ed. Wiley, New York, 1960.

9.8. T. A. Shankoff, "Phase Holograms in Dichromated Gelatin," *Appl. Opt.* **7**, 2101 (1968).

9.9. N. George and J. W. Matthews, "Holographic Diffraction Gratings," *Appl. Phys. Lett.* **9**, 212 (1966).

9.10. L. H. Lin, "Hologram Formation in Hardened Dichromated Gelatin Films," *Appl. Opt.* **8**, 963 (1969).

9.11. L. H. Lin and C. V. LoBianco, "Experimental Techniques in Making Multicolor White Light Reconstructed Holograms," *Appl. Opt.* **6**, 1255 (1967).

9.12. W. R. Klein, "Theoretical Efficiency of Bragg Devices," *Proc. IEEE* **54**, 803 (1966).
9.13. T. A. Shankoff and R. K. Curran, "Efficient, High Resolution, Phase Diffraction Gratings," *Appl. Phys. Lett.* **13**, 239 (1968).
9.14. K. S. Pennington and L. H. Lin, "Multicolor Wavefront Reconstruction," *Appl. Phys. Lett.* **7**, 56 (1965).
9.15. G. W. Stroke and A. E. Labeyrie, "White Light Reconstruction of Holographic Images Using the Lippman–Bragg Diffraction Effect," *Phys. Lett.* **20**, 368 (1966).
9.16. L. H. Lin, K. S. Pennington, G. W. Stroke, and A. E. Labeyrie, "Multicolor Holographic Image Reconstruction with White-Light Illumination," *Bell Syst. Tech. J.* **45**, 659 (1966).

Chapter 10

HOLOGRAM RECORDING MATERIALS

Practical off-axis holography requires high-resolution recording materials. In Section 8.4.1 we show that recording a hologram in a material with too low a cutoff spatial frequency can result in loss of image resolution and limitation of image field. The relatively high photosensitivity of silver-halide emulsion is attractive, but few commercial emulsions possess the required resolution. With the development of high-power coherent light sources it has become feasible to record holograms in materials other than photographic emulsions. Many of these materials have high resolution, but all are relatively insensitive. However, when they are exposed with light from a high-power laser, exposure times can be reduced to the range required for good hologram formation.

The very nature of the holographic process invites the use of materials other than photographic emulsions. An illuminated hologram can reconstruct the wavefront it has recorded by spatially modulating either the amplitude or the phase of the illuminating light. Photographic films which modulate the amplitude are, of course, suitable for forming photographs as well as holograms. On the other hand, phase modulating materials, while of little use in photography, are ideal for holography.

In this chapter we shall discuss some of the general requirements of hologram recording materials. The concept of an ideal exposure characteristic leading to perfect hologram recording and wavefront reconstruction is first advanced. When the characteristic of a representative real material is compared with the ideal, the limitations of real recording material become apparent. Nonetheless, properties suitable for most holography can be selected from among these materials: silver-halide emulsions, dichromated gelatin films, photoconductor–thermoplastic films, photochromic materials, and ferroelectric crystals.

We assume here that a continuous-wave laser is used for hologram formation and reconstruction. Aspects of silver-halide emulsions, magnetic film, and other material pertinent to recording holograms with a pulsed laser are discussed in Chapter 11.

10.1 Optical Changes in Photosensitive Materials

Exposure and (if necessary) development of a photosensitive material must alter the optical transmission properties of the material if it is to be useful for forming holograms. Let us briefly consider wave propagation in a homogeneous lossy medium. As in Section 9.4 a plane wave, having propagated a distance T through a medium, emerges with a complex amplitude

$$\begin{aligned} \mathbf{f} &= A\exp(-\gamma T) \\ &= A\exp(-i\beta T)\exp(-\alpha T) \end{aligned}$$

where $\beta = 2\pi n_0/\lambda_a$ is the propagation constant, α is the absorption constant, n_0 is the index of refraction, and $\alpha \ll \beta$. The exponential factors can thus be written

$$\mathbf{f}/A = \exp(-\alpha T)\exp[-i(2\pi n_0 T/\lambda_a)] \tag{10.1}$$

If a material is to be considered for holographic recording, it must respond to exposure and development with a change in at least one of the parameters α, n_0, or T. It is often the case that only one parameter changes significantly. We can therefore categorize most useful photosensitive materials as (1) amplitude modulating or *absorption* materials if α is exposure-dependent and (2) phase modulating or *phase* materials if either n_0 or T is exposure dependent.

10.1.1 Absorption Materials

Holographic interference fringes are recorded in an absorption material as a spatial variation of its light absorption. Silver halide photographic emulsion and photochromic glasses and plastics are examples of this type of material. In the case of a homogeneous absorption material whose absorption constant is α and whose thickness in the direction of an incident plane wave is T, we can define the following quantities (see Section 9.6.2):

Amplitude transmittance: $t = \exp(-\alpha T)$

Intensity transmittance: $\mathcal{T} = t^2 = \exp(-2\alpha T)$

Optical density: $D = -\log_{10}\mathcal{T} = 2\alpha T \log_{10}(e) = 0.869\ \alpha T.$

To obtain maximum efficiency from an elementary *plane transmission hologram* with sinusoidal fringes, t should vary sinusoidally between 0 and 1 (see Section 8.5). In practice this is difficult to achieve since the corresponding range in optical density would be $\infty > D > 0$. A maximum density of $D = 2$ yielding a transmittance $t = 0.1$ is sufficient and easily achieved with photographic emulsion (but not with some photochromics). The theoretically optimum *average* density of a plane hologram should be $D = 0.6$, corresponding to $t_{E_0} = \frac{1}{2}$ in Eq. (8.30), Section 8.5. This value is close to the optimum obtained experimentally with photographic emulsion.

According to the analysis of an elementary *volume transmission hologram* in Section 9.6.2, maximum efficiency corresponds to the case where the amplitude α_1 of the sinusoidal variation in the absorption coefficient equals the average absorption constant α. Then for $\alpha_1 = \alpha$, the maximum efficiency is obtained when $\alpha T/\cos\theta_0 = \ln 3$, where θ_0 is the Bragg angle. If we take $\theta_0 = 30°$ as a typical value and substitute $\alpha T = \ln 3 \cos 30°$ into our definition of optical density, we obtain for the average optimum density $D = 0.869 \ln 3 \cos 30° = 0.955 \cos 30° = 0.827$ and for the maximum density range $0 < D < 2\alpha = 1.65$.

The experimental value of the optimum average density for thick photographic emulsions is usually lower than 0.827 and not much higher than 0.6.

For *volume reflection holograms*, the theory of Section 9.7.2 predicts that efficiency approaches a maximum value asymptotically with αT. For $\alpha T = 2$, the average optical density is $D \approx 1.7$ and the efficiency is reasonably close to the maximum. When reflection holograms are recorded in thick photographic emulsions, the experimental optimum average density is about 2.

10.1.2 Phase Materials

Holographic fringes are recorded in a phase material as a spatial modulation of either refractive index n or the thickness T. Phase material is usually nearly transparent so that the absorption constant $\alpha \approx 0$. The difference in the phase of light transmitted through a dielectric slab of thickness T and index n_0 compared to that transmitted through an equal thickness of air is given by Eq. (10.1) as

$$\varphi = 2\pi(n_0 - 1)T/\lambda_a . \tag{10.2}$$

Since λ_a is usually very small compared to $n_0 T$, a small variation of n_0 or T produced by the exposing light can lead to a large phase change $\Delta\varphi$. However, we note that maximum diffraction efficiency for either plane phase holograms (see Section 8.5) or volume (thick) phase holograms [see Section 9.6.1, Eq. (9.80)] requires a phase change $\Delta\varphi$ of approximately π radians. We therefore set

$$\Delta\varphi = \Delta\left[\frac{2\pi(n_0 - 1)T}{\lambda_a}\right] = \pi$$

or

$$\Delta\left[\frac{2(n_0 - 1)T}{\lambda_a}\right] = \frac{2(n_0 - 1)T}{\lambda_a}\left(\frac{\Delta n_0}{n_0 - 1} + \frac{\Delta T}{T}\right) = 1. \tag{10.3}$$

For thin or *plane holograms* T approaches λ_a. Since attainable values of index change Δn_0 are generally much less than $n_0 - 1$, we can in this case neglect the term containing Δn_0 in Eq. (10.3). We are left with the condition

$$\Delta T \approx \lambda_a / 2(n_0 - 1) \tag{10.4}$$

for maximum efficiency. Taking typical values $\lambda_a = 0.5\ \mu\text{m}$ and $n_0 = 1.5$ we obtain $\Delta T = \lambda_a = 0.5\ \mu\text{m}$. Diffraction efficiency can be as high as 33.9% for sinusoidal gratings and 40.4% for square wave gratings (see Section 8.5). Examples of thin phase materials include photoconductor–thermoplastic films, etched photoresist films, photopolymers, and photographic emulsions bleached to have only a surface relief.

For a *thick phase hologram*, spatial variations in both refractive index and thickness of the material can coexist. The thickness variation can be a major source of noise in the reconstructed wave and should be minimized by immersing the hologram in an index-matching liquid. Elimination of this undesirable effect implies that $\Delta T/T \ll \Delta n_0/(n_0 - 1)$. We see from Eq. (10.3) that for a maximum efficiency volume hologram

$$\Delta n_0 = \lambda_a / 2T . \tag{10.5}$$

Again taking $\lambda = 0.5\ \mu\text{m}$, we obtain $\Delta n_0 = 2.5 \times 10^{-4}$/mm thickness of material. A maximum efficiency of 100% is possible. Although the maximum variation Δn_0 needed may seem small, only a few materials have the desired photosensitivity. Examples of thick phase materials include dichromated gelatin film, ferroelectric crystals and bleached photographic emulsion.

10.2 Exposure and Sensitivity

One hologram recording material is considered more sensitive than another if, for fixed illumination conditions, less exposure is required to produce a hologram of specified diffraction efficiency. A precise definition of sensitivity is given in Section 10.6. Here we shall merely consider some practical limits to the minimum sensitivity or maximum exposure required by hologram recording materials.

If we assume that the exposing light intensity averaged over an area containing many fringes is uniform over the hologram plate, then the average exposure value E_0 is given by

$$E_0 = \eta_t \tau_e P/A \tag{10.6}$$

where η_t is the net power transfer efficiency from laser to recording material, P is the output power of the laser, τ_e is the exposure time, and A the cross-sectional area of the beam illuminating the hologram plate. In typical hologram recording, only a fraction of the total laser output power that is usable actually exposes the hologram plate. Some 30 to 50% of the power is sacrificed to obtain a degree of uniformity of subject and hologram-plate illumination (see Section 7.3). If the subject is a diffuse reflector of light, then perhaps only 10% of the light incident on it will be received by the hologram plate. In addition there is approximately a 4% loss of the light incident on each air-to-glass interface of the beam-forming optics. Thus if we use Eq. (10.6) to calculate a limit to practical exposure, we should take η_t to be no more than 5%. As discussed in Section 7.5, instabilities and airborne disturbances place a limit on practical exposure time. As a typical upper limit we take $\tau_e = 300$ sec. Then if $A = 100\ \text{cm}^2$, we have that $E_0/P = 0.05 \times 300/100 = 0.15\ \text{mJ/cm}^2/\text{mW}$ of laser output power. Materials requiring exposure values beyond the limit E_0/P are, on this basis, impractical for general holography. Since the maximum output power of a 1-m He–Ne laser is about 50 to 100 mW at 6328 Å, the upper limit of exposure possible with that laser is $E_0 \approx 15\ \text{mJ/cm}^2$ (under the conditions assumed in the calculation of E_0/P). Above this value the more powerful argon ion laser must be used. If a recording material is so insensitive that it requires much more than $15\ \text{mJ/cm}^2$ exposure, instability problems due to the intense illumination become significant. For example, expansion of the subject due to heat absorbed from the laser light may cause fringes to appear in the reconstruction (see Chapter 15).

The sensitivities of photographic emulsion and photoconductor–thermoplastic film are high enough to allow formation of holograms with exposures well below the limit of 15 mJ/cm^2, but the sensitivities of most other materials are only marginally satisfactory. Furthermore most other materials are sensitive only in the blue and green spectral regions.

10.3 Recording Resolution

If a hologram is to be recorded by a photosensitive material, the interference fringe spacing must be resolved in the recording. To estimate the requirements for the resolution capability of a hologram recording material, we consider formation of an *elementary transmission* hologram as in Section 9.1 and Fig. 9.1. Two plane waves form an interference fringe pattern characterized by planes of crest intensity. The wave directions make angles with the normal to the hologram plate of Ω_S and Ω_R in air and ψ_S and ψ_R in the recording material.

A *plane hologram* records the *traces* which the interference (fringe) planes make on it. According to Eq. (9.4) the spacing between crest fringe traces is

$$d_y = \frac{\lambda}{\sin \psi_S - \sin \psi_R} = \frac{\lambda_a}{\sin \Omega_S - \sin \Omega_R} \qquad [9.4]$$

and the fringe frequency $\nu_y = 1/d_y$. A *volume hologram* records the interference planes themselves, and we must consider the spacing d between planes to be taken along the normal to the planes. From Eq. (9.9) we have

$$d = d_y \cos \phi , \qquad [9.9]$$

where ϕ is the angle the interference planes make with the normal to the hologram plate. The planes bisect the wave directions in the medium. We may write

$$1/d_y = (1/d) \cos \phi$$

or

$$\nu_y = \nu \cos \phi \qquad (10.7)$$

where ν is the number of fringe cycles per millimeter in the direction normal to the crest fringe planes. By assigning extreme angles to the forming beams and taking $\phi = 0$, we can determine the maximum range of resolution required of material used to record *transmission holograms*. Thus when $\Omega_S = \Omega_R$, $\nu = 0$. When $\Omega_S = -\Omega_R = \pi/2$, $\nu = 2/\lambda_a$ and

$$0 < \nu < 2/\lambda_a . \qquad (10.8)$$

If we take λ_a to be 0.5 μm, we have

$$0 < \nu < 4000 \text{ cycles/mm} \qquad \text{(transmission)}.$$

For *reflection holograms*, the maximum range of resolution can be obtained by considering holograms formed with two plane waves *oppositely* directed with respect to the z axis but having the same y component of spatial frequency η. As in Section 9.7 the conditions $\eta = \eta_S = \eta_R$ and $\zeta = \zeta_S = -\zeta_R$ lead to crest interference planes oriented parallel to the surface of the hologram plate and spaced by

$$d_z = \frac{1}{2\zeta} = \frac{\lambda}{2(1 - \lambda^2\eta^2)^{1/2}}.$$

If we substitute $\lambda = \lambda_a/n_0$, n_0 being the index of refraction of the recording medium, $(\sin\psi)/\lambda = \eta$, and (from Snell's law) $(\sin\Omega)/n_0 = \sin\psi$, we obtain

$$d_z = \frac{\lambda_a/n_0}{2[1 - (\sin^2\Omega)/n_0^{\,2}]^{1/2}} = \frac{\lambda_a}{2(n_0^{\,2} - \sin^2\Omega)^{1/2}}. \tag{10.9}$$

When $\Omega = \pi/2$, $d_z = \lambda_a/2(n_0^2 - 1)^{1/2}$ and when $\Omega = 0$, $d_z = \lambda_a/2n_0$. We can again take λ_a to be 0.5 μm and the index n_0 for gelatin to be 1.5. In terms of the fringe frequency $\nu_z = 1/d_z$, the range of resolution is

$$4500 < \nu_z < 6000 \text{ cycles/mm} \qquad \text{(reflection)}.$$

From these simple calculations we see that the resolution required of recording materials for holography is much higher than that for photography where 200 cycles/mm is considered high. The fringe frequencies capable of being recorded by various hologram recording materials are given in Sections 10.8.2 and 10.10.4.

10.4 Persistence of the Hologram and Erasability

A hologram need persist only over the period of time required for reconstruction. Holograms formed in photosensitive liquids generally have too little persistence to be useful in holography. Photographic emulsion, on the other hand, forms a relatively permanent record which cannot be erased and reused. Holograms recorded in photochromic materials or in ferroelectric crystals have an intermediate degree of persistence and fade in a relatively brief period. Fading can be caused by thermal relaxation (dark reaction) or by exposure to the reconstructing beam. Thermal relaxation times at room temperature range from minutes to weeks. Fading or decay of a hologram implies an undesirable effect. However, fading can be caused deliberately, either by applying heat or light to the photochromic or ferroelectric, in which case the process is called erasure.

Two hologram recording materials which more suitably combine the qualities of persistence, erasability, and reusability are photoconductor–thermoplastic film and magnetic film. Holograms recorded in these materials are stable at room temperature and yet can be erased in one case by elevated temperature and in the other by an applied magnetic field. Once erased, new holograms can be recorded over the erased areas. (Discussion of magnetic film is reserved for Chapter 11.)

Obviously a given task largely determines the persistence and erasability required of the recording material. However, the sacrifices in efficiency, sensitivity, and resolution which may unavoidably accompany a desired degree of erasability must also be weighed.

10.5 Noise and Recording Linearity

Noise is unwanted light diffracted or scattered from the hologram in the same general direction as the reconstructed wave. We attribute noise to the following sources:

1. Random scattering of both signal and reference beams during exposure due to the granularity of the recording material.
2. Random scattering of the reconstructing beam and the reconstructed wave due to the granularity of the material in the processed hologram.
3. Spatial modulations of the reference and reconstructing waves.
4. Nonlinear recording of the signal wave.
5. Inhomogeneities and surface deformations of the recording material.

Sources 1 and 2 are usually unimportant except in the case of multiple recording of many holograms in the same material and in special cases of hologram formation with incoherent light. Source 3 is unrelated to the properties of the recording material and is important only when a spatially modulated reference beam is used to form the hologram. If the complex amplitude of the reconstructing wave at the plane of the hologram is linearly proportional to that of the original reference wave, then this noise represents a diffusion of the subject wave [see Section 14.5, Eq. (14.23)].

Perhaps the most important source of noise is Source 4, the nonlinear recording of the signal wave. The effect of nonlinear recording is analyzed in Chapter 12. In this chapter we shall discuss experimental methods which ensure linear hologram recording in a given material.

Noise in Source 5 derives from either preparation of the recording material or processing of the material after exposure. If there are defects within the material (as is common with large ferroelectric crystals) or if the relevant surfaces of the material are not sufficiently flat, the wavefronts are distorted in passing through the material. However, these undesirable defects may often be detected prior to exposure and bad hologram "plates" can be rejected. Wet development of photographic emulsion (especially the bleached emulsion) and dichromated gelatin can distort the gelatin surface or the gelatin in bulk. The surface effect is generally the more serious but is easier to remedy.

10.6 Ideal Wavefront Reconstruction and Ideal Recording Material

Having summarized some of the attributes of hologram recording materials which are necessary, desirable, or unavoidable, we now consider the characteristics of a material which would yield ideal wavefront reconstruction. Suppose that a hologram has been formed in a recording material and, after processing, returned to the position it occupied during formation. The hologram is illuminated by the original reference wave. We wish to specify the complex amplitude of an ideal reconstruction of the subject wave in a form independent of whether the hologram is plane or volume, absorption or phase, transmission or reflection.

To do so, we first consider the interference of subject and reference waves in the absence of the recording material. Let us fix our attention on a particular plane xy in the region of interference and call the complex amplitude of the subject wave in that plane $\mathbf{a}(x, y)$. The complex amplitude of the reference wave in the same plane is $\mathbf{r}(x, y)$. As in Eq. (1.15), we can write the intensity of the interference pattern in the xy plane as

$$I(x, y) = \mathbf{aa}^* + \mathbf{rr}^* + \mathbf{ra}^* + \mathbf{r}^*\mathbf{a}. \tag{10.10}$$

To form a hologram, recording material is inserted into a volume of the interference region which is adjacent to, but does not include the plane xy. The material is located so that, on reconstruction, the reconstructed wave emerges from the material and then passes through the xy plane. Thus subject, reference, and reconstructed wave intensities in the plane xy can all be easily measured. We shall refer the illuminating wave to the same xy plane and call its complex amplitude $\mathbf{r}(x, y)$, the value it has in the absence of the hologram.

The fourth term on the right in Eq. (10.10) leads to a reconstruction of the subject wave. In the case of a plane absorption hologram, the reconstructed wave complex amplitude in the xy plane of the hologram is a function of the product of the illuminating wave complex amplitude $\mathbf{r}(x, y)$ and the exposure $\tau_e \mathbf{r}^* \mathbf{a}$ [see Eq. (7.32)]. Since our specification of an ideal reconstruction must hold for plane absorption as well as other hologram types, we find it useful to define the *ideal reconstructed wave complex amplitude* as

$$\mathbf{w}_0(x, y) = 2\mathbf{S}k_1\tau_e \mathbf{r}\mathbf{r}^* \mathbf{a} \tag{10.11}$$

where τ_e is the exposure time, k_1 is the constant of proportionality between the intensities I_p and I (see Section 1.3) and $\mathbf{S}$ is a complex constant. A factor 2 has been introduced to enable a further reformulation in terms of fringe visibility V. A material is said to be *ideal* when the complex amplitudes of waves holographically reconstructed from the material are always given by Eq. (10.11). When $\mathbf{r}\mathbf{r}^*$ is constant over the entire xy plane, the hologram recording is *linear*, i.e., the reconstructed wave complex amplitude $\mathbf{w}_0$ is linearly proportional to the subject wave complex amplitude $\mathbf{a}$.

Let us now consider Eq. (10.11) for the case where $\mathbf{r}(x, y)$ and $\mathbf{a}(x, y)$ are any two *plane waves*

$$\mathbf{r} = r \exp(i2\pi\xi_r x), \qquad \mathbf{a} = a \exp(i2\pi\xi_a x)$$

and where r, a, ξ_r, and ξ_a are real constants. Equation (10.11) then becomes

$$\begin{aligned} \mathbf{w}_0(x) &= \mathbf{S}r(2k_1\tau_e ra) \exp(i2\pi\xi_a x) \\ &= \mathbf{S}rE_0V \exp(i2\pi\xi_a x) \end{aligned} \tag{10.12}$$

where we have substituted

$$2k_1\tau_e ra = k_1(a^2 + r^2)\tau_e V = E_0 V$$

from Eq. (7.31) (for $|\mu| = 1$, $\Omega = 0$) and Eq. (7.33). We see from Eq. (10.12) that under the condition of ideal reconstruction:

$$\mathbf{S} = S \exp(i\sigma) = \text{constant}, \tag{10.13}$$

linear recording is achieved, and the reconstructed wave complex amplitude

$$\mathbf{w}_0(x) = w_0 \exp[i(2\pi\xi_a x + \sigma)] \tag{10.14}$$

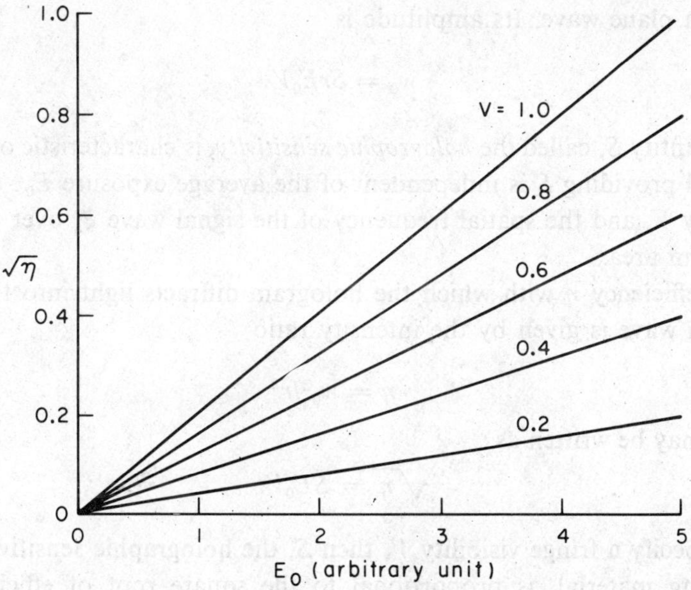

FIG. 10.1. $\sqrt{\eta}$–E_0 characteristics for the ideal recording material. η, diffraction efficiency; E_0, average exposure (arbitrary units); and V, fringe visibility.

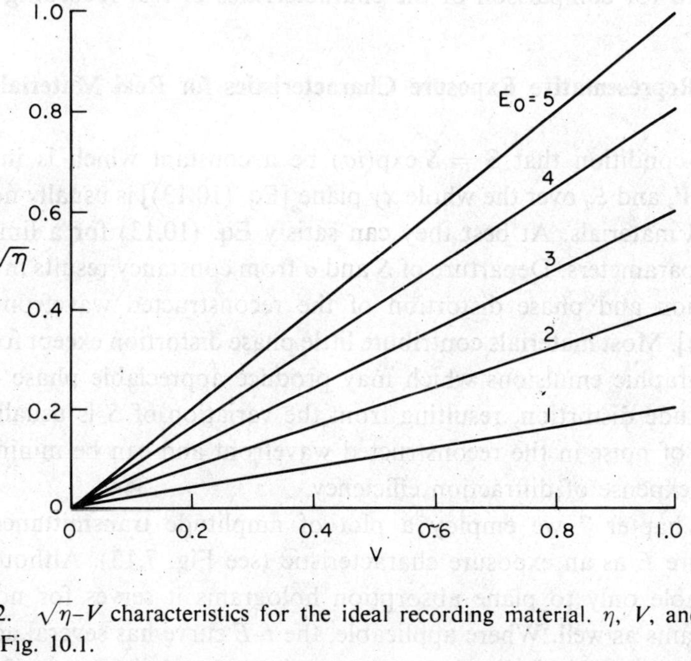

FIG. 10.2. $\sqrt{\eta}$–V characteristics for the ideal recording material. η, V, and E_0 are defined in Fig. 10.1.

is also a plane wave. Its amplitude is

$$w_0 = SrE_0V. \tag{10.15}$$

The quantity S, called the *holographic sensitivity*, is characteristic of an ideal material providing S is independent of the average exposure E_0, the fringe visibility V, and the spatial frequency of the signal wave ξ_a over the entire hologram area.

The efficiency η with which the hologram diffracts light into the reconstructed wave is given by the intensity ratio

$$\eta = w_0^2/r^2$$

which may be written as

$$\sqrt{\eta} = SE_0V. \tag{10.16}$$

If we specify a fringe visibility V, then S, the holographic sensitivity of the recording material, is proportional to the square root of efficiency and inversely proportional to the average exposure. We can plot $\sqrt{\eta}$ versus E_0, with V as a parameter, or plot $\sqrt{\eta}$ versus V, with E_0 the parameter, and obtain families of straight lines as in Figs. 10.1 and 10.2. These represent exposure characteristics of ideal recording material and may be used as a standard for comparison of the characteristics of real recording materials.

10.7 Representative Exposure Characteristics for Real Materials

The condition that $\mathbf{S} = S\exp(i\sigma)$ be a constant which is independent of E_0, V, and ξ_a over the whole xy plane [Eq. (10.13)] is usually not satisfied by real materials. At best they can satisfy Eq. (10.13) for a limited range of the parameters. Departure of S and σ from constancy results in amplitude distortion and phase distortion of the reconstructed wavefront [see Eq. (10.14)]. Most materials contribute little phase distortion except for bleached photographic emulsions which may produce appreciable phase distortion. Amplitude distortion, resulting from the variation of S is usually a major source of noise in the reconstructed wavefront and can be minimized only at the expense of diffraction efficiency.

In Chapter 7 we employ a plot of amplitude transmittance t versus exposure E as an exposure characteristic (see Fig. 7.13). Although strictly applicable only to plane absorption holograms it serves for not-so-plane holograms as well. Where applicable, the t–E curve has several advantages: (1) The curve is obtained experimentally with relative ease. (2) Its form makes the choice of optimum average exposure apparent. (3) It allows a ready estimate of fringe visibility. However, a serious deficiency of this plot is the failure to give explicit information on diffraction efficiency. In fact a t–E plot for a given material does not tell whether the material has even sufficient resolution capability to form a hologram.

A more informative display of the exposure characteristics of hologram recording materials are the curves $\sqrt{\eta}$ versus V (with E_0 the parameter) and $\sqrt{\eta}$ versus E_0 (with V the parameter). Curves in Figs. 10.3 and 10.4 exhibit features representative of real recording material and are to be compared with the corresponding curves for the ideal material (Figs. 10.1 and 10.2). In forming holograms, often the exposure E_0 is nearly constant over the whole hologram plane. In that case the exposure characteristics of Fig. 10.4 are the pertinent ones. In certain cases of hologram formation in incoherent light, where the signal wavefront is made to interfere with itself, the fringe visibility V is nearly constant over the hologram plane (see Chapter 20). Under these relatively rare circumstances Fig. 10.3 is more applicable. We shall confine our discussion to the exposure characteristics shown in Fig. 10.4.

To obtain a similar plot for a given material one must employ two plane waves to form a number of holograms in the material. A unique combination of average exposure E_0 and beam intensity ratio R are associated with each hologram. From the beam ratio R we may derive the fringe visibility $V = 2\sqrt{R}/(1 + R)$ [see Eq. (7.31) for $\mu = 1$, $\Omega = 0$]. If the purpose of plotting the exposure characteristics is to guide one in forming a hologram with some desired angle between subject and reference beam, the plane wave holograms providing the data should be formed with the same mean interbeam angle. After all the plane-wave holograms have been formed and processed, their diffraction efficiencies η are measured. A set of curves $\sqrt{\eta}$ versus V for various values of the parameter E_0 can then be plotted on a linear scale as in Fig. 10.4. For convenience we also show η and R on nonlinear scales.

Several benefits accrue from this method of presenting exposure characteristics. First, the method holds for any material, thin or thick, absorption or phase, and for transmission or reflection holograms. It therefore permits a direct comparison to be made between the exposure characteristics of various materials being considered for a given hologram recording. Furthermore, the departure of the characteristics of any real material from those of the ideal are readily appreciated by comparing the measured curves with those of Fig. 10.2. The following pertinent information is obtained directly from the graph:

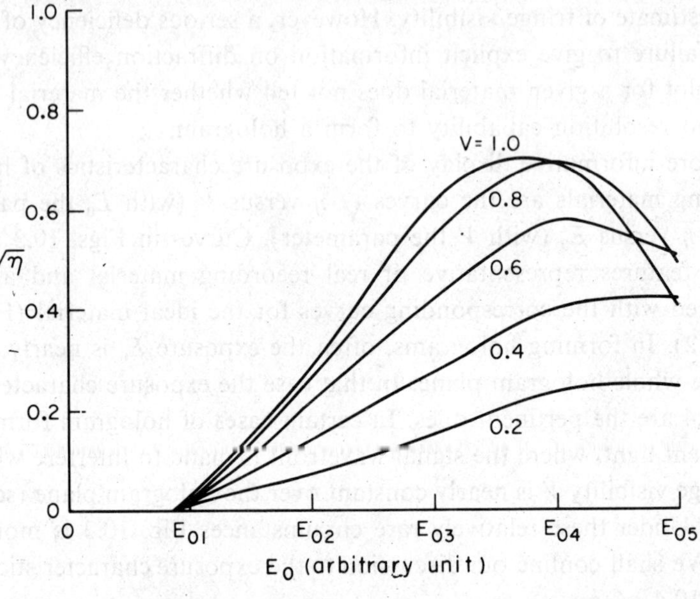

FIG. 10.3. $\sqrt{\eta}$–E_0 characteristics representative of a real recording material.

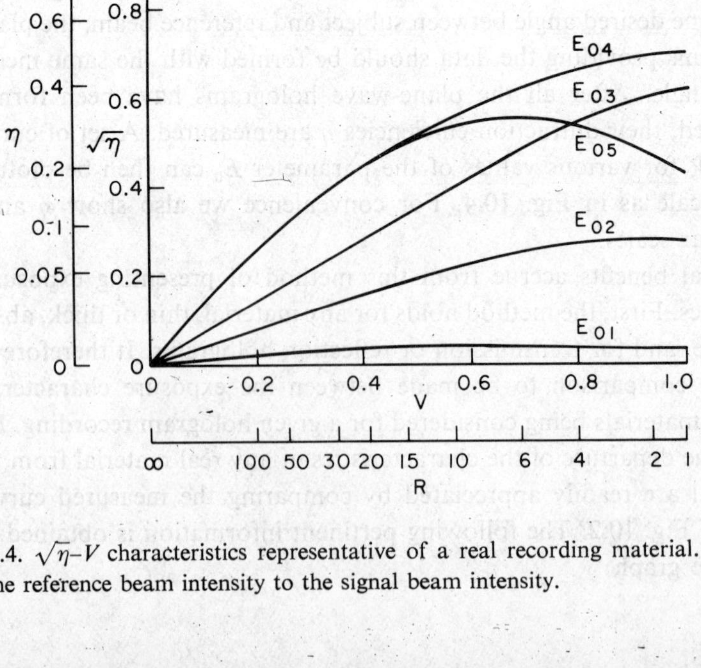

FIG. 10.4. $\sqrt{\eta}$–V characteristics representative of a real recording material. R is the ratio of the reference beam intensity to the signal beam intensity.

1. Range of visibility V or beam ratio R for which the hologram recording is linear (indicated by the straight-line portion of the $\sqrt{\eta}$–V curve at constant E_0).
2. Maximum achievable diffraction efficiency.
3. Exposure necessary to obtain a given diffraction efficiency (V held constant), a measure of sensitivity.
4. Optimum value of average exposure, representing the best compromise between linearity and efficiency.

In Fig. 10.4 exposure value E_{01} represents the minimum exposure to which the material characterized by the curves will respond. E_{02}, E_{03}, E_{04}, E_{05} represent successively higher values of exposure. Value E_{02} is an example of underexposure yielding low efficiency. When the material is exposed to E_{03}, the curve $\sqrt{\eta}$ versus V has its longest straight-line portion; with this exposure the largest variation in V can be linearly recorded. With the exposure E_{04}, maximum efficiency is reached. A trade-off between linearity and efficiency is possible in the exposure range between E_{03} and E_{04}. Finally, when the exposure reaches E_{05}, the material becomes overexposed and efficiency drops below the maximum.

Although the efficiency given by Fig. 10.4 applies to elementary holograms formed with two plane waves, we find that in certain cases the curves can be used to estimate an average efficiency η_{av} for more general holograms. Consider a hologram formed with a diffuse subject beam and a plane reference wave. We must determine an average visibility V_{av} or average beam ratio R_{av} which will allow us to pick off η_{av} from the $\sqrt{\eta}$ versus V plots. The total intensity at the hologram plane due to subject and reference beams is measured with a detector of area large enough to integrate over many interference fringes. Since the intensity component contributed by the diffuse subject beam is itself a high-spatial-frequency interference or speckle pattern (see Section 8.2.2), only its spatially averaged value will be detected. This average value is nearly constant across the hologram plane. Thus we can select a suitable value of the average exposure E_0, but can measure only an average value for the beam ratio R. The associated plot of $\sqrt{\eta}$ versus V determines the choice of V or R. Since we can set the beam ratio only to an average value R_{av}, a reasonable compromise between linearity and efficiency is achieved by selecting a value corresponding to the center of the straight-line portion of the $\sqrt{\eta}$ versus V curve. The average efficiency η_{av} is then read from the vertical scale of the graph.

Holographic sensitivity of a real recording material is defined as in Eq.

(10.16) for the ideal material

$$S = \sqrt{\eta}/VE_0. \tag{10.17}$$

However, in view of the curves of Fig. 10.4, S is no longer strictly constant. For real materials therefore we take S to be the largest of the slopes of the straight-line portions of the $\sqrt{\eta}$–V curves divided by the corresponding value of E_0. The sensitivity S can be considered as a relative figure-of-merit of a material.

10.8 Silver Halide Photographic Emulsions

In the light of the preceding general discussion we now begin to examine the specific properties and characteristics of several important or potentially important materials. Silver halide photographic emulsion was used by Gabor to form the first holograms. Because it has relatively high sensitivity and can be readily purchased, it remains the most common recording material. However, the emulsion which is widely used (type 649F manufactured by Eastman Kodak Co.) was not originally intended for laser light recording. Recently Agfa-Gevaert Inc. has made available silver halide emulsions specially sensitized for laser light photography and holography. At present the following types of emulsions are suitable for holographic recording:

By Eastman Kodak Co.:	649F, 649GH, and High Resolution Plates (HRP)
By Agfa-Gevaert Inc.:	8E56, 8E70, 8E75, 10E56, 10E70, 10E75, 14C70, and 14C75.

10.8.1 General Optical Properties of Emulsion and the Recorded Hologram

Photographic emulsion primarily consists of extremely fine grains of silver halide compounds dispersed in gelatin. Also present in the gelatin are certain sensitizing agents. The emulsion is coated over a transparent substrate which is either a glass plate or flexible acetate film. For hologram recording with a continuous-wave laser, the rigid glass is preferred over the flexible film. Approximate thicknesses and intensity transmittances of some common emulsions are listed in Table 10.1. When a two-beam interference pattern exposes a photographic plate and the plate is developed,

TABLE 10.1

Approximate Thicknesses and Intensity Transmittances of Some Silver Halide Photographic Emulsions

Emulsion	Thickness (μm)	Intensity transmittance		
		6328 Å	5145 Å	4880 Å
649F plate	15	0.42	0.40	0.33
649F film	6	0.81	0.61	0.62
High res. plate	6	0.81	0.58	0.58
10E56 plate	6	0.67	0.35	0.29
10E70 plate	6	0.47	0.37	0.29
10E75 plate	6	0.61	0.50	0.40

an absorption hologram is formed. During the development process the exposed silver halide grains are converted to metallic silver. A bleach process may be used after development to convert the silver into a transparent compound; this in turn converts the absorption hologram to a phase hologram (see Section 10.8.5). Both transmission and reflection holograms can be recorded in photographic emulsion. For absorption transmission holograms the optimum average optical density to be achieved is about 0.6 while for absorption reflection holograms the density should be approximately 2. Bleaching has been successfully applied only to transmission holograms. If the hologram is to be bleached, an exposure considerably greater than that for the unbleached case is required.

The silver halide grains found in emulsions suitable for holography are typically less than 0.1 μm in diameter. They not only absorb light but scatter as well. Figure 10.5 shows that they tend to scatter more light at the blue end of the spectrum than at the red. Forward and back scattering were measured at 45° and 125° relative to normally incident light. Although we state in Section 10.5 that scattering noise is usually unimportant, it becomes a noticeable factor in reducing image contrast when holography is employed for high-quality microimaging.

A word of caution might be of use to those who would form reflection holograms with commercial photographic plates or films. One has the option of purchasing them with an antihalation coating on the nonemulsion side. The coating, black or red, minimizes light reflected from the back surface of the substrate but precludes recording a reflection hologram. It can however be removed in the dark with an appropriate solvent such as alcohol.

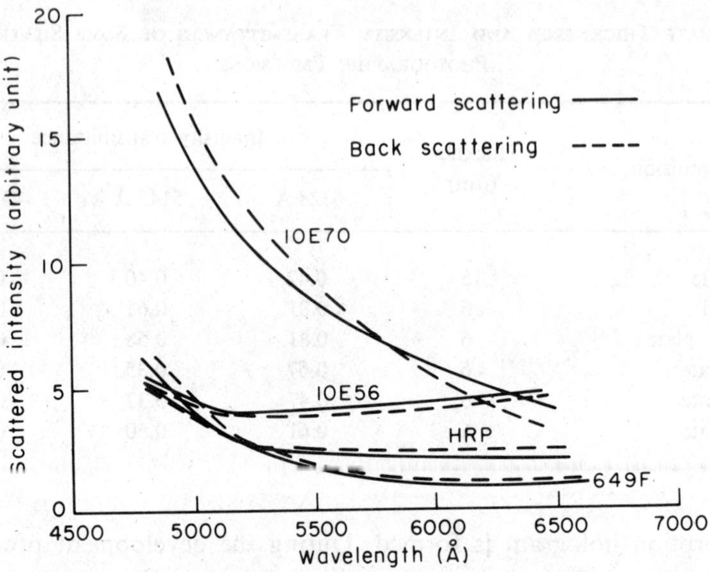

FIG. 10.5. Scattering of normally incident light from photographic emulsions.

10.8.2 SPECTRAL SENSITIVITY AND RESOLUTION

Once a laser has been chosen and the wavelength of the light to be used in recording a hologram has been determined, it is desirable to select the emulsion most sensitive to that wavelength. Figure 10.6 displays approximate spectral sensitivity curves for five emulsions. Since the curves are

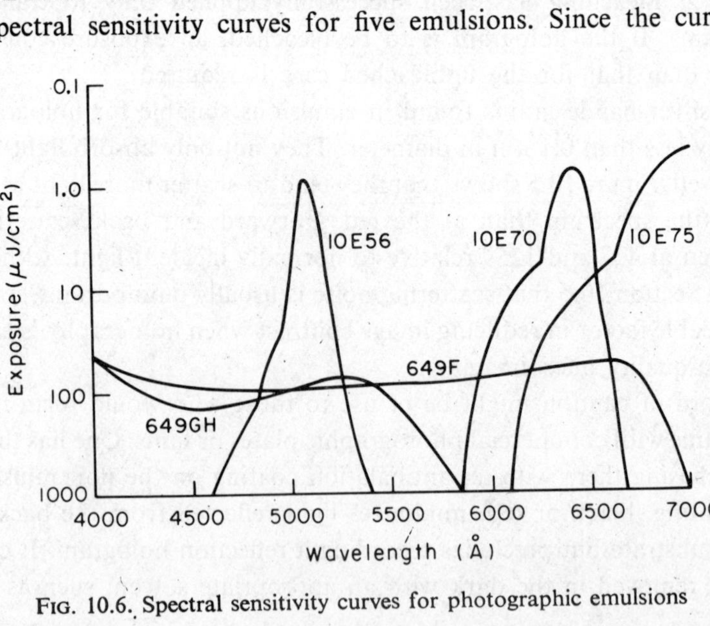

FIG. 10.6. Spectral sensitivity curves for photographic emulsions

based on *photographic* data supplied by the manufacturers, we here employ the term sensitivity or *speed* of the emulsion in its photographic sense, a quantity inversely proportional to exposure and unrelated to diffraction efficiency or fringe visibility. Note that on the vertical axis of Fig. 10.6, exposure in microjoules per square centimeter increases from top to bottom. The curves illustrate that it is possible to obtain emulsions which are uniformly sensitive over the visible spectral range (panchromatic), sensitive only in the blue-green end of the spectrum (orthochromatic), or selectively sensitive at common laser wavelengths.

The highest spatial frequency which can be usefully recorded by a photographic emulsion is often called its resolving power. Table 10.2 lists the

TABLE 10.2

MANUFACTURER-SPECIFIED RESOLUTIONS OF SILVER HALIDE PHOTOGRAPHIC EMULSIONS SUITABLE FOR HOLOGRAPHY

Resolution[a]	Emulsion
1500	14C70, 10C75
2800	10E56, 10E70, 10E75
3000	8E56, 8E70, 8E75
over 2000	649F, 649GH, HRP

[a] Specified by manufacturers as "resolving power" in "lines/mm."

resolving powers of a number of emulsions as advertised by their manufacturers. (In practice the useful resolution of an emulsion for holography, as formulated in Section 10.3, is higher than the advertised value.) It is well known that high speed and high resolution are not obtained simultaneously in a silver halide emulsion. In fact photographic speed decreases very rapidly with increasing resolution. As an example, Eastman Kodak Co. specifies the resolving powers of its 103F and 649F emulsions (having similarly shaped spectral sensitivity curves) as "between 56 and 68 lines/mm" for the former and "2000 lines/mm or more" for the latter [*10.1*]. At the same time, the relative sensitivities of 103F and 649F emulsions are given as the ratio of 60 : 0.005. Thus an improvement in resolution by a factor of 30 is accompanied by a four-orders-of-magnitude loss of speed. In normal photography a resolution of 100 cycles/mm is usually more than adequate, while a resolution of several thousand cycles/mm (Section 10.3) is required for general holography. It is thus not unexpected that the speeds of the emulsions suitable for holography are several orders of magnitude lower than those for ordinary photography. In spite of this, photographic emulsions turn out to be more sensitive than most other hologram recording materials.

10.8.3 Exposure Characteristics

We note in Sections 2.6.2, 7.2.2, and 10.7 that the transmittance versus exposure curve (t–E curve) is a meaningful, though limited, way of characterizing the response of hologram recording materials. Figures 10.7 and 10.8

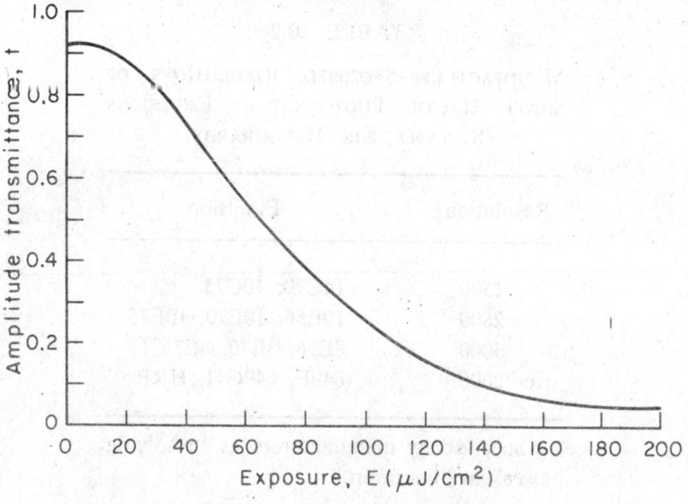

FIG. 10.7. The t–E curve for Kodak 649F emulsion.

show two such curves for Kodak 649F emulsion and Agfa-Gevaert 10E70 emulsion, each coated on glass plates. The process of developing the holograms (from which the data was obtained) is one of the factors determining the shape of the curves. It is described in Section 10.8.4. A wavelength of 6328 Å was used to form the holograms. However, the form of the curve does not depend strongly on the wavelength. If one wishes to have the characteristic for some other wavelength, he need merely change the horizontal scale of exposure E in accord with the spectral sensitivity curve of Fig. 10.6.

As mentioned in Section 10.7 a more informative and useful graphical presentation of the exposure characteristics for any hologram recording material is obtained by plotting $\sqrt{\eta}$–V curves with constant E_0 as a parameter. Two such graphs are shown in Figs. 10.9 and 10.10 for Kodak 649F and Agfa-Gevaert 10E70 emulsions, respectively. The holograms, from which these curves were obtained, were formed at 6328 Å by two plane waves each making an angle of 22.5° to the normal of the hologram plate and an angle of 45° to one another. We see that the optimum exposure value E_0 is about 70 μJ/cm² for 649F emulsion and 2.5 μJ/cm² for 10E70 emulsion. These agree well with values given by the centers of the straight-line portions of the t–E curves in Figs. 10.7 and 10.8, indicating that the applicability of the latter curves extends to "not-so-thin" holograms. From Fig. 10.9 we find the maximum diffraction efficiency to be in the range of 4%. Below

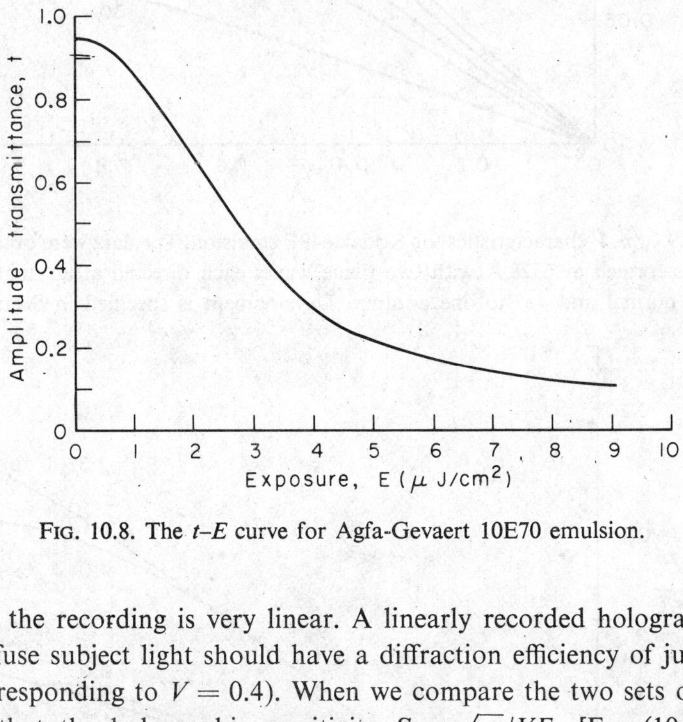

FIG. 10.8. The t–E curve for Agfa-Gevaert 10E70 emulsion.

$V \approx 0.8$ the recording is very linear. A linearly recorded hologram made with diffuse subject light should have a diffraction efficiency of just under 1% (corresponding to $V = 0.4$). When we compare the two sets of curves we see that the holographic sensitivity $S = \sqrt{\eta}/VE_0$ [Eq. (10.17)] for 10E70 emulsion is approximately 7.3 m²/J, while for 649F emulsion it is 0.26 m²/J. The 10E70 emulsion is 28 times more sensitive.

10.8.4 Processing

Processing photographic emulsion exposed to a holographic interference pattern differs very little from processing emulsion exposed to a high-resolution photographic image. Following is a procedure, essentially that recommended by Eastman Kodak Co. for processing scientific plates and

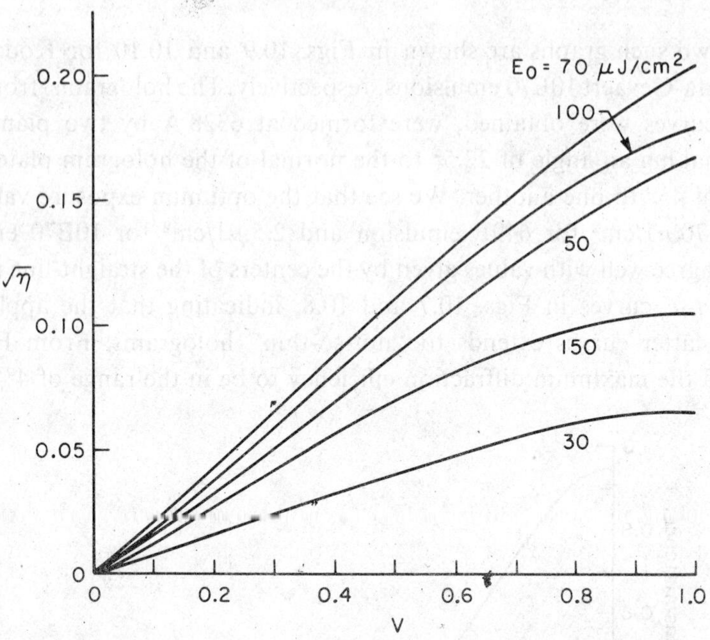

FIG. 10.9. $\sqrt{\eta}$–V characteristics for Kodak 649F emulsion. The data were obtained from holograms formed at 6328 Å with two plane waves each directed at an angle 22.5° to the plate normal and 45° to one another. Development is specified in Section 10.8.4.

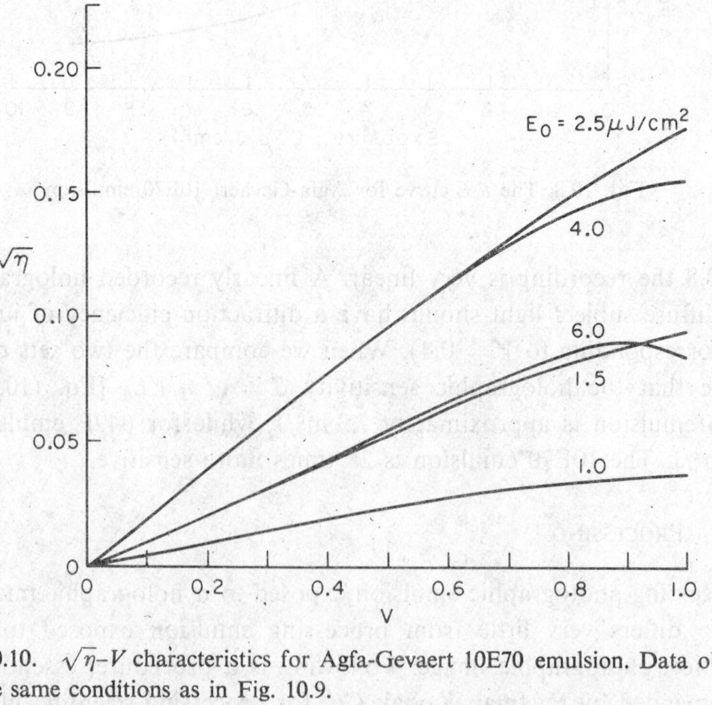

FIG. 10.10. $\sqrt{\eta}$–V characteristics for Agfa-Gevaert 10E70 emulsion. Data obtained under the same conditions as in Fig. 10.9.

films [*10.1*], which has yielded satisfactory results for emulsions manufactured by both Kodak and Agfa-Gevaert:

(1) DEVELOPMENT: 5 min in Kodak D-19 developer bath with continuous agitation. Follow immediately with a 30-sec rinse in water.
(2) FIXATION: 5 min in Kodak Rapid Fixer bath (which contains an emulsion hardener) with continuous agitation. Follow with a 30-sec rinse in water.
(3) RESIDUAL FIXER REMOVAL: $1\frac{1}{2}$ min in Kodak Hypo Clearing Agent bath.
(4) WASH: 5 min in flowing water.
(5) RESIDUAL DYE-SENSITIZER REMOVAL: 5 min in methanol. Follow with a 1-min rinse in water.
(6) DRYING.

The recommended temperature of the developer bath is 20°C or 68°F. Temperatures of other chemical baths are not critical but are usually maintained at the same temperature as the developer bath.

Drying of the processed emulsion is much more important in holography than in photography. As implied in Section 10.1.2, shrinkage of the emulsion thickness by approximately a wavelength can cause a 180° phase reversal of the reconstructed wavefront. If the hologram recording can be considered thin or if the phase of the reconstructed wavefront is not of prime importance, the normal drying method for the photographic emulsion can be followed. For example, it can be drip dried in air after soaking in a diluted Kodak Photo-Flo Solution for about 2 min. If the hologram is to be used in applications such as interferometry, where phase distortion of the reconstructed wavefront must be minimized, a special emulsion-drying method should be considered. A drying method which can result in a more uniform emulsion thickness is the following: After Step 5, the wet emulsion is soaked in denatured alcohol for $1\frac{1}{2}$ min with continuous agitation, and then in isopropanol ($CH_3CHOHCH_3$) for $1\frac{1}{2}$ min. The emulsion is then pulled out from the propanol bath and the excess propanol blown off immediately with a jet of dry air. This is to be done before the alcohol has a chance to evaporate and cause moisture from the air to condense on the emulsion.

Normal processing of a photographic emulsion results in a shrinkage of the emulsion thickness by about 15%. Removal of unexposed silver halide grains from the bulk gelatin base during fixing is the main cause. Shrinkage generates several undesirable effects for holography. Obviously the phase of the reconstructed wavefront is changed from that of the original signal wavefront. Unfortunately this change may not be uniform across the hologram plate. A more serious problem is that the condition for Bragg diffraction of the reconstructed wave, $2d \sin \theta = \lambda$, may no longer be that determined by the original signal and reference waves. Unless the fringe planes are normal to the hologram surface, their spacings d are all reduced. If the hologram were formed with two plane waves (an elementary hologram) with a single original value of d, then the new value d' might be compensated for by changing the wavelength of the reconstructing beam in the ratio d'/d. One might also change the sine of the angle which the reconstructing beam makes with the planes in the ratio d/d'. For the case of a general hologram formed with a signal beam containing a range of spatial frequencies, there will be a range of grating spacings d. Compensation for shrinkage by means of a change in the wavelength or angle of incidence of the reconstructing beam cannot be exact for more than one spatial-frequency component. Failure to satisfy Bragg's law for hologram grating components formed by the other spatial frequencies of the signal results in some attenuation of these frequencies in the reconstruction and some loss of resolution in the image.

When white light is used to reconstruct a wave from a reflection hologram, the most noticeable effect due to emulsion shrinkage is the color shift of the reconstructed wave. When the image is viewed from the same direction as one would view the original subject, it appears in a color corresponding to

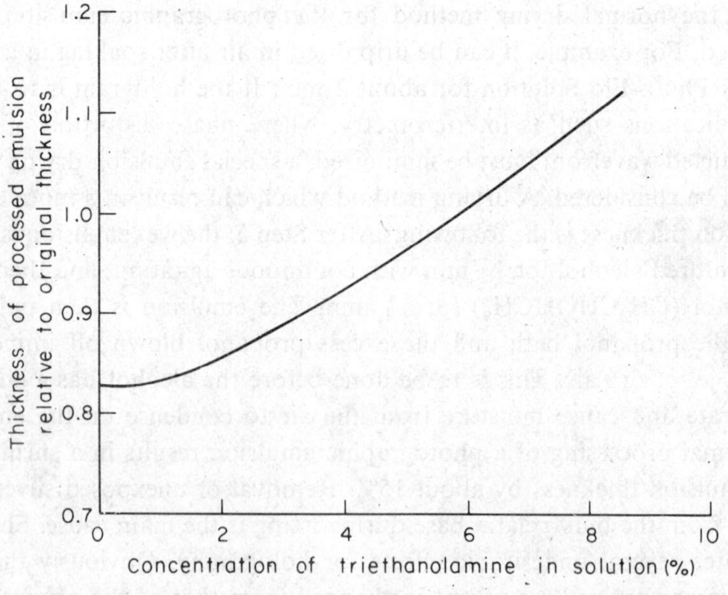

FIG. 10.11. Concentration of triethanolamine required to correct for the shrinkage of processed Kodak 649F emulsion.

a shorter wavelength than that of the original subject illumination. (The new wavelength is given by $\lambda' \approx \lambda d'/d$.) Thus a hologram recorded at 6328 Å wavelength (red) may reconstruct at a wavelength near 5300 Å (green). This, of course, presents problems in color holography (Chapter 17).

By omitting the fixation step in processing the emulsion, shrinkage is reduced. However, an unfixed hologram is unstable (becomes darker with time) and the reconstruction is noisy due to scattering by the unremoved silver halide grains. It is possible to swell the processed emulsion back to its original thickness by soaking it in an aqueous solution of triethanolamine $(CH_2OHCH_2)_3N$ and following with a slow, careful drying in air [*10.3*]. Figure 10.11 shows the relative thickness of the processed Kodak 649F emulsion (developed to an optical density of two) as a function of the concentration of triethanolamine in solution. The concentration needed for the exact correction differs slightly with optical density and can be determined after repeated trials with the same emulsion.

10.8.5 Bleached Holograms

An absorption hologram recorded in a photographic emulsion can be converted into a phase hologram with higher diffraction efficiency. The conversion involves a chemical bleaching of the emulsion. There are basically two kinds of bleach processes. In one process [*10.4–10.6*] the spatial variation of absorption constant is converted into a corresponding thickness variation of the bleached emulsion. This thickness variation is also known as surface relief. In the second process [*10.7–10.11*] the spatial variation of the absorption constant is converted into a corresponding refractive index variation in the emulsion volume. A bleached hologram generally has both thickness and refractive index variations and is not absorptionless; its maximum diffraction efficiency is less than 70%.

A phase hologram with emulsion thickness variation is obtained by bleaching a hologram in a tanning bleach bath. In this process the silver grains developed in the emulsion are removed. Gelatin in the vicinity of the removed silver grains becomes *tanned* or *hardened* relative to gelatin in other areas of the emulsion. [Tanning is the result of a higher-than-normal degree of cross linking of the gelatin molecules (see Section 10.9.1).] Upon drying, that area of the gelatin receiving a greater degree of tanning becomes thicker than the less tanned area. Thus the final record of the interference fringes of a hologram formed in emulsion and subsequently bleached with a tanning bleach is a spatial modulation of the gelatin thickness. The fringe frequency corresponding to *maximum* modulation amplitude is inversely proportional to the average emulsion thickness [*10.6*] and for 15-μm-thick Kodak 649F plate is only about 10 cycles/mm. Consequently one should expect the process to be ineffective for spatial frequencies above a few hundred cycles per millimeter. However, the effect can be observed for spatial frequencies as high as 1200 cycles/mm [*10.5*].

An effective bleach process for obtaining a high resolution phase hologram is one in which the developed silver in the absorption hologram emulsion is replaced by a transparent silver salt with refractive index generally greater than that of the gelatin. When this is done, one finds the scattering of incident light to be substantially greater than from silver grains in the unbleached hologram. Consequently, the noise level in the reconstruction from a hologram of this type is usually higher than either that from an unbleached hologram or from a bleached hologram recorded as pure surface relief. While the resolution to be expected is theoretically limited only by the size of the original silver grains, the process has not been very successful when applied to reflection holograms. The highest diffraction efficiency reported for a bleached reflection hologram is only 10% [*10.12*]. In the following we limit our discussion to transmission holograms.

Bleach baths for converting the silver to a transparent silver salt usually contain one of the following oxidation agents: mercuric chloride $HgCl_2$, ferric chloride $FeCl_3$, potassium ferricyanide $K_3Fe(CN)_6$, ammonium dichromate $(NH_4)_2Cr_2O_7$, or cupric bromide $CuBr_2$. Oxidation of silver $Ag \rightarrow Ag^+$ is accomplished by a chemical reaction in which a multiple-valence metallic ion is reduced from a higher to a lower valence: $Hg^{2+} \rightarrow Hg^{1+}$, $Fe^{3+} \rightarrow Fe^{2+}$, $Cr^{6+} \rightarrow Cr^{3+}$, or $Cu^{2+} \rightarrow Cu^{1+}$. The resulting silver salt is usually a silver halide AgCl or AgBr, silver mercurous chloride $AgHgCl_2$, or silver ferrocyanide $Ag_4Fe(CN)_6$.

Bleaches containing potassium ferricyanide are among the best; they produce stable holograms exhibiting high diffraction efficiency and relatively low noise [*10.10*]. One relatively simple way to bleach a hologram using this oxidation agent is given below. It involves a slight modification of the processing steps described in Section 10.8.4.

(1) DEVELOPMENT: Agitate plate 5 min in Kodak D-76 developer. Follow with 30-sec rinse in water.

(2) FIXATION: Agitate plate 5 min in Kodak Rapid Fixer. Follow with 30-sec rinse in water.

(3) FIXER REMOVAL: $1\frac{1}{2}$ min in Kodak Hypo Clearing Agent.

(4) WASH: 5 min in flowing water.

(5) SENSITIZER REMOVAL: 5 min in methanol. Follow with 1-min rinse in water.

(6) BLEACH: 3 min in 5% aqueous solution of $K_3Fe(CN)_6$.

(7) WASH: 5 min in flowing water.

(8) WATER ABSORPTION: Agitate wet emulsion in denatured alcohol for $1\frac{1}{2}$ min; then in isopropanol for $1\frac{1}{2}$ min.

(9) DRY: Remove plate from propanol and immediately blow off the excess with a jet of dry air.

Exposure characteristics ($\sqrt{\eta}$–V curves) for a bleached hologram recorded in Kodak 649F emulsion are shown in Fig. 10.12.

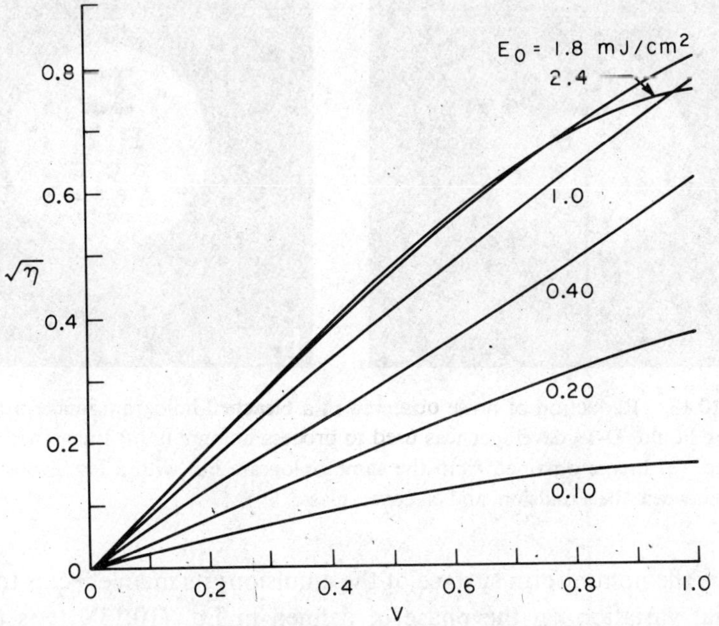

FIG. 10.12. $\sqrt{\eta}$–V characteristics for bleached Kodak 649F emulsion. Data obtained from holograms formed as in Fig. 10.9. Developer Kodak D-76 was used instead of D-19.

Comparing Fig. 10.12 with 10.9, we see that the maximum diffraction efficiency increases from about 4% to 65% after the bleach process. There is also a slight decrease in the length of the straight-line portion of the $\sqrt{\eta}$–V curve, indicating a slight loss in the recording linearity. Optimum exposure is about 1 mJ/cm² instead of 70 μJ/cm² for the unbleached case.

As may be seen from Fig. 10.12, overexposure of the bleached hologram does not cause much reduction in the diffraction efficiency. This contrasts with a rather large drop in efficiency observed for unbleached holograms. The holographic sensitivity S of the bleached hologram, as defined in Eq. (10.17), is about 0.04 m^2/J, as compared to 0.26 m^2/J for the unbleached hologram recorded in the same Kodak 649F emulsion.

The wave reconstructed from a bleached hologram is usually noisier than that reconstructed from an unbleached hologram. Substitution of Kodak D-76 developer for the usual D-19 developer substantially reduces this noise [*10.10*], although D-19 development results in higher holographic sensitivity (about 0.08 m^2/J). An important noise source in bleached holo-

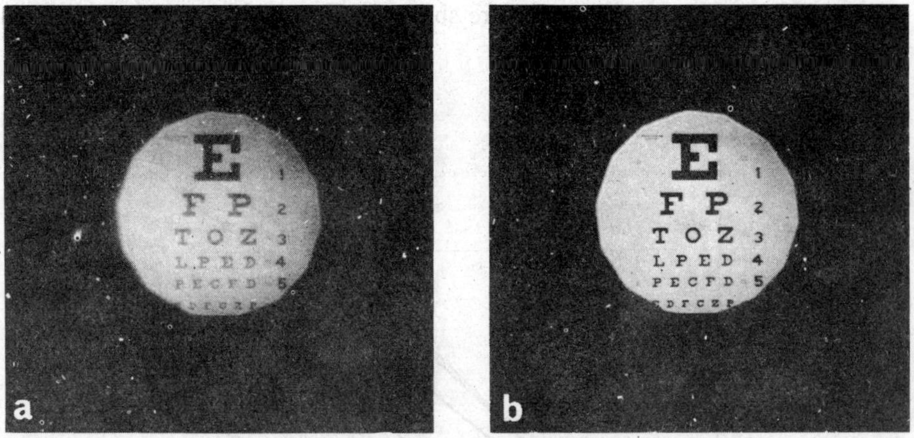

FIG. 10.13. Reduction of noise observed in a bleached-hologram image using index-matching liquid. D-19 developer was used to process hologram. (a) Image without index matching. (b) Image obtained from the same hologram but with a few drops of xylene placed between the emulsion and a cover glass.

grams is the nonuniform surface of the emulsion; its main effect is to impose a spatial variation on the phase σ, defined in Eq. (10.13). It is possible, however, to substantially reduce the noise by immersing the hologram in an index-matching liquid gate or by coating the hologram surface with an index matching film. Figure 10.13a shows a photograph of an image formed by the reconstructed wavefront from a bleached hologram processed with Kodak D-19 developer. Figure 10.13b shows a photograph of the same image from the same hologram, except with a few drops of xylene placed between the emulsion and a cover glass. The refractive indices of gelatin, glass, and xylene are approximately 1.53, 1.54, and 1.49, respectively.

10.9 Dichromated Gelatin Films

The optical properties of holograms recorded in dichromated gelatin films are very close to ideal. A properly recorded and processed hologram has so little light absorption and scattering that under ordinary room light the hologram plate appears almost indistinguishable from clear glass. When volume holograms are formed in dichromated gelatin, their diffraction efficiency can approach the theoretical limit of 100%. Either a transmission or a reflection hologram can be recorded in this material with both high diffraction efficiency and low noise.

10.9.1 Method of Hologram Recording

It has been known since the 1830s that ultraviolet or blue light can cause gelatin molecules to cross link when there is a small amount of dichromate present in the gelatin. Cross linking of gelatin molecules is also known as hardening or tanning of the gelatin. Sufficiently hardened gelatin is insoluble in water. Therefore, when a layer of gelatin containing a small amount of a dichromate is exposed to a light pattern for a sufficient length of time, the illuminated portion of the gelatin becomes insoluble in water while the unilluminated portion can be readily washed away in water. The ability of dichromated gelatin to act in this way as a photoresist has been used for many years in printing processes [*10.13*]. Of course, if the illuminating light is a holographic interference pattern produced by two coherent wavefronts, then a phase hologram is recorded.

There are several methods [*10.14*, *10.15*] of recording holograms in dichromated gelatin film which are superior to the one described above. We shall describe in the following sections a method which can produce very high-quality holograms, despite the fact that some of the chemical reactions and physical mechanisms involved are not fully understood. First let us briefly summarize the process and consider the mechanism whereby information is recorded. The dichromated gelatin film is initially chemically hardened to a sufficient degree that the unexposed film is made insoluble in water. After exposure to a holographic interference pattern, the film (hologram) is washed in water to remove the unexposed dichromate. Subsequently, it is rapidly dehydrated in an isopropanol bath, to complete the development of the film. A hologram is thus recorded on the surface and in the volume of the film as well.

Because the gelatin film is prehardened before the development, no gelatin is dissolved away in water during the washing step. Instead, a large amount of water is absorbed by the gelatin, causing the latter to expand in volume. This expansion, however, is not uniform. It is found that the amount of water absorption, and hence the volume expansion, decreases with increased hardness of the gelatin. Since gelatin exposed to high intensity light is hardened to a greater degree than that receiving a lower exposure, the spatial variation of the light intensity of an exposing interference pattern produces a corresponding spatial variation in gelatin expansion. If the rapid dehydration in isopropanol removes water from the gelatin at a faster rate than the normal shrinkage rate of the gelatin, then there is a strong tendency for the spatial variation of expansion to be retained.

It is not difficult to relate this localized differential expansion of the dehydrated gelatin to a corresponding ripple of the gelatin surface and consequently to a hologram recorded as surface relief. However, it is the differential density variation in the volume of the gelatin which forms the important volume-hologram-recording component. The physical mechanism responsible for the volume hologram behavior is not absolutely determined. There is little doubt that the rapid dehydration and simultaneous shrinkage of the gelatin induce strong, internal, localized stresses. They, in turn, are relieved by a redistribution of gelatin material which may manifest itself in a splitting of the gelatin along the planes of the crest fringes. In that case Bragg diffraction would take place as a result of scattering from the series of air–gelatin interfaces. On the other hand stress relief may cause voids to appear in the gelatin much as in a sponge, the voids more prominent in the less exposed or softer gelatin. If this be the case, a variation in the effective index of refraction of the bulk gelatin is the source of diffraction. When either the splitting or the voids become comparable in size to the fringe spacing, the interference pattern is imperfectly recorded and the reconstruction is noisy. Since the effect is controlled by the hardness of the gelatin, the *initial degree of hardening becomes an important parameter in film preparation.*

10.9.2 Film Preparation

Dichromated gelatin film is not commercially available. We describe here a method of film preparation which can be easily followed in a photographic or holographic laboratory. The first step is to procure a thin layer of pure gelatin on a glass plate or other suitable substrate. This gelatin film should be hardened to a degree where it does not dissolve readily in room temperature water. Numerous techniques for hardening gelatin films are described in the literature dealing with photographic emulsions [*10.16*].

A simple method of preparing a hardened gelatin film is the following: A 7% (by weight) solution of gelatin is poured evenly over a glass plate, and the excess solution is allowed to run off the plate. The glass plate requires no special coating or cleaning; it can be an ordinary glass slide or a photographic cover glass. After the film is dry, it is hardened by soaking it in a photographic fixer containing hardener (such as Kodak Rapid Fixer) for 3 to 5 min. Hardening is followed by a washing in water. Gelatin film obtained by this means is about 3 μm thick and adheres very well to the glass. By varying the concentration of the gelatin solution and by multiple coatings, film thicknesses varying from less than 1 μm to about 20 μm can be obtained. If the film is much thicker than 20 μm, the rapid dehydration in the development becomes less effective. Unless stated otherwise, a film thickness of approximately 12 μm is assumed throughout the remaining sections on dichromated gelatin.

Another convenient source of good quality gelatin film is the commercially available photographic plate.[1] To obtain a clear film of gelatin on a glass plate, one simply dissolves in a fixing bath the silver halides found in the emulsion of an undeveloped photographic plate. The gelatin film is then washed in water and methanol.

Sensitization of the gelatin film is achieved by soaking the film in 5% aqueous solution of ammonium dichromate for 5 min at room temperature. The film is then allowed to dry slowly in darkness. It may be used after 4 to 6 h of drying. No appreciable deterioration due to dark reaction has been observed for a period of several days at room temperature and several months under refrigeration (10°C).

10.9.3 Sensitivity and Resolution

The spectral sensitivity of dichromated gelatin film hardened for hologram recording has not been precisely measured. It is reasonable to expect that the holographic spectral sensitivity of the film should correspond to the spectral sensitivity of the soft dichromated gelatin used as a photoresist. (Much information on the sensitivity of dichromated colloids can be found in [*10.13*].) In this case the holographic sensitivity should have a maximum near 3550 Å, falling off slowly to zero near 5800 Å. The relative holographic sensitivity for the argon ion laser wavelengths 4880 and 5145 Å is in the

[1] Kodak 649F plates have been used successfully for this purpose. A few other types have been tried and found less suitable. The critical parameter is the hardness which as yet has not been adequately specified.

ratio of approximately 5 : 1. At 6328 Å, the wavelength of the He–Ne laser, the material is completely insensitive.

To some extent the sensitivity of the dichromated gelatin film depends on the particular dichromate salt used as the sensitizer. Among the common salts, ammonium dichromate $(NH_4)_2Cr_2O_7$ gives higher sensitivity than potassium dichromate $K_2Cr_2O_7$ or sodium dichromate $Na_2Cr_2O_7$. The sensitivity is considerably increased with pyridine dichromate $Cr_2O_7H_2 \cdot C_5H_5N \times 11H_2O$, perhaps by a factor of three, but this salt is difficult to prepare and the sensitized film has only a few hours of shelf life. We shall limit our discussion to the use of ammonium dichromate.

Holographic sensitivity of dichromated gelatin film is also a function of the degree of hardening of the gelatin. For high sensitivity, the gelatin should not be too hard. On the other hand, if the gelatin is too soft, the rapid dehydration in the development can cause large pores or voids in the gelatin, leaving a frostlike appearance on the hologram. Reconstruction from such a hologram is naturally noisy. Therefore a compromise must be made between high holographic sensitivity and the elimination of frost formation. There appears to be no upper limit on the resolution of this material for hologram recording; sensitized film is grainless. Unlike the bleached hologram recorded in photographic emulsion, the dichromated gelatin film yields high diffraction efficiency for either transmission or reflection holograms.

10.9.4 Exposure Characteristics

When forming a hologram in dichromated gelatin film, signal beam and reference beam can illuminate the same side of the film (for a transmission hologram) or they can illuminate opposite sides of the film (for a reflection hologram). Holographic sensitivity for the reflection hologram is slightly less than for the transmission hologram. Typical exposure characteristics of the dichromated gelatin film are exhibited in Fig. 10.14 in the form of $\sqrt{\eta}$–V curves. The conditions for obtaining the curves were identical to those for obtaining Fig. 10.9 for 649F emulsion and Fig. 10.12 for bleached 649F emulsion except that laser light at 4880 Å wavelength was used instead of 6328 Å. We see that in this example the maximum diffraction efficiency is about 80% (obtained with an exposure value of 20 mJ/cm²); even higher values are possible. The holographic sensitivity S (or the figure of merit) for the dichromated gelatin as defined in Eq. (10.17) is about 0.65×10^{-2} m²/J compared to 0.26 m²/J for Kodak 649F photographic emulsion. Somewhat higher sensitivity is possible if we allow a certain amount of frost. It is apparent from the exposure characteristics of Fig. 10.14 that linear recording is relatively difficult to achieve if very high diffraction efficiency is also demanded. Unlike photographic emulsion, there is no minimum exposure necessary before a response is observed.

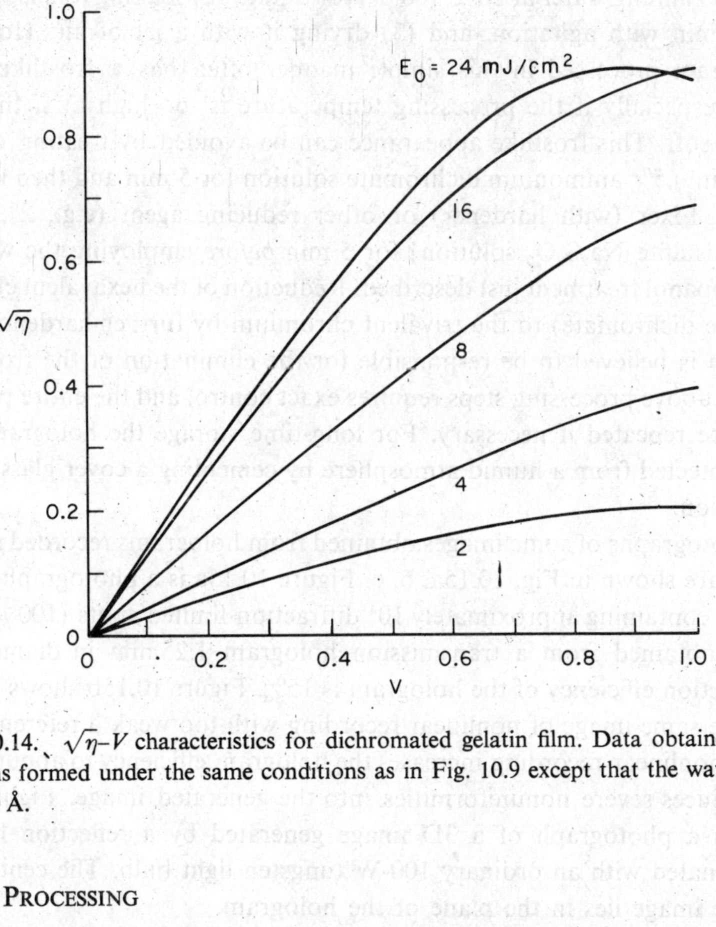

FIG. 10.14. $\sqrt{\eta}$–V characteristics for dichromated gelatin film. Data obtained from holograms formed under the same conditions as in Fig. 10.9 except that the wavelength was 4880 Å.

10.9.5 PROCESSING

Maximum diffraction efficiencies obtained without processing have been observed to be as high as 60% for holograms formed in thick (0.5 mm) dichromated gelatin layers. However, the exposure required for the hologram recording of an ordinary subject can be impractically high. Furthermore, the continued sensitivity of the exposed film to the reconstructing light and the ambient light can eventually deteriorate the reconstruction. In certain applications, e.g., hologram interferometry (see Chapter 15), these disadvantages may be considered secondary to the advantage of not having to remove the exposed film for processing.

In general, a sensitized gelatin film need only be exposed to the extent indicated in Fig. 10.14 and then processed after exposure. Before processing, the maximum efficiency is only a fraction of a percent, but it is greatly improved by the processing. Successful processing of dichromated gelatin of the proper hardness can be accomplished by (1) washing the exposed film in running water at 20°C (68°F) for 5 min, (2) soaking it in isopropanol for 2 min with agitation, and (3) drying it with a jet of air. However, a hologram processed in this simple manner often has a frostlike appearance, especially if the processing temperature is too high or if the gelatin is too soft. This frostlike appearance can be avoided by treating the hologram in 0.5% ammonium dichromate solution for 5 min and then in Kodak Rapid Fixer (with hardener) or other reducing agent (e.g., 2% sodium metabisulfite $Na_2S_2O_5$ solution) for 5 min *before* employing the water and isopropanol treatment just described. Reduction of the hexavalent chromium (of the dichromate) to the trivalent chromium by further hardening of the gelatin is believed to be responsible for the elimination of the frost. None of the above processing steps requires exact control and the entire procedure may be repeated if necessary. For long-time storage the hologram should be protected from a humid atmosphere by cementing a cover glass over the emulsion.

Photographs of some images obtained from holograms recorded in gelatin films are shown in Fig. 10.15a, b, c. Figure 10.15a is a photograph of a real image containing approximately 10^4 diffraction-limited spots (100 μm diameter) obtained from a transmission hologram 1.25 mm in diameter. The diffraction efficiency of the hologram is 15%. Figure 10.15b shows the effect on the same image of nonlinear recording with too weak a reference beam. This nonlinear recording increases the hologram efficiency to about 50% but introduces severe nonuniformities into the generated image. Figure 10.15c shows a photograph of a 3D image generated by a reflection hologram illuminated with an ordinary 100-W tungsten light bulb. The central plane of the image lies in the plane of the hologram.

10.10 Photoconductor–Thermoplastic Films

Holograms can be recorded in a thermoplastic film by causing its surface to deform in accordance with the light intensity variations of holographic interference patterns. Thermoplastics are usually not photosensitive and must be combined with a photoconductor in a film structure which can respond to light. Several unique and desirable properties make the photo-

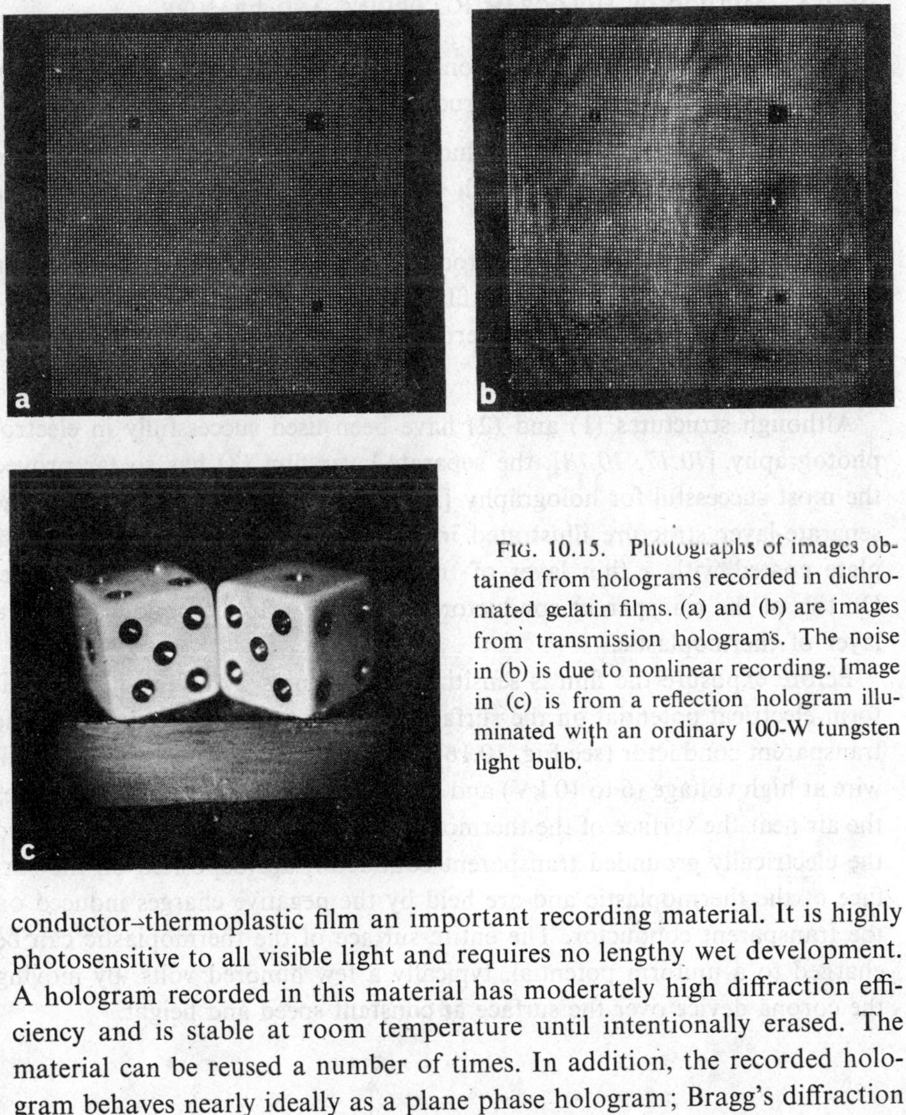

FIG. 10.15. Photographs of images obtained from holograms recorded in dichromated gelatin films. (a) and (b) are images from transmission holograms. The noise in (b) is due to nonlinear recording. Image in (c) is from a reflection hologram illuminated with an ordinary 100-W tungsten light bulb.

conductor–thermoplastic film an important recording material. It is highly photosensitive to all visible light and requires no lengthy wet development. A hologram recorded in this material has moderately high diffraction efficiency and is stable at room temperature until intentionally erased. The material can be reused a number of times. In addition, the recorded hologram behaves nearly ideally as a plane phase hologram; Bragg's diffraction phenomenon is absent. As discussed in Chapter 16 this property is useful when applied to a holographic optical memory.

Photoconductor–thermoplastic film does have several undesirable characteristics. Band-limited spatial frequency response is the most important; the film can record only those interference fringes whose spatial frequencies lie within a limited spatial frequency bandwidth. The maximum resolution obtainable with this material is not much greater than 1000 cycles/mm. In addition high quality film is difficult to prepare (none available commercially), and the thermal development of the exposed film is critical.

10.10.1 Method of Hologram Recording and Erasure

Depending on how the photoconductor is combined with the thermoplastic, the following three film structures are possible:

(1) Homogeneous: Photoconductor and thermoplastic are codissolved or copolymerized into each other (or the thermoplastic is itself a photoconductor).
(2) Heterogeneous: The photoconductor forms a very fine suspension in the thermoplastic. The film is therefore an emulsion.
(3) Separate layer: The thermoplastic is coated over the photoconductor.

Although structures (1) and (2) have been used successfully in electrophotography [*10.17*, *10.18*], the separate-layer film (3) has so far proved the most successful for holography [*10.19*]. We limit our discussion to the separate-layer structure illustrated in Fig. 10.16. The substrate is a glass plate coated with a thin layer of transparent conductor, e.g., tin oxide. On top of the transparent conductor are a layer of photoconductor and a layer of thermoplastic.

Before exposure the film is sensitized in darkness by establishing a uniform electrical potential on the surface of the thermoplastic relative to the transparent conductor (see Fig. 10.16). A corona device, consisting of a thin wire at high voltage (6 to 10 kV) and an electrically grounded shield, ionizes the air near the surface of the thermoplastic. Positive ions, attracted toward the electrically grounded transparent conductor, are deposited on the surface of the thermoplastic and are held by the negative charges induced on the transparent conductor. The entire surface of the thermoplastic can be charged to a uniform potential, typically a few hundred volts, by moving the corona device over the surface at constant speed and height.

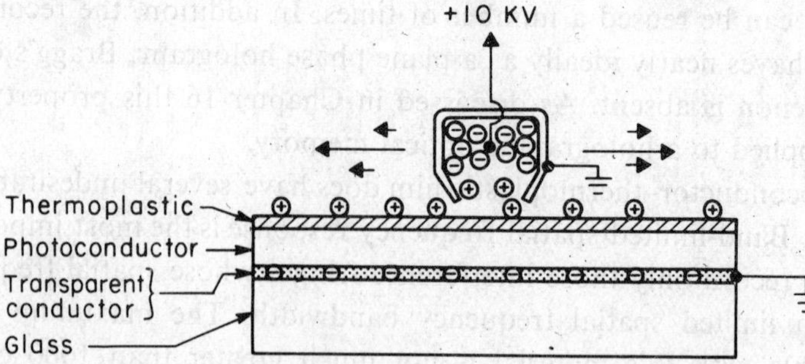

Fig. 10.16. Film structure of a photoconductor–thermoplastic layer system. Also shown is a corona charging device.

The sequence of a complete hologram recording–erasure cycle is illustrated in Fig. 10.17. For simplicity we show only the photoconductor and the thermoplastic, omitting the grounded transparent conductor and the glass substrate. In the *first step*, the *first charging*, a uniform electrostatic potential is applied to the surface of the thermoplastic as described above. In the *second step*, the *exposure*, a holographic interference pattern illuminates the system and causes the photoconductor to discharge in a spatial

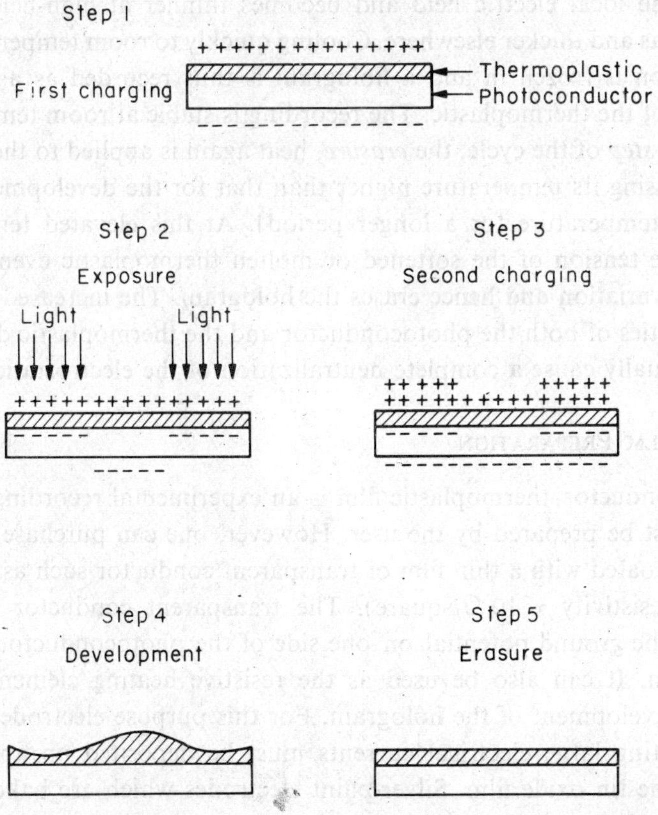

FIG. 10.17. Recording–erasure cycle of the thermoplastic hologram.

pattern dependent on the exposure. In the discharge areas, electrons fall through the potential increment corresponding to the thickness of the photoconductor and deposit on the bottom surface of the thermoplastic. Wherever this happens, the potential measured at the upper surface of the thermoplastic is reduced. However, the electric field in the thermoplastic, which is proportional to the surface charge density, remains unchanged. In the *third step*, the *second charging* (using the same procedure as the first charging), the potential everywhere on the surface of the thermoplastic is recharged to the original value, as determined by the corona device. Wherever the original exposure had caused the photoconductor to discharge, the electric field increases, producing a spatially varying field pattern, while once again on the surface the potential is everywhere uniform. In the *fourth step*, the *development*, heat is applied momentarily to the thermoplastic, raising its temperature to near the softening or melting point, typically between 60° and 100°C. At this temperature the thermoplastic film deforms under the force of the local electric field and becomes thinner at high-field (illuminated) areas and thicker elsewhere. Cooling quickly to room temperature, the deformation is frozen in and a hologram is thus recorded as a thickness variation of the thermoplastic. The recording is stable at room temperature. In the *last step* of the cycle, the *erasure*, heat again is applied to the thermoplastic, raising its temperature higher than that for the development (or at the same temperature for a longer period). At this elevated temperature the surface tension of the softened or molten thermoplastic evens out the thickness variation and hence erases the hologram. The increased electrical conductivities of both the photoconductor and the thermoplastic during the erasure usually cause a complete neutralization of the electrostatic charges.

10.10.2 Film Preparation

Photoconductor–thermoplastic film is an experimental recording medium which must be prepared by the user. However, one can purchase the glass substrate coated with a thin film of transparent conductor such as tin oxide (surface resistivity $\approx 10\ \Omega$/square). The transparent conductor serves to establish the ground potential on one side of the photoconductor–thermoplastic film. It can also be used as the resistive heating element for the thermal development of the hologram. For this purpose electrodes capable of conducting large electrical currents must be deposited on appropriate edges of the tin oxide film. Silver-paint electrodes which are baked on are convenient for experiment. They resist common cleaning solvents, and connections can be soldered to them.

A photoconductor which works well and can be coated on the glass plate in a suitably thin layer is the polymer poly-*N*-vinyl carbazole (PVK) into which is added a small amount of electron donor 2,4,7-trinitro-9-fluorenone[2] (TNF) and a red-sensitizing dye, *brilliant green*. The solvent for the above solids is an equal-volume mixture of dichloromethane and *p*-dioxane. To form a solution from which the glass substrate can be pulled and thus receive

[2] This chemical may be carcinogenic. Care should be taken to avoid breathing vapor or powder.

its coating of photoconductor, solids and solvents are mixed as follows:

PVK	20 gm
TNF	2 gm
Dye	0.2 gm
Solvent	300 cc

Pulling the substrate at a rate of 7.5 cm/min from this solution of photoconductor produces a coating of approximately 2.5 μm.

Of the thermoplastics that are available, a natural tree-resin Staybelite, supplied by Hercules Inc., Wilmington, Delaware, seems to be the most useful. It dissolves in a "super hi-flash naphtha," supplied by American Mineral and Spirits Co., New York. A solution is prepared from 75 gm of the resin per 300 cc of solvent. As soon as the photoconductor coating is dry, the plate is pulled from this solution at a rate of 5 cm/min for a 1-μm-thick coating. The final photoconductor–thermoplastic film is then baked at 60°C for about 1 h.

10.10.3 Hologram Exposure, Development, and Erasure

With the transparent conductor grounded, the film is first electrostatically charged in the dark with a corona device as described previously. Since the photoconductor–thermoplastic film is a thin phase material, both the signal beam and the reference beam usually illuminate the same side (either one) of the film. Any laser wavelength in the visible spectrum can be used and an optimum average exposure is about 100 μJ/cm^2 at 6328 Å. After exposure and the second electrostatic charging, the film need no longer be kept in darkness. Thermal development of the film can be achieved by any of several methods, e.g., convection heating with a stream of hot air, radiation heating by passing the film near an electric heater, or conductive heating by passing an electric current through the transparent conductor (see Section 16.5). The last method is most convenient if the total area of the film is not much larger than 100 cm^2. Conductive heating allows the thermoplastic to be brought to the development temperature very rapidly, typically in a fraction of a second. Erasure usually occurs when the same heating current is applied for twice the development time.

10.10.4 Resolution Limits, Hysteresis, and Reusability

As stated earlier, for a given photoconductor–thermoplastic film, the spatial frequencies of the interference fringes that can be recorded are

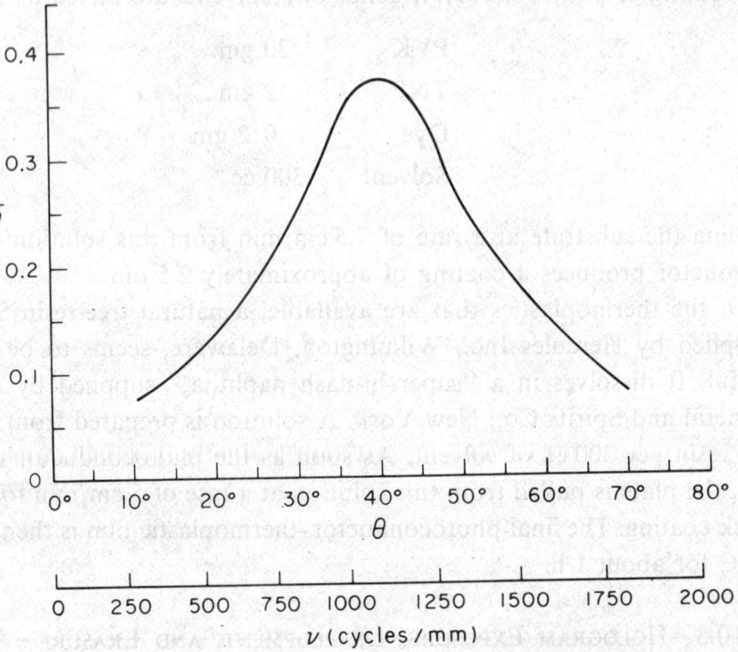

FIG. 10.18. Band-limited character of the spatial frequency response of a thermoplastic hologram. The hologram was formed with plane waves making equal angles to the hologram normal. η, diffraction efficiency; θ, interbeam angle; ν, fringe frequency.

limited to a *band* of spatial frequencies. Surface relief or deformation at spatial frequencies near zero or at several thousand cycles per millimeter is not observed. Depending on the thickness of film and to a lesser degree on the electrostatic field across the thickness prior to development, the peak of the recordable spatial frequency band can range from tens of cycles/mm to over 1000 cycles/mm; the bandwidth is usually about 50% of the peak frequency. Figure 10.18 is a typical curve of the square root of diffraction efficiency versus the spatial frequency of the recorded fringes. It may be regarded as a representative coherent transfer function of the hologram recorded in a particular film. The center frequency can be increased by reducing the film thickness. However, at 1000 cycles/mm, the thickness is only about 1 μm. Such a thin film of uniform thickness is difficult to prepare and is very susceptible to dust and to defects of the substrate surface.

In Chapter 16 we show how erasure and reusability of photoconductor–thermoplastic film can be used to form *write-read-and-erase* optical memories. In such applications one must beware of the possibility of hysteresis, i.e., the possibility that images from earlier holograms, thought to have been erased, reappear in the output of later holograms. Hysteresis is serious if the heat for erasure is insufficient. However, if the method of conduction heating is used, heating can be well controlled and hysteresis need not be a problem. Fig. 10.19, exhibiting images obtained from a hologram after (a), 5; (b), 105; and (c), 106 recording–erasure cycles, is testimony. Insufficient exposure or insufficient heat for development can also leave the hologram with a frostlike appearance, making the reconstruction noisy. On the other hand, overheating can permanently damage the material and terminate the reusability of the film.

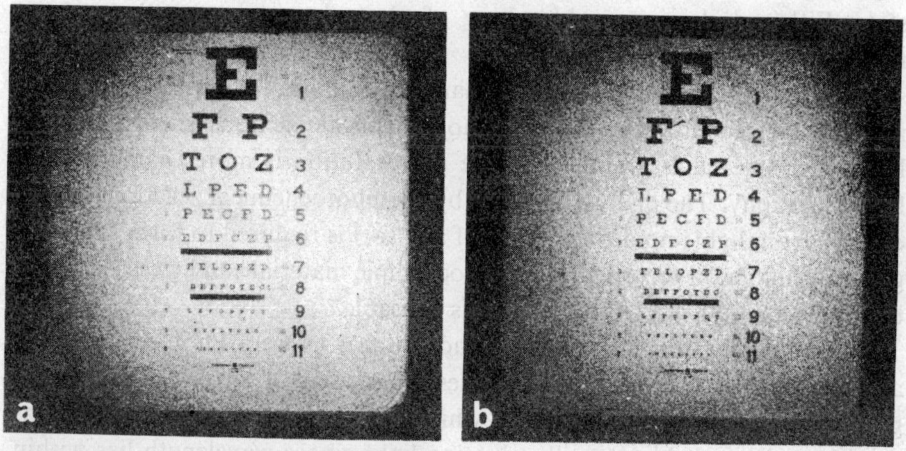

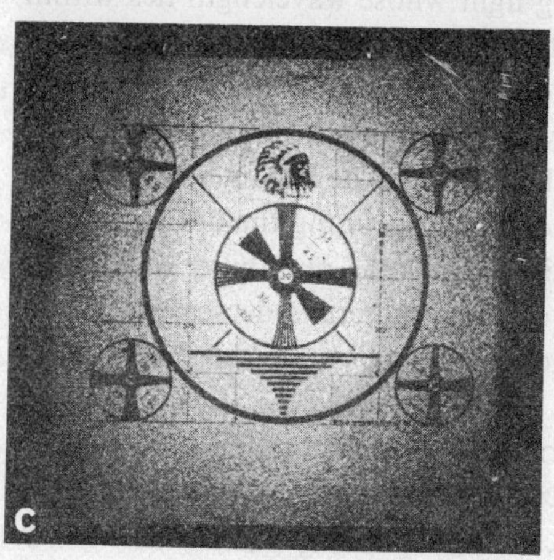

FIG. 10.19. Photographs of images obtained from a thermoplastic hologram after (a) 5, (b) 105, and (c) 106 recording erasure cycles.

10.11 Photochromic Materials

Reversible color change in a material after its exposure to light is called photochromism. Photochromic materials darken after exposure to ultraviolet or blue light. A simple model for crystalline photochromics considers that two classes of impurity centers exist in the photochromic material

[*10.20*]. In an initial state the material absorbs only short-wavelength light and so appears rather transparent in visible light. Absorbed light excites electrons from one class of impurity centers into the conduction band after which they may be trapped by the second class of impurity centers. In this second state the material absorbs in the visible range and appears colored. A return to the original state may be accomplished through natural or stimulated thermal relaxation or by optical bleaching. Bleaching is carried out by illuminating the material with long-wavelength light, which is now absorbed. Electrons in the second class of impurity traps are excited and captured by the traps of the first class, and the material returns to its original state. Either darkening or bleaching of photochromic material can be a mode of hologram recording.

Photochromic materials possess many advantages for hologram recording. Since photochromism involves photochemical reactions at the molecular level, the material is grainless; hence its resolution is more than sufficient for holography. The energy required by the photochemical reactions comes directly from light; no development of any sort is required to form the hologram. A hologram recorded in a photochromic material can be erased either thermally or optically; the material is reusable.

Unfortunately, there are many disadvantages as well. Photosensitivity is at least three orders of magnitude less than that of silver halide photographic emulsion. A photochromic which has recorded a hologram remains sensitive to any ambient or illuminating light whose wavelength lies within the absorption band of the material. It is often possible to choose a wavelength for the reconstructing beam which is substantially different from the wavelength used for recording. However, this leads to aberrations in the reconstructed wavefront. Eventually the recorded hologram is erased by thermal relaxation (dark reaction) although refrigerating the hologram can retard this erasure very effectively. Fatigue limits the reusability of organic photochromics; the photosensitivities of these materials decrease with increasing number of record–erase cycles until finally they become totally insensitive to light.

Photochromism occurs in a wide variety of solid materials which may be organic or inorganic, in solution or in crystalline structures. Given sufficient laser power at the proper wavelength, hologram recording in almost any photochromic material can be at least demonstrated. Diffraction efficiency is a few percent or less. Holograms have been recorded in organic films of spiropyrane derivative [*10.21*] in a darkening mode at 6328 Å, in silver halide photochromic glasses in either a darkening mode at 4880 Å [*10.22*] or a bleaching mode at 6328 Å [*10.23*], in hydrogenated KBr crystals in a bleaching mode at 6328 Å [*10.24*], in CaF_2 crystals doped with La or Ce, and in $SrTiO_3$ crystals doped with Fe–Mo or Ni–Mo, all in a bleaching mode at 6328 Å [*10.25*]. The approximate, room-temperature thermal relaxation times of the four types of CaF_2 and $SrTiO_3$ crystals are reported to be 24 h, 1 week, 10 min, and 15 min, respectively. In most experiments involving photochromic materials, the emphasis has been on the potential for large-capacity optical information storage in thick materials as first proposed by van Heerden (see Chapter 16).

10.12 Ferroelectric Crystals

Certain ferroelectric crystals exhibit small changes in refractive index (typically 10^{-5}) after an intense exposure to light. This photoinduced refractive index change has been called (with good reason) "optical damage" by those working with electro-optic materials. To a certain extent the phenomenon is similar to photochromism; the index change can be reversed with the application of heat and light. Holograms have been recorded [*10.26–10.28*] in the ferroelectrics listed in Table 10.3, which also shows their approximate thermal relaxation times (at room temperature). Most successful results have been obtained with lithium niobate ($LiNbO_3$) crystals.

TABLE 10.3

APPROXIMATE THERMAL RELAXATION TIMES OF HOLOGRAMS RECORDED IN VARIOUS FERROELECTRICS AT ROOM TEMPERATURE

Ferroelectric	Relaxation time	Ferroelectric	Relaxation time
$LiNbO_3$	2 weeks	$BaTiO_3$	1 sec
$LiTaO_3$	2 weeks	KTN[a]	1 sec
$BaCaNaNbO_3$	10 min	$Bi_4Ti_3O_{12}$	50 sec

[a] With 2 kV/cm external electric field applied.

A model [*10.29*] for the mechanism of hologram formation in lithium niobate crystals proposes that the electro-optic effect is responsible for changes in index of refraction. Assume that the holographic interference pattern fringes are a set of planes oriented transverse to the direction of an internal electric field present in the crystal. (The electric field is parallel to the c axis of the crystal.) In the region of the crest fringes, electrons are photoexcited from traps into the conduction band. They drift a short distance under the influence of the internal field and when they reach a region of low light intensity, intermediate to the bright fringes, they are retrapped. This redistribution of charge produces a localized electric field between the retrapped electrons and their ionized donors. As a consequence of the electro-optic effect, the local field in turn effects a change in the local index of refraction. About 1000 V/cm is required to produce an index change of 2×10^{-5}.

The maximum diffraction efficiency observed for a hologram formed in $LiNbO_3$ with fringe planes more or less transverse to the c axis is 42% [*10.26*]. In agreement with the model, only weakly diffracting holograms are formed for parallel orientation. Photosensitivity depends on the direction of polarization of the incident laser light; it is different for ordinary and extraordinary waves. Advantages of hologram formation in ferroelectric crystals parallel those listed for photochromics. They include high resolution in the recording. In addition, no fatigue phenomenon has been observed after many recording–erasure cycles. Since *phase* holograms are formed, their diffraction efficiency can be an order of magnitude better than photochromic holograms.

Ferroelectric crystals have many of the disadvantages of photochromic materials. Specific problems relating to multiple storage of holograms are discussed in Section 16.3.4. The root cause of these problems is that the recording is not "fixed." Another source of difficulty lies in the low holographic sensitivity S. For a 1-cm-thick $LiNbO_3$ crystal illuminated with light of wavelength $\lambda = 4880$ Å, $S \approx 0.3 \times 10^{-6}$ m²/J, a factor 10^6 less than

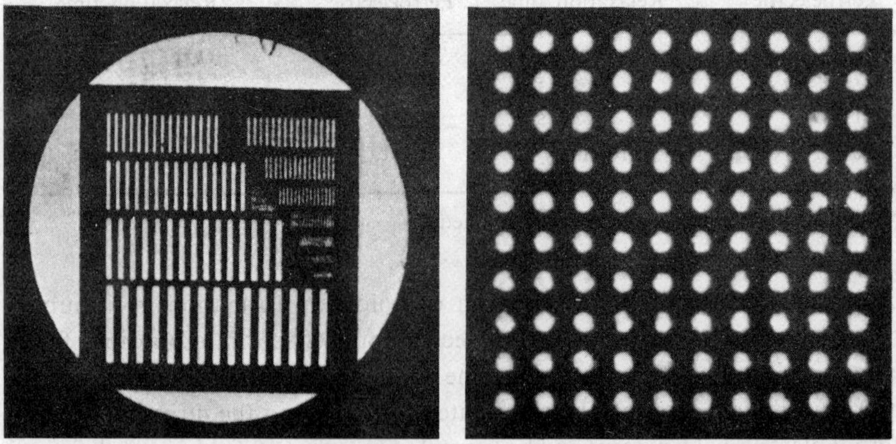

FIG. 10.20. Photographs of images obtained from holograms recorded in a lithium niobate crystal. (Courtesy J. T. LaMacchia, Bell Telephone Labs, Inc.)

the sensitivity of Kodak 649F emulsion. Sensitivity is even less at longer wavelengths. Photographs in Fig. 10.20 show images produced from holograms recorded in a $LiNbO_3$ crystal. (The subjects were transparencies.)

REFERENCES

10.1. *Kodak Plates and Films for Science and Industry*, third printing, Eastman Kodak Co., Rochester, New York, 1967.

10.2. J. C. Wyant and M. P. Givens, "Effects of the Photographic Gamma on the Luminance of Hologram Reconstructions," *J. Opt. Soc. Amer.* **58**, 357 (1968).

10.3. L. H. Lin and C. V. LoBianco, "Experimental Techniques in Making Multicolor White Light Reconstructed Holograms," *Appl. Opt.* **6**, 1255 (1967).

10.4. J. H. Altman, "Pure Relief Images on Type 649-F Plates," *Appl. Opt.* **5**, 1689 (1966).

10.5. V. Russo and S. Sottini, "Bleached Holograms," *Appl. Opt.* **7**, 202 (1968).

10.6. H. M. Smith, "Photographic Relief Images," *J. Opt. Soc. Amer.* **58**, 533 (1968).

10.7. W. J. Cathey, Jr., "Three-Dimensional Wavefront Reconstruction Using a Phase Hologram," *J. Opt. Soc. Amer.* **55**, 457 (1965).

10.8. J. N. Latta, "The Bleaching of Holographic Diffraction Gratings for Maximum Efficiency," *Appl. Opt.* **7**, 2409 (1968).

10.9. J. Upatnieks and C. Leonard, "Diffraction Efficiency of Bleached Photographically Recorded Interference Patterns," *Appl. Opt.* **8**, 85 (1969).

10.10. C. B. Burckhardt and E. T. Doherty, "A Bleach Process for High-Efficiency, Low-Noise Holograms," *Appl. Opt.* **8**, 2479 (1969).

10.11. K. S. Pennington and J. S. Harper, "Techniques for Producing Low-Noise, Improved-Efficiency Holograms," *Appl. Opt.* **9**, 1643 (1970).

10.12. H. Kiemle and W. Kreiner, "High-Efficiency Lippmann–Bragg Phase Hologram," *Phys. Lett.* **28A**, 425 (1968).

10.13. J. Kosar, *Light-Sensitive Systems*, Chapter 2, Wiley, New York, 1965.

10.14. T. A. Shankoff, "Phase Holograms in Dichromated Gelatin," *Appl. Opt.* **7**, 2101 (1968).

10.15. L. H. Lin, "Hologram Formation in Hardened Dichromated Gelatin Films," *Appl. Opt.* **8**, 963 (1969).

10.16. C. E. K. Mees and T. H. James, ed., *The Theory of the Photographic Process*, 3rd ed., Chapter 3. Macmillan, New York, 1966.

10.17. R. F. Kopczewski and H. S. Cole, "Photoconduction in Homogeneous Photoplastic Recording Film," *Appl. Opt. Suppl.* **3**, 156 (1969).

10.18. S. Aftergut, J. J. Bartfai, and B. C. Wagner, "Photoplastic Recording Film Made with CdS," *Appl. Opt. Suppl.* **3**, 101 (1969).

10.19. J. C. Urbach and R. W. Meier, "Thermoplastic Xerographic Holography," *Appl. Opt.* **5**, 666 (1966).

10.20. J. J. Amodei and D. R. Bosomworth, "Hologram Storage and Retrieval in Photochromic Strontium Titanate Crystals," *Appl. Opt.* **8**, 2473 (1969).

10.21. A. L. Mikaeliane, A. P. Axenchikov, V. I. Bobrinev, E. H. Gulaniane, and V. V. Shatun, "Holograms on Photochromic Films," *IEEE J. Quantum Electron.* QE–**4**, 757 (1968).

10.22. J. P. Kirk, "Hologram on Photochromic Glass," *Appl. Opt.* **5**, 1684 (1966).

10.23. W. J. Baldwin, "Determination of the Information Storage Capacity of Photochromic Glass with Holography," *Appl. Opt.* **6**, 1428 (1967).

10.24. G. Kalman, "Holography in Thick Media," *Application of Lasers to Photography and Information Handling* (R. D. Murray, ed.), p. 99. Soc. Photographic Scientists and Engineers, 1968.

10.25. D. R. Bosomworth and H. J. Gerritsen, "Thick Holograms in Photochromic Materials," *Appl. Opt.* **7**, 95 (1968).

10.26. F. S. Chen, J. T. LaMacchia, and D. B. Fraser, "Holographic Storage in Lithium Niobate," *Appl. Phys. Lett.* **13**, 223 (1968).

10.27. J. T. LaMacchia, "Holographic Storage in Ferroelectrics," *Joint IEEE–CUA Symp. Applications of Ferroelectrics* (Oct., 1968). Abst.

10.28. L. H. Lin, "Holographic Measurements of Optically Induced Refractive Index Inhomogeneities in Bismuth Titanate," *Proc. IEEE* **57**, 252 (1969).

10.29. F. S. Chen, "Optically Induced Change of Refractive Indices in Lithium Niobate and Lithium Tantalate," *J. Appl. Phys.* **40**, 3389 (1969).

Chapter 11

PULSED-LASER HOLOGRAPHY

When a hologram is made with a continuous-wave laser, the subject is usually fixed in position so that it cannot move so much as a fraction of the wavelength of light. This high degree of stability, essential for recording high spatial frequency holograms, precludes the use of live or moving bodies as hologram subjects. We noted in Section 7.5 that all problems of subject stability and mechanical or airborne disturbance are lessened as the hologram exposure time is reduced. In this section we discuss pulsed-laser sources whose output energy is sufficient to expose a hologram plate in tens of nanoseconds. With proper technique and suitable precautions, formation of a hologram with a pulsed laser brings a variety of hitherto forbidden subjects, e.g., projectiles, liquid jets, aerosol particles, flying insects (Fig. 11.1), and humans, into the range of holography. Reconstruction of the subject wave is generally carried out with a CW He–Ne laser.

Since 1965, back-illuminated scenes, e.g., Fig. 11.1, have been subjects for in-line Fraunhofer holograms [*11.1*] and off-axis Fresnel holograms [*11.2*] formed with coherent light from pulsed ruby lasers. Brooks *et al.* [*11.2*] have applied pulsed-laser holographic methods to the interferometry of transient events with great success (Fig. 11.2) (see Section 15.2). However, it has been only recently that pulsed-laser holograms of front-lighted scenes (i.e., of reflecting 3D objects) have achieved an image quality comparable to those made with CW lasers. The most significant problem has been the short coherence length of those pulsed lasers with sufficient output power. Working toward a common goal, laser designers have succeeded in obtaining several joules of energy from *Q*-switched, single-frequency, pulsed lasers, while a manufacturer of photographic emulsion has developed plates with

FIG. 11.1. Photograph of the virtual image of a fruit fly in flight. Hologram was exposed with a Q-switched ruby laser pulse of 100-nsec duration. (Courtesy R. F. Wuerker, TRW Inc.)

high sensitivity at the ruby laser emission wavelength. As a result, holograms equal in quality to any made with CW lasers can be obtained with the pulsed laser. What is more, the massive tables and clamps so vital for CW laser holography may be dispensed with. Thus, forming a hologram with a pulsed laser can, in some respects, approach the simplicity of making a snapshot. On the other hand, there is only one color, red, available from the ruby laser so that at present multicolor holographic images are not possible. This deterrent is minor, however, in comparison to that due to the present high cost of pulsed-laser apparatus.

11.1 The Multimode Ruby Laser

The ruby laser emitting red light at 6943-Å wavelength is the only practical pulsed laser currently suitable for optical holography. Still in the future, so far as holography is concerned, are the yttrium aluminum garnet (YAG) and the glass lasers whose high-power infrared emissions can generate intense green light when passed through a frequency doubler. We shall limit our discussion to those properties of the ruby laser [*11.3*] relevant to holography.

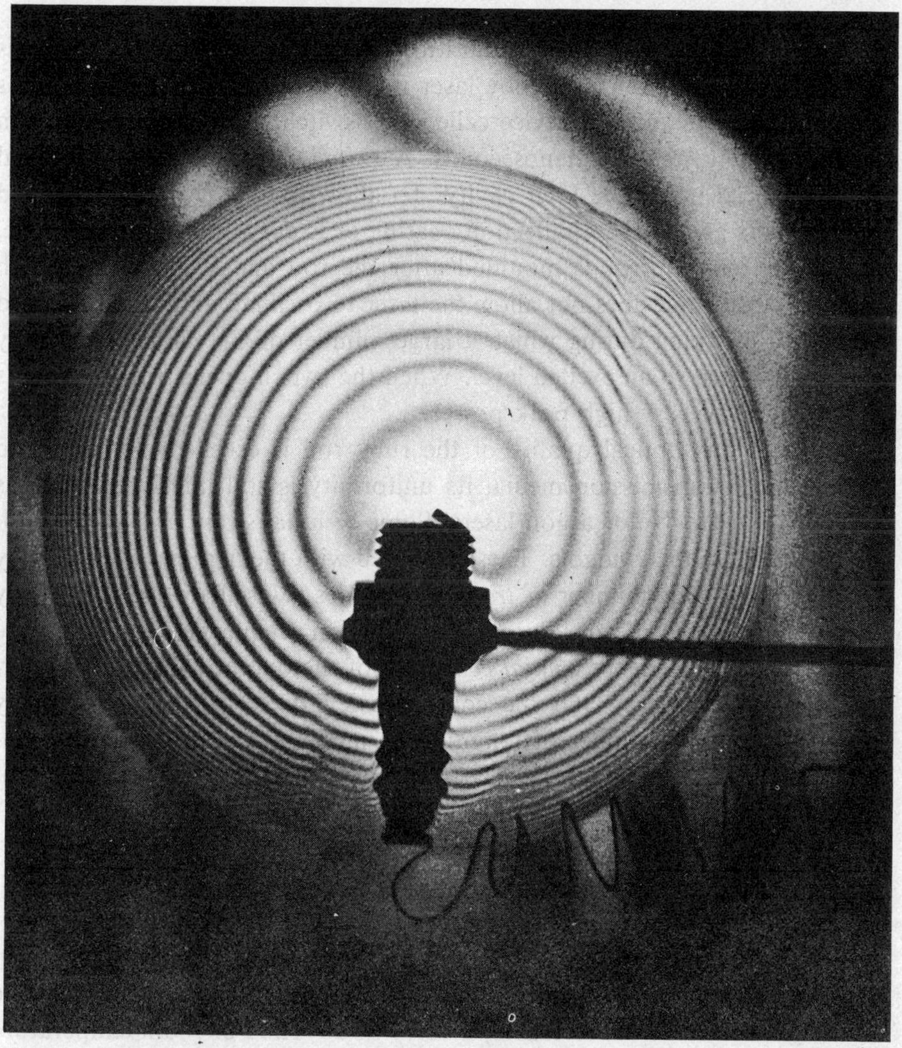

FIG. 11.2. Photograph of the holographic image of a small spark plug igniting an acetylene–oxygen mixture. (Courtesy R. F. Wuerker, TRW Inc.)

Laser action results from stimulation of Cr^{3+} ions in pink ruby (Al_2O_3 containing 0.05% by weight of Cr_2O_3). The ruby is usually cut in the shape of a cylindrical rod; typical dimensions are 5 mm in diameter and 4 cm in length. Both end faces are polished parallel. With a suitable coating of silver or dielectric, one end is made totally reflecting and the other partially reflecting. When the rod is optically pumped with light energy from a xenon flash lamp, approximately 1 J of red, linearly polarized, laser light energy is emitted in a pulse whose duration is ~0.25 msec. The direction of polarization is normal to the plane formed by the rod axis and the crystal c axis.

As we shall learn, the pulse duration of 0.25 msec resulting from this conventional operation of a ruby laser is too long for forming holograms of many subjects. With a device called a Q switch, the pulse duration can be shortened to tens of nanoseconds. A Q switch is a fast-acting optical shutter placed inside the laser cavity. To allow for the Q switch, one end of the rod is given an antireflection coating or is cut at the Brewster angle, and an external mirror is mounted after the rod and switch. The switch remains closed and does not allow the laser to oscillate until after a period of optical pumping during which a large number of Cr^{3+} ions in the ruby rod are elevated to an excited level. When the switch opens, the laser emits a brief burst of very high peak power.

Although the optical quality of the ruby rod is excellent compared to most optical transmission media, its uniformity is still inferior to the gas or plasma in He–Ne or argon lasers. Some residual strain and some inhomogeneities are unavoidable. These, coupled with the high gain of the ruby laser, cause a large number of *transverse* modes to oscillate simultaneously across the rod diameter. Since the oscillation frequencies of transverse modes are unrelated, the *spatial coherence* of the output light is very poor.

Temporal coherence and the coherence length of a laser are strongly dependent on the number of longitudinal modes that can oscillate simultaneously (see Section 7.1.3). The frequency separation of two adjacent longitudinal modes is given by Eq. (7.13) as

$$\Delta f = f_{n+1} - f_n = \frac{c}{2l}$$

where c is the speed of light in air and l is the total *optical* path length in the ruby and air between the mirrors of the laser cavity. A typical value of l might be $l = 20$ cm so that $\Delta f = 0.75 \times 10^9$ Hz. For room temperature the full line width of the ruby fluorescence at 6943 Å is $\Delta f_L = 420 \times 10^9$ Hz. Assuming that all modes whose frequencies $f_n = nc/2l$ lie within Δf_L have sufficient gain to sustain oscillation, we estimate the number of oscillating modes to be $\Delta f_L/\Delta f = 560$. Consequently the coherence length of the laser cannot be much longer than that of the fluorescence. Suppose that the fluorescence line shape is Lorentzian with a width $\Delta f_L = 420 \times 10^9$ Hz. Equation (7.12) indicates that the coherence length of the fluorescent light is $\Delta L_H = 0.11c/\Delta f_L \approx 80\ \mu$m. Coherence lengths as short as this make the multimode laser unsuitable for general hologram formation. When used to form holograms of back-illuminated scenes, the optical paths taken by subject and reference beams must be closely matched at every point on the hologram plane.

11.2 The Single-Frequency Ruby Laser

As in Section 7.1.3, we shall call a ruby laser oscillating in a single transverse and single longitudinal mode, a single-frequency laser. Although the spatial and temporal coherence of light from a single-frequency laser are the highest possible, the energy in an output pulse is far below that delivered by a multimode laser. It is possible, however, to amplify the output of a single-frequency, pulsed ruby laser with ruby amplifiers and to obtain higher energy levels comparable to those achieved with multimode operation of a laser oscillator. The amplified light has essentially the same degree of spatial and temporal coherence as the light from the single-frequency oscillator.

Techniques for constructing a single-frequency laser vary [*11.4–11.7*]. A simple, practical configuration is that shown in Fig. 11.3. (Both the pumping flash lamp and the cooling system for lamp and laser are omitted from the figure.) Ruby rods used for such lasers must be of extremely high optical quality, and the end faces of the rod must be polished optically flat. When the rod is examined in an interferometer, only one or two fringes should be observed across the rod diameter. The end faces of the rod and the surfaces of other components inside the laser cavity are usually antireflection coated. It is often necessary to use an intracavity aperture, typically 2 mm in diameter, to limit the width of the laser beam to the most homogeneous portion of the ruby rod and thereby control the transverse mode oscillation. Longitudinal mode control is achieved with a passive Q switch [*11.8*]. An intracavity etalon (see Section 7.1.4) can be used in addition. The Q switch is simply a liquid cell containing a saturable dye such as cryptocyanine dissolved in methane. It plays the following role: As the ruby is optically pumped, creating an inverted population of Cr^{3+} ions, the dye absorbs radiant energy emitted by the rod and prevents oscillation. This absorption is nonlinear. At a sufficiently high emission rate, the populations of the dye

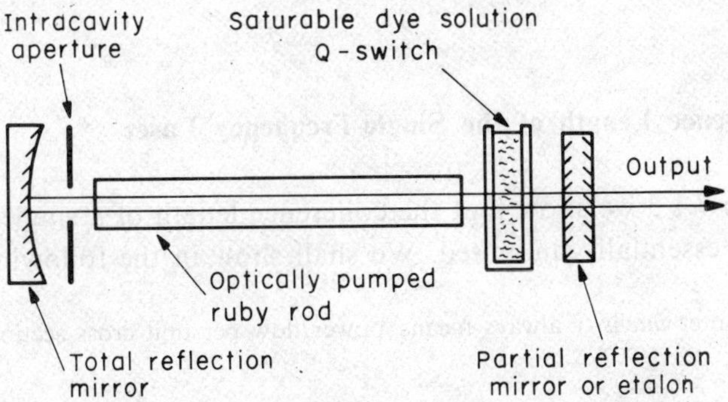

FIG. 11.3. Single-frequency ruby laser configuration.

energy levels involved in the absorption process equalize, and the dye is said to saturate. It is now transparent to the emission from the ruby rod, and laser oscillation occurs in a narrow pulse. Before oscillation begins, light waves corresponding to the various cavity modes make multiple passes through the absorbing dye. Since the absorption by the dye is nonlinear, the process enhances the differences in the loss characteristics of the various longitudinal and transverse modes. The level of optical pumping can be adjusted to permit the gain of only one mode to exceed the threshold for oscillation. When this occurs, all the energy stored in the ruby pours forth at a single frequency.

The pulse shape (variation of intensity with time) of the output of a single-frequency ruby laser is approximately Gaussian. If the laser operates in the lowest-order transverse mode, the TEM_{00}, the variation of intensity with radius is also Gaussian. We may express the intensity[1] as a function of radius r and time t in the form

$$I_P(r, t) = \left(\frac{2}{\pi}\right)^{3/2} \frac{H}{w^2 \Delta t} \exp\left(-\frac{2r^2}{w^2}\right) \exp\left[-\frac{2t^2}{(\Delta t)^2}\right] \qquad (11.1)$$

where $H = \int_0^\infty \int_{-\infty}^\infty I(r, t) 2\pi r \, dr \, dt$ is the total energy per pulse, w is the half-width, and Δt the half-duration of the light pulse. At a radius $r = w$, the intensity is e^{-2} of its maximum value at $r = 0$. Similarly, at $t = \Delta t$ the pulse intensity is e^{-2} of its maximum value at $t = 0$. Some typical values for the measurable quantities are $H = 25$ mJ, $w = 0.5$ mm, and $\Delta t = 15 \times 10^{-9}$ sec $= 15$ nsec.

Ideally, the oscillation frequency of a single frequency laser should remain constant during the pulse. Spatial coherence should be nearly perfect and temporal coherence should depend only on the pulse length. In practice it has been discovered that there can be significant frequency modulation during the pulse [*11.9*, *11.10*]. As we shall learn in the next section, this reduces the temporal coherence. In addition the frequency modulation may not be uniform across the beam diameter thus reducing spatial coherence as well [*11.11*].

11.3 Coherence Length of the Single-Frequency Laser

In Section 7.1.3 we show that the coherence length of a single-frequency CW laser is essentially unlimited. We shall show in the following that the

[1] In this chapter *intensity* always means power flow per unit cross section labeled I_P in Section 1.3.

coherence length of a typical single-frequency pulsed laser is only about 1 m. For many applications this coherence length is quite sufficient.

To determine the coherence length of a light source, we must calculate the complex degree of temporal coherence $\boldsymbol{\mu}_T(\tau)$ defined in Eq. (7.5) as

$$\boldsymbol{\mu}_T(\tau) = \frac{\int_{-\infty}^{\infty} \mathbf{v}(t+\tau)\mathbf{v}^*(t)\, dt}{\int_{-\infty}^{\infty} \mathbf{v}(t)\mathbf{v}^*(t)\, dt} \tag{11.2}$$

where $\mathbf{v}(t)$ and $\mathbf{v}(t+\tau)$ are the complex electric fields of light coming from a single source and arriving via different paths at a point P, and where τ is the transit time difference. In order to carry out the integration indicated in Eq. (11.2) we must express $\mathbf{v}$ in terms which are appropriate to a single-frequency pulsed laser and which also take into account the observed frequency modulation. When the duration of the pulse and the period of the frequency modulation are long compared to the period of the light oscillation, the amplitude of the electric field is proportional to the square root of the intensity. We may then take the square root of $I(r, t)$ in Eq. (11.1) and write

$$\mathbf{v} = a \exp[-t^2/(\Delta t)^2] \exp[i\varphi(t)] \tag{11.3}$$

where a is independent of the time t (but may depend on the spatial coordinates of P) and where $\varphi(t)$ is the phase associated with the modulated frequency $f(t)$. The instantaneous laser light frequency $f(t)$ is defined as

$$f(t) = (1/2\pi)\, d\varphi/dt$$

so that $\varphi(t)$ may be written as

$$\varphi(t) = \int 2\pi f(t)\, dt. \tag{11.4}$$

Let us assume as a first-order approximation that frequency varies linearly with time, as shown in Fig. 11.4, so that

$$f(t) = f_0 + (\Delta f/\Delta t)t \tag{11.5}$$

where f_0 is the mean frequency (at the peak of the pulse) and Δf is the frequency deviation at the half-duration of the pulse Δt. Equation (11.4) can now be written as

$$\begin{aligned}\varphi(t) &= \int 2\pi[f_0 + (\Delta f/\Delta t)t]\, dt \\ &= \varphi_0 + 2\pi f_0 t + \pi(\Delta f/\Delta t)t^2\end{aligned} \tag{11.6}$$

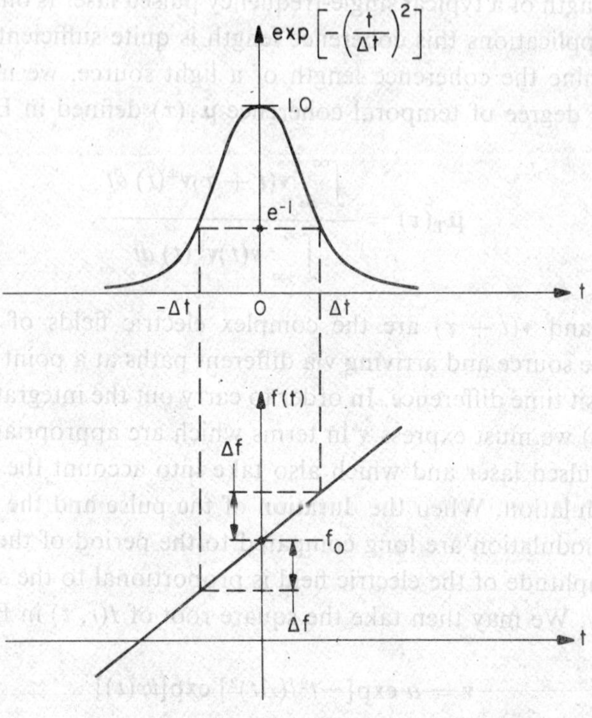

FIG. 11.4. Pulse amplitude and frequency as a function of time.

where φ_0 is a constant of the integration. Equation (11.2) may now be integrated using Eqs. (11.3) and (11.6). A somewhat lengthy computation yields the degree of temporal coherence

$$|\mu_T(\tau)| = \exp\left\{-\left(\frac{\tau}{\Delta t}\right)^2\left[\frac{1+(\pi\,\Delta f\,\Delta t)^2}{2}\right]\right\}. \tag{11.7}$$

From the definition of coherence length given in Section 7.1.2, viz.

$$\Delta L_H = c\tau_H$$

where c is the speed of light and τ_H is the transit time difference which makes $|\mu(\tau_H)| = 1/\sqrt{2}$, we obtain for the single-frequency pulsed laser

$$\Delta L_H = \sqrt{\ln 2}\,\frac{c\,\Delta t}{[1+(\pi\,\Delta f\,\Delta t)^2]^{1/2}}. \tag{11.8}$$

If $\Delta f = 0$, the coherence length is limited only by the pulse duration. Taking $\Delta t = 15 \times 10^{-9}$ sec, a typical value, we find that in the absence of frequency modulation $\Delta L_H = 3.7$ m. Frequency modulation reduces this by a factor $[1 + (\pi \Delta f \Delta t)^2]^{1/2}$. Observed values of Δf for a Q-switched ruby laser are about 50×10^6 Hz [*11.10*], making $\Delta L_H = 0.39 \times 3.7 = 1.4$ m. The value agrees fairly well with that obtained experimentally [*11.12*].

11.4 The Ruby Amplifier

Ruby rods may be used as optical amplifiers to boost the output from single-frequency pulsed lasers up to the level of energy produced by multimode ruby lasers. Both spatial and temporal coherence are essentially preserved in the process. Optical amplification is more practical for ruby lasers than for gas lasers. Whereas the power gain in active gas media is typically a few percent per meter (for visible-light laser media), the power gain in ruby can be as high as 1.1 dB/cm [*11.13*].

Figure 11.5 indicates a practical method of employing an optically pumped ruby rod as an amplifier [*11.7*, *11.12*]. As an amplifier, the ruby can be of lesser quality than that required in single-frequency oscillators and therefore can be made large enough to support a number of long transits by the light

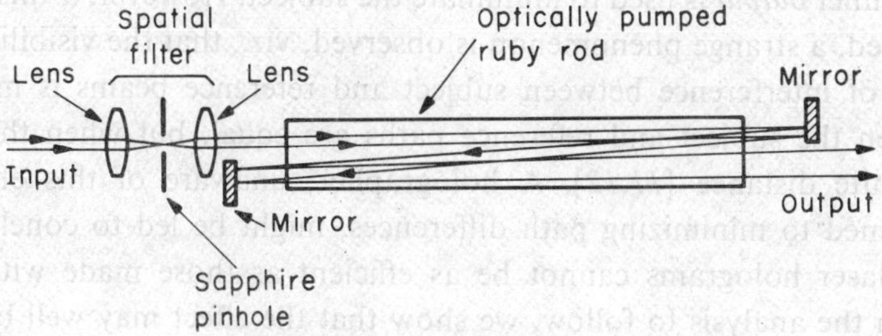

FIG. 11.5. Pulsed ruby amplifier configuration.

beam in the manner shown. For maximum gain, oscillator and amplifier rods are rotated so that the *c* axes of the crystals are coplanar. When the gain is high, precautions must be taken to avoid regenerative oscillation which may damage either of the rods or the mirrors. Of course, it is essential to apply efficient antireflection coatings to both end faces of the amplifier rod. However, even the residue reflection must not be allowed to fall on the output mirror of the oscillator. Optical isolation between oscillator and amplifier is achieved with a simple spatial filter consisting of a sapphire pinhole placed at the common focal point of the two lenses in Fig. 11.5. Input light passes through the pinhole but the reflected light, which can be given a slight off-axis direction, does not. A judicial choice of focal lengths is required to avoid damaging the pinhole or breaking down the air with the focused light. It is also desirable to make the distance between the pinhole and the second lens slightly smaller than the lens focal length. Then the beam expands as it passes through the amplifier rod, and the power density remains below the saturation or damage level for the rod. For an input beam diameter of 1 mm, the output beam diameter should be typically 6 to 7 mm. A typical lens focal length might be 3.3 cm; the pinhole diameter is approximately 30 μm. It is possible to obtain upwards of 10 J from the amplifier output, but the risk of damage to the rod or to other optical components is relatively high. An output of 1 to 3 J is sufficient for general holography and is a reasonably safe level.

As a consequence of inhomogeneities in large-diameter ruby amplifier rods, the radial intensity distribution of the output beam may exhibit large fluctuations superimposed on the Gaussian distribution of the TEM_{00} mode. Although unimportant for illumination of diffusely scattering subjects, these fluctuations can be objectionable when the light is used as a reference beam. Since the energy needed for the reference beam is considerably less than that needed to illuminate the subject, a relatively uniform reference beam may be obtained by tapping part of the *input* to the amplifier. All of the amplifier *output* is used to illuminate the subject. However, if this method is adopted, a strange phenomenon is observed, viz., that the visibility of the pattern of interference between subject and reference beams is maximum not when the subject and reference paths are equal, but when they differ by a finite distance [*11.12*]. A holographer, unaware of this effect and accustomed to minimizing path differences, might be led to conclude that pulsed laser holograms cannot be as efficient as those made with a CW laser. In the analysis to follow, we show that the effect may well be due to a slight shift in the frequency of the laser light as it passes through the amplifier [*11.12*].

Consider the geometry shown in Fig. 11.6. The output and a portion of the input light of a ruby amplifier are directed to a point P via different paths ABP and AP respectively. We assume that the transit time via path ABP exceeds that via AP by an amount τ and that the mean frequency of the light traveling the path ABP through the amplifier is shifted from f_0 to $(1+\delta)f_0$. We shall now compute the *mutual* complex degree of temporal coherence, expressed in a form which distinguishes the two paths and which is analogous to Eq. (7.1):

$$\boldsymbol{\mu}_{\mathrm{T}}(\tau) = \frac{\int_{-\infty}^{\infty} \mathbf{v}_1(t+\tau)\mathbf{v}_2^*(t)\,dt}{\left[\int_{-\infty}^{\infty} \mathbf{v}_1(t)\mathbf{v}_1^*(t)\,dt \int_{-\infty}^{\infty} \mathbf{v}_2(t)\mathbf{v}_2^*(t)\,dt\right]^{1/2}}. \tag{11.9}$$

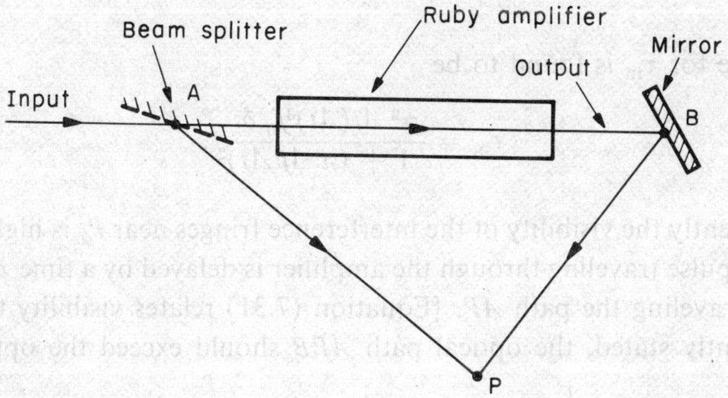

FIG. 11.6. Geometry for the discussion of temporal coherence between input and output of a ruby amplifier.

Here, the complex electric field at P due to the light traveling the path AP is

$$\mathbf{v}_1(t+\tau) = a_1 \exp\{-[(t+\tau)/\Delta t]^2\} \exp[i\varphi_1(t+\tau)] \tag{11.10}$$

where

$$\varphi_1(t+\tau) = \varphi_0 + 2\pi f_0(t+\tau) + \pi(\Delta f/\Delta t)(t+\tau)^2, \tag{11.11}$$

and the complex electric field at P due to light traveling the path ABP is

$$\mathbf{v}_2(t) = a_2 \exp[-(t/\Delta t)^2] \exp[i\varphi_2(t)] \tag{11.12}$$

where

$$\varphi_2(t) = \varphi_0 + 2\pi(1+\delta)f_0 t + \pi(\Delta f/\Delta t)t^2. \tag{11.13}$$

A laborious computation of $\mu_T(\tau)$ yields

$$|\mu_T(\tau)| = \exp[Q(\tau)] \tag{11.14}$$

where $Q(\tau)$ is a quadratic function in τ:

$$Q(\tau) = -\left(\frac{\tau}{\Delta t}\right)^2 \left[\frac{1+(\pi\,\Delta f\,\Delta t)^2}{2}\right] + \pi^2\,\Delta f\,\Delta t\,f_0\,\delta\tau - \frac{\pi^2}{2}(\Delta t)^2 f_0^2\,\delta^2. \tag{11.15}$$

Equation (11.14) reduces to Eq. (11.7) when $\delta = 0$ as it should. Unlike $|\mu_T(\tau)|$ in Eq. (11.7), however, the maximum value of $|\mu_T(\tau)|$ computed here does not occur when $\tau = 0$ but when $\tau = \tau_m$. To maximize $|\mu_T(\tau)|$, we set

$$\left.\frac{dQ(\tau)}{d\tau}\right|_{\tau=\tau_m} = 0.$$

The value for τ_m is found to be

$$\tau_m = \frac{\pi^2 \Delta f (\Delta t)^3 f_0 \delta}{1 + (\pi \Delta f \Delta t)^2}. \tag{11.16}$$

Consequently the visibility of the interference fringes near P_1 is highest when the light pulse traveling through the amplifier is delayed by a time τ_m relative to that traveling the path *AP*. [Equation (7.31) relates visibility to $\mu_T(\tau)$.] Equivalently stated, the optical path *APB* should exceed the optical path *AP* by

$$\Delta L_m = c\tau_m$$

or by

$$\Delta L_m = \frac{\pi^2 c \Delta f (\Delta t)^3 f_0 \delta}{1 + (\pi \Delta f \Delta t)^2}. \tag{11.17}$$

Again setting $\Delta f = 5 \times 10^7$ Hz and $\Delta t = 15 \times 10^{-9}$ sec, the observed values, we obtain

$$\Delta L_m / f_0 \delta = 7.6 \times 10^{-2} \text{ m/MHz}.$$

We see that a frequency shift δf_0 of only 15 MHz can produce the observed value $\Delta L_m \approx 1.2$ m [*11.12*].

11.5 Protection of Optical Components

The peak light intensity in the pulsed beam emitted by a *Q*-switched ruby laser is high enough to destroy some of the optical components used in guiding the light. Proper precautions must be taken to prevent such occurrences. In fact the ruby laser itself may be destroyed if too high an output intensity is demanded. Life expectancy for the ruby rod in a *multimode* laser producing an output intensity[2] of 2×10^8 W/cm² is approximately 1000 pulses but drops to about 1 pulse at 4×10^9 W/cm² [*11.14*]. Multilayer dielectric-coated mirrors inside the laser cavity survive 1 to 5 multimode pulses at 4×10^8 W/cm²; they survive over 50 such pulses when the mirror is immersed in nitrobenzene [*11.15*]. In multimode operation the various transverse modes create high-intensity spots ("hot spots") which are many times the peak intensity value. These hot spots, of course, initiate the damage process. In single-frequency operation, the Gaussian radial distribution of the TEM_{00} mode is smooth so that intensities well over

[2] The intensity values which are quoted in this section were measured as spatial averages.

3×10^8 W/cm^2 can be tolerated without damage to the ruby rod and dielectric mirrors.

Damage thresholds for rods and mirrors outside the cavity are considerably higher. Deployment of lenses for spatial filtering or expanding a pulsed laser beam must be carefully considered. Optical glass reportedly damages at thresholds varying from 2×10^9 W/cm^2 for lanthanium borate glass to 7×10^{10} W/cm^2 for borosilicate crown glass [*11.16*]. (Again, these values may be conservative in the case of single-frequency operation in the TEM_{00} mode.) Multiple-element lenses cemented together with Canada balsam are damaged by lesser intensities. The cement is darkened by a laser pulse of peak intensity 3 to 4×10^7 W/cm^2 [*11.17*]. On the other hand, when the elements are air-spaced, damage may occur when the reflection from one of the elements is focused on another [*11.17*].

Positive lenses are used together with pinholes in metal sheet to expand and spatially filter the output of CW lasers (see Fig. 7.15). Only when it is certain that the peak intensity at the focal point is below the breakdown threshold for air $I_b \approx 7 \times 10^{11}$ W/cm^2 [*11.18*] should positive lenses be used to expand pulsed-laser beams. Knowing the peak output intensity I_1 of a pulsed laser and the half-width of the beam w_1 at the front focal plane of a positive lens, one can calculate the smallest focal length allowed. Assuming a Gaussian beam and a diffraction-limited lens, Eq. (7.46) gives the half-width of the beam in the back focal plane as

$$w_2 = \lambda f/\pi w_1 \tag{11.18}$$

where f is the focal length and λ the wavelength. If all the power flowing through a cross section of the beam in the front focal plane also flows through the back focal plane of the lens, the peak intensity in the back focal plane is given by

$$I_2 = (w_1/w_2)^2 I_1 = (\pi w_1^2/\lambda f)^2 I_1,$$

where we have used Eq. (11.18). If we set $I_2 = I_b = 7 \times 10^{11}$ W/cm^2 and solve for the minimum focal length f_m, we obtain

$$f_m = (\pi w_1^2/\lambda)(I_1/I_b)^{1/2}. \tag{11.19}$$

I_1, the peak intensity for a single-frequency laser, is given by Eq. (11.1) as

$$I_1 = (2/\pi)^{3/2} H/(w_1^2\, \Delta t). \tag{11.20}$$

With the values $H = 25$ mJ, $w_1 = 0.5$ mm, and $\Delta t = 15 \times 10^{-9}$ sec corresponding to the output of a typical single-frequency laser, we find that $I_1 = 3.3 \times 10^8$ W/cm² and $f_m = 24.5$ mm. Substituting the value of f_m for f in Eq. (11.18), we see that the minimum spot size is $2w_2 = 22$ μm. A beam of this diameter will pass through the 30 μm sapphire pinhole shown in Fig. 11.5. (Sapphire is damaged at about the same intensity level as ruby.) To avoid air breakdown at the *output* of a ruby amplifier, where the peak intensity is higher, a single-element negative lens is commonly used to expand the beam.

Mirrors and beam splitters used for hologram formation must also be protected against damage due to high light intensity. The metalized mirrors and beam splitters have a threshold of only about 10^6 W/cm² and can be damaged by an unexpanded beam in one pulse. As mentioned previously, dielectric mirrors have a much higher threshold and can withstand the direct beam from the laser. It is very important, however, to keep dielectric mirrors clean. Dust and other organic contamination on a dielectric mirror can initiate damage at intensities below 10^6 W/cm².

Colored-glass filters or dielectric interference filters have damage thresholds similar to optical glasses and to dielectric mirrors. However, the neutral density filters used with CW lasers are usually composed of organic materials and are generally unsuitable for pulsed-laser holography. Recently, solid glass neutral density filters, which can withstand intensities of approximately 3×10^5 W/cm², have become commercially available.

11.6 Arrangements for Forming Pulsed-Laser Holograms

Although the basic arrangement for hologram formation is not affected by replacement of a CW laser with a pulsed laser, there are some unique aspects arising from the lesser coherence length of the latter and from the greater range of subjects made possible with the short exposure pulse.

11.6.1 Multimode Pulsed-Laser Source

Multimode ruby lasers, conventional and Q switched, were the light sources first used for pulsed-laser holography and are still satisfactory for many back-lighted or transilluminated scenes. Transparent cells containing gases or fluids, transparent bodies, or opaque bodies as silhouettes are typical of such scenes. In each case, scattering of the illuminating beam by the subject scene is confined to small angles. If, in the absence of the subject, the subject illuminating beam and the reference beam can be arranged to have a high degree of mutual coherence everywhere on the hologram plane, then introduction of such subjects will not disturb the mutual coherence of the beams to any significant extent. An ingenious arrangement developed by Brooks *et al.* [*11.19*] satisfies the coherence requirement, even for scenes illuminated through a diffusing screen (Fig. 11.7). As discussed in Section 8.2, diffuse illumination is desirable if the virtual image is to be directly observed by eye (see Fig. 8.13).

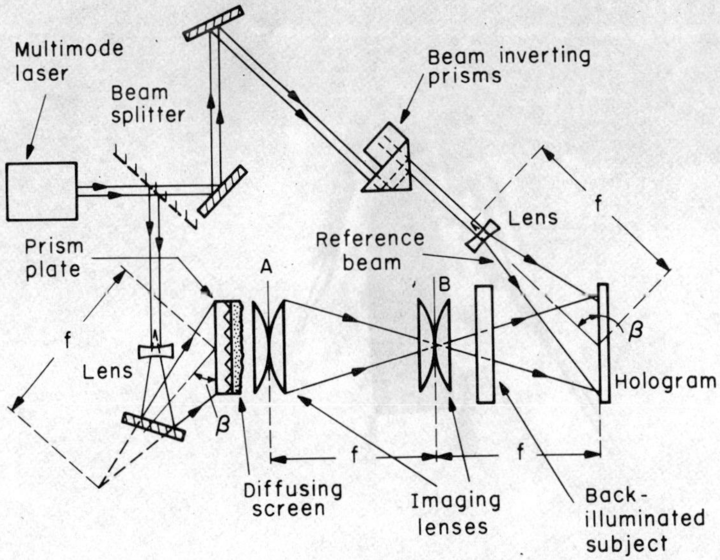

FIG. 11.7. An arrangement for forming a hologram of a subject receiving diffuse back-illumination from a multimode laser. (After Brooks *et al.* [*11.19*].)

Let us examine the optical arrangement of Fig. 11.7, ignoring the subject which we assume to be a small-angle scatterer. At the beam splitter in the upper left of the figure, the wavefront of a multimode laser beam is amplitude divided into two components. It is the function of the remainder of the system to ensure that each pair of rays at corresponding positions on the divided wavefronts travel equal though different optical paths to the hologram plate. There each pair of rays is caused to interfere with each other but with no other rays. In this way high-visibility fringes are recorded on the hologram despite poor spatial or temporal coherence of the laser light.

The diffusing screen is included in the subject beam, and it is imaged onto the hologram plate. If the imaging system does not distort the ray paths, the plane of the diffuser can be considered to reside at the hologram plane. It is then necessary only to illuminate the diffuser at the same angle as the reference beam illuminates the hologram, thus matching ray paths point by point. Of course the lens system inverts the image, and compensating inverting prisms must be introduced in the reference beam. The imaging is carried out by the plano-convex lens pair *B*. Each component of the pair has a focal length *f*. Convex pair *A* acts as a condenser and images the virtual point source illuminating the diffuser onto the central plane of the lens pair *B*.

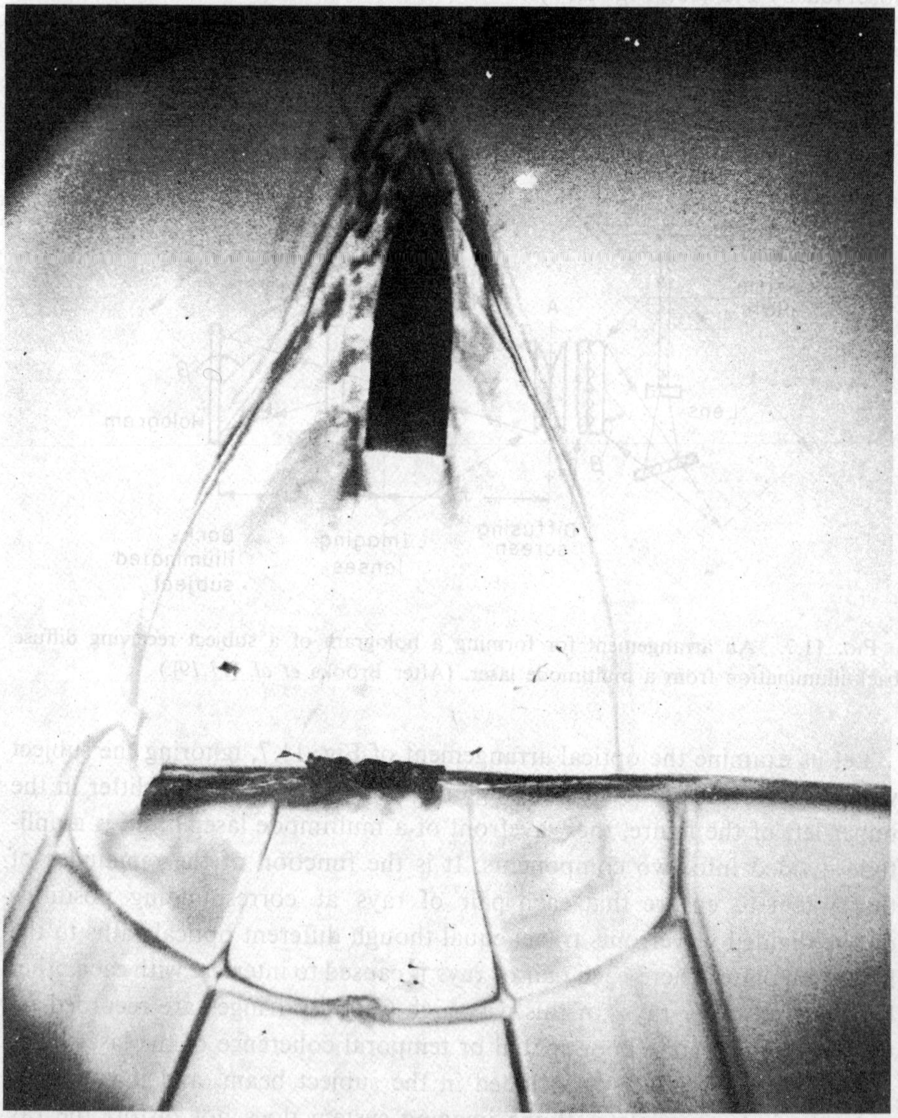

FIG. 11.8. Photograph of an image from a hologram formed with the arrangement of Fig. 11.7. A 22-caliber bullet has passed through a thin brass plate producing shock waves. (Courtesy R. E. Brooks, TRW Inc.)

Lens pairs *A* and *B*, in combination, serve to maintain the optical path length from any point on the screen to the corresponding point on the image more or less independent of position on the screen. Each component of the plano-convex pair also tends to compensate for the other's aberrations. To utilize the diffusing screen efficiently, one should illuminate it in the direction of its normal. An array of thin prisms adjacent to the diffusing screen changes the angle of the illumination for this purpose. Each prism is so small that the optical path difference it introduces is small compared to the coherence length of the laser. Figure 11.8 is a photograph of an image obtained from a hologram formed with the arrangement of Fig. 11.7.

11.6.2 Single-Frequency Pulsed-Laser Sources

Single-frequency pulsed-laser sources, with approximately 1 m of coherence length, do not require special hologram-forming geometry. Because of the short exposure time, mechanical stability requirements are relaxed and it becomes relatively easy to make high-quality transmission holograms [*11.20*] and reflection holograms [*11.21*]. When a ruby amplifier is employed and the reference beam is derived from the input to the amplifier, then the

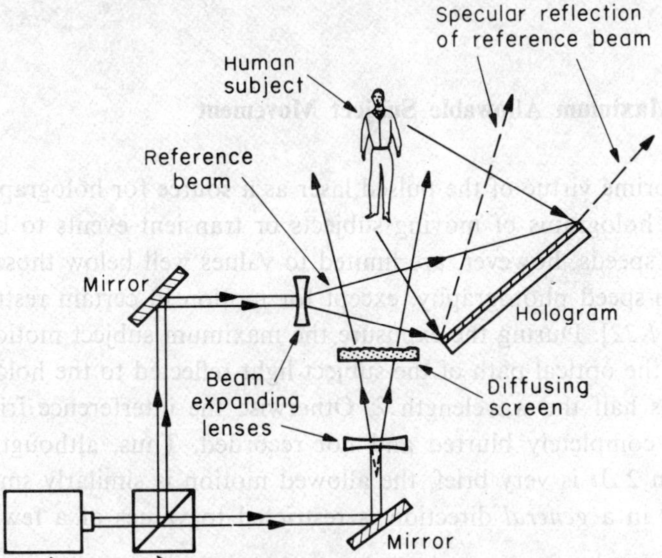

FIG. 11.9. Arrangement for forming a hologram of a human subject using a *Q*-switched single frequency ruby laser. (After Ansley [*11.12*].)

subject beam path must be lengthened by the distance ΔL_m of Eq. (11.17). It is best to determine experimentally the value of ΔL_m which leads to highest visibility.

When a human (or for that matter, any living creature) is part of the subject scene, the subject illumination energy entering the eyes must be at a safe level. Figure 11.9 shows how pulsed-laser holograms of human subjects can be safely made. Instead of using direct light from the expanded laser beam to illuminate the subject, the illumination passes first through a diffusing screen. [Safety is further promoted by the use of sensitive recording emulsion (see Section 11.9).] An arrangement for forming a transmission hologram is shown in the figure. However, the reference beam can illuminate the hologram plate from the side opposite to the subject so as to form a reflection hologram as well. As we shall learn, it is with the reference beam that the greatest care should be taken. Its peak intensity is usually greater than the level thought safe for the eye. Stray reflections or transmissions must not be allowed to reach the eye, since this light can be focused to a high intensity small spot on the retina. To detect the stray light, it is essential to simulate the pulsed-laser light beams with low-power CW laser beams. Figure 11.10 is a photograph of a 3D holographic image of L. D. Siebert.

11.7 Maximum Allowable Subject Movement

The prime virtue of the pulsed laser as a source for holography is that it enables holograms of moving subjects or transient events to be recorded. Subject speeds, however, are limited to values well below those acceptable for high-speed photography, except for motion in certain restricted directions [*11.22*]. During the exposure the maximum subject motion must not change the optical path of the subject light reflected to the hologram by as much as half the wavelength λ. Otherwise the interference fringe pattern will be completely blurred and not recorded. Thus, althougth the pulse duration $2\,\Delta t$ is very brief, the allowed motion is similarly small, and the velocity in a *general* direction is restricted to values of a few meters per second.

Let us consider the maximum speed of a subject point allowed in the hologram-forming geometry of Fig. 11.11. A laser beam, focused to a point at L, provides both illumination for the subject S along the path LS and a reference beam in the direction LH. The reference light proceeding along the path LH and the light reflected by S interfere at a point H on

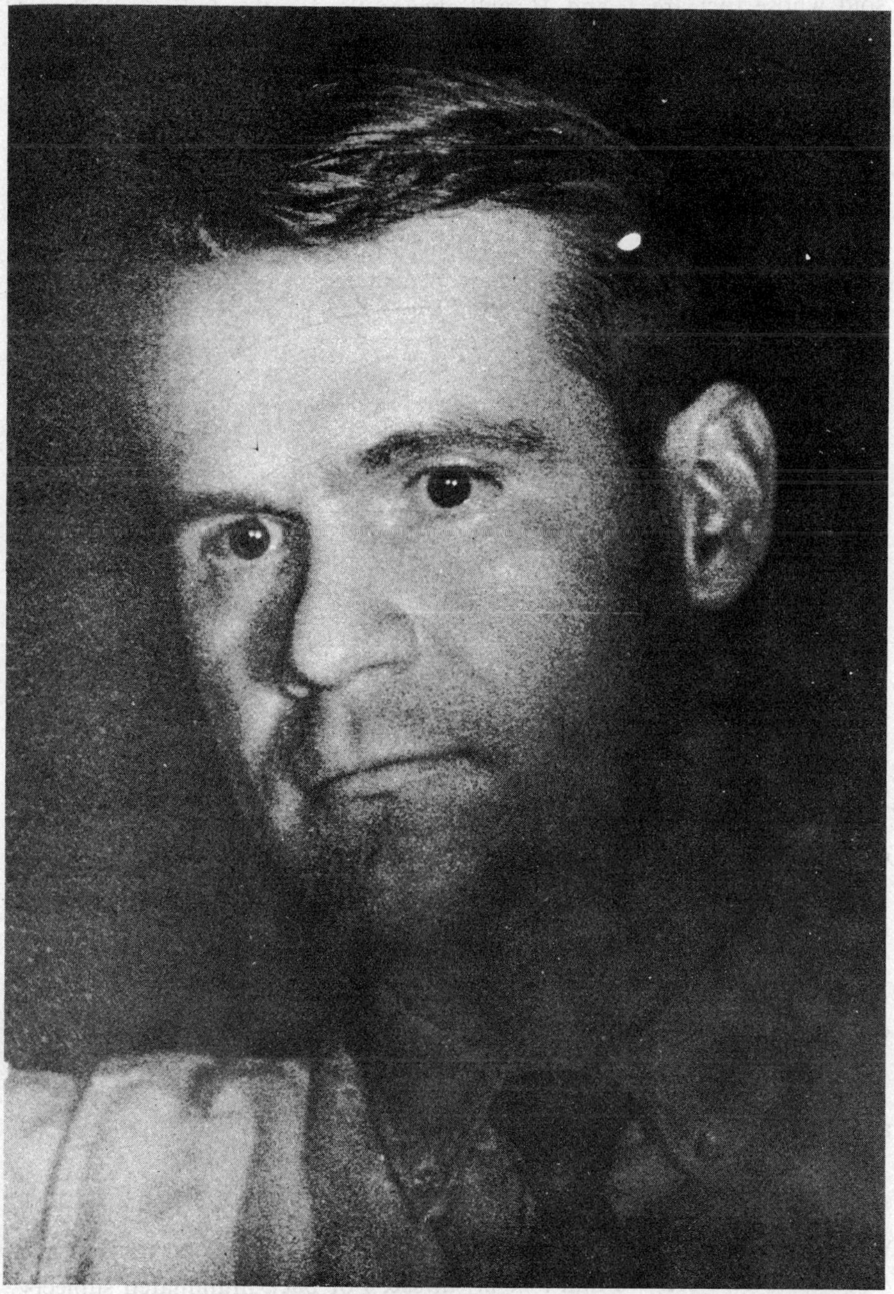

FIG. 11.10. Photograph of a three-dimensional, holographic image of L. D. Siebert Courtesy L. D. Siebert, Conductron Corp.)

the hologram. Any motion of S, other than that on the surface of the ellipsoid whose foci are L and H, will change the length of the optical path LSH and alter the interference fringe pattern at H [11.12]. When the point S does move along the surface of the ellipsoid its speed is limited only by the desired image resolution, as in ordinary photography. A worst-case motion of S is that in a direction normal to the ellipsoid surface. Suppose we define the maximum allowed displacement s in that direction during an

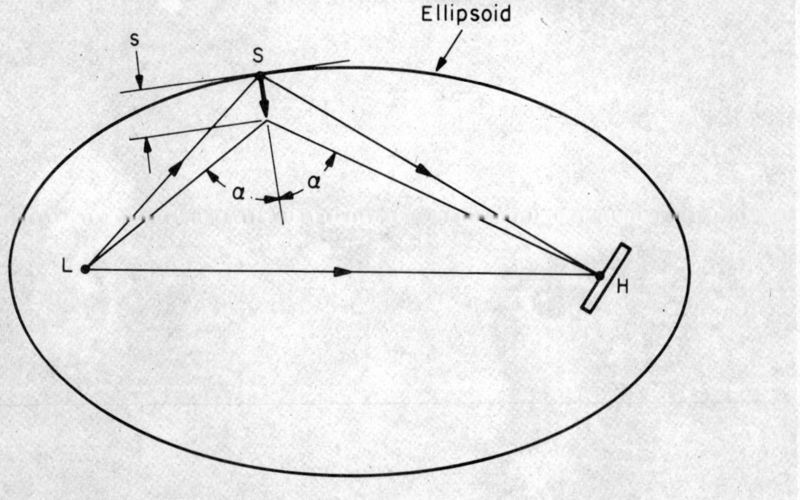

FIG. 11.11. Geometry for computing maximum allowable speed of a point subject.

exposure $2\,\Delta t$ to be that producing a change $\Delta L = \lambda/2$ in the optical path LSH. The maximum allowed speed v is then $v = s/2\,\Delta t$. From Fig. 11.11 it is seen that

$$\Delta L = 2s \cos \alpha = \lambda/2$$

or

$$s = \lambda/(4 \cos \alpha)$$

and

$$v = \lambda/(8\,\Delta t \cos \alpha)$$

where 2α is the angle LSH. Taking α to be 30°, $\lambda \approx 7 \times 10^{-7}$ m, and $\Delta t = 15 \times 10^{-9}$ sec, we find $v \approx 6.7$ m/sec. For back-illuminated subjects, e.g., the bullet in Fig. 11.8, where L, S, and H are essentially in line and $\alpha = 90°$, the maximum allowable speed is determined only by the desired image resolution.

11.8 Safe Illumination of Human Subjects

Because the lens of the eye can focus laser light to a small high-intensity spot on the retina, it is important to know the surface energy density threshold for retinal damage and to ensure that the level is never exceeded. The American Conference of Governmental Industrial Hygienists (ACGIH) [*11.23*] lists the following energy densities as maximum safe levels *on the retina*:[2]

Q-switched laser (pulse duration, 30×10^{-9} sec):	0.07 J/cm^2
Conventional laser (pulse duration, 200×10^{-6} sec):	0.85 J/cm^2.

Since a single-frequency Q-switched laser is the most probable source when humans are to be the subjects of holograms, we shall be concerned with the first-listed value of 0.07 J/cm^2. We regard this as the peak value of energy density that can be tolerated by the retina.

Consider direct or specularly reflected laser light from a Q-switched single-frequency laser to be incident with uniform intensity I_0 over the cornea. Let R be the radius of the pupil and f the focal length of the eye. The intensity distribution $I(r)$ at the retina in the back focal plane of the lens of the eye is given by substituting Eqs. (6.63) and (6.64) into Eq. (6.24) and multiplying Eq. (6.24) by its complex conjugate. Thus

$$I(r) = \left[\frac{\pi R^2}{\lambda f}\right]^2 I_0 \frac{J_1{}^2[2\pi Rr/\lambda f]}{[\pi Rr/\lambda f]^2}$$

where we have set $a_l{}^2 = I_0$ and where r is the radial distance (in the plane of the retina) from the center of the focused spot. When $r = 0$, $I(r)$ has its maximum value

$$I(0) = [\pi R^2/\lambda f]^2 I_0.$$

If we integrate over the duration of the laser pulse, we may rewrite the intensity relation in terms of energy per unit area or energy density. In particular, if E_{ret} is the maximum surface energy density tolerable on the retina and E_{cor} is the corresponding energy density at the cornea, then the maximum allowed energy per unit area of the cornea is

$$E_{\text{cor}} \approx [\lambda f/\pi R^2]^2 E_{\text{ret}}.$$

Taking $\lambda = 7 \times 10^{-7}$ m, $f = 15$ mm, $2R = 7$ mm (maximum), and $E_{\text{ret}} =$

[2] The ACGIH has set 10^{-3} J/cm^2 as the energy density limit for the skin.

0.07 J/cm², we obtain for the maximum safe level at the cornea $E_{cor} = 5.2 \times 10^{-9}$ J/cm². Even the most sensitive photographic emulsions require an energy density of 10^{-6} J/cm² for hologram recording (see Section 11.9). Since the reference beam carries most of this energy density, it is obvious that prevention of retinal damage requires careful control of reference beam reflections.

Let us assume that proper precautions are taken and that the subject receives illumination coming only from the diffusing screen shown in Fig. 11.9. Normally the distance s separating the subject's eye from the screen is much greater than the focal length f of the eye. A small area A_s of the screen will then be imaged to a smaller area A_r on the retina at the back focal plane of the eye. Simple ray geometry gives the ratio of the areas as $A_s/A_r = (s/f)^2$. Now let the surface energy density at the screen be E_s, and let us suppose that the screen area A_s diffuses the energy E_sA_s incident on it into a solid angle Ω_s. If the diameter of the pupil of the eye is D, the solid angle it subtends at the screen is $\pi D^2/4s^2$ and the fraction of the energy it receives is $\pi D^2/4s^2\Omega_s$. Consequently the energy received by the retina of the eye, E_rA_r, where E_r is the surface energy density at the retina is given by

$$E_rA_r = (E_sA_s)\frac{\pi D^2}{4s^2\Omega_s},$$

or

$$E_r = \frac{\pi D^2}{4s^2\Omega_s}\frac{A_s}{A_r}E_s = \frac{\pi D^2}{4f^2\Omega_s}E_s.$$

Note that the relation between E_r and E_s is independent of s. Inserting nominal values $D = 7$ mm, $f = 15$ mm, and $\Omega_s = \pi/10$ steradian, we obtain $E_r = 0.54E_s$. Since the maximum safe value for E_r is $E_r = 0.07$ J/cm², the energy density of the laser beam incident on the diffusing screen must be kept below $E_s = 0.07/0.54 = 0.13$ J/cm². For example, when the total energy per pulse uniformly illuminating the screen is 1 J, the minimum area of the screen is $1/0.13 = 7.7$ cm².

11.9 Hologram Recording Materials

Many of the recording materials suitable for CW lasers (Chapter 10) are found to respond poorly to the short pulse of 6943-Å wavelength light emitted by the ruby laser. In this section we consider materials especially made for the ruby laser or materials which show promise of being useful. Some of their characteristics are summarized in Table 11.1.

TABLE 11.1

RECORDING MATERIALS FOR PULSED LASER HOLOGRAPHY

Recording material	Required energy density ($\mu J/cm^2$)	Maximum diffraction efficiency (%)	Resolution	Reference	Remark
Photographic emulsion Agfa 8E75	11–21[a]	4[b]	"3000 lines/mm"[c]	*11.24*	Can be used for reflection hologram
Photographic emulsion Agfa 10E75	3–6[a]	3[b]	"2800 lines/mm"[c]	*11.24*	
Photographic emulsion Agfa 14C75	0.5–1[a]	0.4[b]	"1500 lines/mm"[c]	*11.24*	
Photographic emulsion Kodak 649F	2000–8000[a]	4.4[b]	"over 2000 lines/mm"[c]	*11.25*	Higher sensitivity with 8-min development time
Magnetic film MnBi	30,000	0.001 Faraday effect 0.01 Kerr effect	"2 μm resolution"[d]	*11.26*	Erasable. Curie temp. 633° K
Magnetic film EuO	100–1000	0.01	"~100 lines/mm"[d]	*11.27*	Erasable. Curie temp. 69.5° K
Metallic film Bi	50,000	6	"close to 1000 lines/mm"[d]	*11.29*	

[a] Effect of reciprocity failure at 10–50 nsec included. For optical density 0.5.
[b] At 45° separation between two interfering plane waves.
[c] Specified by manufacturer.
[d] Quotation from adjoining reference.

Use of a pulsed laser does not lessen the need for materials with sensitivity, resolution, and persistence. Just as with the CW laser we would like the holograms recorded with the ruby laser to be linear, low-noise, and erasable. Thus, much of the basic discussion in Chapter 10 applies here. We note that the expression Eq. (10.6) for the energy density E_0 exposing a hologram plate should now be written

$$E_0 = \eta_t H/A\,,$$

where η_t is the net energy transfer efficiency from laser to recording material, H is the output energy per pulse of the laser, and A is the cross-sectional area of the beam illuminating the hologram. For a typical front-lighted subject η_t is only about 5% but can be much higher for back-lighted scenes.

When exposed for the very brief period of a few tens of nanoseconds by a single pulse from a ruby laser, silver halide photographic emulsions exhibit a so-called *reciprocity failure*. That is, for a given value of exposure $E_0 = I_P \tau_e$ and a given development process, an exposure time τ_e in the nanosecond range produces less optical density in the developed emulsion than longer exposure times. Reciprocity failure is taken into account in the Table 11.1 listing of energy densities required to produce an optical density of 0.5 to 0.6. Note that the Afga-Gevaert 8E75, 10E75, and 14C75 emulsions are much more sensitive than any other materials. For these emulsions, reciprocity failure associated with exposure times in the range of 10 to 50 nsec requires an increase in energy density by a factor of 2 to 4 [*11.24*]. Their high sensitivity helps to make hologram formation with human subjects safer. Kodak 649F emulsion is relatively insensitive to the Q-switched ruby laser. Its spectral sensitivity drops by a factor of 10 when the wavelength changes from 6328 to 6943 Å. Reciprocity failure accounts for another factor of 10. However the latter loss can be diminished somewhat by increasing the development time [*11.25*].

The high peak intensity (1 to 100 MW/cm^2) achieved in a 10- to 50-nsec pulse from the Q-switched ruby laser permits hologram recording in materials which are completely unsuitable for use with CW lasers. Magnetic thin films can be locally heated by laser illumination to near their Curie points, thus allowing their magnetic and magneto-optical properties to be altered. If these properties can be spatially varied at a high enough spatial frequency, holograms can be formed on the film. Suppose initially the magnetic film is uniformly magnetized in a given direction with a relatively strong magnetic field. A small, uniform, reverse-biased magnetic field is now applied to the film; it does nothing in the absence of the pulsed laser light. When the holographic interference pattern produced by the pulsed laser illuminates the film, the temperatures of those areas of the film exposed to the bright fringes of the pattern are raised close to or above the Curie temperature. At such temperatures the reverse-biased magnetic field reverses the orientation of the magnetic domains in the film. After cooling, the orientation reversal remains, even when the magnetic field is removed; a persistent hologram is thus recorded. The spatial variation of the domain orientation gives rise to a spatial variation in the Faraday effect (on transmitted light) or the Kerr effect (on reflected light). Since these effects are concerned with rotation of polarization, the vector nature of the incident light wave must be considered in the reconstruction. One finds that the hologram diffracts as a phase hologram. An attractive feature is that the recorded hologram can be erased with a strong magnetic field and the film can be reused.

One example of such a material is a thin film of MnBi whose Curie temperature is 360°C or 633°K [*11.26*]. Simple interference of two plane waves with fringe spacings of 2 μm has been recorded using an energy density of 0.3 J/cm². Rather low diffraction efficiencies, 0.01% for the Kerr effect and 0.001% for the Faraday, are obtained. Erasure is achieved with a magnetic field of 4000 Oe. Another example is EuO thin film [*11.27*]. Its Curie temperature is only 69°K, requiring cryogenic apparatus for recording and storing the hologram. This material does not seem practical for general holography.

Given sufficient peak power, holograms can be thermally engraved on thin metal films such as Cu, Si [*11.28*], and Bi [*11.29*]. The bright fringes of the interference pattern vaporize the metal on which they are incident. Although the required energy density is high, the resolution is quite good (1000 cycles/mm) and the resulting diffraction efficiency is satisfactory.

REFERENCES

11.1. B. J. Thompson, J. H. Ward, and W. Zinky, "Application of Hologram Techniques for Particle Size Analysis," *J. Opt. Soc. Amer.* **55**, 1566A (1965); *Appl. Opt.* **6**, 519 (1967).

11.2. R. E. Brooks, L. O. Heflinger, R. F. Wuerker, and R. A. Briones, "Holographic Photography of High-Speed Phenomena with Conventional and *Q*-Switched Ruby Lasers," *Appl. Phys. Lett.* **7**, 92 (1965).

11.3. E. L. Steele, *Optical Lasers in Electronics*, Wiley, New York, 1968.

11.4. F. J. McClung and D. Weiner, "Longitudinal Mode Control in Giant Pulse Lasers," *IEEE J. Quantum Electron.* **QE-1**, 94 (1965).

11.5. V. Daneu, C. A. Sacchi, and O. Svelto, "Single Transverse and Longitudinal Mode *Q*-Switched Ruby Laser," *IEEE J. Quantum. Electron.* **QE-2**, 290 (1966).

11.6. J. E. Bjorkholm and R. H. Stolen, "A Simple Single-Mode Giant-Pulse Ruby Laser," *J. Appl. Phys.* **39**, 4043 (1968).

11.7. L. D. Siebert, "Front-Lighted Pulse Laser Holography," *Appl. Phys. Lett.* **11**, 326 (1967).

11.8. W. R. Sooy, "The Natural Selection of Modes in a Passive *Q*-Switched Laser," *Appl. Phys. Lett.* **7**, 36 (1965).

11.9. D. J. Bradley, G. Magyar, and M. C. Richardson, "Intensity Dependent Frequency Shift in Ruby Laser Giant Pulses," *Nature* **212**, 63 (1966).

11.10. A. Flamholz and G. J. Wolga, "Transient Interference Studies of Passively *Q*-Switched Ruby-Laser Emission," *J. Appl. Phys.* **39**, 2723 (1968).

11.11. L. D. Siebert, private communication 1970.

11.12. D. A. Ansley, "Techniques for Pulsed Laser Holography of People," *Appl. Opt.* **9**, 815 (1970).

11.13. J. E. Geusic and H. E. D. Scovil, "A Undirectional Traveling-Wave Optical Maser," *Bell Syst. Tech. J.* **41**, 1371 (1962).

11.14. F. P. Burns, "High-Power Lasers – Their Performance, Limitations, and Future," *IEEE Spectrum* **4**, 115 (1967).

11.15. D. W. Gregg and S. J. Thomas, "Liquid Immersion for Reducing Damaging Effect of Laser Giant Pulses to Dielectric Mirror Coating," *Appl. Phys. Lett.* **8**, 316 (1966).

11.16. J. H. Cullom and R. W. Waynant, "Determination of Laser Damage Threshold for Various Glasses," *Appl. Opt.* **3**, 989 (1964).

11.17. S. D. Rasberry, B. F. Scribner, and M. Margoshes, "Laser Probe Excitation in Spectrochemical Analysis," *Appl. Opt.* **6**, 81 (1967).

11.18. R. W. Minck, "Optical Frequency Electrical Discharges in Gases," *J. Appl. Phys.* **35**, 252 (1964).

11.19. R. E. Brooks, L. O. Heflinger, and R. F. Wuerker, "Pulsed Laser Holograms," *IEEE J. Quantum. Electron.* **QE-2**, 275 (1966).

11.20. L. D. Siebert, "Large-Scene Front-Lighted Hologram of a Human Subject," *Proc. IEEE* **56**, 1242 (1968).

11.21. R. G. Zech and L. D. Siebert, "Pulsed Laser Reflection Holograms," *Appl. Phys. Lett.* **13**, 417 (1968).

11.22. D. B. Neumann, "Holography of Moving Scenes," *J. Opt. Soc. Amer.* **58**, 447 (1968).

11.23. The American Conference of Governmental Industrial Hygienists, "A Guide for Uniform Industrial Hygiene Codes and Regulations" (1968).

11.24. H. Nassenstein, H. Dedden, H. J. Metz, H. E. Rieck, and D. Schultze, "Physical Properties of Holographic Materials," *Photogr. Sci. Eng.* **13**, 194 (1969).

11.25. M. Hercher and B. Ruff, "High-Intensity Reciprocity Failure in Kodak 649-F Plates at 6943 Å," *J. Opt. Soc. Amer.* **57**, 103 (1967).

11.26. R. S. Mezrich, "Curie-Point Writing of Magnetic Holograms on MnBi," *Appl. Phys. Lett.* **14**, 132 (1969).

11.27. G. Fan, K. Pennington, and J. H. Greiner, "Magneto-Optic Hologram," *J. Appl. Phys.* **40**, 974 (1969).

11.28. H. J. Gerritsen and M. E. Heller, "Thermally Engraved Gratings Using a Giant-Pulse Laser," *J. Appl. Phys.* **38**, 2054 (1967).

11.29. J. J. Amodei and R. S. Mezrich, "Holograms in Thin Bismuth Films," *Appl. Phys. Lett.* **15**, 45 (1969).

Chapter 12

NONLINEAR RECORDING, SPECKLE, AND FILM GRAIN NOISE

We state in Section 10.6 that an ideal holographic recording reconstructs a wavefront whose complex amplitude is linearly proportional to that of the original signal wavefront. Properties of the recording causing light other than that proportional to the signal to diffract into the direction of the signal wave may be regarded as sources of noise. Several such sources were identified in Section 10.5, e.g., *film grain* (or granularity) and *nonlinear recording*. An additional potential noise source is *speckle* modulation superimposed on coherent-light images of diffusely reflecting or transmitting surfaces (see Section 8.2.2). In this chapter we discuss in greater detail the adverse effects (on holographic images) engendered by these sources. Methods which can be used to reduce the effect of speckle pattern are also considered.

12.1 Effects of Nonlinear Recording

Our analyses of plane holograms in Chapters 3 and 8 generally assume a linear relation between the amplitude transmittance t and exposure E of the hologram [as, e.g., in Eq. (8.1): $t = t_0 - kI$ where the intensity I is proportional to the exposure]. The t–E curves of actual recording media are always nonlinear to some degree (see Fig. 10.7). However, if the range of exposure fluctuations given by Eq. (7.40), $E_0(1-V_{\max}) < E < E_0(1+V_{\max})$, is kept within the portion of the t–E curve which is approximately linear, adverse effects on image quality can be minimized. In Eq. (7.40) E_0 is the average exposure and V_{max} is the maximum fringe visibility in the exposing interference pattern. Limiting the visibility, on the one hand, reduces the effects of nonlinear recording. On the other hand it restricts the diffraction efficiency η [see Eq. (10.16), $\sqrt{\eta} = SE_0V$ where S is the sensitivity]. Thus in forming holograms one must always compromise between maximizing diffraction efficiency and minimizing nonlinear effects. The compromise is effected by making a suitable choice of the beam ratio R, since R is related to the visibility V by Eq. (7.31).

Analyses of the effects of nonlinear recording have to date been confined to the plane absorption hologram [*12.1–12.5*]. We shall here summarize the important results obtained for plane absorption holograms using a polynomial approximation to the t–E curve. In practice, nonlinear effects are observed in phase as well as absorption holograms and in thick as well as thin media.

Figure 12.1 displays the t–E curve for a Kodak 649F plate. Also shown is a polynomial approximation,

$$t = 0.92 - (0.575 \times 10^{-3})E - (0.137 \times 10^{-3})E^2 + (0.735 \times 10^{-6})E^3, \tag{12.1}$$

where E is the exposure in microjoules per centimeter squared and t is the amplitude transmittance. Since this cubic approximation fits the experimental curve quite well over a considerable range, we adopt as the general

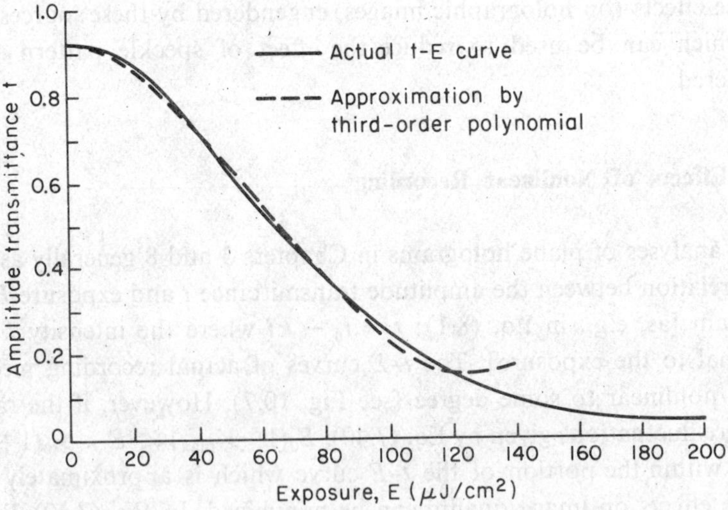

FIG. 12.1. Amplitude transmittance–exposure curve of a Kodak 649F plate and its approximation by the third-order polynomial of Eq. (12.1).

form of the amplitude transmittance of a plane absorption hologram

$$t = c_0 + c_1E + c_2E^2 + c_3E^3. \tag{12.2}$$

For the usual range of exposure, nonlinear effects produced by the quadratic term are the most important. As in Eq. (7.32) the exposure is given by

$$E = k_1I\tau_e \tag{12.3}$$

where k_1 is the constant of proportionality between the intensities I_P and I (see Section 1.3) and τ_e is the exposure time. The first term on the right in Eq. (12.2) is independent of exposure while the second term expresses a linear relation between transmittance and exposure. Nonlinear recording is represented by the third and fourth terms. Let us now investigate the contribution of the third term

$$t_2 = c_2E^2. \tag{12.4}$$

We assume that a subject wave, whose complex amplitude at the recording plate is $\mathbf{a} = a \exp(i\varphi_a)$, and a reference wave, whose complex amplitude at the plate is $\mathbf{r} = r \exp(i\varphi_r)$, interfere for a time τ_e producing an exposure

$$E = k_1I\tau_e = k_1(\mathbf{aa}^* + \mathbf{rr}^* + \mathbf{ar}^* + \mathbf{a}^*\mathbf{r})\tau_e. \tag{12.5}$$

Substituting E into Eq. (12.4) we obtain

$$t_2 = c_2k_1^2(\mathbf{aa}^* + \mathbf{rr}^* + \mathbf{ar}^* + \mathbf{a}^*\mathbf{r})^2\tau_e^2. \tag{12.6}$$

To simplify our analysis let us assume that the reference wave is unmodulated so that $\mathbf{rr}^* =$ constant. We then separate out all terms in Eq. (12.6) which are linearly proportional to any of the terms of Eq. (12.5) and write

$$t_2 = \text{linear terms} + c_2k_1^2\tau_e^2[(\mathbf{aa}^*)^2 + \mathbf{a}^2\mathbf{r}^{*2} + \mathbf{a}^{*2}\mathbf{r}^2 + 2\mathbf{aa}^*\mathbf{ar}^* + 2\mathbf{aa}^*\mathbf{a}^*\mathbf{r}]. \tag{12.7}$$

The cubic term in Eq. (12.2) produces transmittance terms some of which have the same form as in Eq. (12.7). Addition of these terms should be included in a quantitative account of quadratic nonlinear effects. Since we are primarily interested in qualitative analysis, we can neglect their addition. Now consider the significance of the diffracted light $\mathbf{r}t_2$ resulting from illumination of the hologram with the original reference wave $\mathbf{r}$. (In the discussion to follow we neglect the linear terms.) As seen from Eq. (12.7), the diffracted wave that contributes a component in the direction of the original subject wave and thus affects the linear reconstruction of that subject wave is given by the fourth term in the brackets. The complex amplitude of this nonlinear component is $\mathbf{r}(2\mathbf{aa}^*\mathbf{ar}^*) \propto (2\mathbf{aa}^*)\mathbf{a}$. According to the analysis given in Section 8.1.1, the factor $2\mathbf{aa}^*$ represents light with a spread of spatial frequencies centered about zero spatial frequency. Since this factor multiplies the subject wave complex amplitude $\mathbf{a}$, some of the light will travel in the direction of the subject wave.

12.1.1 The Squared Terms

Before considering the contribution of this $(2\mathbf{aa}^*)\mathbf{a}$ term in greater detail, we briefly discuss contributions of the first three terms in the brackets of Eq. (12.7) to the diffracted complex amplitude $\mathbf{r}t_2$. To do this let us assume that $\mathbf{r}$ is an axial plane wave of unit amplitude $\mathbf{r} = r\exp(i\varphi_1) = 1$, so that the terms of interest become

$$\mathbf{r}[(\mathbf{aa}^*)^2 + \mathbf{a}^2\mathbf{r}^{*2} + \mathbf{a}^{*2}\mathbf{r}^2] = (\mathbf{aa}^*)^2 + \mathbf{a}^2 + \mathbf{a}^{*2}. \tag{12.8}$$

As shown in Fig. 8.3 the spectrum of $\mathbf{aa}^*$ has a width twice the spatial frequency bandwidth of the subject itself. (This becomes apparent by Fourier transforming the product $\mathbf{aa}^*$ or, equivalently, autocorrelating the subject Fourier transform $\mathbf{A}$.) Similar arguments apply to the first term on the right in Eq. (12.8). They show that the spatial frequency spectrum of $(\mathbf{aa}^*)^2$, obtained by convolving the spectrum of $\mathbf{aa}^*$ with itself, is four times the width of the subject spectrum and also centered at $\xi = 0$. Thus the offset of the center spatial frequency of the subject wave ξ_a from the reference wave frequency ξ_r must be larger than that required in linear recording if angular overlap of zero-order terms onto the reconstructed subject wave is to be avoided.

The second term on the right in Eq. (12.8), $\mathbf{a}^2$, is the square of the subject wave complex amplitude at the hologram. To illustrate its significance, consider as subject a single point source located at $(x_1 = 0,\ y_1 = -b)$ in an input plane $z_1 = -d$ units from the hologram plane x_2y_2 (Fig. 12.2). According to Eq. (3.3) the phase of the subject wave at the hologram is given by

$$\varphi = -(2\pi/\lambda)(1/2d)(x_2{}^2 + y_2{}^2 + 2y_2b). \tag{12.9}$$

We can then express $\mathbf{a}^2$ as

$$\mathbf{a}^2 = a^2\exp(2i\varphi) \propto \exp[-i(2\pi/\lambda)(1/d)(x_2{}^2 + y_2{}^2 + 2y_2b)], \tag{12.10}$$

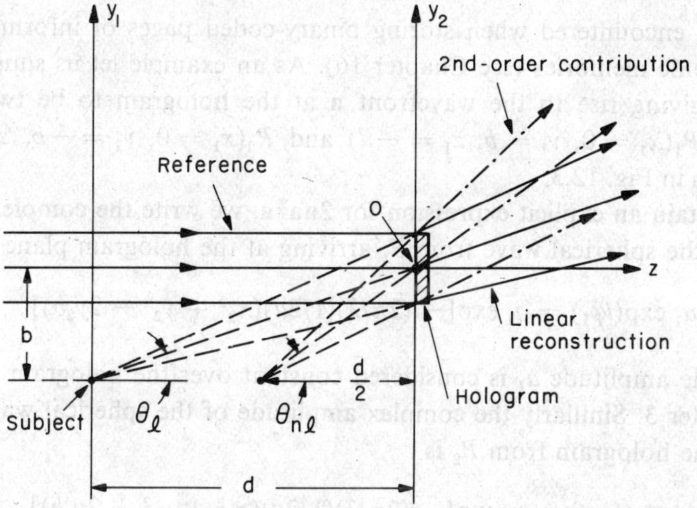

FIG. 12.2. First- and second-order diffraction yielding virtual images of a point source.

where we consider the amplitude a constant across the hologram plane as in Chapter 3. The exponential on the right-hand side of Eq. (12.10) is a spherical phase factor representing the wavefront of a spherical wave diverging from a point source located at ($x_1 = 0$, $y = -b$, $z_1 = -d/2$). It has a curvature twice that of the original subject wavefront. From Fig. 12.2 we see that the mean direction of the wave diffracted by the nonlinear term $\mathbf{a}^2$ makes an angle θ_{nl} to the z axis which is approximately twice that made by the linearly reconstructed subject wave. If the subject is small enough and sufficiently offset from the origin, the angular overlap of the nonlinear contribution onto the linear reconstruction of the subject wave can be avoided. The third term on the right of Eq. (12.8) can be interpreted in terms of the conjugate subject wavefront in a similar manner. Some applications for these squared (and higher-order) wavefront terms have been suggested. The increased curvature of the second- and higher-order diffracted wavefronts may be used to emphasize optical path changes in interferometric measurements [*12.6*] and to sharpen the fringes [*12.7*].

12.1.2 INTERMODULATION

Let us now turn to the fourth term in Eq. (12.7). It produces the diffracted wavefront $\mathbf{r}(2\mathbf{aa}^*\mathbf{ar}^*) \propto 2\mathbf{aa}^*\mathbf{a}$ when the nonlinear recording is illuminated with the original reference wave $\mathbf{r}$. Its effect is of particular importance when the subject of the hologram is a discrete set of point sources. This is the situation encountered when storing binary-coded pages of information in holographic memories (see Chapter 16). As an example let us suppose the subject giving rise to the wavefront **a** at the hologram to be two point sources $P_1(x_1 = 0, y_1 = b, z_1 = -d)$ and $P_2(x_1 = 0, y_1 = -b, z_1 = -d$ as shown in Fig. 12.3.

To obtain an explicit expression for $2\mathbf{a}\mathbf{a}^*\mathbf{a}$, we write the complex amplitude of the spherical wave from P_1 arriving at the hologram plane as

$$\mathbf{a}_1 = a_1 \exp(i\varphi_1) = a_1 \exp[-i(2\pi/\lambda)(1/2d)(x_2{}^2 + y_2{}^2 - 2y_2 b)], \quad (12.11)$$

where the amplitude a_1 is considered constant over the hologram plane as in Chapter 3. Similarly the complex amplitude of the spherical wave arriving at the hologram from P_2 is

$$\mathbf{a}_2 = a_2 \exp(i\varphi_2) = a_2 \exp[-i(2\pi/\lambda)(1/2d)(x_2{}^2 + y_2{}^2 + 2y_2 b)]. \quad (12.12)$$

If we consider $a_1 = a_2 \approx 1$, then the net complex amplitude of subject light at the hologram is

$$\mathbf{a} = \mathbf{a}_1 + \mathbf{a}_2 = \exp(i\varphi_1) + \exp(i\varphi_2). \quad (12.13)$$

Expressing the nonlinear term $2\mathbf{a}\mathbf{a}^*\mathbf{a}$ in terms of φ_1 and φ_2, we have

$$2\mathbf{a}\mathbf{a}^*\mathbf{a} = 2[\exp(i\varphi_1) + \exp(i\varphi_2)][\exp(-i\varphi_1) + \exp(-i\varphi_2)] \times [\exp(i\varphi_1) + \exp(i\varphi_2)].$$

When the multiplication is carried through, we find that

$$2\mathbf{a}\mathbf{a}^*\mathbf{a} = 2\{3 \exp(i\varphi_1) + 3 \exp(i\varphi_2) + \exp[i(2\varphi_1 - \varphi_2)] + \exp[i(2\varphi_2 - \varphi_1)]\}. \quad (12.14)$$

The first two terms on the right of Eq. (12.14) are proportional to the original wavefronts from P_1 and P_2 and thus are reconstructed waves which diverge from virtual images of P_1 and P_2. To reveal the effect of the third and fourth terms we must substitute for φ_1 and φ_2 from Eqs. (12.11) and (12.12). When this is done, $2\varphi_1 - \varphi_2$ becomes

$$2\varphi_1 - \varphi_2 = -(2\pi/\lambda)(1/2d)[x_2{}^2 + y_2{}^2 - 2y_2(3b)]. \quad (12.15)$$

As may be seen from Fig. 12.3, the spherical wave associated with the phase $2\varphi_1 - \varphi_2$ diverges from a false virtual image of a point source P_3 located at $y_1 = 3b$. The false image P_3 and the true image of P_2 are symmetrically

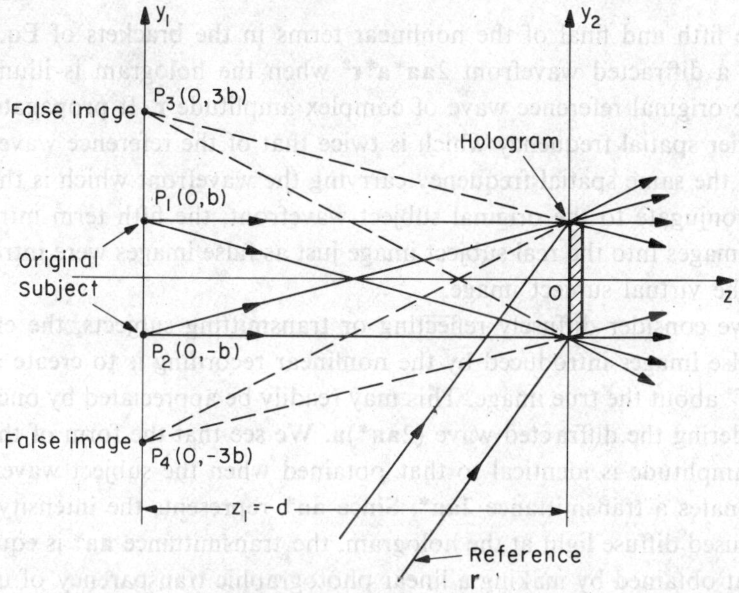

FIG. 12.3. Illustrating the generation of false images of a subject consisting of two point sources. The false images are due to the intermodulation term.

disposed about P_1. Similarly $2\varphi_2 - \varphi_1$ may be shown to represent a spherical wave diverging from a false virtual image of a point source P_4 located at $y_1 = -3b$. Figure 12.4 illustrates false images appearing in the reconstruction of two point sources [*12.2*]. Suppose now that the subject of the hologram is a binary-coded input page, where the presence of point sources on the sites of a rectangular array of spacing $2b$ indicates logical "ones" and their absence indicates logical zeros. It is clear that false logical "ones" are introduced by nonlinear recording.

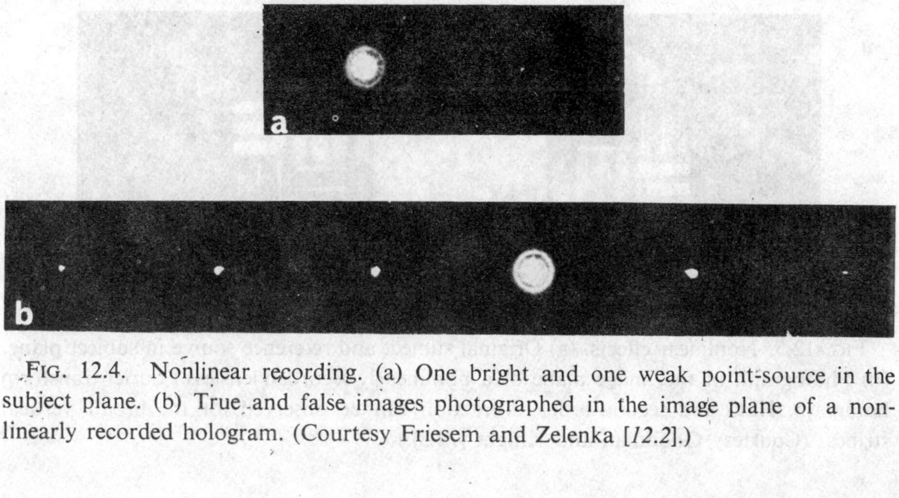

FIG. 12.4. Nonlinear recording. (a) One bright and one weak point-source in the subject plane. (b) True and false images photographed in the image plane of a nonlinearly recorded hologram. (Courtesy Friesem and Zelenka [*12.2*].)

The fifth and final of the nonlinear terms in the brackets of Eq. (12.7) yields a diffracted wavefront $2\mathbf{aa}^*\mathbf{a}^*\mathbf{r}^2$ when the hologram is illuminated by the original reference wave of complex amplitude **r**. It propagates with a carrier spatial frequency which is twice that of the reference wave. Since this is the same spatial frequency carrying the wavefront which is the complex conjugate to the original subject wavefront, the fifth term introduces false images into the real subject image just as false images were introduced into the virtual subject image.

If we consider diffusely reflecting or transmitting subjects, the effect of the false images introduced by the nonlinear recording is to create a noise "halo" about the true image. This may readily be appreciated by once more considering the diffracted wave $(2\mathbf{aa}^*)\mathbf{a}$. We see that the form of the complex amplitude is identical to that obtained when the subject wavefront **a** illuminates a transmittance $2\mathbf{aa}^*$. Since $\mathbf{aa}^*$ represents the intensity of the unfocused diffuse light at the hologram, the transmittance $\mathbf{aa}^*$ is equivalent to that obtained by making a linear photographic transparency of uniform

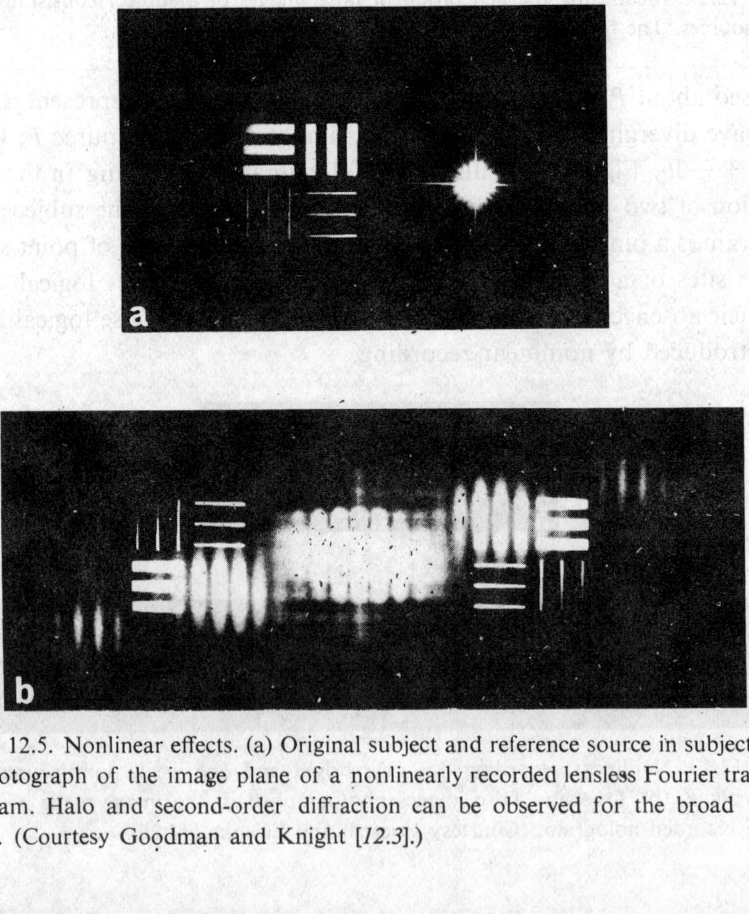

FIG. 12.5. Nonlinear effects. (a) Original subject and reference source in subject plane. (b) Photograph of the image plane of a nonlinearly recorded lensless Fourier transform hologram. Halo and second-order diffraction can be observed for the broad vertical stripes. (Courtesy Goodman and Knight [12.3].)

light with high frequency speckle modulation. If we adopt this representation, then the Fourier components of the transmittance **aa*** must uniformly scatter each ray of the incident subject wave **a** through a band of spatial frequencies given by $\mathscr{F}[\mathbf{aa}^*]$. The band has a width twice the frequency bandwidth of the subject wave. Diffuse noise derived from a highly nonlinearly recorded hologram is thus superimposed over the true image and spread out halolike around it as in Fig. 12.5. In the figure the $\pm$1st-order and $\pm$2nd-order diffraction images generated by a nonlinearly recorded, lensless Fourier transform hologram are symmetrically arrayed about the undiffracted or zero-order light.

Our analysis has been confined to only the quadratic nonlinear terms. Cubic and higher-order nonlinearity can be similarly analyzed. For example, the cubic term leads to a wave diffracted through approximately three times the diffraction angle of the linearly reconstructed image wave, and additional false images are introduced in and around the true images.

12.2 Speckle Pattern

As indicated in Section 8.2.2, the problem of speckle pattern is not unique to holography but common to the observation of any laser-illuminated, diffusely scattering object [*12.8*]. Before noting some of the methods employed to circumvent viewing problems caused by speckle, we present a simple analysis of the phenomenon. (For a more complete analysis see, e.g., Enloe [*12.9*].)

Suppose we consider a ground-glass diffusing screen transilluminated by a plane wave of coherent light. At the scattering surface of the screen, a random phase modulation is impressed on the incident plane wave. At some distance from the screen, the directions or spatial frequencies of the transmitted light are found to be distributed over a wide range. One way to visualize the process is to consider the screen to be a superposition of randomly oriented phase gratings with randomly assigned grating spacings. An incident plane wave is then diffracted by the screen into as many directions as there are (Fourier) phase-grating components required to characterize the screen. While the complex amplitude of the light may be a pure phase function at the surface of the screen, at any other plane the amplitude as well as the phase will fluctuate as a complicated function of the spatial coordinates. This, of course, is due to the interference of the coherent, but randomly directed, diffracted plane waves. A photographic plate exposed to the scattered light will record the corresponding intensity fluctuations.

However, if the screen is finely ground, it will take a high-resolution plate and a microscope to observe the fluctuations of the interference pattern. Figure 12.6 is a hologram of a diffused coherent light pattern magnified to show detail of dimensions the order of the wavelength of light. Although here the subject waves interfere with an additional reference wave, the resulting intensity pattern is similar in appearance to that with no reference wave.

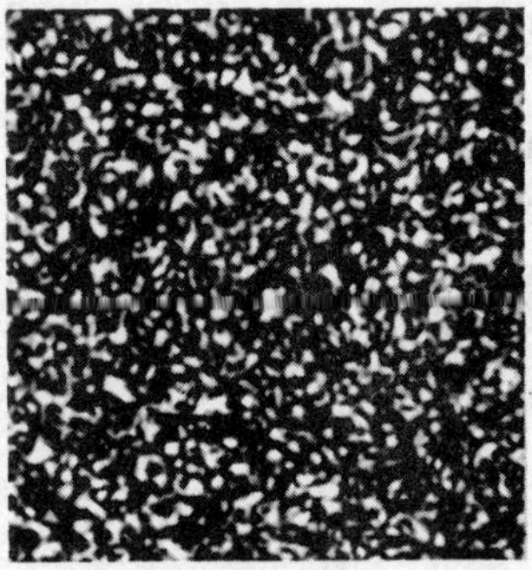

FIG. 12.6. Magnified photograph of a hologram of diffused subject light. (Courtesy J. Upatnieks, University of Michigan.)

In a diffused but coherent light pattern the complex amplitude **a** at any point is the sum of the complex amplitudes of all the interfering plane waves. It can thus be regarded as the sum of a large number of complex amplitudes with random phases,

$$\mathbf{a} = \mathbf{a}_1 + \mathbf{a}_2 + \cdots + \mathbf{a}_n. \tag{12.16}$$

It is sometimes useful to represent complex amplitudes by vectors called phasors. Phasor length is proportional to the amplitude and phasor angle gives the phase of the corresponding complex amplitude. Figure 12.7 graph-

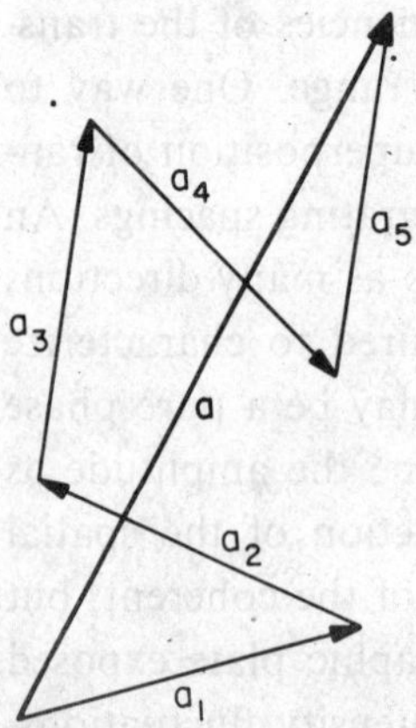

FIG. 12.7. Addition of randomly directed phasors of equal length.

ically illustrates the case for five complex amplitudes. Suppose that the screen diffracts uniformly in all directions so that the amplitudes of the complex amplitudes are all equal (and the length of each phasor in Fig. 12.7 is the same). Then the variation in the amplitude a of the *resultant* complex amplitude $\mathbf{a} = a \exp(i\theta)$, as a function of position in the light pattern, depends only on the phases of the complex amplitudes or the angles of the individual phasors summed at each point. The amplitude a is thus a random function of position and leads to the intensity speckles in Fig. 12.6.

Although the amplitude modulation of the coherent light diffused by the screen is of very high spatial frequency, *observation* of the pattern with the eye or any imaging system of moderate aperture detects a relatively low spatial frequency modulation or *speckle pattern*. Speckle pattern can seriously impede the recognition of whatever information may be present in the image. It does not matter whether one focuses the eye at infinity, on the screen, or attempts to focus on speckle spots at some closer plane; the speckle pattern remains the same. However, if the limiting aperture of the imaging system is increased in diameter, the size of the speckle spots becomes smaller. Thus it is the limited resolution of the imaging system which is responsible for the *observed* speckle pattern.

Consider that the speckle pattern is observed by an eye or optical system focused at infinity. An optical system with a finite aperture c cannot assign a unique point on its back focal-plane (spatial-frequency plane) to each spatial-frequency component diffracted to the system by the diffusing screen. Instead, as we showed in Section 6.4.1, light in a *band* of spatial frequencies of width $\nu = 0.61/c$ is focused to a minimum-resolved spot of diameter $\Delta = 2(0.61)\lambda f/2c$ where f is the focal length and $f/2c$ the f/number of the imaging system. Since the waves in the frequency band are coherent, their amplitudes add as the phasors of Fig. 12.7. Again the random individual phases determine the resultant amplitude and the intensity at any given spot in the focal plane.

From the above discussion we see that speckle pattern spot size is diffraction limited and proportional to the ratio $f/2c$, the f/number of the imaging system. This is illustrated in Fig. 12.8a,b,c, where Fig. 12.8a is the image of a diffusely reflecting surface photographed with a relative aperture of $f/280$ and where Fig. 12.8b displays the same surface photographed with an aperture of $f/93$. It is apparent that speckle spot size is smaller in the latter case. When the pattern of Fig. 12.8b is magnified by the ratio of the f/numbers used for Fig. 12.8a,b, the resulting pattern, shown in Fig. 12.8c, is similar to that in Fig. 12.8a.

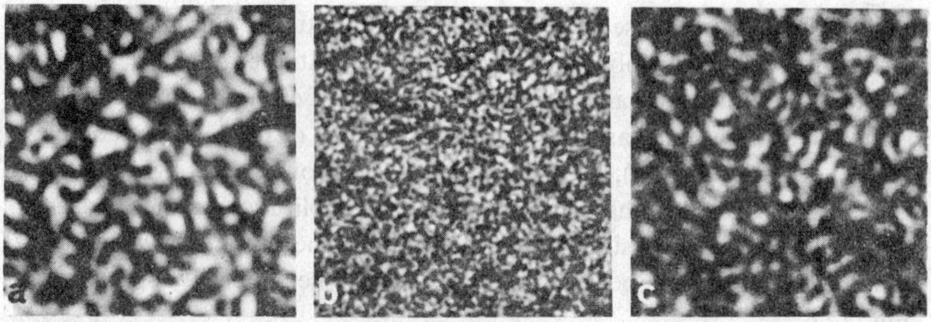

FIG. 12.8. Speckle pattern as a function of f/number. (a) Photograph of speckle pattern with an $f/280$ setting. (b) Photograph of speckle pattern with an $f/93$ setting. (c) Same as (b) but magnified three times.

If the imaging system is so large or the spread of light from the diffuser so small that all of the subject light passes through the imaging system, the image quality is not limited by diffraction from lens boundaries. The imaging system then does not impose speckle on the image. Since even the highest spatial frequencies present at the diffuser plane are resolved in the image of the diffuser, a perfect mapping of the complex amplitude at the diffuser onto the image plane can be achieved. The pure phase modulation and uniform intensity distribution at the diffuser are transferred to the image plane and no intensity modulation or speckle is observed.

Let us now investigate the *magnitude* of the intensity fluctuation in a speckle pattern relative to the mean intensity of the light. We assume that Eq. (12.16) holds for the complex amplitude $\mathbf{a}$ at any point in the light distribution. Each complex amplitude is assumed to have the same amplitude a_k, but a random phase $0 \leq \varphi_k \leq 2\pi$. Our analysis applies to the high spatial frequency fluctuations in the unfocused light from the scattering source as well as to the lower spatial frequency speckle pattern observed with an imaging system. We wish to calculate the ratio of the root mean square of the intensity fluctuations N to the mean intensity $\bar{I}$ where

$$I = a^2, \tag{12.17}$$

$$\bar{I} = \overline{a^2}, \tag{12.18}$$

and

$$N = \left[\overline{(I - \bar{I})^2}\right]^{1/2} = [\overline{I^2} - \bar{I}^2]^{1/2}. \tag{12.19}$$

The statistical distribution of the resultant complex amplitude $\mathbf{a} = a\exp(i\theta)$ in Eq. (12.16) is a solution of the well-known "random-walk" problem and has a probability density (the Rayleigh distribution, Middleton [*12.10*], p. 366)

$$p(a, \theta) = (a/\pi\sigma^2)\exp(-a^2/\sigma^2) \tag{12.20}$$

where σ^2 is a constant whose physical meaning is to be revealed. The mean value of a function $f(a, \theta)$ is defined as

$$\overline{f(a, \theta)} = \int_0^\infty \int_0^{2\pi} f(a, \theta)p(a, \theta)\, d\theta\, da. \tag{12.21}$$

We therefore calculate $\bar{I} = \overline{a^2}$ to be

$$\bar{I} = \frac{2}{\sigma^2}\int_0^\infty a^3 \exp\left(-\frac{a^2}{\sigma^2}\right) da$$

or

$$\bar{I} = \sigma^2. \tag{12.22}$$

To determine N in Eq. (12.19) we must also calculate $\overline{I^2}$. Since $\overline{I^2} = \overline{a^4}$ we can again employ Eq. (12.21) to obtain

$$\overline{I^2} = \overline{a^4} = \frac{2}{\sigma^2}\int_0^\infty a^5 \exp\left(-\frac{a^2}{\sigma^2}\right) da$$

or

$$\overline{I^2} = 2\sigma^4. \tag{12.23}$$

Substituting Eqs. (12.22) and (12.23) into Eq. (12.19) we find that

$$N = [2\sigma^4 - \sigma^4]^{1/2} = \sigma^2 = \bar{I}. \tag{12.24}$$

Therefore *the ratio of the root-mean-square of the intensity fluctuations to the mean intensity is one.*

12.2.1 Avoiding or Reducing Speckle Problems

It is clear from the foregoing that speckle pattern is a serious problem for holographic display. A number of methods for avoiding or reducing the problem apply to the case where the subject is a two-dimensional transparency. If the transparency is illuminated diffusely, one gains the advantages associated with redundant recording noted in Section 8.2, viz., ease of viewing the image and high immunity to dust and scratches on the hologram. A diffusing screen, e.g., a ground-glass plate, is usually the source of the diffuse illumination, and its scattering centers are the sources of the speckle.

It is possible to circumvent the speckle problem and still preserve substantial immunity to dust and blemishes by eliminating the diffusing screen and forming a hologram of the Fourier transform of the subject transparency. A plane wave illuminates the subject and a lens performs the transformation as in Fig. 8.25. The recording is redundant since each point of the Fourier transform represents the total complex amplitude of light from all parts of the subject propagating with a given spatial frequency. However, as pointed out in Section 8.3.3, the large intensity variations in the Fourier transforms of most subjects preclude linear recording of holograms with high diffraction efficiency. One must settle for a compromise between high redundancy in the recording on the one hand and high hologram diffraction efficiency on the other. A means for achieving a satisfactory compromise is to form the hologram a small distance away from the exact Fourier transform plane (see Section 16.4.3).

Another method of distributing subject information over the hologram plane, while avoiding the use of a ground glass screen, and thus avoiding the worst aspects of the speckle problem, depends on the use of a "phase modulator" plate [*12.11*]. The latter might take the form of a glass plate having one flat surface and one with shallow, random, low-spatial-frequency undulations. The phase modulating surface is placed up against the emulsion of the subject transparency, so that the two are essentially in the same plane. Coherent light illuminates the combination. We can regard formation of a hologram and subsequent wavefront reconstruction as equivalent to the action of a diffraction limited lens system which images the plane of the modulator and transparency to a conjugate plane. If the surface variations of the phase modulator are of sufficiently low spatial frequency so that all spatial frequencies of the light diffracted by the modulator–transparency pair pass through the aperture of the imaging system, a perfect mapping of the input plane onto the image plane will be obtained. The uniform intensity at the surface of the phase modulator is preserved in the image and there should be no speckle. In practice there are difficulties in obtaining the desired random phase modulator, and a working solution is to replace the random modulator with periodic phase gratings [*12.12*, *12.13*].

Speckle must be considered when an application requires the projection of a real image from a hologram onto a photosensitive surface. It is possible to retain the diffuser in the formation of the hologram and still limit the effect of speckle if the intensity of each resolved spot in the image is an average of a large number of intensity values. A simple method of achieving this is to form a hologram large enough (with small enough f/number) to produce a resolution in the projected image greater than required in the recording of that image on the photosensitive surface. If the size of a speckle spot is small compared to the smallest recordable spot, then each recorded spot is proportional to a mean intensity averaged over many speckles. The fine-grained speckle modulation is therefore not resolved and speckle intensity fluctuations are circumvented.

Averaging may be carried out in another way: Suppose a large number of small holograms are formed side-by-side on a photographic plate. For each hologram the subject transparency remains the same, but the diffusing screen is altered between exposures. After development and fixing, each hologram in succession is caused to project a real image of the subject onto a photosensitive surface [*12.14*]. This superposition (through multiple exposure) of the intensities of a large number of images, each with identical subject information but different speckle pattern, accomplishes the averaging. However, speckle reduction by this means is relatively inefficient. To obtain an improvement in the ratio of signal to the root mean square of the intensity fluctuations by a factor n, an average over n^2 images must be taken [*12.14*]. It follows that the area needed for recording the holograms is n^2 times the minimum area required for recording the nondiffuse subject.

In Section 8.3.2 we show that an image hologram of a two-dimensional subject can produce an image of good resolution even when illuminated with an extended thermal source. Suppose that the hologram was formed with a ground-glass screen in the illuminating beam so that, when illuminated with coherent light, speckle is observed in the image. If the hologram is instead illuminated with a source whose spatial coherence extends over a lateral distance substantially less than the smallest resolved spot in the image, then the light phasors contributing to a single spot in the image no longer can add coherently. Only their intensities add, and speckle interference is eliminated. Illumination with a low degree of spatial coherence is less useful when the subject is three-dimensional and much of the image lies outside the plane of the hologram. In that case some loss of resolution must be tolerated. Viewers of 3D images formed in laser light can of course resort to continuous movements of the head to average out of the speckle. Certainly better methods, yet to be invented, are desirable if holographic display of 3D images is to be optimized.

12.3 Film Grain Noise

Scattering of light from the individual grains of the developed photosensitive medium in which a hologram is formed is a noise source which becomes important when the diffraction efficiency of the hologram is low. Low diffraction efficiency is unavoidable when the hologram plate must be exposed briefly to a weakly reflecting subject or when a large number of absorption holograms are superimposed on the same plate (see Section 17.5.5). Film grain noise often determines the upper limit to the number of holograms that can be superimposed. An incoherent-light hologram (Section 20.3) is, in effect, a superposition of low efficiency holograms, and film grain noise represents a dominant limitation to its utility. We shall confine the following discussion to absorption holograms, but the general conclusions also apply to phase holograms.

12.3.1 FILM GRAIN NOISE IN FOURIER TRANSFORM HOLOGRAMS

We begin by studying the effect of film grain noise on Fourier transform holograms. Suppose a plane wave traveling normal to a photographic plate uniformly exposes the plate. After exposure, the plate is developed, fixed, placed adjacent to a lens as indicated in Fig. 12.9, and once more illuminated

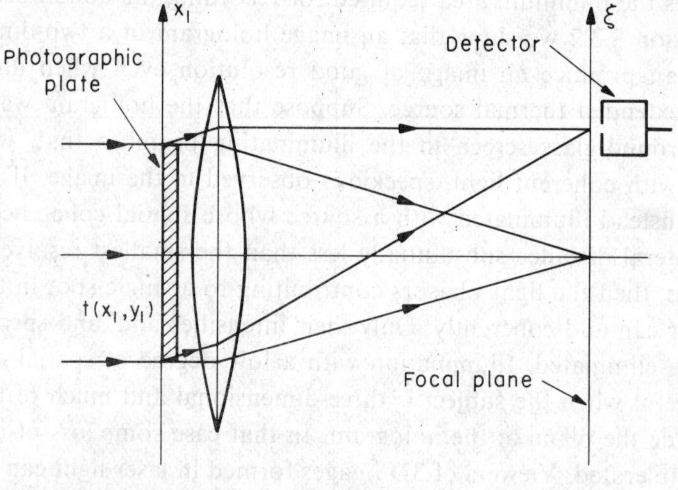

FIG. 12.9. Arrangement for measuring the Wiener spectrum.

with the plane wave. Scattered light or noise intensity is then measured in the back focal plane. Such measurements are relevant to the Fourier transform hologram since the arrangement of Fig. 12.9 is equivalent to that used to illuminate a Fourier transform hologram (see Fig. 8.23), and the usual observation or image plane of the Fourier transform hologram is the back focal plane of the lens. Let us now express the intensity of the light scattered to the back focal plane by the uniformly exposed plate in terms of the spatial frequency coordinates (ξ, η) of that plane.

To do this we write the noise amplitude in the back focal plane of the lens in a form similar to Eq. (8.16) as

$$\mathbf{a}(\xi, \eta) = \frac{ia}{\lambda f} \exp\left(-\frac{i\pi}{\lambda f}(x_2^2 + y_2^2)\right)\mathbf{T}(\xi, \eta) \tag{12.25}$$

where a is the uniform amplitude illuminating the plate transmittance $t(x_1, y_1)$ and where $\mathbf{T}(\xi, \eta) \subset t(x_1, y_1)$. The noise intensity at a point in the spectral plane then becomes

$$\begin{aligned} I_N(\xi, \eta) &= \mathbf{a}(\xi, \eta)\mathbf{a}^*(\xi, \eta) \\ &= (a^2/\lambda^2 f^2)\mathbf{T}(\xi, \eta)\mathbf{T}^*(\xi, \eta) \\ &= (a^2/\lambda^2 f^2)F(\xi, \eta) \end{aligned} \tag{12.26}$$

where

$$F(\xi, \eta) = \mathbf{T}(\xi, \eta)\mathbf{T}^*(\xi, \eta) \subset t(x_1, y_1) \star t(x_1, y_1).$$

Since the light distributed over the spectral plane is a high frequency speckle pattern, we shall henceforth indicate our interest in a local light intensity averaged over many speckles by writing

$$\overline{I_N(\xi, \eta)} = (a^2/\lambda^2 f^2)\,\overline{F(\xi, \eta)}. \tag{12.27}$$

We can write Eq. (12.27) in terms of the *Wiener spectrum* $\Phi(\xi, \eta)$, the ratio of noise power scattered per unit two-dimensional spatial frequency bandwidth P_N/B to the power P_i illuminating the plate. Since bandwidth $B = \xi\eta$ is related to area $x_2 y_2$ in the focal plane by $x_2 y_2 = \lambda^2 f^2 \xi\eta$ (see p. 117), a unit of bandwidth is equivalent to $\lambda^2 f^2$ units of area. If we let A_H be the area of the plate and k_1 the intensity constant given in Eq. (7.32), we have from Eq. (12.27)

$$P_N/B = k_1 \bar{I}_N \lambda^2 f^2 = k_1 a^2 \bar{F} = (k_1 a^2 A_H)\bar{F}/A_H = P_i \bar{F}/A_H,$$

$$\frac{P_N/B}{P_i} = \frac{\overline{F(\xi,\eta)}}{A_H} = \Phi(\xi,\eta), \tag{12.28}$$

$$\overline{I_N(\xi,\eta)} = \left(\frac{P_i}{k_1}\right)\frac{\Phi(\xi, \eta)}{\lambda^2 f^2}. \tag{12.29}$$

The Wiener spectrum depends on the type of photographic plate (i.e., on the grain size) and on the average amplitude transmittance of the plate after exposure and development. Measurement of the noise power in the Fourier transform plane, and thus the Wiener spectrum, is made through an aperture defining a center spatial frequency and a small bandwidth about the center (Fig. 12.9). To obtain the Wiener spectrum as a function of spatial frequency, noise power is plotted as a function of aperture location in the Fourier transform plane. The aperture should be large enough to include many speckles in order to get a reliable average for $\overline{I_N(\xi, \eta)}$ [*12.15*]. Wiener spectrum curves for Kodak 649F plates are plotted in Fig. 12.10 [*12.16*]. The Wiener spectrum was measured as a function of the radial frequency $\nu = (\xi^2 + \eta^2)^{1/2}$, and it was assumed that $\Phi(\xi, \eta)$ was axially symmetric. Further measurements of the Wiener spectrum have been reported [*12.17*].

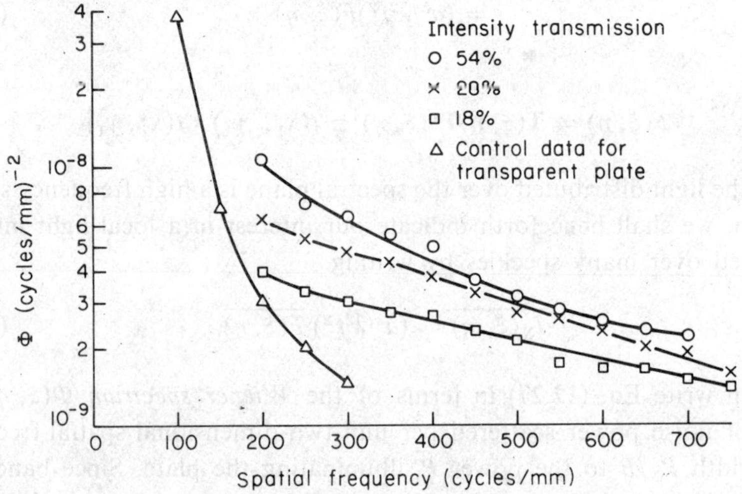

FIG. 12.10. Wiener spectra of Kodak 649F plates measured for several values of intensity transmittance.

12.3.2 FILM GRAIN NOISE IN FRESNEL HOLOGRAMS

In most cases the image plane of a Fresnel hologram is in the *far field* of the individual grains of the photographic plate. Since the far-field pattern given by Eq. (8.22) has the same form as Eq. (12.25), the formal description of film grain noise for Fresnel holograms is similar to that of Fourier transform holograms. The difference is in the definition of spatial frequency used in Eq. (12.28). In the case of the Fourier transform hologram, each point of the observation plane (the Fourier transform plane) receives only the light diffracted by the plate into a single direction characterized by a single pair of spatial frequency coordinates ξ, η. Measurement of the noise power at any such point allows one to plot the Wiener spectrum for one pair of values for ξ and η. However, light rays traveling from various points on a Fresnel hologram to a single point on the image plane arrive at various angles, i.e., with a *range* of spatial frequencies. If noise power is measured in the manner illustrated in Fig. 12.9, we must assign to each point of the image plane of a Fresnel hologram an average value of ξ and η (ξ_A, η_A). For example we may choose ξ_A and η_A to be the spatial frequencies associated with scattered light rays traveling from the center of the plate to the observation point in the image plane. We then obtain for the average value of the noise intensity $\bar{I}_N$ at a point in the image plane

$$\bar{I}_N = \left(\frac{P_i}{k_1}\right) \frac{\Phi(\xi_A, \eta_A)}{\lambda^2 d^2}, \tag{12.30}$$

where d is the distance from the plate to the observation point.

12.3.3 Signal-to-Noise Ratio

In computing the ratio of the signal to the film-grain noise in the image generated by a hologram, one must take into account that signal and noise are coherent and that their amplitudes add [*12.18*]. As in the case of speckle modulation (Section 12.2) it is reasonable to define the noise N as the root-mean-square of the resulting intensity fluctuations:

$$N = \left[\overline{(I - \bar{I})^2}\right]^{1/2} = [\overline{I^2} - \bar{I}^2]^{1/2}. \tag{12.31}$$

The random-walk statistics of Section 12.2 apply here, and the general expression for finding the mean value is given in Eq. (12.21). Suppose that at the image plane the signal intensity is I_i and the signal complex amplitude is $\mathbf{a}_i$. In the same plane the noise intensity contributed by film grain is I_N with a complex amplitude $\mathbf{a}_N$. Let us now evaluate Eq. (12.31). We assume nondiffuse subject light so that the image amplitude $\mathbf{a}_i$ is purely deterministic. The intensity in the image plane due to the interference of signal and noise is given by

$$\begin{aligned} I &= (\mathbf{a}_i + \mathbf{a}_N)(\mathbf{a}_i + \mathbf{a}_N)^* \\ &= I_i + I_N + \mathbf{a}_i\mathbf{a}_N^* + \mathbf{a}_i^*\mathbf{a}_N. \end{aligned} \tag{12.32}$$

Squaring the above yields

$$\begin{aligned} I^2 = I_i^2 + I_N^2 + \mathbf{a}_i^2\mathbf{a}_N^{*2} + \mathbf{a}_i^{*2}\mathbf{a}_N^2 \\ + 2I_iI_N + 2I_i\mathbf{a}_i\mathbf{a}_N^* + 2I_i\mathbf{a}_i^*\mathbf{a}_N \\ + 2I_N\mathbf{a}_i\mathbf{a}_N^* + 2I_N\mathbf{a}_i^*\mathbf{a}_N + 2I_iI_N. \end{aligned} \tag{12.33}$$

To evaluate Eq. (12.31), we must compute the mean value of I^2. Since $\mathbf{a}_N$ is the sum of many complex amplitudes with random phases, its probability density is again given by Eq. (12.20). We may use Eq. (12.21) to show that the mean value of any term in Eq. (12.33) containing a power of $\mathbf{a}_N$ (or $\mathbf{a}_N{}^*$) vanishes. Thus the mean value of $\mathbf{a}_N{}^m$, where m is an integer and where $\mathbf{a}_N = a\exp(i\theta)$ has a probability density given by Eq. (12.20), is

$$\overline{\mathbf{a}_N{}^m} = \overline{a^m \exp(im\theta)} = \int_0^\infty \frac{a^{m+1}}{\pi\sigma^2} \exp\left(-\frac{a^2}{\sigma^2}\right) da \int_0^{2\pi} \exp(im\theta)\, d\theta = 0,$$

since the integral over θ vanishes. Therefore, when the mean of I^2 in Eq. (12.33) is taken, the only nonzero terms are given by

$$\overline{I^2} = I_i{}^2 + \overline{I_N{}^2} + 4I_i\bar{I}_N. \tag{12.34}$$

Returning again to Section 12.2, comparison of Eq. (12.22) with Eq. (12.23) yields

$$\overline{I_N{}^2} = 2\bar{I}_N{}^2, \tag{12.35}$$

where we have substituted I_N for I in the equations cited. Inserting Eq. (12.35) into Eq. (12.34) we obtain

$$\overline{I^2} = I_i{}^2 + 2\bar{I}_N{}^2 + 4I_i\bar{I}_N. \tag{12.36}$$

The other quantity needed to evaluate N in Eq. (12.31) is the square of the mean intensity $\bar{I}^2$. From Eq. (12.32)

$$\bar{I} = I_i + \bar{I}_N$$

and

$$\bar{I}^2 = I_i{}^2 + 2I_i\bar{I}_N + \bar{I}_N{}^2. \tag{12.37}$$

Using Eqs. (12.36) and (12.37) in Eq. (12.31), we can write for the noise N

$$N = [\overline{I^2} - \bar{I}^2]^{1/2} = [\bar{I}_N{}^2 + 2I_i\bar{I}_N]^{1/2}$$
$$= \bar{I}_N\left[1 + \frac{2I_i}{\bar{I}_N}\right]^{1/2}, \tag{12.38}$$

and for the signal-to-noise ratio I_i/N

$$\frac{I_i}{N} = \frac{I_i}{\bar{I}_N[1 + (2I_i/\bar{I}_N)]^{1/2}}. \tag{12.39}$$

Signal-to-noise ratio as a function of $I_i/\bar{I}_N$ is plotted in Fig. 12.11. There are two limiting cases:

$$\text{(1) When} \quad I_i/\bar{I}_N \ll 1, \qquad I_i/N = I_i/\bar{I}_N. \tag{12.40}$$

$$\text{(2) When} \quad I_i/\bar{I}_N \gg 1, \qquad I_i/N = (1/\sqrt{2}) \cdot [I_i/\bar{I}_N]^{1/2}. \tag{12.41}$$

It is the second case which is generally of interest, and we find that, with coherent illumination, the signal-to-noise ratio is proportional to the ratio of the square roots of signal and mean noise intensities, i.e., to the ratio of signal and noise *amplitudes.*

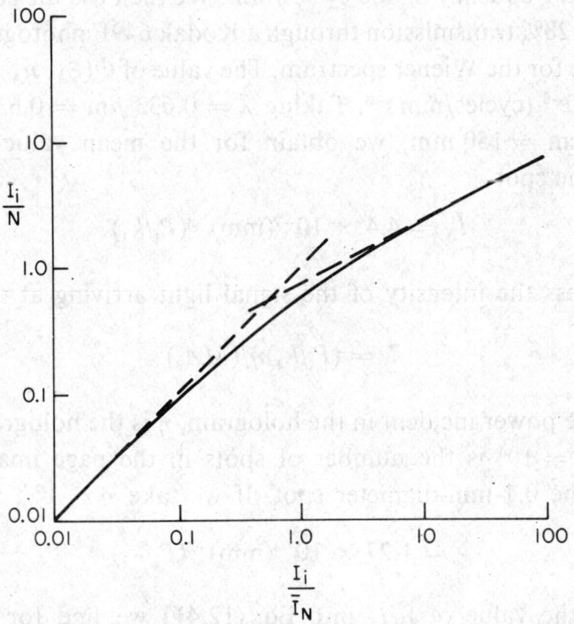

FIG. 12.11. Signal-to-noise ratio I_i/N as a function of the ratio of signal intensity to mean noise intensity $I_i/\bar{I}_N$.

12.3.4 AN ILLUSTRATIVE EXAMPLE

To illustrate the import of Eq. (12.41), we consider the effect of film grain noise on the quality of the image of a page of binary-coded data displayed by a holographic memory (see Section 16.4). Suppose that a small hologram, 1.25 mm in diameter, is illuminated with the conjugate to the reference beam used to form it. It images a 2.5 × 2.5-cm binary page on a plane 15 cm from the hologram. (The small hologram is one of an array of holograms making up the memory.) With the relative aperture of the hologram so defined, one finds that an array of 100×100 bright spots, each 0.1 mm in diameter, can constitute a page of 10^4 logical "ones." We wish to calculate the ratio of the signal intensity I_i in one of these spots to N, the root mean square of the intensity fluctuation from spot to spot, under the conditions $I_i \gg \bar{I}_N$ and $I_i/N = (1/\sqrt{2})[I_i/\bar{I}_N]^{1/2}$ specified in Eq. (12.41). The mean noise intensity $\bar{I}_N$ is given in Eq. (12.30) as

$$\bar{I}_N = \left(\frac{P_i}{k_1}\right) \frac{\Phi(\xi_A, \eta_A)}{\lambda^2 d^2}.$$

Let us consider a spot whose location in the image plane corresponds to an average spatial frequency of 400 cycles/mm. We then use the curve plotted in Fig. 12.10 for 28% transmission through a Kodak 649F photographic plate to obtain a value for the Wiener spectrum. The value of $\Phi(\xi_A, \eta_A)$ is found to be nearly 4×10^{-9} (cycles/mm)$^{-2}$. Taking $\lambda = 0.633\ \mu\text{m} = 0.633 \times 10^{-3}$ mm and $d = 15$ cm $= 150$ mm, we obtain for the mean value of the noise intensity at the spot

$$\bar{I}_N = 4.4 \times 10^{-7} (\text{mm})^{-2} (P_i/k_1).$$

We can express the intensity of the signal light arriving at the spot as

$$I_i = (P_i/k_1)\eta/(MA_s) \tag{12.42}$$

where P_i is the power incident in the hologram, η is the hologram diffraction efficiency, $M = 10^4$ is the number of spots in the page image, and A_s is the area of the 0.1-mm-diameter spot. If we take $\eta = 1\%$, we have

$$I_i = 1.27 \times 10^{-4} (\text{mm})^{-2} (P_i/k_1).$$

Substituting the value of $I_i/\bar{I}_N$ into Eq. (12.41) we find for the signal-to-noise ratio

$$I_i/N = 11.9.$$

Thus the intensity at a bright spot in the binary-coded page image fluctuates (from spot to spot) by about 8% due to film grain noise alone. If, on the other hand, we compare the value of $\bar{I}_N$ at a point in the imaged page where a bright spot is missing (e.g., at the site of a logical zero) with the value of the intensity I_i at a bright spot, we find the noise intensity to be only 0.35% of the bright spot intensity. Our example illustrates that a rather small mean noise intensity leads to a relatively large fluctuation. This is due to the interference of signal and noise amplitudes when coherent illumination is employed.

REFERENCES

12.1. A. Kozma, "Photographic Recording of Spatially Modulated Coherent Light," *J. Opt. Soc. Amer.* **56**, 428 (1966).

12.2. A. A. Friesem and J. S. Zelenka, "Effects of Film Nonlinearities in Holography," *Appl. Opt.* **6**, 1755 (1967).

12.3. J. W. Goodman and G. R. Knight, "Effects of Film Nonlinearities on Wavefront-Reconstruction Images of Diffuse Objects," *J. Opt. Soc. Amer.* **58**, 1276 (1968).

12.4. O. Bryngdahl and A. Lohmann, "Nonlinear Effects in Holography," *J. Opt. Soc. Amer.* **58**, 1325 (1968).

12.5. A. Kozma, "Analysis of the Film Nonlinearities in Hologram Recording," *Opt. Acta* **15**, 527 (1968).

12.6. O. Bryngdahl and A. W. Lohmann, "Interferograms are Image Holograms," *J. Opt. Soc. Amer.* **58**, 141 (1968).

12.7. K. Matsumoto, "Holographic Multiple-Beam Interferometry," *J. Opt. Soc. Amer.* **59**, 777 (1969).

12.8. J. D. Rigden and E. I. Gordon, "The Granularity of Scattered Optical Maser Light," *Proc. IRE* **50**, 2367 (1962).

12.9. L. H. Enloe, "Noise-Like Structure in the Image of Diffusely Reflecting Objects in Coherent Illumination," *Bell. Syst. Tech. J. 46*, 1479 (1967).

12.10. D. Middleton, *An Introduction to Statistical Communication Theory*, McGraw-Hill, New York, 1960.

12.11. J. Upatnieks, "Improvement of Two-Dimensional Image Quality in Coherent Optical Systems," *Appl. Opt.* **6**, 1905 (1967).

12.12. E. N. Leith and J. Upatnieks, "Imagery with Pseudo-Randomly Diffused Coherent Illumination," *Appl. Opt.* **7**, 2085 (1968).

12.13. H. J. Gerritsen, W. J. Hannan and E. G. Ramberg, "Elimination of Speckle Noise in Holograms with Redunancy," *Appl. Opt.* **7**, 2301 (1968).

12.14. W. Martienssen and S. Spiller, "Holographic Reconstruction Without Granulation," *Phys. Lett.* **24A**, 126 (1967).

12.15. A. Kozma, "Effects of Film-Grain Noise in Holography," *J. Opt. Soc. Amer.* **58**, 436 (1968).

12.16. C. B. Burckhardt, "Storage Capacity of an Optically Formed Spatial Filter for Character Recognition," *Appl. Opt.* **6**, 1359 (1967).

12.17. D. H. R. Vilkomerson, "Measurements of the Noise Spectral Power Density of Photosensitive Materials at High Spatial Frequencies," *Appl. Opt.* **9**, 2080 (1970).

12.18. J. W. Goodman, "Film-Grain Noise in Wavefront-Reconstruction Imaging," *J. Opt. Soc. Amer.* **57**, 493 (1967).

Chapter 13
REAL-IMAGE APPLICATIONS

In Chapters 10 and 12 we identified noise sources and other impedimenta contributing to the difficulties of achieving high-resolution, high-fidelity holographic images. Solutions to many of these problems will come about when better photosensitive materials are found. In the meantime, a number of clever ways to employ the unique imaging properties of the hologram have been devised. We shall consider cases where there are distinct technical or economic reasons for employing holographic methods rather than conventional optical methods to form real images of a subject. Potential applications range through the fields of microscopy, photolithography, imaging through the earth's atmosphere, cryptography, and optical memories. We shall leave the last for Chapter 16 and shall present here schemes representative of the other fields of application.

13.1 Microscopy

Microscopy, the application for which holography was invented, while no longer the main focus, nevertheless offers unique employment for the hologram. Gabor had intended to change the wavelength between formation of the hologram and the reconstruction step and obtain a magnification in the ratio of the wavelengths [see Eq. (3.33)]. If it is desired to use visible light in both steps of the holographic process, an alternative is to change the curvature (the reciprocal of the radius) of the reconstructing wave over that of the reference. When the curvature of the reconstructing wave is less than that of the reference wave, magnification will occur (see Section 3.3.1).

The magnification is obtained without need for a lens. However, as noted in Chapter 3, it is accompanied by aberrations. A magnification of ~100× with resolvable detail of several microns in dimension has been reported [*13.1*], but there have been no reports of lensless recording and reconstruction of detail with dimensions very close to the wavelength of light. The accuracy with which the illuminating wavefront must be formed and the need to accurately control the emulsion surface make difficult the attainment of such high resolution with completely lensless imaging.

Resolution of about 1 μm has been obtained from a hologram recorded in the arrangement shown in Fig. 13.1 [*13.2–13.4*]. Here the microscope objective lens forms a magnified real image of the object, as it normally does in the microscope, and the hologram is formed with this image as the subject. On reconstruction, the hologram reproduces the image which is then viewed with a microscope eyepiece. Figure 13.2 is a photograph of the image of fibrous material obtained in this manner. The arrow points to a fiber 1 μm in diameter.

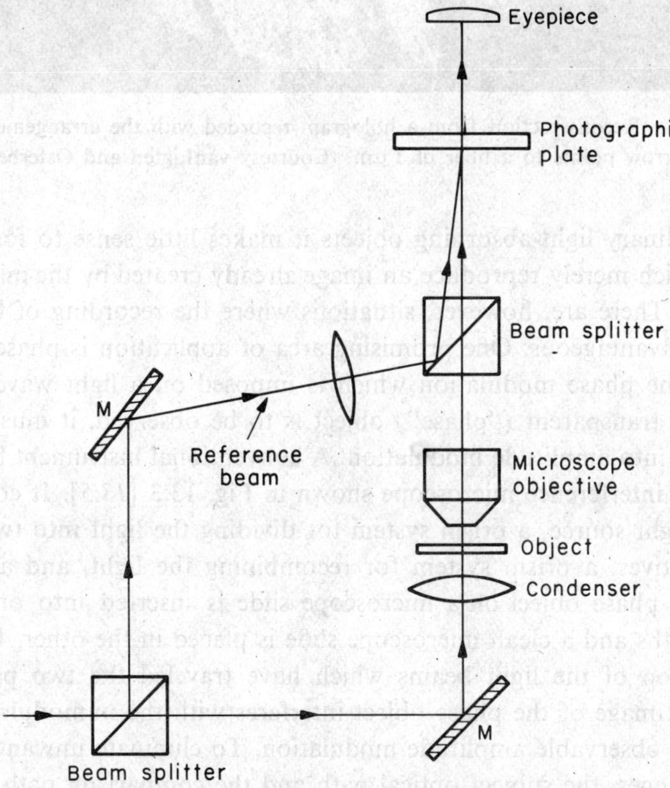

FIG. 13.1. Arrangement for holographic microscopy.

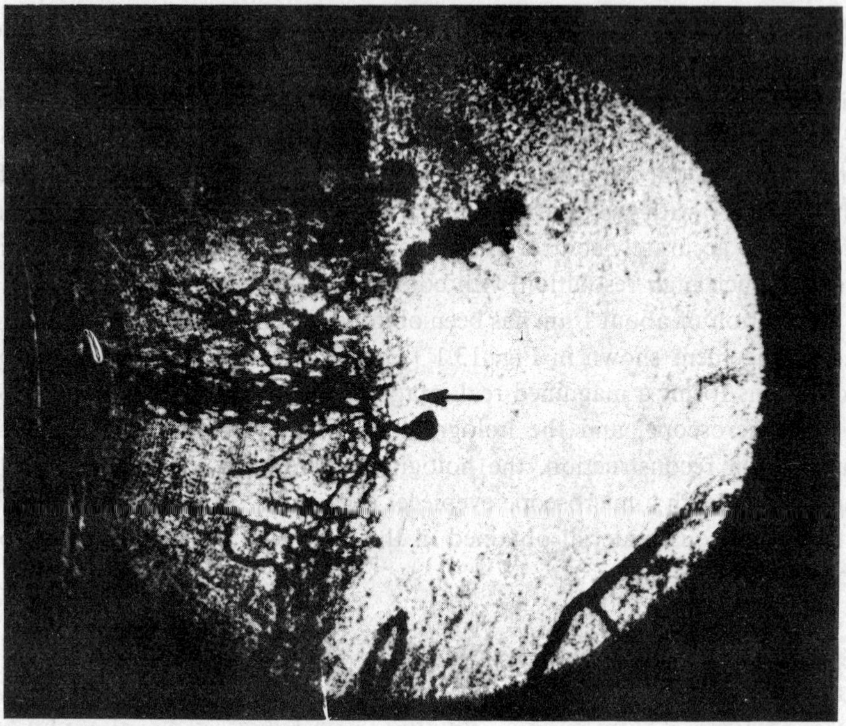

FIG. 13.2. Reconstruction from a hologram recorded with the arrangement of Fig. 13.1. The arrow points to a fiber of 1 μm. (Courtesy vanLigten and Osterberg [*13.2*].)

For ordinary light-absorbing objects it makes little sense to form holograms which merely reproduce an image already created by the microscope objective. There are, however, situations where the recording of the hologram is advantageous. One promising area of application is phase microscopy. If the phase modulation which is imposed on a light wave passing through a transparent ("phase") object is to be observed, it must be first converted into amplitude modulation. A conventional instrument for doing this is the interference microscope shown in Fig. 13.3 [*13.5*]. It consists of a single light source, a prism system for dividing the light into two paths, two objectives, a prism system for recombining the light, and a viewing system. A phase object on a microscope slide is inserted into one of the optical paths and a clean microscope slide is placed in the other. Upon recombination of the light beams which have traveled the two paths, the magnified image of the phase object interferes with the unmodulated light producing observable amplitude modulation. To eliminate unwanted interference fringes, the subject optical path and the comparison path must be carefully matched with respect to index of refraction and thickness of the

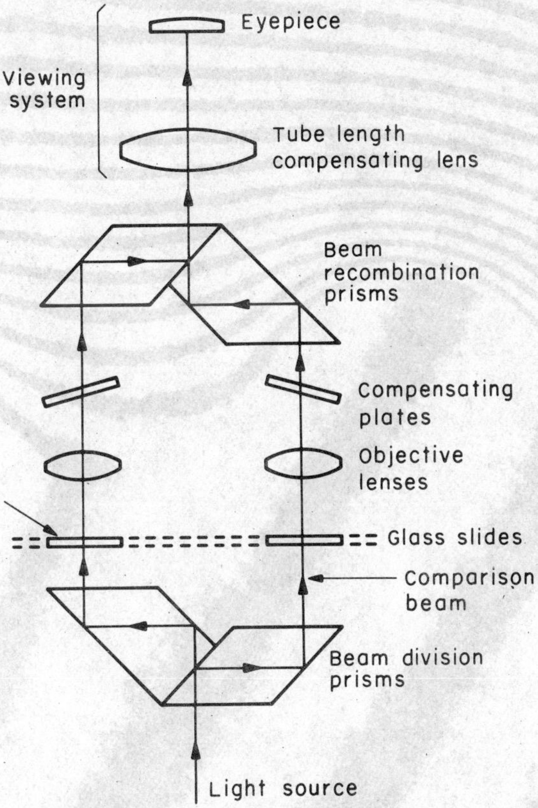

FIG. 13.3. Conventional interference microscope. (Courtesy Snow and Vandewarker [*13.5*].)

glass components and magnification. As a consequence, the interference microscope is an expensive instrument.

The holographic version of the interference microscope can be described with the help of Fig. 13.1. A clean microscope slide is first placed in the subject–beam path, and a hologram is formed. After processing, the hologram is returned to its original position and illuminated by the original reference wave. The hologram then reconstructs the wavefront characteristic of the clean slide. If next a phase object is placed on the slide, the diffracted wave will interfere with the wave reconstructed by the hologram. An observer uses the eyepiece to focus on the image plane of the objective. A drop of oil smeared on a slide would appear as in Fig. 13.4. The function of the hologram is to provide the unmodulated comparison beam at a lesser cost than in the conventional instrument. Unlike the conventional interference microscope, there is no need to match optical components in separate paths.

FIG. 13.4. Interference pattern due to an oil smear seen in a holographic interference microscope. (Courtesy Snow and Vandewarker [13.5].)

Instead, a common optical path is used at separate times. This is possible because of the ability of a hologram to store a wavefront and to reconstruct it at some later time when needed. Formation of the hologram and storage of the comparison wavefront need be carried out only once, presumably by the manufacturer of the microscope. The microscope user never has to concern himself with holography; he uses the hologram merely as another optical element of the microscope. He must, however, use optically flat slides.

Other possibilities arise when the hologram records not the clean microscope slide but the phase object of interest. By placing the processed hologram back in position and reconstructing the subject wave with the reference beam, we can cause the reconstructed wavefront at the image plane of the objective to interfere with the original wavefront at that plane. Suppose the hologram plate is positioned so that the image wavefronts produced by the hologram and by the microscope objective are in phase over the image plane, thereby producing uniform intensity. If one of the images is displaced laterally (the displacement should be less than the resolution limit of the microscope), then intensity changes are observed over those portions of the image where the phase varies rapidly. Where the phase varies slowly, the two images are still essentially in phase. Where the phase is a rapid function of the coordinates of the image plane, e.g., at the edges of the phase object, there will be some phase cancellation and consequently a reduction in the observed light intensity. The technique is the holographic version of *differential interference* and is used for edge enhancement.

In another application a hologram is formed of a microscopic biological specimen which contains a very slowly flowing region in a static environment [*13.6*]. Upon reconstruction, the area of flow stands out prominently as a dark area of the image. Only the subject portions which are sufficiently static during the exposure give rise to a stable, high-visibility interference pattern and a bright reconstruction.

13.2 Analysis of Aerosol Particles

The analysis of aerosol particles was one of the earliest applications of holography and is illustrative of an important advantage that holographic recording has over photographic recording or direct observation of moving microscopic objects. When the object to be observed is small, high magnification is required in the microscope; the depth of field of the microscope becomes extremely small. Suppose that at a given instant of time we wish to record moving microscopic objects distributed throughout a volume; or suppose that we wish to record a single transient event known to occur at a given time within a given volume, but whose exact location within the volume is uncertain. In such transient cases it is advantageous to be able to record the entire volume of interest at the desired instant, as a hologram, and later, at leisure, microscopically examine the three-dimensional image throughout its volume. Since holography does not require the recording plate to be placed in a specified image plane conjugate to a plane in the object, the hologram can record the entire object volume at a resolution which is finally limited only by the hologram aperture.

Aerosol particles are quite small, typically 5 to 50 μm in diameter. A hologram plate placed at almost any distance from the volume containing the particles is in the far field of the individual particle. A lensless Fraunhofer hologram (Section 8.3.6) can be recorded by illuminating the volume of particles with a pulsed ruby laser (Chapter 11). Coherence requirements on the pulsed laser are minimized with in-line geometry, and the far-field diffraction pattern of the "twin" or virtual image overlapping the real image has negligible disturbing effect. Figure 13.5 is a diagram of the arrangement

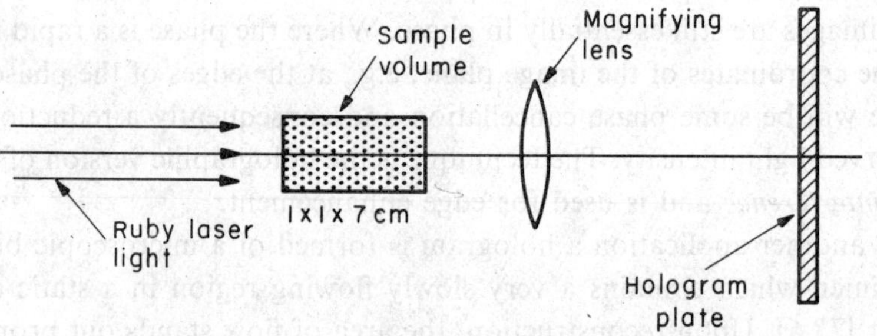

FIG. 13.5. Arrangement for recording holograms of aerosol particles.

used to record the hologram. (See Thompson *et al.* [*13.7*] for a list of early references to this work.) For the brief exposures associated with pulsed laser holograms, high sensitivity emulsion is required. Usually this means low resolution. It is possible to reduce the fringe frequency in the interference pattern to be recorded by first magnifying the sample volume with a lens, as shown in the figure, and then following the magnified image with the recording plate. The effect of magnifying the size of the particles is to decrease the range of spatial frequencies in the Fourier transform and hence in the Fraunhofer pattern [see correspondence (4.22) and Eq. (8.22)]. With a magnification of 5, the resolution required by the hologram plate does not exceed 100 cycles/mm. Kodak SO-243 film, a more sensitive film than 649F, can then be used to record the hologram.

A convenient, safe way to observe the image is indicated in Fig. 13.6 [*13.7*]. A parallel beam of He–Ne laser light illuminates the hologram, producing to the right of the hologram a real image of the particles in the

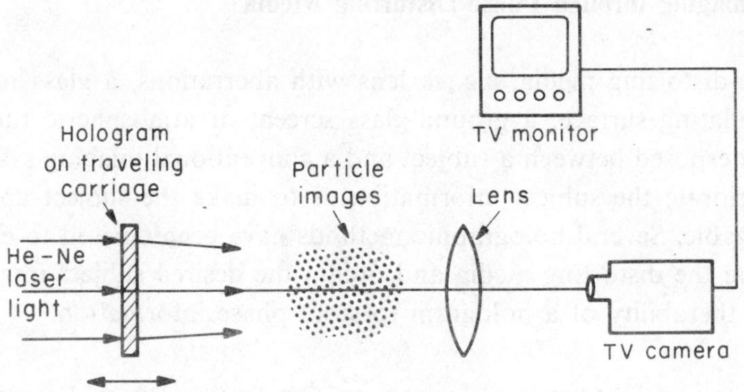

FIG. 13.6. Reconstruction arrangement for holograms of aerosol particles.

sample volume. Lenses magnify one plane of the image and relay it to a TV camera tube where it is displayed on a TV monitor. In the process the particles are magnified 300×. Any plane of the image can be examined by merely shifting the axial position of the hologram with respect to the lens and TV camera. A photograph of particles displayed on the monitor is shown in Fig. 13.7.

FIG. 13.7. Images of aerosol particles displayed on a TV monitor. (Courtesy Thompson *et al.* [*13.7*].)

13.3 Imaging through Phase-Distorting Media

Phase-distorting media, e.g., a lens with aberrations, a glass plate with an undulating surface, a ground-glass screen, or atmospheric turbulence, when interposed between a subject and a conventional imaging system, can so deteriorate the subject information as to make the subject appear unrecognizable. Several holographic methods have been devised to effectively penetrate the distorting media and obtain the desired subject image. They rely on the ability of a hologram to store phase information.

13.3.1 Image Hologram Method of Canceling Phase Distortion

Figure 13.8 indicates how to form a holographic optical system which cancels phase distortions imposed on waves passing through a thin phase distorter. The distortions may be mild, as in the case of aberrations produced

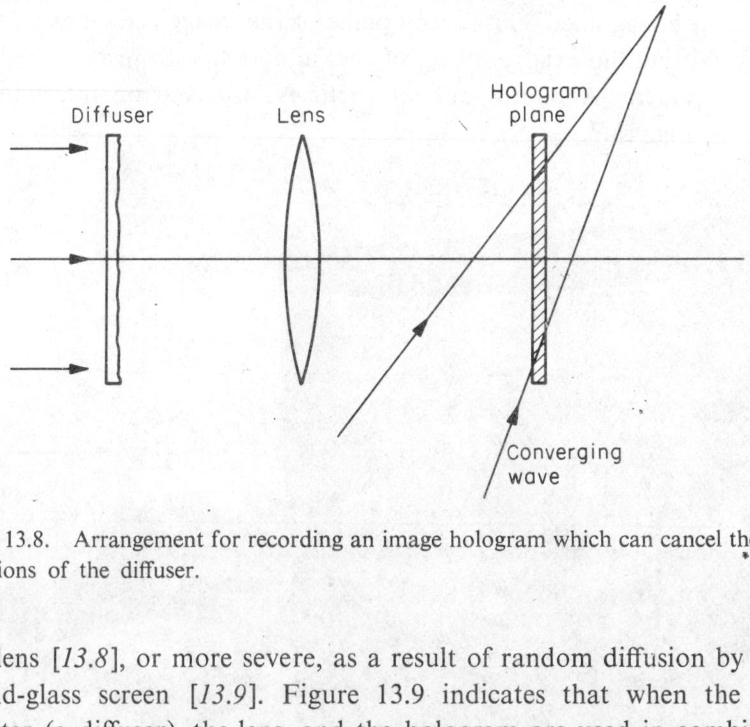

FIG. 13.8. Arrangement for recording an image hologram which can cancel the phase distortions of the diffuser.

by a lens [*13.8*], or more severe, as a result of random diffusion by a thin ground-glass screen [*13.9*]. Figure 13.9 indicates that when the phase distorter (a diffuser), the lens, and the hologram are used in combination they can form a real image of an object located behind the diffuser.

For the method to work, the diffuser must closely approximate a planar phase object, and its transmittance, a spatially varying exponential phase factor, must be imaged onto the hologram plane. If, in Fig. 13.8, the diffuser and hologram plane are located in conjugate planes (object and image planes of the lens) and a plane wave illuminates the diffuser, then the complex amplitude $\mathbf{a}_1$ emerging from the diffuser is also the complex amplitude at the hologram plane. A converging reference wave of complex amplitude $\mathbf{a}_2$ at the hologram interferes with $\mathbf{a}_1$ to form a hologram. We shall be interested in the hologram transmittance term $\mathbf{a}_1{}^* \mathbf{a}_2$ [see Eq. (1.15)]. Suppose the illumination, the lens, and the diffuser remain fixed while the hologram is processed and returned to its original position. The converging wave is removed so that the hologram is illuminated only by $\mathbf{a}_1$. In this case a diffracted wave of complex amplitude $\mathbf{a}_1\mathbf{a}_1{}^* \mathbf{a}_2$ emerges from the hologram on the right. Since $\mathbf{a}_1$ is a pure phase function of the coordinates of the hologram plane, the product $\mathbf{a}_1\mathbf{a}_1{}^*$ is a constant I_1. As in Eq. (1.16), a wave is reconstructed whose complex amplitude $I_1\mathbf{a}_2 \propto \mathbf{a}_2$ where $\mathbf{a}_2$ is the complex amplitude of the original converging wave. Now let a transmission object be placed behind the diffuser and illuminated by a plane wave as shown in Fig. 13.9. Incident on the diffuser is the complex amplitude $\mathbf{a}_0$

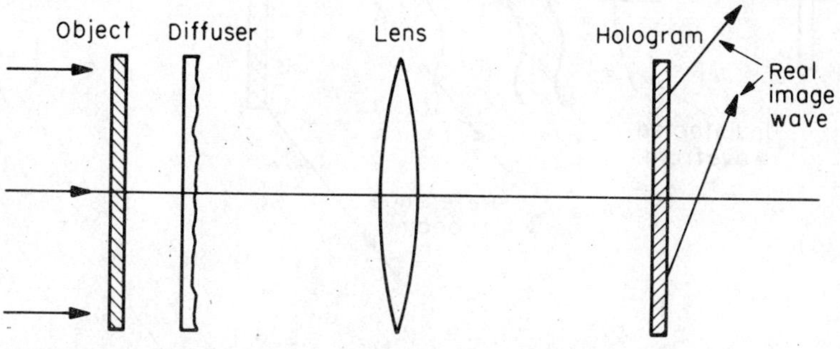

FIG. 13.9. Arrangement using the image hologram to obtain a real or virtual image of an object behind the diffuser.

and imaged onto the hologram is the complex amplitude $\mathbf{a}_0\mathbf{a}_1$. With this illumination the hologram diffracts a wave of complex amplitude $\mathbf{a}_0\mathbf{a}_1\mathbf{a}_1{}^* \mathbf{a}_2$ $\propto \mathbf{a}_0\mathbf{a}_2$. The product $\mathbf{a}_0\mathbf{a}_2$ is similar to that obtained when light transmitted by the object falls on a converging lens. There is no evidence of the diffuser. We may simply regard the hologram as a thin converging lens optically located at the plane of the diffuser. It has an off-axis focus. If we call d_0 the distance between object and diffuser planes and f the distance between the hologram and the point of convergence of the original reference beam (as measured along its central ray), then the thin lens formula written as $1/f = 1/d_0 + 1/d_i$ determines the image distance d_i. When d_0 exceeds f, a real image of the object is formed in a plane to the right of the hologram along an axis defined by the mean direction of the original converging wave. Absence of the wavefront $\mathbf{a}_1$ in the diffracted wave complex amplitude implies an image free of the distortions normally imposed by the diffuser.

13.3.2 Conjugate Reference Wave Method

A second technique for forming images through badly warped glass or ground-glass screens is closely related to the image hologram method. Here, however, the hologram not only removes phase distortions but also stores subject information. Fig. 13.10a indicates a subject separated from a hologram plate by a wavy, phase-distorting, glass surface. An off-axis reference wave interferes with the distorted subject wavefront to form a

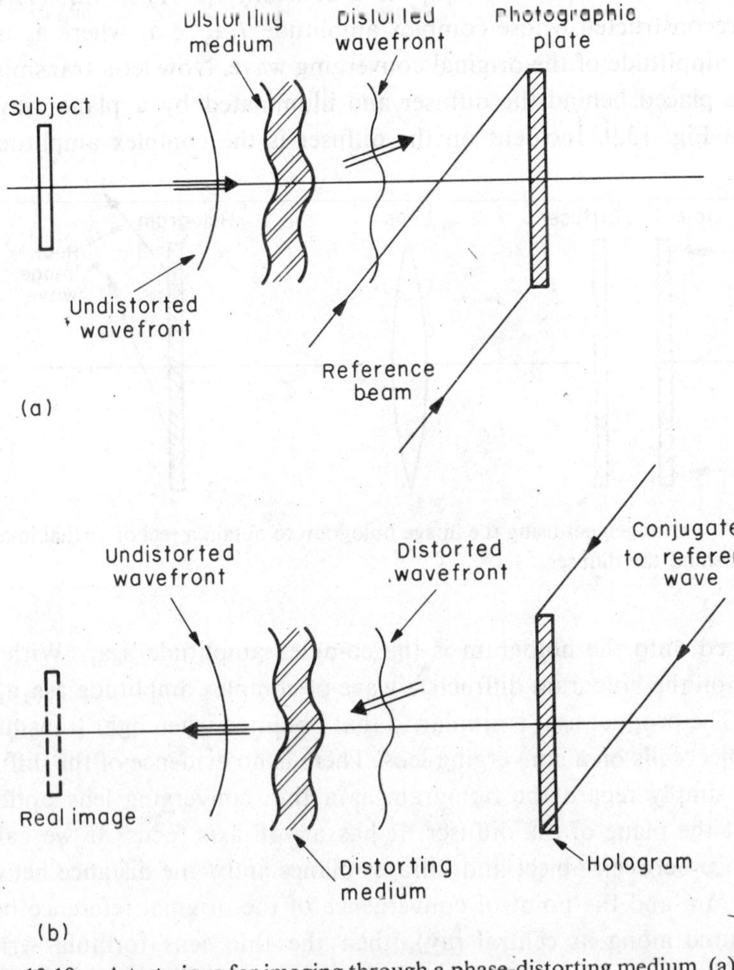

FIG. 13.10. A technique for imaging through a phase-distorting medium. (a) Recording of the hologram. (b) Reconstruction.

hologram. When the hologram is illuminated, as in Fig. 13.10b, with the conjugate to the original reference wave, the conjugate subject wavefront is reconstructed. The wave propagates with ray directions antiparallel to that of the original subject wave and with wavefronts everywhere conjugate to the original wavefronts (see Fig. 8.9). As the wave retraces the subject path back through the glass, the phase factor imposed by the distorting glass is multiplied by its conjugate. The result is elimination of these phase factors in the product, leaving only the undistorted subject wave to form a real image at the original subject position as in Fig. 8.9 and Fig. 13.10b [*13.10*, *13.11*]. If the phase-distorting medium is removed in the reconstruction step, the conjugate wave will not form an image of the original subject.

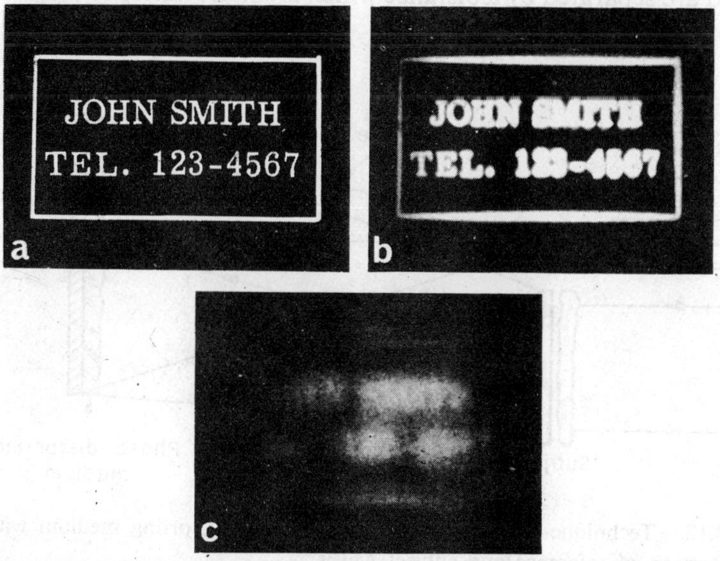

FIG. 13.11. (a) Subject. (b) Image formed in the arrangement of Fig. 13.10b. (c) Image obtained when the phase-distorting medium is removed from the reconstruction arrangement. (Courtesy Kogelnik [*13.10*]. Copyright 1965 by American Telephone and Telegraph Company. Reproduced by permission.)

Evidence that the method works is provided in Fig. 13.11. Figure 13.11a shows the original subject, Fig. 13.11b displays the image observed when the method is properly applied, and Fig. 13.11c indicates the results of removing the distorting glass from the reconstruction arrangement. Direction of the illuminating beam and positioning of the hologram and the glass are critical to the observation of the image. This may be turned to advantage if, as has been proposed, the method is employed for encoding secret messages. Only the possessor of the exact replica of the distorting glass and the knowledge of the reconstruction arrangement can decode the hologram. It should be noted that the technique just described makes it possible to compensate the phase distortions of a three-dimensional transparent medium throughout its depth.

13.3.3 Equal Distortion of Subject and Reference Waves

An experimental arrangement which might simulate conditions producing distortion in vertical observations through the earth's atmosphere is indicated in Fig. 13.12 [*13.12*]. Subject and reference point source lie in a plane and are separated by a distance which is small relative to their distance

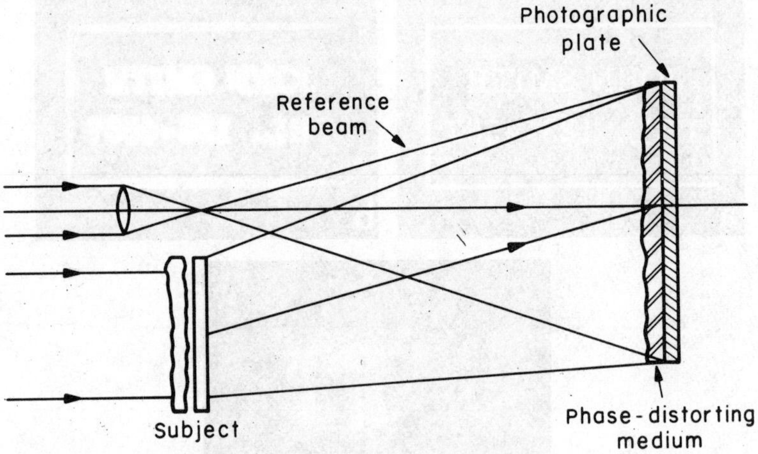

Fig. 13.12. Technique for imaging through a phase-distorting medium with equal phase distortion of reference and subject waves.

from the hologram plate. The distorting medium lies close to the plate. If the medium is thin relative to the smallest spatial period of its optical thickness variation, we can consider subject and reference rays to traverse essentially the same optical path through the medium and suffer the same phase modulation. In this case we can approximate the transmittance of the distorting medium by the phase factor $\exp[i\varphi(x, y)]$. Let us designate the subject complex amplitude incident on the medium as $\mathbf{a}(x, y)$. The subject complex amplitude at the hologram plane is then $\mathbf{a}(x, y) \exp[i\varphi(x, y)]$. Similarly, the reference complex amplitude at the hologram plane is $\mathbf{r}(x, y) \exp[i\varphi(x, y)]$. As has been customary, we assume the hologram transmittance t to be proportional to the intensity of the interference pattern at the plate, with the result that

$$\begin{aligned} t &\propto \{\mathbf{a}(x, y) \exp[i\varphi(x, y)] + \mathbf{r}(x, y) \exp[i\varphi(x, y)]\} \\ &\quad \times \{\mathbf{a}^*(x, y) \exp[-i\varphi(x, y)] + \mathbf{r}^*(x, y) \exp[-i\varphi(x, y)]\} \\ &\propto \mathbf{aa}^* + \mathbf{rr}^* + \mathbf{ar}^* + \mathbf{a}^*\mathbf{r}. \end{aligned} \tag{13.1}$$

We find the hologram transmittance to contain no evidence of the phase deviation $\varphi(x, y)$, and consequently an undistorted subject wave is reconstructed when the hologram is illuminated by the reference wave $\mathbf{r}(x, y)$.

If $\exp[i\varphi(x, y)]$ is to be the same for subject and reference wave the subject and reference source separation must not subtend too great an

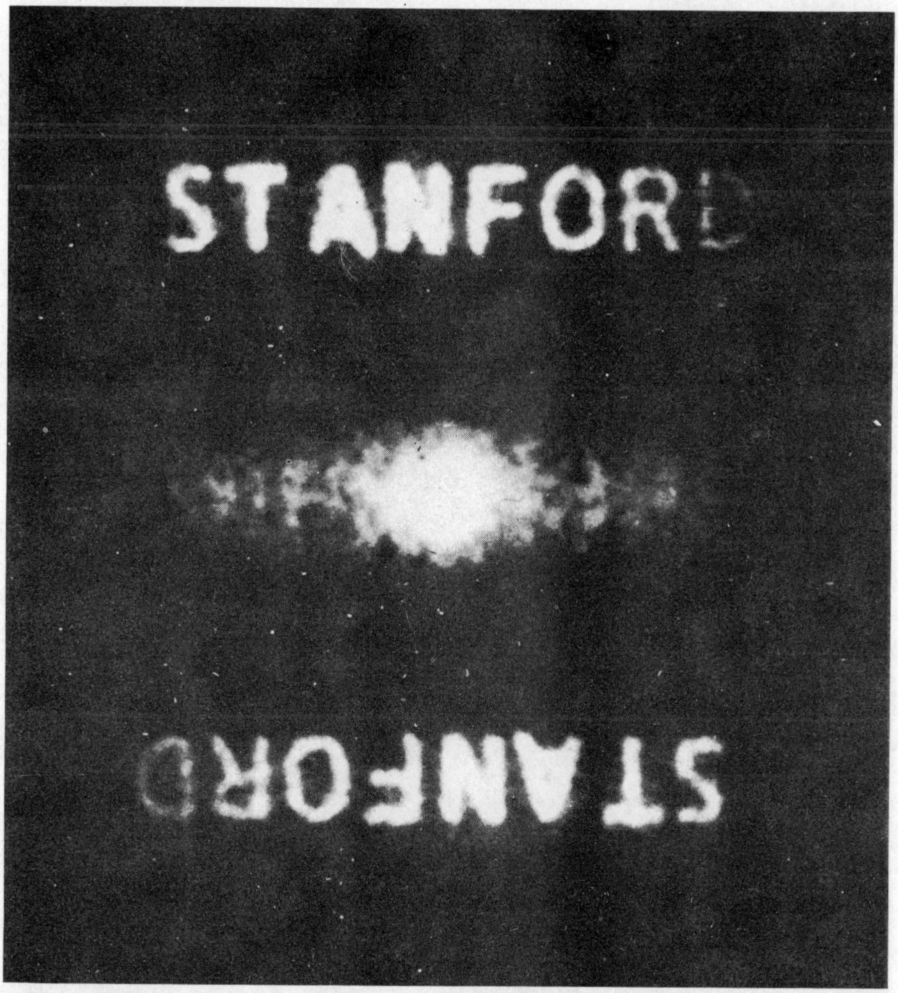

FIG. 13.13. Reconstruction from a hologram recorded in the arrangement of Fig. 13.12. (Courtesy Goodman *et al.* [*13.12*].)

angle at the hologram. We note that the hologram-forming arrangement is that of the lensless Fourier transform hologram and that this restriction on field of view is analagous to that imposed on such arrangements by limited film resolution (see Section 8.4.1). In each case the resolution in the image is not restricted, but the field of view is.

Figure 13.13 is a photograph of the image from a hologram recorded with the method described above. As a comparison the image formed by a lens placed behind the phase-distorting medium is shown in Fig. 13.14. It is evident that conventional imaging is incapable of transmitting information through the medium in intelligible form. Field experiments intended to

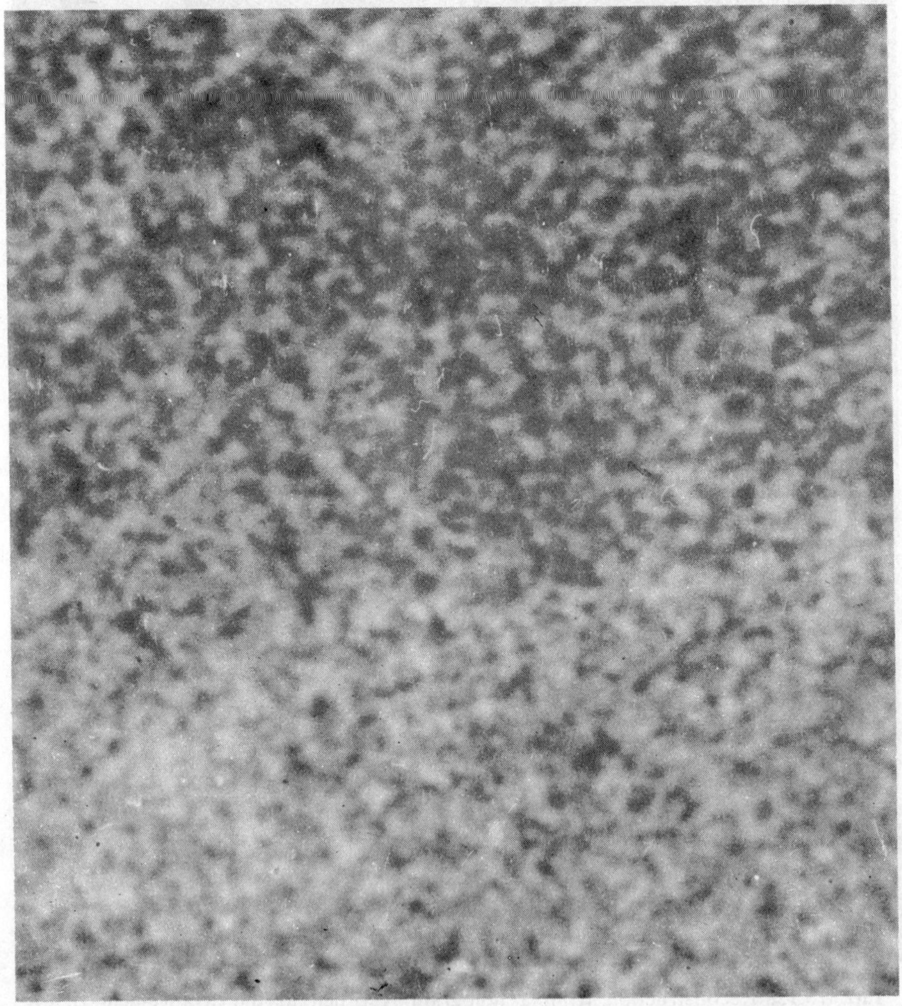

FIG. 13.14. Image formed by a lens behind the phase-distorting medium. (Courtesy Goodman *et al.* [*13.12*].)

explore the holographic method for imaging through turbulent atmosphere are reported by Goodman *et al.* [*13.13*].

13.3.4 Imaging through Moving Scatterers

The holographic technique for imaging through an aggregate of *moving* scatterers, such as fog, is remarkably simple [*13.14*, *13.15*]. We suppose that there is a static subject immersed in the fog. A hologram is recorded of the interference pattern formed by subject light which passes through the fog and a reference wave which does not. Since the scatterers are moving, only subject light passing *unscattered* through the fog is coherent with the reference wave, forms stationary fringes, and is recorded in a hologram. Spatially incoherent light scattered by the fog adds only a constant exposure to the hologram plate which, however, does decrease diffraction efficiency and the signal-to-noise ratio. An analysis of signal-to-noise ratio when the signal contains a strong component of incoherent light is given by Hamasaki [*13.16*].

13.4 High-Resolution Projection Imaging

A hologram illuminated with the conjugate to the reference beam used to form it can, in theory, reconstruct an unaberrated, undistorted subject wave. Resolution in the projected real image should be limited only by diffraction from the boundaries of the hologram. As shown in Sections 6.4.3 and 8.4.2, resolution increases with increasing hologram size. Since holograms can be made large, one might expect to be able to image spatial frequencies as high as 1000 cycles/mm over fields as large as 5×5 cm. Although the magnification in this case would be unity, one-to-one high-resolution projection imaging has an important potential application in the photolithography of integrated circuits. Precision imaging of photolithographic masks onto semiconductor wafers is a task currently performed by contact printing, a process which soon deteriorates the mask. Transfer of the image to the wafer by projection is a desirable alternative, but only the very best and very expensive lenses are capable of projecting the mask image with the desired resolution and field.

Holography would seem to have much to offer in this area, and yet relatively little work has been reported. Image line-widths of 4 μm have been obtained over a field 2.5 cm in diameter [*13.17*]. However, due to use of a diffusing screen in the subject beam during the formation of the hologram, the image has a granular appearance. Many of the difficulties which stand in the way of achieving the optimum from holographic projection imaging can be ascribed to the less-than-ideal properties of available recording materials (see Chapter 10). Shrinkage, surface distortion, nonlinearities, and other noise sources associated with photographic emulsion are of special concern when the illumination is coherent and the record is a hologram. They result in a mottled image with low contrast and fuzzy edges, conditions not tolerable for semiconductor circuit photolithography. New, more stable materials may be the answer to these problems.

When maximum possible resolution is desired, alignment of the hologram surface with respect to the illuminating beam becomes extremely critical [*13.18*]. We also know [from Eqs. (7.52) and (7.53)] that the finite size of the illuminating source and the degree to which the source is monochromatic play a role in determining image resolution. These effects are all minimized when the subject is located close to the photographic plate during formation of the hologram. However, this configuration creates a problem in introducing a reference wave which does not intercept the subject. An image hologram of a two-dimensional mask would be a good solution (see Section 8.3.2) were it not for the need for an excellent, high-resolution, wide-field lens to form the image. An arrangement which overcomes the reference wave problem and requires no lens is shown in Fig. 13.15 [*13.19*]. Here the reference wave passes through a prism and a film of index-matching

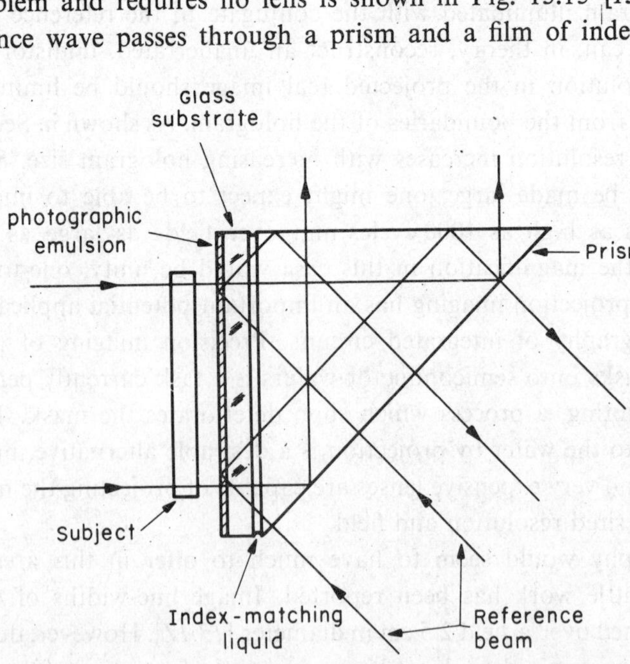

FIG. 13.15. Hologram formation using total internal reflection.

liquid and enters the photographic plate through the back surface of the substrate. When the wave arrives at the air–emulsion boundary, its angle of incidence is sufficiently large to cause it to *totally reflect*. Interference of the subject wave with the reflected reference wave forms a transmission hologram, while interference with the incident reference wave forms a reflection hologram. To obtain a real image of the subject, the hologram is illuminated with a wave traveling antiparallel to the reference wave. Assuming the same arrangement of prism and plate as in hologram formation, both the transmission and reflection holograms recorded in the emulsion project a real image at the original subject position. Note that the undiffracted illuminating light is again totally reflected at the air–emulsion interface and does not appear in the real image.

Preliminary investigations have yielded a maximum image resolution of 640 cycles/mm, although the image is somewhat noisy [*13.20*]. When the hologram is illuminated with incoherent light, the noise is reduced. The reduction is accompanied by only a slightly lesser resolution (500 cycles/mm). Further work with the total internal reflection method or with the holographic recording of evanescent waves [*13.21*, *13.22*] can be expected to increase the quality of projection imaging.

13.5 Multiple Imaging

The mask whose image we might project by the means discussed in the previous section consists of an array of identical images. Conventionally the array is produced by a "step-and-repeat" camera or possibly by imaging through a fly's-eye lens (an array of lenslets). Plane holograms can also be used to form a multiplicity of identical images [*13.23–13.26*]. We shall describe one variant of the general method [*13.25*]. To form the required hologram, one must record the interference of an array of point sources

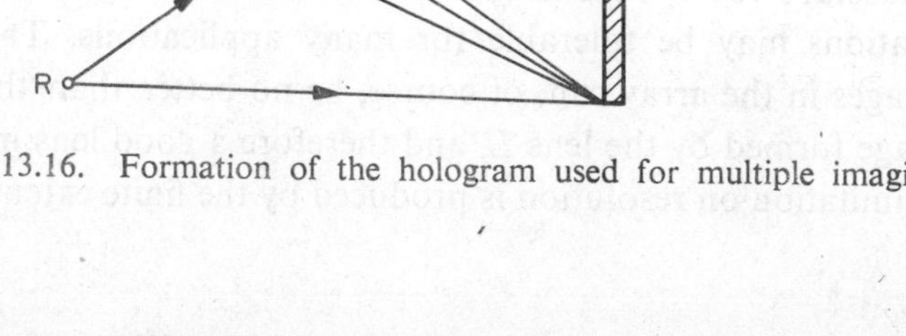

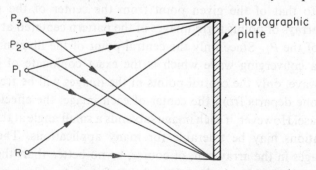

FIG. 13.16. Formation of the hologram used for multiple imaging.

$P_1, \ldots, P_n$ with a point reference source R as in Fig. 13.16. To use the hologram as a multiple-imaging device or fly's-eye lens, a lens is placed near the back of the hologram as shown in Fig. 13.17. A subject pattern illuminated with laser light is positioned so that light passing through the lens and undiffracted by the hologram forms a real image of the pattern in a plane centered at R. This image is called the zero-order image. An array of identical real images is also generated by first-order diffraction from the hologram, and these are focused and centered on the positions of the original point sources P_i.

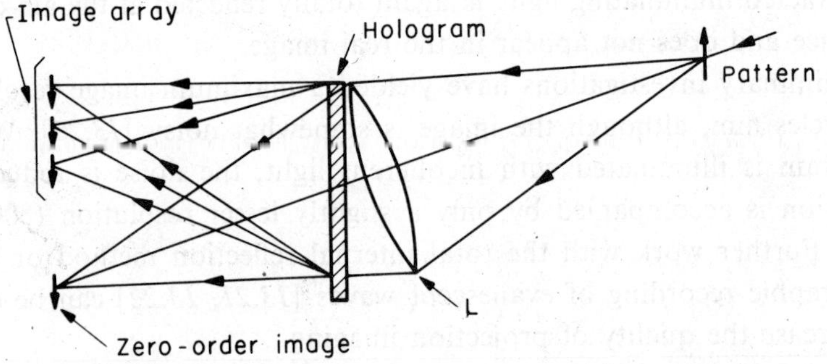

FIG. 13.17. Multiple imaging with the hologram formed according to Fig. 13.16.

The process is understood by first considering the subject pattern to the right of the lens to be merely a point source. In this case the spherical wave passing through the hologram and converging on the point R is the conjugate to the original reference wave. As a consequence, real images of the sources P_i are generated at the original positions of P_i. Ideally these are images free of aberrations. Suppose now that an arbitrary extended pattern replaces the point source to the right of the lens. Each point of the pattern generates a displaced image of the array of points P_i. Since the displacement is proportional to that of the given point from the center of the pattern, the displaced arrays of points form images of the pattern centered at the original positions of the P_i. Since only the central point of the illuminated pattern produces a converging wave which is the exact conjugate of the original reference wave, only the central points of the images will be free of aberrations. As one departs from the center of each image, the effects of aberrations increase. However, if each image subtends a small angle at the hologram, the aberrations may be tolerable for many applications. The resolution of the images in the array can, of course, be no better than the resolution of the image formed by the lens L, and therefore a good lens must be used. Another limitation on resolution is produced by the finite extent of practical "point sources" P_i. With the method described, a resolution of 50 cycles/mm has been reported, while with a slighly different approach, lines of 3-μm width were resolved over a field 5×5 cm [*13.24*].

REFERENCES

13.1. E. N. Leith and J. Upatnieks, "Microscopy by Wavefront Reconstruction," *J. Opt. Soc. Amer.* **55**, 569 (1965).

13.2. R. V. vanLigten and H. Osterberg, "Holographic Microscopy," *Nature* **211**, 282 (1966).

13.3. G. W. Ellis, "Holomicrography: Transformation of Image during Reconstruction a posteriori," *Science* **154**, 1195 (1966).

13.4. W. H. Carter, "Polarization Selection for Reconstructed Wavefronts and Application to Polarizing Microholography," *IEEE J. Quantum Electron.* **2**, 44 (1966).

13.5. K. Snow and R. Vandewarker, "An Application of Holography to Interference Microscopy," *Appl. Opt.* **7**, 549 (1968).

13.6. E. J. Feleppa, "Biomedical Applications of Holography," *Phys. Today* **22**, 25 (July 1969).

13.7. B. J. Thompson, J. H. Ward, and W. Zinky, "Application of Hologram Techniques for Particle Size Analysis," *Appl. Opt.* **6**, 519 (1967).

13.8. J. Upatnieks, A. Vander Lugt, and E. Leith, "Correction of Lens Aberrations by Means of Holograms," *Appl. Opt.* **5**, 589 (1966).

13.9. H. Kogelnik and K. S. Pennington, "Holographic Imaging Through a Random Medium," *J. Opt. Soc. Amer.* **58**, 273 (1968).

13.10. H. Kogelnik, "Holographic Image Projection Through Inhomogeneous Media," *Bell Syst. Tech. J.* **44**, 2451 (1965).

13.11. E. N. Leith and J. Upatnieks, "Holographic Imagery Through Diffusing Media," *J. Opt. Soc. Amer.* **56**, 523 (1966).

13.12. J. W. Goodman, W. H. Huntley, D. W. Jackson, and M. Lehmann, "Wavefront Reconstruction Imaging Through Random Media," *Appl. Phys. Lett.* **8**, 311 (1966).

13.13. J. W. Goodman, D. W. Jackson, M. Lehmann, and J. Knotts, "Experiments in Long-Distance Holographic Imagery," *Appl. Opt.* **8**, 1581 (1969).

13.14. K.A. Stetson, "Holographic Fog Penetration," *J. Opt. Soc. Amer.* **57**, 1060 (1967).

13.15. E. Spitz, "Reconstitution holographique des objets à travers un milieu diffusant en mouvement," *Compt. Rend. Acad. Sci. Paris* **264B**, 1449 (1967).

13.16. J. Hamasaki, "Signal-to-Noise Ratios for Hologram Images of Subjects in Strong Incoherent Light," *Appl. Opt.* **7**, 1613 (1968).

13.17. M. J. Beesley, H. Foster, and K. G. Hambleton, "Holographic Projection of Microcircuit Patterns," *Electron. Lett.* **4**, 49 (1968).

13.18. E. B. Champagne and N. G. Massey, "Resolution in Holography," *Appl. Opt.* **8**, 1879 (1969).

13.19. K. A. Stetson, "Holography with Total Internally Reflected Light," *Appl. Phys. Lett.* **11**, 225 (1967).

13.20. K. A. Stetson, "Improved Resolution and Signal-to-Noise Ratios in Total Internal Reflection Holograms," *Appl. Phys. Lett.* **12**, 362 (1968).

13.21. H. Nassenstein, "Interference, Diffraction, and Holography with Surface Waves ("Subwaves") I., *Optik* **29**, 597 (1969); II., *ibid.* **30**, 44 (1969).

13.22. O. Bryngdahl, "Holography with Evanescent Waves," *J. Opt. Soc. Amer.* **59**, 1645 (1969).

13.23. L. H. Lin and E. O. Schulz-DuBois, "Apparatus for Producing a Fly's Eye Lens," U. S. patent 3, 405, 614, filed 1 December 1965, issued 15 October 1968.

13.24. S. Lu, "Generating Multiple Images for Integrated Circuits by Fourier-Transform Holograms," *Proc. IEEE* **56**, 116 (1968).

13.25. G. Groh, "Multiple Imaging by Means of Point Holograms," *Appl. Opt.* **7**, 1643 (1968).

13.26. S. Lowenthal, A. Werts, and M. Rembault, "Formation de réseaux d'images à l'aide d'un hologramme multiplicateur éclairé en lumiére spatialement incohérente," *Compt. Rend.* **267**, ser. B, 120 (1968).

Chapter 14

HOLOGRAMS AND HOLOGRAM SPATIAL FILTERS FORMED WITH SPATIALLY MODULATED REFERENCE WAVES

Holograms are recordings of the intensities of interference patterns said to be formed with a "subject wave" and a "reference wave." Yet in the mathematical representation of the interference pattern intensity there is nothing except arbitrary notation to distinguish one wave from the other. We find the hologram transmittance t of a plane hologram to be symmetric in the complex amplitudes of the two forming waves. Suppose the complex amplitude (in the hologram plane) of one wave is designated by $\mathbf{a} = a \exp(i\varphi_{\mathrm{a}})$ and that of the other by $\mathbf{r} = r \exp(i\varphi_{\mathrm{r}})$. If the hologram is properly exposed and developed

$$t \propto I = \mathbf{aa}^* + \mathbf{rr}^* + \mathbf{ar}^* + \mathbf{a}^*\mathbf{r} \tag{14.1}$$

as in Eq. (1.15). It is common to choose as illumination for the hologram one of the waves used to form it. Assuming that the original waveforms have certain desirable properties (their nature will be examined in this chapter), illumination with one such wave reconstructs the other. The latter deserves the label "subject wave" since its partner has merely served as a phase reference for it.

Holograms, which respond to one waveform by reconstructing another, may be used as spatial filters to sift data. Let us suppose, for example, that data in the form of coherent light patterns sequentially illuminate the hologram filter. When a pattern identical to the complex amplitude of the reference wave used to form the hologram is a member of the data set, it is recognized by the appearance of the subject wave which it reconstructs. Hologram spatial filters are most effective when formed as plane Fourier transform holograms (see Section 8.3.3). A lens system is used to define the input, spatial frequency, and output planes. The hologram in the spatial frequency plane performs a filtering operation on the spatial frequency components of the input function.

14.1 Associative Storage

Consider illumination of a hologram of transmittance t, as in Eq. (14.1), by a wave of complex amplitude $\mathbf{c}$. The complex amplitude of the light diffracted by the hologram, in the plane of the hologram, is

$$\mathbf{w} = \mathbf{c}t \propto \mathbf{c}I = \mathbf{caa}^* + \mathbf{crr}^* + \mathbf{car}^* + \mathbf{ca}^*\mathbf{r}. \tag{14.2}$$

With proper choice of $\mathbf{c}$, $\mathbf{a}$, and $\mathbf{r}$, the third and fourth terms of Eq. (14.2) can be reconstructions of the forming waves $\mathbf{a}$ and $\mathbf{r}$ respectively. Suppose $\mathbf{a}$ and $\mathbf{r}$ are such that the corresponding intensities over the hologram plane $\mathbf{aa}^* = a^2$ and $\mathbf{rr}^* = r^2$ are constants independent of the xy coordinates of that plane. Then when $\mathbf{c}$ is chosen to be identical to $\mathbf{r}$, the third term in Eq. (14.2), $\mathbf{car}^* = r^2\mathbf{a} \propto \mathbf{a}$, represents reconstruction of the original forming wavefront whose complex amplitude at the hologram plane is $\mathbf{a}$. Similarly, for $\mathbf{c} = \mathbf{a}$, the fourth term represents reconstruction of the original forming wavefront whose complex amplitude at the hologram plane is $\mathbf{r}$. We may summarize these properties by saying that the complex amplitudes $\mathbf{a}$ and $\mathbf{r}$ are stored holographically *in association*. When $\mathbf{r}$ illuminates the hologram, $\mathbf{a}$ is reconstructed; when $\mathbf{a}$ illuminates the hologram, the other member of the associated pair, $\mathbf{r}$, is reconstructed.

Holograms are *usually* formed with *subject* waves whose complex amplitudes are arbitrary spatial functions and with *reference* waves whose complex amplitudes are pure phase factors. If $\mathbf{c} = \mathbf{r} = b \exp(i\varphi_r)$ is the reference wave, where b is a constant, the third term of Eq. (14.2) yields

$$\mathbf{car}^* = b^2\mathbf{a} \exp(i\varphi_r) \exp(-i\varphi_r) \propto \mathbf{a},$$

a reconstruction of the subject wave. The result holds true for any phase function $\varphi_r(x, y)$ of the hologram plane coordinates x and y. It is necessary, however, in the case of a spatially modulated phase $\varphi_r(x, y)$ to ensure that the reconstructing wave $\mathbf{c}$ is properly registered with and identical to $\mathbf{r}$ if an undistorted reconstruction of the subject wave is to be obtained. A more common choice of the reference wave used to form the hologram is one with a phase front unmodulated with information, such as a plane or spherical wave. Then, as we have seen in Chapter 3, the hologram may be illuminated with a plane or spherical wave of different direction or curvature, the first-order consequence being only a change in location or magnification of the subject image.

When the reference source is a reflector (or transmitter) of diffuse light (as many normal objects are), **car*** can be approximately proportional to **a**. For this case the amplitude as well as the phase of the complex amplitude $\mathbf{r}(x, y)$ at the hologram will be spatially modulated. However, apart from speckle pattern (Section 8.2.2 and Chapter 12), the intensity of unfocused light coming from a diffuser is observed to be macroscopically uniform. To this extent we can regard **rr*** as a constant independent of the hologram coordinates. For $\mathbf{c} = \mathbf{r}$

$$\mathbf{car}^* = \mathbf{rr}^*\mathbf{a} \propto \mathbf{a}.$$

14.1.1 A Ghost Image Experiment

An experiment carried out under conditions where both subject and reference sources reflect diffusely [*14.1*] is illustrated in Fig. 14.1A. A slotted metal bar lying on its side served as the subject, and the bar with the series of holes, standing vertically on end, was the reference source. Their mat surfaces were simultaneously illuminated by a laser beam of limited circular cross section, and those portions of the bars covered by the laser beam appear as brighter areas in the upper left photograph of Fig. 14.1B. This photograph of the scene viewed from the hologram plane was taken with room light plus laser light to identify more clearly the subject and reference sources. Because each bar reflects light over a wide range of angles, overlap of the two waves reflected to the hologram plate is ensured. While the distinction between subject and reference sources or subject and reference waves is useful for describing the hologram reconstruction process, a hologram so formed is simply the photographic recording of the Fresnel (near field) diffraction pattern of a coherently illuminated scene.

After proper exposure to this diffraction pattern and after proper development, the hologram was set back precisely into its original position and the horizontal subject bar removed from the scene. The scene is shown in the photograph on the right in Fig. 14.1B. Laser light reflected from the vertical bar continued to illuminate the hologram in exactly the manner it did during the formation of the hologram. Recorded in the bottom photograph of Fig. 14.1B is a view looking through the illuminated hologram. An image

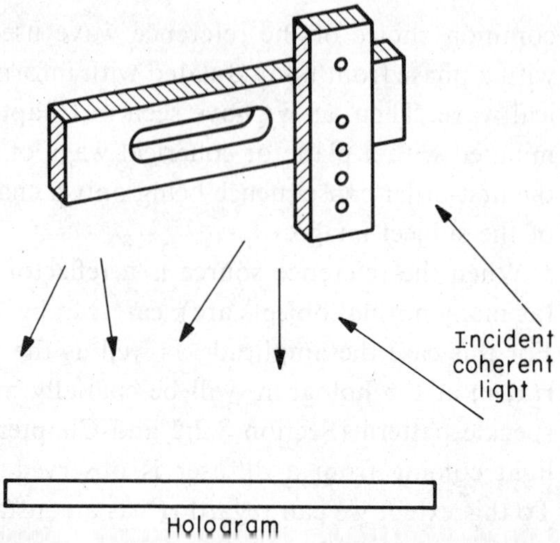

FIG. 14.1A. Recording arrangement for ghost-image experiments.

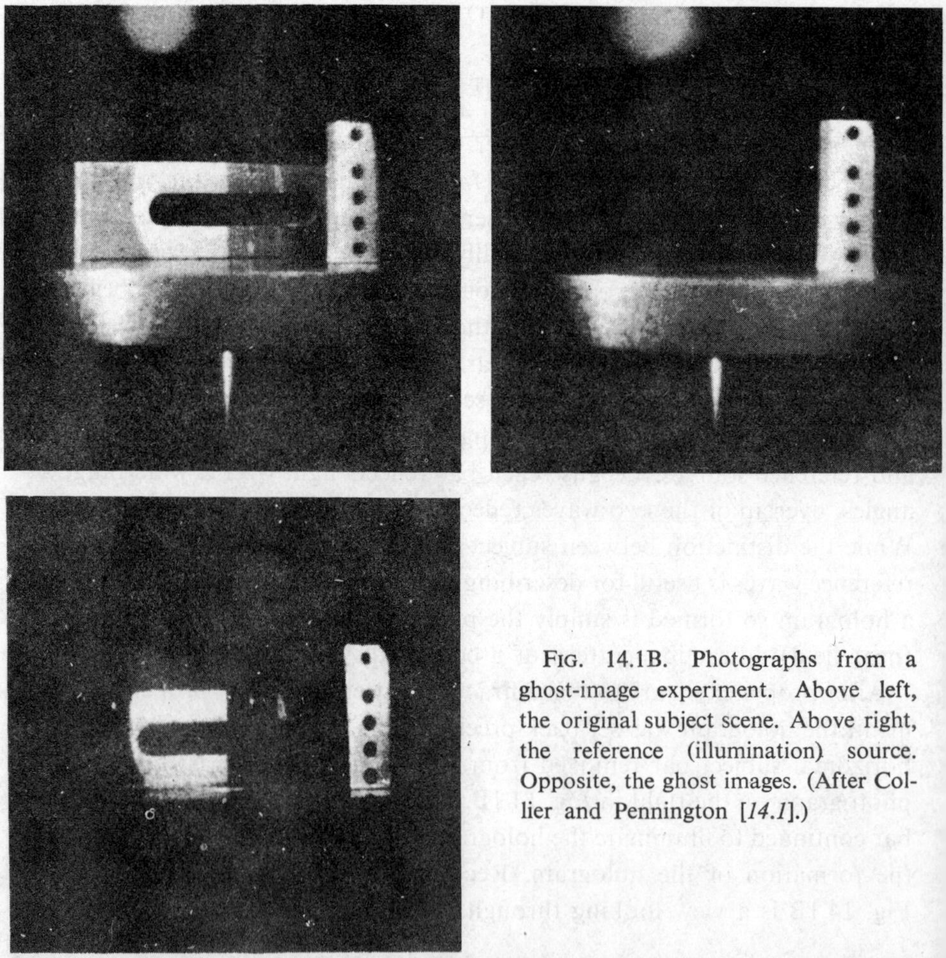

FIG. 14.1B. Photographs from a ghost-image experiment. Above left, the original subject scene. Above right, the reference (illumination) source. Opposite, the ghost images. (After Collier and Pennington [*14.1*].)

of that portion of the horizontal, slotted bar illuminated by the laser during formation is clearly visible along with the reference vertical bar. The image which appears when a hologram is illuminated by its spatially modulated reference wave is often called a *ghost image*. Had the vertical bar been removed and the horizontal bar left to act as the illuminating source for the hologram, then the ghost image of the vertical bar would have been produced.

14.1.2 Allowed Displacement of the Reference Source

To obtain a good reconstruction with a spatially modulated reference wave, the hologram must be returned precisely to its original position and the laser illumination proceeding from the reference source must be precisely the same as during formation. If the vertical bar (reference source) or the hologram, in the experiment just described, is moved ever so slightly, the ghost image of the subject disappears leaving a diffuse "noise" in its place. An estimate of the allowed displacement of the reference source can be obtained from the following considerations: Suppose a hologram is formed with a diffusing reference source. The hologram is processed, replaced in its original position, and illuminated with the same reference source slightly displaced. Let the displacement result in a shift of the complex amplitude of the reference wave at the hologram by a distance c in the y direction. (See Fig. 14.2. For simplicity the analysis is confined to the yz plane.) Shown in the figure is a large diffusing reference source of extent P_1P_2

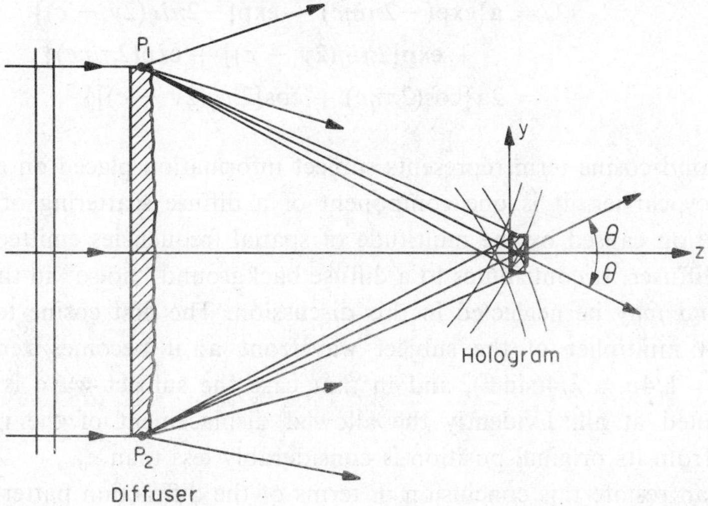

FIG. 14.2. A diffusing reference source illuminating a small hologram.

illuminated by a plane wave. The diffuser emits light over a wide range of angles. However the distance from the diffuser to the (small) hologram limits the directions of the rays accepted by the hologram to a range of only 2θ. (Here *hologram* can be interpreted as any small area of a hologram of arbitrary size.) The bundle of rays from P_1, making a mean angle to the z axis of $-\theta$, and those from P_2, making a mean angle $+\theta$, define the limits. For the purpose of illustrating the effects of shifting the diffusing reference source, we restrict consideration only to these limiting ray directions.

The light from P_1 and P_2 falling on the hologram area may be approximated by plane waves, $\exp(2\pi i\eta y)$ and $\exp(-2\pi i\eta y)$, where $\eta = \sin\theta/\lambda$. (The initial phases of the waves are arbitrary. However, inclusion of the phase constants in the following calculation adds nothing.) If we consider a hologram to be formed with a reference wavefront $\mathbf{r}(y) = \exp(2\pi i\eta y) + \exp(-2\pi i\eta y)$ and a subject wavefront $\mathbf{a}$, then the third term in Eq. (14.1), the virtual-image term, is given by

$$\mathbf{ar}^* = \mathbf{a}[\exp(-2\pi i\eta y) + \exp(2\pi i\eta y)].$$

Now assume the hologram to be illuminated by a reference source displaced so that the illuminating wavefront is $\mathbf{r}(y - c)$. The reconstructed wavefront [the third term in Eq. (14.2)] becomes

$$\begin{aligned}\mathbf{r}(y-c)\mathbf{ar}^* &= \{\exp[2\pi i\eta(y-c)] + \exp[-2\pi i\eta(y-c)]\}\\ &\quad\times \mathbf{a}[\exp(-2\pi i\eta y) + \exp(2\pi i\eta y)]\\ &= \mathbf{a}\{\exp(-2\pi i\eta c) + \exp[-2\pi i\eta(2y-c)]\\ &\qquad + \exp[2\pi i\eta(2y-c)] + \exp(2\pi i\eta c)\}\\ &= 2\mathbf{a}\{\cos(2\pi\eta c) + \cos[2\pi\eta(2y-c)]\}.\end{aligned}$$

The second cosine term represents subject information placed on a spatial frequency carrier. It is one component of a diffuse scattering of subject information caused by the multitude of spatial frequencies emitted by the actual diffuser. It contributes to a diffuse background "noise" in the image plane and may be neglected in this discussion. The first cosine term is a constant multiplier of the subject wavefront $\mathbf{a}$; it becomes zero when $c = c_0 = 1/4\eta = \lambda/4(\sin\theta)$, and in that case the subject wave is not reconstructed at all. Evidently the allowed displacement of the reference source from its original position is considerably less than c_0.

We can restate this conclusion in terms of the diffraction pattern of the reference source alone. At the hologram the reference source diffraction pattern intensity due to the two plane waves from P_1 and P_2 is a system of linear fringes whose periodic spacing $d = \lambda/2(\sin\theta)$ [Section 9.1, Eq. (9.6)]. Our calculation then tells us that the allowed displacement c must be considerably less than half the fringe spacing of the reference source diffraction pattern at the hologram. (Although our calculation, with a single pair of plane waves, implies periodic reconstruction of **a** with linear motion of the diffuser, this will not be true for the real diffuser and the variously oriented fringes it produces at the hologram.)

A more general reference wave is one whose amplitude and phase at the hologram are both spatially modulated. If we use the reference to illuminate the processed hologram, the "virtual image" reconstructed wave amplitude at the hologram is again **rr*a**, where **rr*** is now a function of x and y. We wish to examine the form of the wavefront **rr*a** as it arrives at the plane where the subject should be imaged. Its form determines the relation between the amplitude function of the reference source and the quality of the subject image. To obtain useful results which are easily interpreted, we must consider hologram formation in the Fourier transform plane of the subject and reference source rather than in the more general Fresnel diffraction region. However if one excludes results which derive from the special properties of the Fourier transform hologram, e.g., translation invariance, and if the reference light is well diffused, the analysis serves for Fresnel holograms as well.

14.2 The Fourier Transform Hologram with a Spatially Modulated Reference Wave

A hologram formed in the back focal plane of a lens as a consequence of the interference of light derived from two sources located in the front focal plane of the lens is a Fourier transform hologram. We shall analyze the formation of this type of hologram for the case where the complex amplitude at the hologram plane of each of the forming waves is permitted to be spatially modulated. Restricting the analysis to the Fourier transform hologram produces results which are directly applicable to problems in optical pattern recognition. The analysis is derived from that introduced by Vander Lugt *et al.* [*14.2*] in their work on optical spatial filtering.

Consider the double Fourier transformation or one-to-one imaging system of Fig. 14.3. A focal distance in front of lens L_1 in the input plane x_1y_1 are the input sources. These are two transparencies illuminated by a plane wave. Lens L_1 displays in its back focal plane $\xi\eta$ the Fourier transforms of their complex amplitude transmittances (see Section 6.3.3). (Note the inversion

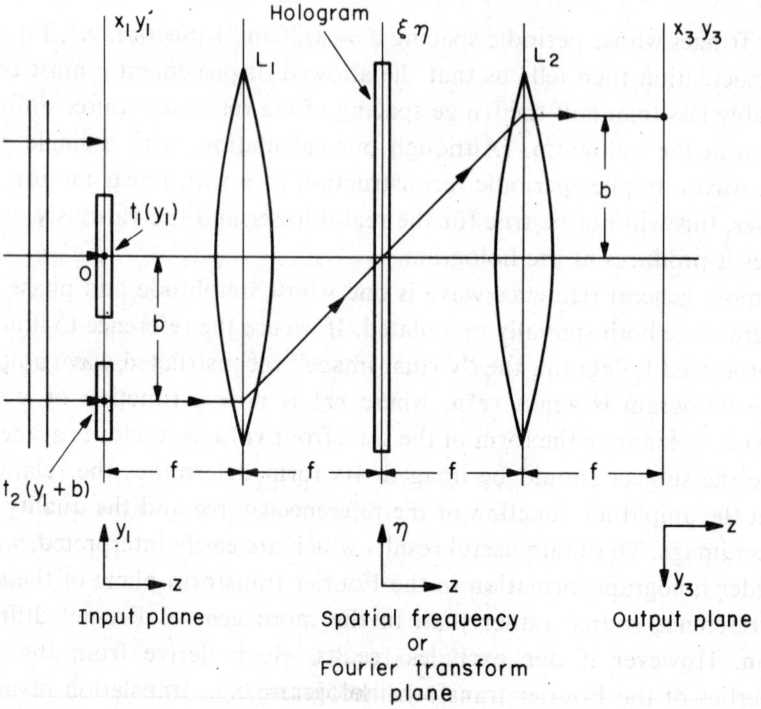

FIG. 14.3. A double Fourier-transformation, one-to-one imaging system.

of the x_3 and y_3 axes of the spatial plane relative to the ξ and η axes of the preceding spatial frequency plane. This is in accord with the discussion of Section 6.3.3.) Lens L_2 performs an inverse Fourier transformation on the complex amplitude distribution in the $\xi\eta$ plane; the resulting transform is displayed in plane x_3y_3 as the image of the input transparencies. We can simplify the notation by considering spatial variation in only the y direction. The complex transmittances of the two transparencies are $\mathbf{t}_1(y_1)$ and $\mathbf{t}_2(y_1 + b)$ where the latter is written as a function whose center has been displaced from an original position at the origin O to a point $y_1 = -b$. As shown in Fig. 14.3, the optical system images the center of $\mathbf{t}_2$ to a point b units above the z axis in the x_3y_3 plane.

Suppose a photographic plate is inserted into plane $\xi\eta$ so as to record the interference of the Fourier transforms of $\mathbf{t}_1$ and $\mathbf{t}_2$ formed by lens L_1. After proper exposure and development the resulting hologram is replaced in its original position in the $\xi\eta$ plane. Transparency $\mathbf{t}_2$ is now removed from the input plane leaving only $\mathbf{t}_1$ illuminated by a plane wave. The complex amplitude transmitted to the hologram is the Fourier transform of $\mathbf{t}_1$, and, if $\mathbf{t}_1$ is suitably chosen, the hologram diffracts a wave whose complex amplitude at the hologram plane is essentially the Fourier transform of $\mathbf{t}_2$. This, passing through the lens L_2 in the same fashion as did the light originating at $\mathbf{t}_2$, is focused to an image of $\mathbf{t}_2$ in the x_3y_3 plane. The image quality will be determined by the form of the transmittance $\mathbf{t}_1$. Again the center of the image of $\mathbf{t}_2$ lies b units above the z axis.

We now proceed to analyze the hologram formation and wavefront reconstruction processes, keeping the notation simple by confining the analysis to the yz or ηz planes. If we consider the plane wave illumination of the input plane to be of unit amplitude, then the transmitted light has a complex amplitude at that plane

$$\mathbf{t}_1(y_1) + \mathbf{t}_2(y_1 + b).$$

At the plane $\xi\eta$, the complex amplitude is

$$\mathbf{T}_1(\eta) + \mathbf{T}_2(\eta)\exp(-2\pi i\eta b)$$

where $\mathbf{T}_1(\eta)$ is the Fourier transform of $\mathbf{t}_1(y_1)$, $\mathbf{T}_2(\eta)$ is the Fourier transform of $\mathbf{t}_2(y_1)$, and where we have used correspondence (4.20). A photographic plate in the $\xi\eta$ plane properly exposed to this light and properly developed has a transmittance t_{H} proportional to the intensity of the light:

$$t_{\mathrm{H}} \propto \mathbf{T}_1\mathbf{T}_1^* + \mathbf{T}_2\mathbf{T}_2^* + \mathbf{T}_1\mathbf{T}_2^*\exp(+2\pi i\eta b) + \mathbf{T}_1^*\mathbf{T}_2\exp(-2\pi i\eta b). \quad (14.3)$$

Illumination of the hologram by $\mathbf{T}_1(\eta)$ produces the transmitted complex amplitude $\mathbf{T}_1 t_{\mathrm{H}}$. The last term in t_{H} is the one of interest since it yields the desired image of $\mathbf{t}_2(y_1 + b)$. From this term alone we obtain a complex amplitude in the plane $\xi\eta$ proportional to

$$\mathbf{T}_1\mathbf{T}_1^*\mathbf{T}_2\exp(-2\pi i\eta b).$$

The complex amplitude $\mathbf{a}(y_3)$ in the output plane due to this term is then proportional to its inverse Fourier transform

$$\mathbf{a}(y_3) \propto \mathscr{F}^{-1}[\mathbf{T}_1\mathbf{T}_1^*\mathbf{T}_2\exp(-2\pi i\eta b)]. \quad (14.4)$$

With the help of correspondences (4.18) and (4.19) we can write the product $\mathbf{T}_1\mathbf{T}_1^*$ as

$$\mathbf{T}_1\mathbf{T}_1^* = \mathscr{F}[\mathbf{t}_1^*(y_3) \star \mathbf{t}_1(y_3)]$$

so that $\mathbf{a}(y_3)$ becomes

$$\mathbf{a}(y_3) \propto \mathscr{F}^{-1}\{\mathscr{F}[\mathbf{t}_1^*(y_3) \star \mathbf{t}_1(y_3)]\mathbf{T}_2\exp(-2\pi i\eta b)\}. \quad (14.5)$$

Since $\mathbf{T}_2 \exp(-2\pi i\eta b)$ is the $\mathscr{F}[\mathbf{t}_2(y_3 + b)]$,

$$\mathbf{a}(y_3) \propto \mathscr{F}^{-1}\{\mathscr{F}[\mathbf{t}_1^*(y_3) \star \mathbf{t}_1(y_3)] \cdot \mathscr{F}[\mathbf{t}_2(y_3 + b)]\}$$
$$\propto [\mathbf{t}_1^*(y_3) \star \mathbf{t}_1(y_3)] * \mathbf{t}_2(y_3 + b), \qquad (14.6)$$

using correspondence (4.11). Equation (14.6) states that the *complex amplitude in the output plane is the convolution of* $\mathbf{t}_2(y_3 + b)$ *with the autocorrelation of* $\mathbf{t}_1(y_3)$. We know from Eq. (4.13e) that the convolution will yield $\mathbf{t}_2(y_3 + b)$, a perfect image of the input transmittance $\mathbf{t}_2(y_1 + b)$, if the autocorrelation function is a delta function. Hence for good imaging of the transmittance $\mathbf{t}_2$, we must seek a transmittance $\mathbf{t}_1$ whose autocorrelation is a peaked function approximating a delta function. If $\mathbf{t}_1$ corresponds to a point source in the input plane, the condition is obviously satisfied.

14.2.1 Autocorrelation of a Diffusing Screen

A ground-glass diffusing screen or scattering plate also has a transmittance which satisfies the criterion for good imaging. Light transmitted through the screen can be regarded as emitted by a collection of secondary point sources. Each source appears to emit a spherical wave of the same amplitude but random phase. Each may be represented by the product of a delta function and a phase factor: $\delta(y - y_m) \exp(+i\varphi_m)$. The Fourier transform of the collection of M such delta functions is, according to correspondence (4.29),

$$\mathbf{D}(\eta) = \sum_{m=1}^{M} \exp(2\pi i y_m \eta) \exp(i\varphi_m).$$

The autocorrelation of the collection of point sources is obtained by forming the product $\mathbf{D}(\eta)\mathbf{D}^*(\eta)$ and then taking the inverse Fourier transform of the product [see correspondence (4.18)]. To this end we form the conjugate

$$\mathbf{D}^*(\eta) = \sum_{n=1}^{M} \exp(-2\pi i y_n \eta) \exp(-i\varphi_n)$$

and the product

$$\mathbf{D}(\eta)\mathbf{D}^*(\eta) = \sum_{m=1}^{M} \sum_{n=1}^{M} \exp[2\pi i(y_m - y_n)\eta] \exp[i(\varphi_m - \varphi_n)]$$
$$= M + \sum_{\substack{m,n=1 \\ m \neq n}}^{M} \sum^{M} \exp[2\pi i(y_m - y_n)\eta] \exp[i(\varphi_m - \varphi_n)].$$

The autocorrelation C is then given by

$$C = \mathscr{F}^{-1}[\mathbf{DD}^*] = M\,\delta(y) + \sum_{\substack{m,n=1\\ m\neq n}}^{M}\sum^{M} \delta[y - (y_m - y_n)] \exp[i(\varphi_m - \varphi_n)]. \quad (14.7)$$

We see that the strength of the first term on the right, the single delta function at $y = 0$, increases linearly with the number of point sources M, while the strength of each delta function in the double summation is independent of M. Hence, for $M \gg 1$, the autocorrelation is peaked at $y = 0$. Assuming a constant spatial density of sources in the diffuser, the greater the extent of the diffuser, the larger M, and the more nearly its autocorrelation function approaches a delta function.

14.3 Some Experiments with Fourier Transform Holograms

Van Heerden [*14.3*] first suggested that a fragment of a subject scene for a Fourier transform hologram can be used to generate a "ghost image" of the remainder. He proposed that this principle be applied to locating information in a library whose entire store of information is recorded as a single Fourier transform hologram. At least in principle, each page of every library book is to be displayed simultaneously in an input plane as collective subject for the hologram. (In practice this would be a formidable microimaging task.) The position of each page in the input plane corresponds to its actual location in the library. Along with the pages of the library, a bright point source is included in the input plane. The hologram is then recorded in the optical system described in Section 14.2 and after development replaced in its original position.

Suppose that a fragment of the total library information, say one page of one book, is presented at the input plane in the same position, in the same format and with the same illumination as in forming the hologram. A ghost image of all of the remaining information should, in theory, be generated by the hologram. Assuming that illumination of the hologram is reduced by the ratio of the fragment size to the size of the entire library, the ghost image intensity will be weak. However, the ghost image of the very bright point source might well be clearly observed in a position in the output plane corresponding to that which it originally occupied in the input plane. If the fragment is presented in a position translated from that which it occupied during formation of that hologram, we shall find that the ghost image of the point source appears in the output plane correspondingly displaced from the position it occupied in the previous case. Thus appearance of the ghost image of the point source signifies that the fragment is in the library, and the vector displacement of the point source from its nominal position reveals the location of the fragment in the library.

That the ghost image is still observed when the fragment is translated from its original position in the input plane is a consequence of the properties of the Fourier transform. This will become evident as we now consider illumination of the hologram transmittance t_H [in Eq. (14.3)] by $\mathbf{T}_1(\eta)\exp(-2\pi i\eta c)$, the Fourier transform of $\mathbf{t}_1(y_1 + c)$. The transmittance $\mathbf{t}_1(y_1 + c)$ represents a translation of the original reference source from $y_1 = 0$ to $y_1 = -c$. At the hologram plane the reconstructed wavefront of interest has a complex amplitude proportional to the product $\mathbf{T}_1(\eta) \times \exp(-2\pi i\eta c)$ times the fourth term in t_H:

$$\mathbf{T}_1\mathbf{T}_1{}^*\mathbf{T}_2 \exp[-2\pi i\eta(c + b)].$$

To obtain the complex amplitude $\mathbf{a}(y_3)$ in the output plane, we perform an inverse Fourier transform operation on the above wavefront and find

$$\mathbf{a}(y_3) \propto [\mathbf{t}_1{}^*(y_3) \star \mathbf{t}_1(y_3)] * \mathbf{t}_2(y_3 + c + b).$$

We can assume for the purposes of this discussion that the autocorrelation function is a sharply peaked one so that $\mathbf{a}(y_3) \approx \mathbf{t}_2(y_3 + c + b)$. The ghost image of $\mathbf{t}_2$ is thus translated the same distance $-c$ in the y_3 direction of the x_3y_3 plane as that undergone by $\mathbf{t}_1(y_1)$ in the x_1y_1 plane.

A verification of the basic ghost-imaging principle of van Heerden was provided by Stroke *et al.* [*14.4*] using an arrangement similar to that of Fig. 14.3. A slide (or transparency) was the original input function, and a fragment of the slide served to illuminate the hologram and reconstruct an image of the remainder. They pointed out that a reference source with a wide range of spatial frequencies, i.e., with a peaked autocorrelation function would convolve with the "subject" input function to produce the best image.

Demonstration of the appearance of the ghost image despite translation of the fragment in the input plane is illustrated in Fig. 14.4 [*14.5*]. Here the subject scene (Fig. 14.4a) is a transparency bearing transparent letters FM and DEMODULATOR on an opaque background. FM may be chosen to correspond to $\mathbf{t}_2(y_1)$ (in Section 14.2) while DEMODULATOR is cast in the role of $\mathbf{t}_1(y_1)$. To form the hologram, an arrangement similar to that of Fig. 14.3 was used. Results of illuminating the hologram with the fragment DEMODULATOR are shown in a series of pictures in Fig. 14.4b taken by a camera looking through the hologram back toward the input plane. As DEMODULATOR was translated to the left in discrete stages, the ghost image

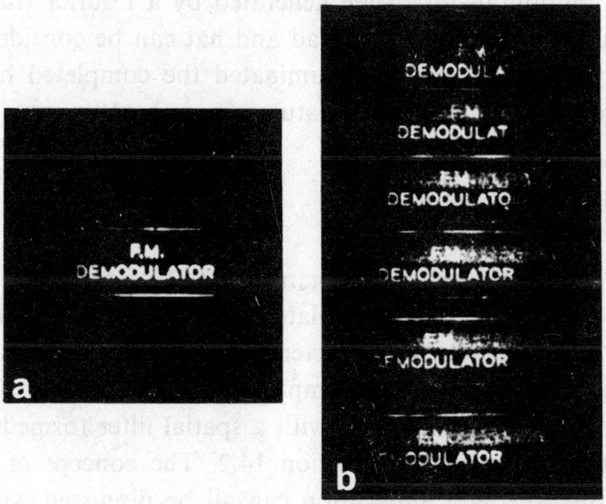

FIG. 14.4. Ghost imaging with the Fourier transform hologram. (a) Original subject scene. (b) Ghost images of FM translated in register with the reference source DEMODULATOR. (After Pennington and Collier [*14.5*].)

of FM was indeed reconstructed and translated in register with the direct image of DEMODULATOR. (The direct image was considerably brighter than the ghost image and had to be attenuated for the photograph. This may not be necessary for recording media which permit high diffraction efficiency.) The coarsely structured background "noise" underlying the image of FM is a consequence of the unsuitable nature of DEMODULATOR as a reference source. When both $\mathbf{t}_1$ and $\mathbf{t}_2$ are made diffusing transmitters, by placing a ground-glass plate adjacent to the transparency (on the source side), the image quality improves, as is evident in Fig. 14.5a. This is to be expected from the peaked autocorrelation function characteristic of a diffusing screen. Figure 14.5b is a photograph of the image of a diffusely

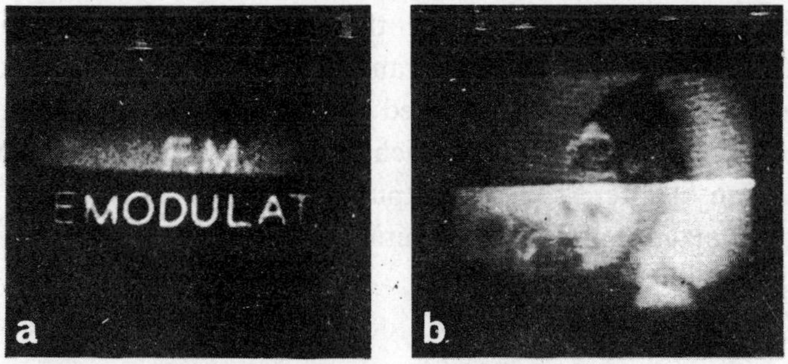

FIG. 14.5. Diffuse-light sources for Fourier transform holograms. (a) Illustration of the improvement in the ghost image of FM. (b) Ghost imaging with a continuous tone slide. (After Pennington and Collier [*14.5*].)

illuminated continuous-tone slide generated by a Fourier transform hologram. The upper portion of the head and hat can be considered to be $\mathbf{t}_1$. When this portion of the slide illuminated the completed hologram, the remainder of the slide, the facial features, was ghost imaged.

14.4 Character Recognition

Many of the properties which characterize the Fourier transform hologram formed with a spatially modulated reference wave are also important features of the approach to character recognition taken by Vander Lugt *et al.* [*14.2*]. They were able to implement the application of "matched filtering" to character recognition with a spatial filter formed by the holographic methods described in Section 14.2. The concept of the matched filter, its formation, and application can all be discussed with the aid of Fig. 14.6. Here the optical arrangement is basically the same double-Fourier transformation, one-to-one imaging system of Fig. 14.3. During formation of a spatial filter for character recognition, the portion of the input plane above the z axis is occupied by a transparency containing M transparent characters on an opaque background and illuminated with a plane wave. We represent the transmittance of this array of one-dimensional characters by

$$\mathbf{t}(y_1 - c_0) = \sum_{j=1}^{M} \mathbf{t}_j(y_1 - c_j) \tag{14.8}$$

where we indicate a symmetric distribution of characters about the point $y_1 = c_0$ and where $\mathbf{t}_j(y_1 - c_j)$ is a typical character in the array, centered at the point $y_1 = c_j$. Also in the input plane, but located at the point $y_1 = -b$, is a bright point source of light of strength $\delta(y_1 + b)$. A Fourier transform hologram is formed in the spatial frequency plane $\xi\eta$. The hologram may be considered to be the result of recording the interference of the Fourier transform of $\mathbf{t}(y_1 - c_0)$ with the plane wave arising from the delta function. When, however, the hologram performs the recognition operation, it is illuminated by light transmitted through only a fragment of $\mathbf{t}$, i.e., through one or more characters which may appear in the input plane. The reconstruction of interest in the output plane is, as we shall see, that of a bright point image signifying recognition.

14.4.1 Matched-Filter Concept

The optical system of Fig. 14.6 can perform a *spatial filtering* operation on unknown transmittances placed in the input plane, providing the complex

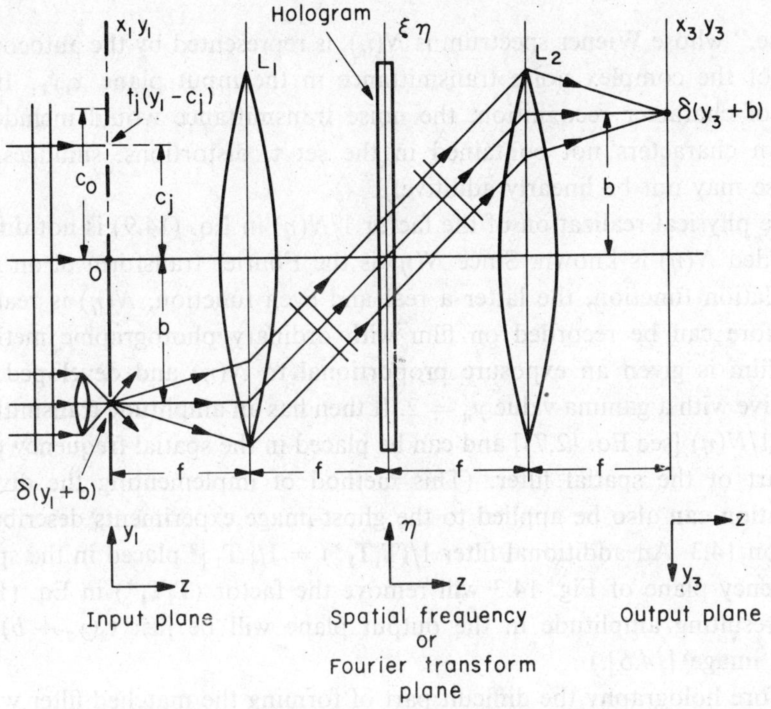

FIG. 14.6. Optical system for character recognition.

transmittance in the spatial frequency plane $\xi\eta$ has the form

$$\mathbf{H}(\eta) = \mathbf{S}^*(\eta)/N(\eta). \tag{14.9}$$

Filter $\mathbf{H}(\eta)$ is an optimum means for detecting a known signal immersed in linearly additive noise (noise not dependent on the form of the signal). Here $\mathbf{S}^*(\eta)$ is the conjugate of the Fourier transform of the signal, and $N(\eta)$ is the Wiener spectrum of the linearly additive noise found in the input to the filter [see Eq. (12.28)]. A filter represented by Eq. (14.9) produces the optimum ratio of peak recognition signal to rms noise [*14.2*]. When $N(\eta)$ is the same for all η (white noise), $\mathbf{H}(\eta)$ is called a *matched filter*.

For recognition of the characters in the array $\mathbf{t}(y_1 - c_0)$, we must set the numerator in Eq. (14.9) equal to the complex conjugate of the Fourier transform of $\mathbf{t}(y_1 - c_0)$, i.e., equal to

$$\mathbf{S}^*(\eta) = \mathbf{T}^*(\eta)\exp(-2\pi i\eta c_0) = \sum_{j=1}^{M} \mathbf{T}_j^* \exp(-2\pi i\eta c_j), \tag{14.10}$$

where $\mathbf{T}(\eta) \subset \mathbf{t}(y_1)$ and $\mathbf{T}_j(\eta) \subset \mathbf{t}_j(y_1)$. As shown in Section 12.3.1, the "noise," whose Wiener spectrum is $N(\eta)$, is represented by the autocorrelation of the complex noise transmittance in the input plane x_1y_1. In the case of character recognition, the noise transmittance would include unknown characters not contained in the set **t**, distortions, smudges, etc. (These may not be linearly additive.)

The physical realization of the factor $1/N(\eta)$ in Eq. (14.9) is not difficult provided $N(\eta)$ is known. Since $N(\eta)$ is the Fourier transform of an autocorrelation function, the latter a real and even function, $N(\eta)$ is real and therefore can be recorded on film with ordinary photographic methods. The film is given an exposure proportional to $N(\eta)$ and developed as a negative with a gamma value $\gamma_n = 2$. It then has an amplitude transmittance $t_n = 1/N(\eta)$ [see Eq. (2.7)] and can be placed in the spatial frequency plane as part of the spatial filter. (This method of implementing the division operation can also be applied to the ghost-image experiments described in Section 14.3. An additional filter $1/(\mathbf{T}_1\mathbf{T}_1{}^*) = 1/|\,\mathbf{T}_1\,|^2$ placed in the spatial frequency plane of Fig. 14.3 will remove the factor $(\mathbf{T}_1\mathbf{T}_1{}^*)$ in Eq. (14.4). The resulting amplitude in the output plane will be just $\mathbf{t}_2(y_3 + b)$, the ghost image [*14.6*].)

Before holography the difficult part of forming the matched filter was in obtaining $\mathbf{S}^*(\eta)$. It is a complex quantity which requires the recording of both phase and amplitude information; van Heerden as well as Vander Lugt realized that holography is perfectly suited for this task. The properly formed Fourier transform hologram recording of $\mathbf{S}^*(\eta)$ and the photographic recording of $1/N(\eta)$ are placed adjacent to each other in the spatial frequency plane. Their combined transmittance $\mathbf{S}^*(\eta)/N(\eta) = \mathbf{H}(\eta)$ is the realization of the filter. In the case of character recognition, the noise spectrum $N(\eta)$ is generally not linearly additive nor is it uniform. Yet optical character recognition experiments seem to suffer little from the practice of merely considering $1/N(\eta)$ a proportionality constant which need not actually be implemented.

14.4.2 The Response of the Filter

Ignoring $N(\eta)$, let us examine in some detail how the filter is formed and used. Depending on the way the filter is illuminated, one can obtain either its impulse response or a recognition signal. The complete set of characters we wish to be able to recognize, $\mathbf{t}(y_1 - c_0)$, is placed in the input plane and recorded holographically in association with the bright point source $\delta(y_1 + b)$. We shall see that the hologram formed in the spatial frequency plane of Fig. 14.6 is the desired filter. It has a transmittance with the general form of Eq. (14.3). This becomes

$$\begin{aligned} t_{\mathrm{H}} &\propto \mathbf{S}(\eta)\mathbf{S}^*(\eta) + 1 + \mathbf{S}(\eta)\exp(+2\pi i\eta b) + \mathbf{S}^*(\eta)\exp(-2\pi i\eta b) \\ &\propto \mathbf{TT}^* + 1 + \mathbf{T}\exp[+2\pi i\eta(c_0 + b)] + \mathbf{T}^*\exp[-2\pi i\eta(c_0 + b)] \end{aligned} \tag{14.11}$$

when we use Eq. (14.10) and the fact that $\delta(y_1 + b) \supset \exp(-2\pi i\eta b)$. Suppose the filter is replaced in the spatial frequency plane and illuminated only with light from the point source (the spatial analog of an impulse) in the input plane. We shall find the impulse response of the filter in the output plane. The complex amplitude at the back side of the hologram in this case is

$$\begin{aligned} [\exp(-2\pi i\eta b)]\cdot t_{\mathrm{H}} \propto{} & \mathbf{TT}^*\exp(-2\pi i\eta b) + \exp(-2\pi i\eta b) \\ & + \mathbf{T}\exp(+2\pi i\eta c_0) + \mathbf{T}^*\exp[-2\pi i\eta(c_0 + 2b)]. \end{aligned} \tag{14.12}$$

Lens L_2 performs an inverse Fourier transformation on the complex amplitude in Eq. (14.12) resulting in a complex amplitude distribution at the output plane

$$\begin{aligned} \mathbf{a}(y_3) \propto{} & [\mathbf{t}^*(y_3) \star \mathbf{t}(y_3)] * \delta(y_3 + b) + \delta(y_3 + b) \\ & + \mathbf{t}(y_3 - c_0) + \mathbf{t}^*(-y_3 + c_0 + 2b). \end{aligned} \tag{14.13}$$

Using Eq. (4.12), we find the first term in $\mathbf{a}(y_3)$ to be the autocorrelation of the collection of characters centered at $y_3 = -b$; the second produces the image of the point source; the third term is the image of the set of characters and is the *impulse response* while the fourth term is the displaced, *conjugate* image of the character set, inverted relative to the normal lens inversion.

Now suppose the point source is removed and the only source of illumination for the hologram is the transmission through a single character of the set $\mathbf{t}$ whose transmittance in the input plane is $\mathbf{t}_l(y_1 - c_l)$. The form of its transmittance indicates that at this juncture the character is located in the original position it occupied during formation of the hologram. If we regard this character as a fragment of the original input information, i.e., a fragment of the character array *plus* the point source, we can expect the hologram to generate ghost images of the remaining characters as well as one of the point source. The images should appear in the output plane in positions which are the normal image locations for the optical system of Fig. 14.6.

The hologram whose transmittance t_{H} is given by Eq. (14.11) is illuminated by the complex amplitude $\mathscr{F}[\mathbf{t}_l(y_1 - c_l)] = \mathbf{T}_l\exp(+2\pi i\eta c_l)$. It diffracts a complex amplitude in the $\xi\eta$ plane,

$$[\mathbf{T}_l \exp(+2\pi i\eta c_l)]\cdot t_{\mathrm{H}} \propto \mathbf{TT}^*\mathbf{T}_l \exp(+2\pi i\eta c_l) + \mathbf{T}_l \exp(+2\pi i\eta c_l) + \mathbf{TT}_l \exp[+2\pi i\eta(c_0 + c_l + b)] + \mathbf{T}^*\mathbf{T}_l \exp[-2\pi i\eta(c_0 - c_l + b)]. \tag{14.14}$$

Inverse Fourier transformation to the output plane now produces the complex amplitude distribution

$$\mathbf{a}(y_3) \propto [\mathbf{t}^*(y_3) \star \mathbf{t}(y_3)] * \mathbf{t}_l(y_3 - c_l) + \mathbf{t}_l(y_3 - c_l) + \mathbf{t}(y_3) * \mathbf{t}_l(y_3 - c_0 - c_l - b) + \mathbf{t}^*(y_3) \star \mathbf{t}_l(y_3 + c_0 - c_l + b). \tag{14.15}$$

The second term on the right is an undistorted image of the input character $\mathbf{t}_l$ centered at $y_3 = c_l$. We may apply theorem (4.12) to the remaining terms. We see that the first term is a distribution also centered at $y_3 = c_l$. Its maximum spatial extent is twice that of the entire array of characters plus the extent of the single character $\mathbf{t}_l$. If, however, the autocorrelation of the array is fairly peaked, we can regard the first term as another ghost image of $\mathbf{t}_l$ coincident with the undiffracted light. The third term is the convolution of the character $\mathbf{t}_l$, with all the members of the array, a distribution centered at $y_3 = c_l + c_0 + b$. Its spatial extent is the sum of the extent of the character array plus the extent of the character $\mathbf{t}_l$. A distribution having the same spatial extent but centered at $y_3 = c_l - c_0 - b$ is represented by the final term in Eq. (14.15). It is the correlation of $\mathbf{t}_l$ with all the characters of the set. To obtain a simple illustration of the spatial relations between the complex amplitude distributions making up $\mathbf{a}(y_3)$, let us arbitrarily choose $\mathbf{t}_l$ to be the center character of the array so that $c_l = c_0 = W/2$, where W is the extent of the array. Let w represent the extent of the character $\mathbf{t}_l$. By plotting the components of $\mathbf{a}(y_3)$, as in Fig. 14.7, we can obtain the minimum separation $d = b + c_0$ between point source and array center in the input plane which will avoid overlap of the output plane distributions. We begin by plotting the first term of Eq. (14.15) centered at $y_3 = c_l = W/2$ and of width $2W + w$. The correlation term [the final term in Eq. (14.15)] is centered at the point $y_3 = -b$ and has a width $W + w$. When the correlation term abuts the distribution centered at $y_3 = c_l$, b has a minimum value given by $b = (2W + w)/2 - c_0 + (W + w)/2$. Thus $d = b + c_0 = (3W/2) + w$. The convolution distribution, the third term, is symmetrically located relative to the center of the first term ($y_3 = c_l$).

A ghost image of the character array is contained in the first term on the right of Eq. (14.15) while a ghost image of the point source is contained

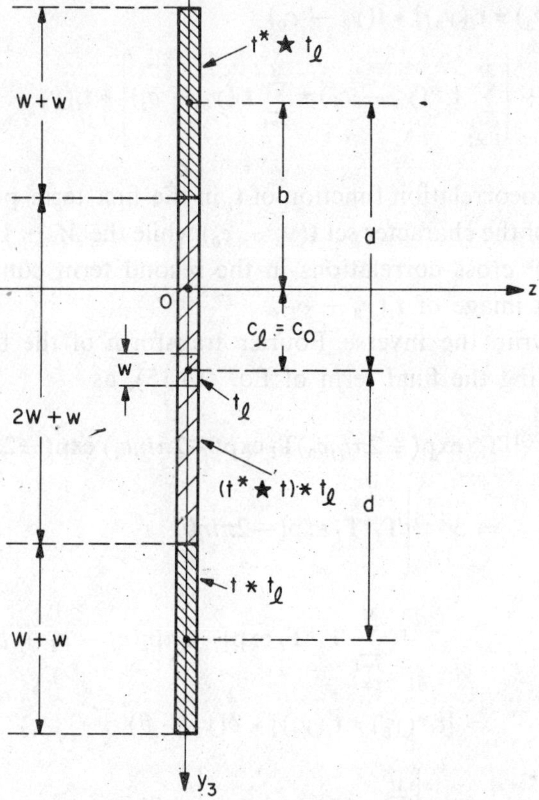

FIG. 14.7. Plot of the complex amplitude distribution in Eq. (14.15.) for the case where $d = 3W/2 + w$. (Positive direction of the y_3 axis is downward.)

in the fourth term. To demonstrate the first of these contentions, let us rewrite the first term on the right of Eq. (14.14) using Eq. (14.10) as

$$\sum_{j=1}^{M} \mathbf{T}_j \exp(+2\pi i\eta c_j) \sum_{k=1}^{M} \mathbf{T}_k{}^* \exp(-2\pi i\eta c_k)\mathbf{T}_l \exp(+2\pi i\eta c_l)$$

$$= \left[\sum_{j=1}^{M} \mathbf{T}_j \exp(+2\pi i\eta c_j)\right] \mathbf{T}_l{}^*\mathbf{T}_l$$

$$+ \sum_{j=1}^{M} \mathbf{T}_j \exp(+2\pi i\eta c_j) \sum_{\substack{k=1\\k\neq l}}^{M} \mathbf{T}_k{}^* \exp(-2\pi i\eta c_k)\mathbf{T}_l \exp(+2\pi i\eta c_l)$$

$$= \mathbf{T} \exp(+2\pi i\eta c_0)\mathbf{T}_l{}^*\mathbf{T}_l$$

$$+ \sum_{j=1}^{M} \mathbf{T}_j \exp(+2\pi i\eta c_j) \sum_{\substack{k=1\\k\neq l}}^{M} \mathbf{T}_k{}^* \exp(-2\pi i\eta c_k)\mathbf{T}_l \exp(+2\pi i\eta c_l).$$

The output plane displays the inverse Fourier transform of the above:

$$[\mathbf{t}_l^*(y_3) \star \mathbf{t}_l(y_3)] * \mathbf{t}(y_3 - c_0)$$

$$+ \left[\sum_{\substack{k=1\\k\neq l}}^{M} \mathbf{t}_k^*(y_3 - c_k) \star \sum_{j=1}^{M} \mathbf{t}_j(y_3 - c_j)\right] * \mathbf{t}_l(y_3 - c_l).$$

A peaked autocorrelation function of $\mathbf{t}_l$ in the first term provides a (weak) ghost image of the character set $\mathbf{t}(y_3 - c_0)$ while the $M - 1$ autocorrelations and $(M - 1)^2$ cross correlations in the second term contribute to an in-register ghost image of $\mathbf{t}_l(y_3 - c_l)$.

We now write the inverse Fourier transform of the final term in Eq. (14.14), yielding the final term of Eq. (14.15), as

$$\mathscr{F}^{-1}[\mathbf{T}^* \exp(-2\pi i\eta c_0)\mathbf{T}_l \exp(+2\pi i\eta c_l) \exp(-2\pi i\eta b)]$$

$$= \mathscr{F}^{-1}\left[\mathbf{T}_l^*\mathbf{T}_l \exp(-2\pi i\eta b)\right.$$

$$\left. + \sum_{\substack{j=1\\j\neq l}}^{M} \mathbf{T}_j^*\mathbf{T}_l \exp[-2\pi i\eta(c_j - c_l + b)]\right]$$

$$= [\mathbf{t}_l^*(y_3) \star \mathbf{t}_l(y_3)] * \delta(y_3 + b)$$

$$+ \left[\sum_{\substack{j=1\\j\neq l}}^{M} \mathbf{t}_j^*(y_3) \star \mathbf{t}_l(y_3)\right] * \delta(y_3 + c_j - c_l + b). \qquad (14.16)$$

We can regard the first term in the result as the ghost image of the point source resulting from the convolution of the delta function with the peaked autocorrelation of $\mathbf{t}_l$. The image $\delta(y_3 + b)$ is located at $y_3 = -b$, its proper position for the imaging system of Fig. 14.6. Perhaps, more naturally, this first term is also the autocorrelation function of $\mathbf{t}_l$ evaluated at $y_3 = -b$, as a result of the sifting property of the delta function. In either case the first term represents a bright spot of light which signifies recognition that $\mathbf{t}_l$ is one of the characters in the hologram filter. For the recognition scheme to work best, the autocorrelation function should be sharply peaked while the cross correlations of the input character with the remaining characters in the array [the second term of Eq. (14.16)] should be broad functions. The latter are distributed in the vicinity of the autocorrelation peak with centers displaced from $y_3 = -b$ by the relative spacing of the input character from the other characters in the original array.

14.4.3 Character Identification

Thus far we have demonstrated only that when a character of the set stored in the hologram is placed in its original position in the input plane, the hologram filter will recognize it as one of the set. It does not identify which character of the set is in the input because the recognition signal, a bright spot at $y_3 = -b$, is the same for any and all characters occupying their original input plane positions. Suppose, however, that the input plane is masked so that there remains only a small window centered at the origin and of spatial extent equal to that of a single character. Unknown characters are to be presented for recognition one at a time at this window. The character $\mathbf{t}_l(y_1 - c_l)$, so placed, is thus displaced from its original position by c_l units in the $-y_1$ direction; its transmittance can be expressed as $\mathbf{t}_l(y_1)$. If in the first line of Eq. (14.16) we substitute $\mathscr{F}[\mathbf{t}_l(y_1)] = \mathbf{T}_l$ for the illuminating amplitude $\mathbf{T}_l \exp(+2\pi i\eta c_l)$, we find that the recognition term in the output [the first term in the last line of Eq. (14.16)] becomes

$$C = [\mathbf{t}_l^*(y_3) \star \mathbf{t}_l(y_3)] * \delta(y_3 + c_l + b). \tag{14.17}$$

The consequence is that the bright recognition spot is also displaced from $y_3 = -b$ by c_l units in the $-y_3$ direction. Since each character in the original array is separated from the origin by a unique distance c_j, the location of the bright-spot recognition signal relative to $y_3 = -b$ identifies which character is in the input.

Before proceeding to a recognition scheme which makes better use of the parallel-processing properties of optical systems, let us summarize the basic recognition technique. We have seen that a matched filter, formed in the spatial frequency plane of Fig. 14.6, displays in an output plane the correlation of an unknown character with those stored in the filter. When the unknown is one of the stored set, recognition is signified by the display of an intense spot of light. This is a consequence of an optical Fourier transformation performed by lens L_2 on the product of the Fourier transform of the input character times the conjugate Fourier transforms of the stored characters. A displacement of the input character from the position it occupied during formation of the filter results in multiplication of that product by a phase factor whose phase is linear in the coordinates of the frequency plane. The inverse Fourier transformation by lens L_2 once again produces the autocorrelation display in the output plane but now shifted from the original position by just the displacement of the input character. Filters formed out of the Fourier transform plane will not recognize a character translated from its original position. On the other hand the tolerance of the

Fourier transform hologram to displacement in the input, stemming from the shift properties of the Fourier transform, is employed to identify the character.

Letter-by-letter transport of a printed page past an input window makes only partial use of the potential for parallel processing of the input information. While the features of a single character are processed simultaneously by the hologram filter and optical system of Fig. 14.6, the N characters on the page are processed in sequence. Vander Lugt *et al.* [*14.2*] have suggested a clever extension of the basic recognition technique which permits simultaneous identification of all the characters on a page.

Suppose that the input plane transmittance, before formation of the hologram filter, is as indicated in Fig. 14.8 [where the black letters and the black

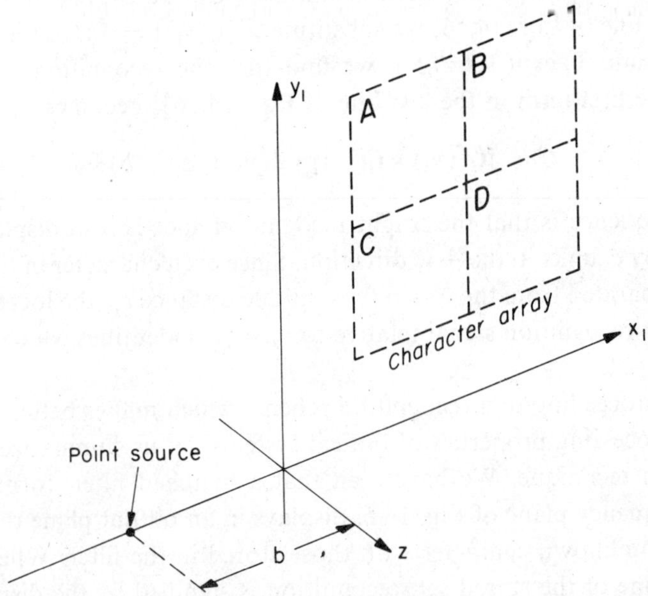

FIG. 14.8. An input plane for forming a filter for character recognition.

dot at $(-b, 0)$ represent the transparent characters and the bright point source, respectively]. The characters in the input are allotted areas on the input plane large enough to contain a projected page of unknown characters to be identified. Figure 14.9a indicates the input plane again, this time opaque except for the input window at the upper left of the first quadrant. A *page* of characters to be identified is in position in the window. We assume that the filter has been formed with the plane-wave-illuminated transmittances of Fig. 14.8, the point source, and the optical system of

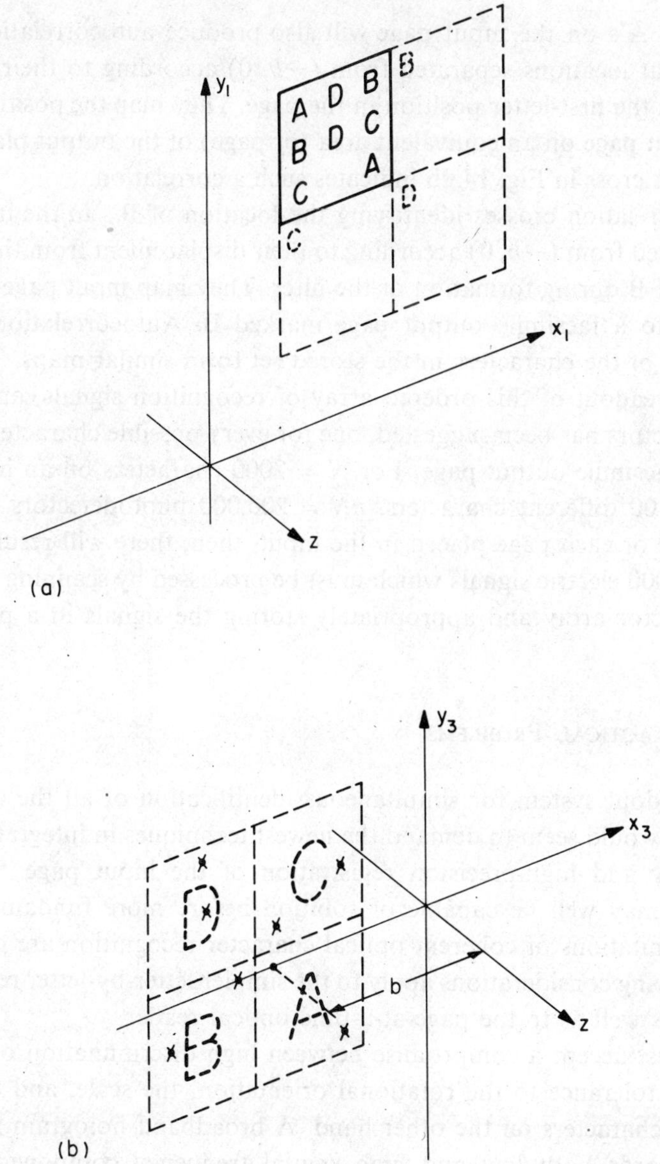

FIG. 14.9. Page-at-a-time character recognition. (a) A page in the input window. (b) The output display.

Fig. 14.6. Figure 14.9b is the output plane. The heavy dot at $(-b, 0)$ is the image of the point source or the autocorrelation spot for a character in the input plane occupying the identical position it had when stored in the filter. With the arrangement of the input plane indicated in Fig. 14.9a, a bright output spot will occur at $(-b, 0)$ only when the first letter of the page is A.

Any other A's on the input page will also produce autocorrelation bright spots but at locations separated from $(-b, 0)$ according to their displacement from the first-letter position in the page. They map the position of A's in the input page on an equivalent area (or page) of the output plane designated A. A cross in Fig. 14.9b indicates such a correlation.

Autocorrelation crosses identifying the location of B's in the input page are displaced from $(-b, 0)$ according to their displacement from the original position of B during formation of the filter. They map input page positions of B's onto a facsimile output page marked B. Autocorrelations for the remainder of the characters in the stored set form similar maps. To implement the readout of this ordered array of recognition signals, an array of photodetectors has been suggested, one for every possible character location in every facsimile output page. For $N = 2000$ characters on an input page and $n = 100$ different characters, $nN = 200{,}000$ photodetectors would be required. For each page placed in the input, then, there will result a maximum of 2000 electric signals which must be processed by scanning the entire photodetector array and appropriately storing the signals in a permanent memory.

14.4.4 Practical Problems

The readout system for simultaneous identification of all the characters on a page would seem to demand the newest techniques in integrated circuit technology and high-precision registration of the input page. Yet these problems may well be capable of solution before more fundamental and general limitations of coherent optical character recognition are overcome. The following considerations apply to the simpler letter-by-letter recognition schemes as well as to the page-at-a-time optical reader.

One must accept a compromise between high discrimination on the one hand and tolerance to the rotational orientation, the scale, and quality of the input characters on the other hand. A broadband hologram filter (one which records both low and high spatial frequency components of the character array Fourier transform) can tolerate typically a $\pm 15\%$ change in the scale of a character and a $\pm 16°$ change in rotational orientation [*14.2*]. The recognition signals remain reasonably bright under such deviations. However, with the broadband filter, cross correlations can have peak intensities which, in certain cases, are the equal of autocorrelation intensities, thereby signaling false recognition. This is certainly true for alphabetic characters whose forms are contained in other members of the set; e.g., O is contained in Q. Attempts to form a broadband filter may often result in overemphasis of the very low spatial frequency components. The latter represent light passing nearly undiffracted through the large open areas of the characters, making up the greatest portion of the transmission. For characters with similar total open area, these strong low-frequency components provide no aid in discrimination; in practice, their recording can become an obstacle to identification. When a reference beam amplitude is chosen large enough to linearly record the low spatial frequency components, the weak high-frequency components are recorded with a large beam ratio R. From Eq. (7.31) we know that when R is large, the visibility of the recorded fringes is proportional to $R^{-1/2}$. Hence the amplitude of light diffracted by these fringes may be so weak as to be indistinguishable from the noise. If, on the other hand, the reference beam intensity is reduced, some of the more important low frequencies will be recorded nonlinearly, introducing additional noise.

Overexposure of the central area of the hologram plate (the area where low spatial frequency components are recorded) can give rise to high-pass spatial filter behavior. The central area becomes highly absorbing and diffracts very little light. Outer areas, which record the higher spatial frequency components, are now better exposed and diffract with greater efficiency. This process emphasizes the importance of the edges and corners of the input characters and thus their differences. At the same time that recognition selectivity increases, however, the tolerance to misorientations, size change, or poor quality decreases.

The high-pass filter concept can be extended to the point where one *records* and *recognizes* the *derivative* of the characters [*14.7*]. If we express a transmittance $\mathbf{t}(y)$ in the input plane in terms of its Fourier transform $\mathbf{T}(\eta)$, we have

$$\mathbf{t}(y) = \int_{-\infty}^{\infty} \mathbf{T}(\eta) \exp(-2\pi i\eta y)\, d\eta.$$

The derivative of the input transmittance is then given by

$$\frac{dt}{dy} = \int_{-\infty}^{\infty} -2\pi i\eta \mathbf{T}(\eta) \exp(-2\pi i\eta y)\, d\eta$$

or

$$\frac{dt}{dy} \supset -2\pi i\eta \mathbf{T}(\eta).$$

To employ the method, the product $\mathbf{T}^*\mathbf{T}_l$ in the recognition term (the last term) of Eq. (14.14) must be replaced by the product

$$[+2\pi i\eta \mathbf{T}^*][-2\pi i\eta \mathbf{T}_l] = 4\pi^2\eta^2\mathbf{T}^*\mathbf{T}_l.$$

This is accomplished by placing another filter adjacent to the standard Fourier transform hologram filter in the spatial frequency plane. The new filter must have a transmittance $4\pi^2\eta^2$ that increases parabolically with distance from the center of the plane and has a null at the center. As one might expect, the technique provides better discrimination between the characters in an array but at the same time impedes recognition of variants of a given character. Figure 14.10 illustrates the signal-to-noise ratio obtainable with the derivative filter and a particular input to the system.

It would seem from our discussion of high-pass filtering that pattern recognition using the hologram filter would be more successful when the input functions are complicated spatial patterns with little completely unmodulated transparent area. Consequently one might expect coherent optical pattern recognition to be well-suited, e.g., to the identification of fingerprints [*14.8*]. The method does tolerate misorientations of $\pm 3^\circ$ and has the ability to identify even a fragment of a print. A problem of some magnitude, however, is caused by superimposed noise (dirt or smear) which obscures the signal. Apparently a human identification expert can tolerate far more of this type of noise than can the optical pattern recognition machine.

A problem which exists for all applications of coherent optical pattern recognition is the need to convert raw input, such as the pages of a book or a fingerprint form, into transparencies. The latter can then be placed in an input window, coherently illuminated, and identified. If this conversion is not made, the diffuse light illuminating the hologram filter will be largely characteristic of the roughness of the input page. In certain applications relief images of input pages can be formed in thermoplastic, the process taking only a few seconds. Where rapid or real-time processing is required, the few seconds may not be brief enough.

14.4.5 Another Method

If one is willing to form a new holographic record each time an input pattern is to be correlated with known patterns, a rather straightforward pattern recognition method can be realized [*14.9*, *14.10*]. The basic idea can be derived from our discussion of Section 14.2 where two transparencies $\mathbf{t}_1(y_1)$ and $\mathbf{t}_2(y_1 + b)$ are the input sources used to form a Fourier transform hologram in the optical system of Fig. 14.3. Suppose $\mathbf{t}_1(y_1)$ is a transparency of the pattern we wish to find in a page of unknowns and $\mathbf{t}_2(y_1 + b)$ is the page transparency which may contain one or more patterns $\mathbf{t}_1(y_1)$. The transmittance of the Fourier transform hologram formed with these sources

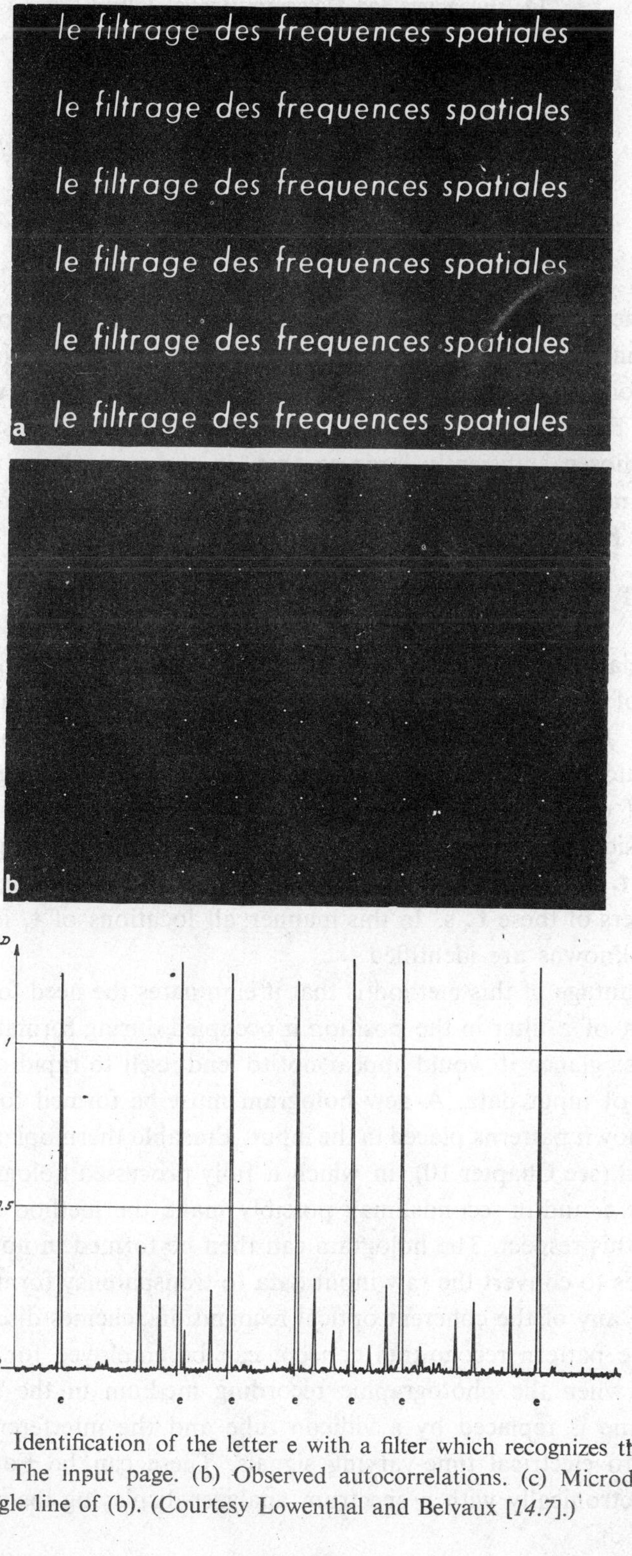

FIG. 14.10. Identification of the letter e with a filter which recognizes the character derivative. (a) The input page. (b) Observed autocorrelations. (c) Microdensitometer traces of a single line of (b). (Courtesy Lowenthal and Belvaux [*14.7*].)

is given by Eq. (14.3) as

$$t_H \propto T_1T_1^* + T_2T_2^* + T_1T_2^* \exp(+2\pi i\eta b) + T_1^*T_2 \exp(-2\pi i\eta b) \quad (14.18)$$

where

$$t_1(y_1) \supset T_1(\eta) \qquad \text{and} \qquad t_2(y_1) \supset T_2(\eta).$$

Consider the hologram to be returned to the frequency plane of Fig. 14.3 and illuminated with an axial plane wave of unit amplitude (the result of placing a point source at the origin of the input plane). Lens L_2 will present the inverse Fourier transform of t_H at the output plane. We assume that b has been chosen sufficiently large so that, in the x_3y_3 plane, the inverse Fourier transforms of the first three terms in Eq. (14.18) do not overlap the inverse Fourier transform of the fourth. The latter

$$\mathscr{F}^{-1}[T_1^*T_2 \exp(-2\pi i\eta b)] = [t_1^*(y_3) \star t_2(y_3)] * \delta(y_3 + b) \qquad (14.19)$$

is the correlation of t_2 with t_1 centered at $y_3 = -b$. The point $y_3 = -b$ is the center of the image of $t_2(y_1 + b)$ as it would normally form in the double lens system. If t_2 happened to be identical to t_1, the peaked autocorrelation $t_1^* \star t_1$ would give rise to a bright spot of light at $y_3 = -b$ signaling recognition. If one or more t_1's are contained in t_2, each of these patterns t_{1i} can be considered to form independent but superimposed holograms with the known t_1; autocorrelation peaks will occur at $y_3 = -b_i$ corresponding to the centers of these t_{1i}'s. In this manner all locations of t_1 in the input page of unknowns are identified.

One advantage of this method is that it eliminates the need for precision replacement of a filter in the position it occupied during formation. However, at first glance it would appear not to lend itself to rapid or real-time processing of input data. A new hologram must be formed for each new set of unknown patterns placed in the input. Erasable thermoplastic recording material (see Chapter 10), in which a fully processed hologram can be formed *in situ* and in seconds, may possibly make the method more competitive in this respect. The hologram can then be formed in no more time than it takes to convert the raw input data to transparency form, a process required of any of the coherent optical recognition schemes discussed.

The same pattern recognition concept can be employed for very rapid processing when the photographic recording medium in the spatial frequency plane is replaced by a vidicon tube and the interference pattern converted to electrical time-varying signals. These can be Fourier-transformed electronically with a spectrum analyzer displaying the correlations

[*14.11*]. There will be some loss in recognition ability due to the one-dimensional nature of the correlation. Of course the vidicon must be able to resolve the interference pattern, but the spatial frequencies of the latter need not be high.

14.5 Multiplexing and Coding

The use of a single Fourier transform hologram filter to identify several different characters in an input page is an example of *spatial frequency multiplexing*. If we regard the hologram in the spatial frequency plane as occupying a band of spatial frequencies, then each character presented to the input window for recognition employs the same band of frequencies in illuminating that hologram. In this sense the hologram spatial frequency channel is being multiplexed. We may consider illumination by a wave complex amplitude proportional to the Fourier transform of any one character's transmittance to represent a *unique coding* of spatial frequencies within the band. Suppose we think of the Vander Lugt hologram filter as a superposition of simultaneously formed holograms where, for each hologram, a particular character serves as the source of a coded reference wave and the point source serves as subject. When a particular character to be recognized is placed in the input and illuminates the hologram filter, only one hologram of the set responds to the unique coding of the illuminating wave. The response is to reconstruct a wave which images the point source.

There is no need to restrict the practice of frequency multiplexing to the case of a single exposure and a single subject. Multiple exposure of the same photographic plate, where both subject and reference code are changed between exposures [*14.12*], has useful application to multicolor holographic imaging [*14.13*] and to digital information storage [*14.14*]. Discussion of the multicolor application will be taken up in Chapter 17. In this section we consider a method of recording point sources of light which can serve as bits of information in a hologram memory.

14.5.1 Digital Storage with Coded Multiple Exposure Holograms

A page of binary information can be represented by an array of bright points of light occupying sites on a regularly spaced two-dimensional lattice. Presence of a bright spot on a given site indicates a binary one while absence of a spot indicates a zero. We wish to record holographically the transmission from such an array in association with a coded reference wave and then to discover how many holograms of similar digital pages can be superimposed by multiple exposure of the same photographic plate. For each exposure both the digital page and the coded reference wave are changed. The latter may be accomplished by using a diffuser or ground-glass screen as the reference source and moving the screen between exposures. Only a small displacement of the screen, perhaps 15 μm, is sufficient to make its transmittance essentially uncorrelated with that for an original position (see Section 14.1.2). An image of any of the digital pages stored in this hologram memory can be produced by illuminating the hologram with the appropriately coded reference wave. Only one of the superimposed holograms will respond to the code and produce the ghost image. The remainder contribute a background of undesirable diffuse light or noise. To determine how many digital pages can usefully be stored, it is necessary to identify the noise sources and calculate the ratio of the signal intensity to the noise intensity at the image plane.

Figure 14.11 is a photograph of the image of a single small source and the surrounding noise produced when a set of 10 superimposed holograms is

FIG. 14.11. Photograph of a 25-μm-diameter spot imaged by one out of a set of 10 superimposed coded holograms. (Courtesy La-Macchia and White [*14.14*].)

illuminated with one of the coded reference waves used to form the set. The same small source was the subject of each hologram. Note that the noise is localized about the bright spot image. Suppose the lattice sites of an input page are spaced sufficiently far apart that the noise surrounding each imaged spot does not overlap that surrounding a neighboring spot.

We can then meaningfully confine our attention to the relation between the number of superimposed holograms of a single point source, the signal intensity, and the noise. The analysis to follow is essentially that given by La Macchia and White [*14:14*]. The superscripts index exposure sequence.

If $\mathbf{a}^{(i)}(y)$ is the complex amplitude of the subject light at the hologram due to the ith point source and $\mathbf{r}^{(i)}(y)$ is that of the associated coded reference wave, then the virtual image term of the hologram amplitude transmittance is proportional to $\mathbf{a}^{(i)}\mathbf{r}^{*(i)}$. For N exposures the total virtual image term is

$$\mathbf{t}_{\mathrm{v}} \propto \sum_{i=1}^{N} \mathbf{a}^{(i)}\mathbf{r}^{*(i)}. \tag{14.20}$$

Suppose the N superimposed holograms are illuminated by the light from the jth reference source so that the diffracted virtual image wave just behind the hologram has the complex amplitude

$$\begin{aligned} \mathbf{w} = \mathbf{r}^{(j)} t_{\mathrm{v}} &\propto \mathbf{r}^{(j)} \sum_{i=1}^{N} \mathbf{a}^{(i)}\mathbf{r}^{*(i)} \\ &\propto \mathbf{r}^{(j)}\mathbf{r}^{*(j)}\mathbf{a}^{(j)} + \mathbf{r}^{(j)} \sum_{\substack{i=1 \\ i\neq j}}^{N} \mathbf{a}^{(i)}\mathbf{r}^{*(i)}. \end{aligned} \tag{14.21}$$

Since all the point source subjects are identical, $\mathbf{a}^{(j)} = \mathbf{a}^{(i)} = \mathbf{a}$. In the practical case the hologram area is small relative to the separation between it and the diffusing reference source. Then, as in Fig. 14.2, we regard the diffuser as a set of M point sources sending M plane waves of unit amplitude, spatial frequency $\eta^{(j)}$ and random initial phase $\varphi^{(j)}$ to the hologram. The reference wave amplitude $\mathbf{r}^{(j)}$ becomes

$$\mathbf{r}^{(j)} = \sum_{m=1}^{M} \exp(+2\pi i \eta_m^{(j)} y) \exp(+i\varphi_{\mathrm{m}}^{(j)}). \tag{14.22}$$

We have for the first term in Eq. (14.21),

$$\begin{aligned} \mathbf{a}\mathbf{r}^{(j)}\mathbf{r}^{*(j)} &= \mathbf{a} \sum_{m=1}^{M} \exp(+2\pi i \eta_m^{(j)} y) \exp(+i\varphi_m^{(j)}) \sum_{n=1}^{M} \exp(-2\pi i \eta_n^{(j)} y) \exp(-\varphi_n^{(j)}) \\ &= M\mathbf{a} + \mathbf{a} \sum_{\substack{m,n=1 \\ m\neq n}}^{M} \sum^{M} \exp[2\pi i(\eta_m^{(j)} - \eta_n^{(j)})y] \exp[i(\varphi_m^{(j)} - \varphi_n^{(j)})]. \end{aligned} \tag{14.23}$$

The first term on the right in Eq. (14.23) leads to the image of the point source and may be considered *the signal*. In the second term of Eq. (14.23) there are $M^2 - M = M(M - 1)$ members each representing the modulation of the subject wavefront $\mathbf{a}$ onto a different spatial carrier frequency with a randomly selected phase. If the sum of these amplitudes is multiplied by its complex conjugate, the product spatially averaged, and the square root taken, we obtain an rms noise amplitude of $[M(M-1)]^{1/2}a$ where $a = |\mathbf{a}|$. When M is large, this noise amplitude component is approximately equal to that of the signal amplitude. The second term in Eq. (14.21) can be written

$$\mathbf{r}^{(j)} \sum_{\substack{i=1 \\ i \neq j}}^{N} \mathbf{a}^{(i)} \mathbf{r}^{*(i)} = \sum_{\substack{i=1 \\ i \neq j}}^{N} \mathbf{a} \left[\sum_{m=1}^{M} \exp(+2\pi i \eta_m^{(j)} y) \exp(+i\varphi_m^{(j)}) \times \sum_{n=1}^{M} \exp(-2\pi i \eta_n^{(i)} y) \exp(-i\varphi_n^{(i)}) \right]. \tag{14.24}$$

For each hologram except the one associated with the illuminating reference wave (the jth), there are M^2 terms in the brackets of Eq. (14.24) each with an amplitude a and a random phase. The total $(N-1)M^2$ terms in Eq. (14.24) produce a rms noise amplitude $[(N-1)M^2]^{1/2}a$ just behind the hologram plate.

We are interested in the ratio of the signal intensity at the image plane to the total noise intensity there. We first compute noise and signal intensities at the hologram plane. The intensity of the noise produced by Eq. (14.23) is $M(M-1)a^2 \approx M^2a^2$ for M large. Equation (14.24) contributes a noise intensity $M^2(N-1)a^2$, and the total noise intensity at the hologram is

$$M^2a^2 + M^2Na^2 - M^2a^2 = NM^2a^2. \tag{14.25}$$

With a signal intensity M^2a^2, obtained from Eq. (14.23), we find for the ratio of the signal intensity at the hologram to the noise intensity at the hologram

$$(I_S/I_N)_H = M^2a^2/NM^2a^2 = 1/N. \tag{14.26}$$

It would seem that the signal is buried in noise when N is large. However it is the ratio of the intensities at the *image* plane which is significant. Since the image is that of a point source, the image power passing the hologram concentrates on a small spot at the image plane producing high intensity. On the other hand the noise power spreads over the image plane and remains low in intensity.

We can compute a ratio ϱ_s of peak signal intensity in the spot image to signal intensity at the hologram plane if we consider the hologram to be small relative to its separation d from the image plane. For this case, illustrated in Fig. 14.12, the reconstructed signal wavefront is essentially plane and uniform over the hologram area. The virtual image it yields is then the far-field pattern of a circular opening of radius h. A general expression for the far-field pattern is given by Eq. (5.39). If we multiply this equation by its complex conjugate, we obtain the desired intensity ratio

$$\varrho_{\mathrm{S}} = \left(\frac{a_2}{a_1}\right)^2 = \frac{1}{\lambda^2 d^2}\,[T(\xi, \eta)_{\mathrm{peak}}]^2 = \frac{(\pi h^2)^2}{\lambda^2 d^2} \tag{14.27}$$

where $T(\xi, \eta)_{\mathrm{peak}} = \pi h^2$, as given by Eq. (4.34) (see Fig. 4.7).

The spread of the noise in the image plane can be estimated by the spread of directions through which the subject wavefront is deflected by the second term in Eq. (14.23). As a result of the double summation, the spread in spatial carrier frequencies is $2(\eta_M - \eta_1)$. In terms of θ_M, θ_1, and θ given in Fig. 14.12, the corresponding spread in angle is $2(\theta_M - \theta_1) = 2\theta$ where we employ the relation $\eta_i \approx \theta_i/\lambda$. Thus the area of the image plane covered by

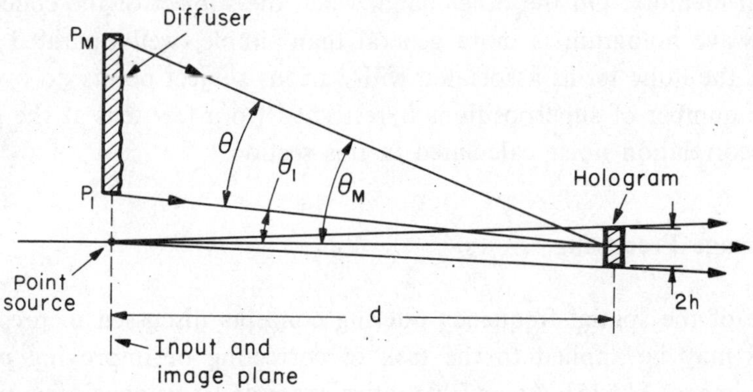

FIG. 14.12. Schematic arrangement for forming and illuminating a coded reference wave hologram of a point source.

the noise is $(\pi/4)(2\theta d)^2$. The ratio of noise intensity in the image plane to noise intensity at the hologram is then given by the inverse ratio of their respective areas:

$$\varrho_{\mathrm{N}} = \frac{\pi h^2}{(\pi/4)(2\theta\, d)^2} = \frac{h^2}{\theta^2 d^2}. \tag{14.28}$$

We may now form the ratio of the signal intensity at the image plane to the noise intensity at the image plane $(I_{\mathrm{S}}/I_{\mathrm{N}})_{\mathrm{I}}$ by multiplying $(I_{\mathrm{S}}/I_{\mathrm{N}})_{\mathrm{H}}$ in Eq. (14.26) by $\varrho_{\mathrm{S}}/\varrho_{\mathrm{N}}$:

$$\left(\frac{I_{\mathrm{S}}}{I_{\mathrm{N}}}\right)_{\mathrm{I}} = \frac{1}{N}\,\frac{\varrho_{\mathrm{S}}}{\varrho_{\mathrm{N}}} = \frac{1}{N}\,\frac{(\pi h^2)^2}{\lambda^2 d^2}\cdot\frac{\theta^2 d^2}{h^2} = \frac{\pi^2\theta^2 h^2}{N\lambda^2}. \tag{14.29}$$

To get the above result we have assumed a uniform noise distribution in the image plane. Actually the noise peaks somewhat at the image location. We note that Eq. (14.29) is not the *signal-to-noise ratio* for coherent illumination. Signal-to-noise ratio is obtained by substituting Eq. (14.29) for $I_i/\bar{I}_N$ in Eq. (12.41) (see also Fig. 12.11).[1]

More than 1000 coded-reference wave holograms of point-source subjects have been superimposed by multiple exposure of a single photographic plate, and this number does not appear to be a limit. However for point-source subjects the practical limitation is the noise produced by film grain or noise introduced in detection rather than the noise intensity derived from the diffuse reference wave. The maximum diffraction efficiency available for a single hologram is reduced by the square of the number of exposures (see Section 17.5.5). With the consequent low signal intensity in any image spot, the detector noise becomes the prime concern of the designer of a practical memory. On the other hand, when the subject of the coded reference wave hologram is more general than simple, well-separated point sources, the noise terms associated with various subject points do overlap, and the number of superpositions is restricted (to a few tens at the most) by the correlation noise calculated in this section.

14.6 Image Processing

Some of the spatial frequency filtering concepts discussed in preceding sections may be applied to the task of correcting or improving photographic images [*14.15*]. As an illustrative example let us consider a photograph deliberately taken out of focus. Suppose that the perfect photographic image of the subject has an amplitude transmittance $f(x, y)$, while its out-of-focus image has the transmittance $g(x, y)$. We determine that the camera, under the out-of-focus conditions, images a point in the subject plane to a spot in the image plane whose amplitude distribution is $h(x, y)$. The distribution $h(x, y)$ is the impulse response of the out-of-focus imaging system. It is possible to record $h(x, y)$ as a transparency if the appropriate gamma is used (see Section 2.5.1). We further suppose that the image $g(x, y)$ can be written

$$g(x, y) = f(x, y) * h(x, y) \tag{14.30}$$

in a form similar to that used in Section 6.4.1 [Eq. (6.66)] to express the point spread due to diffraction from a finite-diameter lens. That is, we as-

[1] The number of exposures N in Eq. (14.29) should not be confused with noise N in Eq. (12.41).

sume in Eq. (14.30) that the amplitude of each point of the perfect-image function $f(x, y)$ is, in the image, multiplied by the same amplitude distribution $h(x, y)$. As shown in Chapter 4, this requires the imaging system to be linear and space-invariant.

Our problem is to discover how to illuminate $g(x, y)$ and spatially filter the light it transmits so as to retrieve only $f(x, y)$. We shall carry out the filtering process in the spatial frequency or Fourier transform plane. To do this, the photograph $g(x, y)$ is placed in the front focal plane of a lens causing its Fourier transform $\mathbf{G}(\xi, \eta)$ to appear in the back focal plane (see Fig. 6.6). From Eq. (14.30) we can write

$$\mathbf{G}(\xi, \eta) = \mathbf{F}(\xi, \eta)\mathbf{H}(\xi, \eta) \tag{14.31}$$

where $\mathbf{F}(\xi, \eta) \subset f(x, y)$ and $\mathbf{H}(\xi, \eta) \subset h(x, y)$. If we are able to multiply

$$\mathbf{G} \cdot \frac{\mathbf{H}^*}{|\,\mathbf{H}\,|^2} = \frac{\mathbf{F} \cdot \mathbf{H} \cdot \mathbf{H}^*}{|\,\mathbf{H}\,|^2} = \mathbf{F}, \tag{14.32}$$

our goal will be achieved. The filter $\mathbf{H}^*/|\,\mathbf{H}\,|^2$ can be generated as two adjacent filters $\mathbf{H}^*$ and $|\,\mathbf{H}\,|^{-2}$ in the manner described in Section 14.4.1 for realizing the filter $\mathbf{S}^*/N$ [*14.2, 14.15*]. Here the transmittance $h(x, y)$ is placed in the front focal plane of Fig. 6.6 and the *intensity* in the back focal plane recorded as a negative *photograph*. The negative is developed with $\gamma = 2$ so that the amplitude transmittance of the record becomes

$$t = [|\,\mathbf{H}\,|^2]^{-\gamma_n/2} = |\,\mathbf{H}\,|^{-2}.$$

With $h(x, y)$ still in the front focal plane of the lens, a new photographic plate is placed in the back focal plane and a Fourier transform hologram is formed there, as in Fig. 8.22 or Fig. 14.6, to record $\mathbf{H}^*$. When the filters $\mathbf{H}^*$ and $|\,\mathbf{H}\,|^{-2}$ are placed together in the spatial frequency plane of Fig. 14.6, and $g(x, y)$ in the input plane is illuminated by an axial plane wave, we find at the spatial frequency plane the product

$$\mathbf{G} \cdot (\mathbf{H}^*/|\,\mathbf{H}\,|^2) = \mathbf{F},$$

and we obtain at the output plane $\mathscr{F}^{-1}[\mathbf{F}] = f(x, y)$ as desired. The method requires careful processing of the photographic plates and good results are difficult to achieve.

Fourier transform holograms (lens or lensless), when illuminated with a spatially modulated wave or formed with a spatially modulated reference wave, can in certain cases be used to process photographic images. Consider the multiplicity of overlapping images produced by a randomly spaced array of pinholes functioning as a multiple-pinhole camera [*14.16*]. The multiple-pinhole camera aperture is the source of the distorting function $h(x, y)$ which operates on each point of the perfect image $f(x, y)$ and produces the garbled image $g(x, y)$. As in Eq. (14.30), we suppose $g(x, y)$ can be expressed as the convolution of $f(x, y)$ with $h(x, y)$. Suppose a Fourier transform hologram is formed as in Fig. 14.6 with $g(x, y)$ replacing the character array in the input plane. The transmittance of the hologram is similar in form to that of Eq. (14.11) with **G** replacing **T**. It is the last term, the product of $\mathbf{G}^*$ times a linear phase factor, which is of interest. If we place the hologram in the back focal plane of the lens, insert the transmittance $h(x, y)$ on axis in the front focal plane, and illuminate the latter with an axial plane wave, we obtain for the amplitude at the spatial frequency plane the product

$$\mathbf{H}\mathbf{G}^* \exp(-2\pi i\eta b) = \mathbf{H}\mathbf{H}^*\mathbf{F}^* \exp(-2\pi i\eta b)$$

from the final term in the hologram transmittance. An inverse Fourier transformation performed with a lens yields

$$[h^* \star h] * f(-x, -y + b).$$

When the autocorrelation of the pinhole array $h(x, y)$ approaches a delta function, the convolution with f yields the inverted sharpened image whose transmittance is essentially f.

If the Fourier transform hologram is formed with $g(x, y)$ and $h(x, y)$ as subject and reference sources in the input plane, and if $g(x, y) = f(x, y) * h(x, y)$, then one of the terms of the hologram transmittance will be

$$\mathbf{G}\mathbf{H}^* = \mathbf{F}\mathbf{H}\mathbf{H}^*.$$

Suppose that $\mathscr{F}^{-1}[\mathbf{H}\mathbf{H}^*] = h^* \star h \rightarrow \delta(x, y)$. When the hologram is illuminated with an axial plane wave and a lens performs an inverse Fourier transformation on $\mathbf{G}\mathbf{H}^*$, we again find in the output plane the transmittance f.

REFERENCES

14.1. R. J. Collier and K. S. Pennington, "Ghost Imaging by Holograms Formed in the Near Field," *Appl. Phys. Lett.* **8**, 44 (1966).

14.2. A. Vander Lugt, F. B. Rotz, and A. Klooster, Jr., "Character Reading by Optical Spatial Filtering," *Optical and Electro-Optical Information Processing* (J. T. Tippett *et al.*, eds.), p. 125, Mass. Inst. Technology Press, Cambridge Massachusetts, 1965).

14.3. P. J. van Heerden "A New Optical Method of Storing and Retrieving Information," *Appl. Opt.* **2**, 387 (1963).

14.4. G. W. Stroke, R. Restrick, A. Funkhouser, and D. Brumm, "Resolution-Retrieving Compensation of Source Effects by Correlation Reconstruction in High Resolution Holography," *Phys. Lett.* **18**, 274 (1965).

14.5. K. S. Pennington and R. J. Collier, "Hologram-Generated Ghost Image Experiments," *Appl. Phys. Lett.* **8**, 14 (1966).

14.6. A. W. Lohmann and H. W. Werlich, "Holographic Production of Spatial Filters for Code Translation and Image Restoration," *Phys. Letters* **25A**, 570 (1967).

14.7. S. Lowenthal and Y. Belvaux, "Reconnaissance des Formes par Filtrage des Fréquences Spatiales," *Opt. Acta* **14**, 245 (1967).

14.8. V. Horvath, J. M. Holeman, and C. Q. Lemmond, "Holographic Technique Recognizes Fingerprints," Laser Focus, p. 18 (June 1967).

14.9. J. E. Rau, "Detection of Differences in Real Distributions," *J. Opt. Soc. Amer.* **56**, 1490 (1966).

14.10. C. S. Weaver and J. W. Goodman, "A Technique for Optically Convolving Two Functions," *Appl. Opt.* **5**, 1248 (1966).

14.11. J. E. Rau, "Real-Time Complex Spatial Modulation," *J. Opt. Soc. Amer.* **57**, 798 (1967).

14.12. E. N. Leith, J. Upatnieks, A. Kozma, and N. Massey, "Hologram Visual Displays," *Laser Focus*, p. 15 (Nov. 1, 1965).

14.13. R. J. Collier and K. S. Pennington, "Multicolor Imaging from Holograms Formed on Two-Dimensional Media," *Appl. Opt.* **6**, 1091 (1967).

14.14. J. T. LaMacchia and D. L. White, "Coded Multiple Exposure Holograms," *Appl. Opt.* **7**, 91 (1968).

14.15. G. Stroke and R. G. Zech, "A Posteriori Image-Correcting 'Deconvolution' By Holographic Fourier-Transform Division," *Phys. Lett.* **25A**, 89 (1967).

14.16. G. W. Stroke, G. S. Hayat, R. B. Hoover, and J. H. Underwood, "X-Ray Imaging With Multiple-Pinhole Cameras Using A Posteriori Holographic Image Synthesis," *Opt. Commun.* **1**, 138 (1969).

Chapter 15

HOLOGRAPHIC INTERFEROMETRY

When a hologram storing more than one wave is illuminated with coherent light, the reconstructed waves can interfere with one another (or with any other phase-related wave derived from the illuminating source). It took some time before the significance of this observation began to be fully appreciated. When it happened, it was with a burst of awareness on the part of many investigators [*15.1–15.6*]; their experiments beginning in 1965 initiated a new branch of interferometry: holographic interferometry. Interferometry has since become the most important application of holography.

A wave recorded in a hologram is effectively *stored* for future reconstruction and use. Holographic interferometry is concerned with the formation and interpretation of the fringe patterns which appear when a wave, generated at some earlier time and stored in a hologram, is later reconstructed and caused to interfere with a comparison wave. It is the storage or time-delay aspect which gives the holographic method a unique advantage over conventional optical interferometry. As we shall see, it permits diffusely reflecting or scattering surfaces which are subjected to stress to be interferometrically compared with their normal state. Thus in Fig. 15.1 (a photograph of the view through an illuminated hologram) the circular fringes superimposed on the image of a thermoelectric cooling module describe the concave distortion of the surface caused by thermal contraction [*15.7*]. The differential displacement of the surface in the direction normal to the surface is approximately one-half the wavelength of the illuminating light per fringe spacing.

Fringes characteristic of small displacements of a subject have relatively low spatial frequencies, as is evident in Fig. 15.1. Since interferometry is

FIG. 15.1. Interference fringes superimposed on the image of a thermoelectric module as seen through a doubly exposed hologram. (Courtesy Wolfe and Doherty [*15.7*].)

primarily concerned with such displacements, we shall restrict the use of the term *fringes* in this chapter to these low-frequency patterns and shall not use it to refer to the high spatial frequency, random-interference, or speckle patterns of diffusely scattered coherent light.

Real-time, *double-exposure*, *time-lapse*, *time-average*, *stroboscopic* are descriptive labels attached to variants of this new interferometric method. We shall first examine the real-time and double-exposure (time-lapse) techniques, as they apply to diffusely reflecting objects, and then turn to the problem of discerning a relation between the observed fringe pattern, its location in space, and the displacements of the subject surface. We show

how time-average and stroboscopic methods apply to vibration analysis and conclude with a discussion of holographic contour mapping.

15.1 Real-Time Interferometry

As discussed in Section 7.2.2, when the interference of a subject wave with a reference wave is properly recorded in a hologram, a wave whose complex amplitude is linearly proportional to that of the original subject wave can be reconstructed. Reconstruction is best achieved by placing the hologram plate back in the position it occupied during its exposure and illuminating it with the original reference beam. If we assume that the hologram plate has been replaced exactly in its original position, then, apart from a constant amplitude and phase factor, the reconstructed wave will be an exact replica of the original subject wave. (The ratio between the amplitudes of the original and reconstructed waves can be made unity by altering the illumination intensity.) Suppose that during the time between exposure of the hologram and its replacement, while the hologram is being processed, the subject stays fixed in position and continues to be illuminated with laser light. We now inquire as to what we observe when we look through the developed and replaced hologram in the direction of the subject. Since most holographic interferometry has been carried out using holograms formed in photographic emulsion, let us assume this to be the case here and recall that the reconstructed wave amplitude will then have a negative sign relative to the original subject wave [Eq. (8.3)]. Provided that (1) processing of the hologram does not distort the emulsion, (2) replacement has been exact, and (3) the illumination has been adjusted to equalize the absolute values of the original and reconstructed subject wave amplitudes, the reconstructed wavefronts will cancel the original subject wavefronts at every point on the viewer's side of the hologram. To the viewer looking through the hologram the subject will therefore not be visible. (In practice at least one broad bright fringe is usually observed because the processing of the hologram inevitably distorts the emulsion to some small degree.)

If the subject of the hologram is a plane mirror surface, the holographic method described above does no more than can be accomplished with a Michelson interferometer illuminated with uniform coherent light (see Fig. 1.16). Because mirror surfaces can be polished so smooth that any surface microstructure is of negligible consequence, one good mirror surface is the equivalent of another. By means of the beam splitter in the Michelson instrument, a given mirror surface can be interferometrically compared with its equivalent. Totally destructive or constructive interference can be observed as with the holographic method.

It is, however, the unique property of the holographic method that an arbitrarily shaped, rough scattering surface with complex microstructure can be interferometrically compared to another similar surface with no more difficulty (in principle) than is encountered with polished mirrors. We may consider the *virtual image* of a diffusely scattering subject, generated by the hologram, to provide one such surface. The other is the actual subject surface itself. If the original reference wave illuminates an exactly replaced hologram, the virtual image will appear to the viewer to be not only a macroscopic replica of the subject, but to lie coincident with it in space and to bear a surface microstructure whose light-scattering properties are identical to those of the subject. It would be a formidable task to attempt to replace the two mirrors of the Michelson interferometer with two such identical physical surfaces. Holography in effect does this, and in the analysis to follow in Section 15.3 we shall treat image and subject surfaces equivalently, much as if they were two *independent*, nearly identical physical surfaces.

Having achieved an effective coincidence of the virtual image with the subject, the latter may be placed under stress and caused to deform, or it may be released from stress and allowed to flow, creep, deflate, or expand. As a result, the optical distance from an observation point to any point of the displaced surface will change relative to its distance to the corresponding point on the comparison surface (that of the virtual image). Distances of surface points from the light source, which in effect illuminates both surfaces, also change. Accordingly, coherent light rays scattering from identical but displaced surface areas (areas with identical microstructure) will acquire a relative phase shift in transit from source to observation plane. The phase shift and the consequent wave-amplitude addition and subtraction which lead to the observed fringe pattern are characteristic of the subject surface deformation. As a change in the subject surface occurs, a change in the fringe pattern can be simultaneously observed—hence the name *real-time* holographic interferometry. It is axiomatic that the stress-induced deformations must not appreciably alter the surface microstructure. Since a rough surface acts as a random scatterer, any significant alteration of the roughness of the surface would impose a new random phase distribution on the light wave reflected from the subject. The subject wave and the reconstructed subject wave would then form a high spatial frequency interference pattern no longer interpretable in terms of the low spatial-frequency surface deformations of interest.

FIG. 15.2. Real-time interference fringes on a metal fork placed under tension. (Courtesy Dudderar [15.8].)

Several factors tend to add to the difficulty of real-time holographic interferometry. If the fringe pattern is to be observed at all or if it is to be characteristic of the intended displacement, the subject position and its illumination at the time when the processed hologram is replaced must be identical to that when the hologram was formed. This generally implies that the subject and the optical components required to form the hologram either must be kept fixed in position during the intervening period between exposure and observation or must be capable of being returned to their original positions with great accuracy. One must register the fringe pattern formed by subject and reference beams at the time of observation with that recorded in the hologram to within a small fraction of a fringe spacing. The task is easier when subject-to-reference beam angles are small. An additional complication is the emulsion shrinkage accompanying the wet processing of photographic plates. This causes some unavoidable distortion of the reconstructed wavefronts and thereby a distortion of the regularity of the observed interference fringes. Moreover, until the plate is thoroughly dry, continuing shrinkage may confuse attempts to register the hologram with the original reference beam.

Despite such impedimenta, it is possible to obtain high-contrast fringes as illustrated in Fig. 15.2. (The real-time fringes displayed in the photograph characterize deformation near the root of a notch machined into a metal fork; the latter had been placed under tension [*15.8*].) Real-time interferometry offers a distinct advantage in providing a capability for continuously monitoring displacements of a rough surface without touching or marking the surface (as, e.g., in moire techniques). A secondary, but useful, advantage over double exposure methods is the freedom allowed by the real-time method to make small adjustments of the hologram plate. These can optimize the fringe display for quantitative interpretation [*15.9*].

15.2 Double-Exposure Interferometry

Continuous comparison of surface displacement relative to an initial state may in certain cases supply more information than is necessary. If it is sufficient to form a permanent record of the relative surface displacement occurring after a fixed interval of time, a method obviating the experimental difficulties of real-time interferometry may be employed. Two exposures of the hologram, once to the initial state of the surface and once to its strained state, are superimposed prior to processing. Each exposure is made with the identical reference wave. With this *double-exposure* or *time-lapse* method, the problems of registering a reconstructed wave with an original are eliminated. After the exposure of the hologram is completed, the subject and the optical components used to illuminate the subject are no longer of concern. Both the comparison wave, characteristic of the surface in its initial state, and the wave representing a later, altered state of the surface are reconstructed in register by illuminating the hologram with a wave similar to the original reference wave. No more care in illumination is necessary than is taken in illuminating any pictorial hologram. Distortion due to emulsion shrinkage is identical for both reconstructed waves and is therefore not a factor in determining the spacing of the fringes formed by the interference of the two waves. Howeliver, mitations on the change of subject surface microstructure apply here as in the real-time method.

Illumination of the doubly exposed hologram not only effects the *simultaneous* reconstruction of two waves which had been scattered from the subject at *different* times but causes them to interfere under ideal conditions. The waves can share the diffraction efficiency of the hologram equally. Thus, their intensities are equal and the interference fringes they produce can have high visibility or contrast. For purposes of analysis we can consider the two virtual images of the subject surface generated by the reconstructed waves to be two slightly different physical surfaces. We imagine these initial-state and final-state surfaces to be simultaneously illuminated with the coherent light originally illuminating the subject. One difference in the interference resulting from the double-exposure method as compared to that obtained from the real-time technique may be observed when initial and final states of the subject surface are identical. We note that each reconstructed wave is negative with respect to the original subject wave and consequently when initial and final states are identical, the reconstructed waves add to give a bright image of the subject.

Figure 15.1 illustrates the fringe contrast obtainable with the double-exposure technique. The first exposure was made with the thermoelectric module at room temperature. Before the second exposure, a current was applied to the module terminals which dropped the temperature of the observed surface by 10°C. The back surface of the module (connected to a heat sink) was warmer, and the resulting thermal gradients gave rise to a contraction of the upper surface and a concave distortion.

The double-exposure method carried out with pulsed lasers has been applied with great success to the interferometry of transient events [*15.10*]. Figure 15.3 is a photograph of the image of a bullet in flight, as observed looking through a doubly exposed hologram. Fringes can be seen in the region of the shock wave. Apart from the problems of obtaining sufficient coherence length from the pulsed laser (see Fig. 11.7), the ease in getting results gives the holographic method an appreciable advantage over difficult-to-align, conventional interferometers. To obtain a photograph such as that shown in the figure, a chamber of gas through which the bullet is

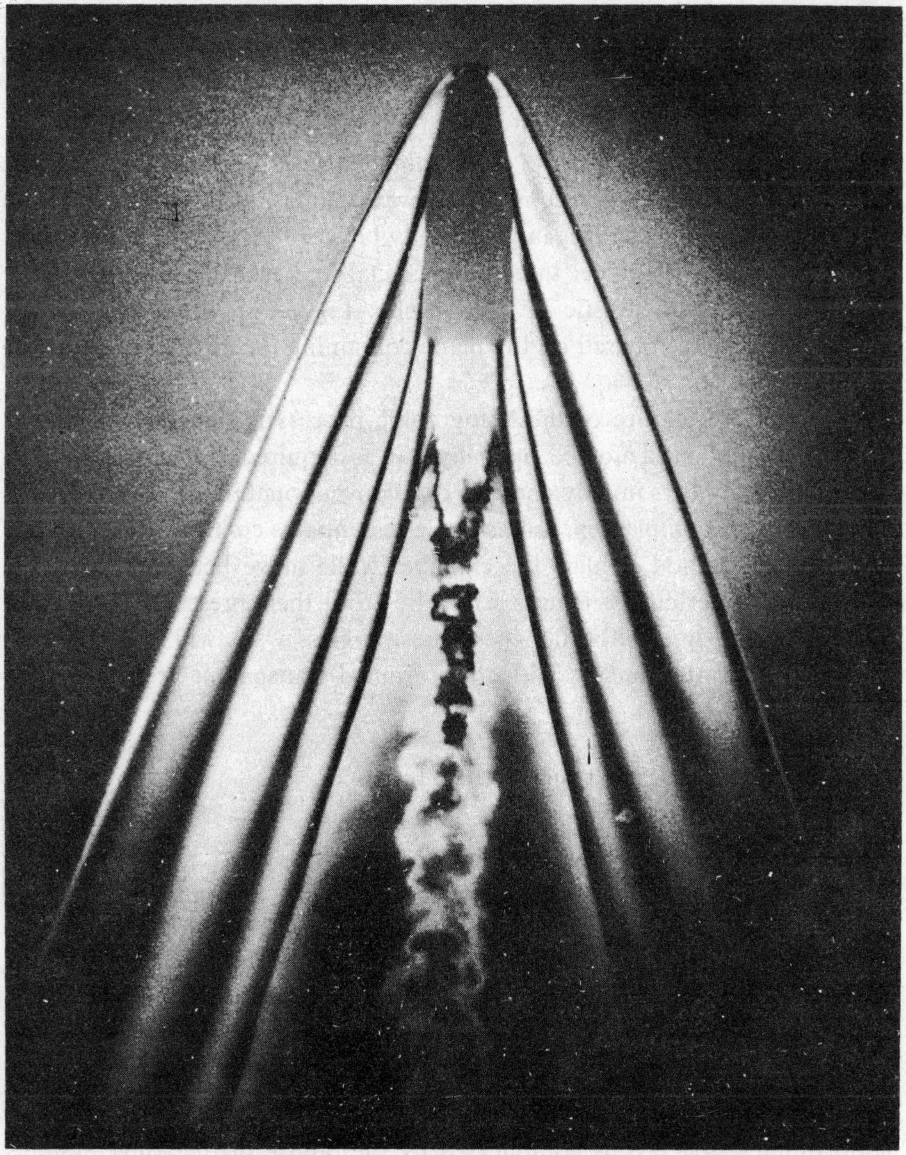

FIG. 15.3. Photograph of the image of a bullet in flight observed through a doubly exposed hologram. (Courtesy R. F. Wuerker, TRW Inc.)

to pass is illuminated from the rear with a pulse of laser light. (When the laser light is diffused before entering the chamber, the reconstructions are readily viewable with the naked eye.) The transmitted light combines with a reference wave outside the chamber to produce a first exposure of a hologram plate. Passage of the bullet through the chamber triggers a second pulse from the laser to illuminate the bullet and (once again) the chamber. The transmitted light together with the original reference beam produces the second exposure. In the region of the bullet shock wave, the density of the chamber gas is altered and with it the optical path through the gas. When the processed hologram is illuminated with a wave similar to the reference wave, the two reconstructed waves interfere to produce fringes characteristic of the changes in gas density between exposures. Both fringes and images are, of course, three dimensional and recorded with the large depth-of-field characteristic of holography. Figure 15.3 is a photograph made by a camera focused on the plane containing the virtual image of the bullet silhouette.

An attractive feature of the holographic process is that the glass walls of the chamber need not be optically flat as required for methods where the chamber appears in only one arm of a conventional interferometer. Both perturbed and comparison beams travel essentially common paths in the holographic method. Unless the chamber walls move between exposures, no spurious interference fringes will arise from their presence. Regardless of their optical nonuniformity the enclosing glass walls will appear in the hologram-generated images with their original transparent properties.

15.3 Fringe Localization and Interpretation

Some applications of holographic interferometry, e.g., nondestructive testing or inspection may require only qualitative information from the fringe pattern. Suppose that the thermoelectric module of Fig. 15.1 contained inadequate connections to semiconductor thermoelements or to heat sinks. Visual inspection for local distortions of the otherwise symmetric fringe pattern would suffice to detect these. For other applications involving quantitative stress–strain analysis, it would be desirable to obtain directly from the fringe pattern a point-by-point map of surface displacement. Unfortunately the task is complicated by the failure of the fringe systems generated by an arbitrary surface deformation to localize on the subject.

A surface on which fringes are said to be *localized* is one for which fringe contrast or visibility is maximum. The surface on which the fringes are localized depends on the subject illumination, the direction of observation, and the nature of the subject deformation. A working definition of localization can be obtained by considering the interference of light rays scattered from pairs of corresponding points on the original and deformed subject surfaces. (As will be explained more fully, scattering by noncorresponding points is randomly phased and can be considered noise.) Suppose the intensity of the scattered light is measured by a light-sensitive detector of small area. The acceptance angle of the detector, though finite, is restricted to receive only the light directed in a narrow cone of rays proceeding from a limited area of the scattering surface to an apex at the detector. We designate the mean direction of the rays as the observation direction. Let δ be the phase difference of light rays scattered to the detector by a pair of corresponding points in the (limited) observed area of the scattering surface. (The original and displaced positions of any point in the area form such a pair.) At some observation distance the variation of δ for all ray pairs which come from corresponding points in the observed area and which enter the finite angular aperture of the detector is zero or nearly zero. We call that position a point of localization. As the detector is translated above the subject surface with its direction of observation kept fixed, it receives light from new areas of the scattering surface. The value of δ (averaged over the small observed area) changes, giving rise to the detected interference fringes characteristic of surface displacement. The distance of the point of localization from the scattering surface may also change.

If the detector is moved so as to always be at a point of localization, the detected fringes will have the best contrast. This derives from the requirement that the variation of δ is nearly zero at such points. Thus the intensity contributed by each pair of corresponding points observed by the detector is nearly the same. As the detector scans over the displaced surface, the value of δ and the value of these intensity contributions changes slowly and smoothly. When, for example, the detected intensity at a point of localization has a minimum value, it is the result of all observed corresponding points producing a near zero of intensity at the detector. Thus the minimum can be nearly zero and the fringe contrast can be nearly maximum. Together, the points of localization form a surface of localization on which the fringes appear focused or localized. By suitably restricting the ray directions received at the detector, fringe contrast can be made as close to maximum as desired. As we shall see in Section 15.3.2, the surface of localization can change with the direction of observation.

Under certain circumstances, one of which we shall examine, the fringes localize on a plane normal to the direction of observation and very close to the subject. It is then easy to photograph both the fringes and the subject and to form a record useful for analysis. In other cases the distance of the surface of localization from the scattering surface varies over a wide range. Then the depth of field which permits the lens system to image both subject and fringes may often preclude the desired resolution.

In seeking to relate deformation to an observed fringe pattern [*15.11–15.13*], we shall follow the approach of Aleksandrov and Bonch-Bruevich [*15.14*] whose analysis focuses on a small area or spot of the surface and its displacement. The small area is observed through a lens system whose restricted aperture provides a large depth of field. Thus subject surface and the surface of fringe localization, under most circumstances, can be kept simultaneously in focus. By (1) sequentially observing the subject spot from four different directions through the hologram and (2) counting the fringes that appear to pass across the spot image as the viewing direction is shifted, three equations are obtained which together yield the displacement of the surface spot.

Before attempting to relate some simple examples of displacement to their associated fringe patterns, let us examine more carefully the underlying concepts. We do this with reference to Fig. 15.4. Shown there are a small area *ABC* of a diffusely reflecting surface, representing the initial position, and an identical but displaced area *A′B′C′*. The surface microstructure in the vicinity of *A′* is identical to that of *A*, etc. For the pairs of points *A*, *A′*, *B*, *B′*, and *C*, *C′*, the displacement vector $\vec{\Delta r}$ is shown to be the same, indicating that we have illustrated the case of pure translation

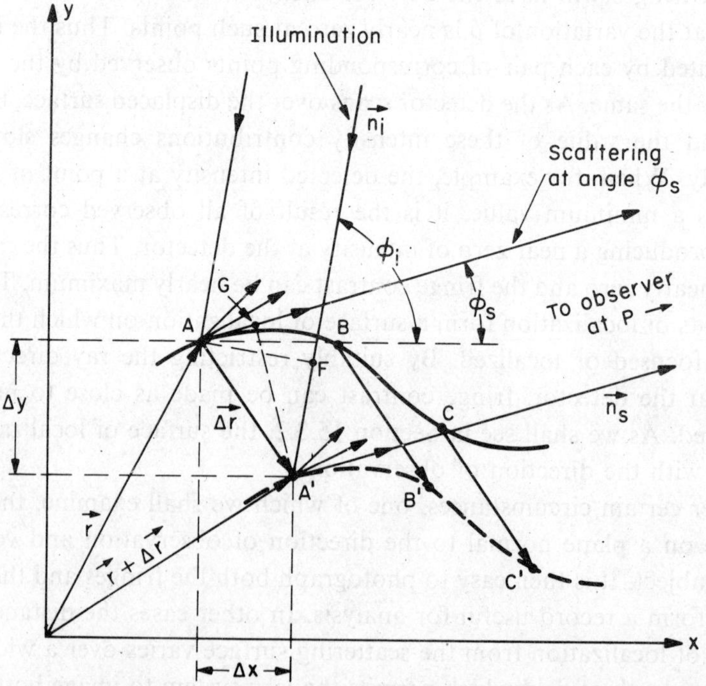

FIG. 15.4. Pure translation of the surface *ABC*.

of ABC. Area ABC represents the virtual image surface in the case of real-time interferometry or the *first-exposure* virtual image surface in the case of the double-exposure method. In either instance ABC may be regarded as a comparison surface. $A'B'C'$ is the actual subject surface when the real-time method is considered or is the *second-exposure* virtual-image surface when the double-exposure method is being analyzed. A plane wave identical to that illuminating the original subject simultaneously illuminates the small areas (in position $\vec{r}$) of both comparison and displaced surfaces.

Upon reflection from the two surfaces the light waves proceed outward at the designated observation angle to the observer (and detector) at P. By translating the detector, keeping the observation direction fixed, we wish to detect a localized fringe pattern characteristic of the surface displacement $\vec{\Delta r}$. In determining a point of localization P, we need only consider interference of light scattered from any point A on the original surface with that scattered from its corresponding point A' on the displaced surface. This may be understood as follows: In the absence of a surface displacement, light rays scattered from A and A' have no phase difference. The displacement, when it occurs, must not alter the surface microstructure. It follows then that the phase difference δ accumulated by light scattering from A' as compared to that scattering from A is strictly determined by the displacement vector $\vec{\Delta r}$. This holds for other pairs of points B and B', etc. In each case, interference of rays from a pair of corresponding points produces an intensity at the detector which depends only on the phase difference δ. However, this is not true for the interference of light scattered from A with that scattered from B' and C'. The random microstructure of the subject surface implies that a random relative phase difference is imposed on the light scattered from noncorresponding points A, B', C', etc. Their interference produces a nonlocalized high-frequency pattern having little effect on the observation of the localized low-frequency fringe system characteristic of $\vec{\Delta r}$.

To understand this, let us consider the interference of light from A, A', B and B' at an arbitrary detector position. Assuming all amplitudes to be unity, the complex amplitude at the detector is given by

$$\mathbf{a} = \{\exp(i\phi_A) + \exp[i(\phi_A + \delta_A)]\} + \{\exp(i\phi_B) + \exp[i(\phi_B + \delta_B)]\}$$

where ϕ_A and ϕ_B are random phases corresponding to the random microstructure of the surface, and δ_A and δ_B are the phase differences derived from surface displacement. To obtain the intensity at the detector we must multiply $\mathbf{a}$ by its complex conjugate. When terms in the first set of brackets (above) are multiplied by their complex conjugates, the result is the intensity at the detector due only to interference of light from A and A'; the value is $2(1 + \cos \delta_A)$. Similarly the intensity due to interference of light from B and B' is obtained from the second set of brackets and given by $2(1 + \cos \delta_B)$. The remaining terms in $\mathbf{aa}^*$ are the cross terms

$$2 \cos(\phi_A - \phi_B) + 2 \cos(\phi_A - \phi_B - \delta_B) + 2 \cos(\phi_A - \phi_B + \delta_A) + 2 \cos(\phi_A - \phi_B + \delta_A - \delta_B).$$

These terms depend on the random phases ϕ_A and ϕ_B at A and B. Suppose we were to include into our intensity computation all the scattering points in the limited area of the scattering surfaces observed by the detector. The resultant of the cross terms at any point on the detector surface will fluctuate very rapidly with detector position. This is due to the addition of a large number of new random phases from new scattering points seen by the detector at each new position. We suppose that the detector area, although small compared to the low frequency fringes characteristic of surface displacement, is sufficiently extended to encompass many such high-frequency fluctuations or speckles. If the detector responds only to the spatial average of the light incident on its photosensitive area, then the cross terms vanish in that spatial average. We are left with intensity terms of the form $N + \cos \delta_A + \cos \delta_B + \cdots + \cos \delta_N$ where N is the number of pairs of points scattering light to the detector and where $\cos \delta_A$, $\cos \delta_B, \ldots,$ are essentially constant over the detector surface.

When the variation in the phases $\delta_A, \delta_B, \ldots,$ is nearly zero, their cosines are essentially in phase and the intensity contributions of each pair of corresponding points AA', BB', etc., are equal. The observation point is then a point of localization. As the detector moves along the surface of localization, it measures an intensity of the form $1 + \cos \delta$ where δ is a slow function of the position of the detector. Depending on what this function is, the detected pattern can be linear fringes of constant spatial frequency or more complicated forms with varying spatial frequencies.

15.3.1 General Expressions for δ and the Condition for Localization

Figure 15.4 illustrates a displacement in the xy plane of the surface ABC to the new position designated $A'B'C'$. Although shown as a pure translation, we shall first consider the displacement to be of general nature. In accordance with the model we have adopted, we consider that both original and displaced surfaces, ABC and $A'B'C'$, are simultaneously present and illuminated by the original subject illumination. The illuminating wave propagates in the direction of the unit vector $\vec{n}_i$. Observation of the light scattered by the surfaces is confined to the direction of the unit vector $\vec{n}_s$. Each of these directions is considered to lie in the xy plane.

Now let us consider the difference in the optical paths taken by light rays traveling from source to observer via point A and point A'. There are two contributions. One stems from the difference in the paths of rays *illuminating* A and A' and the other from the difference in the *scattered* light paths. The first contribution is represented by the projection of the displacement $\vec{\Delta r}$ onto the illumination direction $\vec{n}_i$ while the second is represented by the projection of $\vec{\Delta r}$ onto the scattering direction $\vec{n}_s$. Multiplying $\vec{\Delta r} \cdot \vec{n}_i$ and $\vec{\Delta r} \cdot \vec{n}_s$ by $2\pi/\lambda$, where λ is the wavelength of the illuminating light, gives the corresponding contributions to the difference in phase δ of light traveling from source to observer via the displaced point relative to that traveling via the original point. Consistent with the conventions adopted in Sections 2.4.1, 3.1, and 5.2 with regard to the phase of light as a function of distance from the source, a positive product $\vec{\Delta r} \cdot \vec{n}_i$ implies that the displaced point is farther from the source and the contribution to δ is negative. On the other hand, a positive product $\vec{\Delta r} \cdot \vec{n}_s$ implies that light from the displaced point has a lesser distance to travel to the point of observation than that from the undisplaced point. Consequently its contribution to the phase difference δ is positive. We can therefore write as a general statement of the phase difference

$$\delta = (2\pi/\lambda)\, \vec{\Delta r} \cdot (\vec{n}_s - \vec{n}_i). \tag{15.1}$$

By requiring *zero* variation of δ with the variation in observation angle permitted by the finite acceptance angle of the detector, we can determine where the interference fringes are localized. To do this we must differentiate δ with respect to the angle of observation and set the resulting differential $d\delta = 0$.

15.3.2 Pure Translation of a Surface

Fig. 15.4 illustrates the specific case of pure translation where the products $\vec{\Delta r} \cdot \vec{n}_s = \overline{AG}$ and $\vec{\Delta r} \cdot \vec{n}_i = \overline{A'F}$ are both positive. Let us assume that the illuminating wave is plane so that $\vec{n}_i$ is a constant. For pure translation $\vec{\Delta r}$ is also a constant so that δ in Eq. (15.1) is simply

$$\delta = \vec{c} \cdot \vec{n}_s - c_1 \tag{15.2}$$

where $\vec{c}$ is a constant vector and c_1 is a constant. Of note is the absence in Eq. (15.2) of the position vector $\vec{r}$ so that the equation holds not only for pairs of corresponding points in a restricted area but for all pairs over the entire surface. Substituting $\vec{c} \cdot \vec{n}_s = c_x \cos \phi_s + c_y \sin \phi_s$ into Eq. (15.2), differentiating δ with respect to ϕ_s, and setting $d\delta = 0$ yields the condition for localization: $\phi_s =$ constant. To satisfy this, a detector of finite aperture must be located at infinity, i.e., in the far field of the surface, where the angle the surface subtends at the detector vanishes.

Mere inspection of Eq. (15.2) also tells us that the effect of the interference is simply to impose a constant phase shift on the plane illuminating wave scattered into *any* selected observation direction $\vec{n}_s$. Assume a detector whose area is large enough to average out speckle modulation and whose direction of observation $\vec{n}_s$ is fixed. When translated above the surface it detects no fringes, only the uniform intensity of the plane wave given by the form $1 + \cos \delta =$ constant. Suppose, however, that a lens is placed one focal distance from the surface, and observations are made in its back focal plane.

Each plane wave of direction $\vec{n}_s$ is brought to a discrete focus at this focal plane or spatial frequency plane, optically at an infinite distance from the surface. Since angles are transformed to positions in this plane, we can remove the aperture which had restricted angular reception by the detector. Translation of the detector in the back focal plane will then detect sharp fringes. Because there is no dependence of δ on z, the detector confined to a single value of y and permitted to move only in the z direction will detect a uniform intensity line image parallel to the z axis. This line image represents the localization of the interference pattern for light scattered into a given direction $\vec{n}_s$. Moving the detector along the y direction in the spatial frequency plane is equivalent to varying the direction of observation $\vec{n}_s$ and consequently the phase δ. As δ passes through integral multiples of 2π, a series of bright fringes will be detected. Thus pure translation of a diffusely reflecting surface, under the conditions specified for Fig. 15.4, leads to a system of parallel-line fringes localized at infinity.

15.3.3 Pure Rotation about an Axis in the Surface

At the other extreme from pure translation, so far as fringe localization is concerned, pure rotation about an axis contained in a scattering surface can produce parallel-line fringes localized very *near* that surface. In Fig. 15.5 the axis of rotation is the z axis, normal to the figure plane and the surface is initially considered to be in the xz plane. Rotation of the surface through a small angle α causes all points on the surface to be displaced

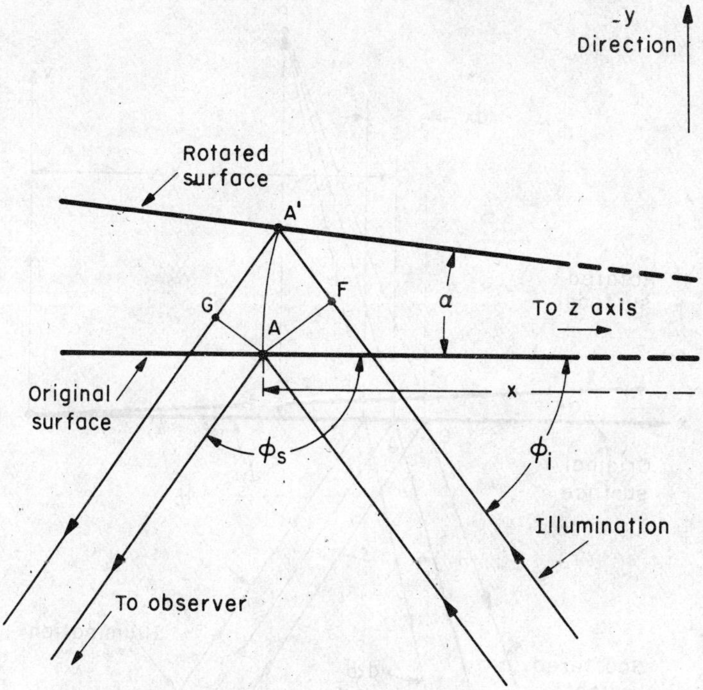

FIG. 15.5. Pure rotation about an axis in the surface.

through small arcs in the xy plane. Illuminating the surface is a plane wave whose wave normal or direction lies in the xy plane and makes an angle ϕ_i with the surface. Observation is restricted to a cone of rays whose mean direction in the xy plane makes an angle ϕ_s with the surface. Our model for analysis assumes, as before, that original and displaced surfaces are simultaneously present in the original subject illuminating beam.

Reference to Fig. 15.5 will show that the phase difference of rays scattered to the observer from the pair of identical but displaced points A, A' is given by

$$\begin{aligned}\delta &= -(2\pi/\lambda)[\overline{FA'} + \overline{A'G}]\\ &\approx -(2\pi/\lambda)(\overline{AA'})[\sin\phi_i + \sin\phi_s]\\ &\approx -(2\pi/\lambda)(x\alpha)[\sin\phi_i + \sin\phi_s].\end{aligned} \tag{15.3}$$

Unlike δ in Eq. (15.2), the phase here varies with the x coordinate of the scattering centers. A detector moving above the surface and observing the wave scattered into a fixed angle ϕ_s will measure a spatially varying intensity. We assume that there exists a surface of localization at some finite distance away from the subject scattering surface, and we wish to compute the finite distance h for an arbitrary point of localization P (Fig. 15.6). With P at a

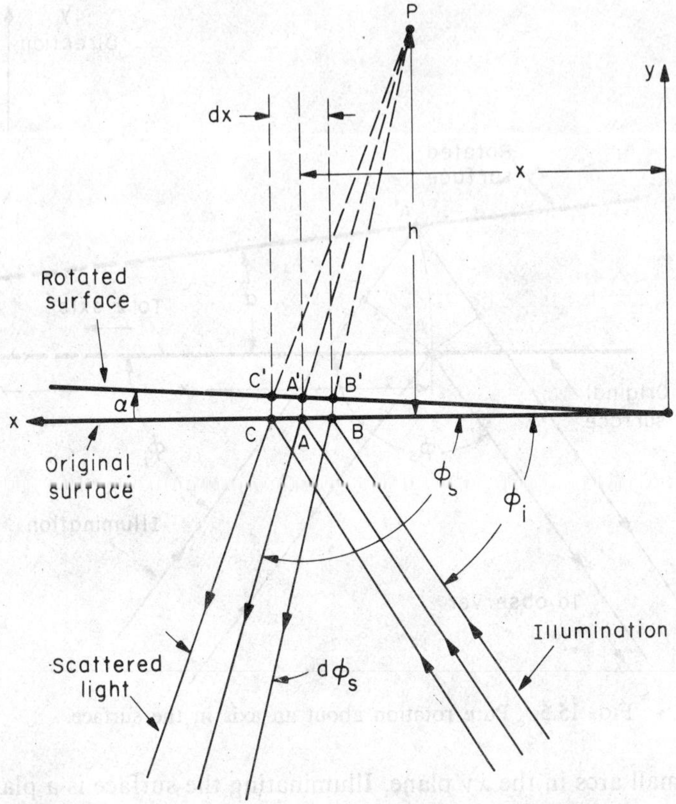

FIG. 15.6. Geometry for calculating h, the distance from the surface to a point of localization of the fringes formed by pure rotation.

finite distance behind the surface, we must consider the divergence from P of light scattered by a small area of the surface between $x - \frac{1}{2}\,dx$ and $x + \frac{1}{2}\,dx$ over a corresponding range of angles $\phi_s \pm \frac{1}{2}\,d\phi_s$. ($P$ can also be a point in front of the surface to which the scattered light converges.) The point P resides on a surface of localization provided that the variation of δ [in Eq. (15.3)] with ϕ_s is nearly zero. Since P in Fig. 15.6 is a virtual point of localization, the detector in this case requires a lens to cause the diverging light to converge into a cone which it can accept. Carrying out the differentiation of δ with respect to ϕ_s and noting that x is a function of ϕ_s while ϕ_i is not, we have for the differential

$$d\delta \approx -\frac{2\pi}{\lambda}\,\alpha(dx)[\sin\phi_i + \sin\phi_s] - \frac{2\pi}{\lambda}\,(x\alpha)\cos\phi_s\,d\phi_s$$

$$\approx -\frac{2\pi\alpha}{\lambda}\left\{\frac{h}{\sin^2\phi_s}\,[\sin\phi_i + \sin\phi_s] + x\cos\phi_s\right\}d\phi_s \qquad (15.4)$$

where we have used the relation

$$dx = \frac{h}{\sin^2 \phi_s} \, d\phi_s$$

obtained from the geometry of Fig. 15.6. Requiring the variation $d\delta$ of the phase difference to be zero and solving $d\delta = 0$ for h, we find

$$h = -\frac{x \cos \phi_s \sin^2 \phi_s}{\sin \phi_i + \sin \phi_s}. \tag{15.5}$$

When the viewing direction is normal to the surface, i.e., when $\phi_s = \pi/2$, then $h = 0$ and the fringes are localized on the scattering surface. Note that in this case corresponding points are essentially in line with the viewing direction. As the viewing angle departs from the normal direction, the surface of localization will lie either in front or behind the surface (depending on whether ϕ_s is $< \pi/2$ or $> \pi/2$). The surface of localization always intersects the scattering surface at the axis of rotation $x = 0$.

At a given point of localization, each pair of scattering points AA', BB', etc., in the coordinate range dx contributes to the detector an equal intensity of the form

$$I = 1 + \cos \delta. \tag{15.6}$$

According to Eq. (15.3) δ is a linear function of x and independent of z. Therefore translation of the detector over the surface of localization detects straight-line fringes cosinusoidal in x with a spatial frequency $\alpha(\sin \phi_i + \sin \phi_s)/\lambda$.

15.3.4 Relating Surface Displacement to Fringe Observation

The preceding calculations of fringe localization for simple, restricted displacements provide some understanding of why *arbitrary* surface deformation (complex translation and rotation) can lead to fringes localized anywhere between the surface and ± infinity. Since exact specification of the surface of localization for arbitrary deformation is difficult, it is perhaps fortunate that the prime task for those who employ interferometric methods is merely to relate fringe observation to surface displacement. To obtain such a relation experimentally, Aleksandrov and Bonch-Bruevich begin by focusing a viewing lens onto the subject surface. Suppose the imaging lens axis is oriented along the extension of the line PA in Fig. 15.6, and its image plane is apertured to receive light scattered from only a small surface area in the vicinity of A. Light appearing to diverge from a point P on the virtual localized fringe system also enters the lens. When the lens is large, resolution in the image of the surface is high. However, a large unrestricted lens may intercept too large a range of ray directions diverging from P and as a consequence the fringe contrast may be reduced to a low value. A compromise is reached by restricting the aperture of the viewing system until both contrast and resolution are acceptable. The increased depth of field will then allow both the surface and the out-of-surface fringes to appear in good focus simultaneously.

It is important to point out that this is not achieved without sacrifice. With the increased depth of field comes a decrease in the ability to resolve not only fine surface detail but to count finely spaced fringes. As noted in Chapter 12, reduction of the aperture increases the grain size of the speckle pattern. When the grain size becomes comparable to the interference fringe spacing, the detector area no longer averages the fluctuations in intensity. Fringe patterns then become more difficult to interpret.

Let us suppose, as will often be the case, that the compromises entailed in employing a small-aperture viewing system are not objectionable. Properly apertured, the viewing system focuses through the hologram onto the nearly coincident original and deformed surfaces and at the same time permits observation of high-contrast interference fringes. The method to be described determines the displacement of one small area or spot A on the subject surface. A series of observations is made on the fringes which appear in the vicinity of A. Suppose initially, when the observation direction is $\vec{n}_{s0}$, a dark (or bright) fringe is observed to be located at A. This corresponds to a phase difference δ_0 between rays scattered from the spot in its displaced and original positions. δ_0 is expressed in general form by Eq. (15.1)

$$\delta_0 = (2\pi/\lambda)\, \vec{\Delta r}\cdot(\vec{n}_{s0} - \vec{n}_i). \tag{15.7}$$

Here the subscript 0 refers to assignment of the counting number 0 to the fringe appearing initially at A. The spot is next sighted from a new viewing direction $\vec{n}_{sk}$. In the process of altering direction, the number of fringes k passing through A is counted. Since the direction of illumination remains unaltered, the phase difference between rays scattered from the spot is now given by

$$\delta_k = (2\pi/\lambda)\, \vec{\Delta r}\cdot(\vec{n}_{sk} - \vec{n}_i). \tag{15.8}$$

Subtracting Eq. (15.7) from (15.8) we have

$$\delta_k - \delta_0 = (2\pi/\lambda)\, \vec{\Delta r}\cdot(\vec{n}_{sk} - \vec{n}_{s0}) = \pm 2\pi k$$

or

$$\vec{\Delta r}\cdot(\vec{n}_{sk} - \vec{n}_{s0}) = \pm k\lambda \tag{15.9}$$

where $2\pi k$ is the number of radians corresponding to the passage of k fringes through A. Equation (15.9) represents a linear equation in the three unknown components of $\vec{\Delta r}$. Two more pairs of sightings provide two additional equations, sufficient to determine all three components of the displacement $\vec{\Delta r}$. (The sign of $\vec{\Delta r}$ is not determined by measuring the phase differences and must be either known *a priori* or obtained by other means.)

Within the limitations pointed out, the method outlined is applicable to any kind of displacement. However, the maximum displacement that can be measured by interferometric techniques is limited by the highest interference fringe frequency which can be resolved by the viewing optics. For the example of pure rotation in Section 15.3.3, the fringe frequency

$$\frac{\alpha(\sin\phi_i + \sin\phi_s)}{\lambda}$$

is proportional to the angle of rotation α. If the fringes are to be resolved, the magnitude of α must be limited.

15.4 Interferometry of Vibrating Surfaces

Powell and Stetson [*15.15*] discovered how to use holographic recording to study vibrating surfaces before the real-time and time-lapse techniques for static surfaces were developed. And yet these latter methods are the fundamental processes. In fact, Powell and Stetson were able to show an analytical equivalence between *time-average* holographic recording of a vibrating surface and simultaneous holographic recording of an ensemble of surfaces whose positions in the limit are all those occupied by the actual vibrating surface over the exposure interval. Later measurements on a sinusoidally vibrating membrane [*15.16*] indicated that, for the case investigated, an even simpler view is reasonably valid. Since the membrane spends more time at the two extreme positions of its displacement than at any other position, its time-average hologram may well be regarded as a double-exposure record. Interference fringes were observed as with the double-exposure method. Predictions of vibration amplitude using the observed spacing of the fringes and this simplified concept were found to agree to within 10% with the more accurate predictions of Powell and Stetson. There are, however, noticeable differences in the visibility of the fringe pattern of the vibrating surface relative to its double-exposure counterpart, and these are explained only by a treatment similar to that of Powell and Stetson.

Figure 15.7 displays photographs of the hologram-generated image of a vibrating guitar surface. Time-average fringes characteristic of the vibration are superimposed on the image.

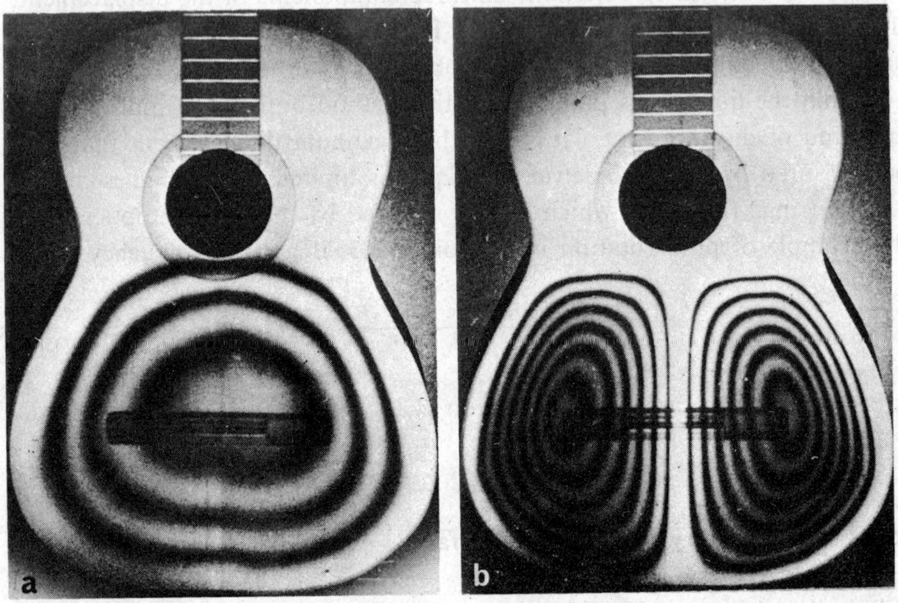

FIG. 15.7. Photographs of images of a vibrating guitar generated by time-average holograms: (1) 185 Hz, (b) 285 Hz. The fringes are characteristic of the first two vibration modes of a guitar made by Georg Bolin, Stockholm. The holograms were recorded by N. E. Molin and K. A. Stetson at the Institute of Optical Research, Stockholm, Sweden.

15.4.1 TIME-AVERAGE HOLOGRAPHIC INTERFEROMETRY

The method for forming a time-average hologram of a vibrating surface is identical to that used for forming a hologram when a surface is stationary. Providing the maximum amplitude of the vibration is limited to some tens of light wavelengths, illumination of the hologram yields an image of the surface on which is superimposed a number of observable interference fringes. It should be emphasized that these fringes are contour lines of equal displacement of the surface. Whereas a few of these may be nodal contours, most are contours of equal displacement plotted on the moving regions of the surface.

To illustrate a relation between vibration amplitude and the intensity of the fringe pattern, we consider a vibrating membrane illuminated with parallel coherent light as in Fig. 15.8. When the membrane surface is static, it is assumed to lie in the xz plane, and its displacement from that plane

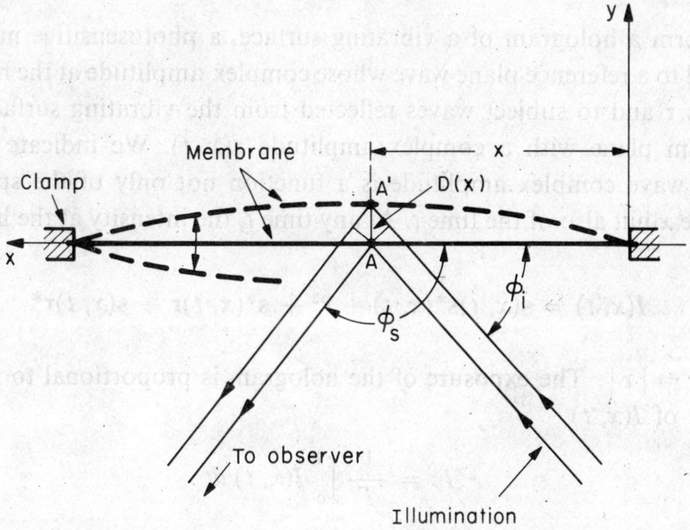

FIG. 15.8. Geometry for calculating the phase shift $\delta(x, t)$ for the time-average method.

during vibration is assumed to be suitably small. A clamp at the outer edges of the surface fixes one nodal contour. The arrangement to be analyzed is similar to the case of pure rotation illustrated in Fig. 15.5, and the phase difference δ of rays scattered by a pair of corresponding points A, A' can be written in a form similar to Eq. (15.3). In the present case we consider the displacement $\overline{AA'} = D(x, t)$ at some instant of time t. If we assume simple harmonic motion, then the displacement of an arbitrary point at time t may be written as

$$D(x, t) = D(x) \cos \omega t, \tag{15.10}$$

where ω is the angular frequency of vibration. By substituting $D(x, t)$ for $x\alpha$ in Eq. (15.3), we obtain for the instantaneous phase shift of the scattered rays

$$\delta(x, t) = -(2\pi/\lambda) D(x) \cos \omega t [\sin \phi_i + \sin \phi_s]. \tag{15.11}$$

Before we can use $\delta(x, t)$ to calculate the intensity of the fringe pattern, we must consider the hologram recording process. We must ensure that the hologram records a time-average of the interference pattern intensity produced by the time-dependent subject complex amplitude at the hologram plane. In practice this means that the exposure time should be long compared to the vibration period; in our calculation however, we can average over exactly one period of vibration.

To form a hologram of a vibrating surface, a photosensitive medium is exposed to a reference plane wave whose complex amplitude at the hologram plane is $\mathbf{r}$ and to subject waves reflected from the vibrating surface to the hologram plane with a complex amplitude $\mathbf{s}(x, t)$. We indicate that the subject wave complex amplitude is a function not only of the spatial coordinate x but also of the time t. At any time t, the intensity at the hologram is

$$I(x, t) = \mathbf{s}(x, t)\mathbf{s}^*(x, t) + r^2 + \mathbf{s}^*(x, t)\mathbf{r} + \mathbf{s}(x, t)\mathbf{r}^* \quad (15.12)$$

where $r = |\mathbf{r}|$. The exposure of the hologram is proportional to the time-average of $I(x, t)$

$$\langle I \rangle = \frac{1}{T}\int_0^T I(x, t)\, dt \quad (15.13)$$

where $T = 2\pi/\omega$ is the vibration period of the surface and ω is the angular frequency. By substituting for $I(x, t)$ from Eq. (15.12), we may write Eq. (15.13) as

$$\langle I \rangle = \frac{1}{T}\int_0^T [\mathbf{s}(x, t)\mathbf{s}^*(x, t) + r^2 + \mathbf{s}^*(x, t)\mathbf{r}]\, dt + \frac{\mathbf{r}^*}{T}\int_0^T \mathbf{s}(x, t)\, dt \quad (15.14)$$

where we have separated out the term of interest, at the far right. This last term leads to a reconstruction of the original subject wave and a virtual image of the subject surface. If a plane absorption hologram is exposed and processed so that it is a linear recording of the subject wave, then its amplitude transmittance is proportional to $\langle I \rangle$. Illumination by the original reference wave reconstructs the time-averaged original subject wave given by the product

$$\mathbf{w} \propto \mathbf{r} \cdot \frac{\mathbf{r}^*}{T}\int_0^T \mathbf{s}(x, t)\, dt$$

$$\propto \frac{1}{T}\int_0^T \mathbf{s}(x, t)\, dt = \frac{1}{2\pi}\int_0^{2\pi} \mathbf{s}(x, t)\, d(\omega t). \quad (15.15)$$

The mathematical form of Eq. (15.15) holds for the light coming from any subject point and for the wavefront of the subject light at an arbitrary distance from the subject. Suppose $\mathbf{a}(x, t)$ represents the instantaneous complex amplitude of the light arriving at an observation point P from a point A on the vibrating surface. We may write $\mathbf{a}(x, t)$ as

$$\mathbf{a}(x, t) = \mathbf{a}(x) \exp[i\delta(x, t)] \quad (15.16)$$

where $\mathbf{a}(x)$ is the complex amplitude at P when the vibrating surface lies in the xz plane and where $\delta(x, t)$ [given in Eq. (15.11)] is the extra phase difference produced by the displacement of the surface $D(x, t)$. If we substitute $\mathbf{a}(x, t)$ for $\mathbf{s}(x, t)$ in Eq. (15.15), the complex amplitude of the reconstruction of the wave originally traveling from A to P may be written as

$$\mathbf{w}_A \propto \mathbf{a}(x) \frac{1}{2\pi} \int_0^{2\pi} \exp[i\, \delta(x, t)]\, d(\omega t). \tag{15.17}$$

The integral in Eq. (15.17) represents interference at the point P of light from the undisplaced point A with that from corresponding points A', A'', etc., where the latter are all the displaced positions of A produced by the vibration. By substituting $\delta(x, t)$ from Eq. (15.11) into Eq. (15.17), we can evaluate the factor modulating the complex amplitude $\mathbf{a}(x)$ of the wave coming from the undisplaced surface. Thus

$$\begin{aligned} &\frac{1}{2\pi} \int_0^{2\pi} \exp[i\, \delta(x, t)]\, d(\omega t) \\ &\quad = \frac{1}{2\pi} \int_0^{2\pi} \exp\left[-i \frac{2\pi}{\lambda} D(x) \cdot (\sin \phi_i + \sin \phi_s) \cos \omega t\right] d(\omega t) \\ &\quad = J_0\left[\frac{2\pi}{\lambda} D(x) \cdot (\sin \phi_i + \sin \phi_s)\right], \end{aligned}$$

where J_0 is the zero-order Bessel function of argument $(2\pi/\lambda)D(x) \cdot (\sin \phi_i + \sin \phi_s)$ [*4.4*, p. 149].[1] The intensity at the observation point is

$$\mathbf{w}_A \mathbf{w}_A{}^* \propto \mathbf{a}(x)\mathbf{a}^*(x)\left\{J_0\left[\frac{2\pi}{\lambda} D(x) \cdot (\sin \phi_i + \sin \phi_s)\right]\right\}^2. \tag{15.18}$$

Equation (15.18) tells us that the dark fringes in the observed interference pattern of a sinusoidally vibrating surface correspond to the zeros of the $J_0{}^2$ function, and the bright fringes correspond to the maxima of the function. These maxima decrease with increasing values of the argument (see, e.g., Fig. 8 of Brown *et al.* [*15.17*]), i.e., with increasing displacement amplitude $D(x)$. The fall-off of intensity limits the number of fringes which have sufficient visibility to be seen by an observer.

Time-average holography enables the vibrational amplitudes of *diffusely reflecting* surfaces to be measured with interferometric precision. It eliminates the need to place sensing elements of any sort in contact with the vibrating surface. Curvature of the surface is not a factor as it is when Chladni patterns are employed. The method has been applied to the study

[1] Note that $J_0(z) = J_0(-z)$.

of a variety of vibrating objects ranging from quartz crystals [*15.18*] to treble-viol plates [*15.19*] and can yield greater detail about the antinodal region than previous techniques. There are, however, limitations which have prompted an investigation of other means to extend holography to the study of vibrating objects. One of these limitations, the diminution in fringe visibility as the vibration amplitude increases has already been cited. Another is that the relative phase of the vibration is not given by the measurement.

15.4.2 Real-Time Interferometry of Vibrating Surfaces

If real-time observation is desirable, this may be achieved by (1) forming a hologram of the static subject, (2) replacing the processed hologram in its original position, and (3) stimulating the subject into vibration. The observer's eye then performs a time average of the intensities of a sequence of interference patterns formed between light appearing to scatter from the static image and that scattered by the time-varying subject. Let us consider a point A on the virtual image of the static subject generated when the hologram is illuminated by the original reference wave. Let $\mathbf{a}(x)$ be the complex amplitude of the light from A arriving at an observation point P. At any instant the observer at P detects the intensity of the interference of light from the virtual image point A with that from the corresponding point A' on the actual moving surface. If we call $-\mathbf{a}(x)\exp[i\,\delta(x,t)]$ the complex amplitude of the light arriving at P from A' where $\delta(x,t)$ the phase difference associated with the displacement $D(x,t)$ is given by Eq. (15.11), then the intensity at time t is

$$\begin{aligned} I(x,t) &= \{\mathbf{a}(x) - \mathbf{a}(x)\exp[i\,\delta(x,t)]\}\{\mathbf{a}^*(x) - \mathbf{a}^*(x)\exp[-i\,\delta(x,t)]\} \\ &= 2\,|\,\mathbf{a}(x)\,|^2[1-\cos\delta(x,t)]. \end{aligned} \quad (15.19)$$

(The negative sign derives from the π phase shift between actual and reconstructed waves.) An average taken over one period of the vibration yields

$$\begin{aligned} \langle I\rangle &= 2\,|\,\mathbf{a}(x)\,|^2\frac{1}{T}\int_0^T[1-\cos\delta(x,t)]\,dt \\ &= 2\,|\,\mathbf{a}(x)\,|^2\frac{1}{2\pi}\int_0^{2\pi}[1-\cos\delta(x,t)]\,d(\omega t) \\ &= 2\,|\,\mathbf{a}(x)\,|^2\left\{1-J_0\left[\frac{2\pi}{\lambda}D(x)\cdot(\sin\phi_{\mathrm{i}}+\sin\phi_{\mathrm{s}})\right]\right\} \end{aligned} \quad (15.20)$$

where we have used Eq. (15.11) and [*4.4*, p. 150]. Because the fringe intensity $\langle I \rangle$ in Eq. (15.20) goes to zero only when $D(x) = 0$, fringe visibility and thus the number of fringes that are observable are reduced in this real-time method.

15.4.3 Stroboscopic Holographic Interferometry

By stroboscopically illuminating the surface of a vibrating subject, fringe systems can be formed holographically which provide information not obtainable from time-average fringes. A variety of methods have proved successful.

Real-time fringes are observed by (1) forming a hologram of the static surface, (2) replacing the processed hologram in its original forming position, (3) setting the surface in vibration, and (4) illuminating the surface once each vibration period with a short pulse of light [*15.20*]. If the pulse is short enough, the method is equivalent to real-time holographic interferometry of static objects (Section 15.1). However, by altering the phase at which the light flash appears, the vibrating surface may be compared with the static image at any phase in its vibration cycle. As with the previous real-time method, one can vary the vibration frequency, in this case keeping the light pulse and surface vibration in synchronism, and so examine the full range of frequency response of the vibrating body. This enables narrowband resonances to be studied with facility. Real-time methods allow observation of the evolution of a fringe pattern as vibration amplitude is increased from zero or as the vibration frequency is brought to the resonance value. From such observations one can count fringes, follow their movement, and detect the positions of nodes. By displacing the surface slightly in the direction of its normal and observing the motion of the fringes, the relative sign of the vibration over the mode pattern can be determined. One major advantage over the time-average system, common to nearly all the stroboscopic techniques, is that the fringes are of equal visibility independent of the amplitude of vibration. [They have an intensity of the form $1 + \cos \delta$ given in Eq. (15.6).] For this reason larger vibration amplitudes can be studied stroboscopically than with the time-average method.

Stroboscopic methods have been applied to double-exposure holographic interferometry as well [*15.21*]. With one technique a first exposure is taken of the static surface using an exposure time half that normally employed to get good reconstruction efficiency. Next, the surface is set into vibration and synchronous pulses of laser light are now provided for subject illumination. In this manner the hologram plate is again exposed for a time calculated to finally give the plate its optimum density (after processing). When illuminated, the hologram displays interference fringes expected from a double-exposure hologram. The method preserves the advantages (over the time-average method) of (1) allowing vibration amplitudes to be measured at any phase and (2) providing uniform fringe visibility.

A second double-exposure stroboscopic technique produces the two exposures with a repetitive *pair* of light pulse synchronized in repetition frequency so that the pair of pulses flash once during the vibrating cycle [*15.22*]. Illumination of the hologram yields fringes which correspond to the relative displacement of the surface in the interval between the first and second light pulse. Of course, the time between the two pulses can be adjusted. This feature is useful when studying strong vibrations. A large vibration amplitude, normally producing fringe patterns too fine to observe, can be examined incrementally and the time between pulses adjusted to give only a limited number of fringes across the field. This method has application in nondestructive testing where interesting events are likely to occur at large amplitudes.

15.5 Contour Generation

Two-beam interference, with at least one of the beams reconstructed from a hologram, can be employed to create a fringe pattern which maps topographical contours onto the surface of a three-dimensional object or its virtual image. Each contour is the locus of all points on the surface that lie at some constant height above a fixed plane. Before considering how holographic contour mapping is implemented, we note that the technique has been used to detect small nonuniformities in optical surfaces. Such errors, incurred in the grinding process, produce irregularities in the contour curves detectable down to one-tenth of a fringe. Because the holographic method works with rough surfaces, areas which must be reshaped can be located and corrected without the need for any intermediate polishing as required for conventional interferometry [*15.23*].

We shall describe two uniquely holographic methods for forming contours. One employs two wavelengths and only one stored wave [*15.24*]. The other method is a double-exposure technique with the refractive medium surrounding the object changed between exposures [*15.25*]. By suitable choice of the wavelengths (Method 1) or indices of refraction (Method 2), the increments of height marked off by the contours can be made to range from microns to millimeters.

In each of the methods, two slightly altered surfaces are compared interferometrically. (We here equate real surfaces and their virtual images.) If the observed interference fringes are to be useful contours, the following conditions should be satisfied:

1. The phase difference of light rays scattered from corresponding points on the two surfaces to the observer must be proportional to the nominal height of the surface above some reference plane.
2. The fringes should localize close to the surface.

15.5.1 Two-Wavelength Method

Suppose a hologram is formed, as shown in Fig. 15.9, with parallel subject illumination introduced on axis and an off-axis parallel reference wave whose direction lies in the *xz* plane. We confine our calculations to that plane. The wavelength of the light used in formation is λ_1. During and

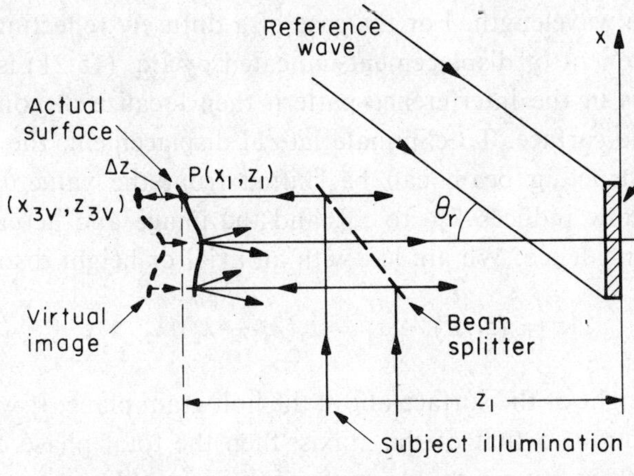

Fig. 15.9. Hologram-forming geometry for two-wavelength contour generation.

after processing of the hologram, the subject is kept fixed in place. Upon replacement of the hologram in its original position, the laser wavelength is shifted to a slightly different wavelength λ_2, but the waveforms and wave direction are assumed, for the moment, to remain the same. Parallel reference light of wavelength λ_2 illuminating the hologram generates a virtual image of the subject. However, the image is displaced from the location of the actual subject which is now illuminated with parallel on-axis light, also of wavelength λ_2. We consider a typical point P on the actual subject with coordinates (x_1, z_1). To find its new coordinates (x_{3V}, z_{3V}) in the image as a result of the change in wavelength, we return to Eq. (3.35) (Section 3.3.2) where we find that for the virtual image

$$x_{3V} = x_1 + z_1[(\theta_c/\mu) - \theta_r], \tag{15.21}$$

$$z_{3V} = z_1/\mu. \tag{15.22}$$

In the above, θ_c and θ_r are the angles the reconstructing and reference waves make with the z axis, and $\mu = \lambda_2/\lambda_1$. Although there is a displacement in the x direction, we note from Eq. (3.29) that the lateral magnification

$$M_{\mathrm{lat,V}} = m\left(1 + \frac{m^2 z_1}{\mu z_c} - \frac{z_1}{z_r}\right)^{-1}$$

is unity as long as reconstructing and reference waves are both parallel (and $m = 1$).

We now wish to compare interferometrically the virtual image surface generated with light of wavelength λ_2 to the actual surface illuminated by light of the same wavelength. For the case of a diffusely reflecting surface, the *lateral* component of displacement indicated by Eq. (15.21) is undesirable since fringes in the interference pattern then localize at considerable distance from the surface. To eliminate lateral displacement, the direction θ_c of the reconstructing beam can be shifted from the value $\theta_c = \theta_r$ to $\theta_c = \mu\theta_r$. This now reduces x_{3V} to x_1, and the image and actual subject are in lateral coincidence. We are left with an axial or height displacement

$$\Delta z = (z_1/\mu) - z_1 = z_1(\lambda_1 - \lambda_2)/\lambda_2 \tag{15.23}$$

where z_1 is the height of the surface above the hologram plane. If we confine the viewing direction to that of the z axis, then the total phase difference of rays passing from source to observer via corresponding points on the two surfaces corresponds to $2\Delta z$ and is given by

$$\delta(x) = (2\pi/\lambda_2)\cdot 2z_1(x)[(\lambda_1 - \lambda_2)/\lambda_2] \tag{15.24}$$

where the height z_1 is a function of x as indicated in Fig. 15.9. Since the actual surface and the virtual-image comparison surface are displaced axially, corresponding points appear *in line* to the observer. Light rays proceeding axially from any pair of corresponding points interfere at the surface, and the variation of their relative phase δ with x gives rise to fringes localized on the surface. According to Eq. (15.24) δ is directly proportional to the height of the surface.

To relate an increment of surface height to a change in δ we can rewrite the linear equation (15.24) in terms of small increments Δz_1 and $\Delta\delta$ and obtain

$$\Delta z_1 = \frac{\lambda_2^2}{\lambda_1 - \lambda_2} \cdot \frac{\Delta\delta}{4\pi}.$$

The increment of height which produces a phase change of $\Delta\delta = 2\pi$ and which corresponds to the observation of a pair of fringes is then

$$\Delta z_1 = \frac{\lambda_2^2}{2(\lambda_1 - \lambda_2)}.$$

This can be written

$$\Delta z_1 = \frac{\lambda^2}{2\Delta\lambda} \tag{15.25}$$

for λ_1 not too different from λ_2. An argon ion laser is a convenient source of light of two wavelengths differing by $\Delta\lambda = 0.0115\ \mu$m and centered at about $\lambda = 0.48\ \mu$m. From Eq. (15.25), the spacing between contours Δz_1 is in this case approximately 10 μm.

15.5.2 Immersion Method

Two-wavelength contour fringes can be realized with a double-exposure technique as well as with the real-time arrangement just described. However, the lateral displacement of the virtual image imposes a complex set of conditions on the reference and reconstructing beams. A simpler contour method which eliminates lateral image shift and hence permits localization of contour fringes on the subject surface is therefore advantageous. By (1) immersing the subject into a cell filled with a transparent medium (gas or liquid) of index of refraction n_1, (2) taking a first hologram exposure, (3) changing the index of the medium in the cell to n_2, and (4) taking a second exposure, a doubly exposed hologram is obtained which serves the desired purpose. Figure 15.10 schematically indicates the arrangement of immersion cell and the subject-illumination and reference beams used to form the superimposed holograms. The forming beams are kept the same for each exposure, and the cell is kept fixed in position while the medium is changed between exposures. After the hologram is processed it is illuminated with the original reference wave.

Before calculating the phase difference δ of rays coming from the virtual images generated by the hologram, let us reflect on the apparent position of a subject immersed in a medium differing from that surrounding the observer. Consider, as in Fig. 15.11, that the subject is enclosed in a cell containing a plane glass window and that in one case the medium inside the cell has an index n_1 identical to the index n_0 outside the cell while in

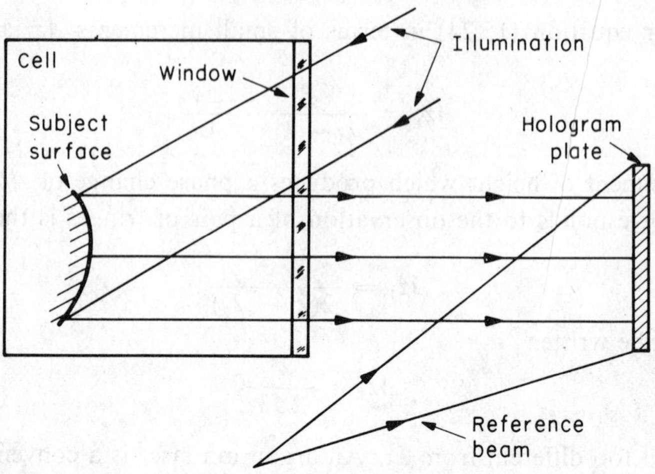

FIG. 15.10. Arrangement for forming the holograms for the immersion method of contour generation.

another case the index $n_1 \neq n_0$. To the viewer whose line of sight makes an angle with the normal to the cell window, a point A on the subject surface is displaced both along the line of sight and transverse to it as the medium is changed. However, to the viewer whose line of sight is normal to the window, the displacement is only in-line. This latter viewing condition is the desirable one and, as indicated previously, leads to fringes localized on the subject surface. As calculations will show, the plane of the window acts as a reference plane from which the subject surface height in the direction normal to the window can be gauged.

We are now ready to calculate the phase difference $\delta(x)$ for rays of light which in their transit from source to observer scatter from corresponding

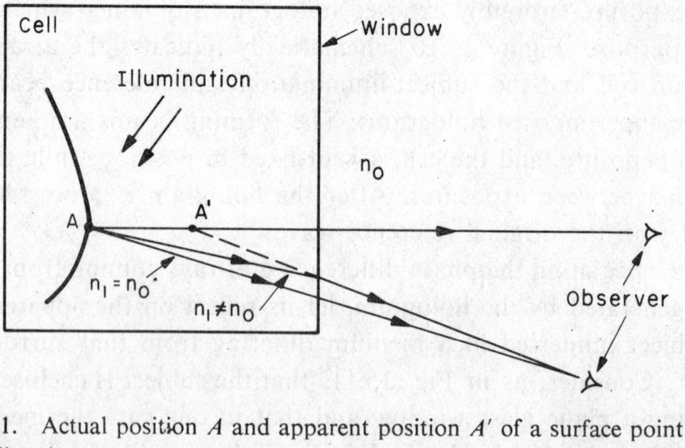

FIG. 15.11. Actual position A and apparent position A' of a surface point as viewed from two directions.

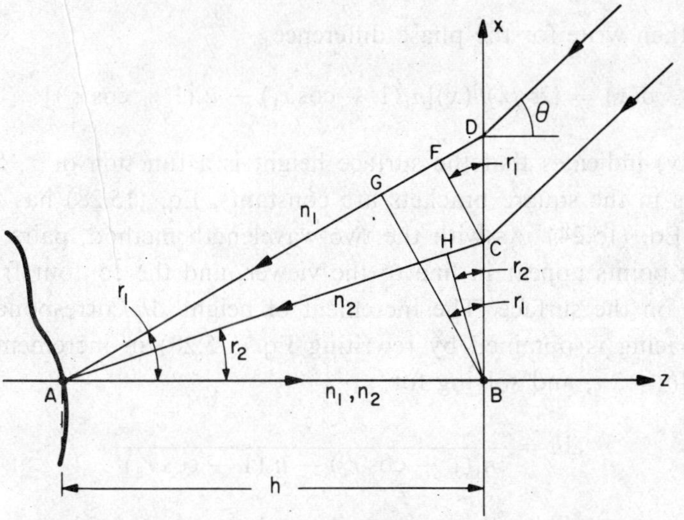

FIG. 15.12. Parameters for calculating phase shift in the immersion method.

points on the virtual images. The parameters of the calculations are indicated in Fig. 15.12 where we have assumed the medium outside the cell to have an index $n_0 = 1$. Inside the cell there are two possible states. In one state the cell is filled with a medium of index n_1 and in the other it is filled with a medium of index n_2, where $n_2 > n_1$. Parallel illuminating light rays proceed to the cell from the right. When the cell medium has the index n_1, a ray refracts at the angle r_1, proceeds to a subject point A, and scatters a ray in the direction of the normal toward B. A corresponding statement is true for the index n_2 and the angle of refraction r_2. The double-exposure hologram records light emerging from the cell at B after traversing different optical paths. Referring to Fig. 15.12 we see that the optical path difference is

$$
\begin{aligned}
l &= n_1[\overline{FGA} + \overline{AB}] - n_2[\overline{CHA} + \overline{AB}] \\
&= n_1[\overline{FGA} + h] - n_2[\overline{CHA} + h]
\end{aligned}
\qquad (15.26)
$$

Snell's law requires $n_1 \sin r_1 = n_2 \sin r_2 = \sin \theta$. This, together with a bit of trigonometry, can be used to prove[2] that $n_1\overline{FG} = n_2\overline{CH}$ and to reduce the optical path difference to

$$
\begin{aligned}
l &= n_1[\overline{GA} + h] - n_2[\overline{HA} + h] \\
&= h[n_1(1 + \cos r_1) - n_2(1 + \cos r_2)].
\end{aligned}
\qquad (15.27)
$$

[2] Note that the phase of the incident wave at B is common to both diffracted wavefronts GB and HB, leading to the same result.

We can then write for the phase difference

$$\delta(x) = (2\pi/\lambda)h(x)[n_1(1 + \cos r_1) - n_2(1 + \cos r_2)], \qquad (15.28)$$

where $h(x)$ indicates that the surface height is a function of x. Since the quantities in the square brackets are constants, Eq. (15.28) has the same form as Eq. (15.24). As with the two-wavelength method, pairs of corresponding points appear in line to the viewer, and the contour fringes are localized on the surface. The increment of height Δh corresponding to a fringe spacing is obtained by rewriting Eq. (15.28) in incremental form, setting $\Delta\delta = 2\pi$, and solving for

$$\Delta h = \frac{\lambda}{n_1(1 + \cos r_1) - n_2(1 + \cos r_2)}. \qquad (15.29)$$

With Freon gas under three atmospheres pressure, the fringes can mark off height differences Δh as small as 100 μm. With liquids, height differences less than 10 μm are measurable.

15.6 Applications and Improvements

Holographic interferometry adds to what can be accomplished with the classical interferometric techniques, e.g., two-beam and multiple beam interferometry, photoelastic displays and moiré methods. The wave-storing property of the hologram enables interference measurements to be extended to diffusely reflecting or diffusely illuminated subjects. As such, it can be put to use by those who are engaged in studying stress–strain relations, fluid mechanics [*15.26*], fracture mechanics [*15.8*], or acoustic resonances; by those who design or produce precisely ground optical glass, precision-machined metallic surfaces [*15.27*], or aerodynamic surfaces; and by those who seek new means for nondestructive testing [*15.28*] (Fig. 15.13). It is evident that the method will become part of the practice of the expert interferometrist.

While most of the basic concepts have been covered in this chapter, it is certain that useful variations will continue to develop as the method matures. Multiple-beam techniques for sharpening fringes [*15.29*, *15.30*], a two-reference beam arrangement for localizing fringes on the subject surface [*15.31*], and an extension of time-average holography using a temporally modulated reference beam [*15.32*, *15.33*] already appear in the literature. Further improvement in technique is to be expected as adaptations are developed to meet specific tasks.

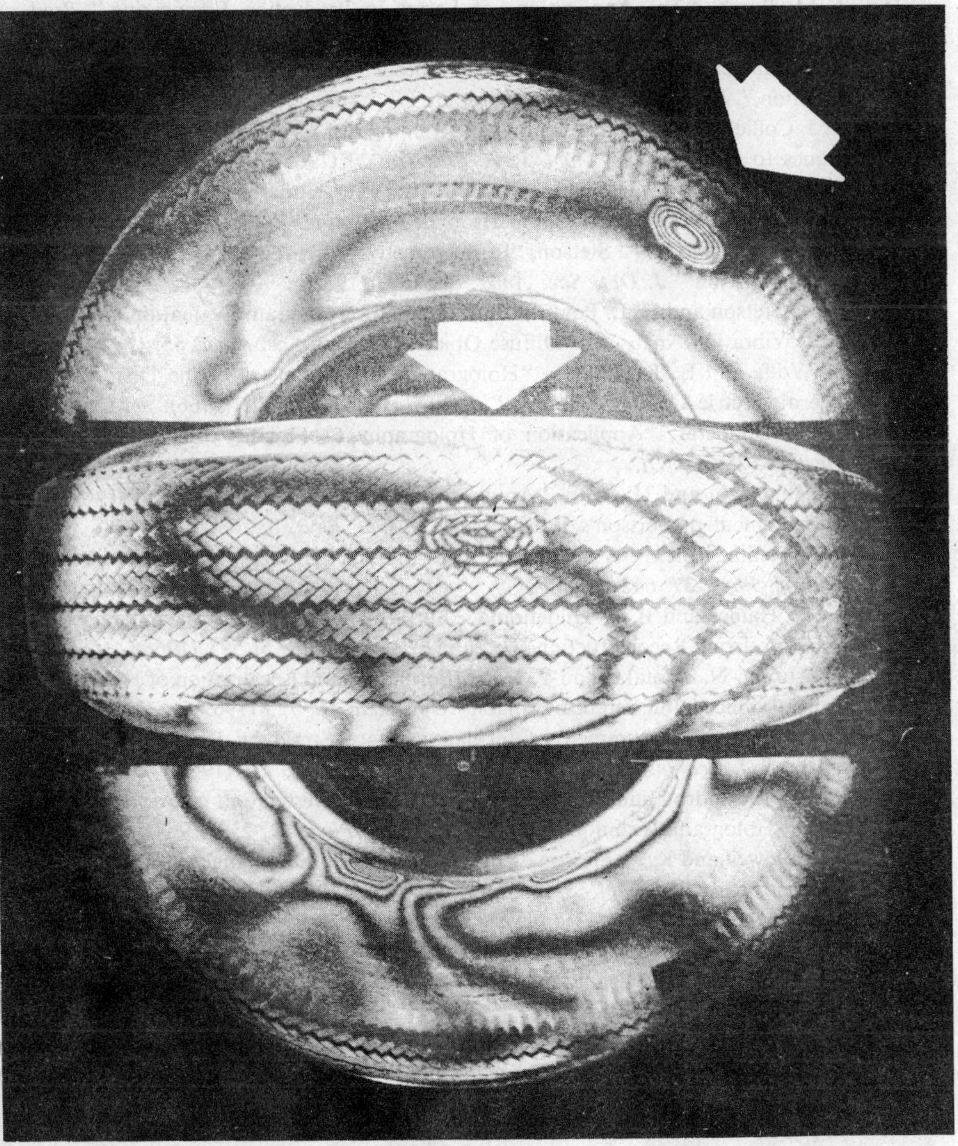

FIG. 15.13. Example of double-exposure holographic interferometry applied to nondestructive testing. The holographic images of a four-ply tire reveal separations between plies at the positions of the localized circular interference patterns. These flaws were brought out by natural creep after inflation to 50 psi. (Courtesy GCO Inc., Ann Arbor, Michigan.)

REFERENCES

15.1. J. M. Burch, "The Applications of Lasers in Production Engineering," *Prod. Eng.* (*London*) **44**, 431 (1965).

15.2. K. A. Haines and B. P. Hildebrand, "Contour Generation by Wavefront Reconstruction," *Phys. Lett.* **19**, 10 (1965).

15.3. R. J. Collier, E. T. Doherty, and K. S. Pennington, "Application of Moiré Techniques to Holography," *Appl. Phys. Lett.* **7**, 223 (1965).

15.4. R. E. Brooks, L. O. Heflinger, and R. F. Wuerker, "Interferometry with a Holographically Reconstructed Comparison Beam," *Appl. Phys. Lett.* **7**, 248 (1965).

15.5. R. L. Powell and K. A. Stetson, "Interferometric Vibration Analysis by Wavefront Reconstruction," *J. Opt. Soc. Amer.* **55**, 1593 (1965).

15.6. K. A. Stetson and R. L. Powell, "Interferometric Hologram Evaluation and Real-Time Vibration Analysis of Diffuse Objects," *J. Opt. Soc. Amer.* **55**, 1694 (1965).

15.7. R. Wolfe and E. T. Doherty, "Holographic Interferometry of the Distortion of Thermoelectric Cooling Modules," *J. Appl. Phys.* **37**, 5008 (1966).

15.8. T. D. Dudderar, "Application of Holography to Fracture Mechanics," *Exp. Mech.* **9**, 281 (1969).

15.9. I. Yamaguchi and H. Saito, "Application of Holographic Interferometry to the Measurement of Poisson's Ratio," *Jap. J. Appl. Phys.* **8**, 768 (1969).

15.10. L. O. Heflinger, R. F. Wuerker, and R. E. Brooks, "Holographic Interferometry," *J. Appl. Phys.* **37**, 642 (1966).

15.11. K. A. Haines and B. P. Hildebrand, "Surface-Deformation Measurement Using the Wavefront Reconstruction Technique," *Appl. Opt.* **5**, 595 (1966).

15.12. T. Tsuruta, N. Shiotake, and Y. Itoh, "Formation and Localization of Holographically Produced Interference Fringes," *Opt. Acta* **16**, 723 (1969).

15.13. A. E. Ennos, "Measurement of In-Plane Surface Strain by Hologram Interferometry," *J. Sci. Instrum.* (*J. Phys. E*) 1968 Series 2, **1**, 731 (1968).

15.14. E. B. Aleksandrov and A. M. Bonch-Bruevich, "Investigation of Surface Strains by the Hologram Technique," *Sov. Phys. Tech. Phys.* **12**, 258 (1967).

15.15. R. L. Powell and K. A. Stetson, "Interferometric Vibration Analysis by Wavefront Reconstruction," *J. Opt. Soc. Amer.* **55**, 1593 (1965).

15.16. M. A. Monahan and K. Bromley, "Vibration Analysis by Holographic Interferometry," *J. Acoust. Soc. Amer.* **44**, 1225 (1968).

15.17. G. M. Brown, R. M. Grant, and G. W. Stroke, "Theory of Holographic Interferometry," *J. Acoust. Soc. Amer.* **45**, 1166 (1969).

15.18. Y. Tsuzuki, Y. Hirose, and K. Iijima, "Holographic Observation of the Parametrically Excited Vibrational Mode of an X-Cut Quartz Plate," *Proc. IEEE* **56**, 1229 (1968).

15.19. C. Ågren and K. A. Stetson, "Measuring the Wood Resonances of Treble-Viol Plates by Hologram Interferometry," *J. Acoust. Soc. Amer.* **46**, 120 (1969) abstract.

15.20. E. Archbold and A. E. Ennos, "Observation of Surface Vibration Modes by Stroboscopic Hologram Interferometry," *Nature* **217**, 942 (1968).

15.21. B. M. Watrasiewicz and P. Spicer, "Vibration Analysis by Stroboscopic Holography," *Nature* **217**, 1142 (1968).

15.22. P. Shajenko and C. D. Johnson, "Stroboscopic Holographic Interferometry," *Appl. Phys. Lett.* **13**, 44 (1968)

15.23. N. Shiotake, T. Tsuruta, and Y. Itoh; J. Tsujiuchi, N. Takeya, and K. Matsuda, "Holographic Generation of Contour Map of Diffusely Reflecting Surface by Using Immersion Method," *Jap. J. Appl. Phys.* **7**, 904 (1968).

15.24. B. P. Hildebrand and K. A. Haines, "Multiple-Wavelength and Multiple-Source Holography Applied to Contour Generation," *J. Opt. Soc. Amer.* **57**, 155 (1967).

15.25. T. Tsuruta and N. Shiotake; J. Tsujiuchi and K. Matsuda, "Holographic Generation of Contour Map of Diffusely Reflecting Surface by Using Immersion Method," *Jap. J. Appl. Phys.* **6**, 661 (1967).

15.26. L. H. Tanner, "The Use of a Ruby Laser with Interferometers Suitable for Work in Fluid Mechanics," *J. Sci. Instrum.* **44**, 1015 (1967).

15.27. E. Archbold, J. M. Burch, and A. E. Ennos, "The Application of Holography to the Comparison of Cylinder Bores," *J. Sci. Instrum.* **44**, 489 (1967).

15.28. Laser Focus, p. 16, Feb. 1969.

15.29. J. M. Burch, A. E. Ennos, and R. J. Wilton, "Dual and Multiple Beam Interferometry by Wavefront Reconstruction," *Nature* **209**, 1015 (1966).

15.30. T. Tsuruta and Y. Itoh, "Holographic Two-Beam Interferometry Using Multiple-Reflected Light," *Appl. Opt.* **8**, 2033 (1969).

15.31. T. Tsuruta, N. Shiotake, and Y. Itoh, "Hologram Interferometry Using Two Reference Beams," *Jap. J. Appl. Phys.* **7**, 1092 (1968).

15.32. C. C. Aleksoff, "Time-Average Holography Extended," *Appl. Phys. Lett.* **14**, 23 (1969).

15.33. F. M. Mottier, "Time-Averaged Holography with Triangular Phase Modulation of the Reference Wave," *Appl. Phys. Lett.* **15**, 285 (1969).

Chapter 16

INFORMATION STORAGE

Holograms can store three-dimensional as well as two-dimensional information. The information may be colored or coded, graphic or alphanumeric. It may be stored on the surface of the hologram or throughout its volume, spatially separated or superimposed, as a permanent record or an erasable one. Components of the record may be unrelated or associated in pairs, may be recognizable images or seemingly unintelligible patterns. Suggested applications for hologram storage range from museum exhibit files to coded credit cards. While recognizing this diversity, we restrict our discussion here primarily to the storage of machine-readable information. The information-storage needs of the computer, in this computer age, seem to warrant the most attention. Moreover, the holographic techniques being proposed for computer memories have a general applicability. We shall be concerned with the assembly of a multitude of holograms into a file memory or storage unit peripheral to the central processor of a computer. Rapid access to the file system is an essential requirement. It is not difficult to show that holographic storage methods have decided advantages over other optical methods such as microimage records, but their real and formidable competitors are the well-entrenched magnetic memories: the core, the tape, the disk and the drum. The last three represent prime targets for the holographic optical memory. With comparable capacity, e.g., 10^8 bits, and cost, e.g., 0.01 cent/bit, the holographic memory can effect information retrieval in microseconds while the magnetic tapes, disks, and drums require tens of milliseconds. As a counter advantage, a magnetic memory is erasable and can be written as fast as it can be read. Fast-write and erasable holographic storage is still in its early experimental stages with only untried possibilities to exhibit.

16.1 Page-Organized Microimage Storage System

Most optical information-storage methods seek to exploit the capability of photographic media to record information at high density. 35-mm-microfilm storage of the pages of books and documents is a working example. By microphotographing one page on every half-frame, 1600 pages are recorded on a reel of 100 ft of film. Little is required in the way of special recording techniques; on the other hand, the density of information stored scarcely taps the potential of high-resolution photographic emulsion. An acme of microphotographic virtuosity was achieved by Emmanuel Goldberg who, in 1926, microphotographed legible copies of a page of 50 lines of print reduced to a height of 100 μm. Individual letters in the microphotograph were but 1 μm high. Storage of information at this density would allow 87,500 pages or 50 Bibles to be imaged onto one square inch [*16.1*]. The feat represents the ultimate in photographic reduction; however, the skill required to form such tiny images, the care needed to exclude dust particles from the image area, and the time spent in trying to focus and retrieve the image with a high-powered microscope combine to make so extreme a reduction prohibitive for practical, rapid-access systems. When alphanumeric information is stored at moderate reduction ratios, so that dust particles of normal size are *not* capable of completely obliterating a character, microfilm storage and microfilm projection systems are found to be well suited to the *human* reader and to the rate at which he can assimilate information.

Today much effort is bent toward storing information that is to be read not by a human but by a computer. For these applications certain properties of the microfilm systems appear as distinct disadvantages. Because machine-read information must be recorded in the form of a binary code, the microimages of the binary encoded pages become more vulnerable to dust, blemishes, or defects in the recording medium. To understand this, consider one form of coding wherein an otherwise opaque slide is modified by the presence of small transparent spots located on selected sites of a regular array. The transparent spots represent logical "ones" while the absence of a spot on an array site constitutes a logical "zero." These *bits* of information can easily be obliterated or misrepresented, e.g., by dust particles alighting on the transparent spots (logical "ones") or by the presence of pinholes in the emulsion at the sites of logical zeros. The easy corruptibility of the binary-coded microimage can be laid to the lack of redundancy in the recording (or to the high degree of localization of information in the microimage). Introduction of redundant storage is offset by increased difficulties in detecting or reading out the information stored. As we shall see in Section 16.4, holography can solve such problems. However, holography requires the use of laser light, and the most desirable high-power lasers are still in the process of being developed into reliable instruments. Thus despite their limitations, versions of a microimage system have at various times been considered for machine-readable storage [*16.2*]. A description of the basic system will serve as foundation of our discussion of the very similar holographic memory in Section 16.4.

Recent concepts in page-organized optical information storage have been directed toward the improvement of overall computer performance by providing high-capacity, rapid-readout, peripheral memories. The goal is to replace the magnetic tape, drum, or disk with a file or catalog memory capable of rapidly supplying the central memory of the computer with blocks or pages of a size it can assimilate. To realize this goal, the information retrieval times must be substantially less than the 10–100 msec characteristic of the magnetic devices. Most certainly the sequential search through a reel of microimages and the mechanical transport of film so as to position a given image in front of a magnifying lens system is not the way to do this. What is needed is a method of displaying to the computer central processor a page of information selected at random. Page selection and display should be carried out in "electronic" times, e.g., in microseconds. A class of optical devices having this capability is the *beam-addressable* or *flying-spot store*.

Figure 16.1 indicates one such beam-addressable system based on an array of microimages formed behind a multiplicity of lenslets called a fly's-eye lens. To the left of the figure is a mask consisting of a number of

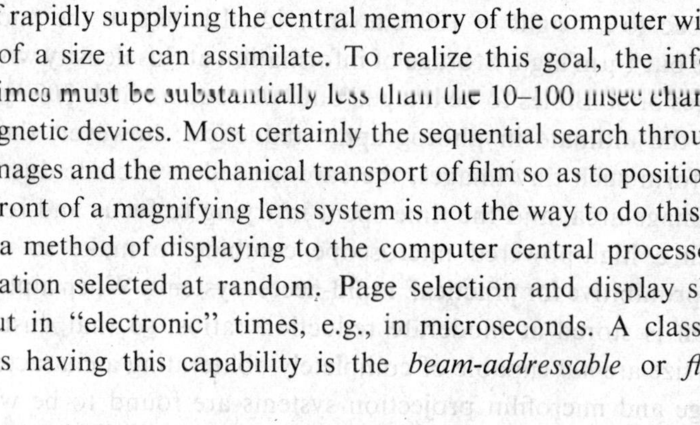

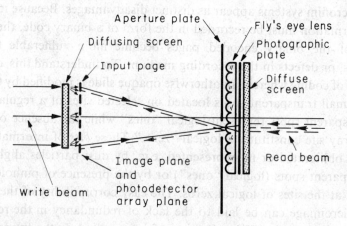

FIG. 16.1. A fly's-eye lens, microimage flying-spot store. When the memory is read, the diffusing screen at the input plane and the aperture plate at the photographic plate are removed and an array of photodetectors replaces the input page.

transparent spots on an opaque background. It represents a binary-coded page of information ready for storage on the memory or storage plane (the photographic plate just to the right of the lenslet array). We call such a mask an *input page*. The input page is illuminated with diffuse light so that information about any bit on the page is distributed by the light rays to every part of the fly's-eye lens. All but one lenslet in the array is masked by a plate with a small aperture in it placed in front of the fly's-eye lens. Light rays received by the exposed lenslet from any bit on the input page are focused to a corresponding small spot on the image plane of the lenslet. By this means a microimage of the input page is recorded on the area of the photographic plate located in the image plane behind the lenslet. When a second input page replaces the first (in the same input position) and the aperture is moved to uncover a new lenslet, a second microimage is recorded. The total number of pages which can be recorded equals the number of lenslets in the array. When the recording is complete, the plate is developed in reversal, fixed, and returned to its original recording position in the image plane of the lenslet array.

Retrieval of the information corresponding to any input page is accomplished by sending light from right to left back through the appropriate microimage and lenslet along ray directions antiparallel to those taken in forming the microimage. The lenslet focuses the transmitted light back to the plane conjugate to the microimage plane, i.e., to a plane at the original input page location. There it forms an image the same size as the original input page. Since each input page was stored from the same input position, each lenslet when interrogated with the read beam will project its microimage onto the identical area of the input plane. Consequently only a single detector area (an array of photodetectors or a vidicon tube face) is required. The readout beam can be laser light deflected to any microimage address at random by an acousto-optic or electro-optic deflector, or it can be an electron beam in a cathode ray tube [*16.2*]. The latter is electrostatically or magnetically deflected to cause a spot of fluorescent light to appear at any desired location on the cathode ray tube screen. The spot is in turn focused on to the corresponding microimage address. A readout scheme which might become more attractive as semiconductor technology improves is to illuminate the array of microimages with an array of electroluminescent diodes, one for each microimage and lenslet. Back at the input plane where each lenslet projects its microimage, one phototransistor per bit site in the original page detects the bright spots, the logical "ones."

We should emphasize here that the microimage–lenslet array system becomes less attractive as one proceeds toward higher storage density. Sensitivity to dust and blemishes or, alternatively, sensitivity to the misregistration of output image and detectors remains the underlying problem, but lenslet quality is also a major difficulty. The resolution, flatness of field, image contrast, and acceptance angle of available lenslet designs and the quality control of lenslets within an array are generally unsatisfactory for the purpose of an optical memory. We shall show that the holographic analog (Section 16.4) is superior in many respects.

16.2 A Standing-Wave Optical Memory

The theory of Chapter 9 predicts the response of volume holograms to illumination as a function of angle and wavelength of the illumination and thickness and spacing of the grating; van Heerden [*16.3*] indicated how these properties may be used to holographically superimpose blocks (pages) of information in the same volume of a thick medium and to retrieve the information, one page at a time, without crosstalk. Precursor to the volume hologram memory (and very closely related) is a so-called *standing-wave memory* concept [*16.4*] based on Gabriel Lippmann's process of color photography [*16.5*]. A brief discussion of the standing-wave memory scheme will preface our consideration of holographic storage in thick media.

Figure 16.2 illustrates how Lippmann-type exposures can be used to form an optically accessible memory. As originally conceived, a collimated source of white light (e.g., the light originating in a zirconium arc lamp and trans-

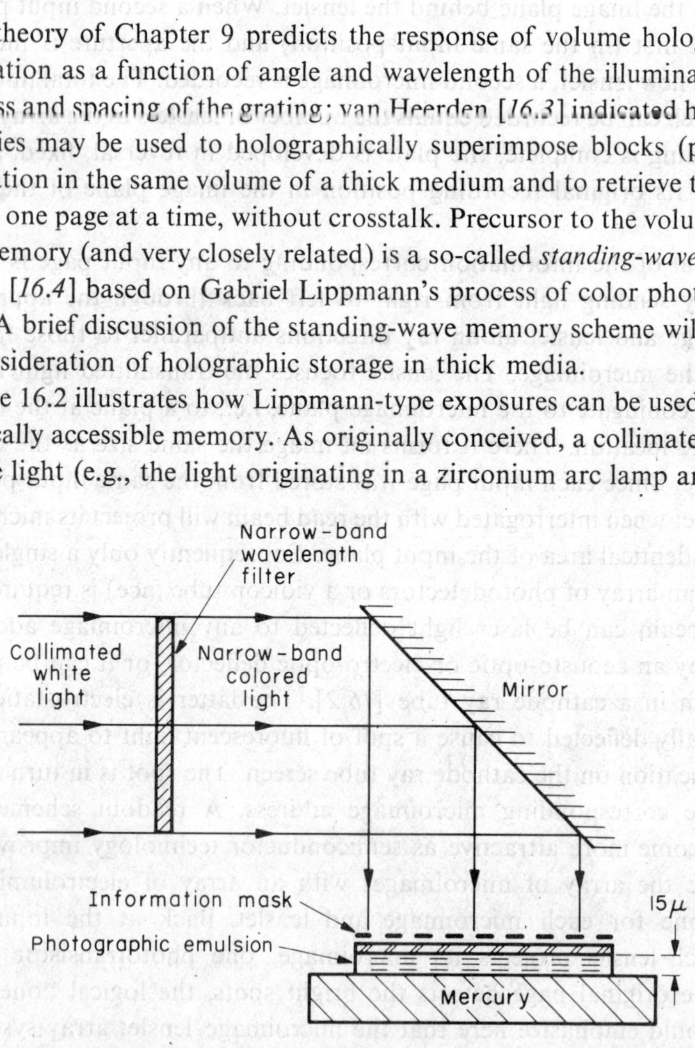

FIG. 16.2. Storing information in a standing-wave memory.

mitted through a collimating lens system) is passed through a narrowband wavelength filter to give the light a useful degree of temporal coherence. The coherence length obtained must be sufficient to allow light transmitted through a 15-μm-thick layer of photographic emulsion to interfere with the same light back-reflected from a pool of mercury in contact with the emulsion. (Laser light may also be used.) A standing-wave pattern is set up with planes of crest intensity oriented parallel to the emulsion surface. If the coherence length is at least 30 μm, then interference fringes are recorded throughout the full thickness of the emulsion. The process of storing information is organised on a *word basis* in the following way: A particular wavelength filter is associated with a particular information mask to be placed over the emulsion before exposure. The otherwise opaque mask is appropriately perforated with small holes occupying certain of the sites in a regular array. Different masks have different distributions of holes. Each hole designates a specific volume of the emulsion (behind the hole) to be reserved for the storage of all the bits of a given word. In the course of recording the total number of words and bits of the memory, a regular array of such volumes is so designated. A logical "one" and its mathematical position within a binary number or word is encoded by exposing the emulsion through the mask to a standing-wave pattern whose periodicity corresponds to the center wavelength transmitted through the wavelength filter. The periodicity $d = \lambda/2$ is given by Eq. (1.10), with $\theta = 90°$ and λ equal to the center wavelength of the light in the emulsion. A given information mask and wavelength-filter combination thus stores a logical "one" in the same binary-code position in all words designated by the mask holes. For a second exposure the wavelength filter and the mask are both changed and a new logical "one" whose binary position is coded with a new grating period is distributed among a new set of word locations. Some of the standing-wave patterns corresponding to the second information bit are stored in volumes previously unexposed (in words where the first bit was a zero) while others are recorded as fringe systems superimposed on fringes from the previous exposure. The process continues until 10 bits/word have been stored.

One mode of information readout is illustrated in Fig. 16.3. It is a method of selecting a word at random and reading out all its bits in parallel. (First note that the mercury reflector has been removed.) A multiwavelength source (a krypton laser might be postulated), emitting at the center wavelengths of the 10 wavelength filters used to encode the bits of each word, directs a beam of light to an electro-optic beam deflector. The deflector selects, illuminates, and thus interrogates any word volume at random. The multi-

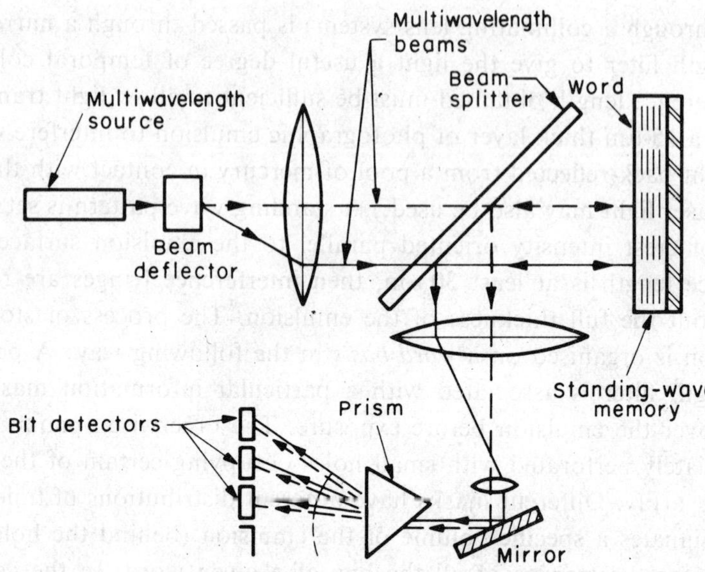

FIG. 16.3. Schematic of a method of random access, parallel-bit readout from a standing-wave memory.

wavelength light diffracts from the multiple gratings in accordance with Bragg's law, travels back from the emulsion along paths antiparallel to the interrogating beam, partially reflects from a beam splitter, and enters a wavelength-dispersing prism. Present in the light incident on the prism is a variety of wavelengths corresponding to the variety of grating periods and bits recorded in the word volume under interrogation. The wavelengths are separated by the prism and detected by individual detectors. All bits in the word are thus read out simultaneously.

A major impediment to successful implementation of the standing-wave memory concept has been the recording medium. The bandwidth of the light reflected from a given volume grating formed in Kodak 649F emulsion has been measured to be 50 Å [*16.4*], and 10 wavelengths or bits have successfully been stored in a single word location. However, the uncontrolled shrinkage of the emulsion during development and fixing causes the grating periods to diminish and causes the response of the gratings to shift toward the blue end of the spectrum. The shrinkage is not uniform; the extent of the wavelength shift, which may be as large as 1000 Å, varies with the number of exposure or bits stored. (A greater number of exposures reduces the amount of unexposed silver halide grains which are removed in fixing and so reduces the collapse or shrinkage of the gelatin.) Uniform swelling cannot properly compensate for the shrinkage and, consequently, bit detection is unreliable. Further shifts in response of the gratings as a function of age must also be contended with.

When one considers the capability of practical beam deflectors and the properties of available recording media, the capacity of the word-organized standing-wave memory cannot be too great. Ten thousand resolvable directions (and thus 10^4 words) are the most that one can expect from today's deflecting devices operating at microsecond addressing times. With only a few tens of bits per word the total capacity would amount to no more than 5×10^5 bits. And yet if optical techniques are to compete favorably with magnetic methods in the development of file and catalog memories, the capacity should be perhaps two to three orders of magnitude higher, or 10^8 bits.

16.3 Holographic Storage in Thick Media

The standing-wave memory we have described stores a bit of information as a standing-wave pattern *distributed* throughout the volume of emulsion assigned to a given word. However, the word-volume itself is *localized* with respect to the emulsion surface and confined to a location behind a small area designated by the holes in the information masks. Should we seek to increase the capacity of the memory by subdividing each original mask hole into a number of smaller diameter holes, i.e., by creating a page of several words centered about each original word location, the memory then suffers from a greater degree of localization and becomes more sensitive to the dust and defect problem and to the directional accuracy of the interrogating beam. van Heerden [*16.3*] has suggested how pages, words, and bits can all be distributed in nonlocalized fashion throughout the volume of a thick medium, resulting in a high-capacity, page-organized memory nearly insensitive to dust or defects. Rapid, random readout of a page at a time is achieved when the medium is illuminated by a properly deflected laser beam. The memory is a superposition of volume holograms, one for each page, formed by multiple exposure of the same volume of a thick photosensitive medium. During the formation of the memory, the wavelength remains the same for all holograms, but the mean angle between subject and reference beams is changed for each exposure. Since it is desirable, in readout, to image each stored page to a single detector area, the input pages are all stored from the same input-plane position (as in the page-organized micro-image memory of Section 16.1). Thus only the reference beam direction and the input page are altered between exposures. Figure 16.4 schematically indicates the method of storage. The reference beams are shown as plane waves. Thirty-five-millimeter transparencies, encoded with binary information in the manner described in Section 16.1, form convenient input pages.

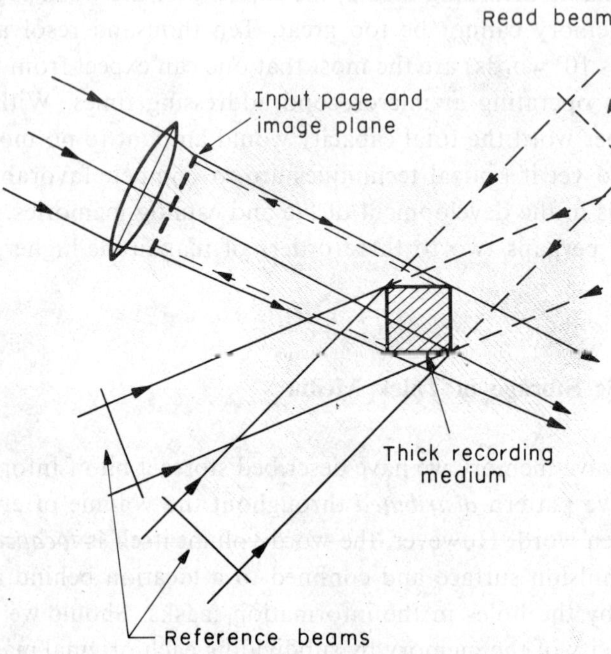

FIG. 16.4. Storage in a thick medium. In formation the reference beam direction is changed with each new input page. In read-only operation, read beams conjugate to the reference beams generate real images on an array of photodetectors replacing the input page.

Each page of stored information is encoded in a volume hologram whose spatial frequencies are characteristic of a unique reference beam as well as the light transmitted through the page. Any one of the superimposed volume holograms will respond selectively to illumination in accordance with Bragg's law, Eq. (1.12). Thus, each diffracts with significant efficiency only when illuminated by a collimated beam whose propagation direction is essentially the original reference wave direction used to form that hologram or its conjugate. (We assume the original wavelength is also used for readout; distortion of the reconstructed wave is minimized in this case.) When the volume containing all the superimposed holograms is illuminated by one such conjugate reference wave, the conjugate of the original subject wave associated with that direction is diffracted and a real image of the original page in its original size is generated at the original input page location.

The hologram itself supplies all the imaging optics necessary. As in Section 16.1, phototransistors arrayed over the image plane detect the presence of logical "ones," i.e., the bright spots in the page image.

The multiply-exposed hologram with its superimposed interference patterns represents distributed or *redundant* storage of all the information in the memory. On the other hand, the image generated from any reconstructed wave represents a selective transformation of a page of information back into the localized form desirable for detection. In this way the holographic process provides selective optical readout while freeing the optical memory from its traditional vulnerability to loss or misrepresentation of information.

16.3.1 Phase Holograms

Volume hologram memories were originally described in terms of absorption holograms formed in alkali halide crystals [*16.3*]. However, as shown in Chapter 9, volume holograms are more efficient when they are recorded as lossless *dielectric* or phase gratings. Consider an essentially lossless recording medium of thickness T whose index of refraction changes as a linear function of its exposure to light. The maximum possible change Δn of the index of refraction corresponds to some limiting exposure beyond which no further change in index will occur. Suppose a hologram is recorded in the medium with an exposure which leads to maximum hologram diffraction efficiency. We inquire as to the number of such holograms which can be stored in a medium offering the recording range Δn. To make the volume hologram analysis of Chapter 9 applicable, we consider only simple holograms formed by the interference of plane waves. As a further simplification, the fringe planes for all holograms superimposed in the medium are assumed to be normal to the surface of the medium. In practice the holograms might be stored so that their fringe planes make different angles ϕ to the surface as shown in Fig. 16.5. Whether the fringes are slanted or not is of little consequence to the calculations we shall undertake here. The results of Chapter 9 which are of particular interest are (1) Eq. (9.81) determining 100% diffraction efficiency:

$$n_1 T/(\cos\theta_0) = \lambda_a/2 \tag{16.1}$$

where n_1 is the amplitude of the sinusoidal index variation produced in the recording medium by the holographic interference pattern, T is the thickness of the medium measured along its normal, θ_0 is the Bragg angle which incident light makes with the fringe planes in the medium, and λ_a is the wavelength in air; and (2) the full angular bandwidth between nulls in

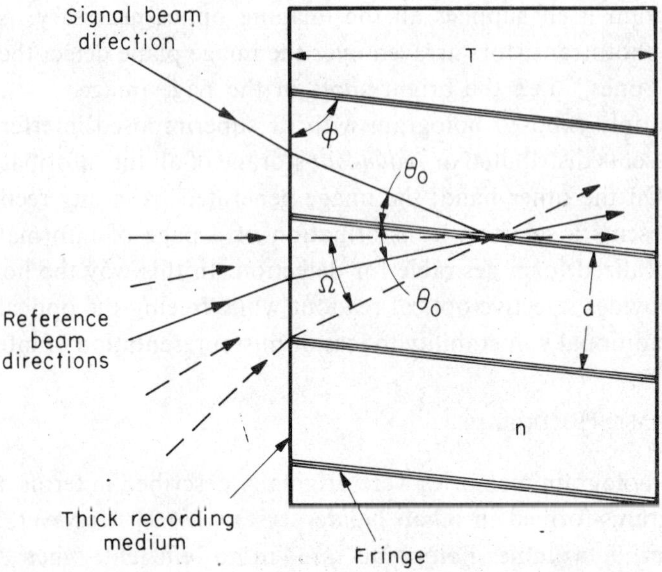

FIG. 16.5. Simple hologram formation in a thick medium.

efficiency:

$$\Phi = 2\,\delta_0 \approx \frac{2d}{T} = \frac{\lambda_a}{n_0 T \sin\theta_0} \tag{16.2}$$

where δ_0 is given in Eqs. (9.86) and (9.87), d is the grating spacing of the hologram, and n_0 is the average index of refraction of the medium. Equation (16.2) expresses the sensitivity of efficiency to deviation of the illuminating beam from the Bragg angle.

16.3.2 STORAGE LIMIT IMPOSED BY RECORDING RANGE

When the recording medium is exposed to a single elementary interference pattern of intensity

$$I = I_0 + I_1 \cos(\vec{K} \cdot \vec{r} + \Delta) \tag{16.3}$$

and when the change in refractive index is linearly proportional to the exposure, we can express the index of refraction as in Eq. (9.38)

$$n = n_0 + n_1 \cos(\vec{K} \cdot \vec{r} + \Delta) \tag{16.4}$$

where n_0 is the average index of the medium after the single exposure, n_1 is the index modulation amplitude, and $\vec{K}$ is the grating vector of magnitude

$K = 2\pi/d$ [see Eq. (9.11)]. Now suppose M exposures of the form of Eq. (16.3) are superimposed sequentially. Each has an arbitrary grating constant $\vec{K}_i$ and arbitrary phase constant Δ_i, and in each case an index change linearly proportional to exposure has been recorded. The index of the medium becomes

$$n = n_0 + \sum_{i=1}^{M} n_i \cos(\vec{K}_i \cdot \vec{r} + \Delta_i) \tag{16.5}$$

where now n_0 is the average index after M exposures. We imagine a worst possible case where, at some point in the medium, all the superimposed sinusoidal index modulations add up exactly in phase to produce a total peak-to-peak modulation equal to $2 \sum_{i=1}^{M} n_i$. Suppose that the minimum of this total peak-to-peak modulation is 0 and the maximum is Δn. Then for the particular point of the medium in question, the total recording range Δn of the material is fully used by the M exposures. This is expressed as

$$2 \sum_{i=1}^{M} n_i = \Delta n. \tag{16.6}$$

[*Note*: Equation (16.6) implies that in each of the M exposures the average intensity I_0 equaled the intensity modulation I_1.]

When each hologram is recorded under conditions satisfying Eq. (16.1), each n_i can be written as

$$n_i/(\cos \theta_{0i}) = \frac{\lambda_a}{2T}. \tag{16.7}$$

Furthermore, any specific illuminating beam must interact with only one hologram of the set, which means that the θ_{0i} must be separated by at least the angle $\Phi = 2\,\delta_0$ given in Eq. (16.2). As a final simplification let us assume that all holograms are formed with Bragg angles θ_{0i} lying in a small range around $\theta_{0i} = 45°$, so that $\cos \theta_{0i} \approx 1/\sqrt{2}$ for all i. From Eq. (16.7)

$$n_1 = n_2 = \cdots = n_M = \frac{\lambda_a}{2\sqrt{2}\,T}. \tag{16.8}$$

We can then write Eq. (16.6) as

$$Mn_i = \Delta n/2, \tag{16.9}$$

from which we determine that, under the conditions specified, the number of holograms that can be recorded in the range of index change Δn is

$$M = (\sqrt{2}\,\Delta n\, T)/\lambda_a. \tag{16.10}$$

For $\lambda_a = 5 \times 10^{-7}$ m, $T = 1$ cm, and a possible index range $\Delta n = 10^{-3}$, $M = 28$, not a very large number. However, if we form our holograms such that each individually can diffract with 50% efficiency, we find from Fig. 9.6 and Eq. (9.76) that the relation

$$n_1 T/(\cos \theta_0) = \lambda_a/4$$

replaces Eq. (16.1) and the value of M increases to 56. Abandoning the requirement on strict linearity between exposure and index change should allow M to reach 100.

16.3.3 Storage Limit Imposed by Angular Bandwidth

To calculate the effect on storage capacity of the finite angular bandwidth Φ [in Eq. (16.2)] over which a hologram will respond to incident light, let us assume the existence of a thick recording material with an unlimited range Δn. When constructing an optical memory, we desire to store each hologram with a unique angle between a fixed subject beam direction and a variable reference beam direction. Any of the reference waves thus defined, when illuminating the completed memory, will diffract only from the grating it helped to form and will reconstruct only the associated subject wave satisfying Bragg's law. In the following discussion we disregard any second-order effect of cross coupling through the reconstructed wave.

Consider an angular range Ω, from which the reference beam direction may be selected, centered about a subject-to-reference-beam angular separation $2\theta_0$ (Fig. 16.5). The angle θ_0 is that which the reference beam makes to the fringe planes in the medium and which appears in Bragg's law:

$$2d \sin \theta_0 = \lambda = \lambda_a/n_0. \tag{16.11}$$

In the above, n_0 is the index of refraction of the medium in which the memory is formed, λ is the wavelength of light in the medium, and λ_a is the wavelength in air. Note that in Fig. 16.5 the range Ω is chosen to lie in a plane. We limit Ω in this way for the following reasons: The Bragg condition [of Eq. (16.11)] for gratings formed by two plane waves, is satisfied by illuminating waves having any of the directions which form the surface of a cone whose axis is normal to the grating planes (see Fig. 16.6). In the practical case of an arbitrary subject wavefront, the hologram is a *mixture* of fringe plane systems each of which has its own cone of Bragg angles. Only the original reference-beam angle satisfies the Bragg condition for all the fringe systems. However partial subject wave reconstruction can result from illumination

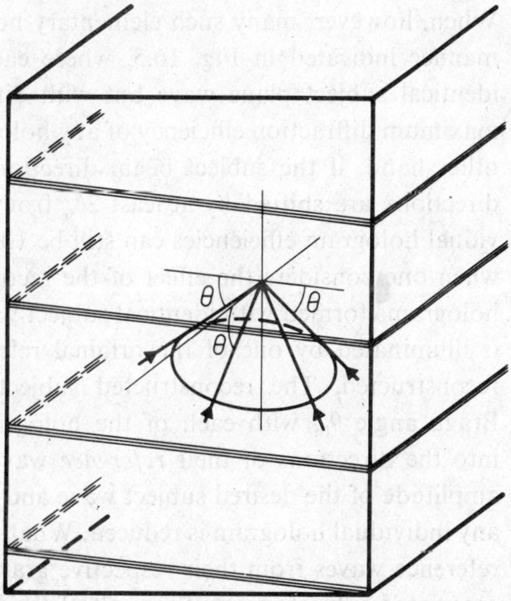

FIG. 16.6. The cone of angles obeying Bragg's law for a simple volume grating.

of the hologram in any of the directions specified by the cones. If cross talk is to be avoided, reference beams for forming additional holograms must not take any of these directions. A simple expedient to ensure this is to confine the choice of reference beam directions to a plane and suitably separate the directions in that plane. As we shall learn, this restriction still allows many more holograms to be formed than permitted by realizable recording ranges (see Sections 16.3.2 and 16.3.5).

With the reference beam direction restricted to a plane, the limit to the number of holograms N which can be stored unambiguously is obtained by dividing the range Ω by the angular bandwidth Φ of each hologram. Thus

$$N \approx \Omega/\Phi \approx (\Omega n_0 T \sin\theta_0)/\lambda_a. \tag{16.12}$$

Substitution of some modest values for the parameters in Eq. (16.12) will illustrate why thick hologram memories are extremely attractive. If we choose $\Omega = 20°$, $n_0 = 2.0$, $\lambda = 0.5$ μm, $\theta_0 = 45°$, and $T = 1$ mm, then a value of 0.02° is obtained for Φ, and $N = 1000$ holograms can be stored in the 1-mm-thick medium without first-order cross talk.

16.3.4 Storage Limit Imposed by Diffraction Efficiency

According to Eq. (16.1) phase holograms which record the interference of two plane waves in a lossless dielectric can diffract with 100% efficiency.

When, however, many such elementary holograms are superimposed in the manner indicated in Fig. 16.5, where each hologram is formed with the identical subject plane wave but with different reference waves, then the maximum diffraction efficiency of any hologram of the set is reduced. On the other hand, if the subject beam directions as well as the reference beam directions are shifted by at least $2\delta_0$ from hologram to hologram the individual hologram efficiencies can still be 100%. This difference is understood when one considers the effect of the reconstructed wave on the remaining holograms formed with identical subject waves. Suppose the storage volume is illuminated by one of the original reference waves and a subject wave reconstructed. The reconstructed subject plane wave forms the correct Bragg angle θ_{0i} with each of the holograms in the volume and diffracts into the directions of their *reference* waves. Because of this coupling, the amplitude of the desired subject wave and hence the diffraction efficiency of any individual hologram is reduced. What is more, diffraction of the coupled reference waves from their respective gratings can in turn reconstruct their associated subject waves. Since all of the latter travel in the same direction overlapping the desired subject wave, they constitute cross talk.

In practice, the effciency of even a single volume phase hologram, formed with a subject wave modulated with information and linearly recorded, cannot be 100%. This is because of the variation in subject amplitude and the corresponding variation in beam ratio R. An efficiency of 15% is representative. When many such holograms are superimposed, a further reduction in efficiency is expected due to the coupling described above. However the reconstruction of the other reference waves by one reconstructed, spatially modulated subject wave depends on the latter's correlation with the other stored subject waves (see, e.g., Chapter 14). In general, the generation of the other reference waves can be expected to be a weak effect and the consequent second-order subject-wave generation or cross talk even weaker.

From the foregoing discussion it is evident that, even in lossless dielectric media, hologram diffraction efficiencies can be expected to be low if they are formed in a manner convenient for readout detection (Fig. 16.4). We shall find that the number B of bits to a stored page of information (i.e., to a hologram) is determined by the available illuminating laser power P_0, the diffraction efficiency of a hologram η, and the detection sensitivity of the photodetectors. If diffraction efficiency is a function of the number of holograms or pages stored, then it not only limits the bits per page but also the total information capacity of the memory.

Suppose $P = \eta P_0$ is the total laser power diffracted by one hologram into the image it generates, and let us assume phototransistors detect the bits or data spots in the image. For a given laser output, the power P_d delivered to one phototransistor is determined by the ratio η/B:

$$P_d = P_0\eta/B = P/B. \tag{16.13}$$

The magnitude of detector signal is thus inversely dependent on B. Either B is sufficiently restricted or other compromises must be made. For example, signal also depends on how long P_d flows, a period fixed by the access rate of the memory. During this time a charge Q is produced by the photons arriving on the sensitive area of the detector. It is desirable that the charge induces the maximum signal voltage $V = Q/C$, where C is the detector capacitance. To promote this, the light spot corresponding to a logical "one" should be as small as possible. The matching sensitive area of the detector along with its associated capacitance C can then be minimized and the signal output maximized. Small image spots require large hologram apertures. If high density storage is desirable, some compromise between image spot size and hologram dimensions and capacity must be reached.

We show in Section 16.4.1 that a 1-mm-diameter hologram can display, on a plane 10 cm away, a 3 × 3 cm page of 10^4 bright spots each 100 μm in diameter (Fig. 16.7). This display can be usefully detected with presently available laser power providing the hologram diffraction efficiency is 10% or more. Our discussion of efficiency in this section leads us to expect that

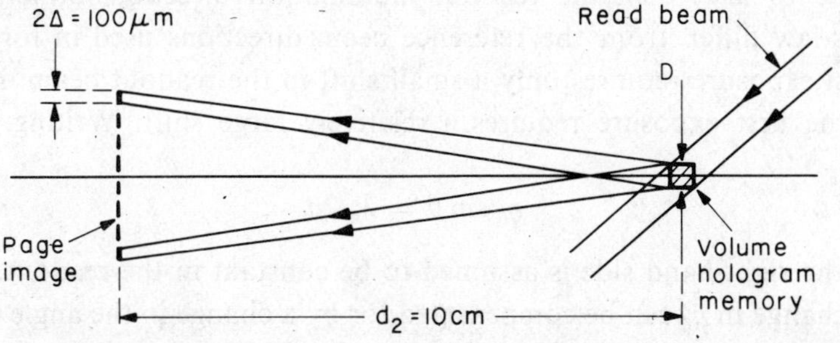

FIG. 16.7. Page-image readout from a volume hologram memory of diameter D.

perhaps only one such phase volume hologram can be recorded if the readout is to be displayed on the same fixed array of detectors. The capacity under these conditions is only 10^4 bits. Of course a smaller number of bits per page and a correspondingly greater number of pages can be allotted. Coming into play then is the limit on the number of pages imposed by the recording range (Section 16.3.2).

16.3.5 Problems Related to Real Recording Media

The storage potential of the ideal volume hologram memory cannot be realized with currently available recording media even when the spatial frequencies of the various subject waves stored in the superimposed phase holograms are completely separated. To begin with, media which have been tested do not exhibit a recording range Δn in excess of 10^{-3}. They do exhibit difficulties which we have not yet discussed. Single-crystal lithium niobate ($LiNbO_3$, see Section 10.12) which has the gross physical attributes of the ideal material will serve as an example illustrating the major problems. $LiNbO_3$ has been shown to be capable of storing phase holograms and can be made as thick as 1 cm [*16.6*]. Despite this large value of T, no exposure can make it satisfy Eq. (16.1) and yield 100% diffraction efficiency. (A very respectable value of 42% efficiency, however, has been achieved with light intensity of 1 W/cm² incident on the crystal for 100 sec. *Note*: Maximum efficiency, e.g., 100% for ideal materials or 42% for $LiNbO_3$, is achieved only when holograms are formed with plane waves and there is no signal modulation.)

Although there are problems with the persistence of holograms recorded in lithium niobate, let us for the moment hypothesize that these have been controlled to the point where many holograms can be superimposed, and let us see what further problems reveal themselves. We first note that if there is a significant change in the refractive index of the medium from first exposure to last, then the readout illumination directions which satisfy Bragg's law differ from the reference beam directions used in formation. The last exposure requires only a small shift in the readout beam direction while the first exposure requires a relatively large shift. Writing Bragg's law as

$$n_0 \sin \theta = \lambda_a / 2d$$

where the right-hand side is assumed to be constant in the readout, we see that a change in n_0 can be compensated for by a change in the angle θ which the illuminating light direction makes with the fringe planes. In so doing, the Bragg condition is maintained. If we assume that the index of refraction increases with exposure, the best procedure to follow in superimposing a number of holograms would be to expose the crystal first to the subject and reference beam pair with the smallest angular separation. The hologram so formed would also require, as the result of subsequent exposures, the greatest shift in readout beam direction (toward smaller values of θ). When the pages of the memory are stored with a minimum reference-beam angular spacing of Φ, this procedure avoids angular overlap and cross talk in the readout. However, the method works only approximately, and a penalty must be accepted. As noted in Section 10.8.4, when the subject wave has a number of spatial frequency components, compensation for changes in the Bragg parameters λ_a or n by means of a change in illumination direction is exact for only one component. The penalty is loss of resolution in the image. If the recording material forms a latent image which can be amplified by development after all exposures are made (as for example in dichromated gelatin), the problem of index change during exposure is eliminated. To date there is no phase material with this property which is also usefully sensitive to visible light and sufficiently stable (both chemically and physically) in the thicknesses required for volume hologram memories.

The holographic recording process in lithium niobate involves the photoexcitation of trapped electrons, a drifting of these electrons in an electric field, and a retrapping of the electrons. Because there is no "fixing" process following a first exposure, the holographic interference pattern incident on the crystal for the second exposure can reexcite some of the retrapped electrons and partially destroy the space charge pattern corresponding to the first hologram. Lack of fixing thus prevents the lithium niobate crystal from storing more than a few tens of holograms. To avoid the same sort of erasure on readout, it is necessary to read with a wavelength different from that used to write the hologram. As noted in Section 3.4, this introduces aberrations.

16.3.6 Absorption Holograms

Lack of persistence of the hologram, slow response to exposure, and the difficulty in obtaining good-quality single crystals are further aspects of lithium niobate which detract from its usefulness. Nevertheless it represents a new species of recording media and perhaps a landmark on the road to the ideal medium. The material which produces the highest-efficiency phase holograms, dichromated gelatin, can be developed and fixed but fails as a volume-hologram memory material due to diffulties in processing very thick gelatin. An alternate class of material for multiple hologram storage consists of media in which absorption holograms can be formed. Such holograms, to start off with, are handicapped by low diffraction efficiency. However, the availability of a variety of thick photosensitive absorption material, e.g., potassium bromide [*16.7*], organic photochromics, silver halide photochromic glass [*16.8*], and doped strontium titanate crystals [*16.9*], has led to a continuing investigation of their storage properties. As many as 100 holograms have been superimposed in a single medium [*16.8*].

Let us consider a thick recording medium whose absorption coefficient increases linearly with exposure up to some limiting value $\Delta\alpha$. We suppose that the medium prior to exposure is lossless. When it is exposed to a single elementary interference pattern it takes on a conductivity

$$\sigma = \sigma_0 + \sigma_1 \cos \vec{K} \cdot \vec{r} \tag{16.14}$$

according to Eq. (9.24). The average conductivity σ_0 and the modulation amplitude σ_1 have been defined in terms of a corresponding average absorption coefficient α and modulation amplitude α_1 by Eqs. (9.30) and (9.32), so that Eq. (16.14) may be written as

$$\alpha_e = \alpha + \alpha_1 \cos \vec{K} \cdot \vec{r}. \tag{16.15}$$

It is shown in Section 9.6.2 that the maximum diffraction efficiency of 3.7% is obtained when the average absorption constant α equals the modulation amplitude α_1 and when the parameter

$$\alpha_1 T/(\cos \theta_0) = \ln 3$$

$$\alpha_1 = \frac{\cos \theta_0}{T} \ln 3. \tag{16.16}$$

As before, θ_0 is the Bragg angle and T is the thickness of the recording medium. Figure 16.8, a duplicate of Fig. 9.7, can serve to indicate the maximum efficiency expected when α/α_1 takes the values 1 through 5. Efficiency is given by the square of the ordinate $|S|$ in Fig. 16.8. We shall use the figure to predict the effect of superposition on the maximum diffraction efficiency of one hologram out of a set of holograms superimposed in absorptive material (see also Section 17.6.3).

Now suppose that M elementary holographic exposures, each of the form of Eq. (16.3), are superimposed on the medium sequentially. Again as in Section 16.3.2 we assume each grating constant $\vec{K}_i$ and phase constant Δ_i is arbitrary, and that in each case the resulting change in the absorption constant is linearly proportional to the exposure. Consequently we may write for the absorption constant after M exposures

$$\alpha_E = \alpha + \sum_{i=1}^{M} \alpha_i \cos(\vec{K}_i \cdot \vec{r} + \Delta_i). \tag{16.17}$$

In the above α is now the total average absorption constant. Since $\vec{K}_i$ and Δ_i are arbitrary, we assume there is a point in the medium where all the super-

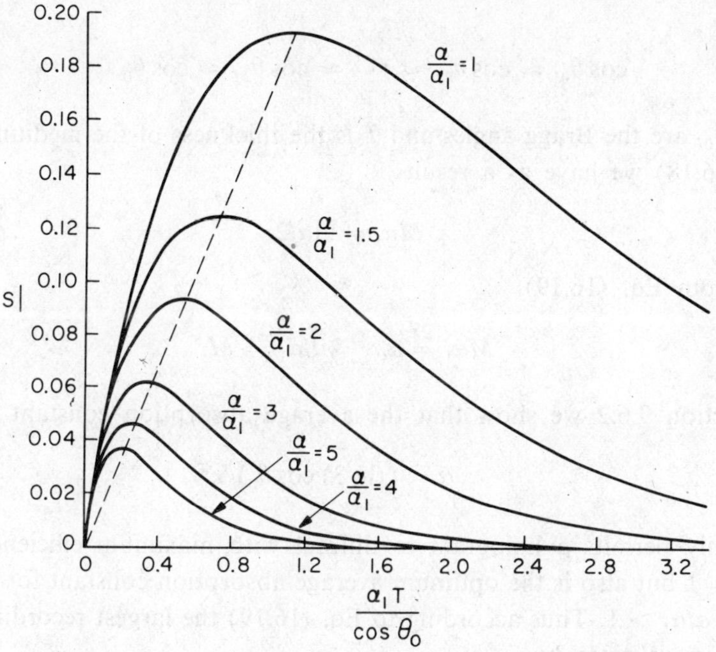

FIG. 16.8. Normalized reconstructed-wave amplitude $|S|$ for an absorption hologram plotted as a function of modulation.

imposed, sinusoidal, absorption-constant modulations add exactly in phase to produce a total peak-to-peak modulation equal to $2\sum_{i=1}^{M}\alpha_i$. As in Section 16.3.2 we assume that the minimum of the total peak-to-peak modulation is 0 and the maximum, $2\sum_{i=i}^{M}\alpha_i$, is equal to the recording range $\Delta\alpha$. Since the medium was initially lossless, this implies that the average absorption constant $\alpha = \Delta\alpha/2$, and that in each of the M exposures the average intensity I_0 equaled the modulation intensity I_1. We therefore may write

$$2\sum_{i=1}^{M}\alpha_i = \Delta\alpha \tag{16.18}$$

and

$$\alpha = \Delta\alpha/2. \tag{16.19}$$

To aid our calculation let us consider the variable $\alpha_i T/(\cos\theta_{0i})$ plotted along the abscissa of Fig. 16.8 (where the subscript i refers to any of the superimposed set of holograms) to be approximately the same for all M holograms. This requirement is satisfied if

$$\alpha_1 = \alpha_2 = \cdots = \alpha_M$$

and

$$\cos\theta_{01} \approx \cos\theta_{02} \approx \cdots \approx \cos\theta_M \approx \cos\theta_0 .$$

Here θ_{0i} are the Bragg angles and T is the thickness of the medium. From Eq. (16.18) we have as a result

$$M\alpha_1 = \Delta\alpha/2 \tag{16.20}$$

and from Eq. (16.19)

$$M\alpha_1 = \alpha, \qquad \alpha/\alpha_1 = M. \tag{16.21}$$

In Section 9.6.2 we show that the average absorption constant

$$\alpha = \big((\ln 3)\cos\theta_0\big)/T$$

not only permits a hologram to diffract with maximum efficiency when $\alpha/\alpha_1 = 1$ but also is the optimum average absorption constant for all cases where $\alpha/\alpha_1 > 1$. Thus according to Eq. (16.19) the largest recording range necessary is given by

$$\Delta\alpha = 2\alpha = \big(2(\ln 3)\cos\theta_0\big)/T. \tag{16.22}$$

For this case the maximum efficiency of a single hologram, when formed as one of set a of 1, 2, 3, 4, or 5 superimposed, equally efficient holograms, is predicated by Fig. 16.8. Maximum efficiency is given by squaring the value of $|S|$ corresponding to points where the dotted line intersects the curves labeled $\alpha/\alpha_1 = 1$, 2, 3, 4, and 5. We see that efficiency drops from 3.7% for a single hologram to approximately 0.1% for 5 holograms. The example serves to illustrate that storage of several hundred holograms in this manner, each with extremely low diffraction efficiency, would pose a very great problem for readout detection. Most probably the low amplitude of the diffracted signal would be submerged in detector noise or noise scattered by the hologram medium. (See also Section 17.6.3, where it is shown that superposition of M holograms reduces the efficiency of an individual hologram by $1/M^2$.)

A practical problem in working with thick absorptive materials is the difficulty in uniformly exposing the medium throughout its thickness. If an initially clear material is darkened by exposure and there is no development after exposure, then the front portion of the medium becomes overexposed before the back receives its optimum density. This is the case for some photochromics. Holograms so formed do not diffract with the same efficiency as those uniformly distributed through the thickness T. The effective thickness appears to the illumination to be less than T, and the angular sensitivity of the grating is correspondingly broadened. As more holograms are stored, the optical density which builds at the front of the medium continues to reduce the relative exposure of the back. Furthermore the diffraction properties of previously recorded holograms noticeably deteriorate. The attenuation of the illuminating beam may become so severe that only a fraction of the original depth of the hologram is usefully interrogated. Noise increases, signal decreases, and the angular sensitivity broadens to the point where images overlap. If the photochromic is initially colored with uv radiation and hologram formation is accomplished by bleaching with visible light, then the first exposure takes place under high attenuation conditions. Bleaching is nonuniform and strongest near the front of the medium. Additional storage eventually overexposes that front portion containing the effective part of the earlier holograms and destroys their diffraction properties. The onset of loss of previously stored information limits the number of absorption holograms that can be superimposed. As noted, approximately 100 have been stored in photochromic glass [*16.8*]. Lack of persistence of the record and the requirement of one wavelength for writing and another for reading pose practical problems. (For nondestructive readout the medium must be insensitive to the read wavelength.)

Photographic emulsions respond to incident light by increasing their optical density as do some photochromics, but the main increase in density comes after the exposure period during development. The problem of nonuniform exposure can thus be avoided. However, the additional problems brought in by wet-processing of the emulsion more than offset this advantage. As a result of the fixing process, the emulsion can shrink (see Section 10.8.4). The degree of shrinkage may not be uniform throughout the depth of the emulsion, and, as a consequence, the originally recorded fringe spacings may be distorted. This reduces angular discrimination and the number of holograms that can be superimposed without cross talk.

Equation (9.88) $\Delta\lambda_0/\lambda_a = (d \cot \theta_0)/T$ expresses the sensitivity of phase (and absorption) gratings to the wavelength of illuminating light. It implies that many holograms can be superimposed in a thick medium and read out unambiguously providing they are each stored with a unique wavelength. (This of course is the principle underlying multicolor volume holography discussed in Chapter 17.) Rapid, random retrieval would require a source whose output wavelength could be tuned electronically. Although there has been some progress in this direction, it would appear that beam deflection making use of the angular sensitivity of volume holograms is a more practical interrogating method. The discussions in Section 16.3.5 and 16.3.6 make it clear that the full potential of volume hologram optical memories will not be realized until a better recording material is developed. The material must be sufficiently sensitive at the wavelength of the light used to form the hologram and must be physically stable after the recording is complete. It is desirable to be able to record phase holograms as latent images which are later developed and fixed.

16.4 A Holographic Flying-Spot Store

Against the preceding background of unfulfilled potential, we now describe a holographic storage system which successfully meets the requirements of high capacity and high speed. The system is a read-only, page-organized flying-spot store [*16.10*, *16.11*]. Any page at random can be retrieved and presented to the central processor of a computer in microseconds. In contrast to the volume hologram memories of Section 16.3, the basic elements of this memory, which are indicated in Fig. 16.9, are

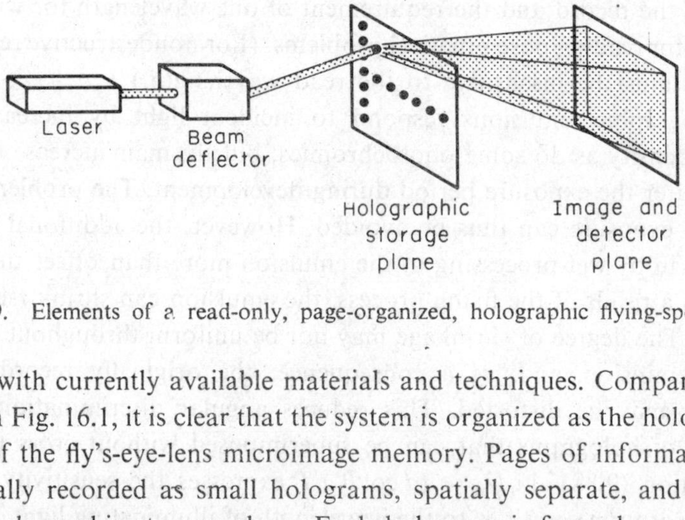

FIG. 16.9. Elements of a read-only, page-organized, holographic flying-spot store.

realized with currently available materials and techniques. Comparing Fig. 16.9 with Fig. 16.1, it is clear that the system is organized as the holographic analog of the fly's-eye-lens microimage memory. Pages of information are individually recorded as small holograms, spatially separate, and arrayed over a holographic storage plane. Each hologram is formed on a unique area of the storage plane with a unique input page as subject but with a commonly directed, collimated reference beam.

16.4.1 CAPACITY AND SPEED

A machine-readable page must be binary coded. We have already indicated (in Section 16.3.4) that the amount of energy received by th photodetectors is a factor determining the parameters of a memory system. To illustrate, consider that the laser in Fig. 16.9 emits a beam whose power is 1 W. Transmission of the beam through two deflectors (for deflection in two orthogonal directions) and a lens system and diffraction at the hologram reduces this power by 20 dB. Suppose we want as many as 10^4 bits per page. Division of the beam into 10^4 images leaves but 1 μW incident per photodetector. A very rapid readout memory might display one page to the detector array every microsecond. One microwatt acting for this interval supplies 3×10^6 photons to a photodetector. Phototransistors able to detect this small energy have sensitive areas of 100 μm diameter.

Accordingly, let us now calculate the diameter of a hologram which can generate a spot 100 μm in diameter on an image plane 10 cm from the hologram. A general expression for the diffraction-limited spot diameter Δ which can be formed with an imaging aperture of diameter D at a distance d_2 is found in Section 6.4.3 and may be written as

$$\Delta = 0.61\, d_2\lambda/r = 1.22\, d_2\lambda/D. \tag{16.23}$$

Here the spot is a spread function of the form given in Eq. (6.74); Δ is the distance from the central maximum of the function to its first null on either side and is generally taken as the nominal value for the spot diameter. A more conservative choice for the spot diameter is 2Δ corresponding to the separation of the nulls of the spread function.

If we set $d_2 = 10$ cm, $\lambda = 0.5$ μm, and $2\Delta = 100$ μm, we obtain $D = 1.22 \times 10^{-1}$ cm. We conclude that a hologram 1.2 mm in diameter illuminated with presently available CW laser power can generate a detectable image of a page containing 10^4 bits.

Bit-to-bit spacing determines the size of a page. The spacing should be large enough to prohibit light intended for one photodetector from overlapping onto the sensitive area of another and large enough to minimize leakage currents between detectors. On the other hand, economic use of the silicon chip on which the phototransistors are formed as integrated circuits dictates that the spacing be made no larger than necessary. If the 100 μm spots are placed on 300-μm-spaced centers, a page 3 cm on a side results. We now proceed to compute the number of pages which can be stored in a fashion permitting rapid retrieval.

The number of pages and therefore the capacity of the memory is determined by the properties of the beam deflector used to interrogate the holograms. Acoustical beam deflectors are well qualified for this interrogation function. Two deflectors, in tandem when combined with a lens system, can displace a laser beam to any of a large number of addresses on an *xy* plane. Figure 16.10A illustrates two orthogonal acoustic deflectors in tandem permitting *x* and *y* deflection. Water is the acousto-optic material. The array of 32×32 resolvable beam positions which they produce is shown in Fig. 16.10B. We note in Appendix III that an acoustic wave traveling through an acousto-optic medium creates a diffraction grating which efficiently diffracts light when illuminated at the Bragg angle [*16.12*]. The grating spacing is the acoustic wavelength. By altering the frequency f of the traveling

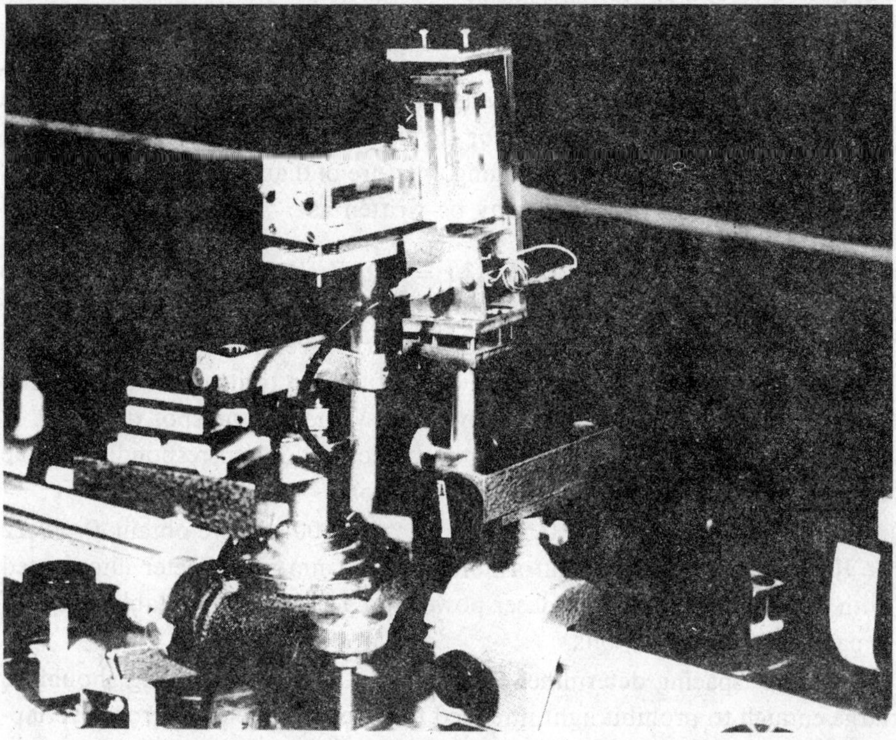

FIG. 16.10A. Acoustic (water-cell) beam deflectors. (Courtesy D. A. Pinnow and S. R. Williamson, Bell Laboratories.)

acoustic wave over some range Δf, the grating spacing and related diffraction angle can be varied. Since the incident beam is fixed in direction, one might expect Bragg diffraction for only one value of f and one grating spacing. As explained in Appendix III, the radiation pattern of the acoustic transducer in effect produces a spread in the *orientation* of the acoustic gratings, thereby allowing Bragg interaction over a range Δf. The effective angular deflection range is considered to be limited to those angles for which the deflected beam intensity exceeds 50% of the beam intensity incident on the deflector. We find in practice that the angular range $\Delta\varphi$ through which the beam is deflected is very small, e.g., 8 mrad. However, with cells of moderate extent it is still possible to distinguish approximately 100 nonoverlapping directions in one plane. When the acoustic cell is a single crystal of a material such as iodic acid or lead molybdate the beam can be directed into any one of these in about 1 μsec (see Appendix III, [*III.1*]).

FIG. 16.10B. Array of 32 × 32 resolvable beam positions produced by deflectors of Fig. 16.10A. (Courtesy Anderson [*16.11*].)

A useful measure of the potential of a *single* acoustic deflector is the capacity–speed product

$$N \cdot (1/\tau) = \Delta f \tag{16.24}$$

derived in Appendix III. Here N is the number of resolvable directions into which the beam can be directed within an access time τ. We see that speed $1/\tau$ can be traded for number of addresses. A frequency bandwidth $\Delta f = 80$ MHz has been employed to obtain random access to any of 64×64 sites in 1.6 μsec. (A safety factor of 2 has been provided. Two deflectors work in tandem to produce x and y deflections.) Controlling the bandwidth is the frequency response of the electroacoustic transducer (lithium niobate) which launches the acoustic wave and that of the acoustooptic material (e.g., alpha-iodic acid [*16.13*] wherein the wave propagates. These allow values of Δf as high as 300 MHz (centered at 350 MHz). A memory plane containing a 100×100 array of holograms any one of which can be interrogated in 1 μsec is therefore entirely reasonable. At 10^4 bits/page the total memory capacity can be 10^8 bits.

16.4.2 Performance

It is desirable to be able to detect the image of any page stored in the memory without recourse to an auxiliary lens. For a large array of holograms, e.g., 100×100, the storage plane might be 15 cm square; an imaging lens converting the virtual images generated by the holograms into real ones would be large and expensive. Instead, real images are obtained directly by illuminating the holograms with a beam conjugate to the original reference. In any case, a beam deflector and collimating lens form beams similar to the original reference in direction and wavefront. A lens system then magnifies their displacements so that they intercept the storage plane at the hologram sites. When the collimated, deflected beams are directed antiparallel to the original reference beam with respect to the hologram plate and thus strike the *backsides* of the holograms, real images form on the detector array. Bright spots in a page image (logical "ones") match the locations of the phototransistors, and the latter convert the optical page of information into an electrical page. Electrical signals resulting from the conversion are stored in a semiconductor buffer memory. The buffer can be scanned for words or commands at electronic speeds, and when the search is complete, a new page of 10^4 bits can be placed in the buffer within microseconds.

Figure 16.11A is a photograph of a 32×32 array of holograms each of diameter 1.2 mm. When illuminated, each hologram produced a test page of the form photographed in Fig. 16.11B. (Note that this was a *test* page containing close to 64×64 logical "ones" arranged in blocks of 64 bits. A "zero" was included in each block so that the signal-to-noise ratio could be checked.) The page format was chosen so that each block of 64 bits fell on a block of 64 phototransistors formed on a single integrated circuit chip as shown in Fig. 16.11C. A total of 64 such chips form the detector array. When the holograms in Fig. 16.11A were formed either on dichromated gelatin or on a photographic emulsion which was bleached after exposure, they diffracted 10–15% of the incident interrogating beam power into the page image.

FIG. 16.11A. 32 × 32 array of holograms. Each hologram is 1.2 mm in diameter. (Courtesy Anderson [*16.11*].)

While an array of spatially separated holograms does not represent as elegant an organization as the superposition of many volume holograms, it has distinct practical advantages over all of the optical memory concepts described thus far. Among the foremost is the evidence that the performance goals set for machine-readable optical memories, 10^8 bits randomly accessible in electronic times (~1 μsec), can be achieved with currently available recording materials. Angular tolerances on the direction of the interrogating beam, moreover, are manageable, and errors are amenable to correction by feedback mechanisms. The storage density of the holographic flying-spot store is far less than that potentially achievable in volume hologram memories but is certainly sufficient for the 10^8-bit memory. For example, a 100×100 array of 1.2-mm-diameter holograms need occupy no more than a square of recording medium 15 cm on a side. A reasonable estimate of the

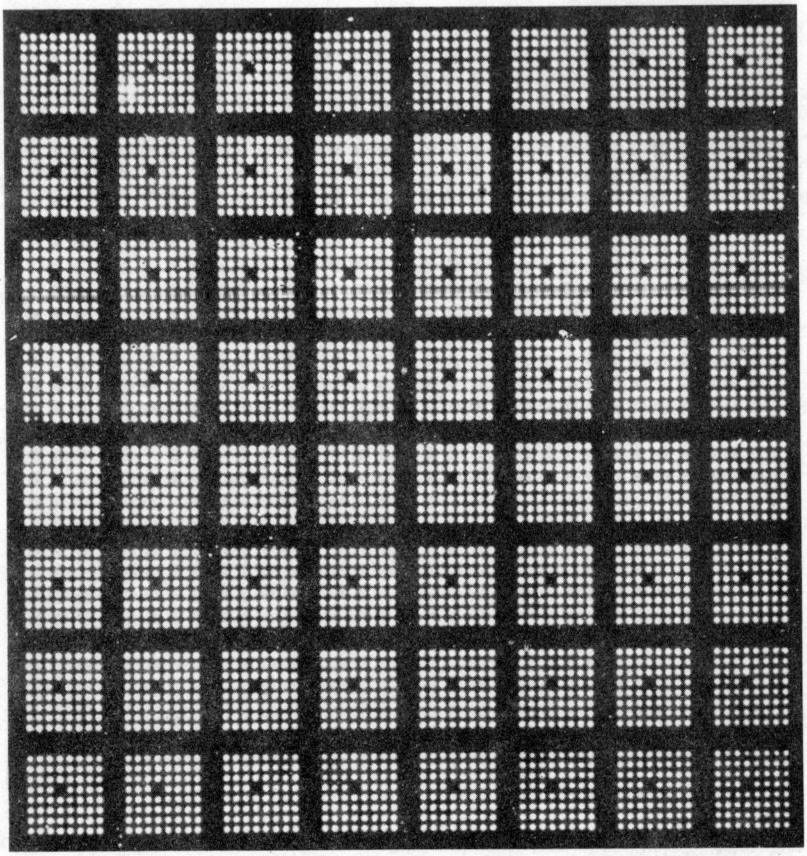

FIG. 16.11B. Test page imaged from a single hologram in the array. There are 4032 logical "ones" and 64 logical "zeros." Each spot is 100 μm, in diameter. (Courtesy Anderson [*16.11*].)

cost is a few hundredths of a cent per bit. Along with its present practical superiority over other holographic memories, the holographic flying-spot store maintains the following advantages over its microimage analog:

1. The imaging optics required to project the page information stored in *each* hologram onto a *common* detector array is built into the hologram record.

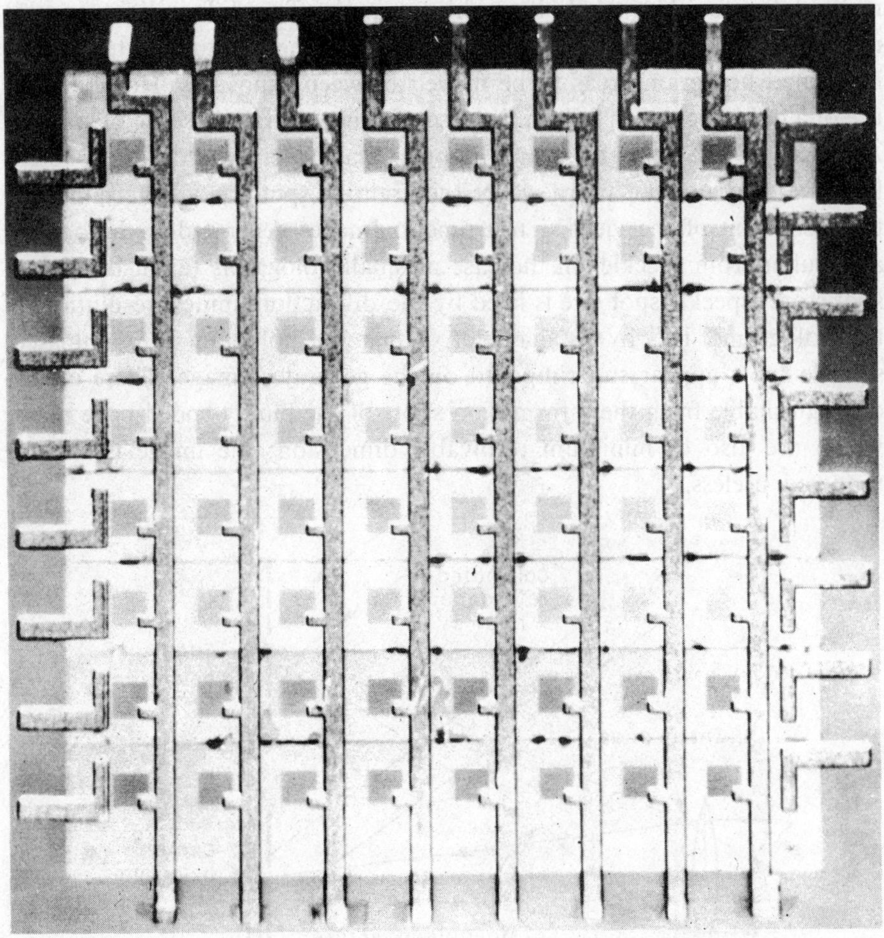

FIG. 16.11C. One block of 64 phototransistor detectors. The light-sensitive area of an individual detector is 100 × 100 μm. The block matches a similar block of 64 data-point sites on the test page. (Courtesy Anderson [*16.11*].)

2. Image resolution can be close to the diffraction limit imposed by the diameter of the hologram.
3. Providing the hologram recording is properly made, the record is redundant and relatively insensitive to dust and blemishes in the recording medium.

16.4.3 FORMING THE HOLOGRAMS

In several respects employment of diffuse light to illuminate the input page, as in the fly's-eye-lens microimage memory of Fig. 16.1, is a desirable method for forming the holographic memory. The diffuse light ensures redundancy in the recording and requires only the aperture plate defining individual hologram areas to be moved between exposures. However, the use of a diffusing screen fails in the case of coherent light because of speckle. As discussed in Section 8.2.2 and Chapter 12, an observer of diffuse coherent light sees a granular pattern where the grain or spot size is determined by the resolution of the viewing instrument. Images generated by holograms also suffer from speckle. In the case of small holograms (as in the flying-spot store) speckle spot size is fixed by the diffraction-limited resolution of the holograms, i.e., by its diameter. When the hologram is illuminated, speckle spots appear superimposed on the generated image. These are indistinguishable from the information spots of the binary-coded page image which are also of minimum resolvable dimension; the image is thereby rendered useless.

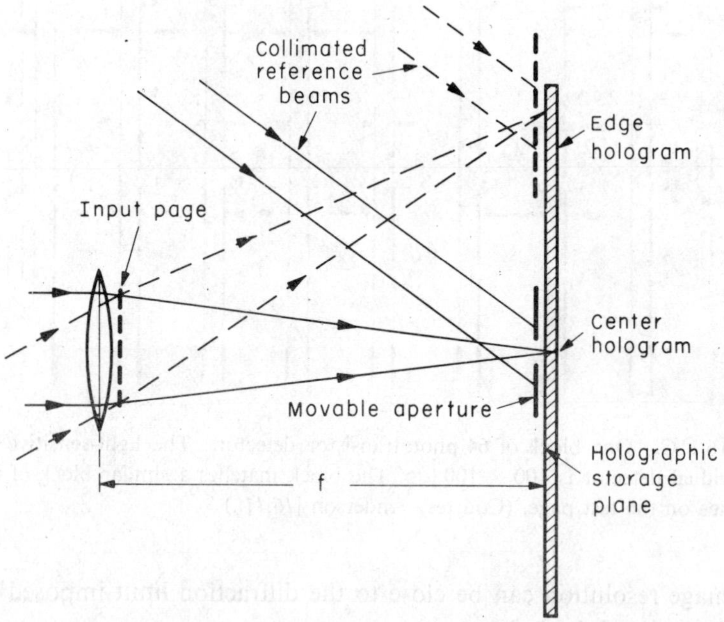

FIG. 16.12. A method of forming Fourier transform holograms for the holographic flying-spot store. The storage plane is approximately in the focal plane of the lens.

A practical method of forming holograms for the storage plane is shown in Fig. 16.12. The basic principle is that enunciated in Section 8.4.2. It is noted there that the Fourier transform arrangement as in Fig. 16.12 makes best use of a limited hologram area. The greatest range of subject spatial frequencies can be recorded on the finite hologram, and consequently the best resolution over the entire image area can be obtained in the reconstruction. Since each point of the Fourier transform plotted on the plane of the hologram represents the total light amplitude of a particular spatial frequency coming from *all* parts of the subject, the Fourier transform hologram record is redundant. One obvious benefit of the Fourier transform method of formation, and one not to be ignored when forming 10^4 holograms, is that all the available laser light is concentrated into the recording area, thus reducing the exposure time to a minimum. On the other hand a major drawback to the Fourier transform hologram is well illustrated in Fig. 16.13. Shown is a magnified photograph of that portion of the Fourier

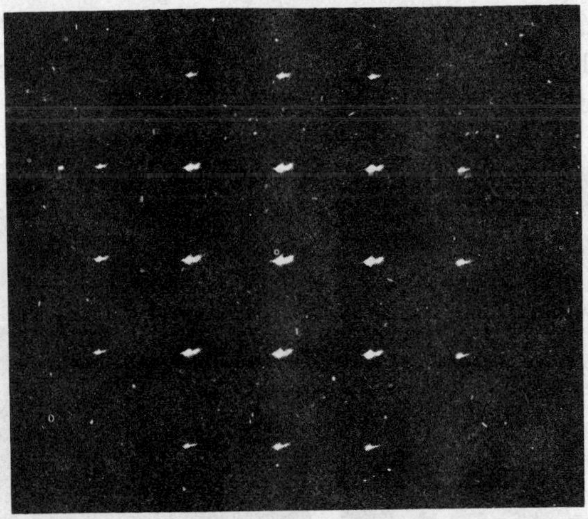

FIG. 16.13. A magnified portion of the Fourier transform of a 10^4-bit input page. The diameter of the area shown is 1.4 mm.

transform of an input page that fits into an individual 1.4-mm hologram area on the Fourier transform plane. [The page itself contains nearly 10^4 spots on a regular array, as shown in Fig. 16.14b.] The Fourier transform of a regular two-dimensional array of transparent spots in an input plane is a similar array in the Fourier transform plane [*16.14*]. From Fig. 16.13 it is evident that much of the hologram area records no holographic interference pattern, and holograms formed with this subject light must be of low diffraction efficiency. The bright areas of the picture in Fig. 16.13 record the basic periodicity of the image-page array. Not recorded by the high-contrast photograph is the vital information on missing spots (zeros) which should appear as weak-intensity light in-between the bright areas. These components are not well recorded by the holographic process either. When the reference beam intensity is set high enough to linearly record the bright areas, it is found to be much too high to adequately record the weak-intensity components. They are reconstructed from the resulting hologram with too little efficiency to yield the necessary signal-to-noise ratio.

An effective compromise between the desirable redundancy and the undesirably large range of subject intensity associated with recording in the exact Fourier transform plane is to place the hologram slightly closer to the lens than the lens focal distance. Figure 16.14a indicates the degree of uniformity in this *near* Fourier transform pattern photographed in a plane displaced from the focal plane by 1/20 of the lens focal length. The diameter of the pattern shown is 1.25 mm. When stored with a collimated reference beam and subsequently illuminated, the hologram generates the image of an input test page 2.5 cm on a side with about 10^4 logical "ones." Figure 16.14b is a photograph of the image. Each bright spot or logical "one" is 100 μm in diameter and the spacing between spots is 250 μm.

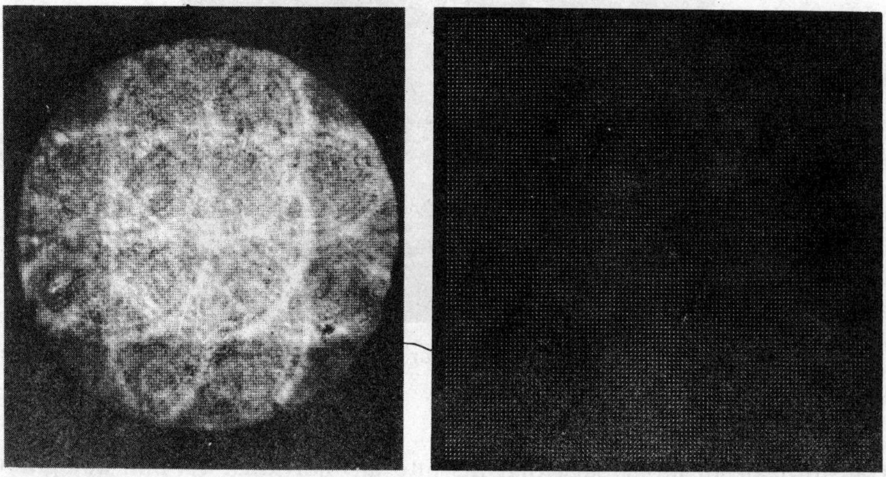

FIG. 16.14. (a) "Near Fourier transform" intensity pattern of an input page. The pattern is 1.25 mm in diameter. (b) Image of a page generated by a hologram formed with pattern of Fig. 16.14a. There are nearly 10^4 bright spots, each 100 μm in diameter.

An alternate method of achieving uniform subject intensity at the hologram while preserving the redundancy attached to recording in the Fourier transform plane is to place a random phase mask in contact with the input page. The effect is to randomize the phase of the light beams passing through the holes of the input page and to randomize their interference at the hologram plane. A suitable phase mask may be prepared by etching a glass plate, which is to be placed adjacent to the input page, such that the glass thickness behind every bit site varies in a pseudorandom manner [*16.15*]. That is, the beamlets transmitted through the page as logical "ones" are given a random distribution of 0 and π phase shifts relative to the phase of light transmitted through the unetched glass. The efficacy of the method is exhibited by the intensity patterns photographed in the exact Fourier transform plane (Fig. 16.15a) and in a plane only 250 μm away from the exact Fourier transform plane (Fig. 16.15b). [The remaining fluctuations in Fig. 16.15a are due to phase errors in the phase mask. The ideal intensity pattern can be shown to be proportional to the intensity of the Fourier transform $[J_1(x)/x]^2$ of a single hole in the input page mask. An aperture of 1.4 mm diameter in the transform plane restricted the pattern to only a central portion of its main lobe.] Results obtained with the two methods of hologram formation are comparable. However, recording in the exact Fourier transform plane with the phase mask may have some practical advantages.

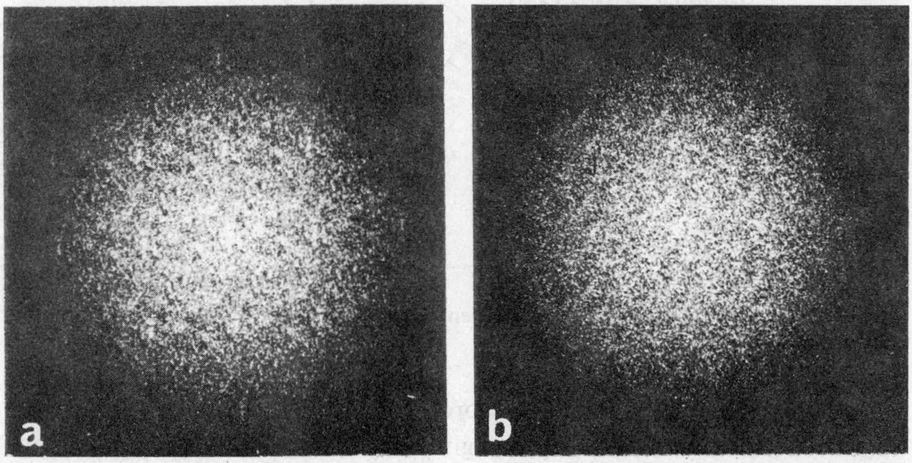

FIG. 16.15. (a) Intensity pattern of the input-page-plus-phase-mask at the holographic storage plane. The latter is in the exact Fourier transform plane. (b) The intensity pattern in a plane only 250 μm from the exact Fourier transform plane.

An arrangement for efficiently forming the hologram array is indicated in Fig. 16.16. Fourier transformation of the input page is carried out by a large lens. After each exposure, the lens and an aperture plate covering the recording medium are shifted in tandem, moving subject beam and hologram aperture in discrete steps. At each step a new input page is inserted in the input plane and a new area of the recording plate is exposed. An experimental model of the arrangement writes an array of 32 × 32 holograms containing 4×10^6 bits in 1.5 h. The same time is required for holograms recorded either on photographic emulsion followed by a bleach process or on dichromated gelatin and includes the post-exposure processing.

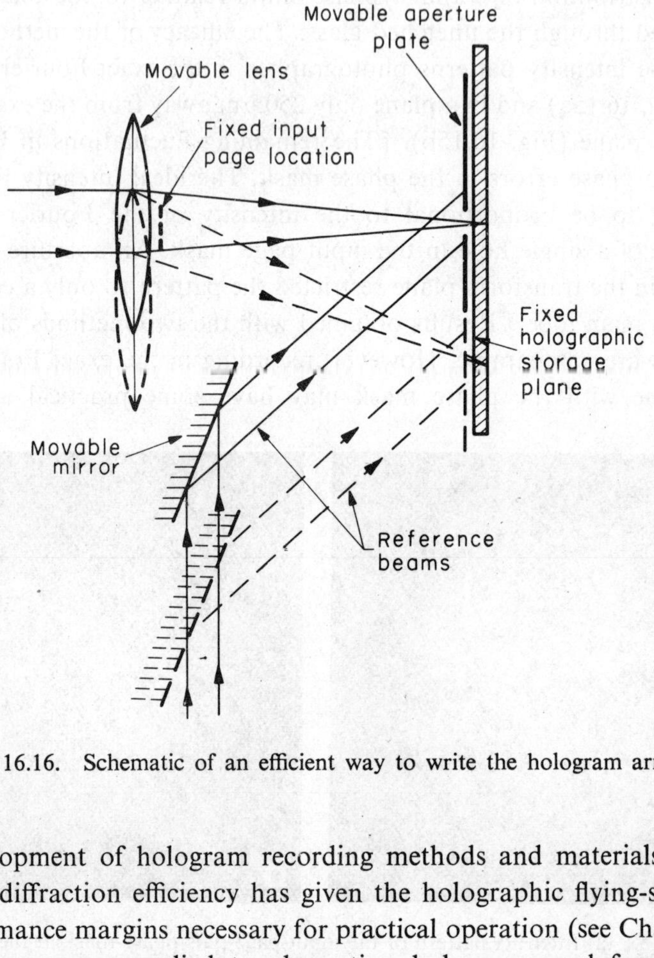

FIG. 16.16. Schematic of an efficient way to write the hologram array.

Development of hologram recording methods and materials yielding improved diffraction efficiency has given the holographic flying-spot store the performance margins necessary for practical operation (see Chapter 10). Bleaching processes applied to absorption holograms and formation of phase holograms in dichromated gelatin have increased the operating hologram diffraction efficiency by one order of magnitude over absorption recordings. Diffraction efficiencies for holograms generating images of binary coded input pages now lie in the range between 10 and 15%. Along with these values, one obtains for the ratio of the signal power at the site of a logical "one" to noise power at the site of a logical zero values exceeding 30 : 1 and for bit-to-bit fluctuations (due to all sources) the values are less than 2 : 1. The numbers imply that the page-organized, holographic flying-spot store is a practical, semipermanent storage technique.

16.5 Write, Read, and Erase *in Situ*

Many applications of high-capacity storage and retrieval systems require updating of the stored information. The semipermanent nature of the holographic flying-spot store is restrictive in that the entire memory record must be discarded and another rewritten when only one page needs alteration. When such a change is made, the new storage plane must be registered with the readout beams and photodetectors. Should updating be a frequent operation, the process would significantly subtract from the useful operating time of the memory. It is therefore desirable to find a way to erase a page of information selected at random and to carry out write, read, and erase operations without having to remove the storage plane from its initially registered position. Whether a successful method will be found depends primarily on the development of adequate recording material. The recording medium and the record it stores should be blessed with all of the properties desired of permanent records, i.e., (1) *stability* of the record with time and temperature under readout conditions and (2) *high* sensitivity, resolution, signal-to-noise ratio, and diffraction efficiency. Beyond these, it must be able to be erased completely, without hysteresis or fatigue, and all operations including erasure and any development or fixing should preferably be accomplished *in situ*. It is clear that the organization of volume hologram memories of the type discussed in Section 16.3 makes selective erasure difficult. All the holograms are superimposed, and it may be impossible to erase one hologram without erasing the rest. On the other hand the spatial array of holograms, which may be plane or volume holograms, lends itself to the purpose.

No presently known medium possesses all the desired attributes, but several warrant further investigation. Photochromic $SrTiO_3$ [*16.9*] can store holograms which diffract with a maximum of 1% efficiency, but if the holograms are to be stored for weeks, they must be maintained at liquid nitrogen temperature. Phase holograms thermomagnetically written on a thin magnetic film of EuO with a giant pulse from a ruby laser can be erased with a small magnetic field. However, only low-resolution holograms with low diffraction efficiency have been formed [*16.16*]. Higher resolution holograms can be stored in nanoseconds on a film of MnBi but the diffraction efficiency remains low [*16.17*]. Perhaps the best compromise candidate to date is *photoconductor-thermoplastic film* which can form a completely processed phase hologram in approximately 1 second. The method whereby electrical charge, light, and heat are combined to expose, develop, and fix the hologram *in situ* has been described in Chapter 10. Useful diffraction efficiencies of 5 to 10% have been observed when the write, read, and erase operations were cycled over 100 times [*16.18*]. In Fig. 10.19 photographs are shown of images at several stages in the cycling test; the last (with a new subject) demonstrates that the images are hysteresis-free and that no loss of resolution is observed after 106 repeated cycles. Properties of thermoplastic holograms which may be considered drawbacks are: (1) the write time is considerably greater than the time in which readout can be carried out and (2) the spatial frequency response is band-limited. A major uncertainty is the aging characteristic of the plastic.

Figure 16.17 displays an electrical circuit for an extremely modest (3 × 3) array of thermoplastic holograms. The electrodes connect rectangular areas

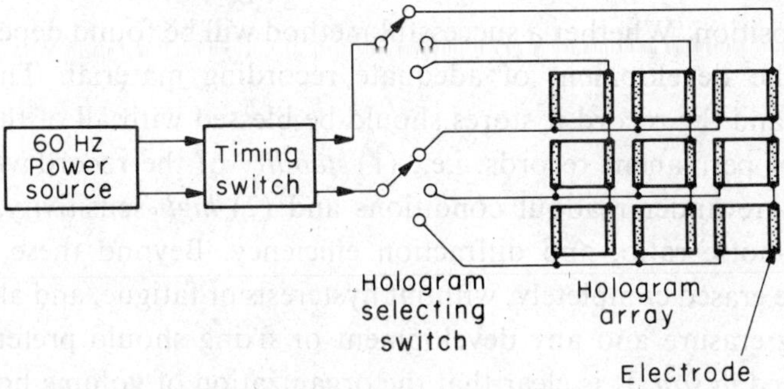

FIG. 16.17. Electrical circuit for development and erasure of thermoplastic holograms in a small array.

of tin oxide to a 60-Hz power source in such a way that any area may be selected to conduct a pulse of current. Current flow through the resistive but transparent tin oxide produces the heat necessary to record or erase a hologram in that area. Figure 16.18 is a photograph of the model storage plane. As explained in Chapter 10 a layer of photoconductor and another of thermoplastic cover the tin oxide. Experiments with the model have demonstrated that the thermoplastic array has the following useful properties [*16.18*]:

1. Many small holograms can be closely packed in an array on a single glass plate. Thermal isolation among the holograms is sufficient to permit the thermal development and erasure of one hologram without affecting its neighbors.
2. Although the circuit of Fig. 16.17 allows some current to pass through tin oxide areas other than the one intended, the thermoplastic responds only to a temperature above some threshold; consequently only that

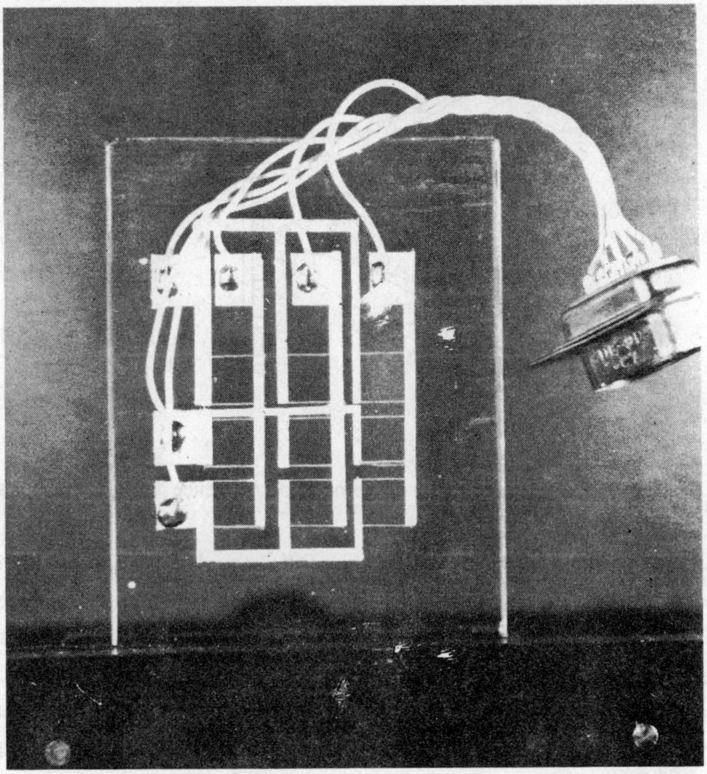

FIG. 16.18. Photograph of a small, model thermoplastic hologram storage plane.

area experiencing a current flow in excess of some threshold current will record or erase.

3. The number of recording–erasure cycles of each hologram can be well in excess of 100. Erasure can be complete and free of hysteresis.
4. Recording resolution exceeds 1000 cycles/mm and diffraction efficiency is ~7%.
5. The behavior of the thermoplastic hologram is close to that of an ideal *plane* hologram. An aberration-free real image can be obtained by locating the collimated reference and readout beams symmetrically about the normal to the storage plane (Fig. 16.19). (See Section 8.1.2.) This permits effective separation of input and output planes.

Figure 16.20 indicates how a thermoplastic hologram array might be employed in a write, read, and erase *in situ* memory. An auxiliary array of holograms acts as a set of beam directors which provides the properly directed input page illumination required to record holograms on the various

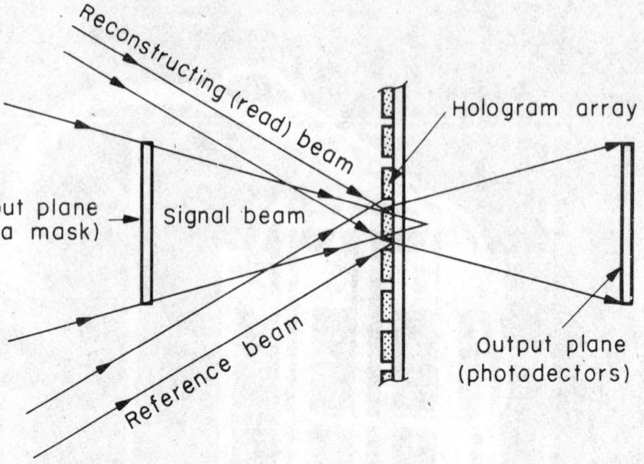

FIG. 16.19. Symmetrical arrangement of reference and read beams.

areas of the storage plane. An array of light valves adjacent to an input page, which consists entirely of logical "ones," would enable new input pages to be composed electronically. At the storage plane there might be spillover of light onto areas other than that intended for a particular page recording. However, a current in excess of the threshold needed to thermally develop a hologram is applied to only one tin oxide area of the storage plane. Only that area will record a persistent record.

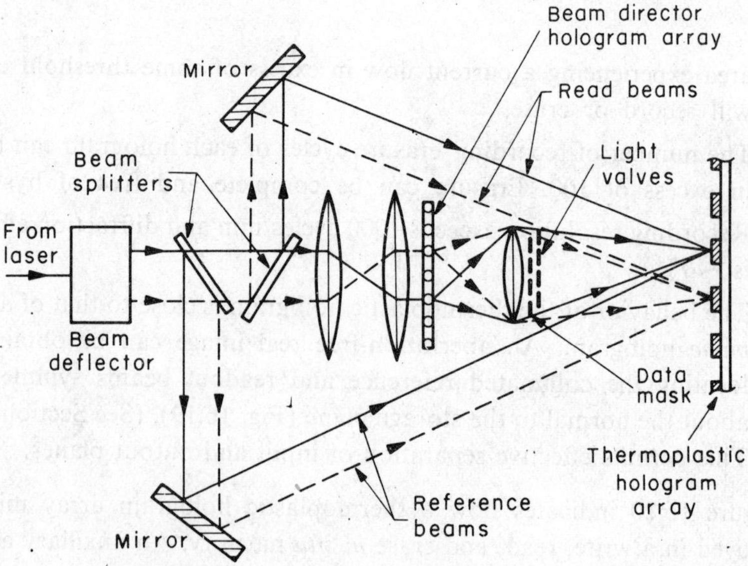

FIG. 16.20. Schematic of a write, read, and erase *in situ* memory. (The shutters required in the read, reference, and subject beams are omitted.)

REFERENCES

16.1. G. W. W. Stevens, *Microphotography*, p. 3. Wiley, New York, 1968.

16.2. A. Reich, "Photochromic, High-Speed, Large-Capacity Semirandom Access Memory," *Optical and Electro-Optical Information Processing* (J. T. Tippett *et al.*, eds.), p. 567. Mass. Inst. Technology Press, Cambridge, Massachusetts, 1965.

16.3. P. J. van Heerden, "Theory of Optical Information Storage in Solids," *Appl. Opt.* **2**, 393 (1963).

16.4. H. Fleisher, P. Pengelly, J. Reynolds, R. Schools, and G. Sincerbox, "An Optically Accessed Memory Using the Lippmann Process for Information Storage," *Optical and Electro-Optical Information Processing*, (J. T. Tippett *et al.* Eds.), p. 1. Mass. Inst. Technology Press, Cambridge, Massachusetts, 1965.

16.5. G. Lippmann, "Sur la Théorie de la Photographie des Couleurs Simples et Composées par la Méthode Interférentielle," *J. Phys.* (*Paris*) **3**, 97 (1894).

16.6. F. S. Chen, J. T. LaMacchia, and D. B. Fraser, "Holographic Storage in Lithium Niobate," *Appl. Phys. Lett.* **13**, 223 (1968).

16.7. G. Kalman, "Holography in Thick Media," *Applications of Lasers to Photography and Information Handling* (R. D. Murray, ed.), p. 99. Soc. of Photographic Scientists and Engineers, 1968.

16.8. A. A. Friesem and J. L. Walker, "Thick Absorption Recording Media," *Appl. Opt.* **9**, 201 (1970).

16.9. J. J. Amodei and D. R. Bosomworth, "Hologram Storage and Retrieval in Photochromic Strontium Titanate Crystals," *Appl. Opt.* **8**, 2473 (1969).

16.10. F. M. Smits and L. E. Gallaher, "Design Considerations for a Semipermanent Optical Memory," *Bell Syst. Tech. J.* **46**, 1267 (1967).

16.11. L. K. Anderson, "Holographic Optical Memory for Bulk Data Storage," *Bell Lab. Rec.* **46**, 318 (1968).

16.12. E. I. Gordon, "A Review of Acousto-Optical Deflection and Modulation Devices," *Proc. IEEE* **54**, 1391 (1966).

16.13. D. A. Pinnow and R. W. Dixon, "Alpha-Iodic Acid: A Solution-Grown Crystal with a High Figure of Merit for Acousto-Optic Device Applications," *Appl. Phys. Lett.* **13**, 15 (1968).

16.14. R. Bracewell, *The Fourier Transform and Its Applications*, p. 86. McGraw-Hill, New York, 1965.

16.15. C. B. Burckhardt, "Use of a Random Phase Mask for the Recording of Fourier Transform Holograms of Data Masks," *Appl. Opt.* **9**, 695 (1970).

16.16. G. Fan, K. Pennington, and J. H. Greiner, "Magneto-Optic Hologram," *J. Appl. Phys.* **40**, 974 (1969).

16.17. R. S. Mezrich, "Curie-Point Writing of Magnetic Holograms on MnBi," *Appl. Phys. Lett.* **14**, 132 (1969).

16.18. L. H. Lin and H. L. Beauchamp, "Write-Read-Erase In Situ Optical Memory Using Thermoplastic Holograms," *Appl. Opt.* **9**, 2088 (1970).

Chapter 17

COLOR HOLOGRAPHY

The goal of color holography is to generate a three-dimensional image whose color closely approximates that of the original subject illuminated with ordinary light. Important steps toward this goal have been taken, but there remain unsolved colorimetric problems as well problems associated with diffraction imaging.

Colorimetric difficulties arise out of the necessity of illuminating the subject with laser light rather than sunlight or incandescent light. To illustrate, consider a particularly difficult subject: a narrowband (100-Å) interference filter centered at a wavelength of 5900 Å. When held in the sunlight, the filter appears orange. Suppose now we illuminate the filter with a combination of 6328-Å light from a He–Ne laser and 5145-Å light from an argon ion laser. When mixed in the proper proportion, these two emissions can approximate the orange color of the filter very closely. No matter what the proportions, however, when the combination actually illuminates the filter, the latter always appears black; the filter only transmits light in the wavelength range 5850 to 5950 Å. Laser emission at many wavelengths covering broad bands of the visible spectrum is required to approach the results of white-light illumination. Although a wide spectrum is indeed presently available, it seems impractical to form holograms with substantially more than three discrete wavelengths.

17.1 Color Mixing with Laser Light

Before entering into a discussion of the problems associated with producing multicolor images from holograms and the several solutions that have been found, let us continue in this section with a brief summary of colorimetric concepts pertinent to color holography [*17.1*]. While it cannot be said to clarify the complicated subject of color, the discussion does at the least provide a rationale for selecting lasers with appropriate wavelengths and relative output powers.

We shall investigate whether a mixture of light at three laser wavelengths can produce any desired color. To reach some quantitative conclusions concerning the proportions, an analytical basis for describing color must be adopted.

17.1.1 Color and Chromaticity

Seeing color involves psychological as well as physical processes, and consequently color is difficult to define. This seems evident from a definition adopted by the Optical Society of America Committee on Colorimetry: "*Color* consists of the characteristics of light other than spatial and temporal inhomogeneities; *light* being that aspect of radiant energy of which a *human observer* is aware through the visual sensation which arises from the stimulation of the retina of the eye [*17.2*]."

Those colors which differ only in brightness are said to have the same chromaticity. For example black, shades of gray, and white are (achromatic) colors differing in brightness but with the same chromaticity. When we are not interested in the relative brightness of a color, we may interchange the terms chromaticity and color. Thus the color of a ground glass screen illuminated with laser light at 6328 Å is red regardless of the illuminating light intensity.

An international body, Comité International d'Eclairage (CIE), has devised a standard method of designating the chromaticity of any color. Each color is assigned a set of chromaticity coordinates (x, y, z) obeying the relation

$$x + y + z = 1. \tag{17.1}$$

If two coordinates are known, the third is uniquely determined by Eq. (17.1). Therefore every color can be represented as a point (x, y) on a plane. The CIE chromaticity diagram shown in Fig. 17.1 is a plot of the area of the plane containing all possible colors. We have labeled regions of the diagram with the name of the prevailing color. There are of course no sharp boundaries.

17.1.2 Colors Obtainable with Laser Light

Points representing monochromatic light (spectral colors) are plotted by wavelength along the horseshoe-shaped outer boundary of the chromaticity

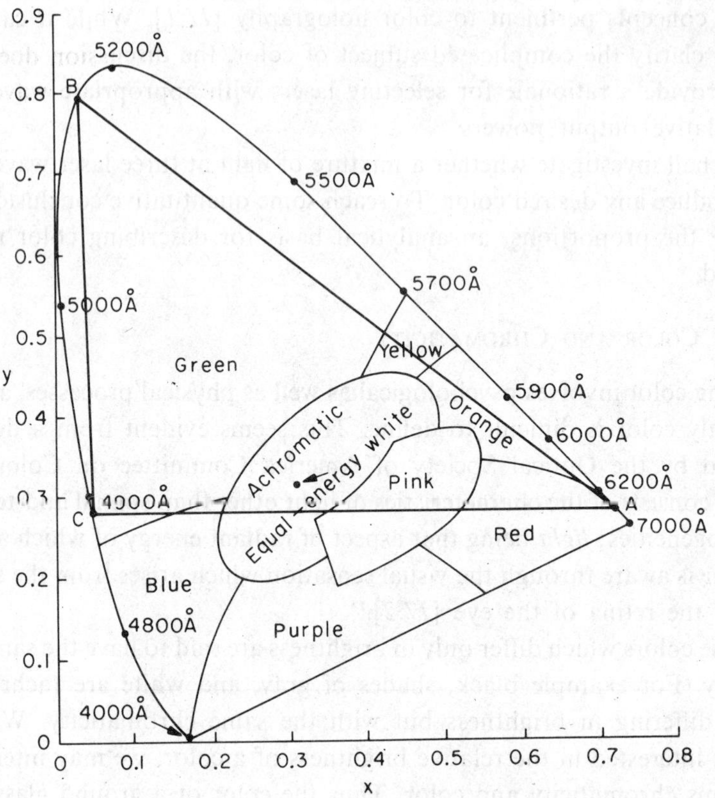

FIG. 17.1. CIE x–y chromaticity diagram.

diagram of Fig. 17.1. The color of the light from a He–Ne laser at 6328 Å wavelength is plotted at point A; argon ion laser emissions at 5145 and 4880 Å are plotted at points B and C respectively. Let us now use the diagram according to the rules developed by the CIE without concerning ourselves with their derivation. We shall delineate the area of the diagram that contains all possible colors resulting from addition of the above spectral or primary colors. According to the rules for use of the diagram, all colors resulting from mixing varying proportions of laser light at the wavelengths 6328 and 5145 Å are represented by points on the straight line AB. When laser light at the wavelength 4880 Å is added to these two components, the possible colors are represented by points on any straight line connecting C with any point on AB. All such points are enclosed in the triangle ABC.

When more than three wavelengths are used, the totality of possible color points on the chromaticity diagram is bounded by a polygon.. A greater area of the diagram can be enclosed and more subject colors can be matched. To match the color of a holographic image with that of the subject, however, requires independent adjustments of the power at each laser wavelength, both in forming the hologram and in the reconstruction process. In practice, successfully adjusting power at three wavelengths is difficult enough. We shall find, furthermore, that the greater the number of wavelengths, the lower the hologram diffraction efficiency in the reconstruction. A better solution to the problem of increasing the range of possible colors is to seek a more optimum set of primary colors *ABC*. Wavelengths of some commonly available, continuous-wave laser emissions are tabulated in Table 17.1. When these are plotted on the chromaticity diagram it is readily seen that the krypton ion laser emissions at the wavelengths 6471 Å (red), 5208 Å (green), and 4762 Å (blue) form a large triangle. On this basis the krypton laser should be of great use in color holography. Unfortunately it is expensive, short-lived, and operates at high current. Consequently the usual choice of sources has been the He–Ne laser for its red output at 6328 Å wavelength and the argon ion laser for its green and blue emissions at the wavelengths 5145 and 4880 Å, respectively. The argon laser also produces another blue at 4765 Å wavelength which allows a greater range of colors in the blue and purple region of the chromaticity diagram. However, this advantage is somewhat offset by the lower power at 4765 Å compared to that at 4880 Å, a factor which is important when the overall output of the laser is low.

TABLE 17.1

WAVELENGTHS OF CW LASER EMISSIONS

Wavelength (Å)	Approximate color	Laser medium	Approximate power[a] (mW)
4416	purplish blue	He–Cd	50
4579	purplish blue	Ar	50
4762	blue	Kr	30
4765	blue	Ar	250
4880	blue-green	Ar	700
4965	bluish green	Ar	150
5017	green	Ar	150
5145	green	Ar	700
5208	green	Kr	60
5682	yellow-green	Kr	60
6328	red	He–Ne	50
6471	red	Kr	150

[a] These values hold for commercial models which are commonly available.

17.1.3 Production of White Color with Laser Light

In practicing color holography, it is generally unnecessary to calculate the proportions of red, green, and blue laser illumination required to give the subject natural color. One empirically adjusts the relative powers of the sources until the desired result is observed or approached. However, as a guide to the selection or purchase of lasers meeting the general requirements of color holography, it is useful to predict the relative powers needed at the primary wavelengths. Since the power relations will differ with the subject, we set as a standard the color *equal-energy white*. Consider, therefore, a mixture of laser light at three wavelengths λ_1, λ_2, and λ_3 illuminating a diffusing screen. Light diffused and transmitted by the screen is observed by a human eye. We wish to determine the ratios of laser powers $P_1 : P_2 : P_3$ at λ_1, λ_2, and λ_3 which cause the observer to see a color which matches an equal-energy white. The latter is assumed to have equal energy at all wavelengths in the visible region of the spectrum; its chromaticity coordinates are $(\frac{1}{3}, \frac{1}{3}, \frac{1}{3})$ as indicated in Fig. 17.1.

Depending upon the wavelength, a given amount of radiant power will provide a different amount of visual stimulus per watt of monochromatic light. A plot of this latter quantity is given in Fig. 17.2. Visual stimulus is

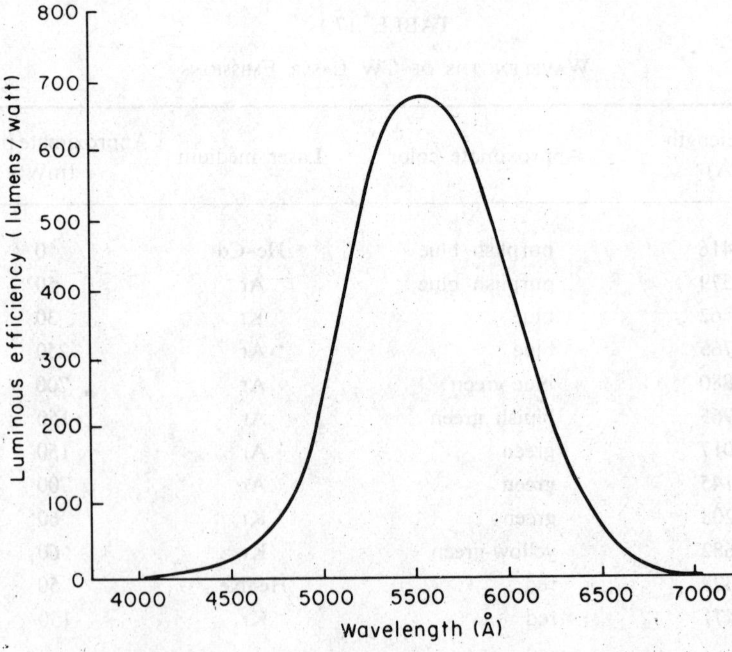

FIG. 17.2. Luminous efficiency of monochromatic light.

measured by a unit called the lumen, and the number of lumens per watt of a light source is called its luminous efficiency. Let η_1 be the luminous efficiency and (x_1, y_1, z_1) be the chromaticity coordinates of the laser light at λ_1 with similar notations for λ_2 and λ_3. We also let (x_0, y_0, z_0) be the chromaticity coordinates of the color observed on the diffusing screen illuminated with laser light of these wavelengths. The coordinate x_0 is to be computed from a linear combination of x_1, x_2, and x_3 with coefficients $\eta_1 P_1$, $\eta_2 P_2$, and $\eta_3 P_3$:

$$x_0 = r(\eta_1 P_1 x_1 + \eta_2 P_2 x_2 + \eta_3 P_3 x_3) \tag{17.2}$$

where r is a constant, η_1, η_2, and η_3 are the luminous efficiencies and P_1, P_2, and P_3 are the powers of the light at λ_1, λ_2, and λ_3 respectively. Similar expressions can be written for the other two coordinates

$$y_0 = r(\eta_1 P_1 y_1 + \eta_2 P_2 y_2 + \eta_3 P_3 y_3) \tag{17.3}$$

and

$$z_0 = r(\eta_1 P_1 z_1 + \eta_2 P_2 z_2 + \eta_3 P_3 z_3). \tag{17.4}$$

The constant r ensures that $x_0 + y_0 + z_0 = 1$. Equations (17.2)–(17.4) contain four unknowns, P_1, P_2, P_3, and r. Let us specify the wavelengths to be $\lambda_1 = 4880$ Å, $\lambda_2 = 5145$ Å, and $\lambda_3 = 6328$ Å; from Figs. 17.1 and 17.2, and Eq.(17.1) we obtain the luminous efficiencies and chromaticity coordinates listed in Table 17.2. When these together with the chromaticity coordinates of equal-energy white, $x_0 = y_0 = z_0 = \frac{1}{3}$, are substituted into Eqs. (17.2)–(17.4) and the equations solved for the ratios $P_1 : P_2 : P_3$, we obtain the desired relation $P_1 : P_2 : P_3 = 1.42 : 0.362 : 1.22$.

TABLE 17.2

LUMINOUS EFFICIENCY AND CHROMATICITY COORDINATES FOR THREE SPECTRAL COLORS

Wavelength (Å)	Luminous efficiency (lumens/W) η	Chromaticity coordinates x	y	z
4880	130	0.06	0.25	0.69
5145	415	0.04	0.81	0.15
6328	163	0.71	0.29	0

17.2 Recording Materials

To achieve a wide range of color we require widely separated laser light wavelengths. This in turn means that we need suitable recording materials responsive to these wavelengths. Some of the more efficient materials, such as dichromated gelatin, are ruled out because of their insensitivity to red light. To date, most of the multicolor holography that has been reported has employed absorption holograms formed in photographic emulsion. Volume holograms are generally best for multicolor imaging; their wavelength sensitivity (Sections 9.6.1 and 9.7.1) prevents color cross modulation. Holograms generating multicolor images can also be formed in photoconductor–thermoplastic film, but they are perforce plane holograms and must rely on other, less convenient techniques to avoid cross modulation.

A major drawback of volume absorption holograms made for multicolor holography is their low diffraction efficiency. They are formed by superimposing three recordings corresponding to three primary wavelengths. As we have noted in Section 16.3.5, the maximum possible efficiency for a single recording is only 3.7%, and efficiency drops off rapidly with the number of holograms superimposed. Reflection volume holograms permit white-light illumination, a great convenience for the viewer. However the hologram diffracts only narrow spectral bands of light representing a small fraction of the total illuminating power. This inefficient use of illumination coupled with poor diffraction efficiency generally means that multicolor holographic images must be viewed in low ambient light. Attempts to increase the diffraction efficiency of reflection holograms by applying the bleaching techniques, so successful with transmission holograms, have thus far been fruitless.

17.3 Monochrome Images

The color of any point on a holographic image is determined by the power spectrum of the light diverging from that point. If we restrict our consideration to some of the more practical methods of hologram formation, we find that holographic images can be classified as (1) *monochrome*, when the spectrum of each image point is a narrow band of wavelengths distributed about a single center wavelength, (2) *achromatic*, when the spectrum of any point contains nearly equal power at all visible wavelengths, and (3) *multicolor*, when the spectrum varies from point to point.

Monochrome images, the most common, are generated from holograms formed with laser light at a single center wavelength. To reconstruct the subject wave, one normally illuminates the hologram with light of the original wavelength. Another wavelength, however, may be used if a change in magnification and some aberrations are tolerable. Where hologram display using a laser is considered hazardous, the illumination laser may be replaced by a less coherent, small-area, narrowband source such as a filtered mercury arc lamp. Some small sacrifice in resolution is incurred but is not usually serious for pictorial display. Reflection holograms formed with monochromatic light can be illuminated with white light. In this case the filtering action of the hologram itself selects the narrow band of wavelengths with which the image is formed.

17.4 Achromatic Images

Holographic images which are called achromatic appear so only under restricted conditions. Either the viewing conditions are restricted or the subject depth is restricted.

17.4.1 Image Holograms

A plane transmission hologram formed with monochromatic light and illuminated with white light can produce an achromatic image. As we note in Section 8.3.2, when the image volume contains the plane of the hologram, then image resolution in that plane is unaffected by the spread in wavelength of the illuminating light. All wavelength components of white light illuminating such image holograms are equally well focused to image points on the hologram plane. For image points out of the hologram plane, color blurring (dispersion) becomes noticeable to a degree determined by the offset angle of the reference beam and the distance of the points from the hologram plane [Eq. (7.53)]. Figure 8.21 shows the variation in image sharpness due to the extent of the source as well as its broad band of wavelengths.

17.4.2 Dispersion Compensation

Achromatic images can be very bright because they use all the white light illuminating the hologram. When the image is entirely out of the hologram plane, it is still possible to cause an off-axis transmission hologram to form an achromatic image. Suppose the hologram is illuminated by white light traveling in the direction of the original reference beam. The hologram, acting as a diffraction grating, disperses the various wavelength components over a range of diffraction angles. We now place close to the hologram a compensating grating which produces an equal but opposite dispersion. When the image is observed from the proper location, it will appear achromatic [*17.3*, *17.4*]. If the observed image is to be very bright, the compensating grating must be highly efficient.

A grating which performs this function to a useful degree is easily formed by replacing the original subject with a source of a plane wave whose direction is identical to the mean subject-light direction. Recording the interference of the plane wave with the original reference wave produces the desired grating. The grating may be employed as in Fig. 17.3.

As a consequence of the double diffraction, the reconstructed wave appears in the direction of the illuminating wave. In effect hologram and grating together diffract as a single hologram formed with subject and reference source centered on the same axis. This means that the image generated by the reconstructed wave must be viewed against a background of undiffracted light, as in the Gabor in-line holograms. In the present case the undiffracted light may be screened out by inserting a venetian-blind light shield between hologram and grating as shown in Fig. 17.3.

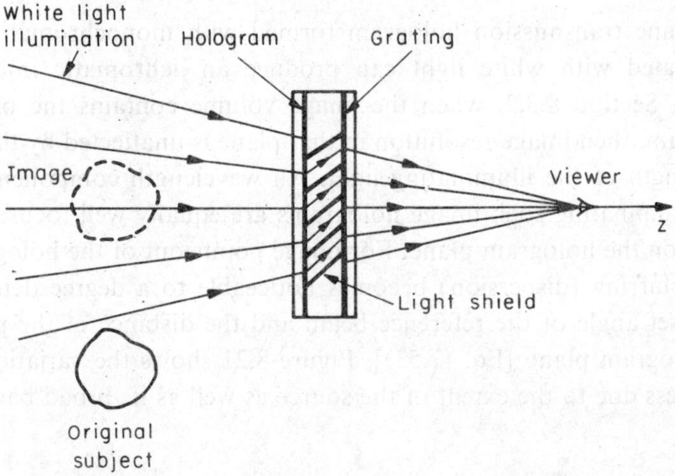

FIG. 17.3. Achromatic image by means of dispersion compensation.

To describe the process analytically, suppose the subject to be located off-axis so that the subject light at the hologram plane during formation of the hologram, $\mathbf{a}(x)$, is equivalent to a complex amplitude $\mathbf{a}_0(x)$ modulated onto a high spatial frequency carrier $\exp(2\pi i\xi x)$. [$\mathbf{a}_0(x)$ is assumed to have a spectrum centered at zero spatial frequency.] Thus

$$\mathbf{a}(x) = \mathbf{a}_0(x) \exp(2\pi i \xi x). \tag{17.5}$$

A spherical reference wave $\mathbf{r} = \exp[i\varphi(x)]$ with a spectrum centered at zero spatial frequency is derived from a source located on axis. Interference of subject and reference light leads to a hologram whose transmittance contains the virtual image term

$$\mathbf{t}_\text{v} \propto \mathbf{r}^*\mathbf{a} = \mathbf{r}^*\mathbf{a}_0(x) \exp(2\pi i \xi x). \tag{17.6}$$

The compensating grating is formed by recording the interference of the plane wave $\exp(2\pi i \xi x)$ with $\mathbf{r}$. We are interested in the transmittance term $\mathbf{r} \exp(-2\pi i \xi x)$. To simplify analysis, let us neglect the venetian blind and consider that grating and hologram are joined and illuminated by $\mathbf{r}$. A wave proportional to

$$\mathbf{r}\mathbf{t}_\text{v}\mathbf{r} \exp(-2\pi i \xi x) = \mathbf{r}\mathbf{a}_0(x) \tag{17.7}$$

is diffracted; its spectrum is centered at zero spatial frequency. We see that the reconstructed subject wave has been removed from its high carrier frequency and now has a mean direction along the axis. This holds true for multiwavelength illumination since the mean spatial frequency of $\mathbf{r}$, $\bar{\xi} = 0$, is independent of wavelength.

When the original reference wave is a converging beam and the illuminating beam is a similar converging beam of white light, a viewer's eye placed at the center of convergence, as in Fig. 17.4, receives a converging beam appearing to come from all points on the virtual image [see Eq. (17.7)]. In

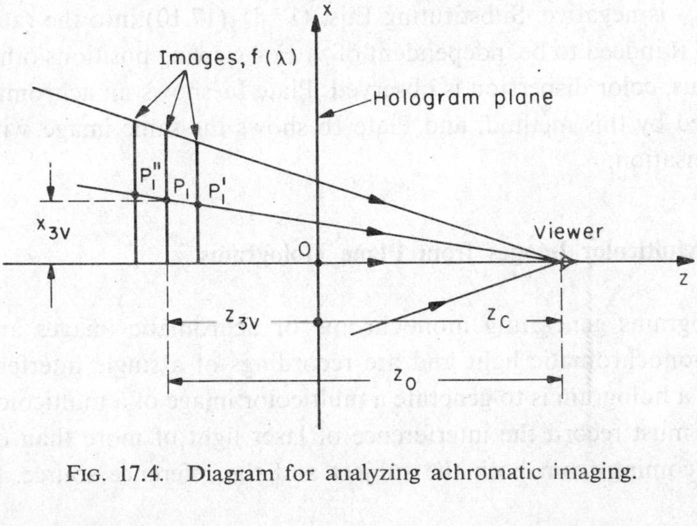

FIG. 17.4. Diagram for analyzing achromatic imaging.

spite of the fact that various wavelength components generate a series of images differing in magnification and in their separation from the eye, the eye perceives an achromatic image. For this to be so, any ray entering the eye must have passed through similar or corresponding points on each of the series of images (see Fig. 17.4). Equal-energy contributions at all wavelengths add at each point imaged onto the retina to give the achromatic impression. This may be achieved if, for any image point, the ratio of the lateral coordinate (the x coordinate) to the distance of the image from the eye is independent of the wavelength of the reconstructing light. We can regard the combined hologram and grating as an on-axis hologram [see Eq. (17.7)] and can predict this result using Eq. (3.27). We consider reference and illuminating sources to be on axis so that $x_r = x_c = 0$. As shown in Fig. 17.4 the axial distances of the reference and illuminating sources from the hologram are positive, i.e., $z_c = z_r > 0$, while the subject distance z_1 and the virtual image distance z_{3V} are negative. With the substitution of $x_r = x_c = 0$ and $z_c = z_r$; the pertinent equations in Eq. (3.27) become

$$x_{3V} = \frac{\mu x_1 z_c^2}{z_1 z_c + \mu z_c^2 - \mu z_c z_1} \tag{17.8}$$

$$z_{3V} = \frac{z_c^2 z_1}{z_1 z_c + \mu z_c^2 - \mu z_c z_1} \tag{17.9}$$

where μ is the ratio of the forming to reconstructing wavelength and where we have set $m = 1$ in Eq. (3.27). The distance from image to eye, z_0, is given by

$$z_0 = z_c - z_{3V}, \tag{17.10}$$

since z_{3V} is negative. Substituting Eqs. (17.8)–(17.10) into the ratio x_{3V}/z_0, we find it indeed to be independent of μ. For viewing positions other than at the focus, color dispersion is observed. Plate Ia shows an achromatic image obtained by this method, and Plate Ib shows the same image without any compensation.

17.5 Multicolor Images from Plane Holograms

Holograms generating monochrome or achromatic images are formed with monochromatic light and are recordings of a single interference pattern. If a hologram is to generate a multicolor image of a multicolor subject, then it must record the interference of laser light of more than one wavelength coming from both the subject and the reference source. However, those wave components of the subject and reference light which are not of the same wavelength cannot produce the standing wave patterns required for hologram recording. Their electric fields vibrate, at different frequencies; when they are added and squared, the resulting intensity shows no interference term (see Section 1.9.1). Those subject and reference wave components which do have the same wavelength, of course, will record holograms. We may regard these as superimposed independent holograms; there are as many holograms as there are wavelength components.

17.5.1 Cross Modulation and False Images

To generate a multicolored image, the superimposed holograms are illuminated with the original reference wave. Each wavelength component diffracts from the hologram formed with that wavelength and reconstructs the original subject wave. If the viewer were to observe only these diffracted waves, he would see a multicolored virtual image of the subject located in the original subject space. We assume here, however, that the recordings are plane holograms; each diffracts not only light of the wavelength used to form it but all components in the illumination. From the discussion in Section 3.3.2 it is clear that as a result of this cross modulation of incident light at one wavelength by a hologram formed in another, false reconstructions of the subject wave will emerge from the hologram at various angles. These can generate false, overlapping virtual images of the subject with angular magnifications dependent on the wavelength ratio μ [see Eq. (3.30)]. (We shall here be concerned only with the virtual images although the same cross-modulation problems apply to real images as well.)

Let us now carry out a simple analysis of the cross modulation problem. An initial simplification is to confine all light rays to the xz plane. We assume the subject is illuminated with a mixture of laser light at M separate wavelengths, $\lambda_1, \lambda_2, \ldots, \lambda_m, \ldots, \lambda_M$. We can write the complex amplitude of the multiwavelength subject wavefront at the hologram in a form

$$\mathbf{s}(x) = \sum_{m=1}^{M} \mathbf{s}_m(x) \exp(2\pi i \xi_{sm} x) \tag{17.11}$$

where ξ_{sm} is the mean spatial frequency for the mth wavelength component and $\mathbf{s}_m(x)$ is assumed to have a spatial frequency spectrum centered at zero frequency [see, e.g., Eq. (17.5)]. If the reference beam consists of M plane waves at the same M wavelengths, then the reference wavefront at the hologram is

$$\mathbf{r}(x) = \sum_{m=1}^{M} r_m \exp(2\pi i \xi_{rm} x) \tag{17.12}$$

where the r_m are constants. Let us suppose the recording of the interference of $\mathbf{s}$ and $\mathbf{r}$ is an absorption hologram with an overall transmittance

$$t(x) = \sum_{m=1}^{M} c_m \{s_m{}^2 + r_m{}^2 + \mathbf{s}_m r_m \exp[2\pi i(\bar{\xi}_{sm} - \xi_{rm})x]$$
$$+ \mathbf{s}_m{}^* r_m \exp[2\pi i(\xi_{rm} - \bar{\xi}_{sm})x]\} \tag{17.13}$$

where c_m is a constant related to the spectral sensitivity of the recording material at the wavelength λ_m and where $s_m = |\mathbf{s}_m|$.

To reconstruct the original subject wave we must illuminate the hologram with a multiwavelength wavefront

$$\mathbf{r}'(x) = \sum_{n=1}^{M} r_n{}' \exp(2\pi i \xi_{rn} x) \tag{17.14}$$

which differs from the original reference wavefront $\mathbf{r}$ only in that r_m is replaced by another constant $r_n{}'$. This allows a correction to the color of the image to be made in the reconstruction process. Diffracted from the hologram will be the virtual image waves

$$\mathbf{w}(x) = \mathbf{r}'\mathbf{t}_v$$
$$= \mathbf{r}' \sum_{m=1}^{M} c_m \mathbf{s}_m r_m \exp[2\pi i(\bar{\xi}_{sm} - \xi_{rm})x],$$

where $\mathbf{t}_v$ is the third term on the right in Eq. (17.13). Substituting from Eq. (17.14) for $\mathbf{r}'$ we have

$$\mathbf{w}(x) = \sum_{m=1}^{M} (c_m r_m r_m{}')\mathbf{s}_m \exp(2\pi i \bar{\xi}_{sm} x)$$
$$+ \sum_{m=1}^{M} \sum_{\substack{n=1 \\ m \neq n}}^{M} (c_m r_m r_n{}')\mathbf{s}_m \exp[2\pi i(\bar{\xi}_{sm} + \xi_{rn} - \xi_{rm})x]. \tag{17.15}$$

The sum of the terms under the *single* summation sign in Eq. (17.15) is proportional to the original subject wavefront $\mathbf{s}$ when

$$c_1 r_1 r_1{}' = c_2 r_2 r_2{}' = \cdots = c_M r_M r_M{}'. \tag{17.16}$$

Equation (17.16) expresses the condition necessary for a true multicolor representation of the original subject as seen in the original laser illumination. Each term under the double summation sign in Eq. (17.15) is a false reconstruction of the subject wave arising out of the diffraction of light of wavelength λ_n by a hologram formed with light of wavelength λ_m. We see from the phase factor that the false image wavefront is placed on a carrier wave of spatial frequency $\bar{\xi}_{sm} + \xi_{rn} - \xi_{rm}$ varying according to the wavelengths λ_n and λ_m. There are $M^2 - M$ such waves generating an equal number of false, displaced virtual images each of a single color. Plate II is a photograph of the true and false images generated by a hologram formed in 12-μm-thick emulsion with only two wavelengths $\lambda_1 = 6328$ Å and $\lambda_2 = 4880$ Å. To ensure plane hologram behavior, the mean subject-to-reference-beam angle was only 15°. Notice the magnification change in the false images overlapping the true, equal-size, red and blue image of the pair of letters BR.

Although several ways have been devised to eliminate or diminish the effect of the false images generated in *plane*-color holograms, they all involve a penalty. Either spatial frequency bandwidth is restricted, image resolution is reduced, or image signal-to-noise ratio is decreased.

17.5.2 Separation of the Spectra

One method of diminishing the effect of false reconstructions is to separate the spectrum of the true image from the spectra of the false images for all the wavelengths used in the reconstruction process. Suppose that the subject spatial frequency bandwidth on either side of the mean frequency $\bar{\xi}_{sm}$ is $\Delta\xi_{sm}$ for any value of m. Reference to Eq. (17.15) shows the mean spatial frequency of a false reconstruction to be $\bar{\xi}_{sm} + \xi_{rn} - \xi_{rm}$. Its bandwidth is also $\Delta\xi_{sm}$ on either side of the mean. If complete separation of true and false spectra is to be achieved, the condition

$$\tfrac{1}{2}\,|\,\xi_{rn} - \xi_{rm}\,| > \Delta\xi_{sm} \qquad (17.17)$$

must be satisfied for all λ_m and λ_n. Since ξ_{rm} can be expressed in terms of the angle θ_{rm} which the plane wave of wavelength λ_m makes with the hologram normal [$\xi_{rm} = (\sin\theta_{rm})/\lambda_m$], the inequality in relation (17.17) can be satisfied by proper choice of the direction and wavelength of each component of the reference beam [*17.5–17.7*].

Suppose we consider the subject spatial frequency bandwidth allowed when only two wavelengths are employed, e.g., $\lambda_1 = 6328$ Å and $\lambda_2 = 4880$ Å. If we set $\theta_{r1} = 30°$ and $\theta_{r2} = 45°$, the maximum allowable subject bandwidth is

$$\Delta\xi_{s1} = \frac{1}{2}\left|\frac{\sin 45°}{\lambda_2} - \frac{\sin 30°}{\lambda_1}\right| = 330 \quad \text{cycles/mm.}$$

This corresponds to an angular viewing range of $\pm 12°$ about the mean direction of the reconstructed subject light. When more than two wavelengths are employed, the value of $|\xi_{rn} - \xi_{rm}|$ will be smaller, requiring a greater spread in reference beam angles to maintain bandwidth. Perhaps the greatest deterrent to the use of this method, or to the use of any of the plane-hologram, multicolor-imaging methods yet to be discussed, is the need in the reconstruction step for several laser beams of different wavelength in accurate relative alignment. Another undesirable feature of this particular method is that the false images may still be seen by the observer, even though their spectra are separated from that of the true image.

17.5.3 Coded Reference Beam Method

A second method of diminishing the effects of false reconstructions is to code uniquely each wavelength component of the reference wavefront [*17.8*]. In the reconstruction step the identically coded wavefront must again illuminate the hologram. It is important, in this step, that the spatial relations between the laser beams, the reference source, and the hologram be maintained as they were in forming the hologram. Figure 17.5 illustrates one way to carry out the coding. Laser light of several wavelengths is directed as a single, mixed beam at an area of a ground glass screen. The light diffused by the screen and transmitted to the hologram plate provides the reference wavefront. Although each wavelength component is scattered by

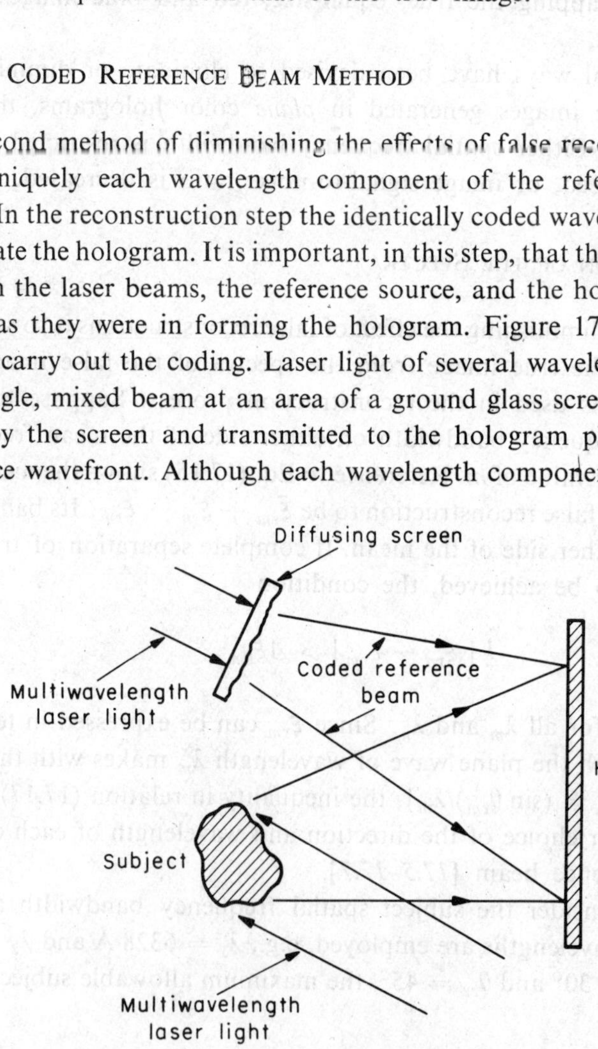

FIG. 17.5. Coding the reference beam.

the same area of the diffusing screen, the complex amplitude of each component of the light arriving at the hologram plate is coded uniquely. This result stems from nothing more than the wavelength dependence of the complex amplitude of diffracted light given in Eq. (5.25). For a diffusing screen we may interpret Eq. (5.25) to represent interference of a great many plane waves propagating in many directions. The distribution of amplitudes and phases with spatial frequency is random, with a different random distribution for each wavelength λ. Interference of this coded reference wavefront with the corresponding wavelength component in the subject beam produces a hologram whose properties are discussed in Section 14.5. Similar holograms are formed and superimposed for each wavelength component of the subject and reference beams.

The result of illuminating any one of the component coded reference wave holograms with its *original* reference wavefront is given in Eq. (14.23) of Section 14.5.1. In addition to the reconstructed subject wave, we find there is a background noise nearly uniform in intensity. Examination of Eq. (14.23) shows the noise to consist of the subject complex amplitude modulated onto a band of carrier waves whose phases are random and whose spatial frequencies range over twice the spatial frequency bandwidth of the diffuse reference wavefront. Similar subject wave reconstructions and noise distributions are produced by each of the other wavelength components in the reference wave illuminating their respective holograms. If we assume that the subject light amplitude is equally divided among the wavelengths and the sensitivity of the recording medium is independent of wavelength, then the background noise contributed by each wavelength component is determined by the intensity of the reference wave of that wavelength [see Eq. (14.23)]. We can adjust the individual reference intensities until the total reference illumination falling on the diffusing screen appears white. In this case the major effect of the background noise is to reduce color saturation and image contrast.

In addition to the above autocorrelation noise produced by a coded wavefront illuminating a hologram formed with the identical coded reference wave, cross-modulation noise is generated when a coded wavefront of wavelength λ_n illuminates a hologram formed with a differently coded wavefront of wavelength λ_m. The form of the noise amplitude is given by Eq. (14.24). [*Note*: The subscripts m and n in Eq. (14.24) refer to the multitude of plane waves diffused by the ground glass. It is the superscripts i and j that correspond to different codes and, in the present case, different wavelengths.] If one regards the interaction of the coded reference wave with the hologram as the diffraction of a set of plane waves from a two-dimensional grating, it is not surprising that the spectrum of the cross-modulation noise of wavelength λ_n has a mean spatial frequency which is shifted from that of the reconstructed subject wave (and the autocorrelation noise) of wavelength λ_m. We can compute the shift by restricting attention to the mean spatial frequencies of the illuminating wave of wavelength λ_n and reference wave of wavelength λ_m. Thus, if the spatial frequencies in the second summation in Eq. (17.15) are taken to be mean values, then the spatial frequency shift is

$$\xi_{rn} - \xi_{rm} = \frac{\sin\theta_{rn}}{\lambda_n} - \frac{\sin\theta_{rm}}{\lambda_m} = \sin\theta_r\left(\frac{1}{\lambda_n} - \frac{1}{\lambda_m}\right), \tag{17.18}$$

since all illuminating and reference beams have the same mean direction. We see that when $\lambda_n > \lambda_m$, the shift is negative, while for $\lambda_n < \lambda_m$ the shift is positive. If the reference beam angle is kept small, this color dispersion, which creates a multicolor background, is minimized.

Plate III is a photograph of a multicolor image generated by a coded reference hologram. Two wavelength components at 6328 and 4880 Å were encoded as in Fig. 17.5. Again the mean subject-to-reference beam angle was 15° as in Plate II. We see that the false images have been transformed into a more tolerable background noise whose uniformity allows us to perceive the desired image information. Notice the color dispersion.

Two conditions make this technique difficult. After processing, the hologram must be replaced in its original position with the high degree of accuracy computed in Section 14.1.2. Then the subject wave will be reconstructed only when the originally coded reference wave illuminates the hologram. Unintentional disturbance of the reference source can be disastrous.

17.5.4 Spatial Sampling

If the hologram is recorded as a composite of many small areas each of which records holographic fringes formed with one wavelength only, false reconstruction and false images can be eliminated entirely [*17.8*]. A random distribution of these areas or elements is preferable, but an interlaced sequence, ordered according to wavelength, is permissible if the elements are small. In the reconstruction step each element must be illuminated only by light of the wavelength used to form it. There then can be no cross-modulation and no false images. Each wavelength component of the subject wave is reconstructed from discrete noncontiguous samples of hologram area. Resolution is sacrificed in the process.

Several ways of implementing this form of sampling are possible. One can overlay the panchromatic recording material with a mosaic of color filters. Each filter allows only one wavelength component to expose the material behind it. After processing, the hologram must be placed back in register with the mosaic. If the spectral bandwidth of the filters is sufficiently narrow, e.g., 50 Å or less, white light can be used to illuminate the hologram and generate a multicolor image. A better alternative would be to make the recording material itself have the filtering property. Figure 17.6

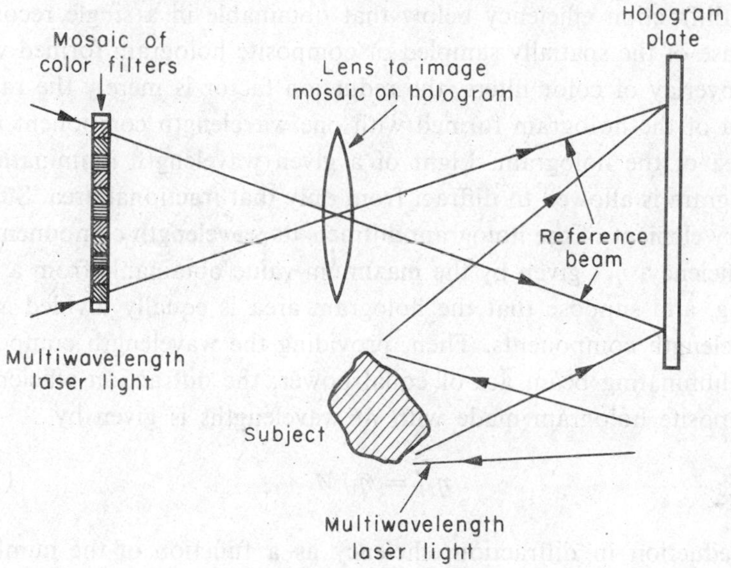

FIG. 17.6. Formation of a composite hologram for multicolor imaging.

illustrates a method which makes use of the fact that a hologram can only be formed on those areas of the recording medium where *both* subject and reference wave of the same wavelength are simultaneously present. A mosaic of color filter strips, one color for each wavelength, is imaged onto the hologram plate with the multiwavelength reference beam. The image, consisting of discrete areas of single-wavelength light, then acts as the reference wavefront. Although no filtering is applied to the subject wavefront, the holograms formed are spatially separated according to wavelength. The multiwavelength subject wavefront is reconstructed when the hologram is placed back in register and illuminated with the reference wavefront mosaic image. Plate IV is a photograph of a multicolor image obtained in this way. Two wavelengths, 6328 and 4880 Å, were used and the reference-to-subject beam angle was again 15°.

17.5.5 Diffraction Efficiency of Plane Holograms Producing Multicolor Images

We define the diffraction efficiency of a hologram formed with light of several wavelengths and illuminated with these same wavelength components as a ratio of the total optical power in the multiwavelength reconstructed subject waves to that in the incident light. The definition is consistent with that applied in Section 8.5 to plane holograms generating monochrome images. However, the multiple recording required for color holography reduces diffraction efficiency below that obtainable in a single recording. In the case of the spatially sampled or composite hologram formed with a mosaic overlay of color filters, the reduction factor is merely the ratio of that area of the hologram formed with one wavelength component to the total area of the hologram. Light of a given wavelength illuminating all the hologram is allowed to diffract from only that fractional area. Suppose that every element of the hologram diffracts its wavelength component with equal efficiency, η_1, given by the maximum value obtainable from a single recording, and suppose that the hologram area is equally divided among the wavelength components. Then, providing the wavelength components in the illuminating beam are of equal power, the diffraction efficiency of the composite hologram made with M wavelengths is given by

$$\eta_M = \eta_1/M. \tag{17.19}$$

The reduction in diffraction efficiency as a function of the number of wavelength components is worse for the case where holograms are superimposed, e.g., as in Section 17.5.2, where the cross-modulation spectra are separated from the subject wave spectrum, or in Section 17.5.3 where a coded reference wave is used. In the following analysis we shall consider M superimposed holograms, each formed with one of M wavelength components and recorded linearly on thin absorptive material. Each hologram, under circumstances identical but for the wavelength, records the simple interference of equal-amplitude signal and reference plane waves. Each diffracts with the same, maximum possible efficiency.

Let us first recall that a single plane hologram formed with two plane waves diffracts with maximum efficiency when its transmittance is given by Eq. (8.31). We rewrite that equation once again as

$$t = t_0 - t_{\mathrm{E}} \tag{17.20}$$

where t is the transmittance of the hologram, t_0 is the transmittance of the unexposed hologram plate and the maximum value that t can have, and

$$t_E = t_{E_0} + t_1 \cos(2\pi\xi x) \tag{17.21}$$

is the exposure-dependent term see Eq. (8.30). The relative amplitude of light diffracted into a first-order wave increases with the modulation amplitude t_1. However, t_1 cannot exceed the average or bias component t_{E_0} without t periodically exceeding the unexposed value t_0. Since this is not allowed, the highest efficiency for a given exposure is obtained when

$$t_1 = t_{E_0}. \tag{17.22}$$

We now consider the superposition of M holographic exposures on the same thin recording material. Each exposure is made with a new wavelength. After the M exposures, the amplitude transmittance t becomes

$$t = t_0 - \sum_{m=1}^{M} [t_m + t_m \cos(2\pi\xi_m x + \Delta_m)], \tag{17.23}$$

where Δ_m is a phase constant, or

$$t = t_0 - \sum_{m=1}^{M} [t_m + \tfrac{1}{2}t_m\{\exp(2\pi i\xi_m x + i\,\Delta_m) + \exp(-2\pi i\xi_m x - i\,\Delta_m)\}] \tag{17.24}$$

where we have applied Eq. (17.22) to each exposure. We assume that $t_m = \bar{t}$, a constant for all m, but that the phase Δ_m is arbitrary. When the hologram of transmittance t given by Eq. (17.24) is illuminated by the M original reference waves each of unit amplitude, then each hologram component of the superimposed set reconstructs its *original* signal wave at its *original* wavelength with an efficiency $(t_m/2)^2 = (\bar{t}/2)^2$. It follows that the set as a whole has a collective efficiency also given by

$$\eta_M = (\bar{t}/2)^2. \tag{17.25}$$

Because Δ_m is arbitrary and the holograms are linear recordings, we assume that in the worst case there is some point in the recording material where the M sinusoidal transmittance modulations add their maximum amplitudes $\bar{t}$. In that case Eq. (17.23) becomes

$$t = t_0 - M\bar{t} - M\bar{t} = t_0 - 2M\bar{t}. \tag{17.26}$$

Since the minimum value of t is 0, we find that the maximum value of $\bar{t}$

is given by

$$t_0 - 2M\bar{t} = 0$$

or

$$\bar{t} = \frac{t_0}{2M}. \tag{17.27}$$

Substituting Eq. (17.27) into (17.25) we have

$$\eta_M = \left(\frac{t_0}{4M}\right)^2. \tag{17.28}$$

For a single exposure $M = 1$ and $\eta_1 = (t_0/4)^2$. Hence

$$\eta_M = \eta_1/M^2. \tag{17.29}$$

Thus the efficiency reduction factor for holograms superimposed on a thin recording medium is M^2. On the basis used for the computation, one would expect a hologram, recorded with three wavelength components, to provide a multicolor image $\frac{1}{9}$ as bright as the comparable monochrome image. In practice the linearity condition is compromised somewhat to achieve higher diffraction efficiency. Nevertheless multicolor images recorded in photographic emulsions are still substantially dimmer than their monochrome counterparts.

17.6 Multicolor Images from Volume Holograms

Volume holograms offer distinct advantages over plane holograms for multicolor imaging. False reconstructions are eliminated without having to incur penalties in resolution, contrast, or spatial frequency bandwidth. No special techniques in formation or reconstruction are necessary. What is required is a recording medium of sufficient thickness.

Consider several holograms superimposed in a thick medium, each formed with a different wavelength. Suppose for the moment that the holograms are elementary ones formed with plane signal and reference waves. If diffraction efficiency is to be a maximum, the holograms should be illuminated with the original reference waves satisfying Bragg's law:

$$2n_0\, d \sin\theta_0 = \lambda_a. \tag{9.10}$$

The parameters in the above equation apply to each hologram and the associated reference wave used to illuminate that hologram. Here n_0 is the average index of refraction of the recording medium, d is the grating spacing, θ_0 is the Bragg angle made by the illuminating beam with the fringe planes, and λ_a is the illuminating wavelength. When the wavelength of the illuminating light deviates from that satisfying Bragg's law, diffraction efficiency drops, as illustrated in Figs. 9.6, 9.8, 9.12, and 9.14 for the several varieties of volume holograms. The abscissa ξ in the first two figures is related to $\Delta\lambda$, the deviation from the Bragg wavelength, by

$$\xi = -(\Delta\lambda/\lambda_a) \tan\theta_0 (2\pi n_0/\lambda_a) T \sin\theta_0 \quad \text{(transmission)} \qquad [9.85]$$

where T is the thickness of the recording medium. In Figs. 9.12 and 9.14 the abscissa is $\xi_r = \Gamma T/(2\cos\psi_0)$, and ξ_r is related to $\Delta\lambda$ by

$$\xi_r = -(\Delta\lambda/\lambda_a)(2\pi n_0/\lambda_a) T \sin\theta_0 \quad \text{(reflection).} \qquad [9.108]$$

All four figures thus indicate the sensitivity of volume holograms to deviation from the Bragg wavelength when illuminated at the Bragg angle. For sufficiently large values of $\Delta\lambda$, diffraction efficiency can become negligible. Superimposed volume holograms illuminated at the Bragg angle by the original reference wave therefore exhibit no cross-modulation when the wavelengths of the illumination components are sufficiently separated.

In Chapter 9 we compute the smallest wavelength deviation $\Delta\lambda$ which reduces the diffraction efficiency of a phase volume hologram to zero. However in the case of reflection absorption holograms recorded for maximum efficiency ($\alpha T/\cos\psi_0 = 2$ in Fig. 9.14), the effciency versus ξ_r curve does not exhibit a zero minimum, and these holograms cannot be characterized in the same way. And yet it is important to include them since reflection holograms formed in photographic emulsion (absorption holograms) are currently the best way to accomplish multicolor holographic imaging. Let us therefore take as a nominal measure of the wavelength response of both transmission and reflection holograms the deviation $\Delta\lambda = \Lambda$ for which the relative diffraction efficiency $\eta/\eta_0 = 0.5$. We shall restrict our concern here to absorption holograms formed in emulsion and to factors which bear on color cross-modulation. We assume that the multiwavelength components used to form the holograms are combined into a single reference beam, and that we can substitute for the Bragg angle θ_0 [in Eq. (9.85) or (9.108)] half the angle between the reference and mean subject beam directions in the medium. Our computations should take into account the fact that Λ varies with the square of the wavelength [see Eqs. (9.85) and (9.108)]; for any pair of wavelength components λ_i and λ_{i+1} used to form and illuminate a multicolor-imaging volume hologram, the computation of Λ should be made at the higher wavelength λ_{i+1}. The criterion for cross-modulation elimination can therefore be expressed as

$$\lambda_{i+1} - \lambda_i \gg \Lambda_{i+1}. \tag{17.30}$$

If the value of Λ is small enough, then the volume hologram can be illuminated by a continuous spectrum of white light. As we note in Section 9.7.1, the wavelength deviation characterizing the response of reflection holograms formed in lossless dielectrics can be less than 100 Å. For white-light illumination the maximum permissible value of Λ is determined not by the relatively widely spaced, laser-light wavelength components used to form the hologram but by the acceptable resolution $\Delta\sigma$ given in Eq. (7.53),

$$\Delta\sigma = \theta_r z_1 \, \Delta\lambda/\lambda$$

where we may substitute Λ for $\Delta\lambda$. In the above expression θ_r is the reference beam angle and z_1 is the distance of the subject or its image from the hologram. It is advantageous to use a white-light source whose area is as small as is consistent with the required light power and to arrange that the subject image appears as close to the hologram plate as possible.

17.6.1 Transmission Holograms

The absence of cross modulation in the multicolor images generated by volume holograms was first demonstrated with a transmission hologram recorded in Kodak 649F photographic emulsion as in Fig. 17.7 [*17.9*]. In the arrangement, the mean subject-beam direction and the plane-wave, reference-beam direction make an angle of 90°. Each beam is incident on the hologram plate at an angle of 45°. Though only two wavelengths, $\lambda_1 = 4880$ Å and $\lambda_2 = 6328$ Å, were used in the original demonstration, the same arrangement can be used for three wavelengths [*17.10*]. Plate V is a photograph of the first multicolor image obtained with the 6328 and 4880 Å laser wavelength components from the He–Ne and argon lasers, respectively. The original subject was a color transparency, and the reconstruction of the subject wave was achieved by illuminating the hologram with the original reference wave.

Let us compute the value of Λ from Eq. (9.85) for the conditions indicated in Fig. 17.7. A nominal value for the emulsion thickness after processing is $T = 12$ μm; the index of refraction is $n_0 = 1.5$. We take the longer of the two wavelengths, 6328 Å, for λ_a the wavelength in air. Figure 9.8 indicates a value for ξ of 1.5 when $\eta/\eta_0 = 0.5$. After the subject and reference beams

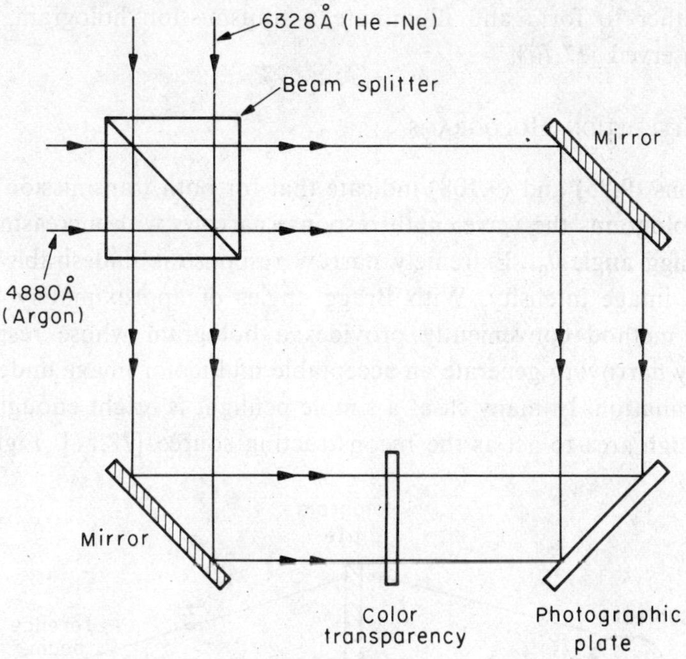

FIG. 17.7. An arrangement for forming a transmission volume hologram with two wavelengths. (After Pennington and Lin [*17.9*].)

refract at the air–emulsion boundary, the angle between them reduces to approximately 56°, making the Bragg angle $\theta_0 \approx 28°$. We now use these parameter values to solve Eq. (9.85) for the wavelength deviation $\Delta\lambda = \Lambda$, corresponding to half the maximum diffraction efficiency, and we obtain

$$\Lambda = 210\ \text{Å}.$$

A glance at Fig. 9.8 shows that the efficiency becomes zero when $\xi = 3$ or $\Delta\lambda = 420$ Å. For higher values of ξ, the efficiency rises again but only to a small sidelobe maximum of $\eta/\eta_0 \approx 0.05$ at $\Delta\lambda = 620$ Å. Other minima and lesser maxima follow. This behavior indicates that the computed value of Λ is sufficiently small relative to $\lambda_2 - \lambda_1 = 6328\ \text{Å} - 4880\ \text{Å} = 1448$ Å for adequate reduction of cross-modulation of 6328 Å light by the hologram formed at 4880 Å.

A similar calculation, when carried out for the green emission from the argon laser at 5145 Å, wavelength, yields a value for Λ of approximately 140 Å. This green argon line is separated from the blue at 4880 Å by only 265 Å. Since $\Delta\lambda = 265$ Å corresponds to $\xi \approx 2.8$ and $\eta/\eta_0 \approx 0.02$, one might perhaps see some small evidence of cross modulation when they are used together to form and illuminate a transmission hologram. This is indeed observed [17.10].

17.6.2 Reflection Holograms

Equations (9.85) and (9.108) indicate that for both transmission and reflection holograms the wavelength response narrows with increasing values of the Bragg angle θ_0. Extremely narrow response is undesirable since it limits the image intensity. With Bragg angles of approximately 80°, the reflection method conveniently provides a hologram whose response is sufficiently narrow to generate an acceptable multicolor image under white-light illumination. In many cases a simple penlight is bright enough and of small enough area to act as the reconstructing source [17.11]. Figure 17.8

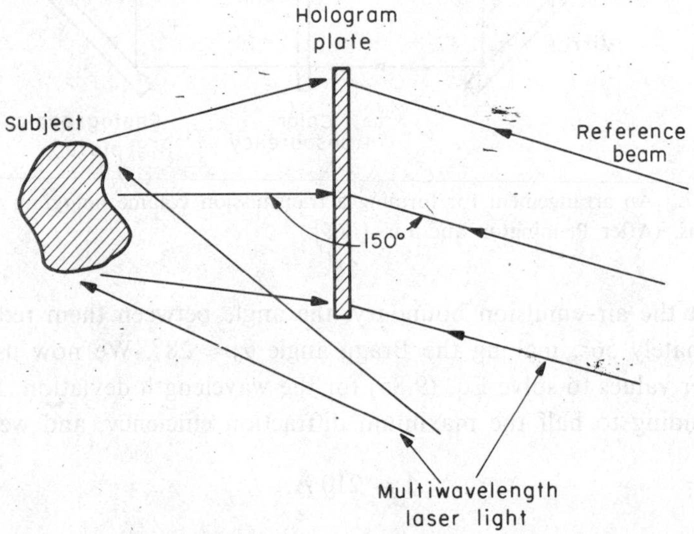

FIG. 17.8. Formation of a reflection hologram.

indicates the arrangement for forming a reflection hologram. The only unique aspect is that the reference wave and subject wave illuminate opposite sides of the hologram plate. Either monochrome or multicolor imaging holograms can be formed in this way. However if more than one wavelength is used to form a volume hologram, the rewards of the reflection hologram technique become multiplied [17.12]. Because white-light illumination is permitted, a potential viewer is saved the necessity of having to provide not one but several laser wavelengths to generate the multicolor image.

Let us now compute Λ for the reflection hologram formed with a longest wavelength of 6328 Å and with the geometrical arrangement of Fig. 17.8. Again using $n_0 = 1.5$ for the index of refraction of the emulsion, we calculate the Bragg angle in the emulsion to be $\theta_0 \approx 80°$. From Fig. 9.14 $\xi_r = 2$ when $\alpha T/(\cos \theta_0) = 2$ and $\eta/\eta_0 = 0.5$. If we insert these values and the thickness of the hologram $T = 12$ μm into Eq. (9.108) we obtain $\Lambda \approx 70$ Å. Similar values are obtained experimentally and prove to be sufficiently small to allow multicolor imaging with white-light illumination [*17.12–17.14*].

Since the interference maxima recorded in reflection holograms are spaced by approximately one-half wavelength in the medium, the resolution required of the recording material must be high (see Section 10.3). For the same reason a high degree of mechanical stability is required of the optical components used to form reflection holograms (Section 7.5). When reflection holograms are formed in photographic emulsion, however, the major problems arise from chemical processing. Distortion or shrinkage of the emulsion is far more significant than for transmission holograms.

During chemical processing and afterward, the thickness of the emulsion undergoes expansion and contraction, and it is difficult to avoid some deformation in drying the plate. The recorded interference fringe surfaces lie nearly parallel to the plate surface in a very thin (10 μm) layer of emulsion. A deformation normally regarded as small may still be an appreciable fraction of the emulsion thickness and may distort the spacing of the many fringes stacked throughout the thickness. Distorted in turn are the reconstructed wavefronts and the color of the image generated by a white-light illuminated hologram. To minimize these undesirable effects, the reference beam, when forming the hologram, should be incident on that side of the emulsion adjacent to the glass plate. Emulsion closest to the glass can be expected to distort the least. On reconstruction, the illuminating wave again illuminates the least-distorted side of the emulsion first. Since there is considerable attenuation through reflection holograms (they have an optical density of about 2) this least-distorted portion of the hologram will contribute most to the reconstruction of the subject waves.

Normal chemical processing of photographic emulsion usually results in shrinkage of the emulsion thickness. At the root is the removal of the undeveloped silver halide compounds from the emulsion during the fixing process and the subsequent collapse of the gelatin. Most available photographic emulsions contain an excess amount of silver halide for holographic purposes and much remains after development of the hologram. Consequently the emulsion shrinkage after fixing is appreciable (20 to 30% for Kodak 649F). The most prominent effect is a change in the color of images generated from reflection holograms illuminated with white light. As an example, consider a reflection hologram recorded in Kodak 649F emulsion with 6328-Å red laser light. After normal processing, the hologram is illuminated with white light incident at the original reference beam angle. The color of the image is observed to be green (~5300 Å). If multicolor images from reflection holograms are to be reasonably faithful color reproductions of the original subjects, emulsion shrinkage must be corrected. Corrective swelling using the agent triethanolamine as described in Section 10.8.4 can do this to a high degree of accuracy. Since the emulsion thickness is also sensitive to the ambient humidity, it is best to place a cover glass over the emulsion and seal the edges so as to isolate the corrected hologram from the atmosphere.

Reflection holograms formed in photographic emulsion generally diffract with less efficiency at the shorter wavelengths. Unfortunately, most white-light sources which might be used to illuminate such holograms emit more power at the longer wavelengths. One should try to compensate when forming the hologram by adjusting the exposures for various wavelength components accordingly. A zirconium-arc lamp is one of the best choices for a white-light illuminating source. It has relatively high output near the blue end of the spectrum, and the source size is small. Plate VI is a photograph of a multicolor holographic image (of a test pattern) generated with white light from a zirconium-arc lamp. The reflection hologram was formed with three wavelengths, 6328, 5145 and 4765 Å. Plate VII is a photograph of a multicolor three-dimensional image obtained in the same way except that only two wavelengths, 6328 and 5145 Å, were used in forming the reflection hologram.

17.6.3 Diffraction Efficiency of Volume Holograms Producing Multicolor Images

In Section 16.3.6 we analyze the effect that superposition has on the maximum efficiency of any one hologram in a set of M superimposed volume holograms. The theory is directly applicable here where now we understand that each hologram is formed with the same beam directions but with a different wavelength. As in Section 16.3.6 we assume the absorption constant α_E is proportional to exposure and that M independent superimposed recordings, identical but for wavelength, lead to the condition

$$\alpha_1 = \alpha_2 = \cdots = \alpha_m = \cdots = \alpha_M = \alpha/M. \qquad (17.31)$$

Here α, the total average absorption constant after M exposures, is half the recording range of the photosensitive material, and α_1 is the modulation amplitude of each hologram [see Eq. (16.21)]. By substituting Eq. (17.31) into the expression for the amplitude of the wave diffracted by an absorptive transmission volume hologram, Eq. (9.94), viz.

$$S(T) = -\exp\left(\frac{-\alpha T}{\cos\theta_0}\right)\sinh\left(\frac{\alpha_1 T}{2\cos\theta_0}\right), \tag{9.94}$$

and by maximizing $|S(T)|$ with respect to $\alpha_1 T/(2\cos\theta_0)$, we obtain $|S_{\max}|$ as a function of M. Squaring the result yields the efficiency

$$\begin{aligned} S^2_{\max} &= \exp\left(\frac{-2\alpha T}{\cos\theta_0}\right)\cdot\frac{1}{4M^2-1} \\ &\approx \exp\left(\frac{-2\alpha T}{\cos\theta_0}\right)\frac{1}{4M^2} \\ &\propto \frac{1}{M^2} \end{aligned} \tag{17.32}$$

for $M > 1$. The same result can be readily obtained graphically from Fig. 9.7. We see that the efficiency of any one of the superimposed volume holograms falls off with square of the number of holograms, just as for plane absorption holograms. A similar result is obtained when Eq. (17.31) is applied to Eq. (9.111), the amplitude of the wave diffracted by an absorption reflection hologram.

REFERENCES

17.1. G. Wyszecki and W. S. Stiles, *Color Science*, Sections 3 and 4. Wiley, New York, 1967.

17.2. The Optical Society of America, Committee on Colorimetry, *The Science of Color*, p. 220. Crowell, New York, 1953.

17.3. C. B. Burckhardt, "Display of Holograms in White Light," *Bell Syst. Tech. J.* **45**, 1841 (1966).

17.4. D. J. DeBitetto, "White-Light Viewing of Surface Holograms by Simple Dispersion Compensation," *Appl. Phys. Lett.* **9**, 417 (1966).

17.5. E. N. Leith and J. Upatnieks, "Wavefront Reconstruction with Diffused Illumination and Three-Dimensional Objects," *J. Opt. Soc. Amer.* **54**, 1295 (1964).

17.6. L. Mandel, "Color Imagery by Wavefront Reconstruction," *J. Opt. Soc. Amer.* **55**, 1697 (1965).

17.7. E. Marom, "Color Imagery by Wavefront Reconstruction," *J. Opt. Soc. Amer.* **57**, 101 (1967).

17.8. R. J. Collier and K. S. Pennington, "Multicolor Imaging from Holograms Formed on Two-Dimensional Media," *Appl. Opt.* **6**, 1091 (1967).

17.9. K. S. Pennington and L. H. Lin, "Multicolor Wavefront Reconstruction," *Appl. Phys. Lett.* **7**, 56 (1965).

17.10. A. A. Friesem and R. J. Fedorowicz, "Multicolor Wavefront Reconstruction," *Appl. Opt.* **6**, 529 (1967).

17.11. G. W. Stroke and A. E. Labeyrie, "White Light Reconstruction of Holographic Images Using the Lippmann–Bragg Diffraction Effect," *Phys. Lett.* **20**, 368 (1966).

17.12. L. H. Lin, K. S. Pennington, G. W. Stroke and A. E. Labeyrie, "Multicolor Holographic Image Reconstruction with White-Light Illumination," *Bell Syst. Tech. J.* **45**, 659 (1966).

17.13. J. Upatnieks, J. Marks and R. J. Fedorowicz, "Color Holograms for White Light Reconstruction," *Appl. Phys. Lett.* **8**, 286 (1966).

17.14. L. H. Lin and C. V. LoBianco, "Experimental Techniques in Making Multicolor White Light Reconstructed Holograms," *Appl. Opt.* **6**, 1255 (1967).

Chapter 18

COMPOSITE HOLOGRAMS

A *composite hologram* is a collection of small holograms arranged in a plane with each individual hologram close to and often contiguous with its neighbor. We shall call the individual holograms *elements* of the composite. The wavefronts *recorded* by neighboring elements are not necessarily continuous or coherent with one another. Nevertheless, when all elements are simultaneously illuminated, the wavefronts they reconstruct can cooperate to produce a desired image.

In Chapter 17 we describe employment of a composite hologram to form a multicolor image. There, each interlaced set of elements is illuminated by a source of a different wavelength. Here we confine the discussion, for the most part, to illumination with a single coherent source. Although the subject waves may not be mutually coherent when recorded, they are coherent on reconstruction.

We shall find that formation of composite holograms and manipulation of their elements lead to results not obtainable with a single hologram. Composite holograms have been employed to reduce information content in a hologram, to exaggerate or diminish stereoscopic effects, to provide a full 360° view of an image, and to permit 3D holographic images to be synthesized from 2D photographs of real scenes or nonexistent scenes generated by a computer.

18.1 Image Resolution and Element Size

All of the images generated by the composite holograms we shall discuss are intended to be viewed directly by the human eye and are therefore virtual images. As a consequence the smallest detail in the image need be no smaller than that which can be resolved by the eye at normal viewing distances. However, a poor choice of the size of a hologram element can degrade the image quality to a degree apparent to the eye. In this case the limit to resolution is imposed by diffraction from the boundaries of the element. It is reasonable to choose the size of an element such that it deteriorates the resolution by the same amount as the finite aperture of the eye. Making the element smaller would only produce a lesser resolution. On the other hand, increasing the element size would be of little benefit, since resolution would then be limited by the eye. Moreover, when a large number of elements are required, too large an element size tends to make the composite hologram impractical.

Let us now establish a rough criterion for selection of the element size. In Fig. 18.1 an eye with an aperture of diameter D_e is shown observing a

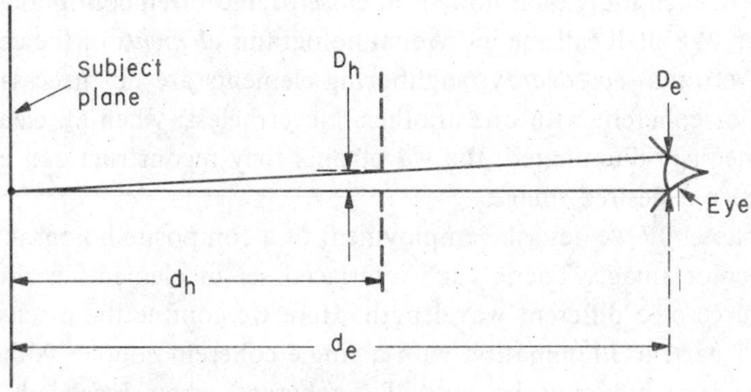

FIG. 18.1. Observation of a subject point.

subject point at a distance d_e. A spread function in the *subject plane* can be defined in the same form as Eq. (16.23) (see also Section 6.4.3). Thus, diffraction from the boundaries of the eye aperture permits the eye to observe a spot on the subject no smaller than

$$\Delta_e = (1.22\lambda d_e)/D_e \tag{18.1}$$

where λ is the wavelength of the light coming from the subject. We assume the resolution of a healthy eye to be diffraction limited. Also shown in Fig. 18.1, in dotted lines, is a mask containing a circular element of the composite hologram whose diameter is D_h and whose separation from the subject is d_h. The element, when properly illuminated, generates a virtual image at the original subject site with a spread function of width

$$\Delta_h = (1.22\lambda d_h)/D_h. \tag{18.2}$$

For equal widths of the two spread functions in the observed image, comparison of Eqs. (18.1) and (18.2) yields the relation

$$d_e/D_e = d_h/D_h. \qquad (18.3)$$

We should then choose the diameter of the element to be

$$D_h = D_e d_h/d_e. \qquad (18.4)$$

Assigning some typical values: $D_e = 3$ mm, $d_e = 60$ cm, and $d_h = 20$ cm, we find the diameter of an element to be $D_h = 1$ mm. Suppose we illuminate the hologram element with light of wavelength $\lambda = 0.5$ μm. Then substitution into Eq. (18.2) yields for the width of the spread function, $\Delta_h = 122$ μm.

When we consider the cooperative effects of more than one element, we find additional factors which can affect the quality of the observed image. As we shall see, certain applications of composite holograms require adjacent elements to reconstruct waves diverging from identical, supposedly coincident, virtual images. When the adjacent elements are imperfectly aligned, their images will not appear coincident, and the overall effect on the observation is blurring. On the other hand, very close alignment can introduce noticeable effects on the image as well. When elements reconstructing nearly identical wavefronts are illuminated coherently, the wavefronts can interfere and produce a set of linear fringes superimposed on the image.

As an example, let us suppose that two identical adjacent elements, whose centers are separated by a distance D_h equal to the element diameters, lie in the Fourier transform or spatial frequency plane of the image. According to Eq. (6.21), a physical separation D_h in the frequency plane corresponds to a spatial frequency shift $\xi = D_h/\lambda d_h$ [where d_h has been substituted for f in Eq. (6.21)]. Now consider the effect of this frequency shift at the image plane. If $\mathbf{a}(x)$ is the complex amplitude of the image generated by one of the elements, then $\mathbf{a}(x)\exp(-2\pi i(D_h/\lambda d_h)x)$ is the complex amplitude generated by the other [see the shift operation, correspondence (4.21)]. The intensity of the interference pattern is given by

$$I = 2|\mathbf{a}|^2[1 + \cos\big(2\pi(D_h/\lambda d_h)x\big)],$$

yielding a sinusoidal fringe pattern modulating the image intensity with a period

$$\delta = \lambda d_h/D_h. \qquad (18.5)$$

Comparison of Eq. (18.5) with Eq. (18.2) shows that an eye which can resolve $\Delta_h/1.22$ will also resolve the fringes.

Thus when the elements are made too small, not only is the resolution poor, but annoying interference fringes are more easily observed [Eq. (18.5)]. In composite holograms applied to information reduction, the elements represent repeated samples of an original hologram (Section 18.2). Since the aim is to reduce the information content in the composite over that in the original hologram, the desire to minimize the sample size is understandable. Equation (18.4) provides a reasonable compromise in element size. If the element has the shape of a long strip rather than a circle, then the strip width can be substituted for D_h in Eq. (18.4). Equation (18.2), modified by removal of the factor 1.22, then gives the resolution Δ_h in the direction transverse to the strip length.

18.2 Hologram Information Reduction

Having established criteria for choosing the size of an element, we now consider the first of several possible applications of composite holograms. Suppose we wish to transmit all the information present on a 10 × 10-cm hologram over a video channel of 5-MHz bandwidth. If the subject of the hologram is a diffusely reflecting object, fringe frequencies perhaps as high as 2000 cycles/mm in the horizontal direction and 1000 cycles/mm in the vertical direction may be recorded. To transmit this information we must scan the hologram with an electro-optic sensor capable of resolving the fringes and then send the results to a receiver at the 5-MHz rate. Since there can be (2000 cycles/mm) × (100 mm) = 2×10^5 cycles of modulation in the horizontal or x direction, the time required for a single horizontal scan is

$$t = \frac{2 \times 10^5}{5 \times 10^6} = 4 \times 10^{-2} \quad \text{sec.}$$

The number of horizontal scans required to cover the 10-cm-high hologram is determined by the maximum spatial frequency in the vertical or y direction and by the sampling theorem (see Section 19.1). According to the latter, we should sample each of the 1×10^5 cycles of vertical spatial modulation at least twice. Hence, the number of scans is $N = 2 \times 10^5$ and the total transmission time is

$$Nt = (2 \times 10^5)(4 \times 10^{-2}) = 8 \times 10^3 \text{ sec} = 2.2 \text{ h}.$$

It is obvious that the information content represented by the hologram must be drastically reduced if video transmission of a hologram is to be practical.

Several methods to effect the reduction have been suggested [*18.1–18.5*], although there are as yet no reports of any being used in actual transmission. All but one [*18.4*] involve composite holograms.

18.2.1 The Screen Effect

Reduction of the area of the hologram is an obvious way to reduce information content. It should be feasible to do this by sacrificing some image resolution, since a large-area hologram generates images with far more resolution than required for normal viewing. However, the large area also provides a wide-angle view of a 3D image, and preservation of this feature is highly desirable. It is the purpose of all the methods referenced to reduce information content at a sacrifice in resolution but without a significant loss in the range of viewing angles. A simple means for accomplishing this is to place adjacent to the hologram a mask with a series of slits. The slits expose corresponding strips of the hologram. These strips are assumed to be electrically transmitted and exactly recreated at a receiving terminal. When illuminated with a laser beam that duplicates the reference beam used to form the original hologram, they generate a 3D image. However, the viewing is highly unsatisfactory in that the image appears to reside behind a picket fence or screen located at the hologram plane (Fig. 18.2a). This screen effect becomes more pronounced and annoying as the opaque bands of the mask are enlarged to gain further reduction in hologram area.

The receiver can eliminate the annoyance at a further sacrifice in resolution by post-processing the recreated hologram. This is done by illuminating the set of spatially separated strip holograms with a laser beam and employing the generated virtual image as a subject for a second hologram [*18.1*]. Suppose that the width of each strip of the first hologram is $1/n$ times the spacing between strips. The photographic plate for the second hologram is in that case exposed n times. Between successive exposures the first hologram is translated the width of a strip and the illuminating beam direction is rotated. It is rotated so that points in the central plane of the virtual image are returned to their original location in the image plane prior to the translation (see Section 3.3.2). (Rotation of the illuminating beam can compensate for image displacement in only one image plane; in other planes some loss of resolution due to blurring must be tolerated.) After n translations and n exposures the first hologram has traversed a distance equal to the strip spacing, and the second hologram plate has formed a record equivalent to that coming from a first hologram with no spacing between strips; the screen effect is thereby eliminated (Fig. 18.2b).

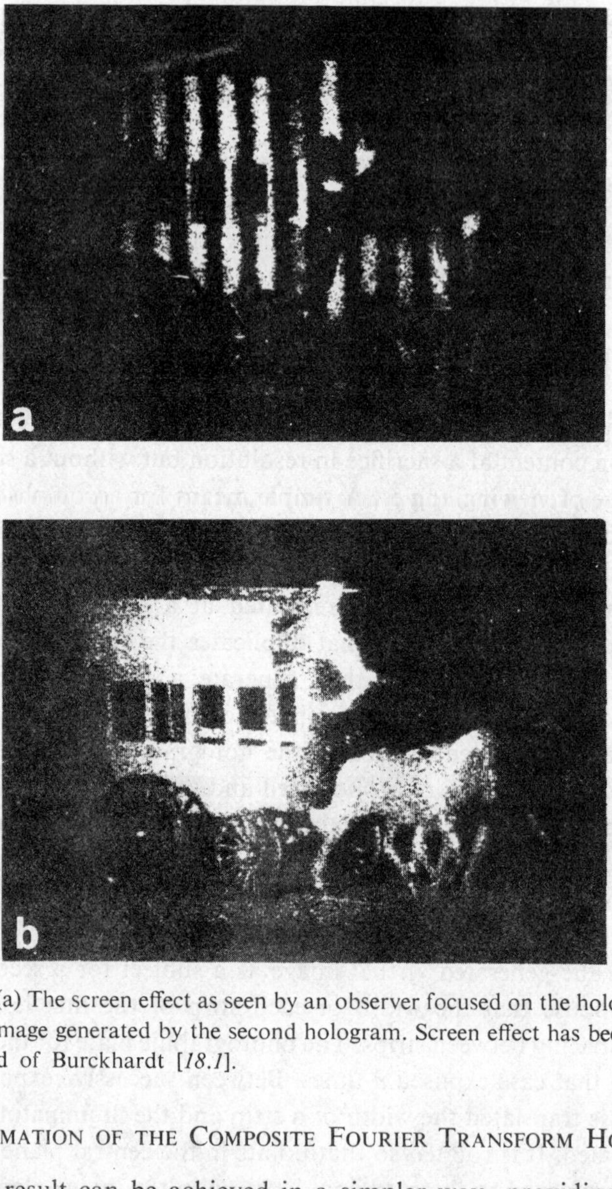

FIG. 18.2. (a) The screen effect as seen by an observer focused on the hologram image. (b) Observed image generated by the second hologram. Screen effect has been eliminated by the method of Burckhardt [*18.1*].

18.2.2 FORMATION OF THE COMPOSITE FOURIER TRANSFORM HOLOGRAM

A similar result can be achieved in a simpler way, providing the transmitted hologram is a Fourier transform hologram [*18.2*]. At the receiving station each space in the recreated hologram can be filled with *n* duplicates of an adjacent transmitted strip. Since each of the *n* duplicates is a Fourier transform hologram, their relative translations do not give rise to corresponding translations of the images they generate. As noted in Section 8.3.5, the image remains stationary with hologram translation, and all elements cooperate to form a single 3D image.

A lensless Fourier transform hologram with substantially reduced area and information content can be formed in the arrangement of Fig. 18.3a. Subject and reference source are equidistant from the hologram plane as required in Section 8.3.5, and a mask allows only the numbered areas of the hologram plate to be exposed. The recorded hologram, in the form indicated in Fig. 18.3b, has a total area and information content no larger than that required for a satisfactory image. As in the previous section, we assume that the hologram is magnified and electro-optically scanned and then electrically transmitted to a receiving terminal. There the process is reversed, and the hologram is rewritten and duplicated in large numbers. The duplicates are the elements to be assembled into a composite hologram.

Each element has the form of Fig. 18.3b with 7 subelements of area $h_x \times h_y$ spaced by a distance L. One row of the composite hologram is formed by interlacing L/h_x elements to fill the gaps in the row in the manner indicated

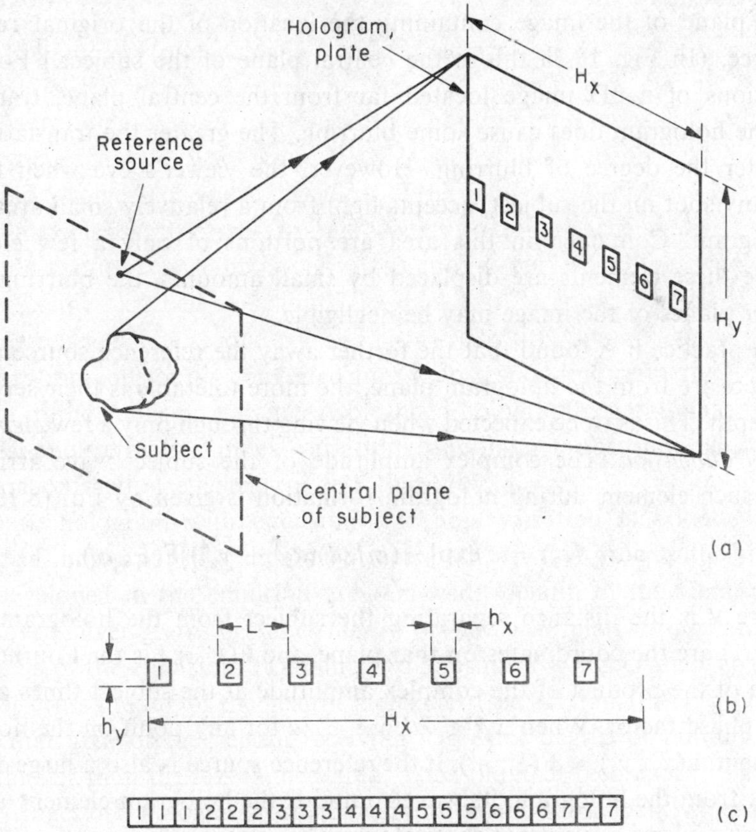

FIG. 18.3. (a) Arrangement for forming an element of a Fourier transform composite hologram. (b) An element. (c) One row of the composite hologram.

in Fig. 18.3c. Repeating the row H_y/h_y times in the vertical direction produces a filled-in composite hologram with an area $H_x \times H_y$. In a practical case this assembly process would be carried out electronically. When the assembled hologram is illuminated by the original reference, all portions diffract light and contribute to the formation of a virtual image of the subject. Comparison of the area of an element with that of the entire composite yields the factor by which the information transmitted is reduced over that normally contained in a hologram $H_x \times H_y$. The reduction factor K is given by

$$K = (L/h_x) \cdot (H_y/h_y). \tag{18.6}$$

Here K is also the number of elements in the composite.

We have noted that the geometry for forming lenless Fourier transform holograms (Fig. 18.3a) is a simple way of bestowing on the hologram image an invariance to element displacement. This invariance holds exactly only for that plane of the image containing the location of the original reference source. (In Fig. 18.3a this is the central plane of the subject.) For those portions of a 3D image located far from the central plane, translation of the hologram does cause some blurring. The greater the translation, the greater the degree of blurring. However, the viewer's eye when focused on any spot on the subject, accepts light from a relatively small area of the hologram. Contained in this area are portions of only a few elements. Since these elements are displaced by small amounts, the blurring of the outer planes of the image may be negligible.

In practice it is found that the farther away the reference source and the subject are from the hologram plane, the more tolerable is their separation in depth. This is to be expected when viewing through only a few *elements of small dimension*. The complex amplitude of the subject wave arriving at any such element during hologram formation is given by Eq. (8.19)

$$\mathbf{a}(x_2, y_2) = c \exp[-(i\pi/\lambda d)(x_2^2 + y_2^2)]\mathbf{F}(\xi', \eta'),$$

where d is the distance separating the subject from the hologram plane, (x_2, y_2) are the coordinates on that plane, and $\mathbf{F}(\xi', \eta')$ is the Fourier transform of the product of the complex amplitude at the subject times a spherical phase factor. When $x_2^2 \ll \lambda d$, $y_2^2 \ll \lambda d$ for any point on the hologram element, $\mathbf{a}(x_2, y_2) \approx \mathbf{F}(\xi', \eta')$. If the reference source is also a large distance away from the hologram, its wavefront over the hologram element approximates a plane wavefront; the hologram transmittance then approximates that of the Fourier transform hologram [see Eq. (8.20)]. Small displacements of the element therefore produce nearly imperceptible image translation.

18.2.3 Loss of Vertical Parallax

The ratio L/h_x contributes no more than a factor of 10 to the value of K in Eq. (18.6), since the dimension h_x must not be so small as to limit resolution and L must not be so large as to introduce abrupt changes in horizontal parallax. However, the factor of information reduction in the vertical direction H_y/h_y can be 100 or more if we present only one angular view of the image in the vertical direction. This large factor of information reduction is achieved at the expense of all vertical parallax [*18.2*, *18.3*, *18.5*]. Fortunately the loss has little effect on the depth perception of a human viewer; his eyes are separated horizontally and his head motion is generally in the horizontal plane. Figure 18.4 compares the quality of an image generated by a 1000-element composite hologram with that of an image from a

Fig. 18.4. Information reduction by a factor of 1000. (a) Photograph of an image generated by a normal hologram of area $H_x \times H_y$. (b) Photograph of an image generated by a composite hologram also of area $H_x \times H_y$. Each element has an area $(H_x \times H_y)/1000$

normal hologram of the same size. The composite was formed with $L/h_x = 10$, $H_y/h_y = 100$, and $K = 1000$.

18.3 Hyper- and Hypostereoscopic Hologram Images

As a second application of the composite hologram, let us consider a means for generating hologram images that exhibit greater-than-normal binocular parallax (hyperstereoscopic images) and less-than-normal binocular parallax (hypostereoscopic images). A hyperstereoscopic 3D image causes the normal dissimilarity of the images received by a viewer's left and right retinas to increase. The effect is to heighten the 3D sensation, especially for images that are far away. On the other hand, hypostereoscopic imaging improves the ease of viewing a nearby 3D image. In this case the normal angular divergence between light rays proceeding from a point on the image to the left and right eye is diminished; the physical effort of focusing both eyes on a nearby object (accommodation) is thereby lessened [*18.6*, p. 85]. These changes in binocular parallax are achieved by redistributing the spectrum of the original image [*18.7*]. The redistribution is necessary only for spatial frequency components in the horizontal direction (parallel to the line separating the eyes). We shall confine our discussion to composite hologram methods for effectively doubling or halving the distance between left and right eyes. It will be evident that the methods may be generalized to achieve any desired effective interocular distance or stereo base.

To produce a hyperstereoscopic image equivalent to that observed with twice-normal eye separation, a Fourier transform hologram of the subject is formed first. Since the hologram is in the spatial frequency plane of the subject, the image it generates is invariant to translation of the hologram in this plane. A spatial frequency redistribution is accomplished by dividing the hologram (the spatial frequency plane) into thin vertical strips of equal width (equal spatial frequency bandwidth in the horizontal direction), discarding every other strip, and reassembling the remainder contiguously in their original order. If the hologram is formed in photographic emulsion carried on an acetate base, frequency redistribution is a simple matter of cutting and regrouping.

When an image equivalent to that observed with half the normal stereo base is desired, two identical Fourier transform holograms are formed; they are cut up into thin vertical strips and interlaced with identical strips adjacent. Illumination of the composite hologram with the original reference beam produces the hypostereoscopic image.

18.4 Wide-Angle Hologram Images

As a viewer scans an ordinary, flat hologram from one side to the other in a horizontal plane, he may be able to vary the angle at which he views the image by perhaps 60°. If we define the *viewing angle* as that subtended by the hologram at the center of the image, it is clear that for a flat hologram the angle cannot exceed 180°. Yet a flat *composite hologram* can provide the viewer with a complete 360° of angular range.

Viewing angles in excess of 180° are also achieved with holograms which are formed on cylindrical surfaces surrounding the subject; these holograms are not composites [*18.8–18.11*]. However, there are certain disadvantages associated with their formation, storage, and viewing. For example, when the subject is large, the area of recording material required to surround it may become formidable. To properly view the image, the hologram should be returned to the cylindrical surface on which it was formed. These problems are absent when the composite hologram approach is taken. The flat, compact format suggests that 360°-view holograms may one day be displayed as a page in a book.

Figure 18.5 indicates the arrangement required to form a composite hologram for wide-angle viewing [*18.12*]. Each element of the composite is formed with the identical reference wave and with standard holographic procedure. However, the subject, mounted on a turntable, is rotated through a small angle between exposures. A mask, an opaque plate with a single

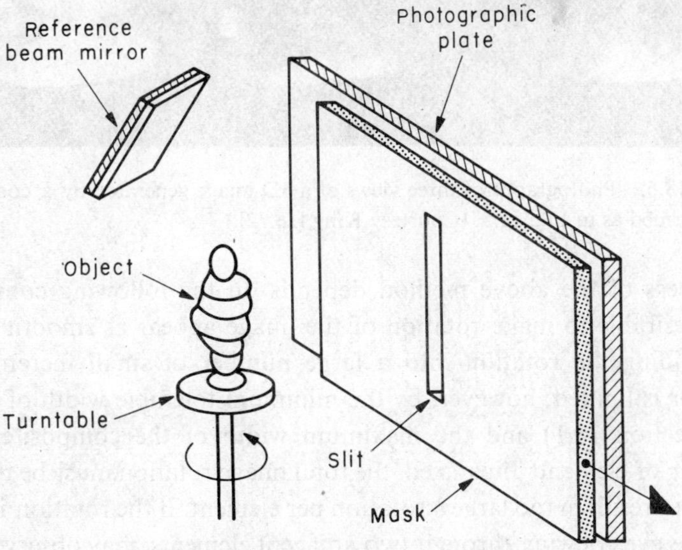

FIG. 18.5. Arrangement for forming a composite hologram for wide-angle viewing.

transparent slit placed over the recording medium during its exposure, defines the width of an element. Between exposures the mask is translated the width of an element in a horizontal direction opposing the direction of subject rotation (see Fig. 18.5). The full width of the recording medium is thus exposed strip by strip in sequence. Illumination of the entire composite hologram with a beam duplicating the original reference allows the viewer to see a virtual image of the subject. As he scans across the hologram from one side to the other, he sees the image rotate in the same direction as did the subject during formation of the hologram. Figure 18.6 displays photographs of three views of a 3D image taken at the center and at the two extreme edges of a composite hologram formed in this fashion. In this case a full 360° rotation is obtained from 80 elements, each 1.25 mm wide.

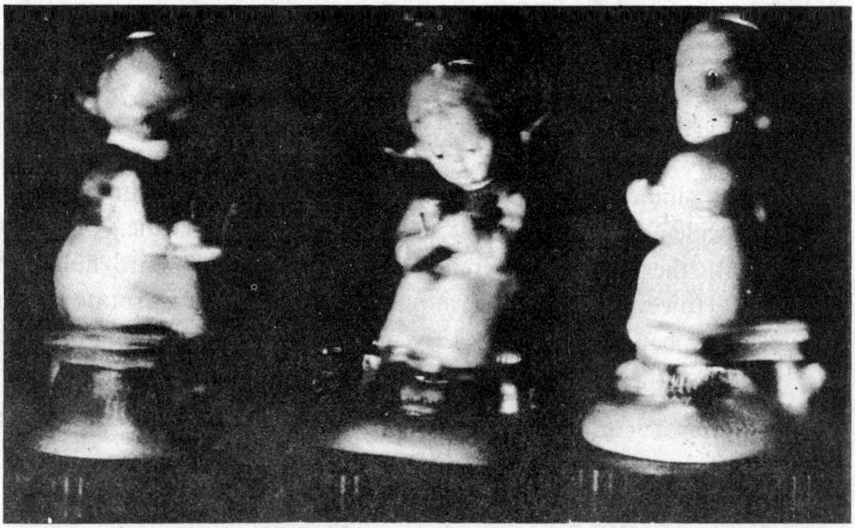

FIG. 18.6. Photographs of three views of a 3D image generated by a composite hologram formed as in Fig. 18.5. (Courtesy King [*18.12*].)

Success of the above method depends on the following considerations. It is desirable to make rotation of the image appear as smooth as possible by dividing the rotation into a large number of small increments. That number is limited, however, by the minimum tolerable width of an element (see Section 18.1) and the maximum width of the composite. With the number of elements thus fixed, the total image rotation must be restricted so as not to result in too large a rotation per element. If the rotation is excessive, a single eye, viewing through two adjacent elements, may observe a blurring or doubling of those portions of the image which are far removed from the axis of rotation and which undergo substantial displacement per element. Moreover when the image is viewed with both eyes, the disparity between the images received by left and right retinas may be so great as to prevent the brain from fusing them into a 3D representation. Double images would again be observed, and again the effect is greatest for image points farthest from the axis of rotation. It is therefore necessary to experimentally determine a satisfactory compromise between image depth and image rotation. For the composite hologram which generated the images of Fig. 18.6, a rotation of 360° was compressed into a hologram width subtending only a 30° angle at the subject; the image depth was 4 cm.

18.5 3D Image Synthesis from Photographs

When a diffusely scattering, three-dimensional body is the subject of a hologram, each small area of the hologram can generate a unique three-dimensional virtual image (see Section 8.2). The images differ according to the location of the generating areas and according to the associated subject aspect. An observer's eye, focused through one such small area of the hologram onto the image, forms a 2D image on the retina. Dissimilar images on the left and right retina are fused by the brain to give the observer an impression of depth. Let us suppose now that we can form a hologram with the following property: Each small area of the hologram through which a single eye is focused generates only a two-dimensional virtual image of the subject. If that image varies suitably from area to area, the observer's brain will still receive dissimilar left and right 2D images, and the observer will still gain the sensation of depth. A composite hologram can be formed to effect this simulation. Each hologram element is formed with light passing through one member of a set of photographs of a subject. A viewer scanning the composite from side to side perceives parallax in normal fashion.

18.5.1 Formation of the Appropriate Perspectives

One method of recording the photographs required for the hologram elements is indicated in Fig. 18.7. Ordinary white light illuminates a 3D subject. An array of lenslets, called a fly's-eye lens (when the array is two-dimensional), forms an array of images on a photographic plate [*18.13*]. We call each recorded image, when developed as a positive, a *perspective*. Each is an image of the subject as seen from a slightly different angle.

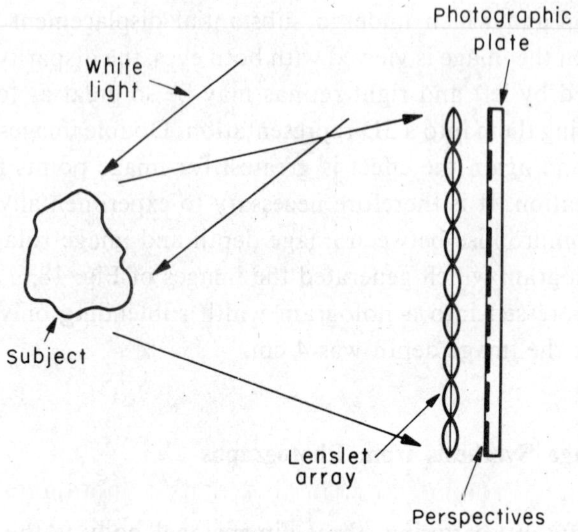

FIG. 18.7. Method of forming perspectives behind an array of lenslets.

Use of an array of lenslets is necessary only when the subject alters in time so that an instantaneous record must be formed. If the subject is stationary, the perspectives may be formed with a single lens system or camera *18.14–18.16*]. The camera can be moved to various positions around the subject corresponding, e.g., to the lenslet positions in Fig. 18.7 and the perspectives recorded sequentially on a roll of film. As with stereo pairs, the perspectives need not be photographs of physical objects. In fact one of the more interesting applications of holographic image synthesis employs perspectives drawn by a computer [*18.17*]. Actual existence of the subject, except as a set of coordinates in the memory of a computer, is not required. A computer, programmed to design a three-dimensional object, can be requested to plot on its microfilm output plotter a series of drawings representing as many views of the object as desired.

Recording the subject initially as a set of perspectives evidently has some clear advantages over direct holography. White-light-illuminated subjects, moving subjects, large-scenes and even nonexistent subjects are all suitable. Let us now consider methods for using perspectives to form the composite hologram.

18.5.2 FORMATION OF THE HOLOGRAM

Suppose that the photographic plate in Fig. 18.7 has been exposed through the array of lenslets, developed in reversal to obtain a positive, and replaced back behind the lenslet array. We now illuminate all the perspectives with diffused laser light incident on the photographic plate from the right, as in Fig. 18.8. With the light passing through each image, the corresponding lenslets project real images of the original subject focused in the original subject space. Together, these images synthesize the complete 3D image of the subject. This synthesized image can now serve as the subject of a hologram [*18.13*]. However, we should note that a real image formed by rays projected back from the perspectives along directions antiparallel to the original subject rays is pseudoscopic (see Section 8.2.1). A hologram formed with a pseudoscopic real image as subject should be illuminated with the conjugate to the reference beam; it then generates an orthoscopic image.

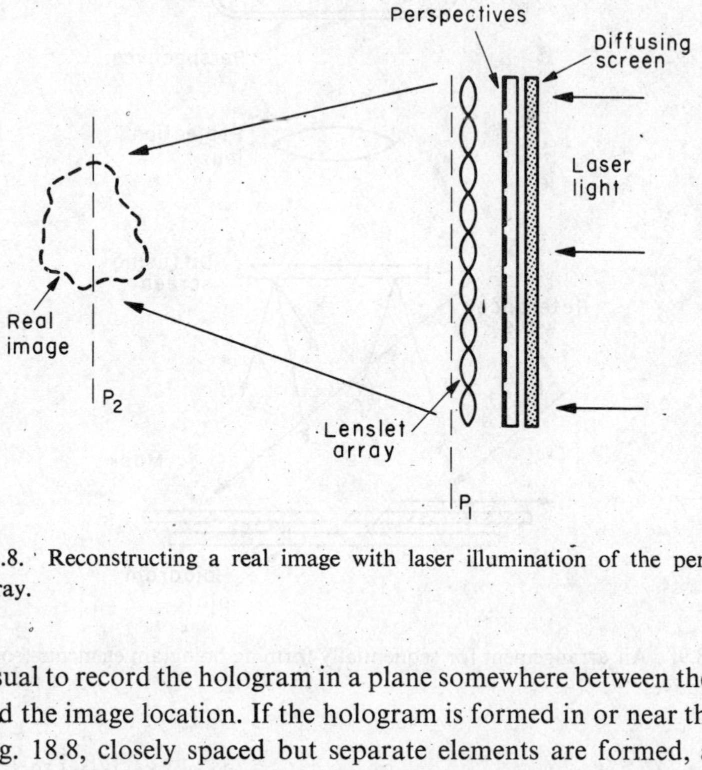

FIG. 18.8. Reconstructing a real image with laser illumination of the perspective-lenslet array.

It is usual to record the hologram in a plane somewhere between the lenslet array and the image location. If the hologram is formed in or near the plane P_1 of Fig. 18.8, closely spaced but separate elements are formed, and the record is a composite hologram in the sense defined in this chapter. If the hologram is recorded in the image plane P_2, the record is a superposition of image holograms [*18.16*]. In the latter case it is preferable to form the hologram with all perspectives simultaneously illuminated. Sequentially exposed, superimposed holograms generally diffract with low efficiency and poor signal-to-noise ratio (see Section 17.5.5). When the elements are spatially separated, as in the case of a hologram formed near plane P_1, either simultaneous or sequential exposure is equally good. Proximity of the hologram plate to the lenslet array requires the reference beam to be incident on the back side of the hologram, forming a reflection hologram.

Much more freedom in generating a composite hologram from a set of perspectives is obtained with the arrangement of Fig. 18.9. The basic components are the perspectives and the hologram plate, a laser source, a single stationary projection lens, a diffusing screen, and a mask which can be

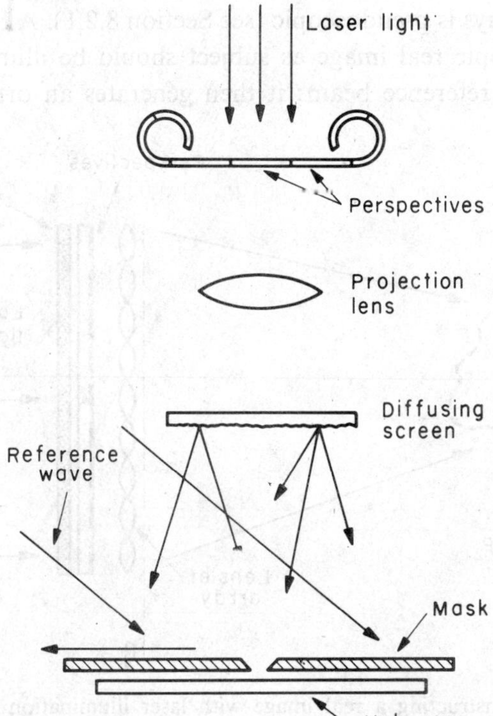

FIG. 18.9. An arrangement for sequentially forming hologram elements from a series of perspectives.

accurately translated over the surface of the hologram plate [*18.17*, *18.18*]. We describe the process for the case where the perspectives differ only in the azimuthal subject angle which they record. In this case the process is the analog of that used to form the wide-angle holograms of Section 18.4.

Suppose that we insert into the laser illumination the perspective representing the subject view to be seen through the extreme right-hand border of the hologram. Its image is projected onto the diffusing screen which, in turn, distributes the light to every part of the hologram plane. The slit in the mask is positioned so as to expose a vertical strip of the hologram plate at the extreme right edge. We may consider the diffuse light proceeding from the screen to the slit equivalent to that coming from the diffusely scattering subject in the wide-angle hologram process. Here, however, there is no vertical parallax at all. A fixed reference beam interferes with the subject light to form the hologram element. A second hologram strip to the left of the first

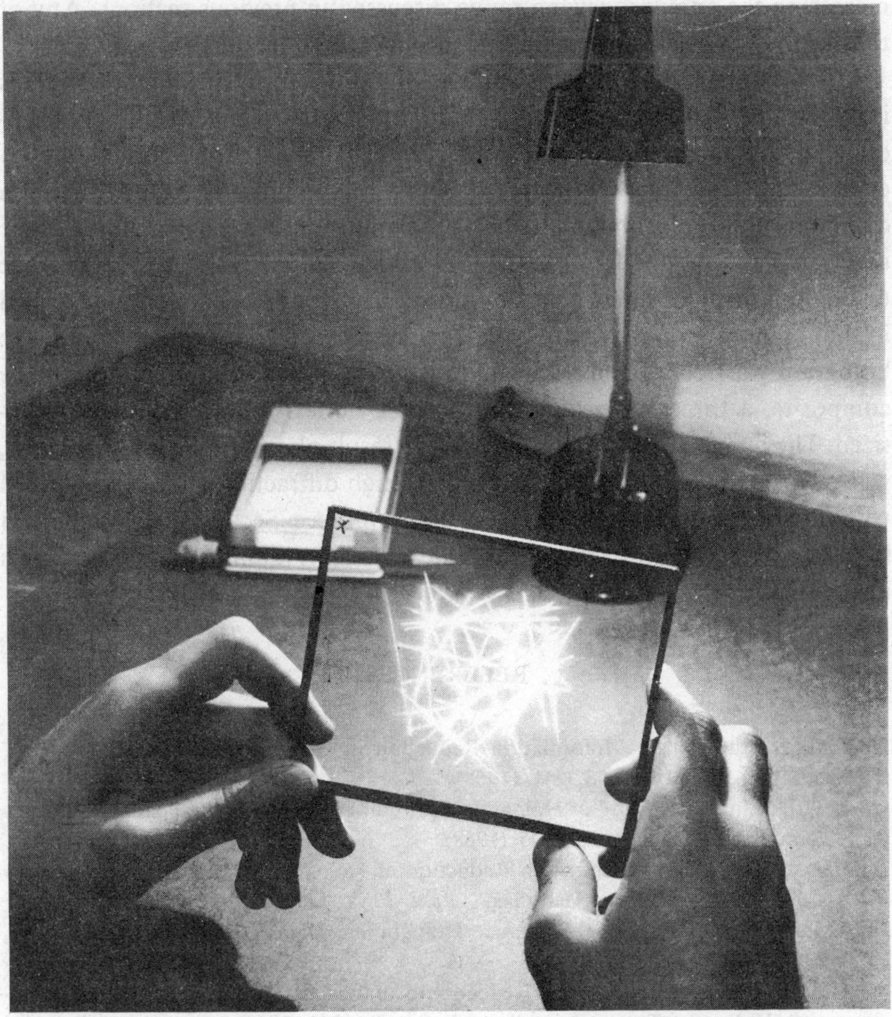

FIG. 18.10. Photograph of a 3D image of a nonexistent subject. The lamp shown in the figure is the illuminating source for the image hologram. Computer-plotted perspectives recorded as a composite hologram enabled the image, a set of randomly oriented bright lines, to be synthesized. (Courtesy King *et al.* [*18.17*].)

and contiguous with it is exposed by replacing the first perspective with the next in order and translating the mask one slit-width to the left. In sequence the entire set of perspectives is stored in the elements of a composite hologram. Illumination of the composite with the single original reference beam used to form all the elements allows the viewer to perceive a 3D virtual image having all the parallax associated with the azimuthal range of the perspectives. The advantage of the method is that it permits one to apply all the tricks for saving bandwidth, altering the degree of stereoscopy, and increasing the angular viewing range discussed in previous sections. Any of the techniques for forming the perspectives are suitable.

It is advantageous to form an image hologram from the composite. This is best done by placing a recording plate in the plane of the real image (generated by illuminating the composite with the conjugate to the original reference wave). Since the 2D perspectives are holographically recorded from a common subject position, all their 2D real images also lie in a common plane of best focus. Figure 18.10 is a photograph of the synthesized 3D image (a random array of lines in space) displayed by an image hologram formed in that plane. A computer drew the perspectives, and the original composite hologram was recorded as indicated in Fig. 18.9. Upon illumination of the composite, a three-dimensional image of a nonexistent object was synthesized. The image hologram, producing the result photographed in Fig. 18.10, was recorded on dichromated gelatin. Its high diffraction efficiency coupled with the properties of the image hologram (see Section 8.3.2) create a bright display requiring only the small, white-light illuminating lamp shown in the figure.

REFERENCES

18.1. C. B. Burckhardt, "Information Reduction in Holograms for Visual Display," *J. Opt. Soc. Amer.* **58**, 241 (1968).

18.2. L. H. Lin, "A Method of Hologram Information Reduction by Spatial Frequency Sampling," *Appl. Opt.* **7**, 545 (1968).

18.3. D. J. De Bitetto, "Bandwidth Reduction of Hologram Transmission Systems by Elimination of Vertical Parallax," *Appl. Phys. Lett.* **12**, 176 (1968).

18.4. K. A. Haines and D. B. Brumm, "Holographic Data Reduction," *Appl. Opt.* **7**, 1185 (1968).

18.5. D. Fritzler and E. Marom, "Reduction of Bandwidth Required for High Resolution Hologram Transmission," *Appl. Opt.* **8**, 1241 (1969).

18.6. L. P. Dudley, "Stereoscopy" *in Applied Optics and Optical Engineering* (R. Kingslake, ed.), Vol. 2, p. 77. Academic Press, New York, 1965.

18.7. L. H. Lin, "Hyperstereoscopic- and Hypostereoscopic-Hologram Images," *J. Opt. Soc. Amer.* **58**, 1539 (1968).

18.8. R. Hioki and T. Suzuki, "Reconstruction of Wavefronts in All Directions," *Jap. J. Appl. Phys.* **4**, 816 (1965).

18.9. E. P. Supertzi and A. K. Rigler, "Wide-Angle Holography," *J. Opt. Soc. Amer.* **56**, 524 (1966).

18.10. T. H. Jeong, P. Rudolf, and A. Luckett, "360° Holography," *J. Opt. Soc. Amer.* **56**, 1263 (1966).

18.11. T. H. Jeong, "Cylindrical Holography and Some Proposed Applications," *J. Opt. Soc. Amer.* **57**, 1396 (1967).

18.12. M. C. King, "Multiple Exposure Hologram Recording of a 3-D Image with a 360° View," *Appl. Opt.* **7**, 1641 (1968).

18.13. R. V. Pole, "3-D Imagery and Holograms of Objects Illuminated in White Light," *Appl. Phys. Lett*, **10**, 20 (1967).

18.14. J. T. McCrickerd and N. George, "Holographic Stereogram from Sequential Component Photographs," *Appl. Phys. Lett*, **12**, 10 (1968).

18.15. N. George, J. T. McCrickerd, and M. M. T. Chang, "Scaling and Resolution of Scenic Holographic Stereograms," *Proc. SPIE Seminar-in-Depth Holography*, p. 117 (1968).

18.16. J. D. Redman, "The Three-Dimensional Reconstruction of People and Outdoor Scenes Using Holographic Multiplexing," *SPIE Seminar-in-Depth Holography*, p. 161 (1968).

18.17. M. C. King, A. M. Noll, and D. H. Berry, "A New Approach to Computer-Generated Holography," *Appl. Opt.* **9**, 471 (1970).

18.18. D. J. De Bitetto, "Holographic Panoramic Stereograms Synthesized from White Light Recordings," *Appl. Opt.* **8**, 1740 (1969).

Chapter 19

COMPUTER-GENERATED HOLOGRAMS

Computer-generated holograms, in their most promising applications, would perform functions similar to those proposed for optically formed holograms. They would recognize characters or features, restore degraded images, and test optical surfaces interferometrically. To be able to realistically appraise their merit relative to holograms formed with laser light, one should understand the principles underlying the computer method. We shall confine the discussion in this chapter to basic principles and refrain from presentation of details relevant to a particular computer or plotter.

Before entering into a mathematical analysis, let us briefly summarize the method. Suppose that a laser-illuminated subject can be adequately described by a limited number of light-scattering points. If the coordinates of these points together with parameters specifying the wavelength and direction of both subject illumination and reference wave are given to a computer, then the amplitude transmittance of the hologram they form can be calculated. With this information stored, the computer is able to control a plotting device, or perhaps the position of an electron beam on the face of a cathode ray tube, and so make a magnified display of the hologram. Magnification is necessary because of the limited resolution of plotting and display devices. The display is next optically demagnified to dimensions appropriate to the wavelength used in the calculation and then photographed as a transparency. When the hologram thus recorded is illuminated with laser light, an image of the subject is generated. It is evident that holograms of subjects which do not exist physically can be computed, since only a mathematical knowledge of the subject is necessary.

The computer methods we shall discuss generate Fourier transform holograms [*19.1*, *19.2*]. To do this, the computer must compute a great many samples of the two-dimensional Fourier transform of the subject. Each sample consists of the amplitude and phase of the transform at an assigned point. Computation time and cost place a limit on the number of samples that can be taken. As might be expected, this limit has its deleterious effect on the imaging properties of the hologram. For example, limiting the spatial rate of sampling restricts the field of view or size of the subject and image. This result is analogous to that obtained when Fourier transform holograms are formed in recording media of limited resolution. On the other hand, a limit on the total number of samples taken at a given rate restricts the size of the hologram and hence the diffraction-limited resolution in the image. A further simplification used in many methods of generating holograms by means of the computer is to consider the amplitude transmittance of the hologram at any point to be either zero or one. Such two-tone holograms are called binary holograms. The simplification makes the display of the hologram compatible with the capability of most plotters and facilitates subsequent photographic recording of the demagnified hologram.

19.1 The Sampling Theorem

It is possible to compute a limited number of samples of a continuous function, e.g., a Fourier transform function, and with these samples reconstruct the continuous function exactly. According to the *sampling theorem*, *if a continuous function $f(x)$ is band-limited, i.e., its spectrum can have nonzero values only over a limited band of frequencies, and if $f(x)$ is sampled at least twice in any increment Δx as large as the spatial period of the highest spatial frequency component in $f(x)$, then $f(x)$ can be exactly reconstructed from the discrete samples.* In the remainder of this section we shall attempt to provide some insight into the application of the sampling theorem. (For further discussion of the sampling theorem and its applications see Bracewell [*4.1*, Chapter 10].)

To begin we need to introduce the comb function illustrated in Fig. 19.1:

$$\mathrm{comb}(x) = \sum_{m=-\infty}^{\infty} \delta(x - m) \tag{19.1}$$

where m is an integral number of units of distance. We see that the function comb(x) is an infinite series of delta functions of unit strength, i.e., of unit area under the delta function [see Eq. (4.13c)], each spaced a unit distance

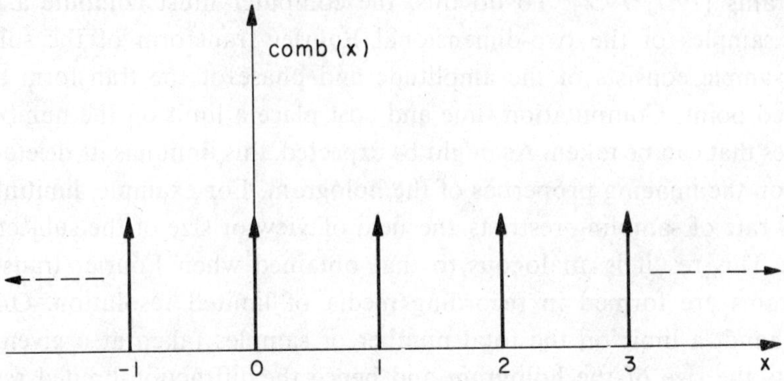

FIG. 19.1. The function comb(x).

apart. It is shown by Bracewell [*4.1*, p. 214] that the comb (or shah function as it is called there) has a Fourier transform which again is a comb function. Thus

$$\mathrm{comb}(x) \supset \mathrm{comb}(\xi). \tag{19.2}$$

The spacing in the series of delta functions represented by the comb function can be set arbitrarily by choosing a scale factor c and inserting cx as the variable in Eq. (19.1). We may then write

$$\mathrm{comb}(cx) = \sum_{m=-\infty}^{\infty} \delta(cx - m)$$

which with the aid of Eq. (4.13d) becomes

$$\mathrm{comb}(cx) = \frac{1}{|c|} \sum_{m=-\infty}^{\infty} \delta\left(x - \frac{m}{c}\right). \tag{19.3}$$

We may also apply the similarity theorem (4.22) to correspondence (19.2) and obtain the transform relation

$$\mathrm{comb}(cx) \supset \frac{1}{|c|}\,\mathrm{comb}\left(\frac{\xi}{c}\right). \tag{19.4}$$

Let us now employ the comb function relations to clarify the meaning of the sampling theorem. Suppose we sample the function $\mathbf{a}(x)$ regularly at a series of points Δx units apart, as shown in Fig. 19.2 for $\mathbf{a}(x)$ real. It is apparent that the mathematical representation of the sampling operation is a multiplication of the continuous function $\mathbf{a}(x)$ by the function $\sum_{m=-\infty}^{\infty} \delta(x - m\,\Delta x)$. We may express the latter in terms of the comb function in Eq. (19.3) by assigning to c the value $c = 1/\Delta x$. Thus

$$\sum_{m=-\infty}^{\infty} \delta(x - m\,\Delta x) = \frac{1}{\Delta x}\,\mathrm{comb}\left(\frac{x}{\Delta x}\right).$$

The sampled function $\mathbf{a}_s(x)$ is then given either by

$$\mathbf{a}_s(x) = \mathbf{a}(x)\,\frac{1}{\Delta x}\,\mathrm{comb}\left(\frac{x}{\Delta x}\right) \tag{19.5a}$$

or by

$$\mathbf{a}_s(x) = \mathbf{a}(x) \sum_{m=-\infty}^{\infty} \delta(x - m\,\Delta x) = \sum_{m=-\infty}^{\infty} \mathbf{a}(m\,\Delta x)\,\delta(x - m\,\Delta x). \tag{19.5b}$$

Each sample of $\mathbf{a}(x)$ in Eq. (19.5b) is a delta function whose strength is given by the value of $\mathbf{a}(x)$ at the position of the delta function. In Fig. 19.2 the delta functions are represented by arrows whose lengths are proportional to the strengths of the delta functions.

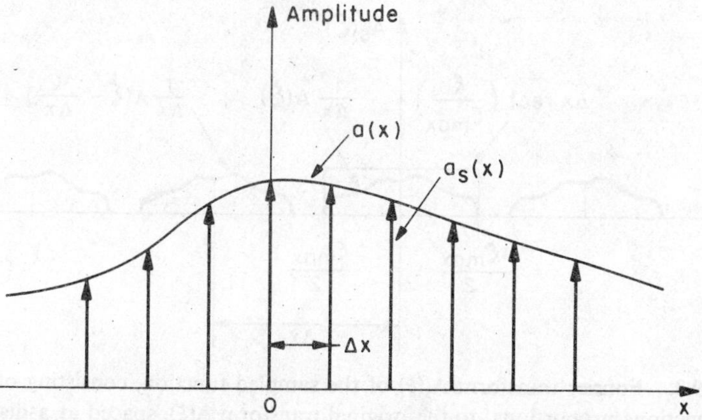

FIG. 19.2. The continuous function $\mathbf{a}(x)$ and the sampled function $\mathbf{a}_s(x)$, consisting of the array of delta functions.

Computer-generated holograms require computation of the Fourier transform $\mathbf{A}_s(\xi)$ of a sampled spatial function $\mathbf{a}_s(x)$. We now show that if the sampling has been carried out in accordance with the sampling theorem, the exact Fourier transform $\mathbf{A}(\xi)$ of the continuous function $\mathbf{a}(x)$ may be obtained from $\mathbf{A}_s(\xi)$. An inverse transformation will then recover $\mathbf{a}(x)$. In transforming Eq. (19.5a), the product of $\mathbf{a}(x)$ and $(1/\Delta x)\,\mathrm{comb}(x/\Delta x)$ in the spatial domain becomes the convolution of their transforms $\mathbf{A}(\xi)$

and comb($\Delta x \xi$) in the frequency domain [see convolution theorem (4.11)] so that

$$\mathbf{A}_s(\xi) = \mathbf{A}(\xi) * \text{comb}(\Delta x\, \xi)$$
$$= \mathbf{A}(\xi) * \left[\frac{1}{\Delta x} \sum_{m=-\infty}^{\infty} \delta\left(\xi - \frac{m}{\Delta x}\right)\right] \tag{19.6}$$

using correspondence (19.4) and substituting ξ for x in Eq. (19.3). With the definition of the convolution operation given in Eq. (4.11) we obtain

$$\mathbf{A}_s(\xi) = \frac{1}{\Delta x} \sum_{m=-\infty}^{\infty} \int_{u=-\infty}^{\infty} \mathbf{A}(u)\, \delta\left(\xi - \frac{m}{\Delta x} - u\right) du$$
$$= \frac{1}{\Delta x} \sum_{m=-\infty}^{\infty} \int_{u=-\infty}^{\infty} \mathbf{A}(u)\, \delta\left(u - \xi + \frac{m}{\Delta x}\right) du$$
$$= \frac{1}{\Delta x} \sum_{m=-\infty}^{\infty} \mathbf{A}\left(\xi - \frac{m}{\Delta x}\right) \tag{19.7}$$

where we have applied the symmetry property of the delta function Eq. (4.13b) and its sifting property Eq. (4.13e). Figure 19.3 shows the Fourier

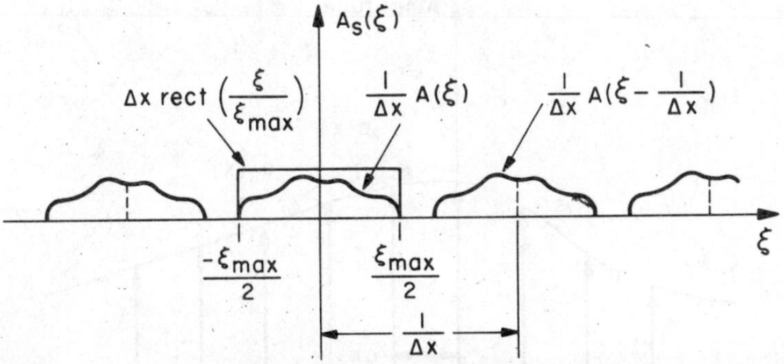

FIG. 19.3. Fourier transform $\mathbf{A}_s(\xi)$ of the sampled function, consisting of an infinite set of functions proportional to the original transform $\mathbf{A}(\xi)$, spaced at a distance $1/\Delta x$. Multiplication of $\mathbf{A}_s(\xi)$ with the rectangular window Δx rect(ξ/ξ_{max}) gives the Fourier transform $\mathbf{A}(\xi)$ of the original function.

transform $\mathbf{A}_s(\xi)$ as a regular series of shifted Fourier transforms $(1/\Delta x) \times \mathbf{A}(\xi - m/\Delta x)$, each proportional to the Fourier transform of the original continuous function $\mathbf{a}(x)$ and spaced by the interval $\xi = 1/\Delta x$. Suppose that the original function $\mathbf{a}(x)$ is band limited so that $\mathbf{A}(\xi)$ has nonzero values only in the range $-\xi_{max}/2 \leq \xi \leq \xi_{max}/2$. Overlap of the shifted transforms $(1/\Delta x)\mathbf{A}(\xi - m/\Delta x)$ is prevented if

$$\frac{1}{\Delta x} \geq \xi_{max} \quad \text{or} \quad \Delta x \leq \frac{1}{\xi_{max}}. \tag{19.8}$$

Inequality (19.8) specifies sampling intervals satisfying the conditions of the sampling theorem. We see from Fig. 19.3 that when inequality (19.8) is satisfied, the Fourier transform $\mathbf{A}(\xi)$ can be recovered from the Fourier transform $\mathbf{A}_s(\xi)$ of the sampled function by simply multiplying $\mathbf{A}_s(\xi)$ with a rectangular window function $\Delta x \operatorname{rect}(\xi/\xi_{\max})$. Thus

$$\mathbf{A}(\xi) = \mathbf{A}_s(\xi)\, \Delta x \operatorname{rect}\left(\frac{\xi}{\xi_{\max}}\right). \tag{19.9}$$

If inequality (19.8) is not satisfied, then overlap of the shifted Fourier transforms $(1/\Delta x)\mathbf{A}(\xi - m/\Delta x)$ makes it impossible to recover $\mathbf{A}(\xi)$. Multiplication of $\mathbf{A}_s(\xi)$ with the rectangular window function would in that case give something other than $\mathbf{A}(\xi)$. The resulting error is called *aliasing*.

Having obtained $\mathbf{A}(\xi)$ from samples of $\mathbf{a}(x)$ let us now proceed to form the inverse Fourier transform of $\mathbf{A}(\xi)$ and so recover the original spatial function $\mathbf{a}(x)$. Inverse Fourier transformation of the product of $\mathbf{A}_s(\xi)$ and $\Delta x \operatorname{rect}(\xi/\xi_{\max})$ in Eq. (19.9) is equivalent to convolving in the spatial domain the inverse Fourier transforms of these functions. From correspondence (4.31) we have for the inverse Fourier transform of the rectangular window

$$\Delta x \operatorname{rect}\left(\frac{\xi}{\xi_{\max}}\right) \subset \Delta x\, \xi_{\max} \cdot \frac{\sin \pi\xi_{\max}x}{\pi\xi_{\max}x}. \tag{19.10}$$

We may therefore write that

$$\begin{aligned}
\mathbf{a}(x) &= \mathscr{F}^{-1}[\mathbf{A}(\xi)] = \mathbf{a}_s(x) * \left[\Delta x\, \xi_{\max} \frac{\sin \pi\xi_{\max}x}{\pi\xi_{\max}x}\right] \\
&= \left[\sum_{m=-\infty}^{\infty} \mathbf{a}(m\,\Delta x)\, \delta(x - m\,\Delta x)\right] * \left[\Delta x\, \xi_{\max} \frac{\sin \pi\xi_{\max}x}{\pi\xi_{\max}x}\right] \\
&= \Delta x\, \xi_{\max} \sum_{m=-\infty}^{\infty} \mathbf{a}(m\,\Delta x) \frac{\sin[\pi\xi_{\max}(x - m\,\Delta x)]}{\pi\xi_{\max}(x - m\,\Delta x)}
\end{aligned} \tag{19.11}$$

where we have used Eq. (19.5b) and the sifting property, Eq. (4.13e). If the maximum permissible sampling interval $\Delta x = 1/\xi_{\max}$ is chosen, Eq. (19.11) simplifies to the following very useful relation:

$$\mathbf{a}(x) = \sum_{m=-\infty}^{\infty} \mathbf{a}\left(\frac{m}{\xi_{\max}}\right) \frac{\sin(\pi\xi_{\max}x - \pi m)}{\pi\xi_{\max}x - \pi m}. \tag{19.12}$$

Figure 19.4 illustrates the reconstruction of $\mathbf{a}(x)$ with the sum of sine functions [of the form $(\sin z)/z$]. Only two terms of the summation in Eq.

(19.12) are shown. Note that

$$\frac{\sin(\pi\xi_{max}x - \pi m)}{\pi\xi_{max}x - \pi m}$$

has the value of one at the mth sampling position, $x = m/\xi_{max}$, and the value zero at all other sampling positions. When account is taken of all terms in Eq. (19.12), each with its positive center lobe and negative as well as positive side lobes, the continuous function $\mathbf{a}(x)$ is exactly reconstructed.

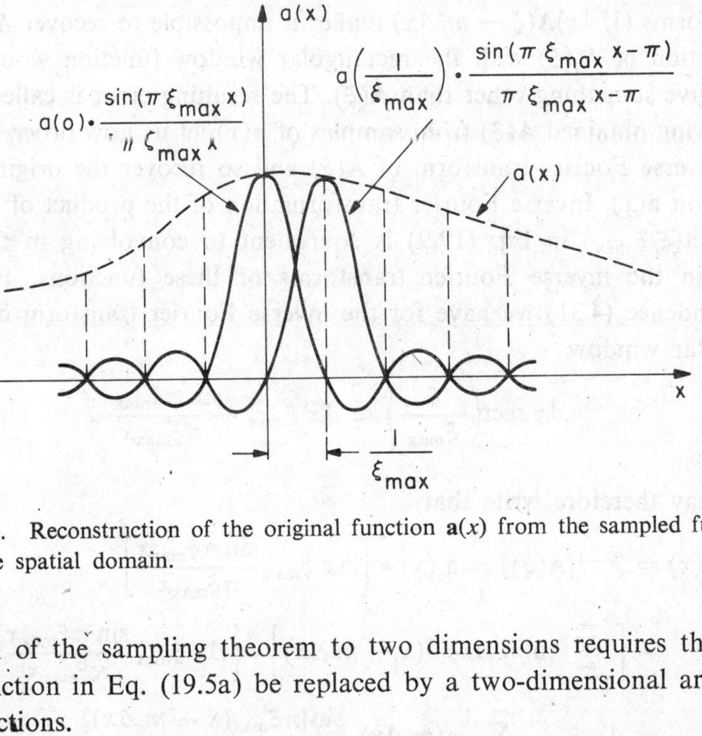

FIG. 19.4. Reconstruction of the original function $\mathbf{a}(x)$ from the sampled function $\mathbf{a}_s(x)$ in the spatial domain.

Extension of the sampling theorem to two dimensions requires that the comb function in Eq. (19.5a) be replaced by a two-dimensional array of delta functions.

19.2 The Discrete Fourier Transform and the Fast Fourier Transform

As we noted earlier, computer methods of generating Fourier transform holograms require the computation of a great many samples of the Fourier transform of a subject function. Since the computer stores the coordinates of only a limited number of samples of the subject function, it is of interest to discover the form taken by the Fourier transform of these samples. We shall confine the analysis to one dimension as before.

Recall that the Fourier transform of a function $\mathbf{a}(x)$ is defined in Eq. (4.9) as

$$\mathbf{A}(\xi) = \int_{-\infty}^{\infty} \mathbf{a}(x) \exp(2\pi i \xi x)\, dx. \tag{19.13}$$

If we substitute for $\mathbf{a}(x)$ the form given in Eq. (19.12), we obtain

$$\mathbf{A}(\xi) = \int_{-\infty}^{\infty} \sum_{m=-\infty}^{\infty} \mathbf{a}\left(\frac{m}{\xi_{\max}}\right) \frac{\sin(\pi\xi_{\max}x - \pi m)}{\pi\xi_{\max}x - \pi m} \exp(2\pi i \xi x)\, dx. \tag{19.14}$$

Correspondences (4.20) and (4.31) tell us that

$$\int_{-\infty}^{\infty} \frac{\sin(\pi\xi_{\max}x - \pi m)}{\pi\xi_{\max}x - \pi m} \exp(2\pi i \xi x)\, dx$$
$$= \frac{1}{\xi_{\max}} \operatorname{rect}\left(\frac{\xi}{\xi_{\max}}\right) \exp\left(2\pi i \frac{m\xi}{\xi_{\max}}\right). \tag{19.15}$$

Substitution of Eq. (19.15) into Eq. (19.14) yields

$$\mathbf{A}(\xi) = \frac{1}{\xi_{\max}} \sum_{m=-\infty}^{\infty} \mathbf{a}\left(\frac{m}{\xi_{\max}}\right) \exp\left(2\pi i \frac{m\xi}{\xi_{\max}}\right) \quad \text{for} \quad \frac{-\xi_{\max}}{2} \le \xi \le \frac{\xi_{\max}}{2}.$$
$$= 0 \qquad \text{otherwise.} \tag{19.16}$$

In obtaining Eq. (19.12) we had assumed that $\mathbf{a}(x)$ is a band-limited function. We now make the additional assumption that it is space limited to the range $-x_{\max}/2 \le x \le x_{\max}/2$. (This is inconsistent because a band-limited function cannot be space limited too, and the inconsistency leads to some error in computation of the Fourier transform $\mathbf{A}(\xi)$. However, when the product $x_{\max}\xi_{\max}$ is much larger than one, as is always the case for holograms, the error is negligible [19.3].) If $\mathbf{a}(x)$ is space limited to the range $\pm x_{\max}/2$, the sampling theorem assures us that $\mathbf{A}(\xi)$, the Fourier transform of $\mathbf{a}(x)$, can be exactly reconstructed from samples of the transform taken at intervals of ξ given by $\Delta\xi = 1/x_{\max}$ [the analog to Eq. (19.8)]. In generating a hologram, computation time is saved by computing $\mathbf{A}(\xi)$ only at the sampling points $\xi = n/x_{\max}$, where n is an integer. We therefore write for one sample of $\mathbf{A}(\xi)$ in Eq. (19.16)

$$\mathbf{A}\left(\frac{n}{x_{\max}}\right) = \frac{1}{\xi_{\max}} \sum_{m=-M/2}^{M/2} \mathbf{a}\left(\frac{m}{\xi_{\max}}\right) \exp\left(2\pi i \frac{mn}{\xi_{\max}x_{\max}}\right) \tag{19.17}$$

where M is the total number of samples of the space-limited function $\mathbf{a}(x)$. Relation (19.17) is called the *discrete Fourier transform*.

At this point let us estimate the number of steps, the time, and the cost required to compute the discrete Fourier transform of some subject function. For each sample $\mathbf{A}(n/x_{\max})$ of the Fourier transform, we must compute M products corresponding to the M samples of the subject spatial function stored in the computer. If, in accordance with the sampling theorem, these latter samples were taken over spatial intervals $\Delta x = 1/\xi_{\max}$ and if the subject spatial function has a length $x_{\max}$, then the total number of stored spatial samples is

$$M = x_{\max}/\Delta x = x_{\max}\xi_{\max}. \tag{19.18}$$

In a similar manner we obtain for the total number of Fourier transform samples

$$N = \xi_{\max}/\Delta\xi = \xi_{\max}x_{\max}. \tag{19.19}$$

We see that $M = N$; the original spatial function and the Fourier transform require the same number of samples. Computation of all the samples of a one-dimensional Fourier transform implies computation of a total of $MN = N^2$ terms. For the two-dimensional Fourier transform represented by an array of $N \times N$ samples, N^4 terms are required.

Suppose we have stored an array of 100×100 samples of the subject spatial function and suppose that 30 μsec are needed to compute and add one term of the sum in Eq. (19.17). (The time is typical for a large modern computer.) Under these conditions the 100×100 samples of the Fourier transform can be calculated in 3000 sec or a little less than 1 h. An array of 100×100 samples represents a rather primitive picture. Since an hour of time on a large computer can cost several hundred dollars, the prospects for computer-generated Fourier transform holograms would appear rather dim. Fortunately a time-saving computational technique, called the *fast Fourier transform*, has recently become well known; its power has markedly changed the outlook.

The fast Fourier transform is an algorithm for computing Fourier transforms. When N is large, it requires far fewer steps than indicated in Eq. (19.17). (Explanation of this algorithm would take us farther afield than we intend. Instead the reader is referred to Cochran [*19.3*].) Computer programs implementing the fast Fourier transform are readily available. To apply the algorithm it is advantageous that N be a power of 2. In this case a total of $4N^2 \log_2 N$ multiplications and additions is required to compute the Fourier transform of an $N \times N$ array [*19.3*] where $\log_2$ denotes the logarithm to the base 2. We shall tabulate some estimates [*19.4*] of the times necessary to compute Fourier transforms of several $N \times N$ arrays via the direct method (N^4 terms) and via the fast Fourier transform. The time-saving advantage of the fast Fourier transform is apparent from Table 19.1.

TABLE 19.1

COMPARISON OF COMPUTATION TIME: DIRECT VERSUS FAST FOURIER TRANSFORM METHOD

Array size	Direct computation	Fast Fourier transform
64 × 64	8 min	3 sec
256 × 256	30 h	1 min
512 × 512	20 days	5 min
1024 × 1024	1 yr	20 min

Fourier transform computations for arrays up to approximately 1000 × 1000 appear feasible when the fast Fourier transform algorithm is used.

19.3 Binary Fourier Transform Holograms

We shall consider here the instructions which must be provided by the computer to a plotting device so that the plotter can write a binary Fourier transform hologram. We suppose that the Fourier transform samples stored in the computer have been taken at all sampling points on a regular square array whose spacing has been specified in accordance with the sampling theorem. The sampling points cover a finite area of a hypothetical $\xi\eta$ plane corresponding to the Fourier transform plane of the subject and to the hologram plane. To display the hologram in binary form, the plotter must represent the sample values on a real surface as an array of either transparent or opaque areas. Because of the limited resolution of the plotter, the sampling intervals and sample areas are scaled up to produce an enlarged hologram. Subsequent optical demagnification results in a hologram with the intended dimensions. Restriction to a binary transmittance makes the hologram easier to write. Many computer-controlled plotters, for example, can produce only the two extreme transmittances, one or zero. Since there are no gray tones, the photographic demagnification can be carried out simply with high-contrast processing. (Computer-generated holograms have also been displayed by a continuous-tone plotter [*19.5*].) Although our discussion is confined to

Fourier transform holograms, it also applies *mutatis mutandis* to Fresnel holograms.

We preface a quantitative treatment of hologram formation with a qualitative picture of the process. In both presentations we neglect the practical necessity of forming an enlarged hologram and then demagnifying it. Figure 19.5 indicates a one-dimensional representation of a hologram located in the Fourier transform plane ξ of a subject and illuminated by an off-axis plane wave. Part of this wave will diffract from the hologram and

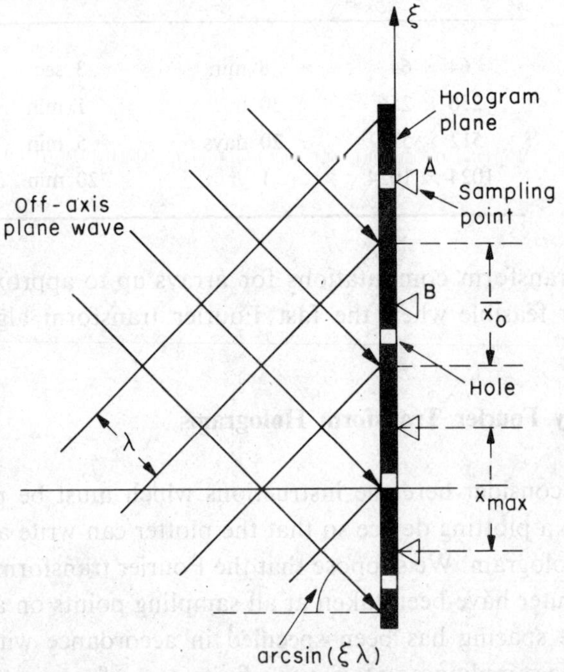

FIG. 19.5. Off-axis plane wave illuminating a one-dimensional, computer-generated, Fourier transform hologram. The hologram is an array of small holes in an opaque plane.

generate the holographic image. The phase of the wave in the plane of the hologram, $2\pi\xi x_0$, varies linearly with ξ and increases by 2π radians in an interval $\Delta\xi = 1/x_0$, where x_0 is the location of the illuminating source in the spatial plane. On the Fourier transform plane or hologram plane of the figure, triangles mark the regularly spaced sampling points at which the Fourier transform has been computed. Their spacing is $1/x_{\max}$, where $x_{\max}$ is the total spatial extent of the subject ($\pm x_{\max}/2$). We have arbitrarily set $1/x_{\max} = 1/x_\sigma$.

Let us assume that the hologram plane ξ is initially opaque. To encode the computed Fourier transform complex amplitudes in a hologram, we perforate the plane with small holes. Each hole is associated with a particular sampling point and has an area proportional to the (real) amplitude of the Fourier transform computed for that point. The location of the hole relative to the sampling point encodes the phase of the Fourier transform sample. Suppose that the phase of the illuminating wave at the sampling point A is exactly that calculated for the subject Fourier transform sample at A. We can then perforate the plane at A with a small hole of the appropriate area. At the sampling point B, a distance $1/x_{\max}$ units away, the phase of the illuminating wave and that of the Fourier transform sample computed for B need not agree. If it is valid to assume that the phase of the *continuous* Fourier transform function varies smoothly and slowly between A and B, the following approximate method can be employed to encode the phase ϕ_B of the Fourier transform sample computed for B. The location of the small hole whose area is appropriate for the sampling point B is translated along the hologram plane away from B until the phase of the illuminating wave is found to be ϕ_B. A suitable perforation is made at that point. In a similar manner all the Fourier transform samples are plotted. When the hologram so formed is illuminated by the off-axis plane wave, the diffracted wavefront will approximate the Fourier transform of the subject transmittance. The degree to which the wavefront faithfully represents the Fourier transform depends, of course, on whether the small holes have been located closely enough to satisfy the sampling theorem and on the errors introduced by the approximate method of encoding phase. Optical Fourier transformation of the diffracted wave yields the subject image.

Let us now consider hologram generation in two dimensions more quantitatively. Suppose the complex amplitude at the subject $\mathbf{a}(x, y)$ is space limited and nonzero for

$$|x| \leq x_{\max}/2, \qquad |y| \leq y_{\max}/2 = x_{\max}/2. \tag{19.20}$$

By illuminating a suitably formed hologram we desire to produce a waveform whose complex amplitude is $\mathbf{A}(\xi, \eta)$, the Fourier transform of $\mathbf{a}(x, y)$. Since $\mathbf{A}(\xi, \eta)$ is to be reconstructed from computed samples of the function, we should indicate this in our representation of $\mathbf{A}(\xi, \eta)$. Equation (19.12) can be altered for this purpose. As written it represents a spatial function whose Fourier transform is band limited. By interchanging ξ for x and η for y, the form can represent a spatial frequency function whose transform in the spatial domain is space limited. Thus, in two dimensions the spatial frequency analog of Eq. 19.12 is

$$\mathbf{A}(\xi, \eta) = \sum_m \sum_n \mathbf{A}\left(\frac{m}{x_{\max}}, \frac{n}{x_{\max}}\right) \frac{\sin(\pi\xi x_{\max} - \pi m)}{\pi\xi x_{\max} - \pi m} \cdot \frac{\sin(\pi\eta x_{\max} - \pi n)}{\pi\eta x_{\max} - \pi n} \tag{19.21}$$

where (m, n) designates a sampling point on the frequency plane. A lens is employed to perform an inverse Fourier transformation of $\mathbf{A}(\xi, \eta)$ and display in its back focal plane the image of the original subject spatial function $\mathbf{a}(x, y)$. With the aid of correspondences (4.31) and (4.21) we obtain for $\mathbf{a}(x, y) = \mathscr{F}^{-1}[\mathbf{A}(\xi, \eta)]$

$$\mathbf{a}(x, y) = \left(\frac{1}{x_{\max}}\right)^2 \operatorname{rect}\left(\frac{x}{x_{\max}}\right) \operatorname{rect}\left(\frac{y}{x_{\max}}\right) \times \sum_m \sum_n \mathbf{A}\left(\frac{m}{x_{\max}}, \frac{n}{x_{\max}}\right) \exp\left[-2\pi i\left(\frac{m}{x_{\max}} x + \frac{n}{x_{\max}} y\right)\right]. \tag{19.22}$$

We may regard $\mathbf{a}(x, y)$ in Eq. (19.22) as the complex amplitude of the subject image obtained from an *ideal* computer-generated, Fourier transform hologram.

The computer-generated hologram, described qualitatively as an opaque plane perforated with a large number of small holes, is represented mathematically by a two-dimensional array of delta functions $H(\xi, \eta)$. Suppose $H(\xi, \eta)$ is illuminated by an off-axis plane wave whose phase varies with ξ only, as shown in Fig. 19.5; let us specify the delta functions in terms of the lattice of sampling points for which the subject Fourier transform was computed. The sampling points on the frequency plane are spaced by equal distances $\Delta\xi = \Delta\eta$ in the ξ and η directions. Since the phase of the illuminating wave varies only in the ξ direction, phase encoding is effective only in the ξ direction and is accomplished by translating a given hole from its associated sampling point $(m\,\Delta\xi, n\,\Delta\xi)$ a small distance $p_{mn}\,\Delta\xi$ in the ξ direction (see Fig. 19.6). We may therefore write for the array of holes $H(\xi, \eta)$

$$H(\xi, \eta) = \sum_m \sum_n B_{mn}\, \delta[\xi - (m + p_{mn})\,\Delta\xi] \cdot \delta(\eta - n\,\Delta\xi). \tag{19.23}$$

We inquire as to the conditions which make the complex amplitude of the image generated by $H(\xi, \eta)$ equivalent to $\mathbf{a}(x, y)$ in Eq. (19.22).

If the holes are small compared to $\Delta\xi$, approximation of their transmittances by delta functions is valid. The strength of the delta function B_{mn} is

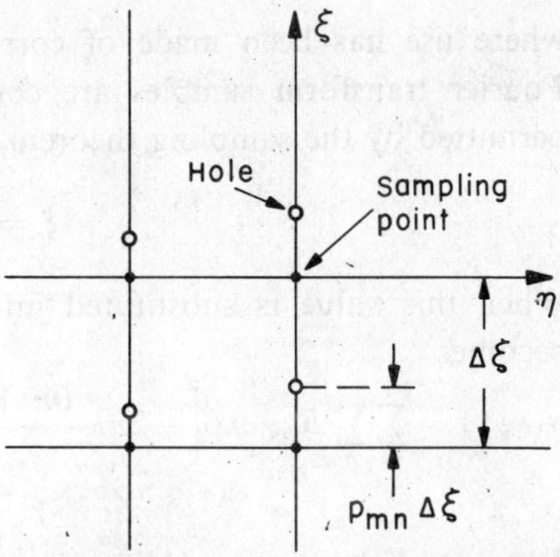

FIG. 19.6. A small section of a computer-generated hologram.

then proportional to the area of the hole. Illumination of the hologram $H(\xi, \eta)$ by an off-axis plane wave derived from a point source located in the spatial plane at $(x = x_0, y = 0)$ (see Fig. 19.7) produces a complex amplitude at the hologram plane

$$\begin{aligned} \mathbf{W}(\xi, \eta) &= \exp(2\pi i x_0 \xi) H(\xi, \eta) \\ &= \exp(2\pi i x_0 \xi) \sum_m \sum_n B_{mn}\, \delta[\xi - (m + p_{mn})\, \Delta\xi]\, \delta(\eta - n\, \Delta\xi). \end{aligned} \tag{19.24}$$

In the back focal plane of lens L_2 in Fig. 19.7 is displayed the inverse Fourier transform of $\mathbf{W}$

$$\begin{aligned} \mathbf{w}(x, y) &= \mathscr{F}^{-1}[\mathbf{W}(\xi, \eta)] \\ &= \sum_m \sum_n B_{mn} \exp[-2\pi i(m + p_{mn})(x - x_0)\, \Delta\xi] \exp(-2\pi i y n\, \Delta\xi) \end{aligned} \tag{19.25}$$

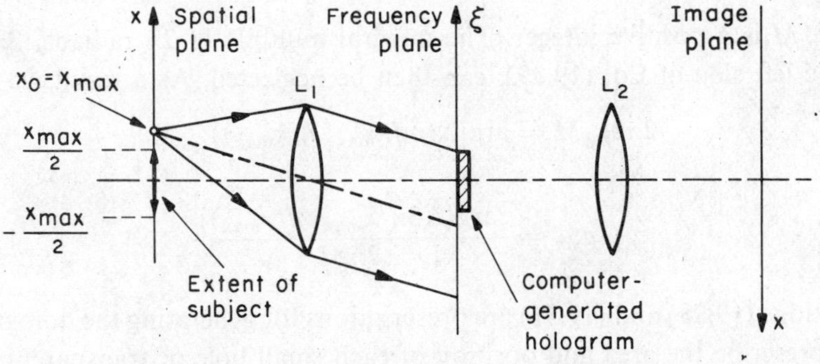

FIG. 19.7. Arrangement for illuminating a computer-generated Fourier transform hologram and displaying the image in the back focal plane of lens L_2.

where use has been made of correspondences (4.20) and (4.30). If the Fourier transform samples are computed at the maximum interval $\Delta\xi$ permitted by the sampling theorem, then we have

$$\Delta\xi = 1/x_{\max}. \tag{19.26}$$

When this value is substituted into Eq. (19.25), the spatial function **w** becomes

$$\mathbf{w}(x, y) = \sum_m \sum_n B_{mn} \exp\left[-2\pi i \frac{(m + p_{mn})(x - x_0)}{x_{\max}}\right] \exp\left(-2\pi i \frac{yn}{x_{\max}}\right). \tag{19.27}$$

A proper hologram has been generated when $\mathbf{w}(x, y)$, the light diffracted by the hologram to the image plane, is identical with the original subject complex amplitude function $\mathbf{a}(x, y)$ in the form of Eq. (19.22).

The magnitude of $\mathbf{w}(x, y)$ in Eq. (19.27) is equal to that of $\mathbf{a}(x, y)$ in Eq. (19.22) when

$$B_{mn} = \left(\frac{1}{x_{\max}}\right)^2 \left|\mathbf{A}\left(\frac{m}{x_{\max}}, \frac{n}{x_{\max}}\right)\right|, \tag{19.28}$$

i.e., when the area of the small hole is proportional to the absolute value of the amplitude of the Fourier transform. Apart from the phase increment $-2\pi p_{mn}(x/x_{\max})$ in Eq. (19.27), the phases of $\mathbf{w}(x, y)$ in Eq. (19.27) and $\mathbf{a}(x, y)$ in Eq. (19.22) are equal when

$$2\pi(m + p_{mn}) \frac{x_0}{x_{\max}} = \arg\left[\mathbf{A}\left(\frac{m}{x_{\max}}, \frac{n}{x_{\max}}\right)\right]. \tag{19.29}$$

We can simplify Eq. (19.29) by choosing

$$x_0/x_{\max} = M \tag{19.30}$$

where M is a positive integer. The integral multiple of 2π radians, $2\pi mM$ on the left side of Eq. (19.29), can then be neglected. As a result we have

$$2\pi p_{mn} M = \arg[\mathbf{A}(m/x_{\max}, n/x_{\max})]$$

or

$$p_{mn} = \frac{\arg[\mathbf{A}(m/x_{\max}, n/x_{\max})]}{2\pi M}. \tag{19.31}$$

Equations (19.28) and (19.31) are prescriptions for generating the hologram; they prescribe the area and position of each small hole or transparent spot of the hologram. If we choose $M = 1$, so that $x_0 = x_{\max}$, then the position of the illuminating point source is also prescribed as shown in Fig. 19.7.

The phase error $-2\pi p_{mn}(x/x_{\max}) = -2\pi(p_{mn}\,\Delta\xi)x$ neglected in Eq. 19.29) is due to the separation (in the frequency plane) $p_{mn}\,\Delta\xi$ of any small hole from its associated sampling point. In the attempt to encode the phase of the Fourier transform sample computed at a given sampling point, we have tacitly assumed that the phase of the continuous Fourier transform function remains independent of ξ between sampling points and then changes abruptly at the sampling points. Errors in the phase of the diffracted wave are produced whenever the true phase behavior of the transform departs from the assumed. A more realistic approach is to assume that the phase of the continuous Fourier transform function varies slowly over a sampling interval. One can then approximate the phase variation of the continuous Fourier transform function between sampling points with a smooth curve based on the phase data at the sampling points. A hole is located where the phase of the interpolation curve equals the phase of the illuminating wave [*19.2*]. Phase errors are reduced by this procedure.

Holograms constructed out of holes small compared to a sampling interval are wasteful of illuminating light, since most of the hologram area is opaque. Efficient holograms are generally about 50% transparent. Suppose, instead of a small hole, we choose a rectangular aperture whose area is proportional to the amplitude of the Fourier transform sample at a given sampling point and whose dimension is no longer small compared to the sampling interval. Each delta function in $\mathbf{W}$ of Eq. (19.24) must be replaced by a rect function; as a consequence, the exponentials appearing in each term of the Fourier transform of $\mathbf{W}$ in Eqs. (19.25) and (19.27) must be multiplied by products of the form $((\sin x)/x)((\sin y)/y)$. The variation of these sinc functions with x or y causes the amplitude of the image $\mathbf{w}(x, y)$ to fall off from the amplitude of the original subject function. This is especially pronounced at large values of x and y, i.e., at the edges of the image. As the size of the aperture increases, the amplitude fall-off becomes more pronounced. However, it can be compensated for in computing the hologram by multiplying the original subject function stored in the computer by reciprocals of sinc functions appropriate to an average-sized aperture. It is therefore possible to choose a size and shape for the aperture which will maximize the hologram efficiency and match the capability of the plotter.

Binary holograms are analogous to holograms recorded on very nonlinear material. Yet the preceding analysis shows that the image complex amplitude can be produced free of the false images normally associated with nonlinear recording (see Section 12.1.2). Sampling, of course, does lead to higher order spectra as shown in Fig. 19.3. However, the spectral orders do not overlap when the sampling theorem is obeyed [see Eq. (19.8)].

Figure 19.8 displays a computer-generated, binary Fourier transform hologram, and Fig. 19.9 is a photograph of the image produced by such a hologram. The image quality is typical of that obtained to date. In this case the hologram was computed for a subject illuminated through a diffusing screen; hence the granular appearance of the image.

FIG. 19.8. A computer-generated, binary Fourier transform hologram. (Courtesy Brown and Lohmann [*19.2*].)

19.4 Applications

Applications for computer-generated holograms are only beginning to emerge, and at present it is not clear where they can best be put to use. It has been suggested that such holograms are suitable for imaging three-dimensional information stored in a computer. Since the subject of a computer-generated hologram need not physically exist, the computed holo-

FIG. 19.9. Photograph of the image produced by a computer-generated, binary Fourier transform hologram. (Courtesy Brown and Lohmann [*19.2*].)

gram might be used to display in 3D the design of an object even before it has been modeled. However, enthusiasm for this use of the computer-generated hologram should be tempered by the realization that a hologram of a diffusely reflecting subject with wide viewing angle requires an enormous amount of computation. In view of other solutions to this problem (see Section 18.5), it would seem unlikely that the computed hologram will emerge the best candidate for the task.

Spatial filtering appears to be a more promising field of application for the computer-generated hologram. In Section 14.6 we discuss some holographic methods of improving degraded images through the use of a filter of the form $\mathbf{H}^*/|\mathbf{H}|^2$ [see, e.g., Eq. (14.32)]. Several precise photographic processes are required to obtain the filter, and the results leave something to be desired. A computer, however, could generate the complete hologram filter in one process and conceivably improve the results. Computed filters for converting phase objects into amplitude objects, e.g., schlieren filters or Zernike phase-contrast filters, have also been demonstrated [*19.6*].

Another promising application of computer-generated holograms is in the testing of aspherical optical surfaces [*19.7*]. It has been known for a long time that the use of aspherical surfaces in an optical system may often be advantageous. Such surfaces are, however, used infrequently because they are difficult to make and test. Computer holograms can help solve the testing problem. A hologram is computed for the wavefront refracted or reflected by a mathematically specified aspherical surface and then illuminated. The diffracted wavefront can be compared interferometrically with the wavefront refracted or reflected by the actual aspherical surface, e.g., as in Section 15.1, thus testing the actual surface.

19.5 The Kinoform

In this section we shall describe a computer-generated, wavefront-reconstructing record which has been called by its originators a *kinoform* [*19.8*]. (A precursor to the kinoform was the *phase Fresnel lens* [*19.9*].) Like the hologram, it can display a three-dimensional image. It differs, however, from the hologram in that it can diffract all the illumination it receives into a single diffraction order. A basic assumption is that a subject wave complex amplitude $\mathbf{a}(x, y)$ can be satisfactorily recorded by considering the wave to have a constant amplitude. Its complex amplitude is then

$$\mathbf{a}(x, y) = \text{constant} \cdot \exp[i\varphi(x, y)]. \tag{19.32}$$

Perhaps the best justification for this assumption lies in the fact that it leads to recognizable images. The computed phase $\varphi(x, y)$ is to be recorded. However before this is done, the computed phase is adjusted so that it varies only between 0 and 2π rad from a phase origin, over the entire xy plane. Integral multiples of 2π rad are subtracted from the relative phase computed for any point (x, y) on the plane. The phase record, the kinoform, is then a thin, transparent sheet whose optical thickness is altered in accordance with the computed phase $\varphi(x, y)$ of the subject wave where $0 \leq \varphi(x, y) < 2\pi$. When illuminated with a plane wave, the kinoform impresses the phase function $\varphi(x, y)$ onto the plane wavefront thereby converting it to a wave of complex amplitude $\mathbf{a}(x, y)$.

The methods used to fabricate the kinoform are made evident by considering a simple spherical subject wave

$$\mathbf{a}(x, y) = \exp[(i\pi/\lambda f)(x^2 + y^2)]. \tag{19.33}$$

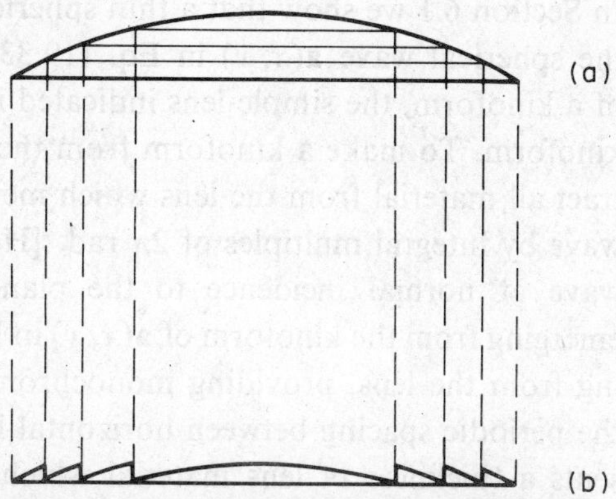

FIG. 19.10. (a) A thin spherical lens. The phase of a plane wave traveling in the vertical direction is shifted 2π radians in the space between horizontal lines. (b) A kinoform representation of the lens in (a) for the case of axial, plane wave illumination.

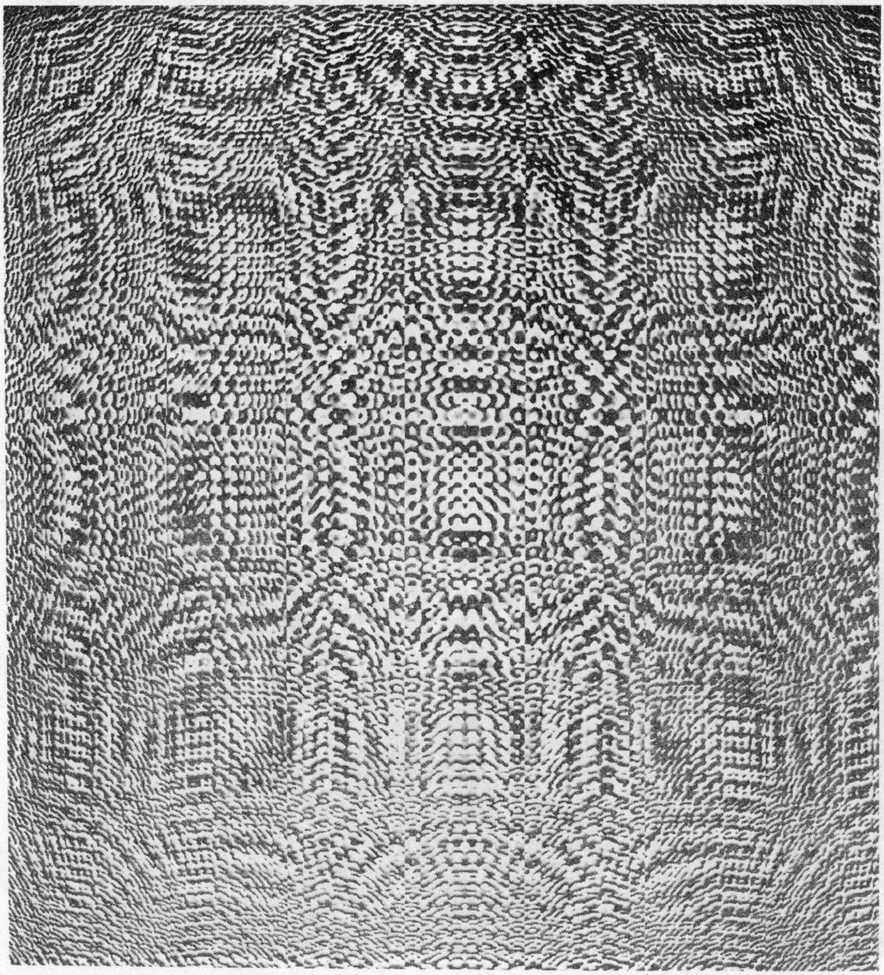

FIG. 19.11. A stage in the kinoform process. The output of the computer-controlled plotter. (Courtesy Lesem *et al.* [*19.8*].)

In Section 6.1 we show that a thin spherical lens converts a plane wave into the spherical wave $\mathbf{a}(x, y)$ in Eq. (19.33). Although it has the attributes of a kinoform, the simple lens indicated in Fig. 19.10a is nevertheless not a kinoform. To make a kinoform from the lens in Fig. 19.10a, we must subtract all material from the lens which merely shifts the phase of an incident wave by integral multiples of 2π rad. [Here we are concerned with a plane wave at normal incidence to the plane surface of the lens. The wave emerging from the kinoform of $\mathbf{a}(x, y)$ in Eq. (19.33) is similar to that emerging from the lens, providing monochromatic light is used.] In Fig. 19.10a the periodic spacing between horizontal lines drawn through the lens represents a thickness of lens material which shifts the phase of a plane wave by 2π rad. Figure 19.10b indicates the contour of the kinoform of $\mathbf{a}(x, y)$ in Eq. (19.33).

In practice the kinoform is generated by first computing samples of the subject wave phase function $\varphi(x, y)$ and then encoding the phase values in

FIG. 19.12. Image produced by a kinoform. Note the residual, zero-order bright spot in the center. (Courtesy Lesem *et al.* [*19.8*].)

a multilevel gray scale. This information is used to control a photographic plotter which exposes a photosensitive surface to the coded pattern. The resulting record is next photoreduced to a size suitable for illumination, and finally the photograph is bleached with a tanning bleach (see Section 10.8.5) to convert the gray levels to the appropriate pattern of optical thickness. If the bleaching is carefully controlled to ensure that the optical thickness of the photoreduced record is altered in proportion to the gray levels, the resulting kinoform can impress the phase function $\varphi(x, y)$ onto an incident monochromatic plane wave. Figure 19.11 shows the output of the computer-controlled plotter before bleaching. An image generated from a kinoform is photographed in Fig. 19.12.

Kinoforms are on-axis devices which have the ability to diffract 100% of incident monochromatic light into an image-forming wave. However, a multilevel gray scale is required and great care in bleaching must be exercised. If the recorded phase shift is not accurate, some light is diffracted into the zero order (see, e.g., the bright spot in the center of Fig. 19.12). This, of course, means less efficiency and a deterioration of the image.

REFERENCES

19.1. A. W. Lohmann and D. P. Paris, "Binary Fraunhofer Holograms, Generated by Computer," *Appl. Opt.* **6**, 1739 (1967).

19.2. B. R. Brown and A. W. Lohmann, "Computer-generated Binary Holograms," *IBM J. Res. Develop.* **13**, 160 (1969).

19.3. W. T. Cochran, J. W. Cooley, D. L. Favin, H. D. Helms, R. A. Kaenel, W. W. Lang, G. C. Maling, Jr., D. E. Nelson, C. M. Rader and P. D. Welch, "What Is the Fast Fourier Transform?," *Proc. IEEE* **55**, 1664 (1967).

19.4. J. W. Goodman, "Digital Image Formation from Detected Holographic Data," in *Acoustical Holography* (A. F. Metherell, H. M. A. El-Sum., and L. Larmore, eds.) Vol. I, pp. 173–185. Plenum Press, New York, 1969.

19.5. W. H. Lee, "Sampled Fourier Transform Hologram Generated by Computer," *Appl. Opt.* **9**, 639 (1970).

19.6. A. W. Lohmann and D. P. Paris, "Computer Generated Spatial Filters for Coherent Optical Data Processing," *Appl. Opt.* **7**, 651 (1968).

19.7. J. Pastor, "Hologram Interferometry and Optical Technology," *Appl. Opt.* **8**, 525 (1969).

19.8. L. B. Lesem, P. M. Hirsch, J. A. Jordan, Jr., "The Kinoform: A New Wavefront Reconstruction Device," *IBM J. Res. Develop.* **13**, 150 (1969).

19.9. K. Miyamoto, "The Phase Fresnel Lens," *J. Opt. Soc. Amer.* **51**, 17 (1961).

Chapter 20

THREE TOPICS IN SEARCH OF A CHAPTER

Replication, TV Transmission, and Incoherent-Light Holograms

20.1 Hologram Replication

When a large number of holograms of the same subject scene are required, one may choose to make replicas of an original hologram or to make many original holograms. Replication is convenient, since the subject is required only for forming the single original hologram. Furthermore, many holograms can be replicated with nonlaser sources of light, few optical components, and without the need for stable, massive optical tables. Proper technique can yield replicas whose outputs differ little from that of the original hologram. We distinguish between replication, which we equate to a process approximating contact printing, and formation of a second hologram where the wave reconstructed from an original hologram interferes with an independently directed reference wave [*20.1*]. We shall not be concerned with this latter method, since it differs little from ordinary hologram formation and requires all of the precautions standard to holography.

20.1.1 Replication of Plane Absorption Holograms with Nonlaser Light

Plane holograms can be replicated by contact-printing the original hologram onto another photosensitive surface [*20.2*]. We assume here that both the original and replicated hologram are recorded in photographic emulsion. For ideal contact printing, the separation between the original hologram emulsion and the replica emulsion should be everywhere less than a wavelength. In this case diffraction of the illuminating light by the hologram can be ignored. [See Eq. (5.25), for $d < \lambda$.] As a result, ordinary (nonlaser) light can be employed to pass through the original hologram and expose the replica emulsion. A negative replica of the original hologram is obtained. It however yields a positive image which appears to be identical to that of the image generated by the original (see Section 2.6.1).

In practice one finds that it is not a simple matter to keep the separation between the original and replica emulsions uniformly less than a wavelength [*20.3*]. When the separation is larger than a wavelength, diffraction effects and the coherence properties of the illumination must be considered. Figure 20.1 indicates the general configuration for hologram replication by "con-

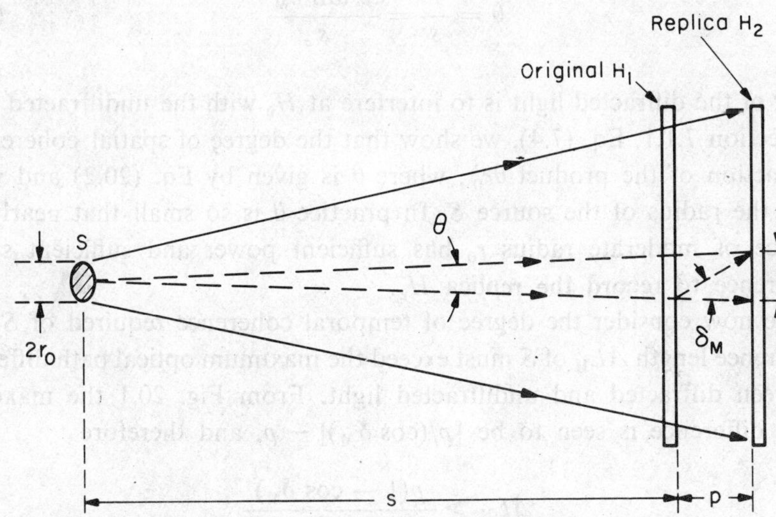

FIG. 20.1. General arrangement for hologram replication.

tact" printing. In the first instance we suppose that the printing light comes from a small thermal source S a distance s from an original hologram H_1. Between H_1 and the replica plate H_2 there is a finite separation $p \ll s$. Waves diffracted from H_1 acquire phase factors as they propagate through the distance separating H_1 and H_2. If the information contained in the original hologram is to be retained in the copy, then the complex amplitudes of the waves must be recorded. This can be done if the degree of coherence of the illuminating light is sufficient to produce high visibility in the fringes of interference between undiffracted and diffracted light over the surface of H_2.

Let us consider the factors which determine the degree of spatial coherence needed in the illumination. Suppose δ_M is the maximum angle at which light is diffracted and suppose the light from S is incident more or less normal to the surface of H_1. According to Eq. (1.11), diffraction of normally incident light ($i = 0$) from the plane hologram grating is described by the relation

$$d \sin \delta_M = \bar{\lambda} \tag{20.1}$$

where d is the smallest fringe spacing recorded in H_1 and $\bar{\lambda}$ is the mean wavelength of the illumination. It is evident from Fig. 20.1 that the light source S must have a high degree of spatial coherence over a cone whose angle is

$$\theta = \frac{r}{s} = \frac{p \tan \delta_M}{s} \tag{20.2}$$

if all of the diffracted light is to interfere at H_2 with the undiffracted light. In Section 7.1.1, Eq. (7.4), we show that the degree of spatial coherence is a function of the product θr_0, where θ is given by Eq. (20.2) and where r_0 is the radius of the source S. In practice θ is so small that nearly any source of moderate radius r_0 has sufficient power and sufficient spatial coherence to record the replica H_2.

We now consider the degree of temporal coherence required of S. The coherence length ΔL_H of S must exceed the maximum optical path difference between diffracted and undiffracted light. From Fig. 20.1 the maximum path difference is seen to be $[p/(\cos \delta_M)] - p$, and therefore

$$\Delta L_H \geq \frac{p(1 - \cos \delta_M)}{\cos \delta_M}$$

or

$$p \leq \frac{\cos \delta_M}{1 - \cos \delta_M} \Delta L_H . \tag{20.3}$$

Holograms formed in photographic emulsion can be considered plane only when δ_M is small, so that we may approximate $\sin \delta_M$ by δ_M and may set $\cos \delta_M \approx 1 - \delta_M^2/2$. Using these approximations first in Eq. (20.1) to obtain $\delta_M \approx \bar{\lambda}/d$ and then in Eq. (20.3), we have

$$p \leq \Delta L_H \, [2(d/\bar{\lambda})^2 - 1]. \tag{20.4}$$

In Section 7.1.2 the coherence length for light at 5461-Å wavelength emitted by a high-pressure mercury arc lamp is calculated to be 8 μm. Since $d/\bar{\lambda}$

for plane holograms is typically 3, we must ensure that $p \leq 17 \times 8\ \mu m = 136\ \mu m$. This is easily satisfied, and therefore a high-pressure mercury arc source is adequate for contact printing.

20.1.2 Double Images from Holograms Replicated with Laser Light

If laser light is used for hologram replication, its high degree of coherence permits H_1 and H_2 to be separated by large distances, e.g., several centimeters. However, we shall see that such large separations are undesirable in that they lead to replicas which generate double images (two virtual and two real images) [*20.4*].

Suppose a hologram H_1 at $z = z_1$ is formed with an off-axis subject and an *axial* reference plane wave of unit amplitude. When illuminated with the same axial plane wave, the hologram diffracts light whose complex amplitude at the hologram is given by

$$\mathbf{w}(x_1, y_1) \propto 1 + \mathbf{a}(x_1, y_1)\mathbf{a}^*(x_1, y_1) + \mathbf{a}(x_1, y_1) + \mathbf{a}^*(x_1, y_1) \quad (20.5)$$

where $\mathbf{a}(x_1, y_1)$ is the complex amplitude of the original subject wave at the hologram H_1 and $\mathbf{a}^*(x_1, y_1)$ is the complex conjugate of that amplitude. (Use of an axial reference wave merely simplifies the notation and does not restrict the results.) We now suppose that the subject is a diffuse scatterer so that $\mathbf{aa}^*$ is essentially constant over the hologram plane or over any other plane. We may now write

$$\mathbf{w}(x_1, y_1) \propto c_1 + \mathbf{a}(x_1, y_1) + \mathbf{a}^*(x_1, y_1) \quad (20.6)$$

where c_1 is a constant. Located in a plane a short distance from H_1, $z = z_2$, is a replica emulsion H_2 (Fig. 20.2).[1] It is to be exposed to the Fresnel transform of $\mathbf{w}(x_1, y_1)$, i.e., to the complex amplitude resulting from the propagation by $\mathbf{w}$ through a distance $p = z_2 - z_1$. We write for the Fresnel transform

$$\mathbf{W}(x_2, y_2) \propto c_2 + \mathbf{A}_V(x_2, y_2) + \mathbf{A}_C(x_2, y_2) \quad (20.7)$$

where c_2 is a constant, $\mathbf{A}_V$ is the Fresnel transform of $\mathbf{a}$, and $\mathbf{A}_C$ is the Fresnel transform of $\mathbf{a}^*$. (Note that $\mathbf{A}_C$ is not equal to $\mathbf{A}_V^*$, except when $z_2 - z_1 = p = 0$.) Both $\mathbf{a}$ and $\mathbf{A}_V$ represent complex amplitudes (in different planes) of the same wave diverging from a virtual image V_1 at $z = v_1$ (see Fig. 20.2). The virtual image V_1 is separated from H_1 by an axial distance $z_1 - v_1$.

[1] Although the photosensitive medium at H_2 can be carried on a plate of glass, it is of course more economic to use an acetate sheet.

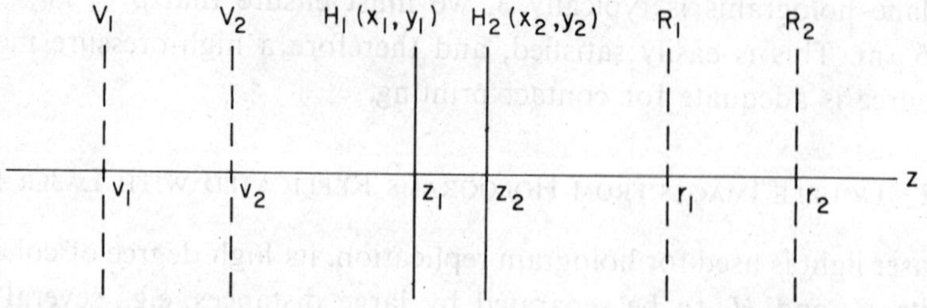

FIG. 20.2. Axial positions of double images obtained from the replica of a plane hologram.

Similarly $\mathbf{a}^*$ and $\mathbf{A}_C$ are the complex amplitudes of a wave converging to a real image R_1 at $z = r_1$, separated from H_1 by $r_1 - z_1$. According to Eq. (3.35), when reference and illuminating waves are plane, the virtual and real images are symmetric about the hologram plane, i.e.,

$$r_1 - z_1 = z_1 - v_1. \tag{20.8}$$

The replica emulsion H_2, if properly exposed to $\mathbf{W}(x_2, y_2)$ and developed, will have a transmittance

$$\begin{aligned} t &\propto (c_2 + \mathbf{A}_V + \mathbf{A}_C)(c_2 + \mathbf{A}_V^* + \mathbf{A}_C^*) \\ &\propto \text{zero-order terms} + c_2(\mathbf{A}_V + \mathbf{A}_V^*) + c_2^*(\mathbf{A}_C + \mathbf{A}_C^*) \\ &\quad + \text{higher order terms.} \end{aligned} \tag{20.9}$$

Equation (20.9) has a form corresponding to two holograms: One is the result of interference of an axial plane wave with the wave designated by subscript V; the other results from interference of the same plane wave with the wave of subscript C. When the replica H_2 is illuminated by the plane wave, the virtual image V_1 is formed at $z = v_1$, a distance $z_2 - v_1$ from H_2, by the wave associated with $\mathbf{A}_V$. We know from Eq. (3.35) that the corresponding real image R_2, generated by the wave associated with $\mathbf{A}_V^*$, must be symmetric about H_2. Therefore its z position, $z = r_2$, is given by

$$z_2 - v_1 = r_2 - z_2. \tag{20.10}$$

Similarly, the wave $\mathbf{A}_C$ yields the real image R_1 at $z = r_1$, a distance $r_1 - z_2$ from H_2. The virtual image V_2, generated by $\mathbf{A}_C^*$, must be symmetric to R_1 about H_2 and its position given by

$$z_2 - v_2 = r_1 - z_2. \tag{20.11}$$

Combining Eqs. (20.11) and (20.8) we obtain

$$v_2 - v_1 = 2(z_2 - z_1) = 2p. \tag{20.12}$$

Combining Eqs. (20.10) and (20.8) yields

$$r_2 - r_1 = 2(z_2 - z_1) = 2p. \tag{20.13}$$

Equation (20.12) indicates that the two virtual images are separated by twice the spacing p between the original hologram and the replica emulsion. We see that the same separation holds for the real images. When $p = 0$, the double images become one. Figure 20.3 is a photograph of double images

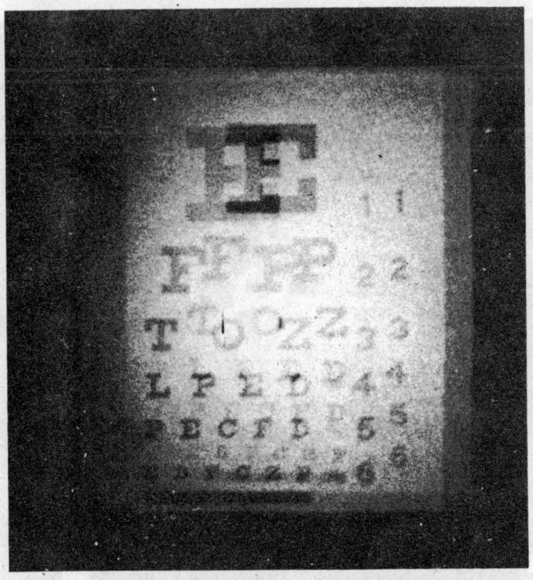

FIG. 20.3. A photograph of double images generated by the replica of a plane hologram.

from a hologram replica. In this case $p = 2.5$ mm. The subject was a transparency backed with a diffusing screen. Contrast is reduced over most of the image due to the superposition of transparent areas onto opaque areas. To the extent that **aa*** is not a constant intensity but only a rapidly varying spatial function whose average is constant, some noise will be added to the image generated by the replica.

20.1.3 Replication of Plane Phase Holograms

A hologram recorded as spatial variations in surface relief, e.g., as in thermoplastic holograms, is a plane phase hologram. Plane-wave light illuminating the phase hologram maintains its unmodulated *intensity* at the surface of the hologram, and only when the light propagates a sufficient distance from the surface can interference and intensity modulation take place. Thus ideal contact printing, i.e., where $p = 0$, is not possible for plane phase holograms, since printing requires intensity variation. Mechanical, rather than optical, methods for replication seem best suited for surface-relief phase holograms. By pressing deformable plastic sheet against the original surface-relief hologram or against a more durable metal master made from the original, replicas can be formed in the same fashion as phonograph records. Figure 20.4 shows an image from an original hologram recorded in thermoplastic film and an image from a replica formed by the mechanical (pressing) method.

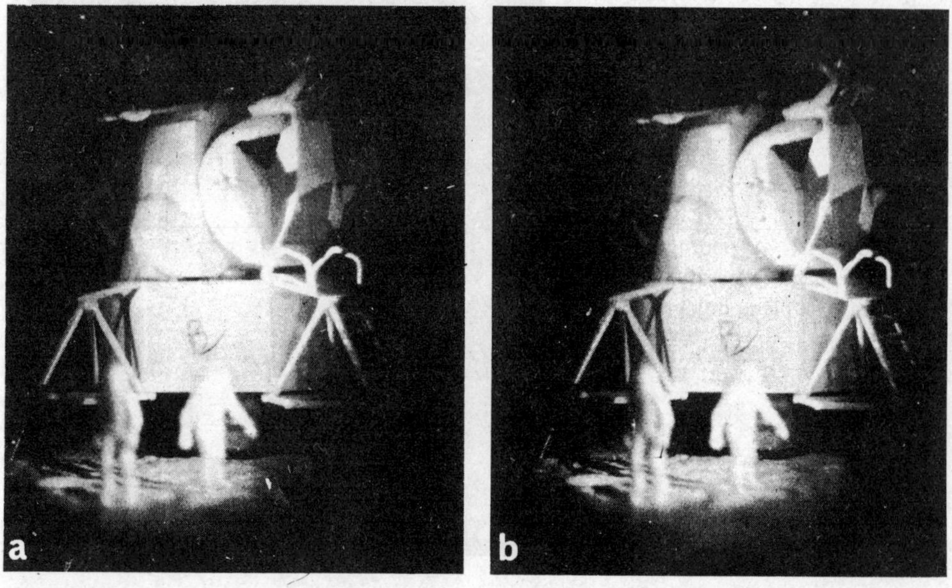

FIG. 20.4. Photographs of images obtained (a) from a thermoplastic hologram and (b) from a replica produced by mechanical methods. (Courtesy J. C. Urbach, W. F. Folger, and R. F. Bergen, Research Lab. Div., Xerox Corp.)

20.1.4 REPLICATION OF VOLUME HOLOGRAMS

Despite their small-fringe spacings, satisfactory replicas of volume *transmission* holograms can be made with an optical arrangement similar to Fig. 20.1. An important difference, however, is that the light illuminating H_1 should duplicate the curvature, direction, and wavelength of the reference wave used to form H_1 [*20.5*]. In this case a maximum-efficiency undistorted reconstruction of the original subject wave is diffracted to H_2. Because of the nature of the volume hologram, a wave with a complex amplitude conjugate to that of the subject wave is not simultaneously diffracted. Hence, not even large values of the separation p will produce double images. Of course the coherence length of the source must exceed the maximum possible optical path differences for diffracted and undiffracted waves passing between H_1 and H_2.

If the replica H_2 is to diffract with high efficiency, the visibility of the fringes it records must be high. High visibility is achieved when the ratio R of the intensities of undiffracted and reconstructed waves from H_1 approaches unity [see Eq. (7.31)]. Let us suppose for the moment that these are plane waves. We assume that H_1 has been optimally exposed (to an intensity ratio $R = 1$) and developed, resulting in an optical density $D \approx 1.0 = \log(1/\mathscr{T})$ and an intensity transmittance $\mathscr{T} = 10\%$ (see Section 9.6.2). Therefore the intensity of the undiffracted wave is approximately 10% of the illuminating beam. As we find in Section 9.6.2, the maximum diffraction efficiency for volume transmission *absorption* holograms is 3.7%, making the reconstructed wave intensity 3.7% of the illuminating beam. With the ratio $R = 10/3.7 = 2.7$, the visibility V of the interference pattern falling on H_2 is given by Eq. (7.31) as $V = 0.89$ (when $|\mu_T| = \cos\Omega = 1$). Diffraction efficiency is proportional to V^2 according to Eq. (7.38); thus the best efficiency η_2 that the replica H_2 can have is given in terms of the efficiency η_1 of H_1 as

$$\eta_2/\eta_1 = V_2^2/V_1^2 = (0.89)^2/1 = 0.8.$$

Figure 20.5 shows photographs of two images one obtained from an absorption transmission hologram and one from its replica. Both holograms were recorded on Kodak 649F photographic plates. During replication, the emulsion of one plate was placed against the uncoated surface of the other. A few drops of xylene were placed between the plates for the purpose of index matching, so that interference of reflections was minimized. The diffraction efficiency of the replica was nearly the same as that of the original.

By reversing the positions of H_1 and H_2 in Fig. 20.1, replicas of *reflection* holograms can be made. Again, as with volume transmission holograms, the illuminating beam should duplicate the curvature, direction, and wavelength of the reference wave used to form H_1. The illuminating beam passes through the replica medium H_2 first and then diffracts from H_1 as a wave traveling back to H_2. Recorded in H_2 is the interference of the incident illuminating wave and the back-diffracted, reconstructed subject wave from

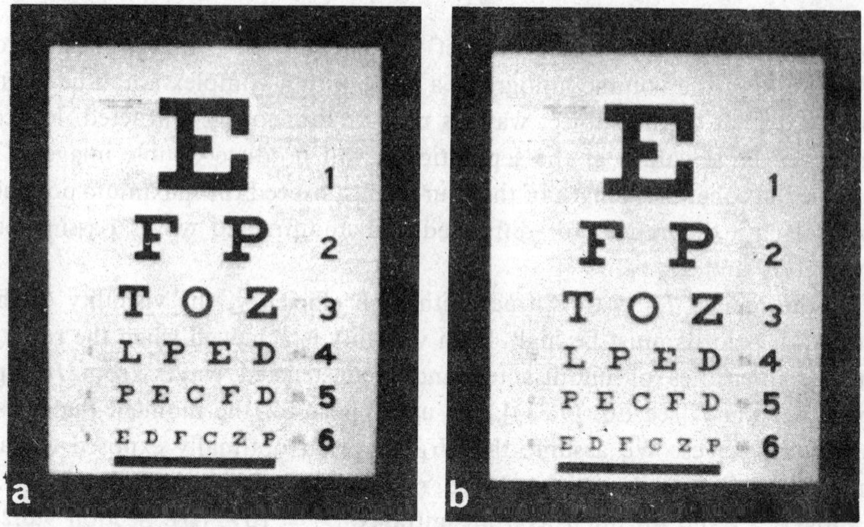

FIG. 20.5. Photographs of images obtained (a) from an original volume absorption transmission hologram and (b) from its replica.

H_1. If H_1 is an absorption hologram, the intensity of the diffracted wave is at best a few percent of the incident intensity (see Table 9.1). Interference of this wave with the incident light produces a low visibility pattern at H_2 and a low efficiency replica. When the original is a phase reflection hologram recorded, e.g., in dichromated gelatin, then the diffracted wave has an intensity comparable to the incident light intensity, and the efficiency and image quality of the replica can be high. Figure 20.6 indicates the quality of an image obtained from a replica of a reflection hologram originally recorded in dichromated gelatin film. The replica, recorded in the same material, was illuminated with laser light to produce the image shown.

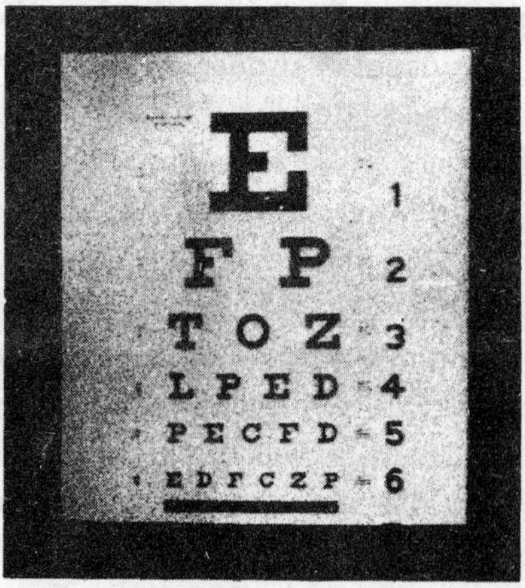

FIG. 20.6. Photograph of the image generated by a replica of a phase reflection hologram. The replica was illuminated with laser light.

20.2 Television Transmission of Holograms

Speculation concerning the feasibility of 3D TV based on hologram transmission grew quite naturally out of the appeal exhibited by holographic 3D images [*20.6*]. In principle one should be able to form a hologram directly on the face of a TV camera tube, transmit the interference intensity pattern over television channels, and replicate the hologram at the receiving station. The replica must be formed on a transparent surface which when illuminated with laser light will allow the reconstructed wave to emerge and generate the desired image. If the process is to add 3D to the viewing of "live" scenes, the formation, transmission, and replication steps must permit a new hologram to be illuminated for the viewer every 1/30 of a second. We note in Section 18.2 that unless information reduction techniques are successful, several hours would be required to transmit the wide-angle hologram of a 3D object over present TV channels. Solution of the transmission problem would still leave much undone at either end of the link. Camera tubes have limited resolution. Either a restricted field of view or reduced image resolution is the consequence (see Section 8.4.1). Coherence limitations of current laser sources limit the spatial extent of the scene. At the receiving end, the hologram must be recreated on a photosensitive material which can be exposed in less than 1/30 of a second and then erased in time for the next frame. Live three-dimensional television employing hologram formation and transmission is indeed a formidable problem.

Less ambitious is the task of transmitting a hologram of a two-dimensional, nondiffusing transparency. One goal of studying such a problem is to discover whether information encoded as a hologram and then transmitted is more immune to channel noise than the same information transmitted in the ordinary mode. The study has led to the invention of transmission schemes which are interesting in their own right.

In the first experiments with hologram transmission, a carrier-frequency (off-axis) hologram was formed directly on the face of the TV camera tube [*20.7*]. Because of the limited resolution capability of the camera tube, the angle between subject and reference beams was made very small. This was accomplished with the arrangement shown in Fig. 8.5 where the subject was a transparency and the photographic plate was replaced by the camera tube face. The hologram so formed was transmitted over closed-circuit television, displayed on the kinescope, and photographed. Since the spatial frequency on which the subject information is carried is very low, the reconstruction step requires a spatial filtering process to isolate the real-image-forming wave from undiffracted light (see Section 8.1.1). This straightforward transmission method has drawbacks. Among these is the need to record on the camera spatial frequencies four times the maximum present in the original subject transparency (see Section 8.1.1). Moreover, the useful subject information occurs near the skirt of the spatial frequency passband.

20.2.1 In-Line Hologram Method

Figure 20.7 indicates an arrangement of components which reduces the resolution required of the camera tube to only that needed to resolve the original subject transparency [*20.8*]. In this case the hologram recorded by

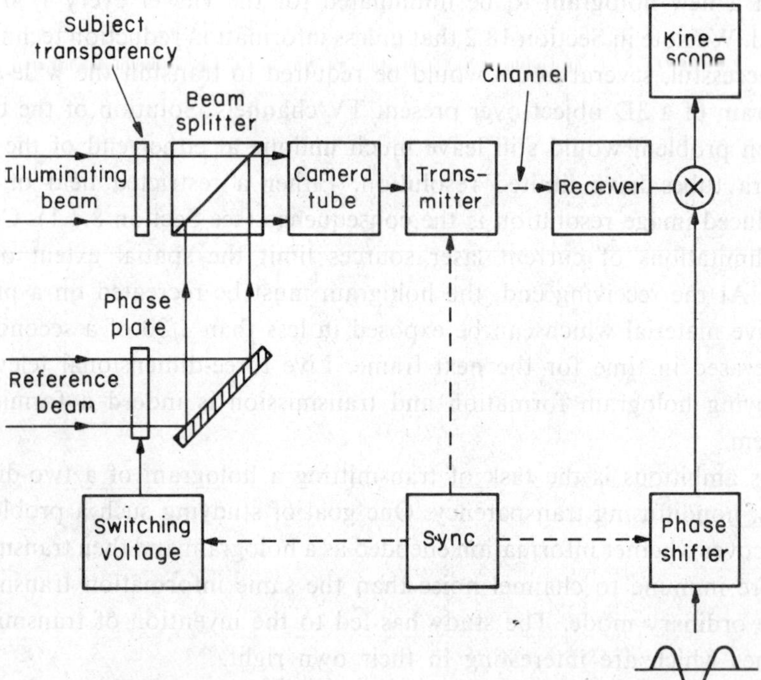

Fig. 20.7. System for television transmission of holograms.

the camera tube is formed as an in-line hologram with the reference beam directed along the mean direction of the subject light and normal to the camera tube face. The maximum spatial frequency of the on-axis or in-line hologram is the maximum spatial frequency of the subject (see Section 8.4.1). Of course in-line holograms have twin-image and other problems noted in Section 2.5. With the arrangement of Fig. 20.7, however, it is possible to display an off-axis or carrier frequency hologram on the receiving station kinescope, while sending only in-line holograms. Three in-line holograms are to be transmitted as follows: A first hologram is formed on the camera with the axial reference plane wave; the light distribution is scanned, converted to electrical signals, and transmitted. At the receiver terminal, the received electrical signal is used to amplitude modulate a sinusoidal electrical carrier and the result displayed on a kinescope. Suppose that the kinescope holds this display for the time necessary to transmit two more in-line holograms. Before forming the second, the phase of the optical reference wave at the transmitter is advanced by 120° from the phase used for the first hologram. Again the hologram is scanned, converted, and transmitted. This time the received electrical signal modulates an electrical carrier advanced in phase by 120° from that used for the first hologram. This second hologram is displayed on the kinescope along with the first. Finally the optical reference phase at the transmitter is advanced another 120°, the sinusoidal electrical carrier at the receiver is once more advanced 120° in phase, and a third hologram is superimposed on the previous two at the kinescope. When one adds the three intensities on the kinescope (assuming linear intensity transmission and recording), he finds a resultant intensity which has the form of a carrier-frequency hologram transmittance. Moreover the undesirable zero-order terms, whose spatial bandwidth is twice that of

FIG. 20.8. Image generated by a hologram transmitted over the system in Fig. 20.7. (Courtesy Berrang [*20.9*].)

the subject, are absent. If the spatial frequencies ξ_s of the subject wave at the camera lie in the range $-\xi_m \leq \xi_s \leq +\xi_m$, then we may set the spatial carrier frequency at the kinescope to be $\xi_r = \xi_m$ making $2\xi_m$ the highest spatial frequency displayed on the kinescope. Only twice the maximum subject spatial frequency must be displayed. Figure 20.8 is a photograph of the image generated by a hologram transmitted in the above manner. Rather than holding the three hologram components on the kinescope, the hologram plate was multiply exposed [*20.9*].

20.2.2 HETERODYNE SCANNER METHOD

Figure 20.9 indicates a method of detecting the intensity pattern of interference between subject light and a scanning reference beam, the latter focused to a spot at the surface of a large-area photodetector [*20.10, 20.11*]. A novel feature of the method is that the temporal angular frequency ω_s of the subject light is not the same as that of the reference beam ω_r. We

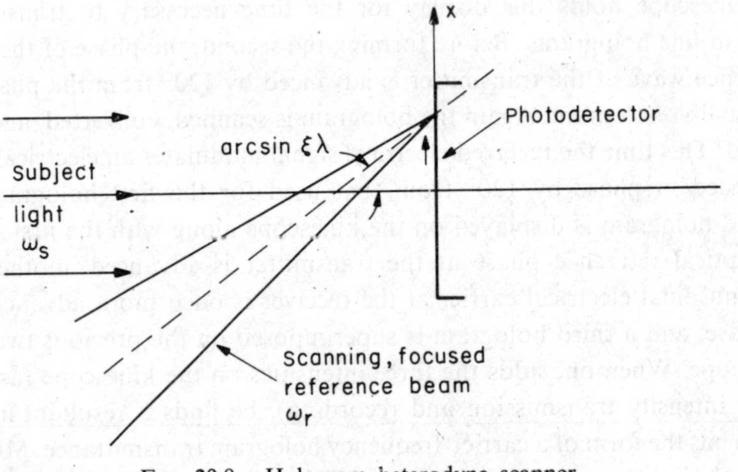

FIG. 20.9. Hologram heterodyne scanner.

designate the complex electric field of the subject wave at the hologram plane (the surface of the photodetector) by

$$\mathbf{v}_s(x, y, t) = a(x, y) \exp\{i[\omega_s t + \varphi_s(x, y)]\}, \tag{20.14}$$

where t is the time and φ_s is a time-independent phase factor, and the complex electric field of the scanning reference wave at the hologram plane by

$$\mathbf{v}_r(x, y, t) = r(x, y, t) \exp\{i[\omega_r t - 2\pi\xi x - \varphi_r]\} \tag{20.15}$$

where ξ is the mean spatial frequency and φ_r is a constant phase factor. In Eq. (20.14) the mean direction of the subject light is assumed to be normal to the detector, i.e., on axis. On the other hand, the mean direction of the reference beam makes an angle $\arcsin(\xi\lambda)$ to the axis (see Fig. 20.9). Interference of $\mathbf{v}_s$ and $\mathbf{v}_r$ produces a pattern whose intensity is

$$\begin{aligned} I &= (\mathbf{v}_s + \mathbf{v}_r)(\mathbf{v}_s + \mathbf{v}_r)^* \\ &= a^2(x, y) + r^2(x, y, t) \\ &\quad + 2a(x, y)r(x, y, t) \cos[(\omega_s - \omega_r)t + \varphi_s(x, y) + \varphi_r + 2\pi\xi x]. \end{aligned} \tag{20.16}$$

If the reference beam is focused to a point on the detector surface and scanned in the x direction with a velocity b, we may write

$$r(x, y, t) = \delta(x - bt)\,\delta(y - y_n) \tag{20.17}$$

where y_n is the vertical height of the scanned line. The time-varying component of the photodetector current $i(t)$ is obtained by integrating the light intensity I over the area of the detector. Substituting Eq. (20.17) into Eq. (20.16), using the sifting property of the delta function [Eq. (4.13e)] in the integration over the photodetector area, and retaining only time-dependent terms, we obtain

$$i(t) = 2Ka(bt, y_n)\cos[(\omega_s - \omega_r + 2\pi\xi b)t + \varphi_s(bt, y_n) + \varphi_r] \tag{20.18}$$

where K is a constant of the recording process. Note that the undesirable zero-order terms do not appear in Eq. (20.18) since their spatial averages over the area of the detector are not functions of the time. Moreover $a(bt, y_n)$ is modulated onto a temporal carrier angular frequency $\omega = \omega_s - \omega_r + 2\pi\xi b$. When the signal is transmitted and then displayed on the kinescope, the carrier frequency ω is converted into a spatial frequency. By choosing $\omega_s \neq \omega_r$ (heterodyning), we can obtain a carrier frequency hologram at the kinescope even though we form an in-line hologram ($\xi = 0$). As it turns out, the heterodyne method increases the resolution of the scanner by a factor 2 over that obtained without heterodyning, i.e., when $\omega_s = \omega_r$, $\xi \neq 0$. A detailed analysis of the heterodyne scanner yields the following interesting results [*20.11*]:

1. The scanning reference beam spot can resolve the phase variations of the subject beam only if the subject beam appears to originate within the cone of the scanning beam (see Fig. 20.10). Thus reference and subject beams must be combined with a beam splitter as shown in Fig. 20.10.
2. The detector need not be located in the focal plane of the reference beam but can be located at any position where it intercepts all the area common to both beams. An out-of-focus position is advantageous because it reduces the peak intensity on the photosensitive surface and minimizes the effect of dust on that surface.

Figure 20.11 shows an image from a hologram transmitted over the system whose essentials we have sketched. Horizontal and vertical deflection of the reference beam were carried out by two scanning mirrors. To

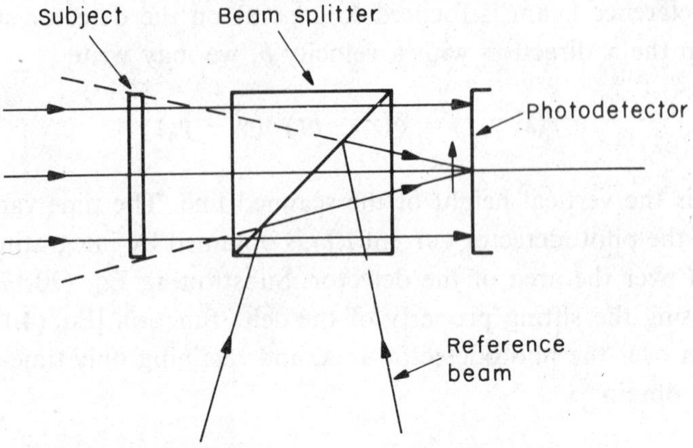

FIG. 20.10. Method of combining the focused scanning reference beam with the subject light to form an in-line hologram.

achieve the angular frequency difference $\omega_s - \omega_r$ the reference beam was reflected from a vibrating mirror. By this means its temporal frequency was Doppler-shifted from that of the subject light.

FIG. 20.11. A photograph of an image from a hologram transmitted by the heterodyne scanner system. (Courtesy Larsen [*20.11*].)

20.3 Holograms Formed with Spatially Incoherent Subject Light

Having begun by extolling the virtues of coherent light, we close this book with a brief consideration of attempts to form holograms with spatially incoherent light. (Coherent light is still used in the reconstruction.) The desired resolution in the image determines the spatial extent over which interference fringes must be produced. This in turn fixes the temporal coherence required to form fringes over that extent. Although a mercury arc lamp has been used as a source, in most experiments the hologram is formed with laser light passed through a moving diffuser. In the latter case the degree of temporal coherence can be high, and a large hologram with a great many fringes can be recorded. Nevertheless, the moving diffuser makes the light in the transmitted laser beam spatially incoherent. Over the time needed to expose a hologram, the phase of the light at any point in the beam exhibits no fixed relation to the phase at a neighboring point. Before proceeding to discuss methods and problems, it is fair to state that incoherent holography has so far had only limited success.

The basic principle of incoherent holography is to amplitude-divide the subject wavefront into two components and to cause two light waves, appearing to come from pairs of corresponding points on the subject surface, to interfere [*20.12*, *20.13*]. Since light from one pair of points is incoherent with that from any other pair, we shall have M independent, superimposed holograms, corresponding to the M subject points. We note that the interference of light coming from the same subject point provides no information on the phase of the light relative to that coming from other points. There is no reference against which phase can be compared. Yet it is possible to coherently illuminate the set of superimposed holograms and diffract waves which focus to M points on a 3D image of the subject.

To illustrate at least one recording technique let us consider formation of a hologram of the transparent letter R illuminated with spatially incoherent light [*20.14*]. An interferometer is devised which makes the light arriving at the hologram plane appear to come not only from the letter R in its normal orientation, but also from a rotated and translated R located in the same subject plane, as shown in Fig. 20.12. Any pair of corresponding points

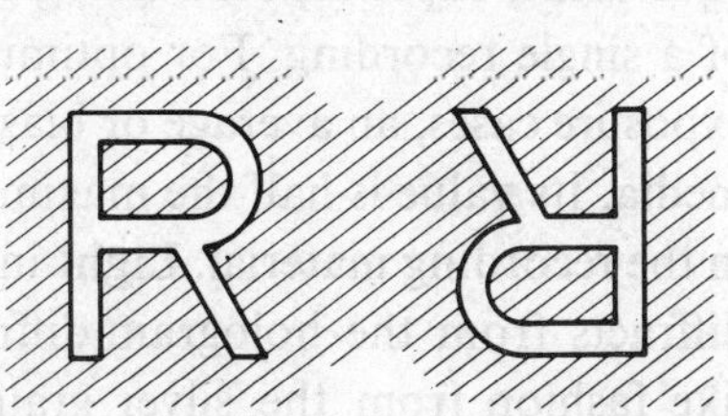

FIG. 20.12. Subject scene as seen from the hologram plane.

can form a lensless Fourier transform hologram. In each case the linear fringes recorded on a plane parallel to the subject plane run at right angles to the line joining the two points, and the fringe frequency is proportional to the separation of the pair (see Section 3.1.3). Examination of Fig. 20.12 reveals that no two pairs of points can form fringes whose orientation and spatial frequency are both identical. Each lenless Fourier transform hologram formed in this way is unique. Each, when placed adjacent to a lens and illuminated with a plane wave of coherent light, generates a real image of the point source forming that hologram. As with any Fourier transform hologram illuminated in the manner of Fig. 8.23, two real images are formed in the back focal plane of the lens. Figure 20.13 shows the image generated by an incoherent Fourier transform hologram [*20.15*]. The subject was illuminated with a mercury arc lamp during formation of the hologram.

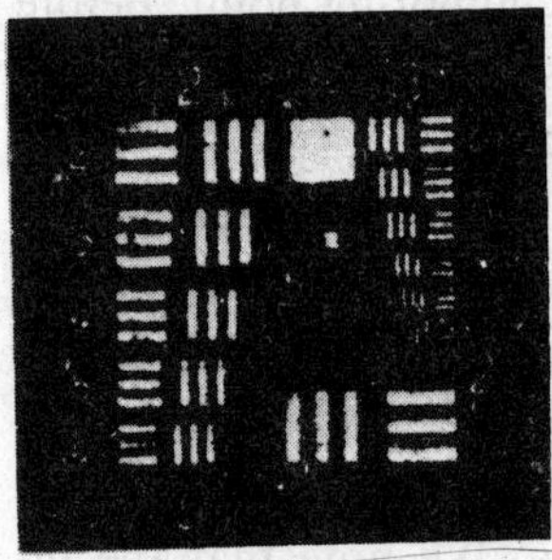

FIG. 20.13. Image generated by a hologram recorded in spatially incoherent light (Courtesy Worthington [*20.15*].)

There are a number of optical geometries or interferometers suitable for forming incoherent-light holograms. In some the subject depth information is preserved; in others it is not. To understand the niceties of these interferometers requires more attention than is warranted by their utility; details are provided by Lohmann [*20.12*], Cochran [*20.13*], Stroke and Restrick [*20.14*], Worthington [*20.15*], and Kozma and Massey [*20.16*].

It is the low diffraction efficiency of each of the superimposed, point-source holograms which is largely responsible for the ineffectiveness of incoherent-light holography. In Section 17.5.5, Eq. (17.29), we compute that the maximum efficiency of an absorption hologram composed of M equivalent, superimposed holograms is $1/M^2$ times the maximum efficiency of a single recording. For optimum diffraction in either multiple or single-exposure cases, an average or bias transmittance is imparted to the recording media. Its value is half the maximum value that the transmittance can attain in the recording material. Light incident on the developed hologram not only diffracts from the hologram diffraction grating but also scatters in noise-like fashion from the silver grains contributing to the bias transmittance.

Since bias and noise are the same for both multiple and single recordings, we can compare signal-to-noise ratios in the two cases by comparing diffraction efficiencies. Let us consider first a single-exposure hologram formed with temporally and spatially coherent light. Its maximum efficiency is η_1. Suppose that the image it generates contains M resolvable points. If the power illuminating the hologram is P_0, then the power delivered to any image point is

$$P_c = P_0\eta_1/M. \qquad (20.19)$$

For the hologram formed with spatially incoherent light, the power diffracted to each image point is determined by the efficiency of each superimposed hologram, given by Eq. (17.29) as

$$\eta_M = \eta_1/M^2,$$

making the power delivered to any image point

$$P_i = P_0\eta_M/M = P_0\eta_1/M^3. \qquad (20.20)$$

Comparison of Eq. (20.19) and (20.20) shows the signal-to-noise ratio for incoherent-light holograms to be reduced by a factor M^2 over that for holograms formed in coherent light. Thus the greater the number of resolvable points, the greater the disparity in signal-to-noise ratio. Even for a picture with the resolution of a television display, the ratio is reduced by 6×10^{10}.

Methods have been proposed to eliminate the noise source by eliminating the average or bias exposure in the hologram recording. Complex electronic circuitry is required by one technique [*20.16*]. Recording materials such as thermoplastics, which show a bandpass response (see Fig. 10.18) and naturally suppress low spatial frequency components of the exposure, might be appropriate. However, any other noise source present in the recording is equally troublesome. Possibly a low-noise phase material such as dichromated gelatin can give better signal-to-noise ratios than photographic emulsion. Another suggestion is to form an image in the vertical direction and a series of one-dimensional holograms in the horizontal direction [*20.17*]. Each one-dimensional hologram receives contributions to its fringe pattern from only the limited number of points in a horizontal strip of the subject, and consequently the signal-to-noise ratio is improved.

REFERENCES

20.1. F. B. Rotz and A. A. Friesem, "Hologram With Nonpseudoscopic Real Images," *Appl. Phys. Lett.* **8**, 146 (1966).

20.2. F. S. Harris, Jr., G. C. Sherman, and B. H. Billings, "Copying Holograms," *Appl. Opt.* **5**, 665 (1965).

20.3. D. B. Brumm, "Copying Holograms," *Appl. Opt.* **5**, 1946 (1966).

20.4. D. B. Brumm, "Double Images in Copy Holograms," *Appl. Opt.* **6**, 588 (1967).

20.5. M. J. Landry, "The Effect of Two Hologram-Copying Parameters on the Quality of Copies," *Appl. Opt.* **6**, 1947 (1967).

20.6. E. N. Leith, J. Upatnieks, B. P. Hildebrand, and K. Haines, "Requirements for a Wavefront Reconstruction Television Facsimile System," *J. Soc. Motion Pict. Telev. Eng.* **74**, 893 (1965).

20.7. L. H. Enloe, J. A. Murphy, and C. B. Rubinstein, "Hologram Transmission via Television," *Bell Syst. Tech. J.* **45**, 335 (1966).

20.8. C. B. Burckhardt and L. H. Enloe, "Television Transmission of Holograms With Reduced Resolution Requirements on the Camera Tube," *Bell Syst. Tech. J.* **48**, 1529 (1969).

20.9. J. E. Berrang, "Television Transmission of Holograms Using a Narrow-Band Video Signal," *Bell Syst. Tech. J.* **49**, 879 (1970).

20.10. L. H. Enloe, W. C. Jakes, Jr., and C. B. Rubinstein, "Hologram Heterodyne Scanners," *Bell Syst. Tech. J.* **47**, 1875 (1968).

20.11. A. B. Larsen, "A Heterodyne Scanning System for Hologram Transmission," *Bell Syst. Tech. J.* **48**, 2507 (1969).

20.12. A. W. Lohmann, "Wavefront Reconstruction for Incoherent Objects," *J. Opt. Soc. Amer.* **55**, 1555 (1965).

20.13. G. Cochran, "New Method of Making Fresnel Transforms with Incoherent Light," *J. Opt. Soc. Amer.* **56**, 1513 (1966).

20.14. G. W. Stroke and R. C. Restrick, III, "Holography with Spatially Noncoherent Light," *Appl. Phys. Lett.* **7**, 229 (1965).

20.15. H. R. Worthington, Jr., "Production of Holograms with Incoherent Illumination," *J. Opt. Soc. Amer.* **56**, 1397 (1966).

20.16. A. Kozma and N. Massey, "Bias Level Reduction of Incoherent Holograms," *Appl. Opt.* **8**, 393 (1969).

20.17. O. Bryngdahl and A. Lohmann, "One-Dimensional Holography with Spatially Incoherent Light," *J. Opt. Soc. Amer.* **58**, 625 (1968).

Appendix I

EQUIVALENCE OF THE FRESNEL–KIRCHHOFF INTEGRAL AND THE DIFFRACTION FORMULA IN THE SPATIAL FREQUENCY DOMAIN

In this appendix we demonstrate the equivalence of the diffraction formula Eq. (5.26) and the Fresnel–Kirchhoff integral Eq. (5.31). Both equations can be used to compute the complex amplitude at a distance d away from a diffracting object. The following material is taken from unpublished work of Rowe [*I.1*]. (Other proofs of the same equivalence are given by Sherman [*I.2*] and Osterberg [*I.3*].[1])

In Eq. (5.26) the spectrum of the complex amplitude distribution at a distance d from the diffracting object is obtained by multiplying the Fourier transform of the amplitude transmittance of the diffracting object by a function

$$\mathbf{H}_1(\xi, \eta) = \exp[-ikd(1 - \lambda^2\xi^2 - \lambda^2\eta^2)^{1/2}], \qquad k = 2\pi/\lambda. \tag{I.1}$$

In the Fresnel–Kirchhoff integral, Eq. (5.31), the complex amplitude distribution at a distance d from the diffracting object is obtained by convolving the amplitude transmittance of the diffracting object with a function $\mathbf{h}_1(x, y)$. From Fig. 5.4 we see that $\cos\theta$ in Eq. (5.31) is

$$\cos\theta = d[d^2 + (x_2 - x_1)^2 + (y_2 - y_1)^2]^{-1/2}. \tag{I.2}$$

[1] Sherman [*I.2*] and Osterberg [*I.3*] show the equivalence of our Eq. (5.26) and Rayleigh's integral. Our Eq. (5.31) follows from Rayleigh's integral by carrying out the differentiation under the Rayleigh integral and assuming that $d \ll \lambda$.

If we insert Eq. (I.2) into Eq. (5.31) we obtain

$$\mathbf{h}_1(x, y) = \frac{i}{\lambda} \frac{\exp[-ik(d^2 + x^2 + y^2)^{1/2}]}{(d^2 + x^2 + y^2)^{1/2}} \cdot \frac{d}{(d^2 + x^2 + y^2)^{1/2}}. \qquad \text{(I.3)}$$

A multiplication in the spatial frequency domain corresponds to a convolution in the spatial domain according to correspondence (4.11). If we are to prove the equivalence of Eqs. (5.26) and (5.31), we must show that $\mathbf{h}_1(x, y)$ is the inverse Fourier transform of $\mathbf{H}_1(\xi, \eta)$.

The inverse Fourier transform of $\mathbf{H}_1(\xi, \eta)$ is obtained as the integral

$$\mathscr{F}^{-1}[\mathbf{H}_1(\xi, \eta)] = \int_{-\infty}^{+\infty}\int_{-\infty}^{+\infty} \exp[-ikd(1 - \lambda^2\xi^2 - \lambda^2\eta^2)^{1/2}] \times \exp(-i2\pi\xi x)\exp(-i2\pi\eta y)\, d\xi\, d\eta. \qquad \text{(I.4)}$$

We now write ξ and η in polar coordinates

$$\xi = \varrho \cos \alpha, \qquad \text{(I.5)}$$

$$\eta = \varrho \sin \alpha. \qquad \text{(I.6)}$$

Equation (I.4) then becomes

$$\mathscr{F}^{-1}[\mathbf{H}_1(\xi, \eta)] = \int_0^{\infty} \exp[-ikd(1 - \lambda^2\varrho^2)^{1/2}]\varrho\, d\varrho \times \int_0^{2\pi} \exp[-i2\pi\varrho(x \cos \alpha + y \sin \alpha)]\, d\alpha. \qquad \text{(I.7)}$$

If, furthermore, we write x and y in polar coordinates

$$x = -r \cos \varphi, \qquad \text{(I.8)}$$

$$y = -r \sin \varphi, \qquad \text{(I.9)}$$

we obtain for Eq. (I.7)

$$\mathscr{F}^{-1}[\mathbf{H}_1(\xi, \eta)] = \int_0^{\infty} \exp[-ikd(1 - \lambda^2\varrho^2)^{1/2}]\varrho\, d\varrho \int_0^{2\pi} \exp[i2\pi\varrho r \cos(\alpha - \varphi)]\, d\alpha$$

$$= 2\pi \int_0^{\infty} \exp[-ikd(1 - \lambda^2\varrho^2)^{1/2}] J_0(2\pi r\varrho)\varrho\, d\varrho. \qquad \text{(I.10)}$$

J_0 stands for the Bessel function of zero order, and reference [*I.4*] has been used to evaluate the integral over α. We define a new variable

$$s = 2\pi\varrho \tag{I.11}$$

and obtain for Eq. (I.10)

$$\mathscr{F}^{-1}[\mathbf{H}_1(\xi, \eta)] = \frac{1}{2\pi} \int_0^\infty \exp[-d(s^2 - k^2)^{1/2}] J_0(rs) s \, ds. \tag{I.12}$$

Consistent with the convention we have adopted for the sign of the phase of a traveling wave, quantities extracted from under the square root of Eq. (I.12) are considered positive. We now make use of Eq. 52 of Erdélyi [*I.5*, p. 95]. It has the form

$$\int_0^\infty J_0(bt) \exp[-a(t^2 - y^2)^{1/2}](t^2 - y^2)^{-1/2} t \, dt$$
$$= \exp[-iy(a^2 + b^2)^{1/2}](a^2 + b^2)^{-1/2};$$
$$\arg(t^2 - y^2)^{1/2} = \pi/2 \qquad \text{if} \quad t < y.$$

Differentiating the above with respect to the parameter a gives us (in the notation of Erdélyi [*I.5*])

$$\int_0^\infty J_0(bt) \exp[-a(t^2 - y^2)^{1/2}] t \, dt$$
$$= \exp[-iy(a^2 + b^2)^{1/2}] \cdot \frac{a}{a^2 + b^2} \left[\frac{1}{(a^2 + b^2)^{1/2}} + iy \right]. \tag{I.13}$$

If we now use Eq. (I.13) in Eq. (I.12) with the appropriate changes of variables we obtain

$$\mathscr{F}^{-1}[H_1(\xi, \eta)] = \frac{ik}{2\pi} \frac{\exp[-ik(d^2 + r^2)^{1/2}]}{(d^2 + r^2)^{1/2}}$$
$$\times \frac{d}{(d^2 + r^2)^{1/2}} \left[1 + \frac{1}{ik(d^2 + r^2)^{1/2}} \right]. \tag{I.14}$$

The second term in the last pair of brackets vanishes for distances that are large compared to the wavelength. When in accord with Eqs. (I.8) and (I.9) we substitute $r^2 = x^2 + y^2$, we find that Eqs. (I.14) and (I.3) are equivalent and

$$\mathscr{F}^{-1}[\mathbf{H}_1(\xi, \eta)] = \mathbf{h}_1(x, y). \tag{I.15}$$

Thus the equivalence of the diffraction formula Eq. (5.26) and the Fresnel–Kirchhoff integral Eq. (5.31) is proven.

REFERENCES

I.1. H. E. Rowe, "Angular Spectrum of Plane Waves and the Fresnel–Kirchhoff Integral for the Scalar Wave Equation," unpublished.

I.2. G. C. Sherman, "Application of the Convolution Theorem to Rayleigh's Integral Formulas," *J. Opt. Soc. Amer.* **57**, 546 (1967).

I.3. H. Osterberg, "Rayleigh's Integral in the Near Fresnel Region," *J. Opt. Soc. Amer.* **55**, 1467 (1965).

I.4. E. Jahnke and F. Emde, *Tables of Functions with Formulae and Curves*, p. 149, 4th ed. Dover, New York, 1945.

I.5. A. Erdélyi, *Higher Transcendental Functions*, Vol. 2. McGraw-Hill, New York, 1953.

Appendix II

COMPLEX REPRESENTATION OF THE ELECTRIC FIELD

In Section 1.3 we note that calculations involving sinusoidal signals associated with monochromatic, perfectly coherent light are simplified when complex notation is introduced. Here we extend the complex notation to wave amplitudes which are not simple sinusoids, such as those characteristic of *partially* coherent light.

Let us first define the *temporal* Fourier transform pair

$$\mathscr{F}^{-1}[\mathbf{g}(t)] = \mathbf{G}(f) = \int_{-\infty}^{\infty} \mathbf{g}(t) \exp(-2\pi i f t)\, dt \tag{II.1}$$

and

$$\mathscr{F}[\mathbf{G}(f)] = \mathbf{g}(t) = \int_{-\infty}^{\infty} \mathbf{G}(f) \exp(2\pi i f t)\, df \tag{II.2}$$

where $\mathbf{g}(t)$ is a function of time t and $\mathbf{G}(f)$, its spectrum, is a function of temporal frequency f. Note that the signs in the exponents of Eqs. (II.1) and (II.2) are interchanged with respect to those in the spatial analogs Eqs. (4.9) and (4.10). Suppose $\mathbf{g}(t)$ in Eq. (*II*.2) is the complex electric field of a light wave. Then $\mathbf{G}(f) \exp(2\pi r f t)$ in Eq. (II.2) represents a Fourier component of $\mathbf{g}(t)$, and we find the choice of signs consistent with Eq. (5.7). We symbolize the temporal transforms in Eqs. (II.1) and (II.2) by

$$\mathbf{g}(t) \subset \mathbf{G}(f). \tag{II.3}$$

Suppose we denote the electric field for monochromatic light of frequency

f_1 by $v_m(t)$ and write it as

$$\begin{aligned} v_m(t) &= a\cos(2\pi f_1 t + \varphi_0) \\ &= (a/2)[\exp(2\pi i f_1 t + i\varphi_0) + \exp(-2\pi i f_1 t - i\varphi_0)]. \end{aligned} \tag{II.4}$$

As in Section 2.4.1, we consider the phase $\varphi = 2\pi ft$ of the light emitted by a continuously oscillating source to be a linearly *increasing* function of increasing time so that the frequency f can assume only positive values. We can find an expression for the *complex electric field* by first interpreting the expanded form of Eq. (II.4) as two complex electric field functions, one of positive frequency f_1, and one of negative frequency $-f_1$. Then, along with the concept of a positive time rate of change of phase we suppress the negative frequency component leaving only $(a/2)\exp(2\pi i f_1 t + i\varphi_0)$. With this choice of phase, we see that the electric field $v_m(t)$ can be given by the real value of the complex electric field $\mathbf{v}_m(t)$. This is achieved by choosing

$$\begin{aligned} \mathbf{v}_m(t) &= 2[(a/2)\exp(2\pi i f_1 t + i\varphi_0)] \\ &= a\exp(2\pi i f_1 t + i\varphi_0). \end{aligned} \tag{II.5}$$

Then

$$v_m(t) = \mathrm{Re}[\mathbf{v}_m(t)] = a\cos(2\pi f_1 t + \varphi_0). \tag{II.6}$$

In a similar way, by restricting frequencies to positive values and by requiring that the electric field $v(t)$ be the real value of the complex electric field $\mathbf{v}(t)$, we can obtain a definition of the complex electric field which holds for the general case of partial coherence [*II.1*, pp. 270–271]. With $H(f)$, the unit step function defined by

$$H(f) = 1 \quad \text{for } f > 0, \qquad H(f) = 0 \quad \text{for } f < 0, \tag{II.7}$$

we define $\mathbf{v}(t)$ in terms of the Fourier transform relation

$$\mathbf{v}(t) \subset \mathbf{V}(f) = 2H(f)\mathbf{V}_R(f) \tag{II.8}$$

where

$$\mathbf{V}_R(f) \supset v(t). \tag{II.9}$$

Equation (II.8) ensures that $\mathbf{v}(t)$ is defined only for positive frequencies. The definition is meaningful if $v(t) = \mathrm{Re}[\mathbf{v}(t)]$. We now show this to be so. Bracewell [*II.1*, p. 270] shows the Fourier transform of $H(f)$ to be

$$\mathbf{h}(t) = \tfrac{1}{2}\,\delta(t) + i/(2\pi t). \tag{II.10}$$

With the aid of the convolution theorem Eq. (4.11) we rewrite Eq. (II.8) in the form

$$\begin{aligned}\mathbf{v}(t) &= 2\mathbf{h}(t) * v(t) \\ &= 2[\tfrac{1}{2}\,\delta(t) + i/(2\pi t)] * v(t).\end{aligned}$$

Using the sifting theorem (4.13e), the above becomes

$$\mathbf{v}(t) = v(t) + i/(\pi t) * v(t) \qquad \text{(II.11)}$$

from which we see that $v(t)$ is indeed the real part of $\mathbf{v}(t)$.

REFERENCE

II.1. R. Bracewell, *The Fourier Transform and Its Applications.* McGraw-Hill, New York, 1965.

Appendix III

CAPACITY–SPEED PRODUCT OF AN ACOUSTIC BEAM DEFLECTOR

An acoustic wave traveling through a cell of acousto-optic material generates a sinusoidal variation in the index of refraction of the medium across the cell, as indicated in Fig. III.1. To light incident on the cell, the variation appears as a slowly moving dielectric volume grating. (The acoustic wave velocity is small compared to the velocity of the light.) When the grating is illuminated at the Bragg angle, light can diffract from the grating with high efficiency (see Chapter 9). As a consequence of Bragg diffraction, an incident light beam is deflected from its original direction through the angle $\varphi = 2\theta_0$ where θ_0 is the Bragg angle given by

$$2\Lambda \sin \theta_0 = \lambda. \tag{III.1}$$

In the above equation the grating spacing Λ is the wavelength of the acoustic wave and λ is the wavelength of light in the cell. Since $\lambda \ll \Lambda$, we may write Eq. (III.1) as

$$\theta_0 = \lambda/2\Lambda$$

and the angle of deflection is therefore

$$\varphi = \lambda/\Lambda = \lambda f/s \tag{III.2}$$

where f is the acoustic frequency and s the acoustic wave speed. By varying the acoustic frequency over a range Δf, an angular deflecting range

$$\Delta\varphi = (\lambda/s)\,\Delta f \tag{III.3}$$

is obtained. With a knowledge of the angular spread of the deflected beam, we can use Eq. (III.3) to calculate the maximum number of resolvable directions into which the beam can be deflected.

Before proceeding to this computation let us examine the deflection process more closely. The laser beam enters the acoustic cell from a *fixed* direction. Suppose the acoustic wave is collimated so that a single train of plane wavefronts progresses across the cell. Then one and only one pair of values of grating spacing and angle of diffraction can satisfy Bragg's law for the fixed wavelength and angle of incidence of the laser beam. Under these circumstances, varying the acoustic frequency from an optimum value f will produce beam *extinction*, not deflection. However, in practice, the acoustic waves are introduced into the cell by an electroacoustic transducer of finite extent (see Fig. III.1). This launches not a single plane wave but,

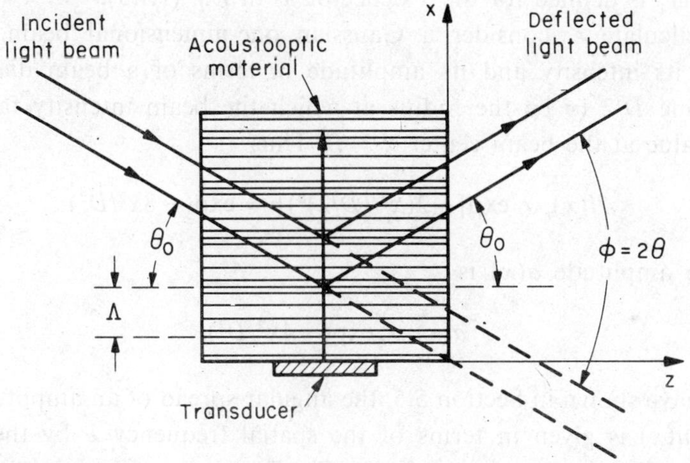

FIG. III.1. Light beam deflected by Bragg diffraction from an acoustic wave.

instead, a sum of waves traveling in a range of directions. Intensities of these waves are specified by the $((\sin x)/x)^2$ acoustic intensity pattern radiated by the transducer. At a given acoustic frequency f, the periodic spacing of the dielectric grating set up by one such wave and the angle that grating makes with the incident light may satisfy Bragg's law. At a second acoustic frequency, another grating with a new spacing and orientation makes a new Bragg angle with the incident light. For each such case there is a new Bragg angle of diffraction and a new angle at which the light beam leaves the cell. Variable beam deflection is thus achieved. The range of angles over which the beam can be deflected with high efficiency is limited by the angular width of the central lobe of the acoustic transducer radiation pattern. Acoustic losses and heat dissipation determine the minimum permissable size of the transducer and limit the width of the lobe.

The direction of the deflected beam is confined to a plane containing the incident light direction and the acoustic wave propagation vector (see Section 9.5). In Fig. III.1 this is the xz plane. Lenses may be used to convert the angular deflection to a linear displacement of the beam in the x direction. To obtain discrete displacements corresponding to the addresses of the holograms on the storage plane, the acoustic frequency is changed in discrete steps. The number N_x of addresses in the x direction to which the beam can be deflected without overlap is determined by the angular beam spread ε due to diffraction. Thus

$$N_x = \Delta\varphi_x/\varepsilon \tag{III.4}$$

where $\Delta\varphi_x$ is defined for the x deflector as in Eq. (III.3).

To calculate ε, consider a Gaussian one-dimensional beam. We can express its intensity and its amplitude in terms of a beam diameter D. We define $D/2$ to be the radius at which the beam intensity falls to e^{-2} of its value at the beam center $x = 0$. Thus

$$I(x) \propto \exp[-2(x^2/(D/2)^2)] = \exp(-8x^2/D^2) \tag{III.5}$$

and the amplitude $a(x)$ is

$$a(x) \propto \exp(-4x^2/D^2). \tag{III.6}$$

As we have shown in Section 5.5, the angular spread of an amplitude distribution $a(x)$ is given in terms of the spatial frequency ξ by the far-field pattern or Fourier transform of $a(x)$. The Fourier transform $A(\xi)$ is according to correspondence (4.27)

$$A(\xi) \propto \exp(-\pi^2 D^2 \xi^2/4). \tag{III.7}$$

The angular spread ε is defined by the range of spatial frequencies between $+\xi_r$ and $-\xi_r$ for which the intensity of the far-field pattern of the beam falls to e^{-2} of the value at beam center $\xi = 0$. In the far field the intensity is

$$[A(\xi)]^2 \propto \exp(-\pi^2 D^2 \xi^2/2)$$

and has the desired value e^{-2} when $\xi_r = \pm 2/\pi D$. For small values of the spread angle θ_1, $\xi_r = (\sin\theta_1)/\lambda \approx \theta_1/\lambda$ so that

$$\theta_1 = \pm 2\lambda/\pi D.$$

We then have for the angular spread

$$\varepsilon = 2\,|\,\theta_1\,| = 4\lambda/\pi D \approx \lambda/D. \tag{III.8}$$

The number of nonoverlapping addresses to which the beam may be sent by assignment of discrete acoustic frequencies in the range Δf becomes

$$N_x = (\lambda\,\Delta f/s)\cdot D/\lambda = (D/s)\,\Delta f. \tag{III.9}$$

Since $D/s = \tau$ is the time taken by the acoustic wave to traverse the diameter of the beam, the expression

$$N_x \cdot 1/\tau = \Delta f \tag{III.10}$$

is called the capacity-speed product for the x deflector. For each address in the x direction a second acoustic cell, in tandem with the first and oriented as a y deflector, produces $N_y = \tau\,\Delta f$ addresses in the y direction. The total number of addresses to which the beam can unambiguously be deflected is the product $N = N_x N_y$. For a constant Δf, Eq. (III.10) states that the access time (determined by τ) may be traded off for more addresses. This is accomplished by increasing D, i.e., by increasing the cross section of the acoustic cell. Large acoustooptic crystals of alpha-iodic acid can be grown and permit the total addresses to increase beyond 10^4 [*16.13*]. Lead molybdate is another excellent material with properties similar to those of iodic acid. Moreover it is insoluble in water, and, unlike the water-soluble iodic acid, its surfaces do not require protection from the normal atmosphere [*III.1*].

REFERENCE

III.1. D. A. Pinnow, L. G. Van Uitert, A. W. Warner, and W. A. Bonner, "Lead Molybdate: A Melt-Grown Crystal with a High Figure of Merit For Acoustic-Optic Device Applications," *Appl. Phys. Lett.* **15**, 83 (1969).

INDEX

A

Abbé, E., 36
Abberations, 78
  freedom from, 78
  third order, 78
Absorption constant, 238
Absorption grating (elementary hologram), 244
Absorption hologram, 15, *see also* Efficiency, plane hologram, volume hologram; Plane hologram; Reflection hologram; Volume transmission hologram
Achromatic images, 500
  by dispersion compensation, 501–504
  from image holograms, 501
Acoustic beam deflector, 477–480, 590–593
  acousto-optic materials, 478–479, 593
  angular deflection range, 479–480
  capacity–speed product, 479, 593
  deflection process, 591–592
  resolvable directions, 592
Aerosol particle analysis, 365–367
Aleksandrov, E. B., 428, 435
Aliasing, 547
Amplitude, 8
  complex, 9
  vector, 9
Amplitude division, 137
Associative storage, 382
Attenuator, beam, 169

B

Background wave, *see* Undiffracted wave
Baez, A. V., 41
Band-limited function, 543
Beam deflector, *see* Acoustic beam deflector
Beam, laser, 164–169
  attenuation of, 169
  division of, 167–169
  expansion of, 165
  intensity, radial distribution of, 164
  spatial frequency filtering, 165
  uniform illumination with, 166
Beam ratio, 153, 157
Beam splitter, 167–169
Binary hologram, 543, *see also* Computer-generated hologram
Bleached hologram, 15, 289–292
  effective process, 290–291
  exposure characteristic, 291
  noise reduction, 292
Bonch-Bruevich, A. M., 428, 435
Bragg, Sir Lawrence, *see* Bragg, W. L.
Bragg, W. L., 14, 36
  Rogers and, 51
Bragg angle, 240, 247–250
  cone of angles, 466–467
Bragg-effect hologram, 21

Bragg's law, 14, 231–233, 241, 466
Brooks, R. E., 325

C

Carrier-frequency hologram, *see* Off-axis hologram
Centrosymmetric distribution, 38
Character recognition, 394–409
  identification of character, 401–404
  page-at-a-time, 402–404
  parallel processing, 402
  problems of optical methods, 404–406
Chromaticity, 495
Chromaticity coordinates, 495
  for three primary wavelengths, 499
Chromaticity diagram, CIE, 495
Circular aperture function, 85
  Fourier transform of, 85
Coded reference wave, 409
Coded reference wave hologram
  for digital information storage, 409–414
  for multicolor imaging, *see* Multicolor images, plane hologram
  multiple exposure, 409–414
  noise, 412–414
Coherence, 26–33, *see also* Spatial coherence, Temporal coherence
  complex degree of, 30–32, 140
  degree of, 32
  partial, 29–33
Coherence length, 26–28, 141
  argon laser, multimode, 174
  He–Ne laser, typical, 158
  for holography, 143
  high pressure discharge, 145
  low pressure discharge, 145
  multimode laser, 147–150
  single-frequency CW laser, 146
  of thermal source, 28
Coherent light, 3, 7
Coherent transfer function, 133, 135
Color, 495
Color holography, 494–522
  achromatic images, 501–504
  choice of primary wavelengths, 498
  guide for laser selection, 498–499
  laser sources for, 497
  monochrome images, 500–501
  multicolor images, 504–521, *see also* Multicolor images
  recording materials, 500
Comb function, 543–544
Composite hologram, 510–511, 523–541
  element size, 523–526
  elements of, 523
  elimination of screen effect, 527–528
  Fourier transform hologram, 528–530
  generation of hyper- and hypostereoscopic images, 532
  for image synthesis, 535–540
  for multicolor imaging, 510–511
  to reduce hologram information content, 526–532
  screen effect, 527–528
  for wide-angle imaging, 533–535
Computer-generated hologram, 542–563
  applications, 558–560
  binary Fourier transform hologram, 551–558
  encoding complex amplitude, 553–558
Conductivity, periodic, 237
Conjugate beam, 17
Conjugate wave, 25, 49
Contact printing of hologram, *see* Replication of plane absorption hologram
Contour generation, 444–450
  immersion method, 447–450
  two wavelength method, 445–447
Contrast, 50
Converging wave, 62, 72
Convolution
  integral, 86
  operation, 86–91
  theorem, 86–87
Correlation, 91
  autocorrelation, 92
  cross correlation, 91
Correspondences, 84–86
  for selected functions, 93–95
Corresponding points, 427, 429
Coupled wave equations, 243
Coupled wave theory, 233
Cross talk, 468

D

Delta functions, 89
  properties of, 89

Density, optical, 47, 267
optimum, 267
Denisyuk, Yu. N., 21
Dichromated gelatin film, 293–298
development and processing, 297–298
exposure characteristics, 296–297
film preparation, 294–295
recording in, 293–294
sensitivity and resolution, 295–296
Dielectric constant, periodic, 236–237
Diffraction, 12–14, 96, 101–110
general, 104
from periodic objects, 101
from volume gratings, 228–264
Diffraction angles, 75
Diffraction gratings, 13–14, 103
plane, 13
volume, 13–14
Diffraction-limited lens, 134
Diffuse subject light, 195
nonlocalized recording of, 196–197
Diffusing reference source, 385
allowed displacement, 385–387
Diffusing screen, 194
autocorrelation of, 390–391
Dirac delta function, *see* Delta function
Distorting media, *see* Imaging through distorting media
Diverging wave, 62, 72
Doppler line-width, 144, 146
Double diffraction, 37, 39
Double-exposure interferometry, *see* Interferometry holographic
Dyson, J., 73

E

Efficiency
of diffraction, definition, 224, 276
maximum values, 261
plane hologram, 223–226
absorption hologram, max., 223–224
absorption Ronchi grating, 224
phase grating, square wave, 226
phase hologram, max., 225–226
volume hologram
absorption reflection hologram, 260–261
absorption transmission hologram, 252–253
phase reflection hologram, 256–258
phase transmission hologram, 247–250
superimposed absorption holograms, 472–474
Electric field of light wave, 5
complex, 9
partially coherent light, 587–589
energy per unit volume, 7
vector, 7, 9
Electron microscopy, 39
resolution, 35
Elementary hologram, 18
El-Sum, H.M.A., 41, 51
Enlarging hologram, *see* Scaling
Erasability, 271–272
Erasable hologram recording
in ferroelectrics, 308
in magnetic thin film, 334–335
in photochromics, 306
in photoconductor–thermoplastic film, 300–302
Erasable holographic memory, 489–492
recording media for, 489
in situ scheme, 491–492
thermoplastic holograms, 489–492
Etalon, 151
Evanescent waves, 104, 377
Exposure, 8, 159, 269
practical limit to, 269
time, 269
Exposure characteristics
ideal material, 275
photographic emulsion, commercial, 284–285
real materials, 276–280

F

Far field
condition, 108
pattern, 108
Ferroelectric crystals, 307–309
Film grain, *see* Noise, film grain
Flying-spot store, 456
holographic, 476–488
bits per page, 477
detector, 481, 483
near Fourier transform hologram, 486

Flying-spot store, holographic (*Cont.*)
Fourier transform hologram with random phase mask, 486–487
hologram diameter, 477
hologram efficiencies, 488
hologram formation, 483–488
page size, 477
pages, number stored, 477–480
system and performance, 480–483
total capacity, 480
microimage, 456–458
Fly's-eye lens, 456
for 3D image synthesis, 537
holographic, 378
Fourier transform, 79
of complex conjugate of function, 92
discrete, 549
fast computing technique, 550
of function with constant multiplier, 92
inverse, 80
of inverted function, 92
one-dimensional, 84
of sum, 92
tables of, 86
temporal, 587
two-dimensional, 80
Fourier transform hologram, 206–210, *see also* Lensless Fourier transform hologram, Quasi-Fourier transform hologram
invariance of image position, 209
with spatially modulated reference wave, 387–394
Fourier transform relations, *see* Correspondences
Fourier transform theorems, *see also* Convolution, Correlation
shift operations, 92
similarity theorem, 92
Fourier transformation, optical, 115–119, 124–127
exact, 119, 126
variable-scale, 127
Fraunhofer diffraction pattern, 108
Fraunhofer hologram, 215–217
Frequency, 8
Frequency transfer functions, 83
Fresnel hologram, 204
image limited by medium, 220–222
Fresnel–Kirchhoff integral, 106
equivalence to diffraction formula in spatial frequency domain, 583–586
Fresnel–Kirchhoff theory, assumptions of, 110
Fringes, 5, 60
bisectors of angle between waves, 11, 231
circular, 34

G

Gabor, D., 1, 20, 35, 36, 39, 49, 72
Gabor hologram, 42–44, 47
Gamma, 50
Gaussian beam, *see* Beam, laser
Gaussian function, 84
Fourier transform of, 85
Ghost image, 383–385
from Fourier transform hologram, 391–394
Grating vector, 234

H

Haine, M. E., 39, 73
Hologram, 2, 17
derivation of word, 14
Hologram formation, 14–16
analysis, elementary holograms, 228–231
coherence requirements, off-axis, 192–194
fringe visibility, optimization of, 152–158
geometries, 19–22, 204–217, *see also* Bragg-effect hologram, Fraunhofer hologram, Fresnel hologram, Fourier transform hologram, Image hologram, In-line hologram, Lensless Fourier transform hologram, Off-axis hologram, Quasi-Fourier transform hologram, Reflection hologram
materials for, *see* Recording materials
with partial temporal coherence, 152–163
polarization direction, preferred, 158
simple technique, 174–177
sources for, 137

space invariant, 223
spatial coherence requirement, 139–142
stable fringes, achievement of, 169–170
temporal coherence requirement, 142–146
Holography, 1, 3, 10
basic equations, 22–26
Hurter and Driffield curve, 47–48
Hyperstereoscopic image, 532
Hypostereoscopic image, 532

I

Illuminating wave, 69
Illumination, 164–167, *see also* Beam, laser
Image hologram, 204–206
for achromatic images, 501
formed from composite hologram, 540
Image processing, 414–416
Imaging
through distorting media, 368–375
conjugate reference method, 370–372
equal distortion method, 372–375
image hologram method, 368–370
with moving scatterers, 375
high resolution
projection, 375–377
noise sources, 376
total internal reflection method, 376–377
multiple, 377–379
Impulse response, 89
Incoherent-light holography, 579–581
signal-to-noise ratio comparison, 581
Index of refraction
average, 238
periodic, 239
Information content, reduction in hologram, 526–532
Information storage, 454–493
nonlocalized, 197
page-organized system, 456
redundant, 223, 463
In-line hologram, 20, 47–52
contrast, 50
In-line point-source hologram, 63–64
fringe frequency, 63–64
fringes, circular, 63
point images generated, 72–74
Intensity, 7–9
contour surfaces of maximum light, 16
square of amplitude, 8
time average of magnitude of Poynting vector, 7
Interference, 3
with light of partial temporal coherence, 153–158
of monochromatic waves, 7, 9
of randomly phased waves, 30
of two waves, 10–12
of waves of different frequency, 30
Interference pattern, 5, 6–12
of plane waves, 18
Interference term, 10, 30
Interferometry, holographic, 418–453
double-exposure, 423–426
pulsed laser, 424–426
real-time, 420–423
vibrating surface, 442–443
stroboscopic, 443–444
surface displacement determination, 435–437
time-average, 438–442
vibrating surfaces, 437–444
Intermodulation, 183
terms, 59
Inverted image, 73
Isoplanatic patch, 82

K

Kellstrom, G., 41
Kinoform, 560–563
Kirkpatrick, P., 41
Klein, W. R., 261
Kogelnik, H., 233, 263

L

Labeyrie, A. E., 22
La Macchia, J. T., 411
Laser, 137, *see also* Ruby lase.
CW gas discharge, 146
argon, 150, 152
He–Ne, 150, 152, 158
single frequency, 146, 149
single frequency operation, achievement of, 150–152
spatial coherence of TEM_{00} mode, 139

Laser emission wavelengths (table), CW lasers, 497
Lateral coherence, *see* Spatial coherence
Leith, E. N., 1, 21, 52, 179
Leith–Upatnieks hologram, *see* Off-axis hologram
Lens
  finite size of, 128–134
    effect on choice of optical system, 129–133
      on Fourier transformation, 128–129
      on resolution, 133–134
  spherical, 112
    imaging condition, 122
  thin, 112
Lens makers' formula, 114
Lensless Fourier transform hologram, 212–215
  field restriction, 220
  high resolution imaging, 220
Lensless Fourier transform point-source hologram, 66–68
  fringe frequency, 67
  point images generated, 77
  resolution, 68
Light waves, 4–6
Lin, L. H., 21
Linear recording, 158–163, 274
  condition for, 161
  plane absorption hologram, 161–162
  plane phase hologram, 162–163
Linear response, 55
Linear space invariant systems, 81–83, 89
Linearity in linear systems, 81, 91
Lippmann, G., 21, 458
Lippmann–Bragg hologram, 22
Lithium niobate, 470–471
  hologram recording in, 471
Localization, fringe, 426–435
  condition for, 431
  point of, 427
  for pure rotation, 432–435
  for pure translation, 431–432
  surface of, 427
Lohmann, A., 51
Lorentz line width, 144
Lukosz, W., 217
Luminous efficiency, 499

M

Magnetic field of light wave, 5
Magnetic thin films, 334
  erasable material, 335
  hologram recording in, 334–335
Magnification
  angular, 72
  factors contributing to, 74, 78
  of hologram, *see* Scaling
  lateral, 71
  with wavelength change, 35, 73
Matched filter, 394–396, 401
  ghost images from, 397–400
  impulse response, 397
  recognition signal, 400
Maxwell, J. C., 5
Maxwell's equations, 235
Meier, R. W., 60, 78, 217
Memory, optical, *see* Erasable holographic memory; Flying-spot store, holographic, microimage; Information storage; Standing-wave memory; Volume hologram memory
Michelson, A., 28
Michelson interferometer, 157, 158
Microimage system, 455–458
Microscopy, holographic, 360–365
  differential interference, 365
  interference, 363–365
  of phase objejcts, 362–365
  resolution obtainable, 361
Modulation transfer function, 135
Monochromatic waves, 8–10
Motion of subject, maximum allowed, 328–330
Multicolor images, 504–521
  plane hologram, 504–514
    coded reference beam method, 508–510
    composite hologram, 510–511
    cross modulation and false images, 505–507
    diffraction efficiency, 512–514
    mosaic color filter, 511
    separation of spectra, 507–508
    spatial sampling method, 510–511
  volume hologram, 514–521
    criterion for cross modulation elimination, 516

cross modulation suppression, 515
diffraction efficiency, 520–521
effect of emulsion shrinkage, 519–520
first multicolor image, 516
reflection hologram, 518–520
from superimposed holograms, 514–516
from transmission holograms, 516–518
with white-light illumination, 516, 518
Multiple-exposure hologram, *see* Superimposed holograms
Mulvey, T., 39

N

Negative hologram, 49, 51
Noise, 272–273
film grain, 351–358
in Fourier transform hologram, 352–354
in Fresnel hologram, 354–355
signal-to-noise ratio, coherent light, 355–358
Wiener spectrum, 353–354
sources of, 272–273
Nonlinear recording, 337–345
false images, 343
halo, 344–345
higher-order diffraction, 341
intermodulation, 341–345

O

Obliquity factor, 106
Kirchhoff, 107
Sommerfeld, 106
Off-axis hologram, 21, 52–56
contrast, 54
Off-axis holography, 180–203
angular separation of diffracted waves, 181–189
diffuse subject light, 194–203
nondiffuse subject light, 180–194
spectra of diffracted waves, 153–156
Off-axis point-source hologram, 64–66
fringe frequency, 65
point images generated, 74–77
On-axis sources, 72
Optical density, 47
Optical holography, 3–4, 42
Optical transfer function, 135
Orthoscopic image, 202

P

Parabolic approximation, 106
Paraxial approximation, 114
Pattern recognition, 406–409, *see also* Character recognition
fingerprint identification, 406
Pennington, K. S., 21
Persistence, 271–272
Perspective, 535
computer-drawn, 540
Phase, 8
Phase grating (elementary hologram), 244
Phase hologram, 15
efficiency maximization, 268
Photochromic materials, 305–307
Photoconductor–thermoplastic film, 298–305
exposure, development, erasure, 303
holographic properties, 303–305
preparation, 302–303
recording and erasure, 300–302
for write, read, erase, *in situ* memory, 489–492
Photographic emulsion, 15, 280–292
bleached, *see* Bleached hologram
development, fixing, drying, 287
exposure characteristics, commercial emulsions, 284–285
interference of plane waves recorded in, 18
properties of commercial emulsion, 280–284
response of, 47–49
shrinkage, 287–289
effect on standing-wave memory, 460
suitable commercial, 280
swelling with triethanolamine, 288–289
Photosensitive medium, 15–16
thick, 16
thin, 16

Plane hologram, 16, 161
absorption, 23
analysis of, 179–227
Plane wave, 11, 97
interference of, 228–231
propagation
diffraction and, 96
in homogeneous lossy medium, 238
Point source, 26, 43
illuminating, 69
images, 71–77
reference, 59
subject, 59
Point-source hologram, 46, 58–78
Polarization
preferred direction, 158
of two interfering waves, 10
Polarization angle, 153
Positive hologram, 49, 50
Powell, R. L., 437
Poynting vector, 7
Projection method, 39, 40, 72
magnification, 73
Propagation
direction of, 98
equation in spatial frequency domain, 105
in homogeneous medium, 238
number, 99
vectors, 240
Pseudoscopic image, 26, 199–203
from lenslet array, 537
Pulsed-laser holography, 311–336
geometry with multimode laser, 324–327
protection of optical components, 322–324
safe geometry for living subjects, 327–328

Q

Quasi-Fourier transform hologram, 210–212

R

Rayleigh distribution, 349
mean value, 349
probability density, 349
Rayleigh rule, 106
Rays, 12
Real image, 17, 50, 72
applications, 360–380
generation of, 189–192
lensless recording of, 191
Real-time interferometry, *see* Interferometry, holographic
Reciprocity failure, 334
Reconstructed wave, 17, 24
ideal, 274
intensity at hologram, 161
Reconstructing source, 170–174
multilongitudinal-mode laser, 173–174
nonlaser
extended, 171–172
of finite bandwidth, 172–173
Reconstructing wave, *see* Illuminating wave
Reconstruction, 49–51
with point source, 68–71
with spatially modulated wave, 385
Recording materials, 265–310
absorption, 266–267
ideal, 273–276
phase, 266, 267–268
for pulsed-laser holography, 332–335
sensitivity, holographic, 276, 279–280
Rectangular function, 90
Redundant storage, *see* Information storage
Reference beam, 15
wavefronts, 15
Reference offset, 56
Reference wave, 47, 381
coded, 409
spatially modulated, 381–383
Reflection hologram, 22, 253–261
absorption hologram, 258–261
angular response, phase hologram, 257–258
diffracted amplitude, absorption hologram, 259–260
effect of emulsion shrinkage on image color, 519–520
efficiency
absorption hologram, 260–261
phase hologram, 256–258
optimum density, absorption hologram, 260–261

phase hologram, 255–258
wavelength sensitivity, phase hologram, 257–258
white-light illumination, 520
Replication
plane absorption hologram, 564–569
coherence required, 565–567
double images generated with laser light, 567–569
with nonlaser light, 564–567
plane phase hologram, 569–570
by pressing, 570
volume holograms, 570–572
Resolution
recording, 270–271
reflection hologram, 271
transmission hologram, 270–271
of recording medium, effect on image resolution, 217–222
Rogers, G. L., 42
Ruby amplifier, 319–322
frequency shift, 320
Ruby laser, 312–316
coherence length, single-frequency, 316–319
frequency modulation, 317
multimode, 312–314
single-frequency, 315–319

S

Safe illumination of humans, 331–332
Sampling theorem, 543
Scalar waves, 11
Scaling, 35, 39
Sensitivity, holographic, *see* Recording material
Shift operation, 92
Similarity theorem, 92
Size of hologram, effect on imaging, 222–223
Snell's law, 228
Source
argon discharge, 145–146
argon ion laser, 150
electron, 40
He–Ne laser, 150
krypton discharge, 145
mercury arc, high pressure, 142, 145
neon discharge, 145–146
point subject, 19
reference, 25
thermal, 28
X-ray, 42
Space invariance, 81, 91
Spatial coherence, 32–33
complex degree of, 140
degree of, 32
nonlaser source, 140, 141
perfect, 26
Spatial domain, 79
Spatial filter, 381, 394
derivative filter, 405–407
high pass, 405
for image processing, 414–416
Spatial frequency, 80, 99
Spatial frequency domain, 79
Spatial frequency multiplexing, 409
Speckle pattern, 203, 345–351
intensity ratio, rms-to-mean, 348–349
reducing problem, 349–351
spot size versus *f*/ number, 347–348
Spectrum, 83
power, 143
high pressure discharge, 144
low pressure discharge, 144
Spherical phase factor, 117
Spherical wave, 61, 71, 115
Spread function, 89, 133
Standing-wave memory 458–461
Standing wave pattern, *see* Interference pattern
Stetson, K. A., 437
Stroke, G. W., 22, 67, 392
Subject wave, 381
Superimposed holograms, 461
deterioration with multiple exposure, 475
formed in multiwavelength light, 515–516
in photochromic material, 474–475
reduced efficiency
absorption, 472–474
plane absorption, 512–514
volume absorption, 520–521

T

Television transmission of holograms, 526, 573–578

Television transmission of holograms (*Cont.*)
information reduction techniques, 527
methods of reducing resolution requirements, 574–578
Temporal coherence, 26–30
complex degree of, 143
degree of
multimode laser, 147–150
nonlaser source, 144
gas laser, 146–150
perfect, 26
Thermally engraved hologram, 325
Thermoplastic film, *see* Photoconductor–thermoplastic film
Thermoplastic hologram, 489–492, *see also* Photoconductor–thermoplastic film
Thompson, B. J., 52
Three-dimensional image, 197–203
Three-dimensional image synthesis, 535–540
forming composite hologram, 536–540
forming perspectives, 535–536
Time-average interferometry, *see* Interferometry, holographic
Total internal reflection, 367–377
Transmission hologram, *see* Plane hologram, Volume transmission hologram
Transmission method, 40–41, 73
Transmittance
amplitude, 23, 49, 55, 267
Taylor series, 160
intensity, 47, 267
Transmittance–exposure curve, 55, 284–285
Twin image, 49, 51–52
separation of, 52
Two-hologram method, 51

U

Undiffracted wave, 39
Upatnieks, J., 1, 21, 52, 179
Urbach, J. C., 217

V

van Cittert–Zernike theorem, 32–33. 139
van Heerden, P. J. 21, 391, 458, 461
Vander Lugt, A., 207, 387, 394, 402
operational notation, 120
Virtual image, 17, 24, 50, 72
Visibility, 29, 32, 153, 156
optimization of, 153–158
Visual stimulus, 498–499
Volume hologram, 16
criterion for, 261
response of elementary, 228–264
Volume hologram memory, 461–475
absorption holograms, superimposed, 471–475
angular bandwidth, phase holograms, 466–467
diffraction efficiency, phase holograms, 467–469
phase holograms, superimposed, 463–471
recording media limitations, 470–471
recording range, phase holograms, 464–466
Volume reflection hologram, *see* Reflection hologram
Volume transmission hologram, 244–253
absorption hologram, 250–253
angular response, phase hologram, 248–250
diffracted amplitude, absorption hologram, 251
efficiency
absorption hologram, 252–253
phase hologram, 247–250
optimum density, absorption hologram, 252
phase hologram, 246
wavelength response, phase hologram, 249–250

W

Wave equation, 97
for periodic inhomogeneities, 233–237
solution satisfying Bragg's law, 239–244
Wave normal, 11
Wavefront, 11
Wavefront reconstruction, 1, 3, 16–19, 23, 39
ideal, 273–276
Wavefront division, 137

2.50